Fundamentals of HEAT TRANSFER

Fundamentals of HEAT TRANSFER

Alan J. Chapman
Rice University

Macmillan Publishing Company
New York
Collier Macmillan Publishers
London

Printed in the United States of America

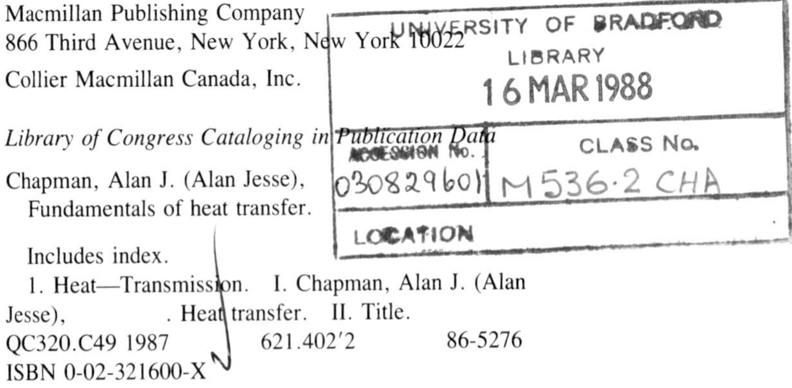

Macmillan Publishing Company
866 Third Avenue, New York, New York 10022

Collier Macmillan Canada, Inc.

Library of Congress Cataloging in Publication Data

Chapman, Alan J. (Alan Jesse),
Fundamentals of heat transfer.

Includes index.
1. Heat—Transmission. I. Chapman, Alan J. (Alan Jesse), . Heat transfer. II. Title.
QC320.C49 1987 621.402′2 86-5276
ISBN 0-02-321600-X

Printing: 1 2 3 4 5 6 7 8 Year: 7 8 9 0 1 2 3 4 5

ISBN 0-02-321600-X

Dedicated to

Marjorie E. Chapman

Preface

This work is an extension of the approach set forth by the author in the earlier four editions of *Heat Transfer*, also published by the Macmillan Company. Since heat transfer is a synthesis of a number of engineering science and mathematics subjects, its study is significantly influenced by the rapidly occurring changes in modern technology. To provide the modern engineer with the flexibility required to work effectively in the variety of emerging fields and areas of application, it is important to teach heat transfer in a way that stresses the fundamentals underlying the subject. This text emphasizes, then, the physical concepts underlying heat transfer and how these fundamental concepts may be applied to obtain answers to real problems of practical interest. The details of mathematical analysis are de-emphasized in favor of physical understanding.

This text, then, is meant to be a teaching text written for the student as a reader. As such, numerous worked-out examples are included throughout the work, and these examples are presented in sufficient detail that the student may understand the way in which a particular problem is being solved. In like fashion, a very large number of problems for student exercises are provided at the end of each chapter, in an order reflecting that in which subject matter is presented in the chapter. The SI system of units is stressed and forms the basic units system for the text. However, since engineers will be required to work also in the English system during the foreseeable future, a number of the worked examples and problems also involve the use of these units.

An applied subject such as heat transfer yields meaningful results only when real problems are solved under the same conditions as experienced in practice. Thus, this text includes an appendix which provides an extensive collection of the thermal properties of a wide variety of gases, liquids, and solids frequently encountered in engineering practice. Most physical properties are tabulated in SI units; however, the data for the important engineering fluids air, water, and steam are given in both SI and English units to avoid extensive conversion calculations.

The approach of the presentation presumes that the subject is being studied at the introductory level by students either late in their junior year or early in the senior year. Knowledge of thermodynamics is presumed and a prior or concurrent course in fluid mechanics is also assumed. Mathematical preparation through advanced calculus and ordinary differential equations is also presumed; however advanced topics such as the use of Fourier series or Bessel's equation are not presented.

The work on conduction treats one-dimensional cases having engineering significance, including extended surfaces and finned arrays which lead naturally into heat exchanger applications in later chapters. Two-dimensional conduction is treated mainly through the presentation of known solutions by use of the conduction shape factor rather than mathematical analysis. The traditional results for transient conduction are presented, without derivation, in graphical form for engineering computations, However, new approximate, closed-form, solutions are also given which permit the accurate computation of most transient problems of practical interest.

The presentation of the study of convective heat transfer begins with a review of the basic concepts of fluid mechanics, interwoven with the notion of the boundary layer. Sample solutions are presented to illustrate the methodology; however, the most useful engineering results are simply quoted. There is a strong emphasis placed on the importance of fluid friction as well as heat transfer for convective problems, and extensive use is made of the analogy between heat and momentum transfer, especially for turbulent flows. Particular attention is paid to the use of similarity parameters and the legitimate use of empiricism. The final correlations recommended for use by engineers in determining convective heat transfer and friction are carefully highlighted along with the applicable limits for their use.

Radiative heat transfer is explained in terms of real surfaces and includes the idealized analysis of black or gray surfaces. Included are treatments of enclosure analysis, solar radiation, and application to industrial and aerospace problems of significance. The often-tedious problem of shape factor algebra is presented in an appendix.

The final two chapters are devoted to the ultimate aim of a heat transfer course: the solution of problems involving the combined modes of conduction, convection, and radiation. Particular emphasis is placed on the necessity for iterative calculations so often required in such combined problems. The work presented on heat exchangers is more extensive than is often found in an undergraduate heat transfer text—with particular attention given to the methodology of carrying out the performance and design calculations so often encountered by practicing engineers. The analysis of thermometric errors incurred by conductive, convective, and radiative interactions is presented, as are certain modern applications such as space radiators, solar collectors, and heat pipes.

Throughout the text, acknowledgment has been made of the numerous authors of books and journal articles that formed the sources from which many data and results were taken. The author would also like to acknowledge the many helpful suggestions of students, reviewers, and users of the earlier texts. Their comments were invaluable in the preparation of this volume.

A.J.C.

Houston, Texas

Contents

CHAPTER 1

Introduction

1.1 INTRODUCTORY REMARKS

In the science of thermodynamics, which deals with energy in its various forms and with its transformation from one form to another, two particularly important *transient* forms are defined, work and heat. These energies are termed transient since, by definition, they exist only when there is an exchange of energy between two systems or between a system and its surroundings, i.e., when an energy form of one system (such as kinetic energy, potential energy, internal energy, flow energy, chemical energy, etc.) is transformed into an energy form of another system or of the surroundings. When such an exchange takes place without the transfer of mass from the system and not by means of a temperature difference, the energy is said to have been transferred through the performance of *work*. If, on the other hand, the exchange of energy between the systems is the result of a temperature difference, the exchange is said to have been accomplished by the transfer of *heat*. It is this energy transfer form, heat, and the basic laws governing its exchange with which this book is concerned. It should be noted that the existence of a temperature difference is the distinguishing feature of the energy form known as heat.

It will prove convenient to speak of the energy transfer in the form of heat as the "flow of heat" from one body to another although there is actually nothing flowing between the two bodies. There is an exchange of the internal energy of one body to that of the other through the mechanism of *heat flow*. The first law of thermodynamics requires that the heat (internal energy) given up by one body must equal that taken up by the other. The second law of thermodynamics requires that the heat flow take place from the hotter body to the colder body.

1.2 THE IMPORTANCE OF HEAT TRANSFER

The importance of a thorough knowledge of the science of heat transfer and the necessity of being able to analyze, quantitatively, problems involving a transfer of heat have become increasingly important as modern technology has become more and more complex. In almost every phase of scientific and engineering work, processes involving the exchange of energy through a flow of heat are encountered.

Mechanical and chemical engineers are particularly concerned with problems of heat transfer. Modern power generation involves the production of work from either a combustible fuel or a nuclear reaction. This energy is converted into useful work by means of boilers, turbines, condensers, air heaters, water preheaters, pumps, etc. All these pieces of apparatus involve a transfer of heat by one means or another, as does almost every piece of apparatus found in a chemical process industry or a petroleum refinery. Certainly, designing the familiar internal combustion engine, gas turbine, and jet engine requires a complete understanding of heat transfer for a thorough analysis of the combustion and cooling processes.

The so-called "thermal barrier" in aerodynamics involves finding means of transferring away from the aircraft the enormous amounts of heat produced by the dissipative effect of the viscosity of the air. Indeed, since all processes in nature have been observed to be irreversible, it follows that all natural processes involve a dissipation of the various forms of mechanical energy into thermal energy with consequent heat transfer processes taking place.

It is this dissipative aspect of the second law of thermodynamics which leads to the much-discussed problem of "thermal pollution," created by the inevitable discharge of waste heat into the environment (air and water). The development of effective solutions to the problems of thermal pollution is perhaps one of the most challenging applications of the knowledge of heat transfer at the present time.

The importance of heat transfer in the production of comfort cooling or comfort heating is readily apparent. This influences the design of building structures of all kinds.

1.3 THE FUNDAMENTAL CONCEPTS AND THE BASIC MODES OF HEAT TRANSFER

It is customary to categorize the various heat transfer processes into three basic types or modes, although, as will become apparent as one studies the subject, it is certainly a rare instance when one encounters a problem of practical importance which does not involve at least two, and sometimes all three, of these modes occurring simultaneously. The three modes are conduction, convection, and radiation.

Heat *conduction* is the term applied to the mechanism of internal energy exchange from one body to another, or from one part of a body to another part, by the exchange of the kinetic energy of motion of the molecules by direct communication or by the drift of free electrons in the case of heat conduction in metals.

This flow of energy or heat passes from the higher energy molecules to the lower energy ones (i.e., from a high-temperature region to a low-temperature region). The distinguishing feature of conduction is that it takes place within the boundaries of a body, or across the boundary of a body to another body placed in contact with the first, without an appreciable displacement of the matter comprising the body.

A metal bar heated on one end will, in time, become hot at its other end. This is an example of conduction. The laws governing conduction can be expressed in concise mathematical terms, and the analysis of the heat flow can be treated analytically in many instances.

Convection is the term applied to the heat transfer mechanism which occurs in a fluid by the mixing of one portion of the fluid with another portion due to gross movements of the mass of fluid. The actual process of energy transfer from one fluid particle or molecule to another is still one of conduction, but the energy may be transported from one point in space to another by the displacement of the fluid itself.

The fluid motion may be caused by external mechanical means (e.g., by a fan, pump, etc.), in which case the process is called *forced convection*. If the fluid motion is caused by density differences which are created by the temperature differences existing in the fluid mass, the process is termed *free convection* or *natural convection*. The circulation of the water in a pan heated on a stove is an example of free convection. The important heat transfer problems of condensing and boiling are also examples of convection—involving the additional complication of a latent heat exchange.

It is virtually impossible to observe pure heat conduction in a fluid because as soon as a temperature difference is imposed on a fluid, natural convection currents will occur as a result of density differences.

The basic laws of heat conduction must be coupled with those of fluid motion in order to describe, mathematically, the process of heat convection. The mathematical analysis of the resulting system of differential equations is perhaps one of the most complex fields of applied mathematics. Thus, for engineering applications, convection analysis will be seen to be a subtle combination of powerful mathematical techniques and the intelligent use of empiricism and experience.

Thermal *radiation* is the term used to describe the electromagnetic radiation which has been observed to be emitted at the surface of a body which has been thermally excited. This electromagnetic radiation is emitted in all directions; and when it strikes another body, part may be reflected, part may be transmitted, and part may be absorbed. If the incident radiation is thermal radiation (i.e., if it is of the proper wavelength), the absorbed radiation will appear as heat within the absorbing body.

Thus, in a manner completely different from the two modes discussed above, heat may pass from one body to another without the need of a medium of transport between them. In some instances there may be a separate medium, such as air, which is unaffected by this passage of energy. The heat of the sun is the most obvious example of thermal radiation.

There will be a continuous interchange of energy between two radiating bodies, with a net exchange of energy from the hotter to the colder. Even in the case of thermal equilibrium, an energy exchange occurs, although the net exchange will be zero.

1.4 THE FUNDAMENTAL LAWS OF CONDUCTION

Thermal Conductivity and Thermal Conductance

The basic law governing heat conduction may best be illustrated by considering the simple, idealized situation shown in Fig. 1.1. Consider a plate of material having a surface area A and a thickness Δx. Let one side be maintained at a temperature t_1, uniformly over the surface, and the other side at temperature t_2. Let q denote the rate of heat flow (i.e., energy per unit time) through the plate, neglecting any edge effects. Experiment has shown that the rate of heat flow is directly proportional to the area A and the temperature difference $(t_1 - t_2)$ but inversely proportional to the thickness Δx. This proportionality is made an equality by the definition of a constant of proportionality k. Thus

$$q = kA\frac{t_1 - t_2}{\Delta x}. \tag{1.1}$$

The constant of proportionality, k is called the *thermal conductivity* of the material of which the plate is composed. It is a property dependent only on the composition of the material, not its geometrical configuration. While more details concerning thermal conductivity are given in Sec. 1.5, it is worth noting here the typical order of magnitude of this property. Values of k may range from 0.008 W/m-°C (0.005 Btu/h-ft-°F) for gases such as Freon-12 to as great as 417 W/m-°C (241 Btu/h-ft-°F) for metals such as silver.

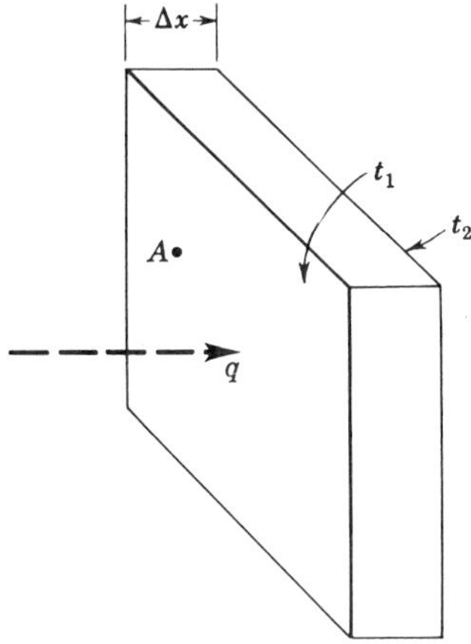

Figure 1.1

Sometimes a gross quantity, *unit thermal conductance,* is used to express the heat conducting capacity of a given physical system, so that if $C = k/\Delta x$ denotes the unit thermal conductance,

$$q = CA(t_1 - t_2)$$

Thus it is seen that thermal conductance is the conductivity of a substance divided by its thickness. It is no longer a physical property but depends, as well, on the geometrical configuration at hand and, thus, is a less general quantity than is thermal conductivity.

Equation (1.1) forms the basis for the fundamental relation of heat conduction. Consider now a homogeneous, isotropic solid as depicted in Fig. 1.2. If the solid is subjected to certain known boundary temperatures, what is the rate at which heat is conducted across a surface, S, within the solid? Selecting a point P on the surface S, one can select a wafer of material having an area δA, which is part of the surface S containing P, and having a thickness δn in the direction of the normal drawn to the surface at P. If the difference between the temperature of the back face of the wafer and its front face is δt, and if δA is chosen small enough so that δt is essentially uniform over it, the rate of heat flow across the wafer, δq is, by Eq. (1.1),

$$\delta q = -k\delta A \frac{\delta t}{\delta n}.$$

The negative sign that has been introduced in the expression above is the result of the usually adopted convention that the heat flow rate is taken to be positive in the direction of a decreasing temperature. That is, if for a positive displacement of the normal coordinate ($\delta n > 0$) the temperature decreases (i.e., $\delta t < 0$), then the flow δq is defined as positive. Forming the ratio $\delta q/\delta A$ and allowing the area

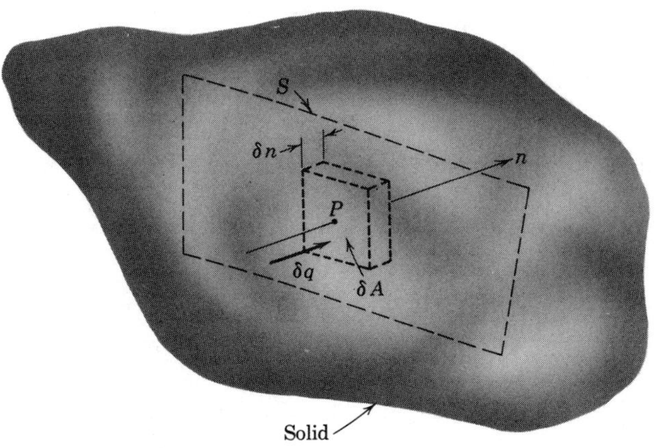

Figure 1.2

$\delta A \to 0$, one obtains what is termed the *flux* of heat conducted through the thickness δn at one point P,

$$f_n = \frac{dq}{dA} = -k\frac{\delta t}{\delta n}.$$

Further, allowing $\delta n \to 0$, one arrives at the flux of heat across S at the point P in terms of the *temperature gradient* at P in the n direction, $\partial t/\partial n$:

$$f_n = -k\frac{\partial t}{\partial n}, \tag{1.2}$$

wherein the notation f_n is used to denote the flux in the n direction. The statement in Eq. (1.2) is called Fourier's conduction law after the French mathematician who first made an extensive analysis of heat conduction. It states that the flux of heat conducted (energy per unit time per unit area) across a surface is proportional to the temperature gradient taken in a direction normal to the surface at the point in question. It should be emphasized that Fourier's law is based on *empirical* observation and is not derived from other physical principles.

Upon returning to the situation pictured in Fig. 1.2, the total rate of heat transferred across the finite surface S would be

$$q = -\int_S k\frac{\partial t}{\partial n}\,dA.$$

Generally speaking, the normal gradient $\partial t/\partial n$ may vary over the surface, but in many instances it is possible to select the surface as one on which the gradient is everywhere the same. This is the situation in the case depicted in Fig. 1.1, in which every plane normal to Δx is such a surface. In the case of a hollow cylinder with uniform outside and inside surface temperatures, every concentric interior cylindrical surface is isothermal with a uniform temperature gradient normal to it. In such cases, then,

$$q = -kA\frac{\partial t}{\partial n}, \tag{1.3}$$

where A is the total area of the finite surface.

The relations above serve to introduce the important physical property of thermal conductivity, which is discussed in more detail in the next section. These relations are also used to develop expressions describing the distribution of temperature and the flow of heat through a heat conducting body as described in some detail in Chapters 2 through 4.

1.5 THERMAL CONDUCTIVITY OF MATERIALS

As discussed in Sec. 1.3, heat conduction is, basically, the transmission of energy by molecular communication. Thermal conductivity is, then, the physical property denoting the ease with which a particular substance accomplishes this energy exchange. The thermal conductivity of a material is found to be dependent on the chemical composition of the substance, or substances, of which it is composed, the phase (i.e., gas, liquid, or solid) in which it exists, its molecular structure, and whether or not it is a homogeneous material. The thermal conducting characteristics of nonhomogeneous materials are better understood after first examining the behavior of homogeneous materials.

The relative order of magnitude of the thermal conductivity of homogeneous materials is best illustrated by examination of Table 1.1, in which some typical values are tabulated. One immediately observes a vast range of possible values of thermal conductivity, with silver having a conductivity almost 50,000 times as great as that of Freon.

One also sees that, generally, a liquid is a better conductor than a gas and that a solid is a better conductor than a liquid. These differences are explained partially by the fact that while in a gaseous state, the molecules of a substance are spaced relatively far apart and their motion is random. This means that energy transfer by molecular impact is much slower than in the case of a liquid in which the motion is still random but in which the molecules are more closely packed. The same is true concerning the difference between the conductivities of the gaseous and solid phases; however, other factors become important when the solid state is formed, producing significant differences between solid and liquid conductivities.

Referring again to Table 1.1, one sees that a solid having a crystalline structure, such as quartz, has a higher thermal conductivity than an amorphous solid, such

Table 1.1. Thermal Conductivity of Various Substances*

Substance	Thermal Conductivity W/m-°C
Gases	
Freon-12 (0°C, 1 atm)	0.0083
Air (0°C, 1 atm)	0.0241
Liquids	
Carbon dioxide (sat. liq., 0°C)	0.105
Glycerine, pure (0°C)	0.282
Water (sat liq., 0°C)	0.562
Solids	
Glass, plate (20°C)	0.76
Ice (0°C)	2.22
Magnesite brick (204°C)	3.81
Quartz (20°C)	7.6
Stainless steel (18% Cr, 8% Ni)(0°C)	16.3
Iron, pure (0°C)	73
Zinc, pure (0°C)	112
Aluminum, pure (0°C)	202
Copper, pure (0°C)	386
Silver, pure (0°C)	417

*Abstracted mainly from tables in Appendix A.

as glass. Also, metals, crystalline in structure, are seen to have greater thermal conductivities than do nonmetals. In the case of the amorphous solids, the irregular molecular arrangement inhibits the transfer of energy by molecular impact and the thermal conductivity is of the same order of magnitude as that for liquids. On the other hand, solids of a crystalline structure exhibit higher conductivities as the result of enhanced energy exchange resulting from vibratory motion of the crystal lattice as a whole. In the case of metallic conduction, still a third mechanism of energy transfer comes into play. In a metal crystal the valence electrons of the constituent atoms (i.e., the outermost electrons) become detached and are free to move within the crystal lattice. It is the drift of these free electrons which makes the metals such better conductors than other solids and accounts for the observed proportionality between the thermal and electrical conductivities of pure metals.

With these brief remarks, the following sections describe the characteristics of the thermal conductivity of various substances, its dependence on certain physical and thermodynamic parameters, and the tabulated data of Appendix A.

The Thermal Conductivity of Homogeneous Solids

Table A.1 of Appendix A tabulates, among other properties, the thermal conductivity of various pure metals and metal alloys. Many factors are known to influence the thermal conductivity of metals, such as chemical composition, atomic structure, phase changes, grain size, temperature, pressure, and deformation. The factors with the greatest influence are the chemical composition, phase changes, and temperature. Usually, the first two of these do not enter a case in which one is interested in a particular material, and, hence, only the temperature effect has to be accounted for.

It is known that the thermal conductivity of metals is directly proportional to the absolute temperature and the mean free path of the molecules. The mean free path tends to decrease with increasing temperature so that the net variation is the result of opposing influences. Pure metals generally have thermal conductivities which decrease with temperature, but the presence of impurities or alloying elements, even in minute amounts, may reverse this trend. These effects may be observed by examination of Table A.1. The data of Table A.1 for chrome steels are plotted in Fig. 1.3, which illustrates typical effects of temperature and composition on the thermal conductivity of a metal.

It is usually possible to represent the temperature dependence of the thermal conductivity of a metal by a linear relation of the form $k = k_0(1 + bt)$. Here k_0 is the thermal conductivity at $t = 0°C$, and b is a constant which represents the rate of change of k with respect to temperature—a positive value of b indicating a material whose conductivity increases with temperature, and a negative value the reverse.

The thermal conductivities of other homogeneous solids are presented in Table A.2, along with many nonhomogeneous solids. Generally speaking, the thermal conductivities of these materials do not vary with pressure, only with temperature.

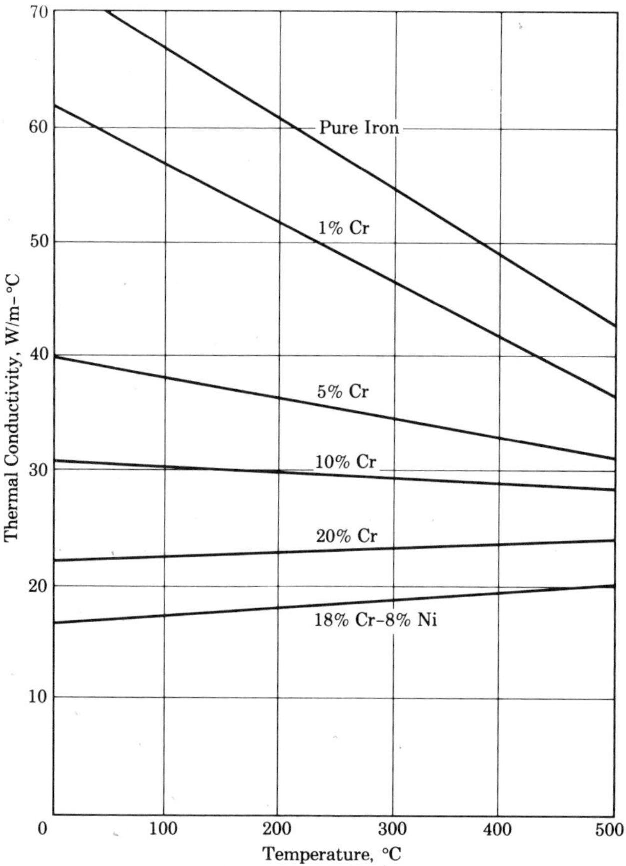

Figure 1.3. The effect of temperature and composition on the thermal conductivity of steel.

The variation with temperature is usually approximately linear, with the conductivity increasing with temperature.

The Apparent Thermal Conductivity of Nonhomogeneous Solids

Many of the solid materials encountered in engineering practice, particularly building materials and insulations, are nonhomogeneous in structure. Some materials may exhibit nonisotropic conductivities that result from a directional preference caused by a fibrous structure, as in the case of wood. Other materials can be discussed only from the point of view of an *apparent thermal conductivity* due to inhomogeneities present because of a porous structure (glass wool, cork, etc.) or because of a structure that is composed of different substances (concrete, brick, etc.). In any of these instances the thermal conductivity may vary from sample to sample, and the values reported in Table A.2 should be looked upon as typical only—with considerable deviation from these values expected in practice.

Several insulating materials are shown in Table A.2 for both low-temperature applications (rockwool, glass wool, etc.) and high-temperature applications (asbestos, magnesia, etc.). These materials exhibit low thermal conductivities due primarily to the air (a poorly conducting gas) that is entrapped in the porous structure of the insulation rather than a low conductivity of the solid substance itself. One observes a strong dependence of the thermal conductivity of many insulating materials on the *apparent bulk density* for this same reason.

The Thermal Conductivity of Liquids and Gases

Tables A.3 and A.4 present the thermal conductivities of various saturated liquids, along with other properties to be used later. In general the conductivities of liquids of importance in engineering applications are relatively insensitive to pressure, particularly at pressures not too close to the critical pressure. For this reason, the temperature variation is usually the only influence taken into account, and the saturated state is the condition at which the conductivity is reported because of the uniqueness of this state.

Most liquids exhibit a decreasing thermal conductivity with temperature, although water is a notable exception. Because of its importance as an engineering fluid, the properties of water are presented separately in a more extensive tabulation in Table A.3 in both SI and English units to avoid excessive conversion calculations. Table A.4 present data for several other nonmetallic liquids, and Table A.9 gives the properties of several liquid metals.

Some of the data of Tables A.3 and A.4 are plotted in Fig. 1.4 for illustrative purposes. Of the nonmetallic liquids, water has the highest thermal conductivity.

The thermal conductivities of several gases are given in Table A.7 for a pressure

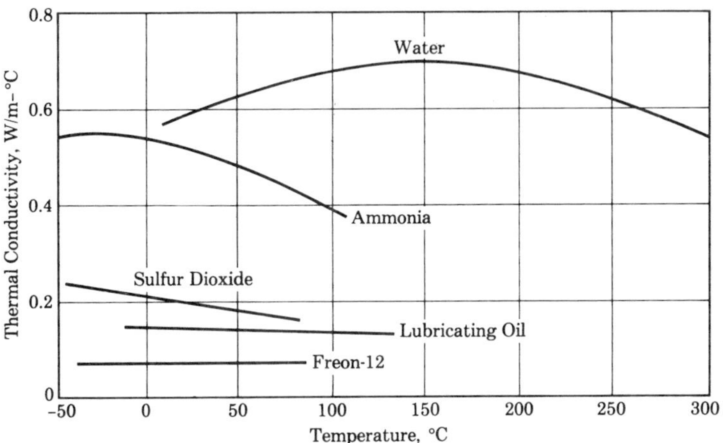

Figure 1.4. The effect of temperature on the thermal conductivity of several saturated liquids.

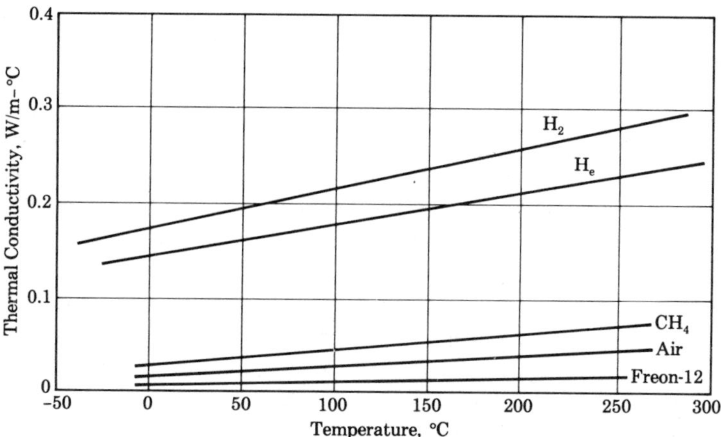

Figure 1.5. The effect of temperature on the thermal conductivity of several gases at atmospheric pressure.

of 1 atm, and Table A.6 presents similar data of a more extensive nature for air. Again, because of its importance as an engineering fluid, the data for air are given in both SI and English units. In general, the thermal conductivity of a gas is relatively independent of pressure as long as neither the saturation state nor the critical state are approached too closely. Hence the data of Tables A.6 and A.7 may be applied up to pressures of about 10 atm without grave error. Figure 1.5 illustrates some of the data of Tables A.6 and A.7. The increase of conductivity with temperature is typical.

As usual, steam is an irregularly behaving gas, showing relatively strong pressure dependence for conductivity as well as a temperature dependence. Table A.5 gives the data for this important gas in both SI and English units.

1.6 CONVECTION HEAT TRANSFER

The Boundary Layer Concept

The discussion of Sec. 1.3 defined "convection" as the term applied to the heat transfer mechanism which takes place in a fluid because of a combination of conduction within the fluid and energy transport which is due to the fluid motion itself—the fluid motion being produced either by artificial means or by density currents.

Since fluid motion is the distinguishing feature of heat convection, it is necessary to understand some of the principles of fluid dynamics in order to describe adequately the processes of convection. When any real fluid moves past a solid surface, it is observed that the fluid velocity varies from a zero value immediately adjacent to the wall to a finite value at a point some distance away.

Considering the simplest case of flow past a plane surface as shown in Fig.

1.6(a), the fluid velocity will vary from a uniform, "freestream" value at points far away from the surface to zero at the surface—somewhat in the way shown in the figure. For fluids of low viscosity, such as air or water, the region near the surface, in which most of the velocity variation occurs, may be quite thin—depending on the freestream velocity of the fluid. In many applications—such as low-speed aerodynamics, hydraulics, etc.—it is possible to obtain satisfactory results by assuming that the fluid is inviscid (i.e., without viscosity). Hence the flow may be treated as though it slips past the surface with no viscous retardation. However, since the process of convection of heat away from the surface (if the surface is at a temperature different from the free stream of the fluid) is intimately concerned with thermal conduction and energy transport due to motion in the fluid layers in the immediate vicinity of the surface, the simplification of assuming the fluid to be inviscid may not be made when analyses of heat convection are undertaken.

Since the region in which the retarding effect of the fluid viscosity plays a dominent role will often be a very thin layer near the surface, it is possible to simplify the description of the convection process by introducing the concept of the *velocity boundary layer*. The velocity boundary layer is defined as the thin layer near the surface in which one assumes that viscous effects are important. Within this region the effect of the surface on the motion of the fluid is significant. Outside the boundary layer it is assumed that the effect of the surface may be neglected. The exact limit of the boundary layer cannot be precisely defined because of the asymptotic nature of the velocity variation. The limit of the boundary layer is usually taken to be at the distance from the surface at which the fluid velocity is equal to a predetermined percentage of the freestream value. This percentage depends on the accuracy desired, 99 or 95% being customary.

Outside the boundary layer region the flow is assumed to be inviscid. Inside the boundary layer the viscous flow may be either *laminar* or *turbulent*. In the case of laminar boundary layer flow, adjacent fluid layers slide relative to one another but do not mix in the direction normal to the fluid streamlines. Thus any heat that flows from the surface to the freestream fluid does so mostly by conduction—although a transport of energy is also accomplished by virtue of the fact that the fluid has a velocity component normal to the surface. This normal velocity component is caused by the fact that the boundary layer must become progressively thicker as it moves along the surface.

In the event that the fluid motion in the boundary layer is turbulent, the mean flow is essentially parallel to the surface, but it has superimposed upon it a fluctuating motion—in directions both parallel and normal to the surface. The transverse fluctuations cause additional fluid mixing, which increases the rate at which heat is transported in the direction perpendicular to the surface. Typical velocity variations through laminar and turbulent boundary layers are illustrated in Fig. 1.6(a).

If the solid surface is maintained at a temperature, say t_s, which is different from the fluid temperature, t_f, measured at a point far removed from the surface, a variation of the temperature of the fluid is observed which is similar to the velocity variation described above. That is, the fluid temperature varies from t_s at the wall

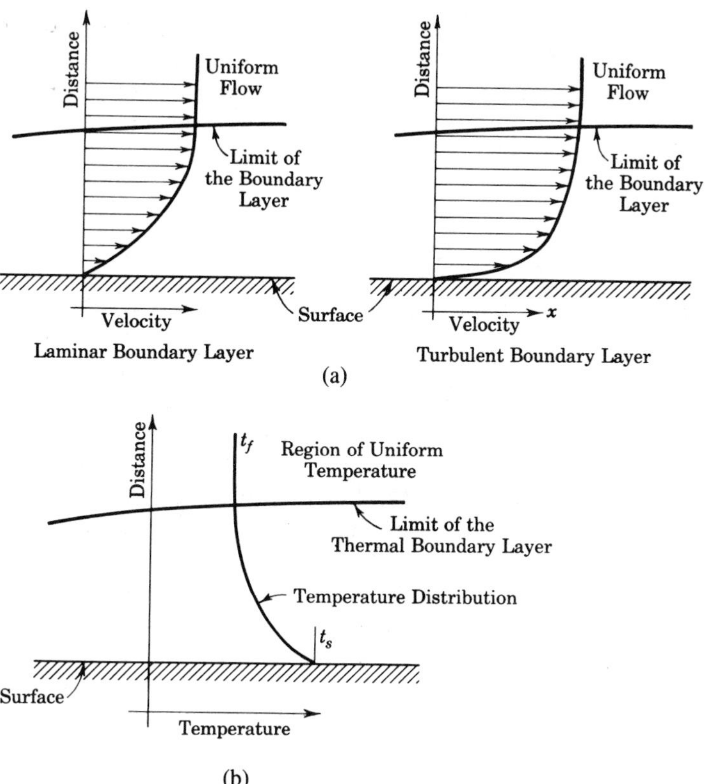

Figure 1.6. Boundary layer flow past a flat surface: (a) the velocity boundary layer; (b) the thermal boundary layer.

to t_f far away from the wall—with most of the variation occurring close to the surface. This is illustrated in Fig. 1.6(b), where it has been assumed that the surface is hotter than the fluid. The fluid temperature approaches t_f asymptotically. However, a *thermal boundary layer* may be defined (in the same sense that the velocity boundary layer was defined above) as the region between the surface and the point at which the fluid temperature has reached a certain percentage of t_f. Outside the thermal boundary layer the fluid is assumed to be a heat sink at a uniform temperature of t_f. The thermal boundary layer is generally not coincident with the velocity boundary layer, although it is certainly dependent on it—i.e., the velocity boundary layer thickness, the variation of the velocity, whether the flow is laminar or turbulent, etc., are all factors which determine the temperature variation in the thermal boundary layer.

Newton's Law of Cooling

Chapters 5 through 8 consider, in some detail, the present state of knowledge of convective heat transfer—from both the theoretical and empirical points of view.

As should be apparent from the discussion, the prediction of the rates at which heat is convected away from a solid surface by an ambient fluid involves a thorough understanding of the principles of heat conduction, fluid dynamics, and boundary layer theory. All the complexities involved in such an analytical approach may be lumped together in terms of a single parameter by introduction of Newton's law of cooling:

$$\frac{q}{A} = h(t_s - t_f). \tag{1.4}$$

The quantity h in this equation is variously known as the *heat transfer coefficient, film coefficient,* or *unit thermal conductance*. It is seen to be (refer to Sec. 1.4) a unit conductance, not a material property as is thermal conductivity. It is a complex function of the composition of the fluid, the geometry of the solid surface, and the hydrodynamics of the fluid motion past the surface. The typical range of values one might expect to encounter for h quoted in Table 1.2 indicates the complexity of the convective process and the difficulty in determining h. Chapters 5 through 8 present detailed analyses of the determination of the heat transfer coefficient.

Before proceeding to the study of the determination of the heat transfer coefficient h, Chapters 2 through 4 concentrate on the solution of the heat conduction equation subjected to various boundary conditions, including the condition of convection to a fluid adjacent to the boundary. In these cases, the boundary condition will be imposed, using Newton's law of cooling, presuming h to be known and given. Following the study of convection in Chapters 5 through 8, some general problems involving the simultaneous solution of the conduction and convection processes will be studied.

The boundary condition to be applied to the solution of a conduction problem when one has a solid bounded by a convecting fluid of known temperature is simply obtained by equating, at the surface, the heat fluxes given by Newton's law in Eq. (1.4) and Fourier's law in Eq. (1.3). Thus, at the convecting surface, the condition to be satisfied by the temperature field in the solid is

$$-\left(\frac{\partial t}{\partial n}\right)_s = \frac{h}{k}(t_s - t_f). \tag{1.5}$$

Table 1.2. Typical Values of the Convective Heat Transfer Coefficient

Situation	h W/m²-°C	h Btu/h-ft²-°F
Free convection in air	5–25	1–5
Free convection in water	500–1000	100–200
Forced convection in air	10–500	2–100
Forced convection in water	100–15,000	20–3000
Boiling water	2500–25,000	500–5000
Condensing steam	5000–100,000	1000–20,000

In Eq. (1.5), n denotes the normal direction to the surface (positive away from the surface), and the subscript s indicates quantities evaluated at the surface. It is important to note that the k in Eq. (1.5) is the thermal conductivity of the *solid,* not the surrounding fluid.

1.7 RADIATION HEAT TRANSFER

As mentioned in Sec. 1.3, it has been observed that a body may gain or lose thermal energy without the need of a physical medium of transport via the mechanism known as thermal radiation. That is, if a body is placed in the presence of cooler surroundings but having no physical contact with them (imagine it suspended in a vacuum), it is observed to lose energy. This energy loss, thermal radiation, is one aspect of a more general phenomenon known as electromagnetic radiation and is electromagnetic emission produced as the result of thermal excitation of the matter comprising the body.

One distinguishing feature of thermal radiation, then, is the lack of a medium of transport as required in the conductive and convective mechanisms. A second distinguishing feature is the effect of the level of the temperature of a thermally radiating body. Ignoring nonlinearities introduced by temperature-dependent properties, the basic rate laws discussed in Sec. 1.4 for conduction (Fourier's law) and Sec. 1.6 for convection (Newton's law of cooling) are *linear* laws. That is, the temperature level does not affect the heat transfer rate by these mechanisms, only the temperature difference. However, the thermal *radiation* emitted by a heated body is proportional to the fourth power of the body temperature. Thus, for a given temperature difference, the heat transfer rate by thermal radiation is much greater at high temperatures than at low temperatures.

While heat exchange by thermal radiation does not depend on a transport medium, the geometric configuration of bodies exchanging heat by this mechanism is quite important. The amount of the radiation emitted by one body that is intercepted by another is highly dependent on the size, shape, and relative orientation of the bodies.

Chapter 9 treats heat exchange by thermal radiation in some detail and all the forementioned effects are discussed in that chapter. It is sufficient, however, for present purposes to quote some of the elementary relations developed in Chap. 9. The basic law of thermal radiation is that the energy emitted by an ideal *blackbody* (a perfectly emitting body) is given by the Stefan–Boltzmann equation:

$$q_r = A\sigma T^4. \tag{1.6}$$

In Eq. (1.6) q_r is the rate of radiant emission, A is the body surface area, T is the absolute temperature of the body surface, and σ is a constant with the value: $\sigma = 5.67 \times 10^{-8}$ W/m^2-°K^4.

Not all radiating bodies behave as the ideal blackbody according to Eq. (1.6). A less-than-perfect emitting surface, called a *gray* surface (described in detail in

Chapter 9), emits according to

$$q_r = A\epsilon\sigma T^4, \tag{1.7}$$

in which ϵ is known as the emissivity of the surface, a physical property less than unity.

The expressions quoted in Eqs. (1.6) and (1.7) for black or gray bodies give only the *emission* of thermal radiation from those bodies. However, in many applications one is interested in the net radiant *exchange* that occurs when there are two or more bodies radiating in the presence of one another. In such instances, the analysis of the radiant exchange is a complex problem involving the geometric shape and location of the bodies as well as their temperatures and surface properties. This is the subject of much of Chapter 9. For purposes of the present discussion it is sufficient to quote one of the results of Chapter 9. If a gray body (temperature T_s, emissivity ϵ_s, surface area A_s) is completely surrounded by another surface which is very large when compared with A_s, the net radiant energy exchanged by the gray body and these surroundings is given by

$$q_r = A_s\epsilon_s\sigma(T_s^4 - T_e^4), \tag{1.8}$$

where T_e represents the absolute temperature of the surrounding surface (i.e., the environment).

Comparable expressions exist for other geometric configurations, but one should not expect such a simple relation in all cases. However, the above will suffice for the present purposes of describing the basic heat transfer mechanisms and some of their general aspects.

1.8 DIMENSIONS AND UNITS

Careful attention to dimensions and units is necessary if one is to obtain correct results in any engineering problem. The English engineering system of units is still widely used in the practice of engineering in the United States. However, the use of SI units (Système International d'Unités) is increasing rapidly in view of the almost universal adoption of these units in the rest of the world. Consequently, the current engineering practitioner must be versatile in both systems. This text will concentrate on the use of SI units, but occasional examples and problems will be given in English units.

The main source of difficulty in dealing with systems of units results from the interrelation of *force* and *mass* through Newton's second law of motion. It is convenient to express Newton's law as

$$F = \frac{1}{g_c} ma,$$

where F denotes force, m denotes mass, a is the acceleration of the mass, and g_c is a dimensional constant. The four physical dimensions: force, mass, length, and time are interrelated in the above law. Presuming, as is always the case, the length and time are fundamental quantities, then one can either make $g_c = 1$ (without dimensions) and define either mass or force in terms of the other, or one can define mass and force as primitive quantities and derive dimensions for g_c. The foregoing possibilities lead to several feasible systems of units. While all of these possibilities have been used in various systems, only those of interest to this text will be discussed: the English engineering system and the SI system.

In addition to the mechanical and kinematic dimensions just discussed, every system of dimensions and units must also include a temperature scale. While not involved in the confusion that often results between the dimensions of mass and force, completeness in the definition of a system of units requires the inclusion of a temperature scale.

The English Engineering System

In the English engineering system of units all four of the dimensions in Newton's law are taken as fundamental, making it necessary to give dimensions to g_c. This, then, is not a consistent system of units. In this system one takes the following fundamental dimensions with the units indicated:

Dimension	Unit	Unit Abbreviation
Force	1 pound force	lb_f
Mass	1 pound mass	lb_m
Length	1 foot	ft
Time	1 second	s
Temperature	degree Rankine	°R

The derived dimensions of g_c are then

$$g_c = \frac{\text{mass} \times \text{length}}{\text{force} \times (\text{time})^2}.$$

If one, in addition, makes the standard arbitrary definition

$$1\ lb_f \text{ will accelerate } 1\ lb_m \text{ at } 32.1739\ ft/s^2,$$

the units of g_c become

$$g_c = 32.1739\,\frac{lb_m\text{-}ft}{lb_f\text{-}s^2}.$$

Usually, a value of $g_c = 32.2(lb_m\text{-}ft)/(lb_f\text{-}s^2)$ is accurate enough. One of the main difficulties encountered in the English system is the fact that when objects are weighed, the weight is a force and the acceleration is that of gravity. At sea level the acceleration of gravity is $g = 32.2\ ft/s^2$. This is the same magnitude as g_c, but certainly does not have the same dimensions. This means that at sea level, 1 lb_m will weigh 1 lb_f—numerically the same, but with different units! This fact produces confusion in the dimensions of certain "specific" quantities such as density, ρ, and specific weight, γ. Density has the dimensions of lb_m/ft^3, whereas specific weight has the dimensions of lb_f/ft^3. Again one has the situation of two quantities being numerically equal but having different dimensions. The relation between the two is

$$\gamma = \frac{g}{g_c}\rho.$$

The same is true for other quantities based on mass or weight, such as specific heat, etc.

In terms of the dimensions and units defined above, the English system adopts a thermal measure of energy. That is, in terms of the above, one can define energy as force × length so that the fundamental measure of energy is the ft-lb_f. However, the English system defines an equivalent thermal unit, the British thermal unit, Btu:

$$1 \text{ Btu} = 778.16 \text{ ft-lb}_f,$$

introducing, then, the additional dimensional constant:

$$J = 778.16 \frac{\text{ft-lb}_f}{\text{Btu}}.$$

Finally, the English system uses the Fahrenheit temperature scale, based on an ice point of 32°F and a steam point of 212°F.

Thus the *derived* quantities in the English system are:

Quantity	Unit	Unit Abbreviation
Energy	1 foot pound force	ft-lb_f
Energy	1 British thermal unit	Btu
g_c	32.2 lb_m-ft/lb_f-s^2	
J	778 ft-lb_f/Btu	
Temperature	degree Fahrenheit = degree Rankine − 459.69	°F

The SI System

In the SI system of units, the dimensions of length, time, and mass are taken as fundamental, the quantity g_c is taken as 1, without dimensions, and force becomes a derived quantity. Further, the Kelvin scale is taken as the measure of temperature. Thus the SI system takes as fundamental the dimensions and units:

Dimension	Unit	Unit Abbreviation
Mass	1 kilogram	kg
Length	1 meter	m
Time	1 second	s
Temperature	degree Kelvin	°K

Mechanical energy is, then, measured in terms of the derived force times length; however, unlike the English system no thermal energy unit is introduced and all energy terms are expressed in mechanical units (i.e., the calorie used in the metric system is not used in the SI system). Thus both force (from Newton's law) and energy (from force × length) are *derived* quantities; although they are given special names for their units. Thus the *derived* units in the SI system are:

Quantity	Unit	Unit Abbreviation
Force	1 newton $= 1 \dfrac{\text{kg-m}}{\text{s}^2}$	N
Energy	1 joule = 1 N-m	J
Power	1 watt = 1 J/s	W
Temperature	degree Celsius = degree Kelvin − 273.15	°C

As may be noted the time rate of energy is given a special unit name, the watt.

The advantages of using the SI system are apparent. No dimensional conversion constants, such as g_c and the mechanical equivalent of heat, are necessary. All units are expressed in terms of the four fundamental units: kilogram, meter, second, and °K.

As noted earlier, the principal emphasis of this text is in the SI system. Because of the still extensive use of English units, some problems and examples will use them. Most of the tables of physical properties, discussed in the next chapter will be given in SI units; however, some of the tabular presentations in the appendices are given in both SI and English units to avoid the necessity of extensive conversions between the systems. Appendix D presents a tabulation of the most commonly needed conversions between these systems of units for quantities commonly encountered in heat transfer calculations.

Following from the foregoing discussion, the units to be used in this text for heat flow rate, q, and heat flux, q/A, are

$$q = \text{W, Btu/s, Btu/h}$$

$$\frac{q}{A} = \text{W/m}^2\text{, Btu/s-ft}^2\text{, Btu/h-ft}^2.$$

Referring to the definition of thermal conductivity, k, in Eq. (1.1) or (1.2) this quantity is expressed in the following units:

$$k = \text{W/m-°C, Btu/h-ft-°F,}$$

while conductance, most notably the heat transfer coefficient h, is expressed as

$$h = \text{W/m}^2\text{-°C, Btu/h-ft}^2\text{-°F.}$$

EXAMPLE 1.1

Figure 1.7 depicts a large slab of material that is 5 cm thick. The inside surface of the slab is maintained at a uniform temperature $t_1 = 150°\text{C}$, and the outside

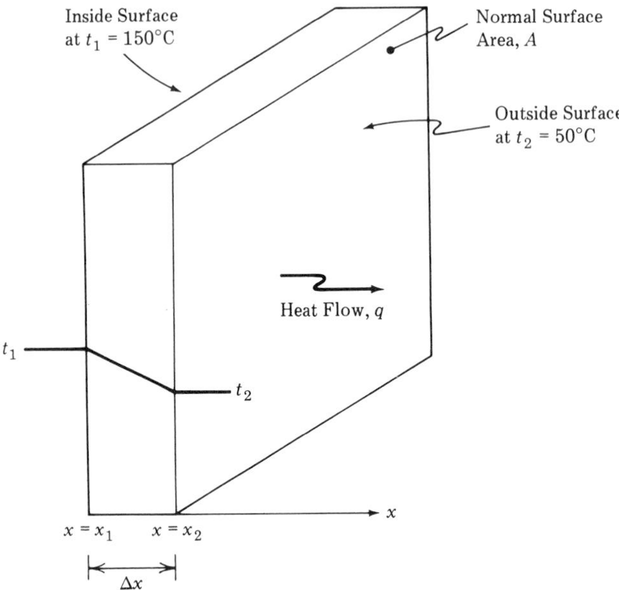

Figure 1.7. The slab of Example 1.1.

surface is maintained at 50°C. Find the heat flow, per unit of normal surface area, if the slab is composed of (a) common brick; (b) wrought iron.

Solution. Assume that the dimensions of the slab are such that the normal surface area is so large that edge effects are negligible and the temperature in the slab may be taken as dependent only on the coordinate x, measured normal to the slab face. In this instance, then, the temperature gradient $\partial t/\partial n$ in Fourier's law [Eq. (1.3)] may be written as a total derivative:

$$\frac{q}{A} = -k\frac{\partial t}{\partial n} = -k\frac{dt}{dx}.$$

For the steady state the heat flux, q/A, is constant, and with the thermal conductivity k also taken as constant, the expression above shows that the gradient of the temperature through the slab, dt/dx, is constant, as the sketch in Fig. 1.7 suggests. The equation above may then be readily integrated from the inside face (where $x = x_1$, $t = t_1$) to the outside face ($x = x_2$, $t = t_2$) to give

$$\frac{q}{A} = -k\frac{t_2 - t_1}{x_2 - x_1} = k\frac{t_1 - t_2}{\Delta x}.$$

This result is the same as that given in Eq. (1.1), used to introduce the concept of thermal conductivity. In the present example $t_1 = 150°\text{C}$, $t_2 = 50°\text{C}$, and $\Delta x = 5\ \text{cm} = 0.05\ \text{m}$.

(a) At an average temperature of the slab material of (150 + 50)/2 = 100°C, Table A.2 gives the thermal conductivity of common brick to be $k = 0.69$ W/m-°C. Thus the heat flux is

$$\frac{q}{A} = 0.69\,\frac{150 - 50}{0.05} = 1380\ \text{W/m}^2\ (437.5\ \text{Btu/h-ft}^2).$$

(b) Similarly, for wrought iron Table A.1 gives $k = 57$ W/m^2-°C, and one obtains

$$\frac{q}{A} = 114{,}000\ \text{W/m}^2\ (36{,}140\ \text{Btu/h-ft}^2).$$

Note the profound effect that the material composition has on the resultant heat flux.

EXAMPLE 1.2

The flat surface of an oven is exposed to ambient air at a temperature of 40°C. The oven surface is covered with a layer of insulation ($k = 1.5$ W/m-°C) 3 cm thick. The exposed surface of the insulation loses heat by convection to the air, and the temperature of the exposed surface is observed to be 45°C. The temperature of the junction between the insulation and the oven surface is measured to be 100°C. What value of the convective heat transfer coefficient must exist at the outer insulation surface?

Solution. Figure 1.8 illustrates the physical situation at hand. The slab of material comprising the insulation is shown with the inside insulation surface at $t_1 = 100$°C, the outside insulation surface at $t_2 = 45$°C, and the ambient air at $t_f = 40$°C. The insulation thickness is $\Delta x = 3$ cm $= 0.03$ m. If, as in Example 1.1, the insulation thickness is considered to be small when compared with the face dimensions of the oven surface, then the conduction through the insulation may be taken to depend only on the normal coordinate, x, with a resultant uniform temperature gradient from t_1 to t_2 as shown in the figure.

The heat flux conducted through the insulation layer is then given by the same relation derived in Example 1.1, or Eq. (1.1),

$$\frac{q}{A} = k\,\frac{t_1 - t_2}{\Delta x}.$$

For the specified values of t_1 and t_2, and the known insulation thermal conductivity $k = 1.5$ W/m-°C, one finds the heat flux to be

$$\frac{q}{A} = 1.5\,\frac{100 - 45}{0.03} = 2750\ \text{W/m}^2.$$

For steady conditions this same heat flux must be dissipated by convection between the exposed surface at $t_2 = 45$°C and the ambient air at $t_f = 40$°C. The

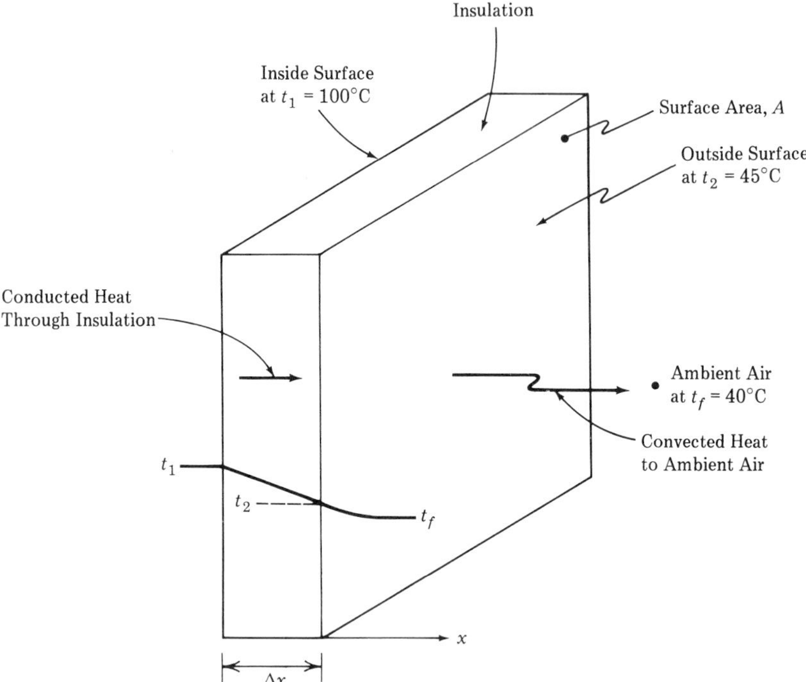

Figure 1.8. The insulation layer of Example 1.2.

temperature in the air adjacent to the insulation surface must vary in some fashion, from t_2 to t_f as suggested in Fig. 1.8. The precise nature of this variation is a complex problem dependent on the motion of the air past the surface. However, Newton's law of cooling [Eq. (1.4)] lumps these hydrodynamic effects into a single parameter, the heat transfer coefficient h:

$$\frac{q}{A} = h(t_s - t_f).$$

In this application the heat flux is known from the earlier calculation for the conduction through the insulation, the surface temperature is known ($t_s = t_2 = 45$°C), as is the ambient air at $t_f = 40$°C. Thus the surface heat transfer coefficient is readily calculated from the equation above:

$$\begin{aligned} 2750 &= h(45 - 40) \\ h &= 550 \text{ W/m}^2\text{-°C (96.9 Btu/h-ft}^2\text{-°F)}. \end{aligned}$$

EXAMPLE 1.3

The flat wall of a furnace, 6 cm thick and composed of a material with $k = 2.5$ W/m-°C, is maintained at 200°C on its inside face. The outside face is exposed to a convective heat loss to ambient air at 35°C through a surface heat transfer coefficient of $h = 50\ \mathrm{W/m^2\text{-}°C}$. The outside face, which is gray with an emissivity $\epsilon = 0.8$, also exchanges radiant heat with the walls of a very large room in which the furnace is located. These walls are at a temperature of 30°C. Find the temperature of the exposed surface of the furnace.

Solution. Figure 1.9 depicts the physical processes taking place. The furnace wall is shown with the inside surface at $t_1 = 200°\mathrm{C}$ and the outside surface at the to-be-determined t_2. The outside surface is exposed to a convective heat exchange with the ambient air at $t_f = 35°\mathrm{C}$ and a simultaneous radiative heat exchange with the room walls at $t_e = 30°\mathrm{C}$.

If, as in Examples 1.1 and 1.2, the thickness of the furnace wall is taken as small when compared with its other dimensions, then the conduction of heat through it may be taken as dependent only on the normal coordinate, x, so that the heat flux by conduction is again given by Eq. (1.1):

$$\left(\frac{q}{A}\right)_{\mathrm{cond}} = k\,\frac{t_1 - t_2}{\Delta x}.$$

The outer surface temperature t_2 is unknown, but it must assume a value so that the surface heat losses by convection and radiation equal that conducted through the wall.

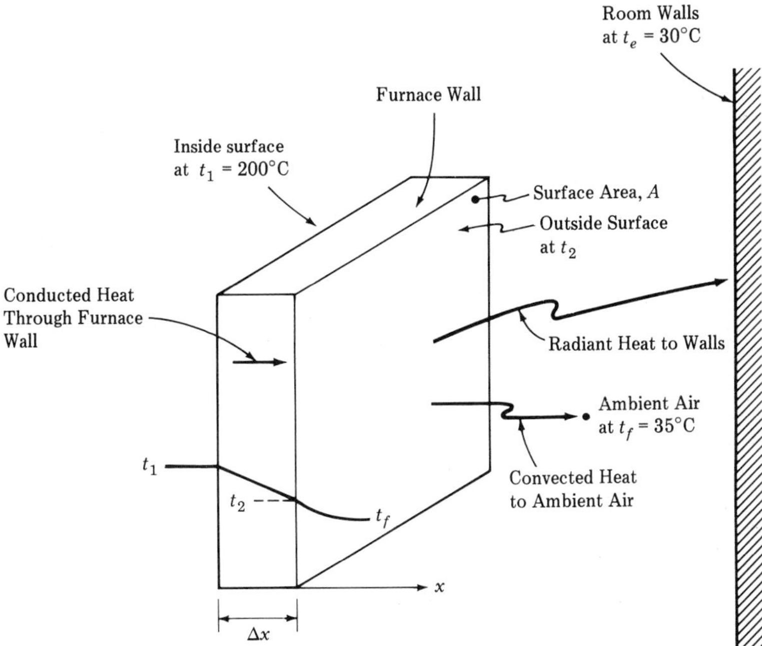

Figure 1.9. The furnace wall of Example 1.3.

The surface heat loss by convection to the ambient air is again expressed by Newton's law of cooling [Eq. (1.4)] in terms of the surface heat transfer coefficient, h, the unknown temperature, t_2, and the ambient air temperature, t_f:

$$\left(\frac{q}{A}\right)_{\text{conv}} = h(t_2 - t_f).$$

If the surrounding room walls are considered as very large when compared to the furnace surface, the radiative flux between the furnace and the walls is given by Eq. (1.8),

$$\left(\frac{q}{A}\right)_{\text{rad}} = \epsilon\sigma(T_2^4 - T_e^4),$$

in which T_2 and T_e represent the absolute values of the surface temperature t_2 and the wall temperature t_e and ϵ is the surface emissivity.

For steady-state conditions, the heat conducted through the wall must be transferred away from the exposed surface by the convective and radiative mechanisms:

$$\left(\frac{q}{A}\right)_{\text{cond}} = \left(\frac{q}{A}\right)_{\text{conv}} + \left(\frac{q}{A}\right)_{\text{rad}}$$

or

$$k\frac{T_1 - T_2}{\Delta x} = h(T_2 - T_f) + \epsilon\sigma(T_2^4 - T_e^4).$$

In the latter expression, absolute temperatures have been used throughout since they *must* be used in calculating the radiant term. In the equation above, all temperatures are known except the surface temperature: $T_1 = 200 + 273 = 473°\text{K}$, $T_f = 35 + 273 = 308°\text{K}$, and $T_e = 30 + 273 = 303°\text{K}$. With the given parameters k, h, ϵ, and Δx as specified, one has

$$2.5\frac{473 - T_2}{0.06} = 50(T_2 - 308) + 0.8 \times 5.67 \times 10^{-8}(T_2^4 - 303^4).$$

This expression is a nonlinear one in the unknown surface temperature T_2. It may be solved by an iterative calculation or by use of the root-finder function available on many pocket calculators. The result is

$$T_2 = 377°\text{K} = 104°\text{C } (219°\text{F}).$$

It is interesting to note the relative magnitudes of the various modes of heat transfer. Using the $T_2 = 377°\text{K}$ just found, application of each of the heat flux relations noted in the foregoing gives

$$\left(\frac{q}{A}\right)_{\text{cond}} = k\frac{T_1 - T_2}{\Delta x} = 2.5\frac{473 - 377}{0.06} = 4000 \text{ W/m}^2,$$

$$\left(\frac{q}{A}\right)_{\text{conv}} = h(T_2 - T_f) = 50(377 - 308) = 3460 \text{ W/m}^2,$$

$$\left(\frac{q}{A}\right)_{rad} = \epsilon\sigma(T_2^4 - T_e^4) = 0.8 \times 5.67 \times 10^{-8}(377^4 - 303^4) = 540 \text{ W/m}^2.$$

It is seen that the radiative mechanism accounts for more than 10% of the heat flux away from the exposed surface.

PROBLEMS

1.1 A plate glass window, 50 cm by 100 cm, is 0.6 cm thick. If the inside and outside surface temperatures are 20°C and 0°C, respectively, find the conduction loss through the window.

1.2 A heat flux of 4000 W/m^2 is conducted through a slab of asbestos insulation, 3 cm thick, that is applied to the walls of a furnace. If the thermal conductivity of the insulation is 0.25 W/m-°C, what is the temperature difference imposed across it?

1.3 The heat conduction rate through a slab of 1.0% carbon steel is observed to be 650 Btu/h-ft^2. If the slab is 10 in. thick and the cooler surface is maintained at 200°F, what is the temperature of the other surface?

1.4 The wall of a furnace is 15 cm thick and has dimensions of 3 m by 2 m. It is composed of fire-clay brick with a thermal conductivity of $k = 1.65$ W/m-°C. Measurement of the surface temperatures shows that the inside surface is at 1120°C and the outside at 875°C. Find the rate of heat loss through the wall.

1.5 A heat flow rate of 5 kW is conducted through a layer of insulating material with a face area of 15 m^2 and a thickness of 3 cm. If the temperature of the hot side of the insulation is 420°C, what is the temperature of the cooler side if the insulation has a thermal conductivity of 0.2 W/m-°C?

1.6 A concrete wall ($k = 0.6$ Btu/h-ft-°F) has a surface area of 300 ft^2 and a thickness of 1 ft. One surface of the wall is maintained at 75°F and the other at 5°F. What is the heat loss through the wall?

1.7 The thermal conductivity of a material is often measured by heating one surface of a slab of the material with an electric heater and cooling the other surface with a plate through which cooling water flows. If the two surface temperatures and the slab thickness are measured, the heat flux may be found from the electrical input and the thermal conductivity calculated. Determine the apparent thermal conductivity of a 5-cm-thick slab of glass wool if one surface is maintained at 30°C, the other at 80°C, and a heat flux of 50 W/m^2 is observed to occur.

1.8 A slab of concrete is tested in the manner described in Prob. 1.7. The slab is 3 in.

thick and a heat flux of 50 Btu/h-ft^2 is measured when one surface is maintained at 80°F and the other at 60°F. What is the thermal conductivity of the concrete?

1.9 The walls of a freezer compartment are to be insulated with hair felt (density: 12.8 lb$_m$/ft^3). It is desired to limit the heat flow into the freezer to a flux of 5 Btu/h-ft^2 when the inner and outer insulation surface temperatures are 15°F and 75°F, respectively. Determine the required thickness of the insulation.

1.10 Determine the instantaneous rate of heat transfer from a 1.5-cm-diameter ball bearing with a surface temperature of 150°C submerged in an oil bath at 75°C if the surface convective heat transfer coefficient is 850 W/m^2-°C.

1.11 Air at 480°F flows over a flat surface 20 in. by 12 in. in size. The surface is maintained at 75°F and the convective heat transfer coefficient at the surface is 45 Btu/h-ft^2-°F. Find the rate of heat transfer to the surface.

1.12 What is the heat transfer coefficient due to free convection between the walls of a room and the ambient air if a heat flux of 20 W/m^2 is observed when the air temperature is 32°C and the wall temperature is 35°C?

1.13 A steam pipe 30 cm in diameter and 5 m long passes through a room where the air temperature is 20°C. If the exposed surface of the pipe is uniformly at 40°C, find the rate of heat loss from the pipe into the air if the surface heat transfer coefficient is 8.5 W/m^2-°C.

1.14 A spherical tank used to store petroleum products has a diameter of 10 ft. The products stored in the tank keep the exposed surface at 75°F. Air at 60°F blows over the surface of the tank, producing a heat transfer coefficient of 10 Btu/h-ft^2-°F at that surface. Determine the rate of heat transfer from the tank.

1.15 The heat transfer coefficient for water flowing normal to a cylinder is to be measured by passing water over a cylindrically shaped electric resistance heater. The heater has a diameter of 3 cm and a length of 0.5 m. When water at a temperature of 25°C flows across the cylinder with a velocity of 1 m/s it is found that 30 W of electrical power are required to maintain the heater surface temperature at 60°C. Estimate the heat transfer coefficient existing at the heater surface. What is the significance of the water velocity in determining the answer?

1.16 A cylindrical electric resistance heater has a diameter of 1 cm and a length of 0.25 m. When water at 30°C flows across the heater a heat transfer coefficient of 15 W/m^2-°C is observed to exist at the surface. If the electrical input to the heater is 5 W, what is the surface temperature of the heater?

1.17 A typical value of the heat transfer coefficient for water boiling on a flat surface is 900 Btu/h-ft^2-°F. Estimate the heat flux on such a surface when the surface is maintained at 222°F and the water is at 212°F.

1.18 A layer of insulation (k = 1.5 W/m-°C) is applied to a plane surface so that the inner surface of the insulation is maintained at 150°C. The insulation is 2.5 cm thick and its outer surface is exposed to ambient air at 35°C to which it loses heat by convection. Find the value of the surface heat transfer coefficient that will ensure that the surface temperature will not exceed 45°C. What is the heat flux through the insulation?

1.19 The wall of a building is 8 in. thick and is composed of common brick. This inside surface of the wall is maintained at 100°F. The outer surface of the wall loses heat to ambient air by convection. If the surface heat transfer coefficient is 5 Btu/h-ft^2-°F at the outer surface and the air temperature is 40°F, find the heat flux through the wall and the temperature of the exposed surface.

1.20 The accompanying figure depicts a case of combined conduction and convection that is frequently encountered in practical heat transfer calculations. A plane wall, such as that of a building, is exposed to convective heat transfer at each of its two surfaces. The wall has a thickness Δx and a thermal conductivity k. The inside surface of the wall is exposed to convective heat transfer with a fluid at temperature t_{f_i} through a heat transfer coefficient h_i. The other side of the wall also experiences convective heat transfer to a second fluid at t_{f_o} through a heat transfer coefficient h_o. Show that, in terms of the overall temperature potential, $t_{f_i} - t_{f_o}$, the heat flux through the wall is given by

$$\frac{q}{A} = \frac{t_{f_i} - t_{f_o}}{1/h_i + \Delta x/k + 1/h_o}.$$

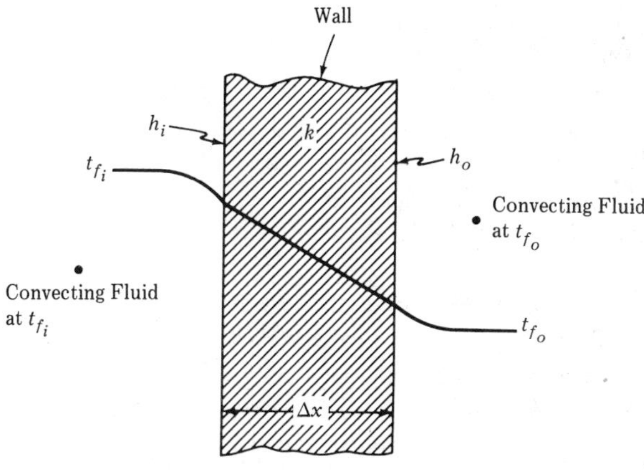

Problem 1.20

1.21 The wall of a building, exposed to air on each side, is an example of the situation described in Prob. 1.20. Consider a wall made of common brick that is 10 cm thick. One side is exposed to air at 40°C through a convective heat transfer coefficient of 30 W/m^2-°C and the other side to air at 32°C through a heat transfer coefficient of 8 W/m^2-°C. Find the heat flux through the wall and the temperature of each surface.

1.22 A slab of material (k = 2.5 W/m-°C) is 12 cm thick. One surface is maintained at 100°C. The outer surface is exposed to a convecting fluid at 70°C through a heat transfer coefficient of 150 W/m^2-°C. What is the temperature of the surface exposed to the fluid? What is the heat flux through the slab?

1.23 The dividing surface in a simple heat exchanger is a stainless steel (18% Cr, 8% Ni) plate 0.6 cm thick. Hot combustion gases flowing past one side of the plate are at 425°C and produce a convective heat transfer coefficient of 48.8 W/m^2-°C at that surface. Air at 60°C flows past the other side of the plate, producing a convective heat transfer coefficient of 36.9 W/m^2-°C at that surface. Find the heat flux exchanged between the two fluids.

1.24 Find the radiant energy emitted by an ideal blackbody, 10 m^2 in area, if it is maintained at **(a)** 50°C, **(b)** 100°C, or **(c)** 500°C.

1.25 A surface is maintained at 100°C and is enclosed by very large surrounding surfaces at 80°C. Find the radiant flux lost by the surface if its emissivity is **(a)** 1.0 or **(b)** 0.8.

1.26 Repeat Prob. 1.25 if the surface temperature is raised to 150°C, all other data remaining unchanged.

1.27 Idealize a 75-W light bulb as a sphere 7 cm in diameter. Further, assume that only 10% of the power input to the bulb is converted to light, the remaining being absorbed into the glass bulb. Estimate the surface temperature of the bulb **(a)** if the only heat loss from the bulb surface is that due to convection to ambient air at 20°C through a surface heat transfer coefficient of 8 W/m^2-°C, and **(b)** if the heat loss from the bulb surface is taken to be convection, as above, *plus* that due to radiation. Assume that the glass emissivity is 0.9 and that the radiant environment is a large room also at 20°C.

1.28 A furnace wall, made of brick with a thermal conductivity of 1.3 W/m-°C, is 15 cm thick. The outside surface of the wall loses heat by convection to ambient air at 25°C through a surface heat transfer coefficient of 25 W/m^2-°C. The surface also loses heat by radiation to a very large room the walls of which are also at 25°C. The emissivity of the outside surface is 0.8 and the temperature of that surface is measured to be 100°C. What is the temperature of the inside surface of the brick?

1.29 The surface of the wall of a house, 30 m^2 in area, maintained at 80°C is exposed to a convective loss to air at 40°C through a heat transfer coefficient of 50 W/m^2-°C. The surface simultaneously exchanges heat by radiation to very large surroundings also at 40°C. If the surface emissivity is 0.75, find the combined heat loss due to convection and radiation.

1.30 Repeat Prob. 1.22 if the side of the wall exposed to the convecting fluid also exchanges radiant heat with surroundings at 50°C. The surface emissivity is 0.8.

CHAPTER 2

Steady-State Heat Conduction in One Dimension

2.1 THE GENERAL HEAT CONDUCTION EQUATION

The expressions of Fourier's conduction law given in Section 1.4 may be used to develop an equation describing the temperature distribution throughout a heat-conducting solid. In general, a heat conduction problem consists of finding the temperature at any time and at any point within a specified solid which has been heated to a known initial temperature distribution and then subjected to known boundary conditions on its surface.

The development that follows will consider the case in which the heat-conducting solid may also have internal sources of heat generation. Such sources may be the result of chemical or nuclear reaction, electrical dissipation, etc., and may be either concentrated at certain locations or distributed throughout the solid.

To develop the differential equation governing this problem, consider a solid, as shown in Fig. 2.1, and select arbitrarily three mutually perpendicular coordinate directions, x, y, and z. Select in the solid a parallelepiped of dimensions Δx, Δy, and Δz. By making an energy balance on this element between the heat conducted in and out of its six faces, the heat stored within, and the heat generated within the element, an expression interrelating the temperatures throughout the solid is obtained. First, consider heat conduction in the x direction only. Let f_x denote the flux of heat conducted into the element through the left yz face and f_x' the flux of heat conducted out the right yz face. If f_x' is written in terms of f_x by use of a Taylor's expansion, then

$$f_x' = f_x + \left(\frac{\partial f_x}{\partial x}\right) \Delta x + \left(\frac{\partial^2 f_x}{\partial x^2}\right) \frac{(\Delta x)^2}{2!} + \cdots .$$

The excess rate at which heat is conducted into the element over that conducted out is $(f_x - f_x') \Delta y \Delta z$. Thus the net energy flow into the element due to conduction

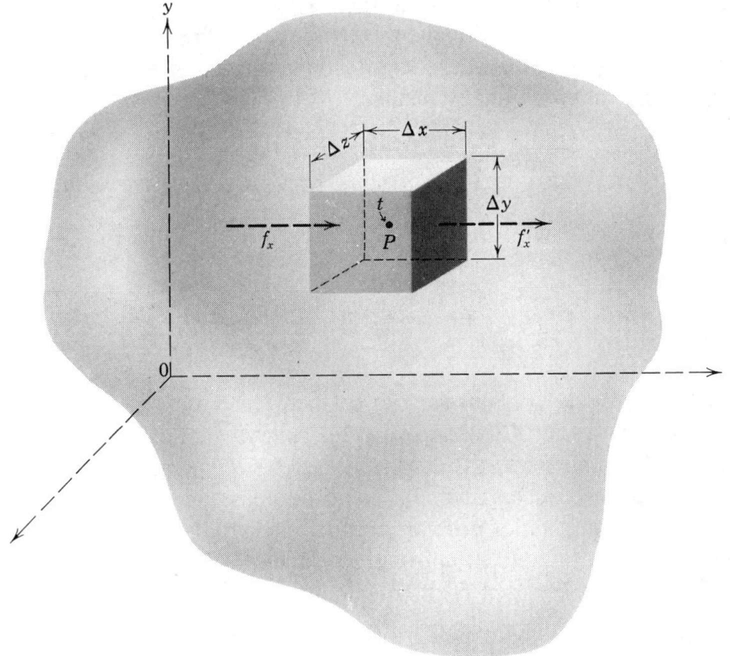

Figure 2.1

in the x direction is

$$-\left[\frac{\partial f_x}{\partial x}+\left(\frac{\partial^2 f_x}{\partial x^2}\right)\frac{\Delta x}{2!}+\cdots\right]\Delta x\,\Delta y\,\Delta z.$$

Similar expressions may be written for conduction in the other two coordinate directions, so that the total *net* rate of conduction *into* the element is

$$-\left[\frac{\partial f_x}{\partial x}+\frac{\partial f_y}{\partial y}+\frac{\partial f_z}{\partial z}+\frac{\partial^2 f_x}{\partial x^2}\frac{\Delta x}{2!}+\frac{\partial^2 f_y}{\partial y^2}\frac{\Delta y}{2!}+\frac{\partial^2 f_z}{\partial z^2}\frac{\Delta z}{2!}+\cdots\right]\Delta x\,\Delta y\,\Delta z. \tag{2.1}$$

If q^* is used to denote the rate at which heat is being internally generated in the element, *per unit volume*, then

$$q^*(\Delta x\,\Delta y\,\Delta z) \tag{2.2}$$

is the total heat released in the element.

The energies introduced in the element as given in Eqs. (2.1) and (2.2) must equal the rate of heat storage which is reflected in the time rate of change of the average temperature of the element t_{av}, as given by

$$\rho c_p\,\frac{\partial t_{\text{av}}}{\partial \tau}\,\Delta x\,\Delta y\,\Delta z. \tag{2.3}$$

Energy conservation requires that Eq. (2.3) equal the sum of Eqs. (2.1) and (2.2). Upon making this equality; dividing by the volume of the element, $\Delta x\ \Delta y\ \Delta z$; allowing Δx, Δy, and $\Delta z \rightarrow 0$; and noting that under these conditions $t_{av} \rightarrow t$ (the temperature at point P), one finally obtains

$$\rho c_p \frac{\partial t}{\partial \tau} = -\left(\frac{\partial f_x}{\partial x} + \frac{\partial f_y}{\partial y} + \frac{\partial f_z}{\partial z}\right) + q^*.$$

According to Fourier's law given in Eq. (1.2), each of the fluxes in the equation above is proportional to the temperature gradient in each particular direction. Thus

$$\rho c_p \frac{\partial t}{\partial \tau} = \frac{\partial}{\partial x}\left(k \frac{\partial t}{\partial x}\right) + \frac{\partial}{\partial y}\left(k \frac{\partial t}{\partial y}\right) + \frac{\partial}{\partial z}\left(k \frac{\partial t}{\partial z}\right) + q^*. \tag{2.4}$$

Equation (2.4) represents a *volumetric* heat balance which must be satisfied at each point in the body. This expression, known as the *general heat conduction equation,* describes, in differential form, the dependence of the temperature in the solid on the spatial coordinates and on time. It should be noted that the heat generation term, q^*, may be a function of position or time, or both.

In the form given in Eq. (2.4) allowance is made for the possible dependence of the thermal conductivity k on position within the solid (or on temperature and, hence, on position). If the thermal conductivity is treated as constant, an assumption the significance of which will be discussed in more detail later, the following special form is obtained:

$$\rho c_p \frac{\partial t}{\partial \tau} = k\left(\frac{\partial^2 t}{\partial x^2} + \frac{\partial^2 t}{\partial y^2} + \frac{\partial^2 t}{\partial z^2}\right) + q^*. \tag{2.5}$$

Frequently, Eq. (2.5) is written in the following form, which, although useful, destroys the volumetric representation of each term:

$$\frac{\partial t}{\partial \tau} = \alpha\left(\frac{\partial^2 t}{\partial x^2} + \frac{\partial^2 t}{\partial y^2} + \frac{\partial^2 t}{\partial z^2}\right) + \frac{q^*}{\rho c_p}. \tag{2.6}$$

The quantity

$$\alpha = \frac{k}{\rho c_p} \tag{2.7}$$

is called the *thermal diffusivity* and is seen to be a physical property of the material of which the solid is composed. This property is discussed in more detail later.

Other Coordinate Systems

The discussion above was carried out in terms of rectangular coordinates. It is often useful to write the equation in cylindrical or spherical coordinates. The results are, for the case of constant k:

Cylindrical coordinates, r (radius), z (axis), θ (longitude):

$$\frac{\partial t}{\partial \tau} = \alpha\left(\frac{\partial^2 t}{\partial r^2} + \frac{1}{r}\frac{\partial t}{\partial r} + \frac{1}{r^2}\frac{\partial^2 t}{\partial \theta^2} + \frac{\partial^2 t}{\partial z^2}\right) + \frac{q^*}{\rho c_p}. \tag{2.8}$$

Spherical coordinates, r (radius), φ (longitude), θ (colatitude):

$$\frac{\partial t}{\partial \tau} = \alpha\left[\frac{1}{r^2}\frac{\partial}{\partial r}\left(r^2\frac{\partial t}{\partial r}\right) + \frac{1}{r^2 \sin\theta}\frac{\partial}{\partial \theta}\left(\sin\theta\frac{\partial t}{\partial \theta}\right) + \frac{1}{r^2\sin^2\theta}\frac{\partial^2 t}{\partial \varphi^2}\right] + \frac{q^*}{\rho c_p}. \tag{2.9}$$

The proof of these equations will be left as exercises for the reader.

The Steady State

A particularly useful special case of Eq. (2.6) is one which has a very wide range of application in engineering. This is the *steady state*, in which there is no dependence on time. The heat conduction equation then reduces to Poisson's equation: In rectangular coordinates this is

$$\alpha\left(\frac{\partial^2 t}{\partial x^2} + \frac{\partial^2 t}{\partial y^2} + \frac{\partial^2 t}{\partial z^2}\right) + \frac{q^*}{\rho c_p} = 0.$$
$$\frac{\partial^2 t}{\partial x^2} + \frac{\partial^2 t}{\partial y^2} + \frac{\partial^2 t}{\partial z^2} + \frac{q^*}{k} = 0. \tag{2.10}$$

In the absence of internal heat generation, Laplace's equation is obtained:

$$\frac{\partial^2 t}{\partial x^2} + \frac{\partial^2 t}{\partial y^2} + \frac{\partial^2 t}{\partial z^2} = 0. \tag{2.11}$$

Thermal Diffusivity

In the preceding development of the general heat conduction equation an important property, the thermal diffusivity, was defined in Eq. (2.7):

$$\alpha = \frac{k}{\rho c_p}$$

The thermal diffusivity expresses the magnitude of the ratio of a body's conducting capability to its heat-storing capability. As such the diffusivity plays an essential role in the analysis of time-dependent conduction problems, discussed in some detail in Chapter 3. As will be seen in Chapter 5, the thermal diffusivity is also an important parameter in the analysis of fluid convection.

Introduction of the dimensions of thermal conductivity discussed in Sec. 1.8 into Eq. (2.7) along with the known dimensions of density and specific heat yields the fact that the dimensions of thermal diffusivity are

$$\alpha = \frac{(\text{length})^2}{\text{time}}$$
$$= \text{m}^2/\text{s},\ \text{ft}^2/\text{h}.$$

The units noted above are those commonly used in the SI and English systems. It is interesting to note that although it is a *thermal* property, diffusivity involves no thermal dimensions or units—only kinematic ones.

Tables A.1 and A.2 tabulate values of the thermal diffusivity of the solids noted. Tables A.3 and A.4, similarly show values of α for liquids, and these values may be taken as being fairly independent of pressure since k, ρ, and c_p are virtually pressure independent for liquids. The thermal diffusivity of gases is necessarily pressure dependent since gas densities are strong functions of pressure. Thus α is not tabulated for the gases given in Tables A.5 through A.7. The tabulated values shown for k and c_p of gases may be taken as virtually pressure independent and α calculated from its definition, $\alpha = k/\rho c_p$, after ρ is found from an appropriate equation of state or tables of thermodynamic properties.

2.2 THE STEADY STATE AND ONE-DIMENSIONAL SIMPLIFICATIONS

The various forms of the general heat conduction equation given in Eqs. (2.4) through (2.9) express the temperature in a heat-conducting body as a function of time and up to three spatial variables. Various simplifications may be made to reduce the number of independent variables involved—thereby reducing the complexity of seeking solutions. One such simplifying assumption that has already been mentioned is that of the *steady state*. In this instance one presumes that the conditions to which the body is subjected are such that the temperature at any location is independent of time. The associated heat flows are then also independent of time. With the assumption of the steady state, the heat conduction equation assumes the simpler forms given in Eqs. (2.10) and (2.11), in which only the spatial coordinates are involved.

The steady-state problem may be simplified further by assuming the existence of *one-dimensional* conduction. This term is applied when only one space coordinate is required to describe the distribution of temperature within the body. Such a situation rarely exists precisely in real problems, but a great number of practical engineering problems may be approximated rather well by such an assumption.

The flow of heat through a plane wall with uniform temperatures imposed on its surfaces may, at points well removed from the edges, be presumed to depend only on the coordinate measured normal to the plane of the wall. Similarly, conduction through a very long hollow cylinder (such as a pipe) maintained at uniform inside and outside temperatures may be considered to depend only on a single coordinate, the radial distance. A very thin rod, or wire, maintained at different temperatures at its ends may be considered to conduct heat one-dimensionally along its length if it is sufficiently thin that its temperature may be taken as uniform over any cross section.

If a system may be taken as both one-dimensional and in the steady state, the conduction equation (2.4) reduces to a considerably simpler, ordinary, differential equation

$$\frac{d}{dx}\left(k\,\frac{dt}{dx}\right) + q^* = 0$$

or its equivalent in other coordinate systems.

The remainder of this chapter is devoted to the solution of steady one-dimensional problems. Cases with and without internal heat generation will be studied, and the boundary conditions of known surface temperature and convection to an ambient fluid will be considered.

2.3 STEADY CONDUCTION IN A PLANE WALL

As mentioned in the foregoing, the simplest one-dimensional steady conduction problem is that depicted in Fig. 2.2. Shown is a plane wall of thickness Δx maintained at known temperatures t_1 and t_2 on its two faces, with t_1 assumed greater than t_2 for purposes of discussion. The wall is taken as very large in extent in the directions other than normal to its face so that the temperature may be taken as dependent only on the normal coordinate, x. Assuming the thermal conductivity, k, to be constant (an assumption to be examined in more detail shortly), the preceding equation becomes

$$\frac{d^2t}{dx^2} = 0. \tag{2.12}$$

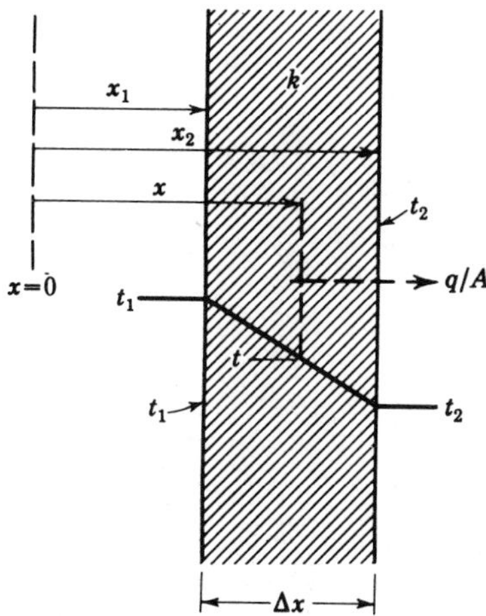

Figure 2.2

This equation may be readily integrated twice to yield the temperature within the wall as a function of the one-dimensional coordinate x and two integration constants:

$$t = B_1 x + B_2. \tag{2.13}$$

Using x_1 and x_2 to denote the values of x at the two faces of the wall, the constants B_1 and B_2 are easily found from applying the specified boundary temperatures at these two locations:

$$t = t_1 \quad \text{at } x = x_1,$$
$$t = t_2 \quad \text{at } x = x_2.$$

When B_1 and B_2 are thus evaluated and eliminated from the Eq. (2.13), one finds the expected linear distribution of temperature in the wall:

$$t = t_1 + (t_2 - t_1)\frac{x - x_1}{x_2 - x_1}. \tag{2.14}$$

Since the one-dimensional representation in this instance presumes that the temperature depends only on the normal coordinate, the temperature gradient, dt/dx, at any location x is also single valued. That is, on any plane within the wall parallel to the bounding surfaces dt/dx is uniform, and Fourier's law in the form of Eq. (1.3) is applicable:

$$q = -kA\frac{dt}{dx}.$$

For the steady state, the heat flow rate q must be the same at any location x, and since the area available for heat flow, A, is the same at every location in a plane wall, the relation above verifies the fact just derived—that the gradient x is not only uniform over every interior plane, it is also the same at every value of x. The gradient dt/dx is readily evaluated from Eq. (2.14),

$$\frac{dt}{dx} = \frac{t_2 - t_1}{x_2 - x_1},$$

so that Fourier's law gives the heat flow rate to be

$$q = -kA\frac{dt}{dx} = -kA\frac{t_2 - t_1}{x_2 - x_1} = kA\frac{t_1 - t_2}{\Delta x}$$

or

$$\frac{q}{A} = k\frac{t_1 - t_2}{\Delta x}. \tag{2.15}$$

The latter form, in terms of the heat flux q/A, is preferred since the one-dimensional assumption actually implies an infinitely large surface area. Thus the heat flux is a more meaningful physical parameter than is the total heat flow, q.

For a given wall of known k and thickness Δx, Eqs. (2.14) and (2.15) allow the determination of the heat flux and the temperature at any interior location when the surface temperatures are specified. These relatively simple results depend on the presumption that the thermal conductivity is constant. The following discussion examines the case when this is not true.

The Effect of Variable Thermal Conductivity

The thermal conductivity, k, was taken to be constant in the discussion just concluded in order to simplify the analysis. However, the discussion of Sec. 1.5 noted the fact that almost all substances show some temperature dependence of thermal conductivity. Examination of the tables of Appendix A, or Figs. 1.3 through 1.5, will show that for most substances, particularly solids, a linear dependence of the thermal conductivity on temperature exists, at least for limited ranges of temperature. Thus one may write k as the following linear function of temperature:

$$k = k_0(1 + bt), \tag{2.16}$$

in which k_0 is the conductivity at $t = 0$ and b is another physical constant indicating the rate of change of k with t.

The physical effects of a linearly varying conductivity are best illustrated by again considering one-dimensional conduction in a plane wall of specified surface temperatures as illustrated in Fig. 2.3. In this instance it is more physically instructive to work directly from Fourier's law rather than integrate the general conduction equation. Writing Eq. (1.3) in terms of the heat flux, q/A, since both q and A are constant,

$$\frac{q}{A} = -k\frac{dt}{dx}.$$

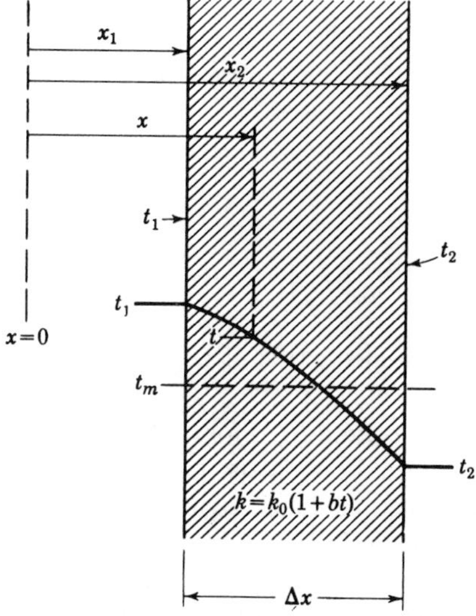

Figure 2.3

Introduction of the linear relation for k gives

$$\frac{q}{A} = -k_0(1 + bt)\frac{dt}{dx}. \tag{2.17}$$

Since q/A is constant in the steady-state plane wall, one may note from Eq. (2.17) that the temperature gradient will not remain constant through the wall as in the case of constant k. If, as illustrated in Fig. 2.3, the temperature decreases through the wall and if the constant b is positive, Eq. (2.16) shows that dt/dx must increase with x. The decreasing temperature t implies a decreasing conductivity, so that dt/dx must increase with x to maintain q/A constant as required by Fourier's law.

Proceeding more formally, Eq. (2.17) may be separated and integrated between the surface at $x = x_1$ and some generic location x:

$$\frac{q}{A}\int_{x_1}^{x} dx = -k_0\int_{t_1}^{t}(1 + bt)\,dt,$$

$$\frac{q}{A}(x - x_1) = -k_0\left[t + \frac{b}{2}t^2\right]_{t_1}^{t}.$$

The result above shows that the temperature distribution is a quadratic function of x instead of linear as in the constant-k case. The heat flux through the wall may be expressed in terms of the two boundary temperatures by evaluating the expression above at $x = x_2$ where the other temperature t_2 is specified:

$$\frac{q}{A}(x_2 - x_1) = -k_0\left[\left(t_2 + \frac{b}{2}t_2^2\right) - \left(t_1 + \frac{b}{2}t_1^2\right)\right],$$

which may be rewritten as

$$\frac{q}{A} = -k_0\left(1 + b\frac{t_1 + t_2}{2}\right)\frac{t_2 - t_1}{x_2 - x_1}.$$

This equation is equivalent to

$$\frac{q}{A} = k_m\frac{t_1 - t_2}{\Delta x},$$
$$k_m = k_0\left(1 + b\frac{t_1 + t_2}{2}\right), \tag{2.18}$$

in which k_m is the thermal conductivity evaluated at the mean of the two imposed surface temperatures.

Equation (2.18) is exactly the same form as Eq. (2.15) for the wall with constant k. Thus, as far as the determination of the heat flow rate is concerned, the solution for a constant k may be applied when the thermal conductivity is a linear function of temperature as long as the conductivity is evaluated at the mean of the imposed surface temperatures. In the case of multilayer walls, to be discussed next, it may not be possible to know the mean temperature a priori, and an iterative calculation could be required.

The Multilayer Wall

Consider now a plane wall composed of layers of materials having different thicknesses and thermal conductivities. Figure 2.4 illustrates a wall of three layers for the purposes of discussion, although any number of layers may be considered.

Denote each junction of two different materials by a number (1, 2, 3, 4 in Fig. 2.4) and adopt the following notational convention:

Δx_{mn} = thickness of material between plane m and plane n,

k_{mn} = thermal conductivity of material between plane m and plane n.

If the outside temperatures are specified (i.e., t_1 and t_4) and if one knows the thickness and conductivities of all the materials between the planes where these temperatures are specified, what is the distribution of the temperature through the wall and what is the rate of heat flow through the wall?

Since the steady state is assumed to exist, Eq. (2.15) may be applied to obtain the following set of equations since the rate of heat flow per unit area, q/A, is the same for each layer:

$$\frac{q}{A} = \frac{t_1 - t_2}{\Delta x_{12}/k_{12}} = \frac{t_2 - t_3}{\Delta x_{23}/k_{23}} = \frac{t_3 - t_4}{\Delta x_{34}/k_{34}}.$$

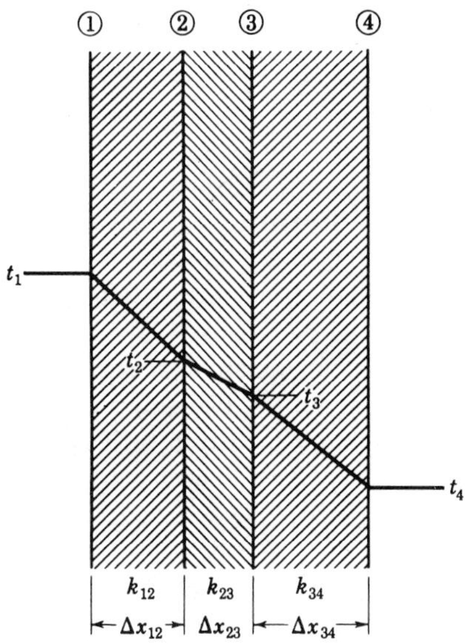

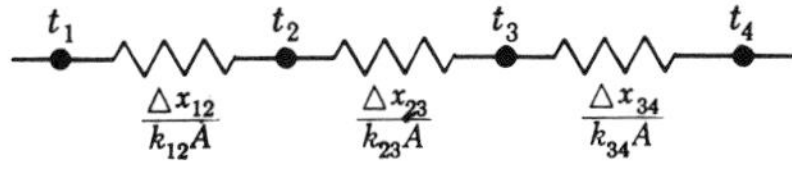

Figure 2.4

Solving for each of the temperature differences, one obtains

$$\begin{aligned} t_1 - t_2 &= \frac{q}{A}\frac{\Delta x_{12}}{k_{12}}, \\ t_2 - t_3 &= \frac{q}{A}\frac{\Delta x_{23}}{k_{23}}. \\ t_3 - t_4 &= \frac{q}{A}\frac{\Delta x_{34}}{k_{34}}. \end{aligned} \tag{2.19}$$

Upon addition of these equations the unknown temperatures are eliminated, and the following expression for the heat flow through the wall in terms of the overall temperature difference may be obtained:

$$\frac{q}{A} = \frac{t_1 - t_4}{\dfrac{\Delta x_{12}}{k_{12}} + \dfrac{\Delta x_{23}}{k_{23}} + \dfrac{\Delta x_{34}}{k_{34}}}. \tag{2.20}$$

Once the heat flow rate is obtained from Eq. (2.20), one may determine the temperatures t_2 and t_3 from Eq. (2.19). The distribution of temperature is then known to be linear between the known values.

The application of Eq. (2.20), or its equivalent for a different number of wall layers, may offer some difficulty if the materials involved have thermal conductivities showing a strong temperature dependence. In general, if the boundary temperatures t_1 and t_4 are specified, the intermediate temperatures t_2 and t_3 needed to evaluate mean values of the conductivities k_{12}, k_{23}, and k_{34} are not known in advance. In such an instance one would have to assume reasonable values for k_{12}, k_{23}, and k_{34} and use these values to find q/A, t_2, and t_3 as noted in the preceding paragraph. Then the mean temperature of each layer may be determined and the correctness of the assumed conductivities ascertained by reference to the tables of Appendix A. The process could be repeated until a satisfactory agreement is obtained. However, unless there is a strong dependence of the thermal conductivity on temperature and the imposed temperature differences are large such a procedure may not be justified—depending on the accuracy of the known conductivity data and the given data. Engineering judgment has to be applied in such instances.

Thermal Resistance and Thermal Conductance

In Sec. 1.4 a unit thermal conductance was defined:

$$C = \frac{k}{\Delta x}.$$

A unit thermal resistance R_k could be defined as the reciprocal of the unit conductance:

$$R_k = \frac{1}{C} = \frac{\Delta x}{k},$$

which would lead to an analogy between electrical and thermal conduction. The subscript k is used to emphasize the fact that the resistance defined is a *conduction* resistance. However, a clearer analogy may be obtained if one defines a *total* thermal resistance

$$\mathcal{R}_k = \frac{1}{CA} = \frac{\Delta x}{kA}.$$

On the basis of this definition, the total heat flow through a multilayer wall is

$$\begin{aligned} q &= \frac{\Delta t_{\text{overall}}}{\mathcal{R}_{k12} + \mathcal{R}_{k23} + \cdots}, \\ \frac{q}{A} &= \frac{\Delta t_{\text{overall}}}{A(\mathcal{R}_{k12} + \mathcal{R}_{k23} + \cdots)}, \\ \frac{q}{A} &= \frac{\Delta t_{\text{overall}}}{R_{k12} + R_{k23} + \cdots}. \end{aligned} \tag{2.21}$$

Introduction of the definition of $\mathcal{R}$ or R will yield Eq. (2.20).

Figure 2.4 depicts the equivalent electrical network implied by the above analysis. One sees that thermal resistance in a series should be added—as in the electrical equivalent. For this plane wall, the area, A, is the same for each layer, and the unit resistance could have been employed. However, for instances in which the areas might not be the same (in the case of parallel circuit, for instance), the formulation given above is necessary.

EXAMPLE 2.1

A house wall consists of an outer layer of common brick 10 cm thick (k = 0.69 W/m-°C), followed by a 1.25-cm layer of Celotex sheathing (k = 0.048 W/m-°C). A 1.25-cm layer of sheetrock (k = 0.744 W/m-°C) forms the inner surface and is separated from the sheathing by 10 cm of air space—as provided by the wall studs. The air space has a *unit conductance* of 6.25 W/m^2-°C. The outside brick temperature is 5°C; the inner wall surface is maintained at 20°C. What is the rate of heat loss, per unit area of wall? What is the temperature at a point midway through the Celotex layer?

Solution. Let the house wall be presumed to be large with respect to its thickness. Then the conduction through it may be treated as one-dimensional and the principles developed in the preceding section may be applied. If the inside surface is denoted by 1, the sheetrock as layer 1–2, the air space as 2–3, the Celotex as 3–4, the brick as 4–5, and the outer surface as 5, then Eq. (2.21) gives the heat flux through the wall in terms of the resistances of each layer:

$$\frac{q}{A} = \frac{t_1 - t_5}{R_{k12} + R_{23} + R_{k34} + R_{k45}}.$$

The conduction resistances are each given by $R_k = \Delta x/k$, while the resistance of the air space is the reciprocal of its conductance, $R_{23} = 1/C_{23}$. In these terms, the heat flux is

$$\frac{q}{A} = \frac{t_1 - t_5}{\dfrac{\Delta x_{12}}{k_{12}} + \dfrac{1}{C_{23}} + \dfrac{\Delta x_{34}}{k_{34}} + \dfrac{\Delta x_{45}}{k_{45}}},$$

and the data given yield

$$\frac{q}{A} = \frac{20 - 5}{\dfrac{0.0125}{0.744} + \dfrac{1}{6.25} + \dfrac{0.0125}{0.048} + \dfrac{0.10}{0.69}}$$

$$= \frac{20 - 5}{0.0168 + 0.1600 + 0.2604 + 0.1449}$$

$$= 25.77 \text{ W/m}^2 \text{ (8.17 Btu/h-ft}^2\text{)}.$$

The thermal resistance of the sheetrock layer is seen to be small compared with those of the other layers.

A point midway through the Celotex layer is separated from the outside brick surface by a layer of Celotex $0.0125/2 = 0.00625$ m thick and the resistance of the brick. Writing the heat flux between these two points in terms of the thermal resistances between them gives

$$\frac{q}{A} = \frac{t - t_5}{\dfrac{\Delta x}{k} + \dfrac{\Delta x_{45}}{k_{45}}}.$$

Since the steady state exists, the heat flux is the same as that already found for the entire wall, and one obtains

$$\frac{q}{A} = 25.77 = \frac{t - 5}{\dfrac{0.00625}{0.048} + \dfrac{0.10}{0.69}}$$

$$t = 12.1°\text{C (53.8°F)}.$$

Thermal Contact Resistance

The discussion of the multilayer wall in the foregoing presumed perfect contact between adjacent layers so that it could be assumed that the temperatures of the layers were the same at the plane of contact. In real systems such is rarely the case, and the contacting surfaces touch only at discrete locations, due to surface roughness, interspersed with void spaces. The void spaces are usually filled with air. Figure 2.5 depicts such a situation in an exaggerated scale. Thus there is not a single plane of contact, and heat may flow across the interface by parallel paths: some through the contact spots by conduction and some across the voids by convection and, perhaps, radiation. Thus an apparent temperature drop may be presumed to occur between the two materials at the interface, as suggested in Fig. 2.5. If t_A and t_B represent the temperatures at the theoretical plane interface obtained by extrapolation of the temperature gradients in the materials on either side, then a *unit thermal contact resistance* may be defined as

$$R_{tc} = \frac{t_A - t_B}{q/A}.$$

Correspondingly, for a given wall area A, a *total* thermal contact resistance

$$\mathcal{R}_{tc} = \frac{R_{tc}}{A}$$

can be defined.

The effect of interfacial contact can be introduced into Eqs. (2.20) and (2.21) by including values of R_{tc} or $\mathcal{R}_{tc}$ in the resistance summations. The utility of the contact resistance concept depends on the availability of reliable numerical values. In most instances such values have to be obtained experimentally; however, in the case of metal-to-metal contact many theories have also been developed to predict values of R_{tc}. As would be expected, metallic contact resistance depends on the metals involved, the surface roughness, the contact pressure, the material occupying the void spaces, and temperature. Reference 1 may be consulted as a source of both theoretical and experimental values of R_{tc}. As a guide to the typical orders of magnitude and the effect of various parameters, sample experimental data are given in Table 2.1.

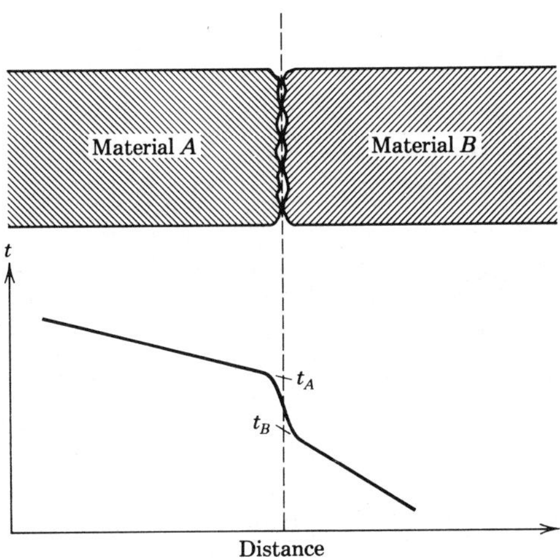

Figure 2.5. Typical temperature distribution due to contact resistance.

Table 2.1. Typical Values of the Unit Contact Resistance for 75S-T6 Aluminum at 93°C

RMS Surface Roughness μm	$R_{tc} \times 10^4$ m^2-°C/W Contact Pressure 100 kN/m^2	Contact Pressure 1000 kN/m^2
3.05	3.5	1.6
1.65	2.0	0.98
0.25	0.98	0.59

The Plane Wall with Convection at the Boundaries

The cases presented thus far have presumed that the exposed boundary temperatures of the walls were known. In many engineering applications the wall surfaces are exposed to ambient fluids and it is the temperatures of these fluids, not the surfaces, that are known.

If fluid motion exists in the ambient fluids, as it invariably will because of either free or forced convection, the resulting thermal and velocity boundary layers cause a temperature difference to exist between the main bulk of the fluid (where the temperature is known) and the surface. This was discussed in Sec. 1.6, and in that section the heat transfer coefficient was defined in Eq. (1.4) to relate the heat flux from the surface to the difference between the surface and fluid temperature:

$$\frac{q}{A} = h(t_s - t_f).$$

Application of this definition will now be made to the plane wall bounded by convecting fluids, but it is important to note that at this point the discussion is directed primarily to the analysis of the conduction problem in the wall. The details of the convection process at the surfaces will not be of interest here and it will be presumed that the heat transfer coefficient h is a known quantity. Chapters 5 through 8 are devoted to a detailed analysis of convection and the determination of the coefficient h. Then in Chapters 10 and 11 the problems of combined conduction and convection will be reconsidered.

Figure 2.6 shows a plane wall (composed of two solid layers for purposes of discussion) bounded on each side by convecting fluids. By denoting the fluid regions of uniform temperature and the various injunctures between different materials by numbers, the subscript notation introduced earlier may be used to denote the thermal conductances and conductivities separating the numbered regions: h_{12}, k_{23}, k_{34}, and h_{45}.

The heat flow per unit wall area may be written for each "layer" by application

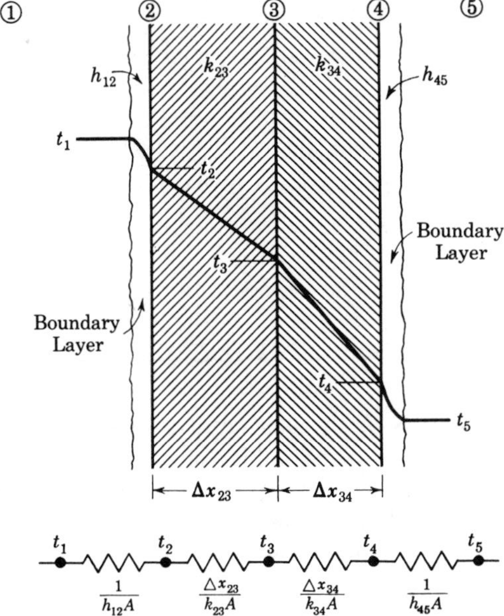

Figure 2.6

of the defining relation for h given above and the equations obtained above for the multilayer wall. Thus,

$$\frac{q}{A} = h_{12}(t_1 - t_2),$$

$$\frac{q}{A} = \frac{t_2 - t_4}{\dfrac{\Delta x_{23}}{k_{23}} + \dfrac{\Delta x_{34}}{k_{34}}},$$

$$\frac{q}{A} = h_{45}(t_4 - t_5).$$

Again, by combining these relations to eliminate t_2 and t_4, the following expression for the heat flux is obtained in terms of the overall temperature difference and the thermal properties of the matter in between:

$$\frac{q}{A} = \frac{t_1 - t_5}{\dfrac{1}{h_{12}} + \dfrac{\Delta x_{23}}{k_{23}} + \dfrac{\Delta x_{34}}{k_{34}} + \dfrac{1}{h_{45}}}. \tag{2.22}$$

In keeping with the thermal resistance defined above,

$$\mathcal{R}_k = \frac{1}{CA},$$

a *convective* thermal resistance may be defined:

$$\mathcal{R}_c = \frac{1}{hA}.$$

The application of this definition, along with that already used for conduction resistances, to the network shown in Fig. 2.6 leads to the result given by Eq. (2.22).

EXAMPLE 2.2

The masonry wall of a building consists of an outer layer of facing brick (k = 1.32 W/m-°C) 10 cm thick, followed by a 15-cm-thick layer of common brick (k = 0.69 W/m-°C), followed by a 1.25-cm layer of gypsum plaster (k = 0.48 W/m-°C). An outside coefficient of 30 W/m²-°C may be expected, and a coefficient of 8 W/m²-°C is a reasonable value to use for the inner surface of a ventilated room. What will be the rate of heat gain, per unit area, when the outside air is at 35°C and the inside air is conditioned to 22°C? What will be the temperature of the surface of the plaster?

Solution. As in Example 2.1, the conduction through the wall may be treated as one-dimensional and the heat flux expressed in terms of the overall temperature difference and the sum of the intervening resistances. Denoting the outside air as region 1 with t_1 = 35°C, the inside air as region 6 with t_6 = 22°C, and the intermediate material surfaces by 2 through 5, an equation in the form of Eq. (2.22) may be written for the heat flux:

$$\frac{q}{A} = \frac{t_1 - t_6}{\dfrac{1}{h_{12}} + \dfrac{\Delta x_{23}}{k_{23}} + \dfrac{\Delta x_{34}}{k_{34}} + \dfrac{\Delta x_{45}}{k_{45}} + \dfrac{1}{h_{56}}}$$

$$= \frac{35 - 22}{\dfrac{1}{30} + \dfrac{0.1}{1.32} + \dfrac{0.15}{0.69} + \dfrac{0.0125}{0.48} + \dfrac{1}{8}}$$

$$= \frac{35 - 22}{0.0333 + 0.0758 + 0.2174 + 0.0260 + 0.1250}$$

$$= \frac{35 - 22}{0.4775} = 27.2 \text{ W/m}^2 \text{ (8.63 Btu/h-ft}^2\text{)}.$$

Note that the thermal resistance of the common brick and the inside heat transfer coefficient (0.1274 + 0.1250 = 0.2524 m²-°C/W) contribute over 70% of the total resistance, 0.4775 m²-°C/W.

Application of Newton's law of cooling [Eq. (1.4)] at the inside surface yields the temperature of that surface since the existence of the steady state demands that the heat flux there be the same as that through the wall:

$$\frac{q}{A} = h_{56}(t_5 - t_6)$$

$$27.2 = 8 \times (t_5 - 22)$$

$$t_5 = 25.4°\text{C } (77.7°\text{F}).$$

2.4 STEADY CONDUCTION IN THE CYLINDRICAL GEOMETRY

A geometrical configuration which is mathematically simple and also of great engineering importance is that of a hollow cylinder, as pictured in Fig. 2.7. Pipes and tubes are used in a great variety of engineering applications and the need to determine the heat flow through such configurations is frequent.

The cylindrical coordinate system shown in Fig. 2.7 is the natural one to use for the given geometry. If the inner surface of the cylinder, of radius r_1, and the outer surface, of radius r_2, are maintained at uniform temperatures t_1 and t_2, respectively, then for a very long cylinder the end effects may be ignored. Thus, one may eliminate any dependence of the temperature within the cylinder wall on either the axial coordinate, z, or the circumferential coordinate, θ. Thus, the problem of determining the temperature is reduced, in the steady state, to a one-dimensional case with r as the single independent variable. If, in addition, internal heat generation is absent and the thermal conductivity is taken as constant, the general conduction equation in cylindrical coordinates [Eq. (2.8)] becomes

$$\frac{d^2t}{dr^2} + \frac{1}{r}\frac{dt}{dr} = 0 \qquad \text{or} \qquad \frac{d}{dr}\left(r\frac{dt}{dr}\right) = 0.$$

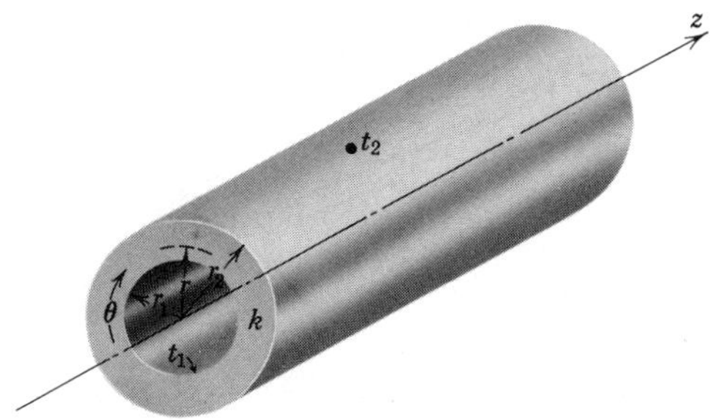

Figure 2.7

This equation may be integrated twice to yield the following logarithmic dependence of the temperature on the radial position:

$$t = B_1 \ln r + B_2. \tag{2.23}$$

The constants B_1 and B_2 are determined by application of the known surface temperatures at the inside and outside: $t = t_1$ at $r = r_1$ and $t = t_2$ at $r = r_2$. Proper evaluation of B_1 and B_2 and substitution into Eq. (2.23) gives the following for the temperature at any interior location r in terms of those at the surfaces:

$$t = t_1 + (t_2 - t_1)\frac{\ln (r/r_1)}{\ln (r_2/r_1)}. \tag{2.24}$$

The necessity of a logarithmically varying temperature, unlike the linearly varying temperature in the plane wall, is best seen by examining Fourier's conduction law. Since symmetry and the one-dimensional simplification demand that the temperature gradient be uniform over any interior cylindric surface, at some radius r, the form of Eq. (1.3) is applicable:

$$q = -kA_r \frac{dt}{dr},$$

where A_r represents the area available for conduction at the surface in question. Clearly, for a cylinder length L,

$$A_r = 2\pi rL,$$

$$q = -k2\pi rL \frac{dt}{dr}.$$

For the steady state q must be the same at all values of r. Since A_r varies directly with r, the gradient, dt/dr, must vary as $1/r$—hence the logarithmic variation of t with r.

Substitution of Eq. (2.24) into the foregoing relation for q gives the heat flow, per unit length, through a cylinder of known geometry and surface temperatures:

$$\frac{q}{L} = \frac{2\pi k(t_1 - t_2)}{\ln (r_2/r_1)}. \tag{2.25}$$

One may proceed to show that if the thermal conductivity has a linear dependence on temperature, as in Eq. (2.16), the same result is obtained as in the case of the plane wall. That is, Eq. (2.25) for the case of constant k will apply if k is evaluated at the mean temperature $(t_1 + t_2)/2$. Proof of this is left as a problem at the end of the chapter.

The Multilayer Cylinder

A pipe covered with insulation is a perfect example of the next configuration to be discussed—that of a long cylinder composed of two or more layers of materials having different thermal conductivities. For purposes of discussion consider the

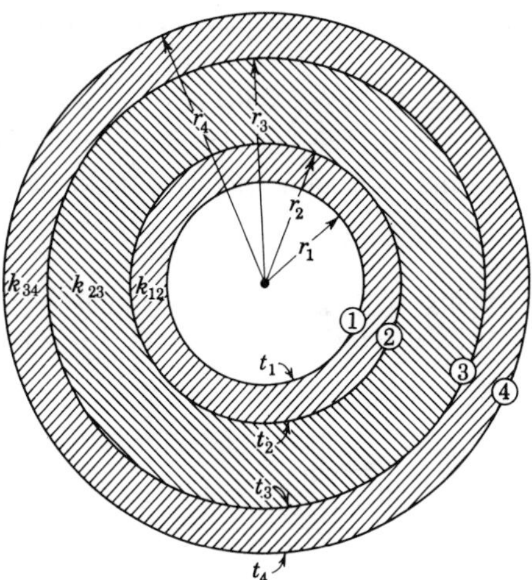

Figure 2.8

three-layer cylinder shown in Fig. 2.8. The same notational scheme employed in the case of the multilayer wall is used here.

The same procedure used for the plane wall may be applied in this case. Writing three expressions for the heat flow through each of the layers, in terms of unspecified temperatures t_2 and t_3:

$$\frac{q}{L} = \frac{2\pi k_{12}(t_1 - t_2)}{\ln (r_2/r_1)} = \frac{2\pi k_{23}(t_2 - t_3)}{\ln (r_3/r_2)} = \frac{2\pi k_{34}(t_3 - t_4)}{\ln (r_4/r_3)}.$$

Elimination of the unknown t_2 and t_3 gives

$$\frac{q}{L} = \frac{2\pi(t_1 - t_4)}{\dfrac{\ln (r_2/r_1)}{k_{12}} + \dfrac{\ln (r_3/r_2)}{k_{23}} + \dfrac{\ln (r_4/r_3)}{k_{34}}} \tag{2.26}$$

as the heat flow per unit length in terms of the overall temperature difference. Once q/L is known, the temperatures t_2 and t_3 may be found from the expressions for the heat flow through each layer.

Equation (2.25) suggests that for a cylindrical wall, a *total* thermal resistance may be defined

$$\mathcal{R}_k = \frac{\ln (r_2/r_1)}{2\pi Lk}. \tag{2.27}$$

Based on this definition, the electrical network analogy

$$q = \frac{\Delta t_{\text{overall}}}{\mathcal{R}_{k12} + \mathcal{R}_{k23} + \cdots}$$

yields the same result as that given by Eq. (2.26).

EXAMPLE 2.3

A 4-in. schedule 40 wrought iron pipe ($k = 55$ W/m-°C) is covered with 2.5 cm of magnesia insulation ($k = 0.071$ W/m-°C). If the inside pipe wall and outer insulation surface temperature are 150°C and 30°C, respectively, find the heat loss per meter of pipe length.

Solution. If the pipe is regarded as being very long, the conduction through it may be taken as dependent only on the radial coordinate and the heat loss may be expressed in the form of Eq. (2.26):

$$\frac{q}{L} = \frac{2\pi(t_1 - t_2)}{\dfrac{\ln(r_2/r_1)}{k_{12}} + \dfrac{\ln(r_3/r_2)}{k_{23}}}$$

with 1 denoting the inside pipe surface, 2 the outside pipe surface, and 3 the outer insulation surface. Appendix B gives the inside and outside radii of a 4-in. schedule 40 pipe to be $r_1 = 10.266/2 = 5.133$ cm, and $r_2 = 11.43/2 = 5.715$ cm. Hence the heat loss is

$$\frac{q}{L} = \frac{2\pi(150 - 30)}{\dfrac{\ln(5.715/5.113)}{55} + \dfrac{\ln[(5.715 + 2.5)/5.715]}{0.071}}$$

$$= \frac{2\pi(150 - 30)}{0.00202 + 5.11081} = 147.5 \text{ W/m } (153.4 \text{ Btu/h-ft}).$$

The thermal resistance of the pipe wall is seen to be negligible compared with that of the insulation, because of the low conductivity of the latter, and might have been ignored with little error.

The Cylinder with Convection at the Boundary

A cylindrical surface separating fluids of different temperatures is one of immense practical importance since fluids are transported, heated, cooled, evaporated, and condensed in cylindrical pipes, tubes, and vessels. Such processes are encountered in almost every phase of engineering work involving fluids or heat transfer.

The notation and numbering scheme introduced earlier is used in Fig. 2.9, which shows a double-layer pipe separating fluids of different fixed temperatures.

The rate of heat flow through the various layers of the configuration shown in Fig. 2.9 may be written by use of the equations just developed for the multilayer cylinder and the definition of the heat transfer coefficient in Eq. (1.4), remembering that the area through which heat is convected is different at the two exposed surfaces. Thus

$$q = 2\pi L r_2 h_{12}(t_1 - t_2) = \frac{2\pi L(t_2 - t_4)}{\dfrac{\ln (r_3/r_2)}{k_{23}} + \dfrac{\ln (r_4/r_3)}{k_{34}}} = 2\pi L r_4(t_4 - t_5)h_{45}.$$

The set of equations above yields the following expression for the rate of heat flow through the cylinder per unit length:

$$\frac{q}{L} = \frac{2\pi(t_1 - t_5)}{\dfrac{1}{r_2 h_{12}} + \dfrac{\ln (r_3/r_2)}{k_{23}} + \dfrac{\ln (r_4/r_3)}{k_{34}} + \dfrac{1}{r_4 h_{45}}}. \tag{2.28}$$

Once again, one may notice in Eq. (2.28) the application of the electrical network analogy. The total conduction resistances of the cylindrical layers are of the form noted in Eq. (2.27) while the convective resistances at the inner and outer surfaces are $1/2\pi L r_2 h_{12}$ and $1/2\pi L r_4 h_{45}$, respectively. Thus the heat flow through the convective boundary layers and the solid layers is the total potential $(t_1 - t_5)$ divided by the sum of the resistances between t_1 and t_5. The common factor $2\pi L$ has been taken to the numerator of Eq. (2.28).

EXAMPLE 2.4

A $\frac{3}{4}$-in. 18-gage brass condenser tube ($k = 115$ W/m-°C) is used to condense steam on its outer surface at 10 kN/m^2 pressure. Cooling water at 18°C flows through the tube. If the inside and outside heat transfer coefficients are 1700 W/m^2-°C and 8500 W/m^2-°C, respectively, find the mass of steam condensed per hour per unit of tube length.

Solution. Treating the heat flow through the condenser tube as one-dimensional, an equation of the form of Eq. (2.28) may be written:

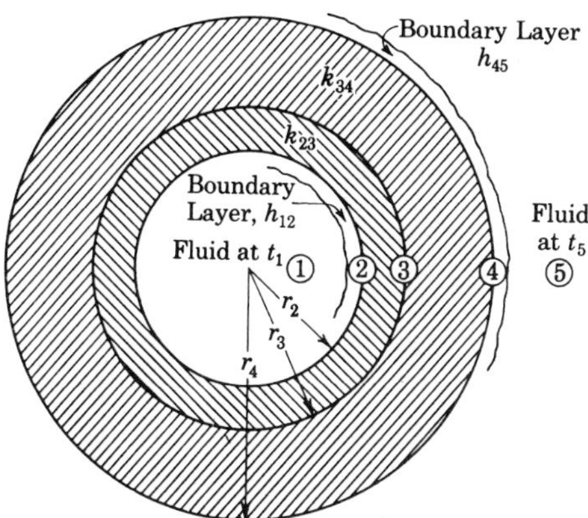

Figure 2.9

$$\frac{q}{L} = \frac{2\pi(t_1 - t_4)}{\dfrac{1}{r_2h_{12}} + \dfrac{\ln(r_3/r_2)}{k_{23}} + \dfrac{1}{r_3h_{34}}},$$

with 1 and 4 denoting the cooling water in the tube and the condensing steam, respectively, and 2 and 3 the inside and outside tube surfaces. Appendix B gives the tube diameters to be $D_2 = 1.656$ cm and $D_3 = 1.905$ cm. The Steam Tables give the saturation temperature of the condensing steam to be $t_4 = 45.8°\text{C}$ and its latent heat of vaporization to be $h_{fg} = 2392.8$ kJ/kg. Thus the heat flow is

$$\frac{q}{L} = \frac{2\pi(18 - 45.8)}{\dfrac{1}{(0.01656/2)1700} + \dfrac{\ln(1.905/1.656)}{115} + \dfrac{1}{(0.01905/2)8500}}$$

$$= \frac{2\pi(18 - 45.8)}{0.07104 + 0.00122 + 0.01235}$$

$$= -2064 \text{ W/m, into the tube.}$$

The rate at which steam is condensed, per unit length of tube, is then

$$\frac{\dot{m}}{L} = \frac{-q/L}{h_{fg}} = \frac{2064}{2392.8 \times 10^3} \times 3600$$

$$= 3.11 \text{ kg/h-m } (2.09 \text{ lb}_\text{m}\text{/h-ft}).$$

2.5 THE CRITICAL THICKNESS OF PIPE INSULATION

An interesting application of the above relations having some practical significance is found in the case of insulation of small pipes or electrical wires. Given a pipe of fixed size, let it be desired to examine the variation in heat loss from the pipe as the thickness of insulation is changed. As insulation is added to the pipe, the outer exposed surface temperature will decrease, but at the same time the surface area available for convective heat dissipation will increase. These two opposing effects produce some interesting optimization effects.

For ease of analysis, some simplifying assumptions will be made. As noted in Fig. 2.10, let the pipe radius be R and the insulation radius r, so that $(r - R)$ represents the insulation thickness. For fixed values of the temperature of the fluid carried by the pipe and the ambient air at t_a, the addition of insulation will alter the pipe surface temperature T. However, the variation of T is generally so slight that one may take it to be constant. If one also assumes that the heat transfer coefficient at the exposed insulation surface, h, is also constant, then the heat lost from the pipe is

$$\frac{q}{L} = \frac{2\pi(T - t_a)}{\dfrac{1}{hr} + \dfrac{\ln(r/R)}{k}}, \tag{2.29}$$

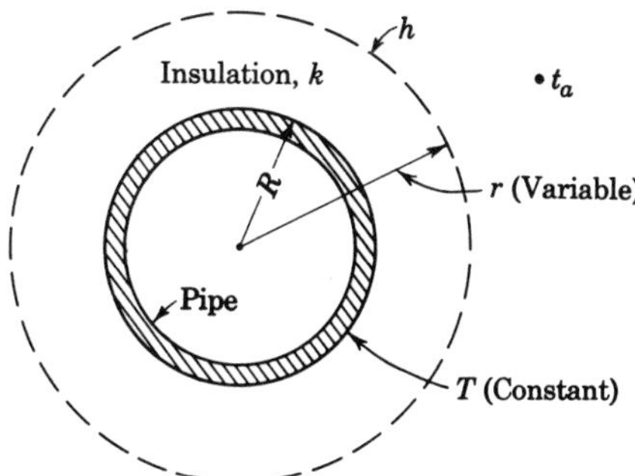

Figure 2.10

in which r is the only variable.

Differentiation of Eq. (2.29) with respect to r will show that the heat loss, q/L, reaches a *maximum* when the insulation radius is equal to

$$r = r_c = \frac{k}{h}. \tag{2.30}$$

The symbol r_c denotes the "critical radius" of the insulation. The fact that q/L attains a maximum at $r = r_c$ is the result of the above-mentioned opposing effects: increasing r increases the thermal resistance of the insulation layer but decreases the thermal resistance of the surface coefficient because of the increasing surface area. At $r = r_c$, the total resistance reaches a minimum.

Thus it appears possible to *increase* the heat loss from a pipe by the addition of insulation. This is illustrated in Fig. 2.11. As shown, the critical radius r_c is a quantity fixed by the thermal properties involved. If the pipe size is such that $R < r_c$, then initial addition of insulation will increase the het loss until $r = r_c$ after

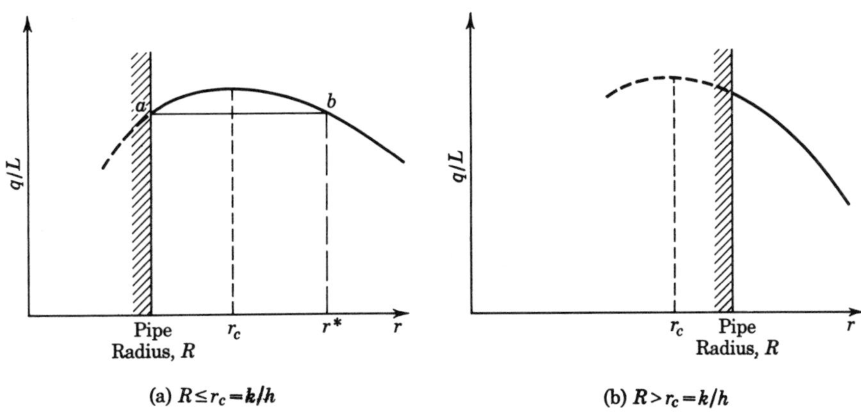

Figure 2.11. The critical thickness of pipe insulation.

which it will begin to decrease. The bare pipe heat loss is again reached at some radius, shown as r^* in Fig. 2.11(a). For a larger pipe, as suggested in Fig. 2.11(b) for which $R > r_c$, any insulation added will decrease the heat loss.

The main purpose of introducing this example is to illustrate the opposing effects a geometry change may introduce in the thermal resistance and available convective area of a given situation. Similar effects will be seen to occur in more subtle ways in other more complex cases to be studied later. This example was not shown as a definitive study of insulation optimization. Indeed, such optimization studies must also include other considerations such as the effect of radiation as well as economic factors such as the cost of the insulation.

EXAMPLE 2.5

A pipe with a surface temperature of 250°C passes through a room where the air temperature is 30°C. Asbestos insulation (density: 577 kg/m^3) is to be applied to the pipe and a heat transfer coefficient of 4 W/m-°C can be expected to exist at the surface of the bare pipe or the surface of the insulation after it is applied.

(a) Find the critical radius of the insulation.
(b) If the pipe diameter is 5 cm and insulation of the radius found in part (a) is applied to it, find the heat loss from the pipe before and after the insulation is applied.
(c) If the pipe diameter is 12 cm and insulation of the same *thickness* as that in part (b) is used, find the heat loss before and after the insulation is applied.

Solution

(a) The inner surface of the insulation is at 250°C. The outer surface temperature is not known; however, for purposes of estimating the thermal conductivity, evaluate it at the mean of the inside surface and the ambient air temperatures. Thus at (250 + 30)/2 = 140°C, Table A.2 gives k = 0.20 W/m-°C. With the given heat transfer coefficient h = 4 W/m^2-°C, Eq. (2.30) gives the critical radius to be

$$r_c = \frac{k}{h} = \frac{0.20}{4.0} = 0.05 \text{ m} = 5.0 \text{ cm}.$$

(b) From the given data one has the pipe surface temperature to be T = 250°C, the air temperature to be t_a = 30°C, and the pipe radius to be R = 5.0/2 = 2.5 cm. Thus the heat loss, per unit length, from the bare pipe is

$$\begin{aligned}\frac{q}{L} &= 2\pi Rh(T - t_a) \\ &= 2\pi \times 0.025 \times 4.0(250 - 30) \\ &= 138.2 \text{ W/m (143.8 Btu/h-ft)}.\end{aligned}$$

When the insulation is applied with a radius equal to the critical radius, $r = r_c = 5.0$ cm, the insulation thickness is $r - R = 5.0 - 2.5 = 2.5$ cm. If the pipe surface temperature is assumed to remain unchanged at $T = 250°C$ and the same heat transfer coefficient $h = 4.0$ W/m^2-°C is assumed to apply at the insulation surface, the heat loss is given by Eq. (2.29):

$$\frac{q}{L} = \frac{2\pi(T - t_a)}{\dfrac{1}{hr} + \dfrac{\ln(r/R)}{k}} = \frac{2\pi(250 - 30)}{\dfrac{1}{4.0 \times 0.050} + \dfrac{\ln(5.0/2.5)}{0.20}}$$

$$= 163.3 \text{ W/m } (169.8 \text{ Btu/h-ft}).$$

The heat loss *with* the insulation applied is seen to be greater than that for the bare pipe. This is because the original pipe radius, $R = 2.5$ cm, is less than the critical radius, $r_c = 5.0$ cm, and any insulation up to r_c increases the heat flow. If insulation greater than r_c is applied, the heat loss would be reduced below 163.3 W/m, and if thick enough, below that of the bare pipe.

(c) Now, suppose that the bare pipe radius is $R = 12/2 = 6.0$ cm. Then the heat loss is

$$\frac{q}{L} = 2\pi Rh(T - t_a)$$

$$= 2\pi \times 0.06 \times 4.0(250 - 30)$$

$$= 331.8 \text{ W/m } (345.0 \text{ Btu/h-ft}).$$

The *thickness* of the insulation applied in part (b) was $5.0 - 2.5 = 2.5$ cm. If the same thickness is applied in this instance, the insulation radius is $r = R$ + thickness $= 6.0 + 2.5 = 8.5$ cm. Now the heat loss given by Eq. (2.29) is

$$\frac{q}{L} = \frac{2\pi(T - t_a)}{\dfrac{1}{hr} + \dfrac{\ln(r/R)}{k}} = \frac{2\pi(250 - 30)}{\dfrac{1}{4.0 \times 0.085} + \dfrac{\ln(8.5/6.0)}{0.20}}$$

$$= 295.2 \text{ W/m } (307.0 \text{ Btu/h-ft}).$$

In this case any insulation reduces the heat loss since the bare pipe radius is already greater than the critical radius.

2.6 THE OVERALL HEAT TRANSFER COEFFICIENT

In many instances it is customary to express the heat flow rate in the cases of the plane wall and cylinder (single or multilayered) with convection at the boundaries in terms of an "overall conductance" or "overall heat transfer coefficient." This overall heat transfer coefficient, symbolized by U, is simply defined as a quantity such that the rate of heat flow through a configuration is given by taking a product of U, the surface area, and the overall temperature difference:

$$q = UA(\Delta t)_{\text{overall}}. \tag{2.31}$$

The dimensions of U are seen to be those of a conductance.

If one utilizes the equivalent resistance concept of the electrical networks introduced earlier, the overall conductance U is simply related to the total resistance between the points at which the overall potential is applied:

$$U = \frac{1}{\mathcal{R}_{\text{total}} A}.$$

The Plane Wall

The case of the plane wall is easily deduced from the discussion of Sec. 2.3. For a wall of n layers bounded on either side by fluids of temperatures t_1 and t_{n+3},

$$q = UA(t_1 - t_{n+3}), \tag{2.32}$$

$$U = \frac{1}{\dfrac{1}{h_{12}} + \dfrac{\Delta x_{23}}{k_{23}} + \dfrac{\Delta x_{34}}{k_{34}} + \cdots + \dfrac{1}{h_{(n+2)(n+3)}}}.$$

This representation is not essentially different from that already discussed, but it is found to be useful in the calculation of heat flow rates in the determination of heating or cooling needs of buildings. In such cases the overall heat transfer coefficient of standard structural walls, floors, roofs, etc., may be tabulated for ready referernce. This has been done, and the results are available in various handbooks.

The Cylinder

Much use is made of the overall het transfer coefficient for the cylindrical case in expressing the heat transfer capacity of heat exchangers. Frequently, heat exchangers use a bundle of cylindrical tubes to provide the surface area for the transfer of heat between two fluids.

In the case of the cylindrical configuration, the overall heat transfer coefficient depends on what surface area (inside or outside) is used in the defining relation given by Eq. (2.31), although the product UA is always the same. It is customary to use the outside surface area, and the associated overall coefficient is sometimes called the "outside overall heat transfer coefficient." This is best illustrated by use of the double-layered cylinder discussed in Sec. 2.4 and pictured in Fig. 2.9. If U_4 is used to denote the overall heat transfer coefficient based on the outside area A_4 of radius r_4, then (see Fig. 2.9)

$$q = U_4 A_4 (t_1 - t_5). \tag{2.33}$$

Since $A_4 = 2\pi r_4 L$, Eqs. (2.28) and (2.33) combine to give the following expression for U_4:

$$U_4 = \frac{1}{\dfrac{r_4}{r_2 h_{12}} + \dfrac{r_4 \ln (r_3/r_2)}{k_{23}} + \dfrac{r_4 \ln (r_4/r_3)}{k_{34}} + \dfrac{1}{h_{45}}}. \qquad (2.34)$$

EXAMPLE 2.6

Find the overall heat transfer coefficient for the building wall described in Example 2.2.

Solution. Using the notation of Example 2.2, Eq. (2.32) gives the overall heat transfer coefficient to be

$$U = \frac{1}{\dfrac{1}{h_{12}} + \dfrac{\Delta x_{23}}{k_{23}} + \dfrac{\Delta x_{34}}{k_{34}} + \dfrac{\Delta x_{45}}{k_{45}} + \dfrac{1}{h_{56}}},$$

so that the given data yield

$$U = \frac{1}{\dfrac{1}{30} + \dfrac{0.1}{1.32} + \dfrac{0.15}{0.69} + \dfrac{0.0125}{0.48} + \dfrac{1}{8}}$$

$$= 2.09 \text{ W/m}^2\text{-°C } (0.369 \text{ Btu/h-ft}^2\text{-°F}).$$

EXAMPLE 2.7

Find the overall heat transfer coefficient, based on the outside exposed surface area, of the condenser tube described in Example 2.4.

Solution. Using the notation of Example 2.4, Eq. (2.34) gives the overall heat transfer coefficient, based on the surface area A_3, to be

$$U_3 = \frac{1}{\dfrac{r_3}{r_2 h_{12}} + \dfrac{r_3 \ln (r_3/r_2)}{k_{23}} + \dfrac{1}{h_{34}}},$$

so that the given data yield

$$U_3 = \frac{1}{\dfrac{0.01905}{0.01656 \times 1700} + \dfrac{(0.01905/2) \ln (1.905/1.656)}{115} + \dfrac{1}{8500}}$$

$$= 1241 \text{ W/m}^2\text{-°C } (218.5 \text{ Btu/h-ft}^2\text{-°F}).$$

2.7 CASES INVOLVING INTERNAL HEAT GENERATION

The cases considered in the foregoing sections have been ones in which the conducting solid is free of internal heat generation. In this section two cases in which there is uniformly distributed heat generation will be treated. Numerous practical instances exist in which such generation must be accounted for. The dissipative processes in current-carrying electrical conductors result in heat generation. Induction heating produces distributed heat additions, as do certain exothermic chemical reactions such as the curing of concrete.

Uniformly Distributed Generation in a Plane Wall

Consider the instance in which a plane wall, for which the surface temperatures are specified, has a uniformly distributed heat generation rate of q^* (per unit volume) throughout its interior. For simplicity of analysis assume that the two surfaces are maintained at the same temperature t_s and it is desired to find how the internally generated heat alters the temperature distribution within the wall from what would otherwise be a uniform temperature. Figure 2.12 indicates the geometry involved. The wall thickness is denoted by Δx, and the origin of the x coordinate is taken as one face. Thus, the temperature of the faces of the wall at $x = 0$ and $x = \Delta x$ are known to be t_s. The temperature distribution though the wall will look somewhat as suggested in Fig. 2.12, rising from t_s at one surface to a maximum at the center, t_c, and falling again to t_s at the other surface.

For steady one-dimensional heat flow, the general heat conduction equation [Eq. (2.5)] becomes

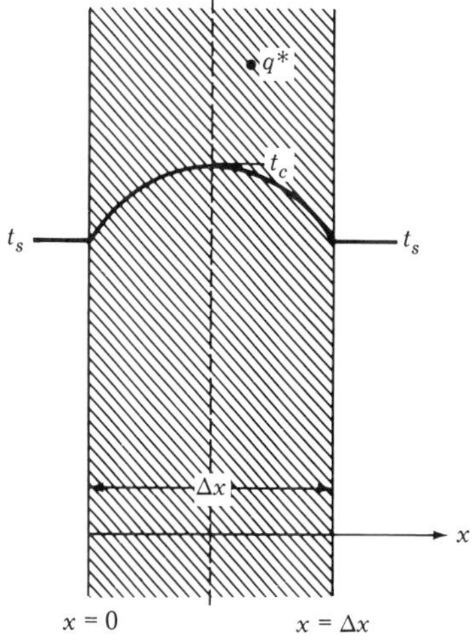

Figure 2.12. The plane wall with internal heat generation.

$$\frac{d^2t}{dx^2} + \frac{q^*}{k} = 0.$$

When integrated twice the equation above becomes

$$t + \frac{q^*}{2k} = Bx + C, \tag{2.35}$$

in which B and C are constants. These constants are readily evaluated from the fact that $t = t_s$ at $x = 0$ and $x = \Delta x$, and the results are

$$B = \frac{q^*\,\Delta x}{2k}, \qquad C = t_s,$$

so that Eq. (2.35) for the temperature through the plane is

$$t = t_s + \frac{q^*\,\Delta x}{2k}\,x - \frac{q^*}{2k}\,x^2. \tag{2.36}$$

Equation (2.36) is of the parabolic form suggested in Fig. 2.12. The maximum temperature at the center of the wall ($x = \Delta x/2$) is then

$$t_c = t_s + \frac{q^*(\Delta x/2)^2}{2k}. \tag{2.37}$$

The heat flow from each surface is readily found from Fourier's law:

$$\left(\frac{q}{A}\right)_s = \left(\frac{q}{A}\right)_{x=\Delta x} = -\left(\frac{q}{A}\right)_{x=0} = -\left(-k\,\frac{dt}{dx}\right)_{x=0} = \frac{q^*\,\Delta x}{2}, \tag{2.38}$$

equal to one-half the total internally generated heat, as should be the case.

The problem just considered applied to a plane wall of thickness Δx, with uniform internal heat generation, when each of its surfaces was maintained at a fixed surface temperature t_s. The results, however, may be applied to other surface conditions. For example, the symmetry of the temperature distribution given by Eq. (2.36) indicates that there is no heat flow at the center of the wall. Thus the results above may be applied to the case of a plane wall with heat generation when one surface is insulated and the other maintained at t_s by simply replacing Δx with twice the wall thickness.

In some applications it may be that the wall surface temperature t_s is not specified, but rather, that each surface is exposed to convective heat transfer to an ambient fluid at t_f through a heat transfer coefficient h—t_f and h being the same at both surfaces. In such a case the surface temperature may be calculated from Newton's law of cooling and the known surface heat flux given in Eq. (2.38):

$$\begin{gathered} h(t_s - t_f) = \left(\frac{q}{A}\right)_s = \frac{q^*\,\Delta x}{2}, \\ t_s = t_f + \frac{q^*\,\Delta x}{2h}. \end{gathered} \tag{2.39}$$

Equation (2.39) could be used to eliminate t_s from Eqs. (2.36) and (2.37) to express those results in terms of the specified fluid temperature t_f.

Uniformly Distributed Generation in a Cylinder

The case of uniformly distributed heat generation in a solid cylinder has application as a model for the process taking place in an electrically conducting wire. By assuming the wire to be quite long, the conduction in it may be taken as one-dimensional in the radial direction. The problem, then, may be solved as much as was just done for the plane wall.

Assume, then, that one has a solid cylinder, outside radius R, maintained at a known surface temperature t_s, in which the internal heat generation is a known, uniform value q^* (per unit volume). For steady one-dimensional conduction in the radial direction only, the general heat conduction equation in cylindrical coordinates, Eq. (2.8) reduces to

$$\frac{d^2t}{dr^2} + \frac{1}{r}\frac{dt}{dr} + \frac{q^*}{k} = 0,$$

$$\frac{d}{dr}\left(r\frac{dt}{dr}\right) + r\frac{q^*}{k} = 0.$$

Integrating twice gives

$$t + \frac{r^2}{4}\frac{q^*}{k} = B \ln r + C, \tag{2.40}$$

where B and C are again constants. Since the cylinder is solid, r may approach zero, and to maintain a finite solution the constant B must be zero. Then the condition that $t = t_s$, known, at $r = R$ yields

$$C = t_s + \frac{R^2}{4}\frac{q^*}{k},$$

so that Eq. (2.40) gives the desired temperature distribution to be

$$t = t_s + \frac{q^*R^2}{4k}\left(1 - \frac{r^2}{R^2}\right). \tag{2.41}$$

The heat flow, per unit of cylinder length, at the surface is

$$\begin{aligned}\left(\frac{q}{L}\right)_s &= -k\left(\frac{A_s}{L}\right)\left(\frac{dt}{dr}\right)_{r=R} \\ &= -k(2\pi R)\frac{q^*R^2}{4k}\left(-\frac{2R}{R^2}\right) \\ &= \pi R^2 q^*,\end{aligned} \tag{2.42}$$

which is, necessarily, the total heat generated in the cylinder. Equation (2.41) readily gives the maximum temperature at the center of the cylinder to be

$$t_c = t_s + \frac{q^*R^2}{4k}. \tag{2.43}$$

As was done for the plane wall case earlier, one may apply the results above to the case in which the heat-generating cylinder, instead of having its surface temperature specified, is exposed to an ambient convecting fluid at a known temperature t_f. In this instance the surface heat flow in Eq. (2.42) determines the surface temperature t_s:

$$h\left(\frac{A_s}{L}\right)(t_s - t_f) = \left(\frac{q}{L}\right)_s,$$

$$h(2\pi R)(t_s - t_f) = \pi R^2 p^*, \tag{2.44}$$

$$t_s = t_f + \frac{q^*R}{2h}.$$

EXAMPLE 2.8

A plane wall of concrete, 7.5 in. thick, is in the process of curing so that internal heat is being generated at the rate of 50 Btu/h-ft^3. Both sides of the wall are exposed to air at 70°F through a surface heat transfer coefficient of 0.8 Btu/h-ft^2-°F. If the thermal conductivity of the concrete is 0.45 Btu/h-ft-°F, find the temperatures of the surfaces and the center of the wall.

Solution. One-half of the internally generated heat must flow out of each surface of the wall. Thus the heat flux there is, by Eq. (2.38),

$$\left(\frac{q}{A}\right)_s = \frac{q^*\,\Delta x}{2} = \frac{50 \times (7.5/12)}{2} = 15.63 \text{ Btu/h-ft}^2.$$

This heat must be transferred to the surrounding fluid, so that

$$\left(\frac{q}{A}\right)_s = h(t_s - t_f)$$

$$15.63 = 0.8(t_s - 70)$$

$$t_s = 89.5°\text{F } (31.9°\text{C}).$$

Application of Eq. (2.39) gives the same result.

The center temperature is given by Eq. (2.37):

$$t_c = t_s + \frac{q^*(\Delta x/2)}{2k}$$

$$= 89.5 + \frac{50 \times (7.5/2 \times 12)^2}{2 \times 0.45}$$

$$= 94.9°\text{F } (34.9°\text{C}).$$

2.8 EXTENDED SURFACES

When it is desired to increase the heat removal between a structure and a surrounding ambient fluid, it is common practice to utilize "extended surfaces" attached to the primary surface. In such instances the extended surfaces are provided to increase artificially the surface area of heat transmission, although the average surface temperature may be decreased by so doing. If the surface is proportioned properly, the net result will be an increase in the heat transmission rate between the structure and the ambient fluid.

The uses of extended surfaces in applications of practical importance are numerous. Examples may be found in the cooling fins of air-cooled engines, the fin extensions to the tubes of radiators and other heat exchangers, the "pins" or "studs" attached to boiler tubes, etc. The extended surface applications noted above are all cases in which one purposely wishes to increase the rate of heat exchange between a source and an ambient fluid. Similar extended surface configurations may occur in other instances where the exchange of heat with the ambient fluid may be a disadvantage. Such instances are encountered in the measurement of temperature—the conduction of heat along thermocouple wires attached to a heated surface being an example.

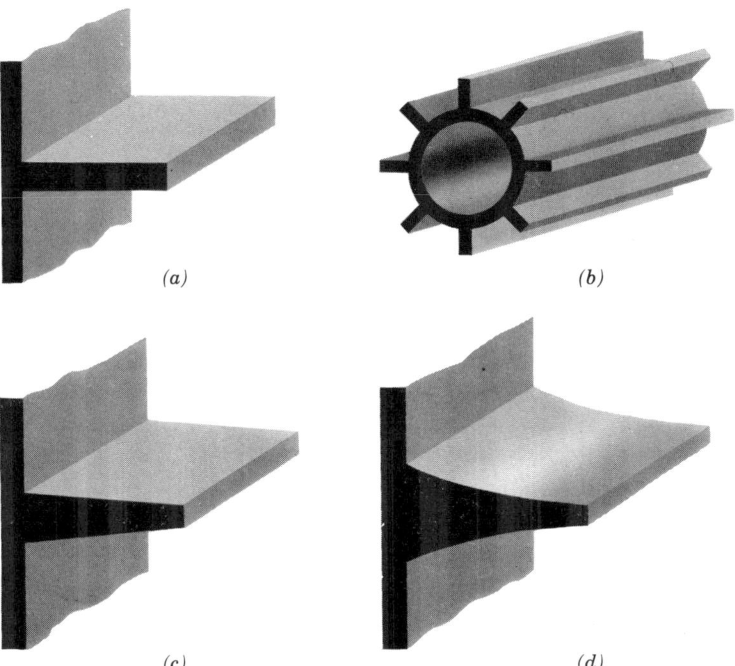

Figure 2.13. Examples of extended surfaces: (a) and (b), straight fins of uniform thickness; (c) and (d), straight fins of nonuniform thickness.

An extended surface configuration is generally classed as a straight fin, an annular fin, or a spine. The term *straight fin* is applied to the extended surface attached to a wall which is otherwise plane, whereas an *annular fin* is one attached, circumferentially, to a cylindrical surface. A *spine* or *pin fin* is an extended surface of cylindrical or conical shape. These definitions are illustrated in Figs. 2.13 and 2.14.

The basic problem with which the designer is faced is: Given a fin of a certain configuration and size attached to a surface of a fixed temperature (i.e., the fin base temperature) and surrounded by a fluid of fixed temperature, what is the rate of heat dissipated by the fin, and what is the variation in the temperature of the fin as one proceeds from the base to the tip?

In most instances the fin proportions used in practice are such that the length of the fin (its dimension measured normal to the primary surface to which it is attached) is large compared to its maximum thickness. When this is the case, one may assume that the temperature of the fin depends on only the single coordinate measured in the direction of the fin's length. For example, if a cylindrical rod protruding from a heated wall is very long compared to its diameter, one may assume that the temperature is uniform over any cross section taken normal to the axis. Hence the temperature in the rod depends only on the axial coordinate measured along the rod. Similarly, if an annular fin is sufficiently slender, the distribution of the temperature may be taken to depend only on the radial coordinate.

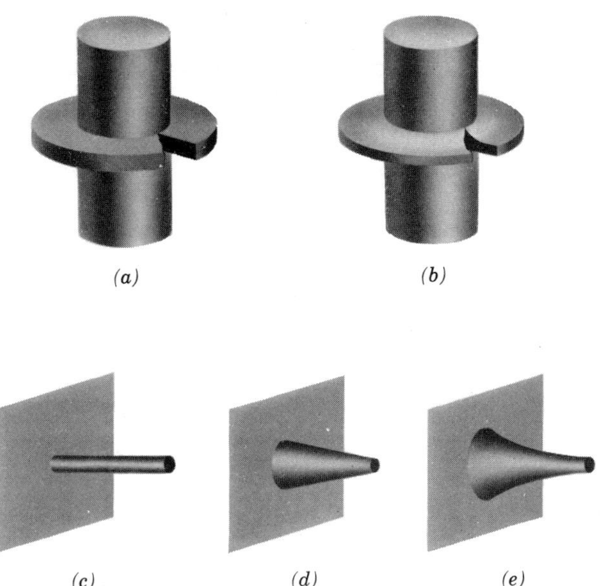

Figure 2.14. Examples of extended surfaces: (a) and (b), annular fins; (c), (d), and (e), spines.

If the simplification above can be made, the problem for the solution of the temperature distribution and heat flux rate becomes a one-dimensional conduction problem. A few shapes of practical importance will be considered in detail in this chapter. Only the steady-state case will be considered.

2.9 THE STRAIGHT FIN OF RECTANGULAR PROFILE AND THE SPINE OF UNIFORM CROSS SECTION

The straight fin of rectangular profile and the spine of uniform cross section may be treated identically since in either case the cross-sectional area for heat flow (normal to the length coordinate) is constant. Also, in either case, the exposed surface area for heat convection is a linear function of the distance measured along the length; i.e., the perimeter of the cross section is constant.

Figure 2.15 illustrates these two configurations and shows the notation to be used in the following analysis. The symbol L will denote the length of the fin, k the thermal conductivity, h the heat transfer coefficient at the exposed surface, t_o the fixed temperature at the fin base, and t_f the temperature of the bulk of the ambient fluid. The coodinate distance along the fin length will be symbolized by x. The symbols A and C will be used to denote the area of a cross section normal to x and the perimeter of this section, respectively.

Usually when dealing with straight fins it is customary to neglect heat losses from the side edges of the fin and to express the heat flow, etc., per unit width. In this case, as illustrated in Fig. 2.15, the perimeter C will be 2 and the cross-sectional area will equal the thickness, w.

If one now considers an element of the fin δx in length (see Fig. 2.15) and defines q_1, to be the heat conducted into the element at x, q_2 the heat conducted out of the element at $x + \delta x$, and q_3 the heat convected out of the surface of the element, then energy conservation requires, for the steady state, that

$$q_1 = q_2 + q_3.$$

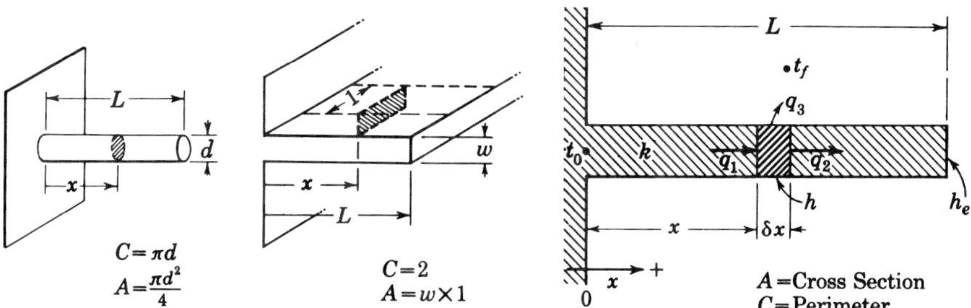

Figure 2.15. The straight fin of rectangular profile or spine of constant cross section.

Fourier's law and the definition of the heat transfer coefficient renders the equation above into the form

$$-kA\left(\frac{dt}{dx}\right)_x = -kA\left(\frac{dt}{dx}\right)_{x+\delta x} + h(C\,\delta x)(t_{av} - t_f),$$

where t_{av} is the average temperature of the element. If the temperature gradient at $x + \delta x$ is written in terms of that at x by a Taylor's expansion and the equation is divided by the element volume, $A\,\delta x$, one has

$$-\frac{k}{\delta x}\left(\frac{dt}{dx}\right)_x = -\frac{k}{\delta x}\left(\frac{dt}{dx}\right)_x - k\left(\frac{d^2t}{dx^2}\right)_x - k\left(\frac{d^3t}{dx^3}\right)_x \frac{\delta x}{2} + \cdots + \frac{hC}{A}(t_{av} - t_f).$$

This energy conservation must be satisfied for all elements, so upon letting $\delta x \to 0$, the governing differential equation for the temperature distribution in the fin is obtained:

$$\frac{d^2t}{dx^2} - \frac{hC}{kA}(t - t_f) = 0.$$

If the thermal conductivity and the heat transfer coefficient are taken as constant, the equation above can easily be solved.

The differential equation may be written more concisely by defining the temperature difference variable

$$\theta = t - t_f$$

and a parameter

$$m = \sqrt{\frac{hC}{kA}}. \tag{2.45}$$

Then the governing equation is

$$\frac{d^2\theta}{dx^2} - m^2\theta = 0. \tag{2.46}$$

Before discussing the solution to Eq. (2.46) it is worth noting special forms of the parameter m defined in Eq. (2.45). The two most commonly encountered geometric shapes conforming to the restrictions of constant C and constant A noted here are the straight fin of rectangular profile, thickness w, and the circular rod of diameter d, both noted in Fig. 2.15. For these two important shapes, the geometries given in Fig. 2.15 show:

$$\text{Rectangular fin, thickness } w\text{:} \quad m = \sqrt{\frac{2h}{kw}}.$$
$$\text{Circular rod, diameter } d\text{:} \quad m = \sqrt{\frac{4h}{kd}}. \tag{2.47}$$

The solution to Eq. (2.46) is

$$\theta = Be^{-mx} + De^{mx}. \tag{2.48}$$

The B and D in Eq. (2.48) are arbitrary constants determined by the boundary conditions imposed at the ends of the fin. At the base of the fin, the temperature is fixed at t_0 if the fin material is integral with the material of the primary surface. If, as is sometimes the case, the fin is a separate piece attached to the primary surface, some contact resistance may exist at the fin base and its temperature may be different from that of the primary surface. In such a case t_0 should be interpreted as the fin base temperature, not the temperature of the primary surface. At the outer end of the fin, convection is taking place. In other instances some other condition may be imposed—such as another fixed temperature if the fin extends between two heat sources.

For the present analysis let the conditions be those indicated by Fig. 2.15—an imposed temperature at the base and convection at the free end. Then the conditions determining B and D are:

$$\text{At } x = 0: \quad t = t_0.$$

$$\text{At } x = L: \quad -k\frac{dt}{dx} = h_e(t - t_f).$$

It should be noted that the boundary condition of convection at the end of the fin as stated above assumes that the heat transfer coefficient at the end, h_e, is different from that at the other surface, h. This is done for two reasons. First, as will be shown later, the convective heat transfer coefficient depends on the orientation of the surface, and it is quite likely that the coefficient at the fin end will differ from that on the other surfaces. Second, this distinction between the heat transfer coefficient will permit easy simplification of further results to the special case in which the heat convected out the end is considered negligible by merely setting $h_e = 0$.

Written in terms of the temperature difference variable, θ, defined above, these boundary conditions are

$$\begin{aligned} &\text{At } x = 0: \quad \theta = \theta_0 = t_0 - t_f. \\ &\text{At } x = L: \quad \frac{d\theta}{dx} = -\frac{h_e}{k}\theta. \end{aligned} \tag{2.49}$$

Application of these boundary conditions to Eq. (2.48) to determine the constants B and D yields

$$\frac{\theta}{\theta_0} = \frac{t - t_f}{t_0 - t_f} = \frac{[e^{m(L-x)} + e^{-m(L-x)}] + \dfrac{h_e}{km}[e^{m(L-x)} - e^{-m(L-x)}]}{(e^{mL} + e^{-mL}) + \dfrac{h_e}{km}(e^{mL} - e^{-mL})}. \tag{2.50}$$

This is, perhaps, more conveniently expressed in terms of the hyperbolic functions:

$$\frac{\theta}{\theta_0} = \frac{t - t_f}{t_0 - t_f} = \frac{\cosh m(L - x) + H \sinh m(L - x)}{\cosh mL + H \sinh mL}, \tag{2.51}$$

where

$$H = \frac{h_e}{km} \quad \text{and} \quad m = \sqrt{\frac{hC}{kA}}.$$

The rate of heat flow from the fin could be evaluated by integrating the convected heat over the fin surface. However, the result is obtained more directly by noting the fact that all the heat dissipated by the fin must be *conducted* past the point where $x = 0$. Thus if q symbolizes the heat flow rate,

$$q = -kA\left(\frac{dt}{dx}\right)_{x=0}.$$

Introduction of the temperature distribution from Eq. (2.51) yields

$$q = kmA\theta_0 \frac{\sinh mL + H \cosh mL}{\cosh mL + HA \sinh mL}. \tag{2.52}$$

Fins with Negligible End Heat Loss

In the development of the relations above for the temperature distribution and heat dissipation, it was assumed that the fin was sufficiently slender that a one-dimensional condition prevailed. If this condition is met, it is quite likely that the heat convected out of the free end at $x = L$ is very small compared with that convected out the lateral surfaces. If this is the case, Eqs. (2.51) and (2.52) may be simplified considerably by setting $h_e = H = 0$:

$$\frac{\theta}{\theta_0} = \frac{t - t_f}{t_0 - t_f} = \frac{\cosh m(L - x)}{\cosh mL}, \tag{2.53}$$

$$q = kmA\theta_0 \tanh mL. \tag{2.54}$$

A case that sometimes arises is that of a fin of constant cross section and perimeter which extends between two heat sources, both at the same temperature θ_0. Equations (2.53) and (2.54) may be applied to such a case, with L being the half-length between the sources, since symmetry indicates zero heat flow at the midpoint.

Very Long Fins

The equations quoted in the foregoing indicate that the product mL is the significant parameter governing the solution. Large values of mL result for fins that are either very long compared to their thickness [refer to the expressions in Eq. (2.47) for m] or for which the surface heat loss is high compared to the conductivity, or both. Thus a large mL product implies that the fin is, thermally speaking, very long. Under such circumstances one may obtain considerably simplified expressions by finding the solution to Eq. (2.48) for a fin of infinite length. For such a case the constant B must approach 0 in order that the fin temperature remain finite. Thus Eq. (2.48) becomes

$$\theta = De^{-mx},$$

for which the condition that $\theta = \theta_0$ at $x = 0$ readily yields

$$\theta = \theta_0 e^{-mx}. \tag{2.55}$$

Then the heat dissipated by the fin is

$$q = k\left(\frac{d\theta}{dx}\right)_{x=0} = kmA\theta_0. \tag{2.56}$$

EXAMPLE 2.9

Three rods, one made of glass ($k = 1.09$ W/m-°C), one of pure aluminum ($k = 228$ W/m-°C), and one of wrought iron ($k = 57$ W/m-°C), all have diameters of 1.25 cm, lengths of 30 cm, and are heated to 120°C at one end. The rods extend into air at 20°C, and the heat transfer coefficient on the surfaces is known to be 9.0 W/m^2-°C. Find (a) the distribution of temperature in the rods if the heat loss from the ends is neglected; (b) the total heat flow from the rods, neglecting the end heat loss; (c) the heat flow from the rods if the end heat loss is not neglected, and the heat transfer coefficient at the ends is also 9.0 W/m^2-°C.

Solution

(a) For a cylindrical rod, the parameter $m = \sqrt{hC/kA}$ is $m = \sqrt{4h/kd}$. Thus

$$m = \sqrt{\frac{4 \times 9.0}{1.09 \times 1.25 \times 100}} = 0.5140 \frac{1}{\text{cm}}, \text{ for glass.}$$

$$m = 0.07108 \frac{1}{\text{cm}}, \text{ for iron.}$$

$$m = 0.03554 \frac{1}{\text{cm}}, \text{ for aluminum.}$$

Then, by Eq. (2.53), the temperature distribution in the rods is given by

$$\frac{t - 20}{120 - 20} = \frac{\cosh m(L - x)}{\cosh mL},$$

in which $L = 30$ cm, and x is allowed to vary from 0 to 30 cm. The temperature distributions given by the expression above are plotted in Fig. 2.16 for comparison.

(b) Equation (2.54) gives the heat flow to be

$$q = kmA\theta_0 \tanh mL.$$

$$q = (1.09)(0.5140) \frac{\pi(1.25)^2}{4 \times 100} (120 - 20) \tanh (0.514 \times 30)$$

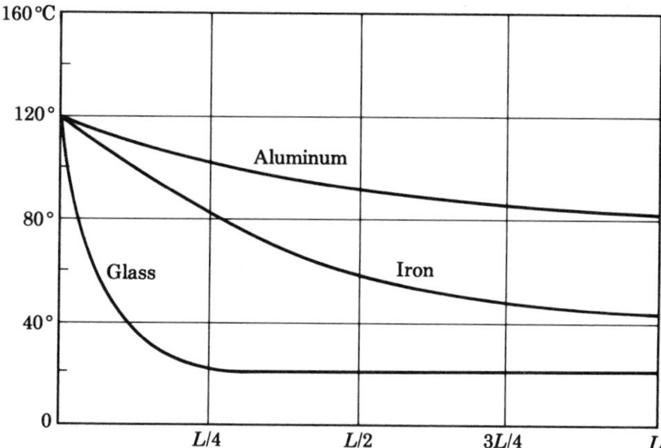

Figure 2.16

$$= 0.688 \text{ W } (2.35 \text{ Btu/h}), \text{ for glass.}$$

$$q = 4.83 \text{ W } (16.49 \text{ Btu/h}), \text{ for iron.}$$

$$q = 7.84 \text{ W } (25.74 \text{ Btu/h}), \text{ for aluminum.}$$

(c) If the heat loss out of the end is to be accounted for, the parameter $H = h_e/km$ is needed. Thus

$$H = \frac{9.0}{1.09 \times 0.514 \times 100} = 0.1606, \text{ for glass.}$$

$$H = 0.0222, \text{ for iron.}$$

$$H = 0.0111, \text{ for aluminum.}$$

Thus, Eq. (2.52) gives the heat flow to be

$$q = kmA\theta_0 \frac{\sinh mL + H \cosh mL}{\cosh mL + H \sinh mL}$$

$$= (1.09)(0.5140)$$

$$\times \frac{\pi(1.25)^2}{4 \times 100} \frac{\sinh (0.514 \times 30) + 0.1606 \cosh (0.514 \times 30)}{\cosh (0.514 \times 30) + 0.1606 \sinh (0.514 \times 30)}$$

$$= 0.688 \text{ W } (2.35 \text{ Btu/h}), \text{ for glass.}$$

$$q = 4.84 \text{ W } (16.5 \text{ Btu/h}), \text{ for iron.}$$

$$q = 7.88 \text{ W } (26.9 \text{ Btu/h}), \text{ for aluminum.}$$

The neglect of the end heat loss is seen to have little effect on the total heat flow.

2.10 FIN EFFICIENCY

The purpose of adding fins to a surface is to increase the area available for convective heat transfer to the surrounding fluid. However, the added surface area is not utilized to the same degree as the original primary surface since the fin surface is at a lower average temperature than that of the primary surface as a result of the temperature gradient within the fin.

In order to express the heat-exchanging capacity of an extended surface relative to that of the primary surface, it is useful to define the *fin efficiency* as the ratio of the heat transfer rate from the fin to the heat transfer rate that would be obtained if the entire fin surface area were maintained at the same temperature as the primary surface. It is usually presumed that no contact resistance exists at the fin base, so that the base temperature and the primary surface temperature are taken to be the same, namely θ_0 in the usual temperature-difference notation. Thus if A_f represents the exposed *surface* area of the fin and q_f is the actual heat transferred by the fin, then the fin efficiency κ, is defined:

$$\kappa = \frac{q_f}{hA_f\theta_0}. \tag{2.57}$$

The fin heat loss, q_f, is found from the relations developed in the preceding section, but before deducing an expression for κ, it is useful to notice another interpretation of the efficiency. The actual heat flow from the fin may also be used to define a *mean surface temperature,* θ_m:

$$\theta_m = \frac{q_f}{hA_f}.$$

Thus Eq. (2.57) gives

$$\kappa = \frac{\theta_m}{\theta_0}.$$

Hence the fin efficiency may also be interpreted as the ratio of the mean fin surface temperature to the base temperature. The efficiency κ must necessarily be less than 1, but hopefully the associated increase in surface area offsets this degradation of the average surface temperature. The latter concept is more accurately expressed in terms of the *total surface effectiveness* discussed in the next section.

For a given fin configuration q_f in Eq. (2.57) may be written in terms of the given fin parameters and equations for κ deduced. Such results are given next.

The Straight Fin of Rectangular Profile

For the straight fin of rectangular profile (or a spine of constant cross section) in which the end heat loss is neglected, Eq. (2.54) gives the fin heat loss to be

$$q_f = kmA\theta_0 \tanh mL,$$

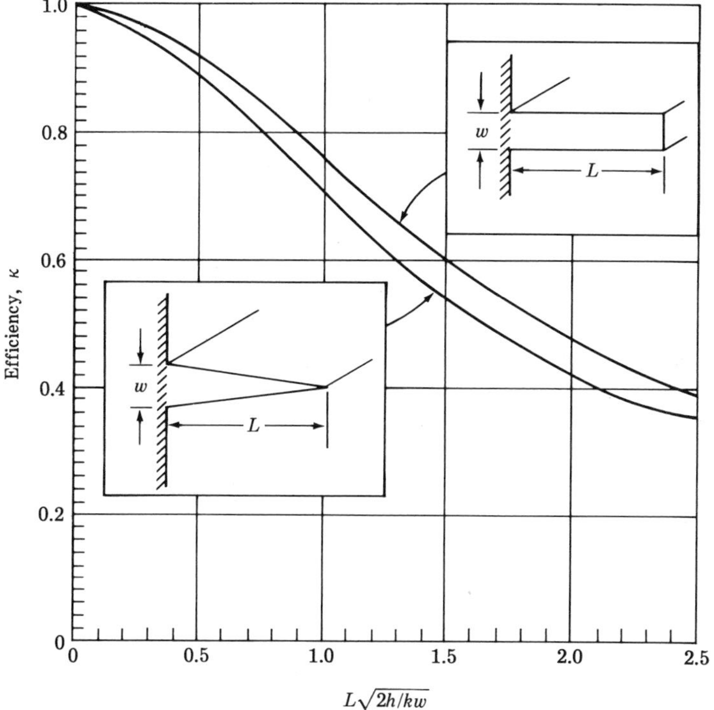

Figure 2.17. Efficiency of straight fins of rectangular and triangular profiles.

where it must be remembered that A is the cross-sectional area for conduction. The exposed fin surface is $A_f = CL$ (C being the exposed perimeter as in Sec. 2.9). Thus Eq. (2.57) for the efficiency is

$$\kappa = \frac{kmA\theta_0 \tanh mL}{hCL\theta_0}.$$

Since $m = \sqrt{hC/kA}$ [Eq. 2.45], one has

$$\kappa = \frac{1}{mL} \tanh mL \tag{2.58}$$

as the effectiveness for the straight fin of rectangular profile. Thus the effectiveness is readily evaluated from the known geometric and thermal parameters of the fin. For the straight fin of rectangular profile, thickness w, it is known that

$$mL = L\sqrt{\frac{2h}{kw}}.$$

While κ is easily evaluated from Eq. (2.58), it is shown as a function of the parameter $L\sqrt{2h/kw}$ in Fig. 2.17 for ease of comparison with the triangular fin to be discussed next.

The Straight Fin of Triangular Profile and the Annular Fin of Rectangular Profile

All of the foregoing analyses of fins, or extended surface, have been limited to those with constant perimeter and constant cross section. As noted in the introductory discussion of Sec. 2.8 and illustrated in Figs. 2.13 and 2.14, many other shapes may be utilized. The analysis of these other shapes becomes more complex since the conduction area may not be constant and the surface area exposed for convection may vary nonlinearly. Consequently, even after invoking the one-dimensional simplification, the resulting differential equations for the fin temperature distribution (from which the heat flow may be deduced) become more complex than that derived for the fin with constant C and A [Eq. (2.46)]. Solutions for many other fin geometries may be found in Refs. 2 through 4.

Two of these other geometries which are frequently encountered in engineering applications are the straight fin of triangular profile and the annular fin of rectangular profile. The triangular straight fin is used since it involves less material than the rectangular one of comparable dimensions. The annular fin is encountered when the primary surface is a cylinder, such as is the case of a finned tube. Of primary interest here is the fin efficiency of these shapes.

The efficiency of the straight fin of triangular profile (base thickness w and length L) may be written (Refs. 2 through 4) in terms of the same parameter developed above for the rectangular straight fin, namely $L\sqrt{2h/kw}$. The results are plotted in Fig. 2.17 along with Eq. (2.58) for the rectangular fin, and this graph may be used to analyze such fins. One should not immediately infer from Fig. 2.17 that a triangular straight fin is always less effectives than the rectangular one. The same w and L may not be used in the two cases, and even if they were, one should remember that the triangular fin uses one-half the material of the rectangular one.

The annular fin of rectangular profile is depicted in Fig. 2.18. The fin thickness is w and r_b and r_e denote the base and end radius of the fin, respectively. The solution for the effectiveness of this geometry (Refs. 2 through 4) is a function of *two* parameters: r_b/r_e and $r_e\sqrt{2h/kw}$. Figure 2.18 presents this functional relation.

Summary

For a given fin configuration Fig. 2.17 or 2.18 may be used [or Eq. (2.58) for the rectangular straight fin] to determine the fin efficiency from the specified geometric parameters (i.e., w, L, r_e, or r_b) and thermal parameters (h and k). Then the heat transferred from the fin to the ambient fluid is readily evaluated from Eq. (2.57), for a given base temperature potential $\theta_0 = t_0 - t_f$, after simply determining the exposed fin surface area:

$$q_f = \kappa h A_f \theta_0.$$

The next two examples illustrate this procedure.

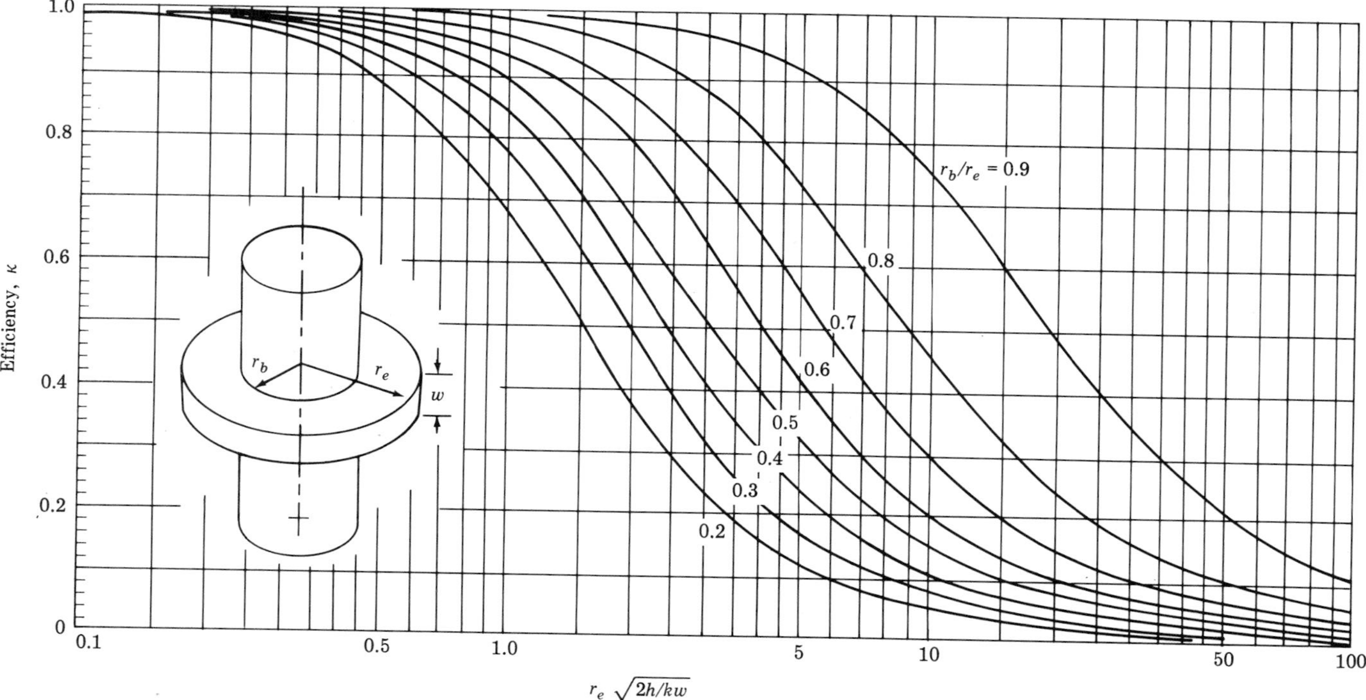

Figure 2.18. Efficiency of the annular fin of rectangular profile.

EXAMPLE 2.10

A fin is to be made of an aluminum alloy ($k = 64$ Btu/h-ft-°F) and is to be used to enhance the heat transfer between a primary surface and an ambient fluid where the surface heat transfer coefficient is 52 Btu/h-ft^2-°F. Different fin shapes are to be considered, but let the fin thickness at the base be 0.05 in. in every case. Find the fin efficiency if the fin is (a) a straight fin of rectangular profile with a length of 1.0 in.; (b) a straight fin of triangular profile with a length of 1.0 in.; (c) an annular fin of rectangular profile with a base radius of 1.0 in. and an end radius of 2.0 in.

Solution

(a) For the straight fin of rectangular profile, Eq. (2.47) gives the parameter m to be $m = \sqrt{2h/kw}$, so that the product mL is

$$mL = \sqrt{\frac{2h}{kw}}\,L = \sqrt{\frac{2 \times 52}{64 \times 0.05/12}}\,\frac{1.0}{12} = 1.646.$$

Thus, by Eq. (2.58), the efficiency is (Fig. 2.17 might be used as well)

$$\begin{aligned}\kappa &= \frac{1}{mL}\tanh mL \\ &= \frac{1}{1.646}\tanh 1.646 = 0.5641.\end{aligned}$$

(b) For the straight fin of triangular profile, since the base thickness and fin length are the same as for the rectangular fin in part (a), the parameter $\sqrt{2h/kw}\,L = 1.646$, as before. Thus Fig. 2.17 gives the efficiency to be

$$\kappa = 0.50.$$

While the straight triangular fin has a significantly lower efficiency than the rectangular one of equal base thickness and length, it should be noted that the triangular fin used one-half the material of the rectangular fin. Thus, under certain conditions, the triangular fin may prove to be more cost-effective.

(c) For the annular fin one has the same base thickness $w = 0.50$ in. The fin has the same length from the base radius $r_b = 1.0$ in. to the end radius $r_e = 2.0$ in. However, its shape, conduction area, and resultant temperature distribution differ significantly from the straight fin of equal dimensions—resulting in a reduced efficiency. From the data given, one has

$$\frac{r_b}{r_e} = \frac{1.0}{2.0} = 0.5,$$

$$r_e\sqrt{\frac{2h}{kw}} = \frac{2.0}{12}\sqrt{\frac{2 \times 52}{64 \times 0.05/12}} = 3.291.$$

Thus Fig. 2.18 gives

$$\kappa = 0.48.$$

EXAMPLE 2.11

If the fins described in Example 2.10 have a base temperature of 200°F and transfer heat with an ambient fluid at 80°F, find the rate of this heat transfer. In the case of the straight fins find the heat flow per unit width of fin.

Solution. According to Eq. (2.57), which defines the fin efficiency, the heat transferred to the ambient fluid is given by

$$q_f = \kappa h A_f \theta_0.$$

With the heat transfer coefficient, h, and the base temperature excess, $\theta_0 = t_0 - t_f$, given, q_f may be found from the efficiency values determined in Example 2.10 simply by evaluating the fin surface area A_f.

(a) For the straight rectangular fin (thickness $w = 0.05$ in., length $L = 1.0$ in.) the fin surface area, per unit width, is

$$A_f = 2L + w$$
$$= \frac{2 \times 1.0 + 0.05}{12} = 0.1708 \text{ ft}^2/\text{ft}.$$

For the given $h = 52$ Btu/h-ft^2-°F and $\theta_0 = t_0 - t_f = 200 - 80 = 120$°F, the value determined in Example 2.10 for $\kappa = 0.5641$ gives the heat flow to be

$$q_f = \kappa h A_f \theta_0$$
$$= 0.5641 \times 52 \times 0.1708 \times 120$$
$$= 601 \text{ Btu/h-ft } (578 \text{ W/m}).$$

(b) For the straight triangular fin the surface is simply

$$A_f = 2L$$
$$= \frac{2 \times 1}{12} = 0.1667 \text{ ft}^2/\text{ft}$$

if the slope of the sides is ignored. Then, with $\kappa = 0.50$ as found in Example 2.10, identical calculations give

$$q_f = \kappa h A_f \theta_0$$
$$= 0.50 \times 52 \times 0.1667 \times 120$$
$$= 520 \text{ Btu/h-ft } (500 \text{ W/m}).$$

(c) For the annular fin the exposed surface area is

$$A_f = 2\pi(r_e^2 - r_b^2) + 2\pi r_e w$$

$$= \frac{2\pi[(2.0^2 - 1.0^2) + 2.0 \times 0.05}{144}$$

$$= 0.1353 \text{ ft}^2.$$

Then, with $\kappa = 0.48$ from Example 2.10, one has the heat flow

$$q = \kappa h A_f \theta_0$$

$$= 0.48 \times 52 \times 0.1353 \times 120$$

$$= 405 \text{ Btu/h (119 W)}.$$

2.11 TOTAL SURFACE EFFECTIVENESS

The fin efficiency discussed in Sec. 2.10 is concerned with describing the performance of a single fin. However, many applications employing extended surfaces involve the use of an array of fins attached to the primary surface. Figure 2.19 depicts such arrays for straight and annular fins. In such applications it is useful to define a total surface temperature effectiveness which gives a measure of the performance of the total exposed surface of the array—that is, both the finned and unfinned surface. Let

A_f = exposed surface area of fins only,

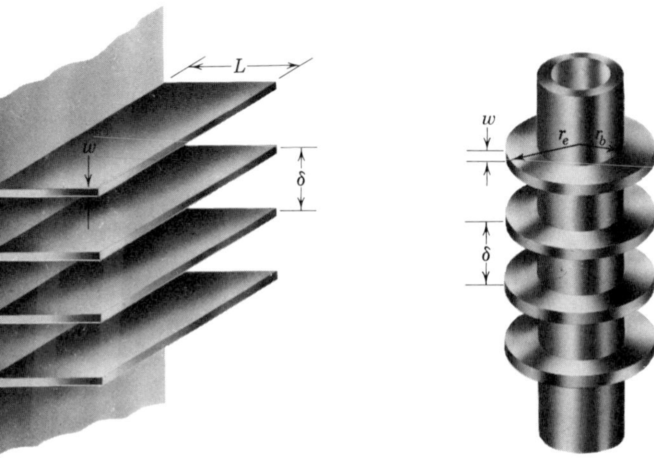

Figure 2.19

A_t = total exposed surface area, including the fins and unfinned surface,
κ = efficiency for the particular fin shape involved.

Then if the total surface temperature effectiveness, η, is defined as the ratio of the actual heat transferred by the array to that it would transfer if its entire surface were maintained at the base temperature,

$$\eta = \frac{q}{A_t h \theta_0}$$

$$\eta = \frac{(A_t - A_f)h\theta_0 + \kappa A_f h \theta_0}{A_t h \theta_0} \tag{2.59}$$

$$\eta = 1 - \frac{A_f}{A_t}(1 - \kappa).$$

Since $A_f/A_t < 1$ and $\kappa \leq 1$, it is apparent that $\eta \leq 1$. The ratio A_f/A_t is readily evaluated from the geometry of the array. For example, for an array of uniform fins of length L, thickness w, spacing on centers δ, as shown in Fig. 2.19(a), one may deduce

$$\left(\frac{A_f}{A_t}\right)_u = \frac{2L + w}{2L + w + (\delta - w)} = \frac{2L + w}{2L + \delta}. \tag{2.60}$$

Similarly, one may show for an array of triangular straight fins:

$$\left(\frac{A_f}{A_t}\right)_t = \frac{2L}{2L + \delta - w}. \tag{2.61}$$

As in the case of the fin efficiency, the total surface effectiveness, η, may also be interpreted as the ratio of the mean temperature of the total exposed surface, A_t, to the base temperature of the primary surface, θ_0. If A_p represents the exposed primary surface *before* the fin array is added, then the original total resistance between the surface and the fluid was

$$\frac{1}{A_p h},$$

while *after* the fins are added the resistance becomes

$$\frac{1}{A_t \eta h}. \tag{2.62}$$

Unless the product $A_t\eta$ is significantly greater than A_p, the addition of the fin array will not enhance the heat dissipation to the ambient fluid materially. This is best illustrated by example, as given next.

Before showing examples, it is worth noting here that the surface effectiveness, η, is used extensively in the representation of the performance of heat exchangers, as will be illustrated in Chapter 10.

EXAMPLE 2.12

A plane surface is provided with an array of straight rectangular fins as depicted in Fig. 2.19. The fin thickness and length are $w = 0.15$ cm, $L = 2.5$ cm. The fins are spaced $\delta = 0.35$ cm on centers. The fins are made of a material with $k = 100$ W/m-°C and the heat transfer coefficient at all exposed surfaces is $h = 300$ W/m^2-°C. If the fin base temperature is $t_0 = 95$°C and the ambient fluid is 25°C, find the total surface effectiveness of the array and the heat transferred to the fluid, per square meter of primary surface area.

Solution. The determination of the total surface effectiveness first requires the finding of the individual fin efficiency for the given shape. For the straight rectangular fin, the required parameter for Eq. (2.58) or Fig. 2.17 is

$$mL = L\sqrt{\frac{2h}{kw}} = 0.025\sqrt{\frac{2 \times 300}{100 \times 0.0015}} = 1.5811.$$

Thus Eq. (2.58), or Fig. 2.17, gives the fin efficiency

$$\kappa = \frac{1}{mL} \tanh mL$$

$$= \frac{1}{1.5811} \tanh 1.5811 = 0.5811.$$

Equation (2.60) shows that for the array of straight rectangular fins in Fig. 2.19, the ratio of the exposed fin area to the total exposed area is

$$\frac{A_f}{A_t} = \frac{2L + w}{2L + w + (\delta - w)} = \frac{2L + w}{2L + \delta}$$

$$= \frac{2 \times 2.5 + 0.15}{2 \times 2.5 + 0.35} = 0.9626.$$

So Eq. (2.59) gives the total surface effectiveness to be

$$\eta = 1 - \frac{A_f}{A_t}(1 - \kappa)$$

$$= 1 - 0.9626(1 - 0.5811) = 0.5968.$$

This indicates that the total exposed surface of the array performs at an average temperature difference that is 59.68% of the imposed θ_0.

For a unit width of 1 m, the total exposed area of a single fin and the accompanying bare primary area is

$$A_t = 2L + w + (\delta - w) = 2L + \delta.$$

Since there are $1/\delta$ fins per unit length down the primary area, the ratio of total exposed area per unit of primary area is

$$\frac{A_t}{A_p} = \frac{2L + \delta}{\delta}$$

$$= \frac{2 \times 2.5 + 0.35}{0.35} = 15.286.$$

Finally, then, the total heat transferred to the fluid from the array is, by the definition of the total effectiveness in Eq. (2.57),

$$q = \eta h A_t \theta_0.$$

With $\theta_0 = t_0 - t_f = 95 - 25 = 70°\text{C}$,

$$\frac{q}{A_p} = \eta h \frac{A_t}{A_p} \theta_0$$
$$= 0.5968 \times 300 \times 15.286 \times 70$$
$$= 191.6 \text{ kW/m}^2 \text{ (60,730 Btu/h-ft}^2\text{)}.$$

Note that if the primary surface had been left bare of fins, the heat flow would be only

$$\frac{q}{A_p} = h\theta_0 = 300 \times 70$$
$$= 21.0 \text{ kW/m}^2.$$

Thus, while the finned array operates with a temperature potential that is 0.5968 of that of the bare surface, the available transfer area is amplified by A_t/A_p = 15.286, so that the heat flow is enhanced by a factor of 0.5968 × 15.286 = 9.12. The effects of such performance on an overall structure is illustrated in the next example.

EXAMPLE 2.13

As depicted in Fig. 2.20, heat is to be transferred through a plane wall, Δx = 1.25 cm thick, composed of a material with k_{23} = 200 W/m-°C. On the left side of the wall is an ambient fluid at t_1 = 120°C and the heat transfer coefficient there for all exposed surface is h_{12} = 450 W/m^2-°C. On the right side of the wall there is another ambient fluid at t_4 = 20°C and the heat transfer coefficient there for all exposed surfaces is h_{34} = 25 W/m^2-°C. It is desired to enhance the heat transfer between the two fluids by using straight, rectangular fins having a length L = 2.5 cm, thickness w = 0.16 cm, and spaced δ = 1.25 cm on centers, and a k the same as the wall. Assuming that a one-dimensional representation may be used, find the heat transfer rate between the two fluids, per unit of primary wall area, if (a) there are no fins; (b) the fins are added to the right side only; (c) the fins are added to the left side only.

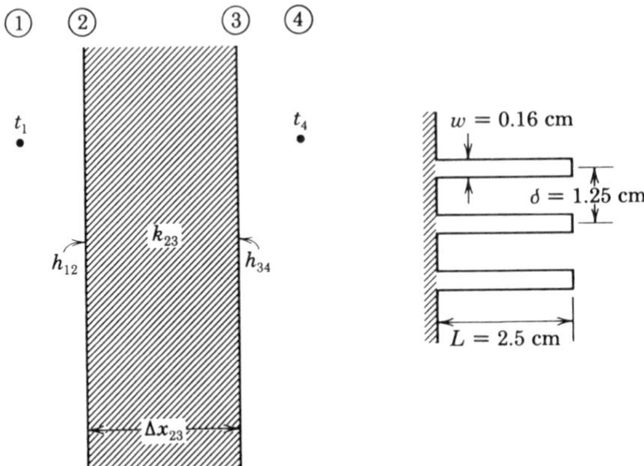

Figure 2.20

Solution

(a) For the bare wall, with primary area A_p, the heat flow is, from Eq. (2.22),

$$\frac{q}{A_p} = \frac{t_1 - t_4}{\dfrac{1}{h_{12}} + \dfrac{\Delta x_{23}}{k_{23}} + \dfrac{1}{h_{34}}} = \frac{120 - 20}{\dfrac{1}{450} + \dfrac{0.0125}{200} + \dfrac{1}{25}}$$

$$= 2364.9 \text{ W/m}^2 \text{ (749.6 Btu/h-ft}^2\text{)}.$$

(b) If the fins are added to the right side, the heat flow is, using Eq. (2.62),

$$\frac{q}{A_p} = \frac{t_1 - t_4}{\dfrac{1}{h_{12}} + \dfrac{\Delta x_{23}}{k_{23}} + \dfrac{1}{(A_t/A_p)_3 \eta_3 h_{34}}}.$$

The effectiveness of the surface 3 is, by Eq. (2.59),

$$\eta_3 = 1 - \left(\frac{A_f}{A_t}\right)_3 (1 - \kappa_3),$$

and κ_3 is found from Eq. (2.58):

$$\kappa_3 = \frac{1}{mL} \tanh mL.$$

For the conditions stated on the right side,

$$mL = \sqrt{\frac{2h}{kw}}\, L = \sqrt{\frac{2 \times 25}{200 \times 0.0016}} \times 0.025 = 0.3125,$$

$$\kappa_3 = 0.9687,$$

$$\frac{A_f}{A_t} = \frac{2L + w}{2L + \delta} = 0.8256,$$

$$\frac{A_t}{A_p} = \frac{2L + \delta}{\delta} = 5.0,$$

$$\eta_3 = 1 - 0.8256(1 - 0.9687) = 0.9742,$$

$$\frac{q}{A} = \frac{120 - 20}{\dfrac{1}{450} + \dfrac{0.0125}{200} + \dfrac{1}{5.0 \times 0.9742 \times 25}}$$

$$= 9526.6 \text{ W/m}^2 \text{ (3019 Btu/h-ft}^2\text{)}.$$

Thus the heat flow is increased by 300%.

(c) Identical calculations for fins attached to the left side, where $h = h_{12} = 450$ W/m^2-°C, yields

$$mL = 1.3258,$$
$$\kappa_2 = 0.6549,$$
$$\eta_2 = 0.7151,$$
$$\frac{q}{A} = \frac{120 - 20}{\dfrac{1}{5.0 \times 0.7151 \times 450} + \dfrac{0.0125}{200} + \dfrac{1}{25}}$$
$$= 2458.0 \text{ W/m}^2 \text{ (779.2 Btu/h-ft}^2\text{)}.$$

Fins applied to the left side increase the heat flow by only 4%. The poor performance of the fins results from the large value of h on the left side, resulting in a relatively poor surface effectiveness.

REFERENCES

1. Rohsenow, W. M., and J. P. Hartnet, eds., *Handbook of Heat Transfer,* New York, McGraw-Hill, 1973, pp. 3–14.
2. Gardner, K. A., "Efficiency of Extended Surface," *Trans ASME,* Vol. 67, No. 8, 1945, pp. 621–631.
3. Chapman, A. J., *Heat Transfer,* 4th ed., New York, Macmillan, 1984.
4. Kern, D. Q., and A. D. Kraus, *Extended Surface Heat Transfer,* New York, McGraw-Hill, 1972.

PROBLEMS

2.1 It is desired to limit the heat loss through a boiler furnace wall to 2000 W/m^2. The wall is composed of fire brick having a thermal conductivity of 1.15 W/m-°C. If the inner surface temperature is 1100°C and the outer surface 370°C, what thickness should be used?

2.2 If the wall of Prob. 2.1 has 5 cm of insulation (k = 0.09 W/m-°C) added to its outer surface and reduces the exposed surface temperature to 180°C, what will be the rate of heat loss through the wall per unit of area?

2.3 The wall of a pressure vessel is so large that it may be treated as plane. If the wall is made of 1% carbon steel and is 0.75 in. thick, find the heat flux through the wall when the inside temperature is maintained at 350°F and the outside at 300°F.

2.4 The basement wall of a building is to be made of concrete (k = 0.95 W/m-°C). The earth on the outside of the wall keeps that surface temperature at 15°C. The interior environment in the basement keeps the inner surface at 26°C. If the heat flux through the wall is not to exceed 25 W/m^2, find the required thickness of the wall.

2.5 To avoid possible burn injury to workers as well as to reduce heat losses, it is desired to apply insulation to the pressure vessel wall described in Prob. 2.3. If a 0.5-in. layer of insulation k = 2.0 Btu/h-ft-°F) is applied to the outer surface of the vessel wall so that the exposed insulation temperature is 140°F, what is the resultant heat flux? Assume that the inside wall surface remains at 350°F.

2.6 The composite wall of a furnace is made of 20 cm of fire-clay brick (burnt at 1330°C), 15 cm of fired diatomaceous earth brick, and an outer layer of common brick 10 cm thick. If the inside surface is at 1050°C and the outside surface is held at 150°C, find **(a)** the heat loss, per unit area; **(b)** the temperatures at the junction between the different layers of brick; and **(c)** the temperature at a point 15 cm from the outer surface into the wall.

2.7 The composite wall of a furnace is made of 7 in. of fire-clay brick (burnt at 2426°F), 5 in. of fired diatomaceous earth brick, and an outer 4-in. layer of common brick. If the inside surface temperature is 1800°F and the outside surface is held at 250°F, find **(a)** the heat loss rate per square foot of wall area; **(b)** the temperature at the junction between the different layers of brick; and **(c)** the temperature at a point 6 in. from the outer surface, into the wall.

2.8 A wall is composed of a 12-cm-thick layer of material A and a 20-cm-thick layer of material B. The temperature of the outer surface of layer A is known to be 260°C when the outer temperature at layer B is at 30°C. A layer of 2.5-cm-thick insulation (k = 0.09 W/m-°C) is added to the outer surface of layer B. Under these conditions it is observed that the surface of layer A rises to 310°C, the junction between layer B and the insulation (formerly the outer surface of layer B) becomes 220°C. If the insulation surface is measured to be 25°C, what is the rate of heat flow, per square meter of wall area, *before* and after the insulation is added?

2.9 A wall of a house is composed of 10 cm of common brick, 1.25 cm of Celotex, a 9-cm air space created by the studs, and 1.25 cm of asbestos cement board. If the outside brick surface is at 30°C, and the inner wall board surface is at 20°C, what is the heat flow rate, per unit area of wall surface, **(a)** if the air space *conductance* is 7.0 W/m^2-°C, and **(b)** if the air space is filled with glass wool at a density of 64 kg/m^2?

2.10 The composite wall of a furnace consists of three layers of different materials. The inner layer has a thickness $\Delta x_{12} = 30$ cm and is composed of a material with $k_{12} = 20$ W/m-°C. The middle layer has a thickness $\Delta x_{23} = 15$ cm but its material is of unknown conductivity. The outer layer has a thickness $\Delta x_{34} = 15$ cm and is composed of a material with $k_{34} = 50$ W/m-°C. Under steady operating conditions the inside surface temperature is observed to be 600°C, the outside surface temperature is 20°C, and the heat flux through the wall is measured to be 5000 W/m^2. What is the thermal conductivity of the middle layer of material?

2.11 A composite wall is formed of a 1-in.-thick copper plate, a 0.1-in.-thick layer of asbestos cement board, and a 2-in.-thick layer of glass wool (64 kg/m^3). If the structure is subjected to an overall temperature difference of 225°F, find the heat flux through the wall.

2.12 A plane wall of material is 12 cm thick. Its inner surface temperature is maintained at 0°C. When its outer surface is maintained at 100°C the heat flux through the wall is measured to be 16,000 W/m^2. When the outer surface is raised to 200°C, the heat flux increases to 40,000 W/m^2. Is the thermal conductivity independent of temperature? If not, estimate the constants in the linear law of Eq. (2.16).

2.13 Find the relation for the rate of heat flow through a single-layered plane wall, the thermal conductivity of which varies quadratically:

$$k = k_0(1 + bt + ct^2).$$

2.14 Two circular bars of 75S-T6 aluminum ($k = 120$ W/m-°C), each 20 cm long, have a surface roughness of 0.25 μm. The cylinders are pressed together on their circular faces and the assembly subjected to an overall temperature difference of 50°C. Find the axial heat flux **(a)** if the contact resistance is neglected, and **(b)** if the contact resistance is accounted for and the contact pressure is (1) 100 kN/m^2 or (2) 1000 kN/m^2.

2.15 Repeat Prob. 2.14 if the surface roughness is 3.05 μm.

2.16 Imagine that the wall described in Prob. 2.6, rather than having its surface temperature specified is surrounded on its fire-brick side by hot gases at 1150°C with a heat transfer coefficient of 32.0 W/m^2-°C and is surrounded on its other side by air at 38°C with a heat transfer coefficient of 8.5 W/m^2-°C. Find **(a)** the rate of heat flow per unit area, and **(b)** the temperature of the two surfaces.

2.17 Imagine that the wall described in Prob. 2.7, rather than having its surface temperatures specified, is surrounded on its fire-brick side by hot gases at 2000°F with a heat transfer coefficient of 5.4 Btu/h-ft^2-°F and is surrounded on its other side by air at 100°F with a heat transfer coefficient of 1.5 Btu/h-ft^2-°F. Find **(a)** the rate of heat flow through each square foot of wall, and **(b)** the temperature of the two surfaces.

2.18 Let the wall of the house in Prob. 2.9 be subjected to an outdoor wind velocity of 16 km/h so that a heat transfer coefficient of 30 W/m^2-°C exists on its outer surface. Assuming that the inside heat transfer coefficient is 10 W/m^2-°C, find the rate of heat flow and the wall surface temperatures for inside and outside air temperatures of 18°C and 38°C, respectively.

2.19 A steel (k = 43 W/m-°C) plate 1.25 cm thick is exposed on one side to steam at 650°C through a heat transfer coefficient of 570 W/m^2-°C. As a safety precaution, it is desired to insulate the outer surface so that the exposed surface of the outer insulation does not exceed 38°C. To minimize cost, an expensive high-temperature insulation (k = 0.26 W/m-°C) is applied to the steel surface, and then a less-expensive insulation (k = 0.09 W/m-°C) is placed on the outside. The maximum allowable temperature of the less-expensive insulation is 315°C. The heat transfer coefficient at the outermost surface is 11.3 W/m^2-°C, and the ambient air there is at 30°C. Find the thickness of the two layers of insulation.

2.20 A house wall consists of 1 in. of plaster (on metal lath), 5 in. of fiberglass insulation (k = 0.02 Btu/h-ft-°F), and 1.5 in. of oak wood. A surface heat transfer coefficient of 2 Btu/h-ft^2-°F exists at the inside plaster surface which is exposed to air at 70°F. The outer, wood, surface is exposed to air at 10°F and a heat transfer coefficient of 5.5 Btu/h-ft^2-°F exists at that surface. Determine the heat flux through the wall and the temperatures of the two surfaces.

2.21 A laboratory furnace has the dimensions: 3 m × 4 m × 5 m. Its wall consists of a 15-cm-thick layer of chrome brick, a 5-cm-thick layer of fiberglass (k = 0.035 W/m-°C), and an outer layer of 1% carbon steel 1 cm thick. The inside brick surface is exposed to hot gases at 475°C, where a surface heat transfer coefficient of 20 W/m^2-°C exists. The outside steel surface is exposed to ambient air at 20°C through a heat transfer coefficient of 25 W/m^2-°C. Assuming that one of the 4 m × 5 m sides rests on an insulated floor, estimate the heat lost through the furnace walls and the temperature at the junction between the brick and the fiber glass.

2.22 Suppose that the maximum allowable temperature for the fiberglass in Prob. 2.21 is 330°C. How thick should the brick be made? What is the resultant heat loss?

2.23 A nominal 5-in. wrought iron schedule 40 pipe is covered with a 5-cm layer of 85% magnesia insulation. If the pipe carries steam so that the inner surface is at 425°C and the outer insulation surface is 40°C, what is the heat loss per hour from 30 m of this pipe?

2.24 If a nominal 8-in. schedule 80 steel steam pipe is covered first with a 7.5-cm layer of 85% magnesia insulation and then a 2.5-cm layer of air-cell insulation (k = 0.064 W/m-°C), find the heat loss, per unit length of pipe, if the inside pipe surface is at 530°C and the outside insulation surface is at 90°C. What error is incurred if the thermal resistance of the pipe wall is neglected?

2.25 Find the temperature at the junction between the two insulations in Prob. 2.24.

2.26 A pipe in a refrigeration system has an outside diameter of 6 cm and a surface temperature of 0°C. It is to be covered with a 2.5-cm thickness of expanded cork insulation and a 2.5-cm thickness of rock wool (64 kg/m^3 density). Which insulation should be placed next to the pipe surface to achieve the maximum insulating effect if the outer insulation surface temperature is 35°C in either instance?

2.27 A nominal-4-in. wrought iron schedule 40 pipe is covered with a 2-in. layer of 85% magnesia insulation. If the pipe carries steam so that the inner surface is at 700°F and the outer insulation surface is 150°F, what is the heat loss per hour from 100 ft of this pipe?

2.28 A steel pipe has an outside diameter of 5 cm. It is covered with a 0.6-cm layer of one insulation (k = 0.17 W/m-°C) followed by a 2.5-cm layer of a second insulation (k = 0.048 W/m-°C). The pipe surface temperature is 315°C and the outer insulation surface temperature is 40°C. Calculate the temperature at the junction of the two insulations.

2.29 A nominal-1-in. schedule 40 wrought iron pipe has an inside surface temperature of 250°F. When a $\frac{1}{2}$-in. layer of magnesia insulation is applied to the pipe the temperature at the outer insulation surface is found to be 100°F. What is the heat loss, per foot of length, from the pipe? Calculate the temperature at the pipe–insulation interface.

2.30 A hollow cylinder of inside and outside radii r_1 and r_2, respectively, is heated such that its inner and outer surfaces are at uniform temperatures t_1 and t_2. If the material of which the cylinder is composed has a thermal conductivity which varies with temperature in the following way:

$$k = k_0(1 + bt),$$

find the rate of heat flow through the cylinder. At what mean temperature should one evaluate k in order to use the constant-thermal-conductivity formula given in Eq. (2.25)?

2.31 A nominal-4-in. schedule 40 wrought iron pipe is covered with 3.8 cm of 85% magnesia insulation. The pipe carries superheated steam at a temperature of 400°C, and the outer insulation surface is exposed to air at 25°C. If the inside and outside heat transfer coefficients are 1400 and 10 W/m^2-°C, respectively, find the rate of heat loss per unit of pipe length, and find the temperatures of the inside pipe surface, outside pipe surface, and outside insulation surface.

2.32 The pipe configuration described in Prob. 2.24 carries steam at 650°C with an inside heat transfer coefficient of 2800 W/m^2-°C. The outer insulation surface is exposed to air at 50°C with a surface coefficient of 10 W/m^2-°C. Find the heat loss per unit length.

2.33 Calculate the heat loss per unit of length of a 4-in. schedule 40 pipe covered with a 1.25-cm layer of insulation (k = 0.09 W/m-°C) if the inside pipe surface temperature is 200°C and the outside air temperature is 20°C. Assume an outside heat transfer coefficient of 170 W/m^2-°C.

2.34 A long pipe (7.5 cm ID, 10 cm OD, k = 35 W/m-°C) is covered with 1.25 cm of insulation (k = 5 W/m-°C). The temperature of a fluid flowing inside the pipe is 20°C, and the temperature of the inner wall of the pipe is 45°C. The temperature of a fluid surrounding the outer surface of the insulation is 150°C, and the heat transfer coefficient at that surface is 85 W/m^2-°C. Find **(a)** the heat flow, per meter of pipe length, between the two fluids; and **(b)** the temperature of the outer insulation surface.

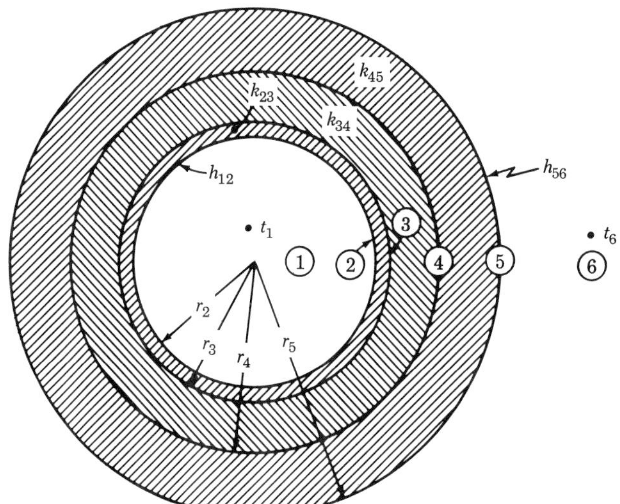

Problem 2.35

2.35 The accompanying sketch depicts a pipe covered with two layers of insulation. The pipe has an inside diameter $D_2 = 6$ in., an outside diameter $D_3 = 8$ in., and a thermal conductivity $k_{23} = 8$ Btu/h-ft-°F. The first layer of insulation is 3 in. thick, but its thermal conductivity is unknown. The second layer of insulation is 4 in. thick and has a thermal conductivity $k_{45} = 0.4$ Btu/h-ft-°F. The pipe carries a fluid of known temperature $t_1 = 1000$°F, and the inner pipe temperature is measured to be 900°F. The heat transfer coefficient at the inner pipe surface is $h_{12} = 5.5$ Btu/h-ft^2-°F. The temperature of the outermost insulation surface is known to be $t_5 = 100$°F, the heat transfer coefficient at that surface is $h_{56} = 10.0$ Btu/h-ft^2-°F, but the ambient fluid temperature, t_6, is not known. Find **(a)** the heat flow from the pipe, per foot of length; and **(b)** the unknown conductivity of the inner insulation, k_{34}.

2.36 A refrigerant at −40°C flows inside a copper pipe (3 cm ID, 3.4 cm OD) and a heat transfer coefficient of 110 W/m^2-°C exists at the inner surface. It is desired to add insulation ($k = 0.04$ W/m-°C) to the outside of the pipe in order to reduce the heat flow into the pipe to 20 W per meter of pipe length. If the ambient air is at 25°C and a heat transfer coefficient of 18 W/m^2-°C can be expected at the insulation surface, determine the required thickness of the insulation.

2.37 Steam at 250°C flows through a 1% carbon steel pipe (7.5 cm OD, 6.0 cm ID) that is covered with 2.5 cm of insulation ($k = 0.05$ W/m-°C). The heat transfer coefficient at the inner pipe surface is 400 W/m^2-°C. The outside insulation surface is exposed to air at 20°C, where it loses heat by convection through a heat transfer coefficient at 20 W/m^2-°C. The outer surface (emissivity 0.8) simultaneously loses radiant heat to the walls of a very large room, also at 20°C. Find the surface temperature of the insulation and the rate of heat loss from the pipe, per meter of length.

2.38 Repeat Prob. 2.33 if, in addition to the convective heat transfer at the outer insulation surface, the pipe also loses heat by radiating to the walls of a very large room at 20°C. The insulation surface emissivity is 0.7.

2.39 A bare 2.5-cm-diameter pipe has a surface temperature of 175°C and is placed in air at 30°C. The convective heat transfer coefficient between the surface and the air is 5.6 W/m^2-°C. It is desired to reduce the heat loss to 50% of its present value by the addition of an insulation with $k = 0.17$ W/m-°C. Assuming that the pipe surface temperature and the exposed surface heat transfer coefficient remain unchanged as insulation is added, find the required thickness of insulation. Is this thickness an economically reasonable value?

2.40 Repeat Prob. 2.39 for a 10-cm-diameter pipe, all other data remaining unchanged.

2.41 Repeat Prob. 2.31 for a nominal-10-in. schedule 40 steel pipe subjected to identical conditions. Take all conductivities and film coefficients to be the same.

2.42 A 6-in. schedule 40 steel pipe carries steam at 200 psia, 400°F. Minimum cost is to determine the selection of the thickness of magnesia insulation to be used in insulating the pipe. The air temperature in the room through which the pipe passes is 90°F and a film coefficient of 1.5 Btu/h-ft^2-°F exists at the exposed insulation surface. The cost of generating the steam is $1.00 per million Btu. The cost of the insulation, installed, depends on the thickness:

1 in.:	$5.00 per foot of length
2 in.:	$8.00
3 in.:	$14.00
4 in.:	$20.00
5 in.:	$30.00

Annual fixed charges for interest, repairs, etc., are 10% of the initial cost. The steam line operates 8000 h per year. Recommend the thickness of insulation to be used based on an estimated life of 10 years.

2.43 A 12-in. schedule 40 steel pipe has a surface temperature (outside) of 480°C. It is desired to insulate the pipe, as a safety precaution, so that the exposed insulation surface does not exceed 40°C. Two insulating materials are to be used. First a high temperature insulation (k = 0.26 W/m-°C) is applied next to the pipe and then a less-expensive insulation (k = 0.073 W/m-°C) is placed on the outside. The maximum temperature of the less-expensive insulation is 315°C. The outermost insulation surface is exposed to ambient air at 24°C through a surface heat transfer coefficient of 9.65 W/m^2-°C. Assuming that the pipe surface temperature remains constant, find the thickness of the two types of insulation.

2.44 An electrical wire has a diameter of 0.32 cm. If it is to be covered with an electrical insulation having a thermal conductivity of 0.09 W/m-°C, find the thickness that will give the maximum rate of heat dissipation for a surface heat transfer coefficient of 22.7 W/m^2-°C.

2.45 A long, electrically heated wire of 0.040 in. diameter has a surface temperature of 400°F. Its surface is exposed to ambient air to 60°F through a surface heat transfer coefficient of 5 Btu/h-ft^2-°F.

(a) Find the heat loss from the wire, per foot of length.

(b) If the wire is covered with insulation (k = 0.025 Btu/h-ft-°F) 0.125 in. thick, find the heat loss if the wire surface temperature, the air temperature, and the heat transfer coefficient remain unchanged.

2.46 Find the overall heat transfer coefficient, U, for the plane walls described in Probs. 2.17 and 2.18.

2.47 Find the overall heat transfer coefficient, U, for the plane walls described in Probs. 2.20 and 2.21.

2.48 Find the overall heat transfer coefficient, U, based on the outside exposed surface area for the pipes described in Probs. 2.31 and 2.32.

2.49 Find the overall heat transfer coefficient, U, based on the outside exposed surface area for the pipes described in Probs. 2.34 and 2.35.

2.50 A $\frac{3}{4}$-in 18-gage brass condenser tube has a heat transfer coefficient at its inner surface of 5400 W/m^2-°C and one of 6800 W/m^2-°C at its outer surface. The cooling water flowing in the tube is at 30°C. Find **(a)** the overall heat transfer coefficient, U and **(b)** the mass of saturated steam at 10 kN/m^2 pressure that will be condensed for each meter of tube length.

2.51 A water-to-water heat exchanger is made of brass tubes, 1 in. OD, 16 gage. The inside and outside heat transfer coefficients are 800 and 1200 Btu/h-ft^2-°F, respectively. Find the overall heat transfer coefficient, U.

2.52 A hollow sphere (inside radius r_1, outside radius r_2) has its inner and outer surfaces maintained at uniform temperatures t_1 and t_2, respectively. Show that the heat flow through the sphere is given by

$$q = \frac{4\pi k(t_1 - t_2)}{\dfrac{1}{r_1} - \dfrac{1}{r_2}}.$$

2.53 Using the results of Prob. 2.52, show that the heat flow through a two-layer hollow sphere with inside and outside convective heat transfer coefficients is given by

$$q = \frac{4\pi k(t_1 - t_5)}{\dfrac{1}{r_2^2 h_{12}} + \dfrac{(1/r_2) - (1/r_3)}{k_{23}} + \dfrac{(1/r_3) - (1/r_4)}{k_{34}} + \dfrac{1}{r_4^2 h_{45}}}.$$

The notation is the same as that used in Fig. 2.9 and Eq. (2.28) for a two-layer cylinder.

2.54 A sphere of fixed outside radius and fixed surface temperature is to be insulated with a material of known thermal conductivity. For fixed values of an ambient fluid temperature and surface film coefficient, find if a "critical thickness" of insulation exists (as in the cylindrical case considered in Sec. 2.5) and, if so, what its value is.

2.55 A hollow stainless steel (18% Cr, 8% Ni) sphere has an inside diameter of 50 cm and an outside diameter of 60 cm. It is placed in an ambient fluid at 15°C. The outer sphere surface is at 50°C and a heat transfer coefficient of 140 W/m^2-°C exists at that surface. What is the temperature of **(a)** the inside sphere surface, and **(b)** a point midway through the sphere wall?

2.56 A sphere with a diameter of 10 cm is electrically heated so that its surface temperature is maintained at 225°C. The surface is exposed to an ambient fluid at 25°C through a heat transfer coefficient 50 W/m²-°C.

(a) Find the heat loss from the sphere surface.

(b) If insulation (k = 0.08 W/m-°C) is added to the sphere, what thickness is required to reduce the heat flow to 20% of its original value, assuming that all other parameters remain unchanged?

2.57 A hot water holding tank consists of a cylindrical section (inside diameter 0.5 m, length 1 m) capped on each end with hemispherical sections. The tank wall is steel (k = 42 W/m-°C) 0.5 cm thick covered with 5 cm of insulation (k = 0.1 W/m-°C). The tank contains water at 100°C, and a heat transfer coefficient of 75 W/m²-°C exists at the inside surface. The outer surface is exposed to ambient air at 20°C with a heat transfer coefficient of 10 W/m²-°C. Determine the rate of heat loss from the tank.

2.58 Show, for a single-layered cylinder, that the equation for the rate of heat flow through a plane wall may be used if one uses for the wall area the log-mean of the inner and outer cylindrical surface areas. That is,

$$A_m = \frac{A_2 - A_1}{\ln (A_2/A_1)}.$$

2.59 Repeat Prob. 2.58 for a single-layered sphere, showing that the geometric mean area should be used:

$$A_m = \sqrt{A_2 A_1}.$$

2.60 A large slab of concrete, 1 m thick, has both surfaces maintained at 20°C. During the curing process a uniform internal heat generation of 60 W/m³ occurs throughout the slab. If the concrete thermal conductivity is 1.13 W/m-°C, find the steady temperature which results at the center of the slab.

2.61 Repeat Prob. 2.60 if instead of having its surface maintained at 20°C, the slab is exposed to a convecting fluid at 20°C through a surface heat transfer coefficient of 2.5 W/m²-°C.

2.62 A copper rod (k = 380 W/m-°C) 0.5 cm in diameter and 30 cm long has its two ends maintained at 20°C. The lateral surface of the rod is perfectly insulated, so conduction may be taken as one-dimensional along the length of the rod. Find the maximum electrical current that the rod may carry if the temperature is not to exceed 120°C at any point and the electrical resistivity (resistance × cross section/length) is 1.73×10^{-6} Ω-cm.

2.63 A bare copper wire 0.3 cm in diameter has its outer surface maintained at 25°C while carrying an electrical current. The electrically generated heat is conducted one-dimensionally in a radial direction. If the centerline temperature of the wire is not to exceed 120°C, find the maximum current the wire can carry. Use the thermal and electrical properties for copper given in Prob. 2.62.

2.64 Repeat Prob. 2.63 if instead of having its surface maintained at 25°C the wire has its surface exposed to air at 25°C through a heat transfer coefficient of 10 W/m^2-°C. Explain the difference in the results of this problem and Prob. 2.63.

2.65 The accompanying figure depicts a plane wall composed of two materials. The inner material, a, has a thickness $\Delta x_a = 5$ cm, a conductivity $k_a = 75$ W/m-°C, and an internal heat generation rate of $q_a^* = 1.5 \times 10^6$ W/m^3. The outer material, b, has a thickness $\Delta x_b = 2$ cm, a conductivity of $k_b = 150$ W/m-°C, but no internal heat generation. The left face of material a is perfectly insulated and the right face of material b is exposed to a fluid at $t_f = 30$°C with $h = 1000$ W/m^2-°C. Determine the temperatures of the insulated surface, the junction of the two materials, and the exposed surface of material b.

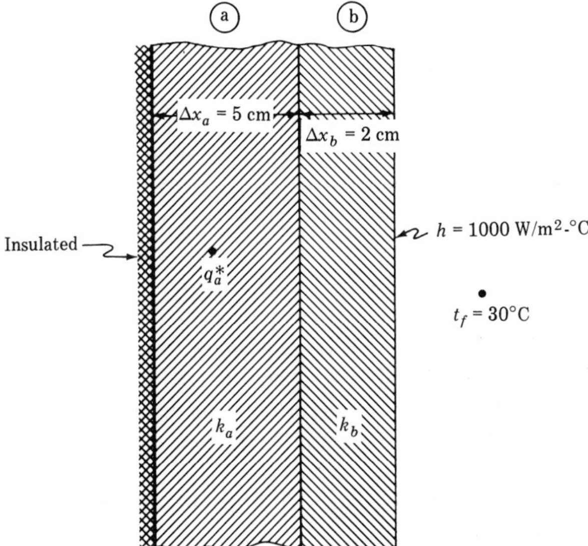

Problem 2.65

2.66 A uranium oxide nuclear reactor fuel element consists of a circular rod 1.2 cm in diameter and has a thermal conductivity of 2 W/m-°C. The fuel element is encased by a layer of cladding material 0.3 cm thick with a conductivity of 25 W/m-°C. The fuel element generates 2.8×10^6 W/m^3 of internal heat and the cladding is exposed to water at 27°C through a heat transfer coefficient of 1000 W/m^2-°C. What is the maximum temperature in the fuel element?

2.67 Consider a solid sphere, radius R, the surface of which is maintained at t_s. If the sphere has a uniform internal heat generation rate of q^*, per unit volume, derive expressions for the temperature distribution in the sphere and the temperature at the sphere center. Take the thermal conductivity to be constant.

2.68 A plane wall with a uniform internal heat generation rate of q^*, per unit volume, has its face at $x = x_1$ maintained at temperature t_1 and the face at $x = x_2$ maintained at t_2. (That is, the wall has a thickness $\Delta x = x_2 - x_1$.) Show that the temperature distribution in the wall is given by

$$t - t_1 = \frac{t_2 - t_1}{x_2 - x_1}(x - x_1) + \frac{q^*}{2k}(x_2 - x_1)^2\left[\frac{x - x_1}{x_2 - x_1} - \left(\frac{x - x_1}{x_2 - x_1}\right)^2\right],$$

where x is any position $x_1 \leq x \leq x_2$.

2.69 A long hollow cylinder with a uniform internal heat generation of q^*, per unit volume, has its inside at radius r_1 maintained at t_1 and the outside surface, radius r_2, maintained at t_2. Show that the temperature at any intermediate radius r is given by

$$t - t_1 = (t_2 - t_1)\frac{\ln(r/r_1)}{\ln(r_2/r_1)} + \frac{q^*}{4k}\left[(r_2^2 - r_1^2)\frac{\ln(r/r_1)}{\ln(r_2/r_1)} - (r^2 - r_1^2)\right].$$

2.70 A 1.25-cm-diameter rod of iron ($k = 45$ W/m-°C) is heated to 260°C at its base and protrudes into air at 38°C where $h = 8.5$ W/m^2-°C. How long must the rod be so that if its end temperature is computed using Eq. (2.55) for the infinitely long fin with $x = L$, the error will be less than 2.5°C as compared to the exact solution given in Eq. (2.53)? What error is made at this length in the total heat flow from the rod?

2.71 A cylindrical rod of 1.9 cm diameter, 24 cm long, protrudes from a casting where its base is maintained at 150°C into air at 32°C. The film coefficient of convective heat transfer is known to be 5.6 W/m^2-°C on all exposed surfaces. Find the following information for the three cases of the rod being composed of copper, cast iron, and glass:

(a) The temperature at points located $\frac{1}{4}$, $\frac{2}{4}$, $\frac{3}{4}$, and $\frac{4}{4}$ of the distance from the source to the rod end. Neglect end heat loss.

(b) The rate of heat flow out of the source if the end heat loss is neglected, and if the end heat loss is not neglected.

2.72 A 0.6-cm-diameter pure copper rod 46 cm long connects two pressure vessels such that it is heated to 90°C at each end. If the air which surrounds the rod is at 25°C and a heat transfer coefficient of 25 W/m^2-°C exists at the surface, find the temperature at the midpoint of the rod and at a point 10 cm from one end. How much heat is dissipated to the surroundings by the first 10 cm of the rod?

2.73 A straight fin of rectangular profile is made of an aluminum alloy ($k = 10$ Btu/h-ft-°F). It has a thickness of 0.01 in. and is attached to a wall so that its base temperature is 150°F. The fin protrudes into an ambient fluid at 75°F, and a heat transfer coefficient of 15 Btu/h-ft²-°F exists on the exposed surface. Find the heat flow, per unit width, from the fin and the temperature of its exposed end if the fin length is **(a)** 0.5 in., **(b)** 1.0 in., or **(c)** 2.0 in.

2.74 A straight fin of uniform cross section A, length L, perimeter C, conductivity k is maintained at a temperature, above the ambient fluid, of θ_0 at the end where $x = 0$ and at θ_L at the end where $x = L$. The heat transfer coefficient at the exposed surface is h. Derive the following expressions for the rate of heat flow at the two ends (positive from $x = 0$ to $x = L$):

$$q_0 = kmA \frac{\theta_0 \cosh mL - \theta_L}{\sinh mL},$$

$$q_L = kmA \frac{\theta_0 - \theta_L \cosh mL}{\sinh mL}.$$

2.75 In Prob. 2.74, let the fin be a circular rod with a diameter of 1.25 cm and a length of 45 cm. The left end is maintained at 175°C, the right end at 50°C, and the ambient fluid is at 20°C. For $h = 11$ W/m²-°C and $k = 43$ W/m-°C, find the heat flow rate out of each source.

2.76 A Chromel–Alumel thermocouple (wire diameters = 0.125 cm) is attached to a surface at 120°C and extends into air at 25°C ($h = 10$ W/m²-°C). Estimate the rate of heat loss from the surface due to the attachment of the thermocouple.

2.77 For free convection coefficients in air of the order of magnitude of 1 Btu/h-ft²-°F, make a reasonable estimate for the minimum length of a wiener-roasting wire made of an old coat hanger in order to avoid an uncomfortably hot temperature on the end held by the user.

2.78 The following technique is sometimes used to measure the thermal conductivity of a substance by comparing it with another material of known conductivity. Consider two rods of identical size and shape that are both supported between two heat sources, each at 100°C. Both rods are exposed to ambient air at 25°C. One rod is known to have a thermal conductivity of 43 W/m-°C and its midpoint temperature is measured to be 49°C. If the midpoint temperature of the other rod is observed to be 75°C, estimate its thermal conductivity.

2.79 For a straight fin of rectangular profile, the parameter m is, for unit width, $m^2 = 2h/kw$. If such a fin, of given k and given width, w, is to be utilized under fixed service conditions (i.e., θ_0, h), the heat loss from the fin may be expressed as a function of the fin length L. Write such an expression, including the loss of heat from the end with $h_e = h$, and show that if $(hw/2k) > 1$, $dq/dL < 0$ for *all* L; and if $hw/2k < 1$, $dq/dL > 0$. This shows that unless $hw/2k < 1$, the application of such fins will produce an insulating effect.

2.80 Consider a straight fin of rectangular profile. Let the "profile area" be that area taken in a plane that is parallel to the fin length and normal to the width, $A_p = w \times L$. For a fixed amount of material (i.e., A_p = constant), show that if the heat loss from the fin end is negligible, the fin dissipates the maximum amount of heat if its length, L, and thickness, w, are related by the following condition:

$$\tanh \xi = 3\xi \operatorname{sech}^2 \xi,$$

$$\xi = L\sqrt{\frac{2h}{kw}}.$$

2.81 An array of straight fins is to be used to enhance the heat exchange between a plane surface of a heat exchanger and a fluid flowing past it. The fins are to be 1 in. in length and made of stainless steel (18% Cr, 8% Ni). The heat transfer coefficient at the surface is expected to be 3 Btu/h-ft^2-°F when the fin extends into a fluid at 120°F. The fin base temperature is 200°F. Find the fin efficiency and the heat flow, per unit width, from the fin if the fin profile is (1) rectangular or (2) triangular when the fin base width is **(a)** 0.1 in. or **(b)** 0.05 in.

2.82 Repeat Prob. 2.81 if the fin material is brass.

2.83 An annular fin of rectangular profile has an inner radius of 7.5 cm and an outer radius of 12.7 cm. The fin thickness is 0.5 cm and it is made of a material with k = 43 W/m-°C. The base of the fin is held at 200°C and the surface is exposed to a fluid at 35°C through a heat transfer coefficient of 56.8 W/m^2-°C. Find the fin efficiency and the heat transferred to the fluid.

2.84 Write an expression for the total surface effectiveness of the annular array shown in Fig. 2.19 in terms of the geometry shown and the efficiency of a single fin.

2.85 Annular aluminum fins of a rectangular profile are attached to the tubes of a heating system steam coil in the manner illustrated in Fig. 2.19. The tube has an outside diameter of 5 cm. The fins have a thickness of 0.2 cm, length of 1.5 cm, thermal conductivity k = 200 W/m-°C, and an on-center spacing δ = 0.8 cm (i.e., 125 fins per meter of tube length). The tube surface is at 200°C and the fins dissipate heat by convection to the ambient air at 30°C through a heat transfer coefficient of 50 W/m^2-°C. Find the heat flow to the air, per meter of tube length.

2.86 The plane wall of a heat exchanger is equipped with an array of straight fins of triangular profile made of steel with k = 43 W/m-°C. The base of the fins is maintained at 120°C and the array is exposed to an ambient fluid at 25°C through a heat transfer coefficient of 20 W/m^2-°C. The fins have a base width of 0.32 cm, a length of 3.8 cm, and are spaced 1.5 cm on centers. Find the total surface effectiveness of the array and the heat transferred to the fluid per square meter of the original primary surface. Compare the heat transfer to that if the primary surface were bare of fins.

2.87 A plane wall of a heat exchanger is equipped with an array of straight fins of rectangular profile, 1.9 cm long, 0.13 cm thick, spaced 1.9 cm on centers. The fins are made of Duralumin and the surface heat transfer coefficient is 142 W/m^2-°C. Find the total surface effectiveness and the heat transfer rate from the array when the base temperature excess is θ_0 = 150°C, per square meter of primary surface.

2.88 A plane heat exchanger surface is provided with an array of aluminum (k = 115 Btu/h-ft-°F) fins of rectangular profile (w = 0.04 in., L = 0.4 in.) spaced 0.2 in. on centers. The surface is maintained at 390°F, and the fins dissipate heat by convection to ambient air at 100°F through a heat transfer coefficient of 10 Btu/h-ft^2-°F. Find the total surface effectiveness and the rate of heat transfer from the array, per square foot of primary surface. Compare this heat transfer rate with that found if the surface was bare of fins.

2.89 An array of straight fins of triangular profile is attached to a plane wall at 460°F. The fins have a base width of 0.08 in., a length of 0.35 in., and are spaced 0.2 in. on centers. The fins are made of aluminum (k = 115 Btu/h-ft-°F) and are exposed to a fluid at 100°F through a heat transfer coefficient of 10 Btu/h-ft^2-°F. Find the total surface effectiveness and the heat transfer rate from the array, per square foot of primary surface area.

2.90 Heat is being transferred through the plane wall of a heat exchanger 1.25 cm thick, composed of a material with k = 17.3 W/m-°C. On the left side of the wall is a fluid of temperature 90°C, and the heat transfer coefficient there is 284 W/m^2-°C. On the right side of the wall is another fluid of temperature 38°C, and the heat transfer coefficient to all exposed surfaces there is 17.0 W/m^2-°C. It is desired to enhance the heat transfer between the two fluids by adding to the right surface either (1) straight rectangular fins, 0.13 cm thick, 2.5 cm long, spaced on 1.25-cm centers; or (2) straight triangular fins, 0.13 cm thick at the base, 2.5 cm long, spaced on 1.25-cm centers. Assume that the fins are made of the same material as the wall, that the heat transfer coefficient is unchanged by adding the fins, and that a one-dimensional representation may be used. Based per unit of area of plane, unfinned, wall, find **(a)** the heat transfer between the fluids when the wall is bare, **(b)** the total surface temperature effectiveness of the right surface and the heat exchange between the two fluids when fins are used as in case 1, and **(c)** same as part (b) when fins are used as in case 2.

2.91 Heat is to be transferred through the plane wall of a heat exchanger, 1.25 cm thick, composed of a material with k = 43 W/m-°C. On the left side of the wall is an ambient fluid of specified temperature, and the heat transfer coefficient there between the fluid and any exposed surface is 182 W/m^2-°C. On the right side of the wall there is also an ambient fluid of known temperature, and the fluid-to-surface heat transfer coefficient is 17 W/m^2-°C. It is contemplated to enhance the heat transfer between the two fluids by using straight, rectangular fins having the same conductivity as the wall. The fins are to be 0.064 cm thick, 2.29 cm long, and spaced 1.5 cm on centers. Presuming that the addition of the fins will not alter the existing heat transfer coefficients, that these coefficients apply to fin surfaces as well as the unfinned surface, and that a one-dimensional representation may be used, find the percent increase in the heat transfer between the two fluids (based per unit of area of the original wall without fins) over the case without fins **(a)** if fins are added to the left side only, **(b)** if fins are added to the right side only, and **(c)** if fins are added to both sides.

2.92 A pipe has an inside diameter of 2.5 in. and an outside diameter of 3.0 in. Flowing inside the pipe is a hot fluid at 200°F and a heat transfer coefficient of 45 Btu/h-ft^2-°F exists at the inner surface. The outer surface is exposed to another fluid at 100°F through a heat transfer coefficient of 30 Btu/h-ft^2-°F. The tube material has a thermal conductivity of 28.125 Btu/h-ft-°F. It is desired to increase the heat exchange rate between the two fluids by adding annular fins of the same material to the outside surface in an array like that pictured in Fig. 2.19. The fins are 0.1 in. thick, 0.75 in. long in the radial direction (from base to end), and spaced 0.4712 in. on centers. Assuming that all temperatures and heat transfer coefficients are the same before and after the addition of the fins, find the rate of heat transfer between the two fluids, per foot of pipe length, **(a)** when the tube is bare, and **(b)** when the fins are in place.

2.93 The accompanying sketch depicts a long tube equipped with straight, longitudinal fins extending in the radial direction. The tube has an inside diameter of 6.3 cm and an outside diameter of 8.9 cm. There are 18 fins, spaced equally around the circumference of the tube, with a base thickness of 0.25 cm and a radial length of 1.9 cm. The tube and the fins have a thermal conductivity of 48.6 W/m-°C. A fluid flowing inside the tube produces a heat transfer coefficient of 285 W/m^2-°C at the inner surface. A fluid outside the tube produces a heat transfer coefficient of 175 W/m^2-°C at all exposed surfaces. Based per meter of tube length, find the percent increase of the heat transfer rate between the inner and outer fluids for the finned tube over a bare tube if the fin profile is **(a)** rectangular or **(b)** triangular.

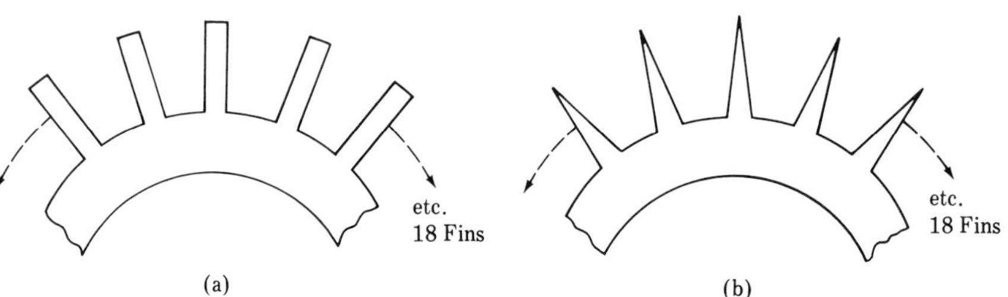

Problem 2.93

2.94 As shown in the accompanying sketch, a straight fin protruding from a plane wall at 200°C consists of a rectangular portion, 12.5 cm long and 1.25 cm thick, capped

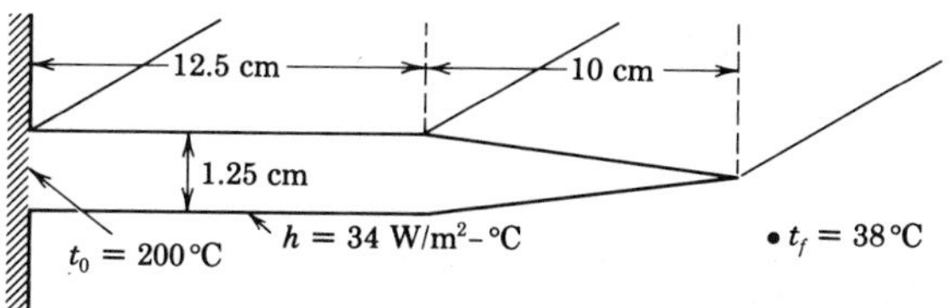

Problem 2.94

by a triangular fin 10 cm long. The fin material has a $k = 86.5$ W/m-°C and the heat transfer coefficient between the fin and an ambient fluid at 38°C is 34 W/m^2-°C. By appropriately combining the solution for a rectangular straight fin maintained at known temperatures at each end (see Prob. 2.74) with the solution for the heat flow in a triangular straight fin as given by its efficiency in Fig. 2.17, find the temperature at the section where the fin changes shape from rectangular to triangular. What is the heat transfer rate, per unit width, from the fin to the fluid?

2.95 The cross section of the core of a plate-fin heat exchanger is shown in the accompanying figure. The two primary heated surfaces are maintained at the same temperature, t_0, as shown. A fluid, of temperature t_f, flows in the open passages between the fin-matrix array shown. The fluid flows in a direction normal to the plane of the paper, and the fins are straight in that direction. The heat transfer coefficient, h, is presumed to be the same for all exposed surfaces.

(a) Find the total surface effectiveness for the primary surfaces maintained at t_0 in terms of the dimensions and parameters shown in the figure.

(b) If $h = 170$ W/m^2-°C, $k = 52$ W/m-°C, $t_0 = 50$°C, $t_f = 30$°C, $d = 1.0$ cm, $L = 1.25$ cm and $w = 0.125$ cm, find the total heat transferred to the fluid for a section of the core 0.5 m in depth (normal to the paper) and 0.5 m in width.

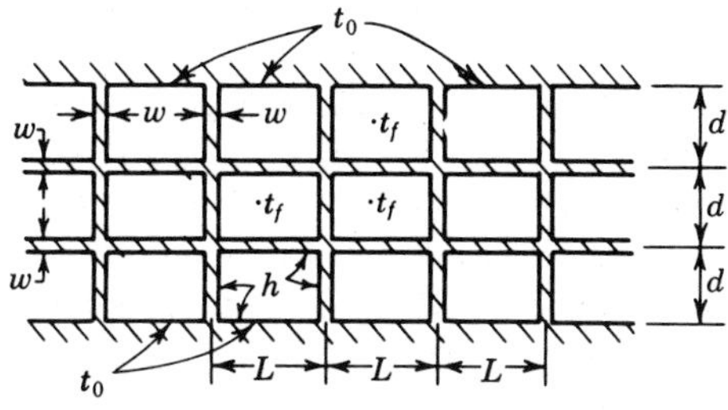

Problem 2.95

2.96 Derive, but do not solve, the governing differential equation for the temperature distribution in an annular fin of rectangular profile. Use the notation shown in the inset in Fig. 2.18.

2.97 Derive, but do not solve, the governing differential equation for the temperature distribution in a straight fin of triangular profile. Use the notation shown in the inset in Fig. 2.17, and establish the origin of the one-dimensional coordinate at the fin tip, measured toward the base.

CHAPTER 3

Heat Conduction in Two or More Independent Variables

3.1 INTRODUCTORY REMARKS

The cases of heat conduction discussed in Chapter 2 were all ones in which the temperature distribution throughout the bodies could be expressed in terms of a single independent variable. This meant that only steady-state, one-dimensional systems could be discussed.

The present chapter is devoted to discussions of conduction problems that have to be described in two or more independent variables. The major portion of the chapter will be used to describe systems in two independent variables. This means either steady conduction in two space dimensions or nonsteady conduction in one dimension (the second variable being time). The two-dimensional steady-state case will be examined first. The particular case to be studied will be conduction in a rectangular plate because of the simplicity in describing the boundary conditions. Then one-dimensional transient conduction in a plane wall will be treated to illustrate nonsteady problems. The solutions presented will involve only the simplest boundary conditions, and even these simple cases will be seen to become mathematically complicated very quickly. Hence no attempt will be made in this text to present a comprehensive coverage of the mathematical theory of multiple variable heat conduction—only the basic concepts will be illustrated. The reader is referred to more comprehensive works, such as Refs. 1 through 3, for further details.

The results obtained by more advanced techniques to problems having engineering significance will be quoted and presented graphically in forms useful for practical applications. In particular, it will be demonstrated how certain nonsteady solutions for one-dimensional problems may be combined to obtain answers for useful two- and three-dimensional cases.

3.2 STEADY-STATE CONDUCTION IN RECTANGULAR PLATES

If attention is now directed to steady-state conduction in rectangular plates, such as shown in Fig. 3.1, it is most convenient to use cartesian coordinates to describe the temperature distribution in the plate. The plate will be considered to be in the x-y plane with the origin of the coordinates at one corner. No conduction will be considered in the z direction normal to the plate. This may be imagined to be the case if the plate has such a great extent in the z direction that no end effects exist or if the x-y faces of the plate are insulated so that no heat will pass in the z direction.

The heat conduction equation for the steady state [Eq. (2.11)], for cartesian coordinates in two dimensions and with constant thermal conductivity, is

$$\frac{\partial^2 t}{\partial x^2} + \frac{\partial^2 t}{\partial y^2} = 0. \tag{3.1}$$

Since the steady-state heat conduction equation is a linear differential equation, the principle of superposition of solutions is applicable. This fact will be used later to build the solutions to more complex situations by adding the solutions of simpler problems.

The solution of Eq. (3.1) is obtained by assuming the temperature distribution to be expressible as the product of two functions, each of which involves only one of the independent variables. That is, if $X(x)$ is a function of x only and if $Y(y)$ is a function of y only, one assumes that the temperature, t, is given by

$$t = X(x)Y(y). \tag{3.2}$$

When this is substituted into Eq. (3.1) and the resulting expression is rearranged,

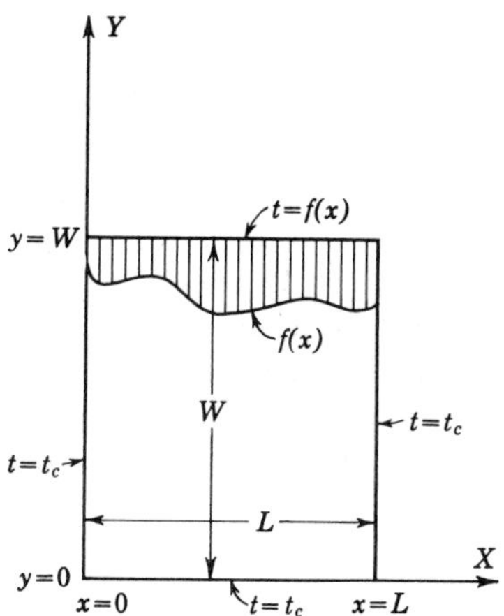

Figure 3.1. Rectangular plate with a specified temperature distribution on one edge, the other edges at constant temperature.

one obtains

$$-\frac{1}{X}\frac{d^2X}{dx^2} = \frac{1}{Y}\frac{d^2Y}{dy^2}.$$

Since each side of this equation involves only one of the independent variables, they may be equal only if they are both equal to the same constant. Calling this constant λ^2, one finds that

$$-\frac{1}{X}\frac{d^2X}{dx^2} = \frac{1}{Y}\frac{d^2Y}{dy^2} = \lambda^2.$$

This equation is equivalent to the following two ordinary differential equations:

$$\frac{d^2X}{dx^2} + \lambda^2 X = 0,$$

$$\frac{d^2Y}{dy^2} - \lambda^2 Y = 0.$$

The solutions of these equations are

$$Y = B_1 \sinh \lambda y + B_2 \cosh \lambda y,$$
$$X = B_3 \sin \lambda x + B_4 \cos \lambda x.$$

When these results are introduced into the assumed form of the solution given in Eq. (3.2), one finally obtains the following general solution of Eq. (3.1):

$$t = (B_1 \sinh \lambda y + B_2 \cosh \lambda y)(B_3 \sin \lambda x + B_4 \cos \lambda x). \tag{3.3}$$

The λ's and B's are constants to be determined by application of boundary conditions. An example of this will now be illustrated.

The Rectangular Plate with a Specified Temperature Distribution on One Edge—The Other Edges at Constant Temperature

Let the configuration under consideration be a rectangular plate of finite width L in the x-coordinate direction and of width W in the y-coordinate direction. This is illustrated in Fig. 3.1.

Let the temperature of the edges of the plate at $x = 0$, $x = L$, and $y = 0$ be maintained at a constant temperature t_c and that of the edge at $y = W$ be maintained at values that vary along that edge. Let this variation be represented as $f(x)(0 \le x \le L)$, remembering that $f(x)$, although unspecified, is to be treated as known.

The differential equation and the boundary conditions to be satisfied here are

$$\frac{\partial^2 t}{\partial x^2} + \frac{\partial^2 t}{\partial y^2} = 0.$$

At $x = 0$: $t = t_c$.

At $x = L$: $t = t_c$.

At $y = 0$: $t = t_c$.

At $y = W$: $t = f(x)$.

The problem is, then, for the system of equations noted above: What is the temperature at any specified point within the plate? That is, find the function $t = t(x, y)$ that gives the distribution of temperature in the plate. This problem may be simplified somewhat by rendering the boundary conditions homogeneous through the introduction of the temperature difference variable.

$$\theta = t - t_c.$$

Then the system to be solved is

$$\frac{\partial^2\theta}{\partial x^2} + \frac{\partial^2\theta}{\partial y^2} = 0.$$

$$\begin{aligned} &\text{At } x = 0: \quad \theta = 0. && (1)\\ &\text{At } x = L: \quad \theta = 0. && (2)\\ &\text{At } y = 0: \quad \theta = 0. && (3)\\ &\text{At } y = W: \quad \theta = f(x) - t_c. && (4) \end{aligned} \qquad (3.4)$$

The solution of the differential equation in Eq. (3.4) is of the same form as Eq. (3.3), which is the solution to Eq. (3.1). Thus

$$\theta = (B_1 \sinh \lambda y + B_2 \cosh \lambda y)(B_3 \sin \lambda x + B_4 \cos \lambda x).$$

Application of condition (3) of Eq. (3.4) shows that $B_2 = 0$, so

$$\theta = B_1 \sinh \lambda y(B_3 \sin \lambda x + B_4 \cos \lambda x).$$

In order to satisfy condition (1) of Eq. (3.4), $B_4 = 0$. Thus

$$\theta = \sinh \lambda y \, B \sin \lambda x, \qquad (3.5)$$

where B replaces the product B_1B_3. Substitution of condition (2) gives

$$0 = \sinh \lambda y \, B \sin \lambda L.$$

The only way that this may be satisfied for *all* values of y is for

$$\sin \lambda L = 0.$$

This expression is satisfied for $\lambda = 0, \pi/L, 2\pi/L, \ldots$, or, in general,

$$\lambda_n = \frac{n\pi}{L}, \quad n = 0, 1, 2, 3, \ldots \qquad (3.6)$$

Each of the λ's of Eq. (3.6) gives rise to a separate solution of Eq. (3.5), and since the general solution will be the sum of these individual solutions, one has

$$\theta = \sum_{n=0}^{\infty} B_n(\sinh \lambda_n y)(\sin \lambda_n x).$$

The symbol B_n represents the constant B for each of the solutions. Since $\lambda_n = 0$ for $n = 0$, no contribution is made by the first term,

$$\theta = \sum_{n=1}^{\infty} B_n(\sinh \lambda_n y)(\sin \lambda_n x). \qquad (3.7)$$

Applying, finally, condition (4) of Eq. (3.4) for $y = W$, one obtains

$$[f(x) - t_c] = \sum_{n=1}^{\infty} B_n \sinh \lambda_n W \sin \lambda_n x,$$
$$\lambda_n = \frac{n\pi}{L}; \quad n = 1, 2, 3, \ldots; \quad 0 \leq x \leq L. \tag{3.8}$$

Equation (3.8) determines the constants B_n. Since $\sinh \lambda_n W$ is constant, Eq. (3.8) basically asks: Is it possible to find the constants C_n:

$$C_n = B_n \sinh \lambda_n W, \qquad \lambda_n = \frac{n\pi}{L}, \qquad n = 1, 2, 3, \ldots \tag{3.9}$$

so that the function $f(x) - t_c$ is expressible as an infinite series of sines? That is, what are the C_n's that give

$$f(x) - t_c = \sum_{n=1}^{\infty} C_n \sin \frac{n\pi x}{L}? \tag{3.10}$$

The mathematically proper answer to this question involves the use of Fourier series and is beyond the scope of this text. Again, the reader is referred to Refs. 1 through 4 for detailed discussions of such mathematical techniques. However, two special cases are discussed below.

One Edge with a Harmonic Temperature Distribution. One solution to the question posed by Eqs. (3.8) and (3.10) that is easily obtained is that when the edge at $y = W$ is maintained at a temperature distribution which is a simple sine function. That is, suppose that the imposed temperature at $y = W$ is such that

$$f(x) = \mathbf{T} \sin \frac{\pi x}{L} + t_c$$

with $\mathbf{T}$ being a constant. Then Eq. (3.10) for C_n is

$$\mathbf{T} \sin \frac{\pi x}{L} = \sum_{n=1}^{\infty} C_n \sin \frac{n\pi x}{L}.$$

Clearly, $C_1 = \mathbf{T}$ for $n = 1$ and $C_n = 0$ for $n \geq 2$. Thus from Eq. (3.9):

$$B_1 = \frac{\mathbf{T}}{\sinh (\pi W/L)} \qquad \text{for } n = 1$$
$$B_n = 0 \qquad \text{for } n \geq 2.$$

Then the solution for the temperature anywhere in the plate is given by Eq. (3.7) to be

$$\frac{\theta}{\mathbf{T}} = \frac{t - t_c}{\mathbf{T}} = \frac{\sinh (\pi y/L)}{\sinh (\pi W/L)} \sin \frac{\pi x}{L}. \tag{3.11}$$

Equation (3.11) permits the calculation of the plate temperature at any values of x and y.

One Edge at a Uniform Temperature. The case in which the plate edge at $y = W$ is maintained at a constant temperature is of some practical interest. That is, let the solution for the temperature at a point within the plate be sought when $f(x) = t_0$, a constant. Such a case is illustrated in Fig. 3.2.

In this instance the C_n's defined by Eqs. (3.8) and (3.9) are to be found from Eq. (3.10):

$$t_0 - t_c = \sum_{n=1}^{\infty} C_n \sin \frac{n\pi x}{L}. \tag{3.12}$$

The theory of Fourier series (Refs. 1 through 4) states that an arbitrary function, call it $g(x)$, may be expressed as an infinite series of sine functions of the form

$$g(x) = \sum_{n=1}^{\infty} C_n \sin \frac{n\pi x}{L} \tag{3.13}$$

if the C_n's are defined as

$$C_n = \frac{2}{L} \int_0^L g(x) \sin \frac{n\pi x}{L}\, dx. \tag{3.14}$$

In this case $g(x) = t_0 - t_c$ constant, so that

$$\begin{aligned} C_n &= \frac{2}{L} \int_0^L (t_0 - t_c) \sin \frac{n\pi x}{L}\, dx \\ &= \frac{2}{L} (t_0 - t_c) \frac{L}{n\pi} \left[-\cos\left(\frac{n\pi x}{L}\right) \right]_0^L \end{aligned} \tag{3.15}$$

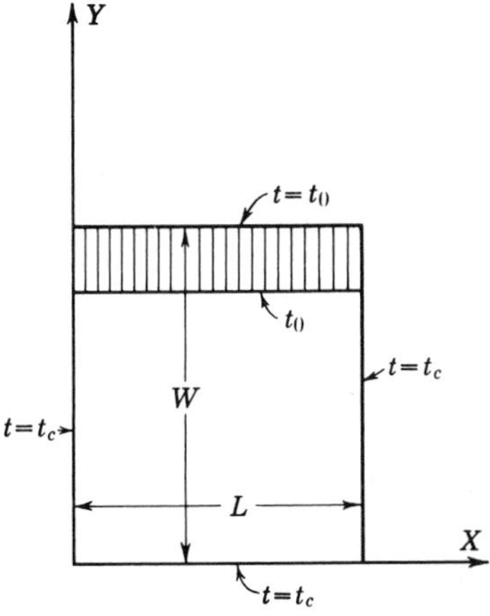

Figure 3.2. Rectangular plate with one edge at a uniform temperature, all other edges at constant temperature.

or

$$\begin{aligned} C_n &= \frac{4}{n\pi}(t_0 - t_c), \qquad && n = 1, 3, 5, \ldots \\ &= 0, && n = 2, 4, 6, \ldots . \end{aligned} \tag{3.16}$$

Since the B_n's are, by Eq. (3.9), $B_n = C_n/\sinh(n\pi W/L)$, Eq. (3.7) for the temperature at any point (x, y) in the plate is

$$\frac{\theta}{t_0 - t_c} = \frac{t - t_c}{t_0 - t_c} = \sum_{n=1,3,5,\ldots} \frac{4}{n\pi} \frac{\sinh(n\pi y/L)}{\sinh(n\pi W/L)} \sin\frac{n\pi x}{L}. \tag{3.17}$$

As an illustration of the use of Eq. (3.17), Fig. 3.3 shows the distribution of temperature in a plate 10 by 6, with one edge held at 100° and all others at 0°.

The temperature distribution given by Eq. (3.17) may also be used to find the heat flow at any of the plate edges. For example, application of Fourier's law at the edge of $y = W$ maintained at t_0 (see Fig. 3.2) shows the heat flow there to be

$$q_{y=W} = \int_0^L -k\left(\frac{\partial t}{\partial y}\right)_{y=W} dx$$

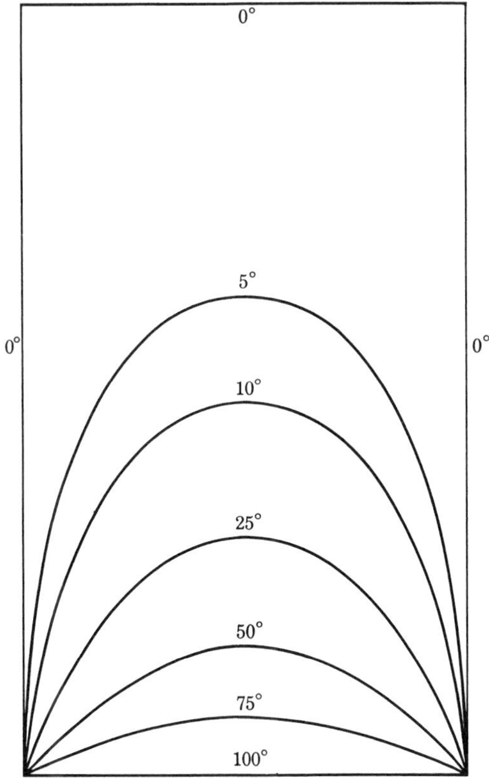

Figure 3.3. Isotherms in a rectangular plate heated on one edge.

for unit depth into the paper. Introduction of Eq. (3.17) gives, after some algebra,

$$q_{y=W} = -k(t_0 - t_c) \sum_{n=1,3,5,\ldots} \frac{8}{n\pi} \frac{1}{\tanh(n\pi W/L)}. \tag{3.18}$$

The negative sign in Eq. (3.18) indicates heat flow into the plate at the face $y = W$.

Note that in the case just discussed only two temperatures were involved: t_0, the temperature at the $y = W$ edge, and t_c, the temperature of the other edges. Equation (3.18) gives the heat flow into the plate as the product of the plate conductivity, k, the overall temperature difference, $(t_0 - t_c)$, and a quantity dependent only on the geometry of the configuration. This representation is explored more fully in Sec. 3.3.

The Rectangular Plate with More than One Edge at a Specified Temperature

The case just considered was that of a rectangular plate with the temperature at $y = W$ held at t_0 and that for all other edges at some other constant value t_c. Figure 3.4 illustrates a case in which the edges at $y = 0$ and $y = W$ are held at two different temperatures, t_{01} and t_{02}, respectively, while the other edges are maintained at t_c'. Since the governing equation for the temperature distribution [Eq. (3.1)] is linear, the additive principle of superposition may be applied. Thus the

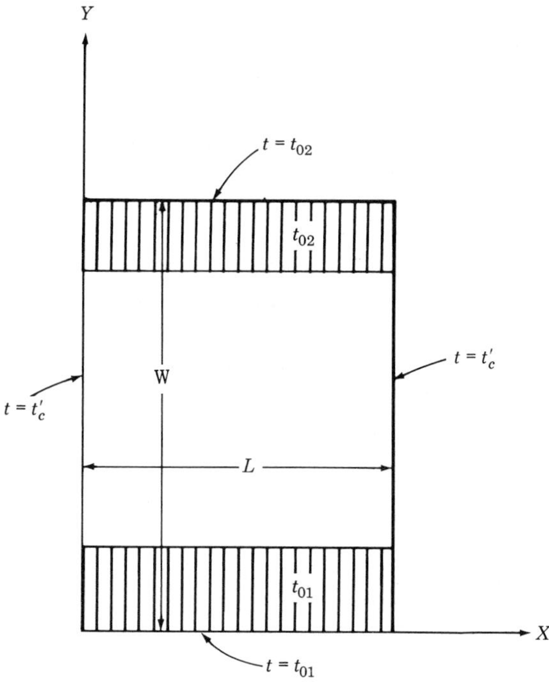

Figure 3.4

temperature at any point in the plane may be obtained by adding the solution of Eq. (3.17) with $t_0 = t_{02}$, $t_c = t'_c$ to that obtained from Eq. (3.17) with $t_0 = t_{01} - t'_c$, $t_c = 0$, and $W - y$ replacing y.

The case when all four edges are maintained at different temperatures may be treated similarly.

3.3. THE CONDUCTION SHAPE FACTOR

The two-dimensional conduction problem considered in the foregoing section, conduction in a rectangular plate, is among the simplest that may be posed. In spite of its simple geometry, the solution was found to be mathematically complex and expressed in the form of an infinite series which is inconvenient to use. The same sort of complexities arise in other geometries having some practical applications. For those cases involving only two specified temperatures, call them t_1 and t_2, the solution for the heat flow between the surfaces maintained at these temperatures may always be written in the form

$$q = kS(t_1 - t_2), \tag{3.19}$$

in which S is a quantity called the *conduction shape factor*. The conduction shape factor is a function of the geometry at hand and attempts to render the solution in a form analogous to that for a plane wall in which S is obviously equal to $A/\Delta x$. For example, for the geometry of a rectangular plate maintained at t_1 on one edge and t_2 on all other edges, the conduction shape factor is given by the summation term in Eq. (3.18). Even this summation is inconvenient to use and one might seek an approximation to it.

It should be apparent from the definition of the conduction shape factor in Eq. (3.19) that it has the dimensions of 1/length (i.e., 1/m or 1/ft).

The conduction shape factor has been worked out for a great number of geometries. References 3 and 5 present extensive summaries of the available shape factors. Table 3.1 presents a number of shape factors for some useful geometries. Since some of these results are approximations to mathematical solutions, restrictions on their applicability are also noted. The results of Table 3.1 may then be used with Eq. (3.19) to find the heat flow for the particular geometry. The first two cases shown in Table 3.1 are included for the sake of completeness; they result from the one-dimensional analyses of Chapter 2.

EXAMPLE 3.1

A long pipe 50 cm in diameter is buried horizontally in the ground so that its centerline is 120 cm below the surface. The pipe carries a hot fluid so that its surface temperature is 90°C, and the earth's surface temperature is 30°C. Find the heat loss, per unit length, from the pipe if the soil conductivity is 1.5 W/m-°C.

Table 3.1. Conduction Shape Factors for Selected Geometries

Configuration	Schematic	Shape Factor	Restrictions
Plane wall	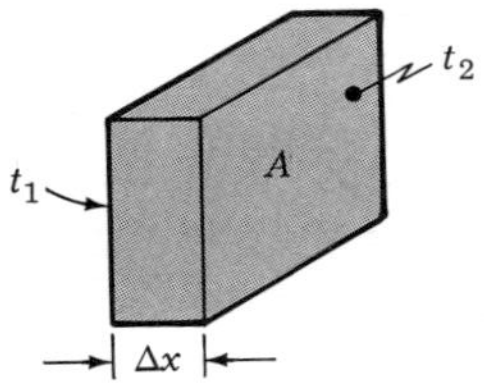	$\frac{A}{\Delta x}$	One-dimensional conduction
Hollow cylinder, length L	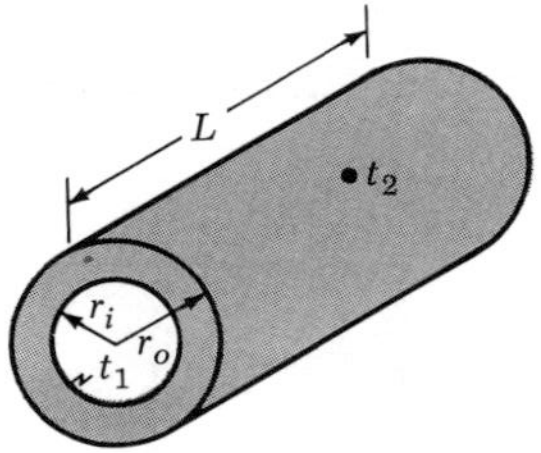	$\frac{2\pi L}{\ln (r_0/r_i)}$	$L >> r$
Conduction through corner of adjoining walls	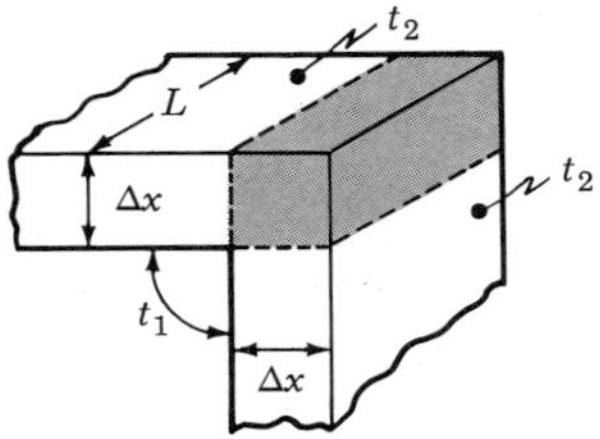	$0.54L$	$L > \frac{\Delta x}{5}$

Table 3.1. (Cont.)

Configuration	Schematic	Shape Factor	Restrictions
Square hole in a square solid, length L	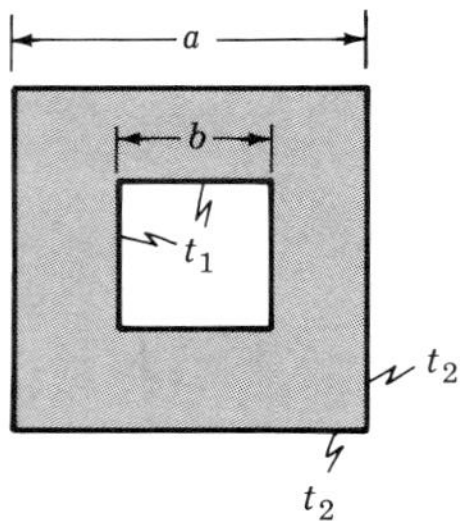	$\dfrac{2\pi L}{0.93 \ln (a/b) - 0.0502}$ $\dfrac{2\pi L}{0.785 \ln (a/b)}$	$\dfrac{a}{b} > 1.41$ $\dfrac{a}{b} < 1.41$
Circular hole in a square solid, length L	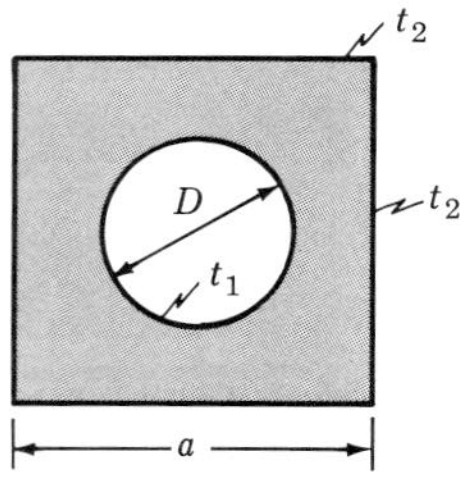	$\dfrac{2\pi L}{\ln (1.08a/D)}$	$a > D$
Sphere buried in a semi-infinite region	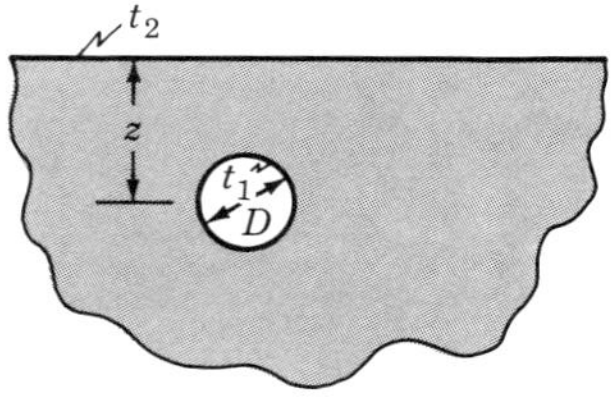	$\dfrac{2\pi D}{1 - D/4z}$	$z > \dfrac{D}{2}$

Table 3.1. (Cont.)

Configuration	Schematic	Shape Factor	Restrictions
Horizontal cylinder length L, buried in a semi-infinite region	t_2, z, t_1, D, L	$\dfrac{2\pi L}{\cosh^{-1}(2z/D)}$	$L >> D$
Vertical cylinder, length L, buried in a semi-infinite region	t_2, t_1, L, D	$\dfrac{2\pi L}{\ln(4L/D)}$	$L >> D$
Thin horizontal disc buried in a semi-infinite region	t_2, z, t_1, D	$4D$ $2D$	$z >> D$ $z = 0$

Table 3.1. (Cont.)

Configuration	Schematic	Shape Factor	Restrictions
Circular cylinder, length L, midway between plates of length L and infinite width	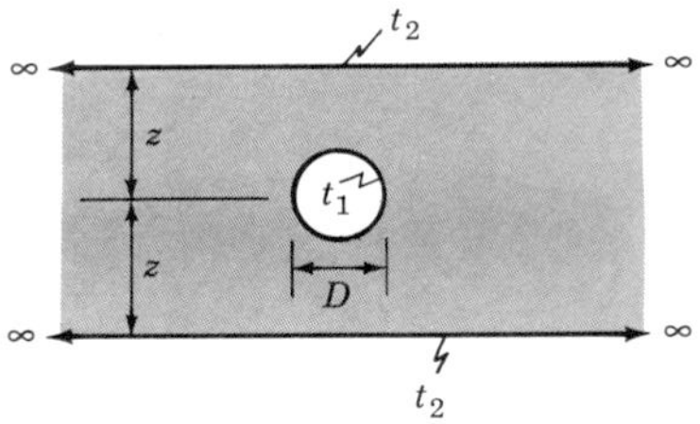 t_2, ∞, z, t_1, D, t_2	$\frac{2\pi L}{\ln (8z/D)}$	$z > \frac{D}{2}$
Two cylinders of equal length L, buried in an infinite medium	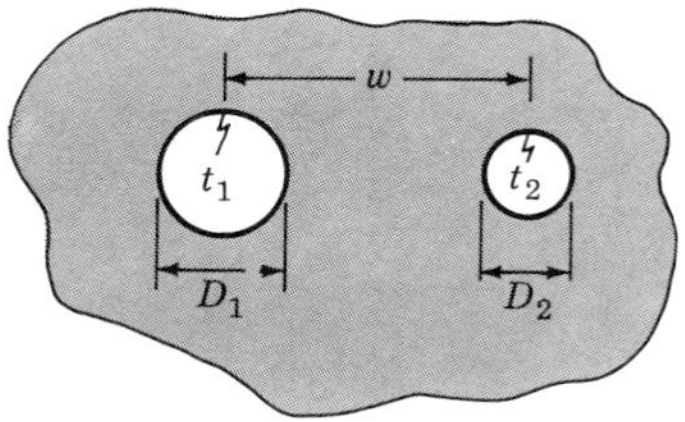 w, t_1, t_2, D_1, D_2	$\frac{2\pi L}{\cosh^{-1} \frac{4w^2 - D_1^2 - D_2^2}{2D_1D_2}}$	$L >> D_1D_2$ $L >> w$

Solution. Table 3.1 gives the conduction shape factor for this configuration, per unit length, to be

$$\frac{S}{L} = \frac{2\pi}{\cosh^{-1}(2z/D)}$$

Thus, for the data given,

$$\frac{S}{L} = \frac{2\pi}{\cosh^{-1}(2 \times 120/50)} = 2.792.$$

[The identity $\cosh x = \ln(x \pm \sqrt{x^2 - 1})$ may be used to evaluate the inverse hyperbolic cosine if necessary.] Equation (3.19) then gives the heat loss:

$$\begin{aligned}\frac{q}{L} &= k\left(\frac{S}{L}\right)(t_1 - t_2)\\ &= 1.5 \times 2.792 \times (90 - 30)\\ &= 251.2 \text{ W/m } (261.3 \text{ Btu/h-ft}).\end{aligned}$$

3.4 NONSTEADY CONDUCTION IN ONE SPACE DIMENSION

The problem of describing the temperature distribution and its variation with time for *nonsteady* heat conduction in only *one space dimension* has much similarity to the two-dimensional steady conduction problems discussed in the foregoing in that both cases involve the determination of the temperature in terms of two independent variables. While many nonsteady cases may be posed, only those cases having some practical engineering application will be described here. The reader is referred to extensive works on conduction such as Refs. 1 through 3, for additional cases.

Some of the most often encountered nonsteady situations are those in which a given body at a uniform temperature suddenly has its boundary conditions altered—either by changing its boundary temperature or by changing the temperature of a surrounding, convecting, fluid. If such is the case, certain generalizations of the form of the expected result may be made before carrying out any particular solution.

For the purposes of discussion, imagine a nonsteady conduction problem, with constant properties, in which the single one-dimensional coordinate is the cartesian variable x. (An analogous discussion could be carried out in any other one-dimensional coordinate system, e.g., the cylindrical coordinates with r as the only variable.) The general conduction equation, for constant properties and no internal heat generation [Eq. (2.6)] is

$$\frac{1}{\alpha}\frac{\partial t}{\partial \tau} = \frac{\partial^2 t}{\partial x^2}. \qquad (3.20)$$

Let the body be initially at a uniform temperature, t_i, and imagine that the temperature of a surrounding fluid (with heat transfer coefficient h) is suddenly changed to some value, t_f. It is then desired to find the temperature within the body at any subsequent time. (The alternative case in which the surface temperature of the body itself is changed to some value will be treated as a special case below.) Thus the boundary conditions on Eq. (3.20) are

$$\text{At } \tau = 0: \quad t = t_i, \text{ for all } x, \qquad (3.21)$$

$$\text{For } \tau > 0: \quad -k\left(\frac{\partial t}{\partial x}\right)_s = h(t - t_f)_s,$$

where the subscript s denotes the body surface.

The solution to the system of Eqs. (3.20) and (3.21) may be generalized by introducing certain nondimensional variables. If l represents some characteristic dimension of the body (e.g., slab thickness, cylinder radius, etc.), then the linear variable may be nondimensionalized by defining

$$\xi = \frac{x}{l}.$$

The temperature may be nondimensionalized by measuring all temperatures above that of the ambient fluid and referencing to the initial uniform temperature:

$$\mathbf{T} = \frac{t - t_f}{t_i - t_f}. \qquad (3.22)$$

Additionally, the time and heat transfer coefficient may be rendered dimensionless by the following definitions:

$$\text{Dimensionless time} = \text{Fourier number} = \text{Fo} = \frac{\alpha\tau}{l^2}. \qquad (3.23)$$

$$\text{Dimensionless coefficient} = \text{Biot number} = \text{Bi} = \frac{hl}{k}. \qquad (3.24)$$

With these definitions, the differential equation and its boundary conditions reduce to

$$\frac{\partial \mathbf{T}}{\partial \text{Fo}} = \frac{\partial^2 \mathbf{T}}{\partial \xi^2},$$

$$\text{At Fo} = 0: \quad \mathbf{T} = 1, \text{ for all } \xi, \qquad (3.25)$$

$$\text{For Fo} > 0: \quad \left(\frac{\partial \mathbf{T}}{\partial \xi}\right)_s = -\text{Bi}\mathbf{T}_s.$$

The solution of Eq. (3.25) will then be of the form

$$\mathbf{T} = \text{fn}(\xi, \text{Fo}, \text{Bi}).$$

Thus all geometrically similar bodies with identical Bi will have the same dimensionless temperature response at geometrically similar points according to time measured as Fo.

Thus the parameters Fo and Bi are important measures as to how a body responds to temperature changes. The Fourier number, Fo, is a measure of time, and its definition in Eq. (3.23) shows that bodies with a high diffusivity respond faster than those with a low diffusivity; large bodies respond more slowly than small bodies. The significance of the Biot number can be more readily seen if it is rewritten

$$\text{Bi} = \frac{l/k}{1/h}.$$

It is seen to be a measure of the ratio of the thermal resistance of the body, l/k, and that of the surface film, $1/h$. Thus the temperature response of bodies with a low Bi is dominated by the surface resistance while those with a large Bi are dominated by the internal resistance. The further significance of these parameters will be seen as specific solutions are obtained.

The case in which the surface temperature is suddenly changed to a fixed value, t_s, was not discussed above. This case may be viewed as that in which the heat transfer coefficient $h \to \infty$, or $\text{Bi} \to \infty$. In such a case the surface temperature, t_s, becomes equal to the fixed fluid temperature, i.e., $t_s \to t_f$. Thus the dimensionless temperature becomes $\mathbf{T} = (t - t_s)/(t_i - t_s)$ and the solution form becomes

$$\mathbf{T} = \frac{t - t_s}{t_i - t_s} = \text{fn}\ (\xi, \text{Fo}).$$

In the solutions presented in the remainder of this chapter, cases for $\text{Bi} \to \infty$ (i.e., fixed surface temperature) will be considered first. Then cases with finite Bi will be studied, and finally, cases for $\text{Bi} \to 0$ will be examined.

3.5 TRANSIENT CONDUCTION IN THE INFINITE SLAB

One of the simplest nonsteady conduction problems in one dimension is that in an infinite slab. This is a plane wall of finite thickness but so large in extent in the other directions that one may ignore edge effects. Thus the only space variable needed to describe the temperature in the slab is that measured normal to its face, x. Thus if initial and boundary conditions are applied that produce time-dependent conditions, the temperature in the slab depends on two independent variables, x and τ (time).

Such a possible situation is depicted in Fig. 3.5. Here a slab of thickness L is subjected to some known initial distribution of temperature, call it $f(x)$. Then at time $\tau = 0$ and all subsequent times, the surface temperatures are reduced to and maintained at a constant temperature t_s. It is desired to find the temperature at any location in the slab as a function of time for $\tau > 0$, i.e., $t(x, \tau)$. The differential equation to be solved is Eq. (2.6) with internal heat generation absent:

$$\frac{\partial t}{\partial \tau} = \alpha \frac{\partial^2 t}{\partial x^2}.$$

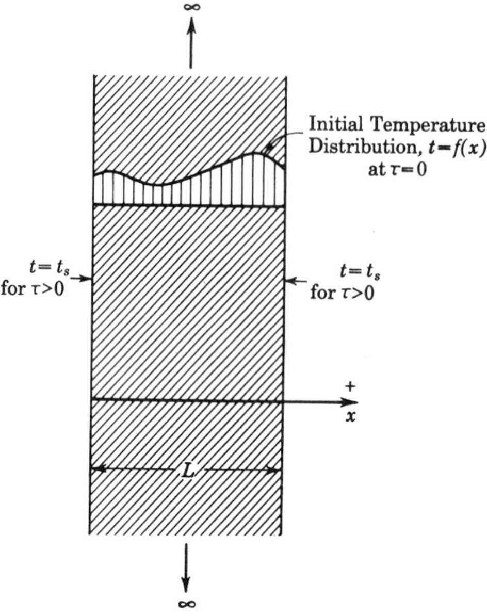

Figure 3.5

As in the cases discussed in Sec. 3.2 and 3.4, it is desirable to work in terms of a temperature difference variable

$$\theta = t - t_s, \tag{3.26}$$

so that one must solve the following equation subject to the initial and boundary conditions noted:

$$\frac{\partial \theta}{\partial \tau} \alpha \frac{\partial^2 \theta}{\partial x^2} \tag{3.27}$$

$$\begin{aligned} \text{For } \tau = 0:&\quad t = f(x) \text{ or } \theta = f(x) - t_s. && (1) \\ \text{For } \tau \geq 0:&\quad \text{at } x = 0,\ t = t_s \text{ or } \theta = 0, && (2) \\ &\quad \text{at } x = L,\ t = t_s \text{ or } \theta = 0. && (3) \end{aligned} \tag{3.28}$$

Seeking, as was done in Sec. 3.2, the existence of a product solution, one assumes that the solution is representable as

$$\theta = X(x) \cdot T(\tau), \tag{3.29}$$

where $X(x)$ and $T(\tau)$ are functions of x and τ only. Introduction of Eq. (3.29) into (3.27) and subsequent rearrangement yields

$$\frac{1}{\alpha T}\frac{dT}{d\tau} = \frac{d^2X}{dx^2}\frac{1}{X} = -\lambda^2, \tag{3.30}$$

where λ is the separation parameter and the negative sign was chosen to assure a negative exponential solution in time. Solution of the two ordinary differential equations implied in Eq. (3.30) yields

$$T(\tau) = B_1' e^{-\alpha\lambda^2\tau},$$
$$X(x) = B_2' \sin \lambda x + B_3' \cos \lambda x,$$

so that Eq. (3.29) gives the general solution to Eq. (3.27) to be

$$\theta = e^{-\lambda^2\alpha\tau}(B_1 \sin \lambda x + B_2 \cos \lambda x). \tag{3.31}$$

The constants λ, B_1, and B_2 are to be determined from the imposed boundary and initial conditions.

Application of conditions (2) and (3) given in Eq. (3.28) yields $B_2 = 0$ and $\sin \lambda L = 0$. Thus, as in the case of the rectangular plate studied in Sec. 3.2, the parameter λ may take on a multiplicity of values and the solution in Eq. (3.31) becomes the sum of the corresponding solutions for each λ:

$$\begin{aligned} \theta &= \sum_{n=1}^{\infty} B_n e^{-\lambda_n^2\alpha\tau} \sin \lambda_n x \\ \lambda_n &= \frac{n\pi}{L}, \qquad n = 1, 2, 3, \ldots . \end{aligned} \tag{3.32}$$

Applying, finally, the initial condition at $\tau = 0$ as given by condition (1) of Eq. (3.28), one has

$$\begin{aligned} f(x) - t_s &= \sum_{n=1}^{\infty} B_n \sin \lambda_n x \\ &= \sum_{n=1}^{\infty} B_n \sin \frac{n\pi x}{L}. \end{aligned} \tag{3.33}$$

Equation (3.33) poses the same question as that given in Eq. (3.10) for the rectangular plate. That is, what are the values of B_n that satisfy Eq. (3.33)? The answer depends on the function $f(x)$ which represents the imposed initial temperature distribution at time zero. As before, the determination of the B_n's requires application of the theory of Fourier series (Refs. 1 through 4). For the special case in which the slab is initially heated to a uniform temperature, call it $f(x) = t_i$, the results given in Eqs. (3.12) through (3.16) apply:

$$\begin{aligned} B_n &= \frac{2}{L}\int_0^L (t_i - t_s) \sin \frac{n\pi x}{L}\, dx \\ &= \frac{4}{n\pi}(t_i - t_s), \qquad n = 1, 3, 5, \ldots \\ &= 0, \qquad n = 2, 4, 6, \ldots . \end{aligned}$$

Equation (3.32) then gives the following solution for the temperature in a slab that is initially heated to a uniform temperature t_i and has its surfaces suddenly reduced to a constant temperature t_s:

$$\frac{\theta}{t_i - t_s} = \frac{t - t_s}{t_i - t_s} = \sum_{n=1,3,5,\ldots} \frac{4}{n\pi} e^{-(n\pi/L)^2\alpha\tau} \sin\frac{n\pi x}{L}. \tag{3.34}$$

At any location x, Eq. (3.34) permits the calculation of the slab temperature at any time τ.

Examination of Eq. (3.34) shows that if the slab thickness L is taken as the characteristic dimension, then the dimensionless parameters of length and time discussed in Sec. 3.4 enter naturally:

$$\text{Distance: } \xi = \frac{x}{L}. \tag{3.35}$$

$$\text{Time} = \text{Fourier number: Fo} = \frac{\alpha\tau}{L^2}.$$

Thus the solution above is

$$\frac{t - t_s}{t_i - t_s} = \sum_{n=1,3,5,\ldots}^{\infty} \frac{4}{n\pi} e^{-(n\pi)^2\text{Fo}} \sin n\pi\xi, \tag{3.36}$$

which is of the form $\mathbf{T} = \text{fn}(\xi, \text{Fo})$ predicted in Sec. 3.4.

The heat flow out of the slab may also be of interest. The rate of heat flow, per unit of slab face area, is

$$\frac{q}{A} = \left[-k\left(\frac{\partial t}{\partial x}\right)_{x=0}\right] 2.$$

The factor 2 appears due to symmetry and the fact that heat flows out *two* faces of the slab, at $x = 0$ and $x = L$. The total heat flow up to time τ is then

$$\frac{Q}{A} = \int_0^\tau \frac{q}{A}\, d\tau.$$

Introduction of Eq. (3.36) gives, since $\alpha = k/\rho c_p$,

$$\frac{Q}{AL\rho c_p(t_i - t_s)} = \sum_{n=1,3,5,\ldots} \frac{8}{(n\pi)^2} [1 - e^{-(n\pi)^2\text{Fo}}].$$

The denominator of the left side of the equation above is recognizable as the total heat stored in the slab initially—measured relative to the fixed boundary temperature. Call this $Q_i = AL\rho c_p(t_i - t_s)$, so the ratio of the total heat flow out of the slab up to time τ to that initially in the slab is

$$\frac{Q}{Q_i} = \sum_{n=1,3,5,\ldots} \frac{8}{(n\pi)^2} [1 - e^{-(n\pi)^2\text{Fo}}]. \tag{3.37}$$

Figure 3.6 shows plots of Eqs. (3.36) and (3.37) so that the temperature–time history and the total heat flow from the slab may readily be determined as functions of time.

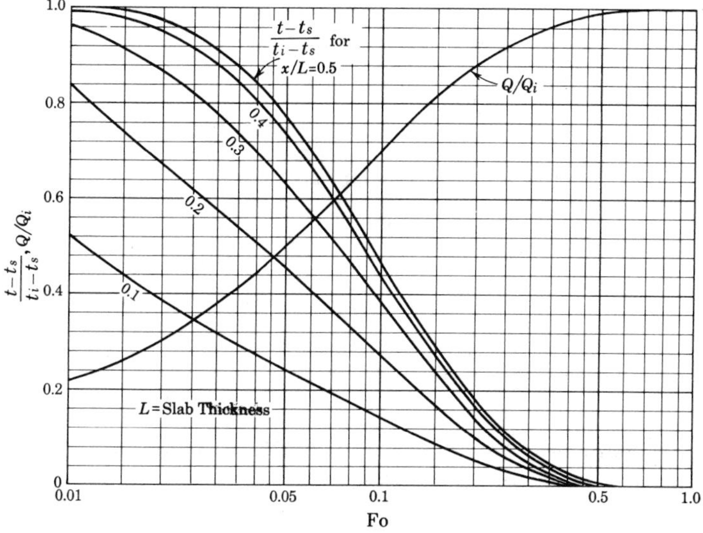

Figure 3.6. Time variation of the temperature distribution and heat flow in an infinite slab initially at a constant temperature which has had its surface temperature suddenly changed.

EXAMPLE 3.2

The wall enclosing a park is 25 cm thick and is made of common brick. Initially, the wall is at a uniform temperature of 35°C. A cold front arrives, suddenly reducing the surface temperatures of the wall to −10°C. Find the temperature at a plane 10 cm from the surface of the wall after 2 h has passed. How much heat has been conducted out of the wall during that time?

Solution. Table A.2 gives the following properties for common brick:

$$c_p = 0.84 \text{ kJ/kg-°C}, \qquad \rho = 1602 \text{ kg/m}^3,$$
$$k = 0.69 \text{ W/m-°C}, \qquad \alpha = 5.2 \times 10^{-7}\text{m}^2\text{/s}.$$

For $\tau = 2$ h and $L = 25$ cm $= 2.5$ m, the Fourier number is

$$\text{Fo} = \frac{\alpha\tau}{L^2} = \frac{5.2 \times 10^{-7} \times 2 \times 3600}{(0.25)^2} = 0.0599.$$

For a plane 10 cm from the surface $\xi = x/L = 10/25 = 0.4$, Fig. 3.6 gives

$$\frac{t - t_s}{t_i - t_s} = 0.669, \qquad \frac{Q}{Q_i} = 0.551.$$

In this instance $t_i = 35°C$ and $t_s = -10°C$, so that the desired temperature is

$$\frac{t - (-10)}{35 - (-10)} = 0.669$$

$$t = 20.1°C\ (68.2°F).$$

The heat conducted out of the wall, per unit area, is

$$\frac{Q}{A} = \frac{Q_i}{A}\frac{Q}{Q_i} = L\rho c_p(t_i - t_s)\frac{Q}{Q_i}$$

$$= 0.25 \times 1602 \times 0.84 \times [35 - (-10)] \times 0.551$$

$$= 8342\ \text{kJ/m}^2\ (734.5\ \text{Btu/ft}^2).$$

3.6 ONE-DIMENSIONAL TRANSIENT CONDUCTION WITH CONVECTIVE BOUNDARY CONDITIONS

The one-dimensional nonsteady conduction problem considered in Sec. 3.5 illustrated the basic procedure to be followed in solving such a problem: The constants resulting from the integration of the conduction equation are determined by application of the known initial temperature distribution at time zero and the imposed conditions at the surface for subsequent times. In the case discussed in Sec. 3.5, the specification of a known surface temperature led to the relatively simple determination of the separation constants, λ_n, as given in Eq. (3.32). Also, the Biot number, $\text{Bi} = hl/k$, did not appear, as predicted in Sec. 3.4, since no boundary convection was present.

Of more interest for engineering applications are nonsteady conduction problems in which the conducting body is exchanging heat at its surface with a fluid of specified temperature through a known heat transfer coefficient, h. In such instances the boundary condition at the surface becomes

$$-k\left(\frac{\partial t}{\partial x}\right)_s = h(t - t_f)_s.$$

where t_f is the ambient fluid temperature and the subscript s denotes the body surface. In such instances the determination of the separation parameter, λ_n, becomes more complex and the subsequent application of the theory of Fourier series is likewise more complex. One does know, however, based on the discussion of Sec. 3.4, that the solution for the temperature within the body will take the form

$$\mathbf{T} = \text{fn}\ (\xi, \text{Fo}, \text{Bi}),$$

in which $\mathbf{T}$ is an appropriately defined dimensionless temperature, ξ a dimensionless position coordinate, and Fo and Bi are dimensionless time and heat transfer coefficients as defined in Eqs. (3.23) and (3.24).

This section will present only the *results* of such analyses applied to the geometrically significant shapes of the infinite slab and the infinitely long cylinder.

The details of how these results are obtained may be found in Refs. 1 through 4. Similar results are also known for another geometry, the sphere, but are not given here because of their relatively limited applicability.

The Infinite Slab

Figure 3.7(a) depicts the infinite slab which is initially heated to a uniform temperature t_i and then has its surface suddenly exposed to a convecting fluid at a constant temperature t_f through a surface heat transfer coefficient h for all times subsequent to $\tau = 0$. It is desired to know how the temperature at any location within the slab varies as a function of time. The coordinate system used is shown in Fig. 3.7, and the one-dimensional coordinate location x is measured from the center of the slab outward. The slab thickness is taken as $2L$, so that $x = L$ denotes

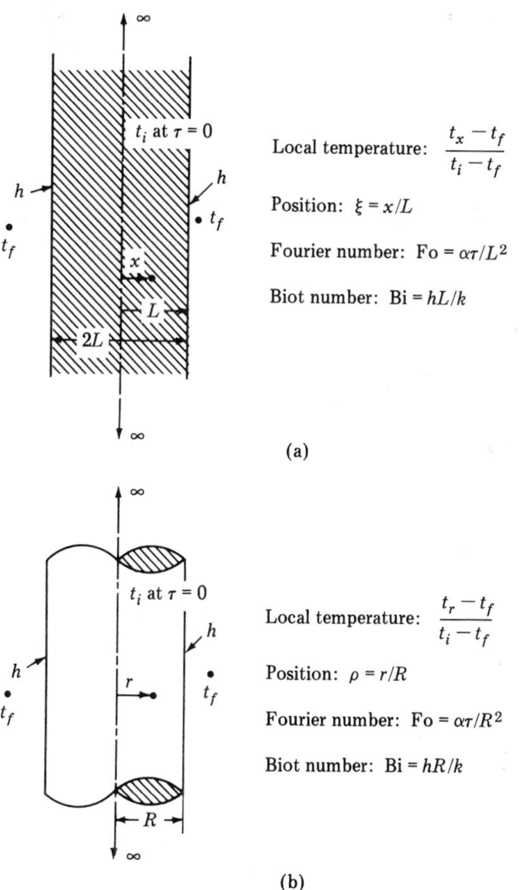

Figure 3.7. Geometry and dimensionless parameters for transient conduction in (a) an infinite slab and (b) an infinite cylinder.

the exposed face of the slab. (Note that this designation of the slab thickness as $2L$ differs from that used in Sec. 3.5, wherein the slab thickness was simply L.)

The dimensionless variables used to express the solution to the problem above are also indicated in Fig. 3.7(a) and are defined by using the half-thickness L as the characteristic length. Thus, consistent with the discussion of Sec. 3.4, one defines:

$$\text{Temperature: } \mathbf{T} = \frac{t_x - t_f}{t_i - t_f}.$$

$$\text{Position: } \xi = \frac{x}{L}.$$

$$\text{Time, Fourier number: Fo} = \frac{\alpha\tau}{L^2}.$$

$$\text{Surface coefficient, Biot number: Bi} = \frac{hL}{k}.$$

In the above the symbol t_x denotes the temperature at the location x and is a function of time as well as of x. Thus one wants the solution given by the function

$$\frac{t_x - t_f}{t_i - t_f} = \text{fn}(\xi, \text{Fo}, \text{Bi}).$$

The function above constitutes a known solution (Refs. 1 through 4); however, it is a function of three parameters and is, hence, difficult to display graphically. Heisler (Ref. 6) found it convenient to recast the solution in an alternate way. If t_c is used to denote the temperature at the center of the slab, $x = 0$, then the ratio

$$\frac{t_c - t_f}{t_i - t_f} = \text{fn}(\text{Fo}, \text{Bi}) \tag{3.38}$$

is clearly a function of only the two parameters Fo and Bi. Then the desired temperature at some other position is

$$\frac{t_x - t_f}{t_i - t_f} = \frac{t_c - t_f}{t_i - t_f}\left(\frac{t_x - t_f}{t_c - t_f}\right), \tag{3.39}$$

in which the last term, $(t_x - t_f)/(t_c - t_f)$, is the ratio of the temperature difference at $x = x$ to that at $x = 0$. One would expect the latter quantity to be a function of all three parameters, ξ, Fo, and Bi. However, a detailed analysis of the exact solution indicates that its dependence on Fo is slight, particularly if Fo is large enough. That is,

$$\frac{t_x - t_f}{t_c - t_f} \approx \text{fn}(\xi, \text{Bi}) \text{ only}. \tag{3.40}$$

Thus one may find the temperature at any x and τ (i.e., any ξ and Fo) for a given h (i.e., Bi) by use of Eq. (3.39) if the functions noted in Eqs. (3.38) and (3.40) are known. The function of Eq. (3.38), i.e., the centerline temperature, is shown in Fig. 3.8. The "position correction" function (3.40) is shown in Fig. 3.9. The

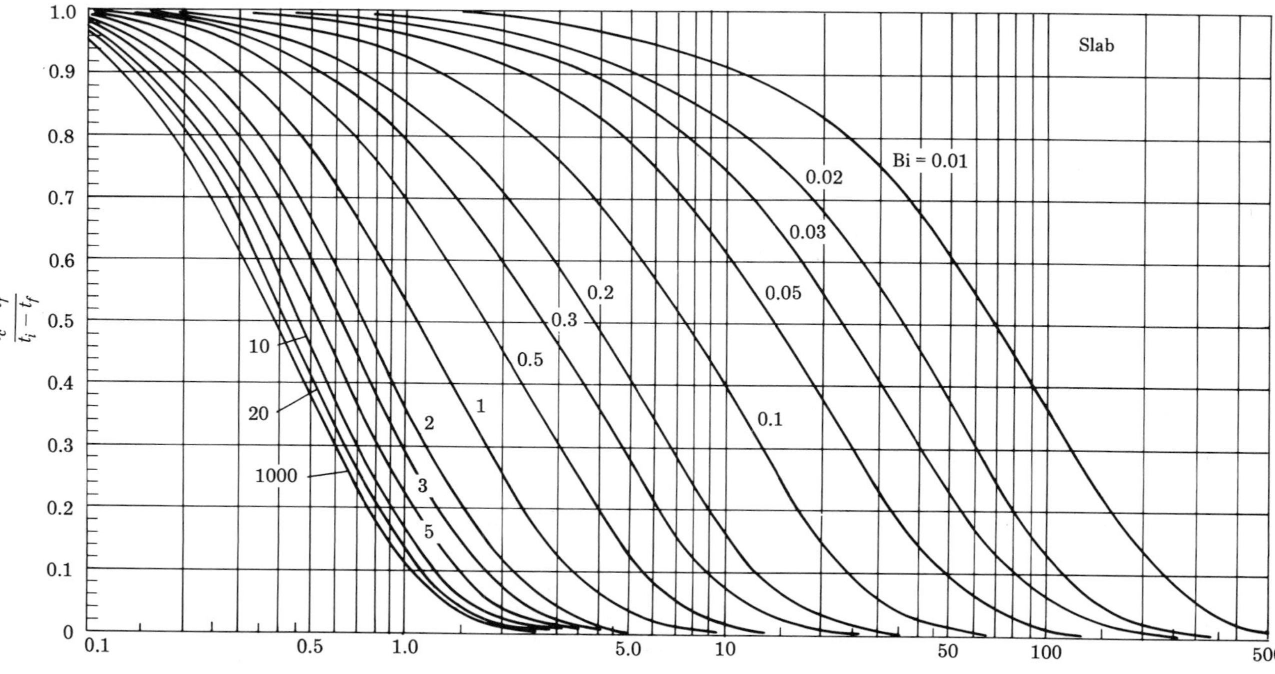

Figure 3.8. Temperature response chart for the center of an infinite slab.

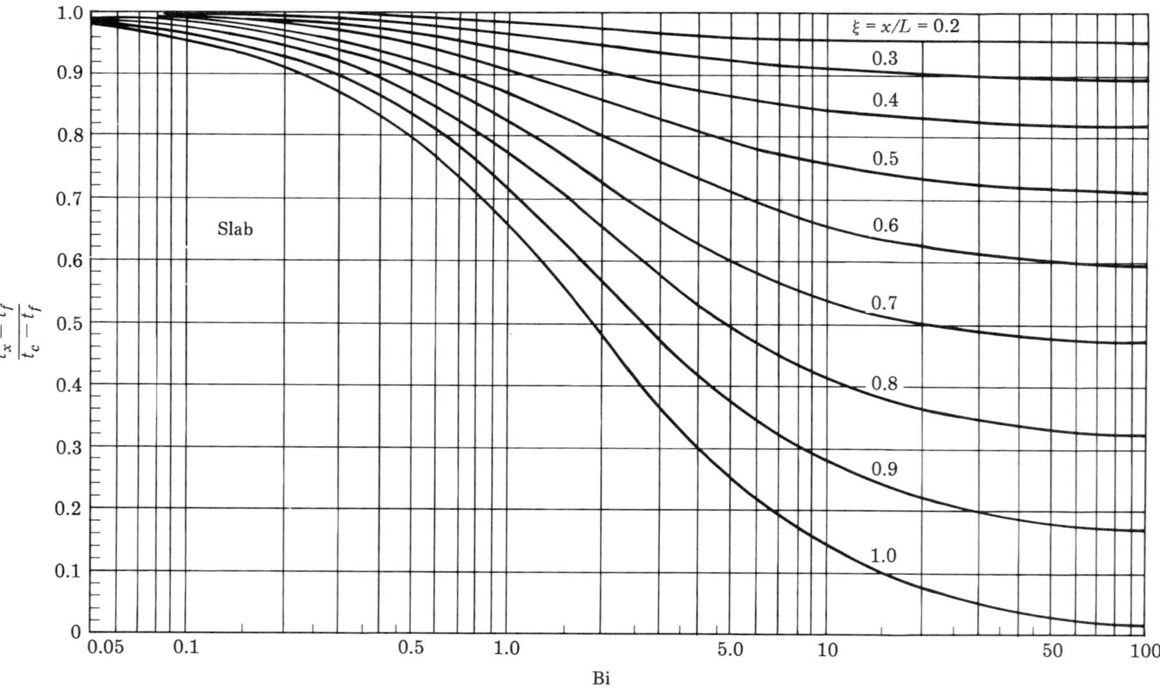

Figure 3.9. Position correction chart for the infinite slab.

data given in Figs. 3.8 and 3.9 are essentially the same as that given earlier by Heisler (Ref. 6); however, following the suggestion of Wolf (Ref. 7) they have been plotted in a different form for easier and more accurate reading. Examples will be presented shortly which will illustrate the use of these figures.

It should be pointed out that Heisler's position correction concept, which implies that the ratio of the temperature difference at a value x to that at the center is independent of time, as stated in Eq. (3.40), is an approximation. It can be applied with confidence for $\text{Fo} > 0.2$, but should probably not be used for Fo below that value.

Another quantity of engineering use is the ratio of the cumulative heat flow out of the slab face from time $\tau = 0$ until some arbitrary time τ. In keeping with the nondimensional representation used in the foregoing, it is useful to express this total heat flow nondimensionally as well. Following the same line of reasoning used in Sec. 3.5 for the slab of specified boundary temperature, let Q_i/A represent the heat initially stored in the slab, per unit of face area. Thus, referenced to the ambient fluid temperature, $Q_i/A = L\rho c_p(t_i - t_f)$. Then from the known solution for the temperature in the slab as a function of position and time, one may determine the heat flow, Q/A, out of the surface at $x = L$, and find the ratio Q/Q_i as a function of h and time:

$$\frac{Q}{Q_i} = \text{fn (Bi, Fo)}. \tag{3.41}$$

The function implied by Eq. (3.41) is shown in Fig. 3.10.

The quantity Q/Q_i may be used for purposes other than the calculation of heat flow. Since Q/Q_i represents the relative heat flow *out* of the slab, $1 - Q/Q_i$ must

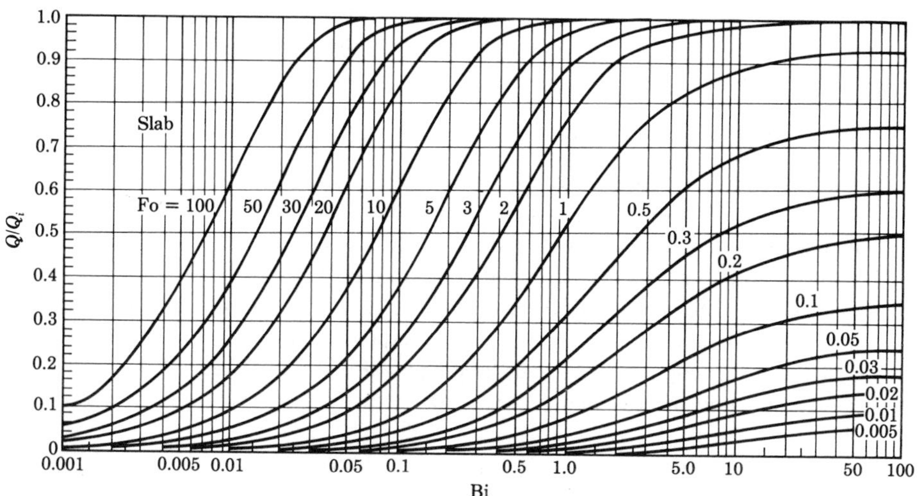

Figure 3.10. Heat flow from an infinite slab as a function of time and thermal resistance.

represent the relative amount of heat remaining *in* the slab. Consequently, $1 - Q/Q_i$ is the dimensionless *average* temperature of the slab at any time. That is,

$$1 - \frac{Q}{Q_i} = \frac{t_{av} - t_f}{t_i - t_f}$$

may be used to determine the average slab temperature.

In summary, then, the transient conduction problem in an infinite slab, half-thickness L, initially at t_i, which has its surface exposed to a fluid at t_f for all times $\tau > 0$, has a solution expressed as:

$$\xi = \frac{x}{L}, \qquad \text{Fo} = \frac{\alpha\tau}{L^2}, \qquad \text{Bi} = \frac{hL}{k}$$

$$\text{At } x = 0: \quad \frac{t_c - t_f}{t_i - t_f} = \text{fn (Fo, Bi)} \quad \text{(Fig. 3.8).}$$

$$\text{At } x = x: \quad \frac{t_x - t_f}{t_i - t_f} = \frac{t_c - t_f}{t_i - t_f}\left(\frac{t_x - t_f}{t_c - t_f}\right) \tag{3.42}$$

$$\frac{t_x - t_f}{t_c - t_f} = \text{fn } (\xi, \text{Bi}) \quad \text{(Fig. 3.9).}$$

$$\frac{Q}{Q_i} = 1 - \frac{t_{av} - t_f}{t_i - t_f} = \text{fn (Fo, Bi)} \quad \text{(Fig. 3.10).}$$

The method of solution for transient conduction in an infinite slab just summarized in Eq. (3.42) requires use of the charts given in Figs. 3.8 through 3.10 and will be referred to in the following as the "temperature response chart" method.

The Infinitely Long Cylinder

Figure 3.7(b) depicts another geometric shape of considerable practical interest. This is the solid circular cylinder so long that end effects may be neglected and conduction treated as depending only on the radial coordinate r. With the outer radius of the cylinder denoted by R, one may pose an analogous problem to that just discussed for the slab. If the cylinder is heated initially to a uniform temperature, t_i, and then has its surface suddenly exposed to an ambient fluid at t_f through a heat transfer coefficient h, find the temperature at any radial location r and time $\tau > 0$ and find the cumulative heat flow from the surface of the cylinder at that time.

Again the solution to the problem above is known (Refs. 1 through 4) and only the results are quoted here, graphically. The cylinder radius, R, is the natural quantity to use as the characteristic length in the definition of the dimensionless position parameter ($\rho = r/R$), Fourier number, and Biot number. Once again, the

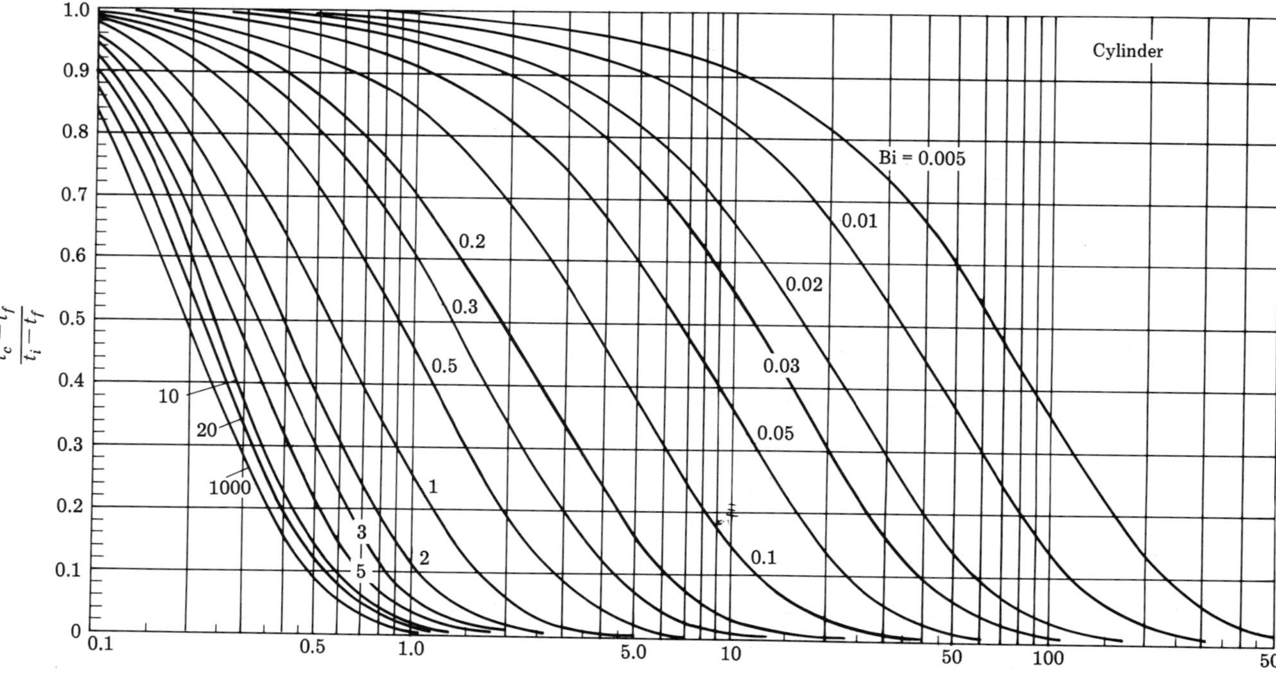

Figure 3.11. Temperature response chart for the center of an infinite cylinder.

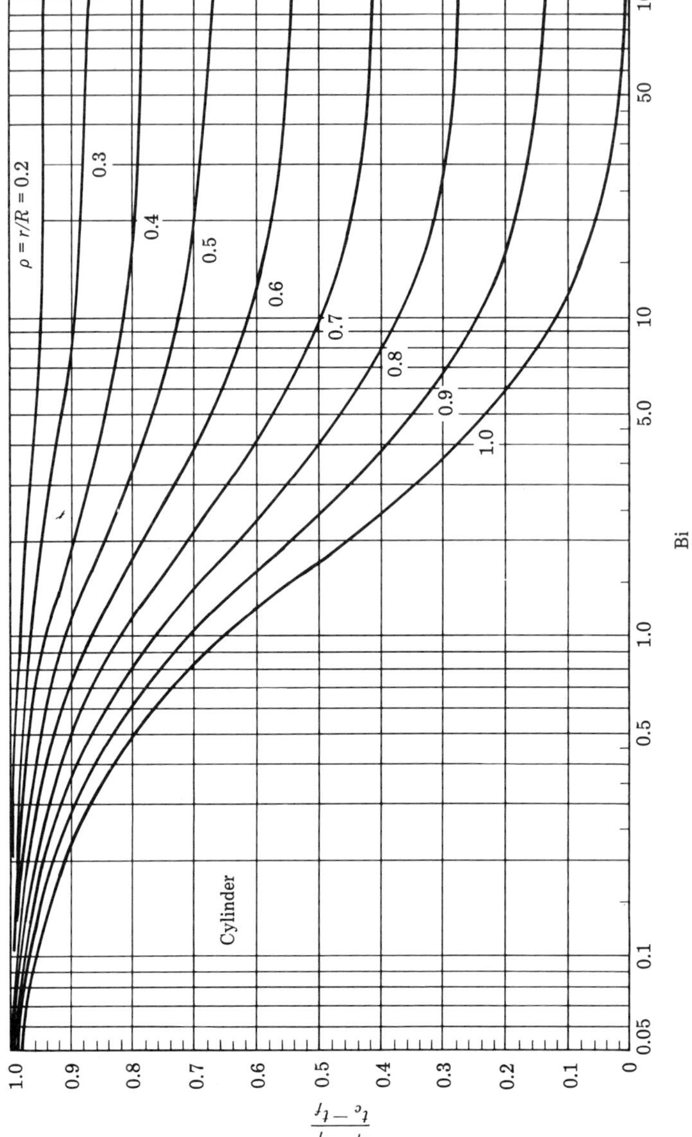

Figure 3.12. Position correction chart for the infinite cylinder.

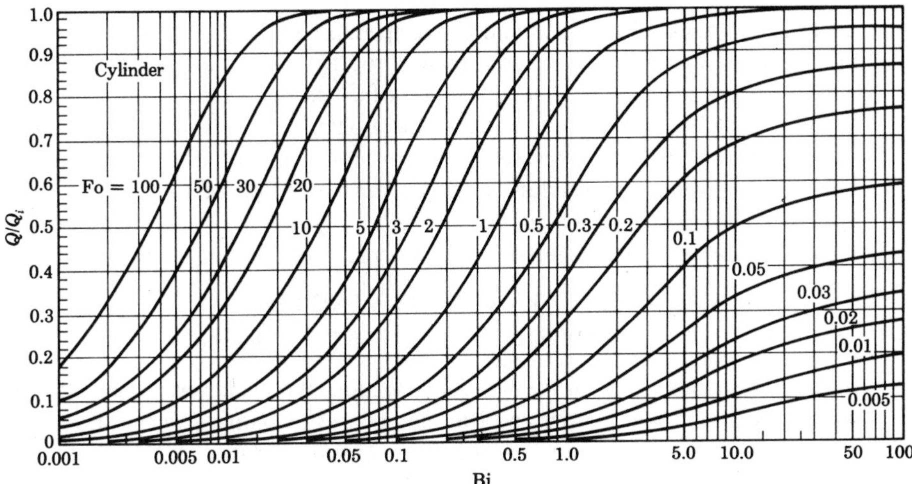

Figure 3.13. Heat flow from an infinitely long cylinder as a function of time and thermal resistance.

solution is more conveniently expressed in terms of the temperature response at the center of the cylinder (i.e., t_c at $r = \rho = 0$) and a position correction factor. The solution is expressed, then, as

$$\rho = \frac{r}{R}, \qquad \text{Fo} = \frac{\alpha\tau}{R^2}, \qquad \text{Bi} = \frac{hR}{k}$$

$$\text{At } r = 0: \quad \frac{t_c - t_f}{t_i - t_f} = \text{fn (Fo, Bi)} \quad \text{(Fig. 3.11).}$$

$$\text{At } r = r: \quad \frac{t_r - t_f}{t_i - t_f} = \frac{t_c - t_f}{t_i - t_f}\left(\frac{t_r - t_f}{t_c - t_f}\right) \tag{3.43}$$

$$\frac{t_r - t_f}{t_c - t_f} = \text{fn } (\rho, \text{Bi}) \quad \text{(Fig. 3.12).}$$

$$\frac{Q}{Q_i} = 1 - \frac{t_{\text{av}} - t_f}{t_i - t_f} = \text{fn (Fo, Bi)} \quad \text{(Fig. 3.13).}$$

The three functions required to apply Eq. (3.43) to the solution of transient problems in the infinite cylinder are displayed in the temperature response charts of Figs. 3.11 through 3.13. Again, the position correction representation should be applied only when Fo > 0.2.

EXAMPLE 3.3

The problem posed in Example 3.2 may be more realistically analyzed by accounting for convective effects at the wall's surface. Thus, consider a wall of common brick, 25 cm thick, at an initial uniform temperature of 35°C. When the cold front moves in, let the ambient air temperature drop to -10°C and presume that the wind velocity past the wall produces a surface heat transfer coefficient of 90 W/m-°C. Find the temperature in the wall at a plane 2.5 cm from the surface after 10 h has elapsed. Also find the heat that has conducted out of the wall during that time and the average temperature in the wall at the end of the 10 h.

Solution. As noted in Example 3.2 the thermal properties of the wall are

$$c_p = 0.84 \text{ kJ/kg-°C}, \qquad \rho = 1602 \text{ kg/m}^3,$$
$$k = 0.69 \text{ W/m-°C}, \qquad \alpha = 5.2 \times 10^{-7} \text{ m}^2/\text{s}.$$

In the formulation noted in the foregoing section, the characteristic length to be used in evaluating the Fourier and Biot numbers is the *half-thickness* of the slab. Thus in this instance $L = 25/2 = 12.5$ cm. The Fourier number (for $\tau = 10$ h), the Biot number, and the dimensionless position parameter ξ (for a depth of 2.5 cm) are found from their definitions given in Eq. (3.42):

$$\text{Fo} = \frac{\alpha\tau}{L^2} = \frac{5.2 \times 10^{-7} \times 10 \times 3600}{(0.125)^2} = 1.20,$$

$$\text{Bi} = \frac{hL}{k} = \frac{90 \times 0.125}{0.69} = 16.3,$$

$$\xi = \frac{x}{L} = \frac{12.5 - 2.5}{12.5} = 0.80.$$

Note that since the coordinate x is measured from the center of the slab, a plane 2.5 cm from the surface is at $x = 12.5 - 2.5$. For these values, the temperature response charts of Figs. 3.8 through 3.10 give

$$\frac{t_c - t_f}{t_i - t_f} = 0.092,$$

$$\frac{t_x - t_f}{t_c - t_f} = 0.38,$$

$$\frac{Q}{Q_i} = 0.94.$$

The temperature response at the desired location is then found by applying the position correction, $(t_x - t_f)/(t_c - t_f)$, to the centerline temperature response, $(t_c - t_f)/(t_i - t_f)$, in accord with Eq. (3.42):

$$\frac{t_x - t_f}{t_i - t_f} = \frac{t_c - t_f}{t_i - t_f}\left(\frac{t_x - t_f}{t_c - t_f}\right)$$
$$= 0.092 \times 0.38 = 0.0350.$$

Then, since $t_i = 35°C$ and $t_f = -10°C$, the temperature at the location x is

$$\frac{t_x - (-10)}{35 - (-10)} = 0.0350$$

$$t_x = -8.4°C\ (16.8°F).$$

The heat flow, per unit area, out of the *half-slab* is found from

$$\frac{Q}{A} = \frac{Q_i}{A}\frac{Q}{Q_i}$$

$$= L\rho c_p(t_i - t_f)\frac{Q}{Q_i},$$

where $Q_i/A = L\rho c_p(t_i - t_f)$ is the heat initially stored in the half-slab, referenced to the ambient fluid temperature. Thus

$$\frac{Q}{A} = 0.125 \times 1602 \times 0.84 \times [35 - (-10)] \times 0.94$$

$$= 7115\ \text{kJ/m}^2 \text{ for one side,}$$

$$= 14{,}230\ \text{kJ/m}^2\ (1253\ \text{Btu/ft}^2) \text{ for both sides.}$$

The average temperature of the slab is represented by the heat remaining:

$$\frac{t_{av} - t_f}{t_i - t_f} = 1 - \frac{Q}{Q_i}$$

$$\frac{t_{av} - (-10)}{35 - (-10)} = 1 - 0.94$$

$$t_{av} = -7.3°C\ (18.9°F).$$

EXAMPLE 3.4

A long cylindrical bar 20 cm in diameter is heated to 980°C and then quenched in an oil bath at 40°C in which the heat transfer coefficient is expected to be 565 W/m^2-°C. How long will it take for the centerline of the cylinder to reach 260°C? The bar is made of 19% Cr, 8% Ni, stainless steel.

Solution. For the stainless steel specified, Table A.1 gives

$$c_p = 0.46\ \text{kJ/kg-°C}, \qquad \rho = 7817\ \text{kg/m}^3,$$
$$k = 16.3\ \text{W/m-°C}, \qquad \alpha = 0.444 \times 10^{-5}\ \text{m}^2\text{/s}.$$

For one-dimensional transient conduction in a cylinder the characteristic length is the radius $R = 20/2 = 10$ cm $= 0.1$ m. Thus the Biot number defined in Eq. (3.43) is

$$\text{Bi} = \frac{hR}{k} = \frac{565 \times 0.1}{16.3} = 3.47.$$

With $t_i = 980°\text{C}$, $t_f = 40°\text{C}$, the desired centerline temperature of $t_c = 260°\text{C}$ yields the dimensionless temperature response at the centerline:

$$\frac{t_c - t_f}{t_i - t_f} = \frac{260 - 40}{980 - 40} = 0.234.$$

Figure 3.11 shows that this temperature is reached, for Bi = 3.47, when the Fourier number is

$$\text{Fo} = 0.53.$$

The definition of the Fourier number in Eq. (3.43) yields the required time:

$$\text{Fo} = \frac{\alpha\tau}{R^2}$$

$$0.53 = \frac{0.444 \times 10^{-5} \times \tau}{(0.1)^2}$$

$$\tau = 1194 \text{ s} = 0.332 \text{ h}.$$

Approximate Equations for the Slab and the Cylinder

The results quoted in the foregoing for transient conduction in a slab or a cylinder heated to a uniform initial temperature and subjected to surface convection for subsequent times were presented graphically in Figs. 3.8 through 3.13. It is difficult to read these temperature response charts with accuracy, and the analytical solutions from which they are derived are complicated infinite series that are difficult to use for computational purposes. Recently, Chen and Kuo (Ref. 8) applied the approximation technique known as the *heat balance integral method* to obtain approximate equations that may be used in lieu of the temperature response charts. These equations may be easily evaluated using the commonly available hand-held calculators and the results are usually as accurate as those obtained from the charts, thus effectively replacing the charts.

The equations found by Chen and Kuo are summarized below and apply in the range of Fourier number for which the Heisler position correction concept is applicable. The expected accuracy is also indicated. Chen and Kuo also quote approximate equations for "short" time (i.e., Fo < 0.25), but these results are not given here, due to their limited practical applicability.

The Infinite Slab. For the infinite slab the functions in Figs. 3.8 through 3.10 needed to apply Eq. (3.42) may be approximated by:

For Fo > 0.25:

$$\frac{t_c - t_f}{t_i - t_f} = \exp\left[\frac{\text{Fo}^* - \text{Fo}}{0.35 + (1/\text{Bi}) + 0.05 \exp(-4/\text{Bi})}\right] \tag{3.44}$$

$$\text{Fo}^* = 0.167 - \frac{0.067}{1 + (6/\text{Bi})}$$

$$\frac{t_x - t_f}{t_c - t_f} = 1 - \frac{(1 + C)\xi^2 - C\xi^3}{1 + (2 - C)/\text{Bi}} \tag{3.45}$$

$$C = \frac{0.315}{1 + (2.5/\text{Bi})}.$$

For Fo > 0.1:

$$\frac{Q}{Q_i} = 1 - \left[\frac{0.63 + (2/\text{Bi})}{1 + (2/\text{Bi})}\right]\frac{t_c - t_f}{t_i - t_f}. \tag{3.46}$$

In Eq. (3.46) the ratio $(t_c - t_f)/(t_i - t_f)$ is found from Eq. (3.44) even when $0.1 < \text{Fo} < 0.25$. In any of Eqs. (3.44) through (3.46), the error in the computed dimensionless ratio will be less than 0.01 when compared with the exact solution.

The Infinite Cylinder. The corresponding approximate equations for the long cylinder are:

For Fo > 0.2:

$$\frac{t_c - t_f}{t_i - t_f} = \exp\left[\frac{\text{Fo}^* - \text{Fo}}{0.13 + (0.5/\text{Bi}) + 0.04 \exp(-2/\text{Bi})}\right] \tag{3.47}$$

$$\text{Fo}^* = 0.14 - \frac{0.056}{1 + (1/\text{Bi})}$$

$$\frac{t_r - t_f}{t_c - t_f} = 1 - \frac{(1 + C)\rho^2 - C\rho^3}{1 + (2 - C)/\text{Bi}} \tag{3.48}$$

$$C = \frac{0.595}{1 + (3/\text{Bi})}.$$

For Fo > Fo*:

$$\frac{Q}{Q_i} = 1 - \left[\frac{0.42 + (2/\text{Bi})}{1 + (2/\text{Bi})}\right]\frac{t_c - t_f}{t_i - t_f}. \tag{3.49}$$

Again, in Eq. (3.49) the centerline temperature ratio from Eq. (3.47) is used even if $\text{Fo}^* < \text{Fo} < 0.2$. Equations (3.47) through (3.49) may, then, be used to replace Figs. 3.11 through 3.13. The error in the centerline temperature ratio [Eq. (3.47)] will be less than 0.02 and less than 0.01 for $(t_r - t_f)/(t_c - t_f)$ and Q/Q_i.

EXAMPLE 3.5

Repeat Examples 3.3 and 3.4 using the approximate equations.

Solution.

(a) For Example 3.3 one has a slab with the following dimensionless parameters:

$$\text{Fo} = 1.20, \quad \text{Bi} = 16.3, \quad \xi = 0.80.$$

Instead of using Figs. 3.8 through 3.10, Eqs. (3.44) through (3.46) give the centerline temperature response, the position correction, and the relative heat flow:

$$\text{Fo}^* = 0.167 - \frac{0.067}{1 + (6/\text{Bi})}$$

$$= 0.167 - \frac{0.067}{1 + (6/16.3)} = 0.1180,$$

$$C = \frac{0.315}{1 + (2.5/\text{Bi})}$$

$$= \frac{0.315}{1 + (2.5/16.3)} = 0.2731,$$

$$\frac{t_c - t_f}{t_i - t_f} = \exp\left[\frac{\text{Fo}^* - \text{Fo}}{0.35 + (1/\text{Bi}) + 0.05\exp(-4/\text{Bi})}\right]$$

$$= \exp\left[\frac{0.1180 - 1.20}{0.35 + (1/16.3) + 0.05\exp(-4/16.3)}\right] = 0.091,$$

$$\frac{t_x - t_f}{t_c - t_f} = 1 - \frac{(1 + C)\xi^2 - C\xi^3}{1 + (2 - C)/\text{Bi}}$$

$$= 1 - \frac{(1 + 0.2731)(0.8)^2 - 0.2731 \times (0.8)^3}{1 + (2 - 0.2731)/16.3} = 0.39,$$

$$\frac{Q}{Q_i} = 1 - \left[\frac{0.63 + (2/\text{Bi})}{1 + (2/\text{Bi})}\right]\frac{t_c - t_f}{t_i - t_f}$$

$$= 1 - \left[\frac{0.63 + (2/16.3)}{1 + (2/16.3)}\right] \times 0.091 = 0.94.$$

The values found for the center temperature response and the position correction differ only slightly from those found in Example 3.3 and that for Q/Q_i is the same. Since Fo = 1.20 is well above the condition Fo > 0.25 for which the approximate equations are applicable, they yield very good results. Performing the same calculations as in Example 3.3 gives identical answers for the desired temperature, heat flow, and average temperature:

$$t_x = -8.4°\text{C } (16.8°\text{F}),$$

$$\frac{Q}{A} = 14{,}230 \text{ kJ/m}^2 \text{ } (1253 \text{ Btu/ft}^2),$$

$$t_{av} = -7.3°\text{C } (18.9°\text{F}).$$

(b) For Example 3.4 one has a cylinder with the Biot number and centerline temperature response:

$$\text{Bi} = 3.47, \qquad \frac{t_c - t_f}{t_i - t_f} = 0.234.$$

Knowing Bi permits calculation of the quantity Fo* given in Eq. (3.47):

$$\text{Fo}^* = 0.14 - \frac{0.056}{1 + (1/\text{Bi})}$$

$$= 0.14 - \frac{0.056}{1 + (1/3.47)} = 0.0965.$$

Then, since the temperature response at the centerline is known, Eq. (3.47) permits the calculation of the Fourier number:

$$\frac{t_c - t_f}{t_i - t_f} = \exp\left[\frac{\text{Fo}^* - \text{Fo}}{0.13 + (0.5/\text{Bi}) + 0.04 \exp(-2/\text{Bi})}\right]$$

$$0.234 = \exp\left[\frac{0.0965 - \text{Fo}}{0.13 + (0.5/3.47) + 0.04 \exp(-2/3.47)}\right]$$

$$\text{Fo} = 0.53.$$

This value of Fo is the same as that found in Example 3.4 using the charts. Thus the time lapse is the same:

$$\text{Fo} = \frac{\alpha\tau}{R^2}$$

$$0.53 = \frac{0.444 \times 10^{-5}\tau}{(0.1)^2}$$

$$\tau = 1194 \text{ s} = 0.332 \text{ h}.$$

The Semi-infinite Solid

For very thick slabs in which one is interested in the temperature response at or near the surface, the solution methods given in Eqs. (3.42) and (3.44) through (3.46) become impossible to apply. This difficulty arises from the fact that the characteristic length (the half-thickness L) becomes so large that the various dimensionless parameters take on values out of the range of applicability of the charts or equations involved. For instance, if the slab in Example 3.3 were 10 times as thick, so that $L = 125$ cm, then one finds Fo $= 0.012$ and $\xi = 0.98$. For these values the centerline temperature response cannot be found from Fig. 3.8, and even if it could, its value would be very close to 1. Heisler's position correction concept is no longer applicable, and even if it were, accurate values are impossible to obtain from Fig. 3.9. Likewise, the approximate relations in Eqs. (3.44) through (3.46) are not applicable for the low value of Fo involved.

For such very thick slabs or for virtually infinite media such as the earth's surface, the problem needs to be recast and resolved. The semi-infinite solid depicted in Fig. 3.14 is the geometry used in analyzing this problem. The solid is presumed to have a plane exposed face but to be of infinite extent into and out of the paper and to the right of the exposed face. The coordinate x is measured from the face into the solid. If the solid is assumed to be of a uniform initial temperature t_i at time $\tau = 0$ and then have its face exposed to an ambient convecting fluid at temperature t_f through a heat transfer coefficient h, it is desired to find the temperature–time response for any location in the solid, x for $\tau > 0$. Since there is no characteristic length on which to base the definition of the Fourier and Biot numbers, it becomes necessary to define *local* values of these numbers based on the local coordinate x:

$$\text{Local Fourier number: } \mathrm{Fo}_x = \frac{\alpha\tau}{x^2}. \qquad (3.50)$$
$$\text{Local Biot number: } \mathrm{Bi}_x = \frac{hx}{k}.$$

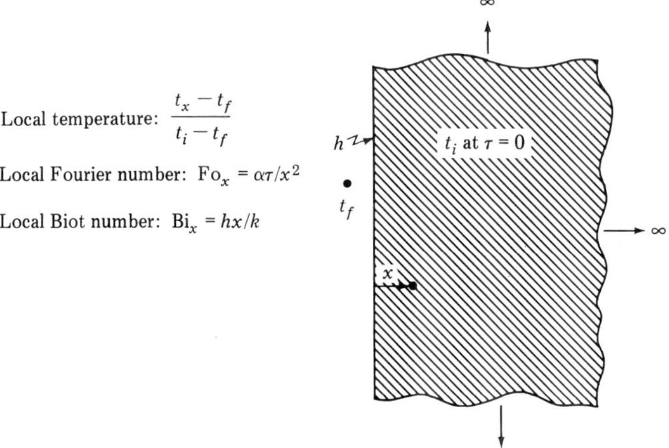

Figure 3.14. Geometry and dimensionless parameters for transient conduction in a semi-infinite solid.

In terms of these parameters the desired temperature response has been obtained by Schneider (Ref. 9) and is

$$\begin{aligned}\frac{t_x - t_f}{t_i - t_f} &= \operatorname{erf}\left(\frac{1}{2\mathrm{Fo}_x^{1/2}}\right) \\ &+ \left\{\left[1 - \operatorname{erf}\left(\frac{1}{2\mathrm{Fo}_x^{1/2}} + \mathrm{Bi}_x\mathrm{Fo}_x^{1/2}\right)\right] \exp\left(\mathrm{Bi}_x + \mathrm{Bi}_x^2\mathrm{Fo}_x\right)\right\}.\end{aligned} \tag{3.51}$$

In Eq. (3.51) erf (z) is known as the *error function* of argument z and is a function of known properties and tabulated values (Ref. 1).

For the solution of practical problems, the results of Eq. (3.51) are plotted in Figs. 3.15 and 3.16. Two plots are presented, as suggested by Wolf (Ref. 7), to avoid iterative calculations depending on which data are given in a particular problem. In both figures the local Fourier number, Fo_x, is used as the abcissa—increasing Fo_x meaning that one is approaching the exposed surface. Figure 3.15 plots the temperature response as a function of Fo_x and Bi_x and is useful when one knows the location x and wishes the temperature there as a function of time or the time at which the temperature reaches a certain value. Figure 3.16 plots the temperature response as a function of Fo_x and the parameter $\mathrm{Bi}_x\mathrm{Fo}_x^{1/2} = h\sqrt{\alpha\tau}/k$. It is useful in finding the value of x at which the temperature reaches a certain value in a known time. In Fig. 3.16 the asymptotic values approached for large Fo_x represent the time dependence of the temperature of the exposed surface.

EXAMPLE 3.6

The soil in a certain locality has $k = 0.85$ Btu/h-ft-°F and $\alpha = 0.03$ ft²/h. Initially, the soil is in equilibrium at 60°F with the atmosphere. The ambient air temperature suddenly drops to 10°F and a heat transfer coefficient of 15 Btu/h-ft²-°F exists at the surface.

(a) For a water pipe buried 6 in. below the surface, how long will it take for the temperature of the soil at that depth to reach the freezing point, 32°F?

(b) How deep should the pipe be buried in order that the temperature there not reach freezing in less than 6 h?

(c) What is the ground surface temperature 2 h after the air temperature drops to 10°F?

Solution

(a) Since the depth is known to be $x = 6$ in., the local Biot number may be found:

$$\mathrm{Bi}_x = \frac{hx}{k} = \frac{15 \times (6/12)}{0.85} = 8.82.$$

For this depth the temperature ratio is known to be

$$\frac{t_x - t_f}{t_i - t_f} = \frac{32 - 10}{60 - 10} = 0.440,$$

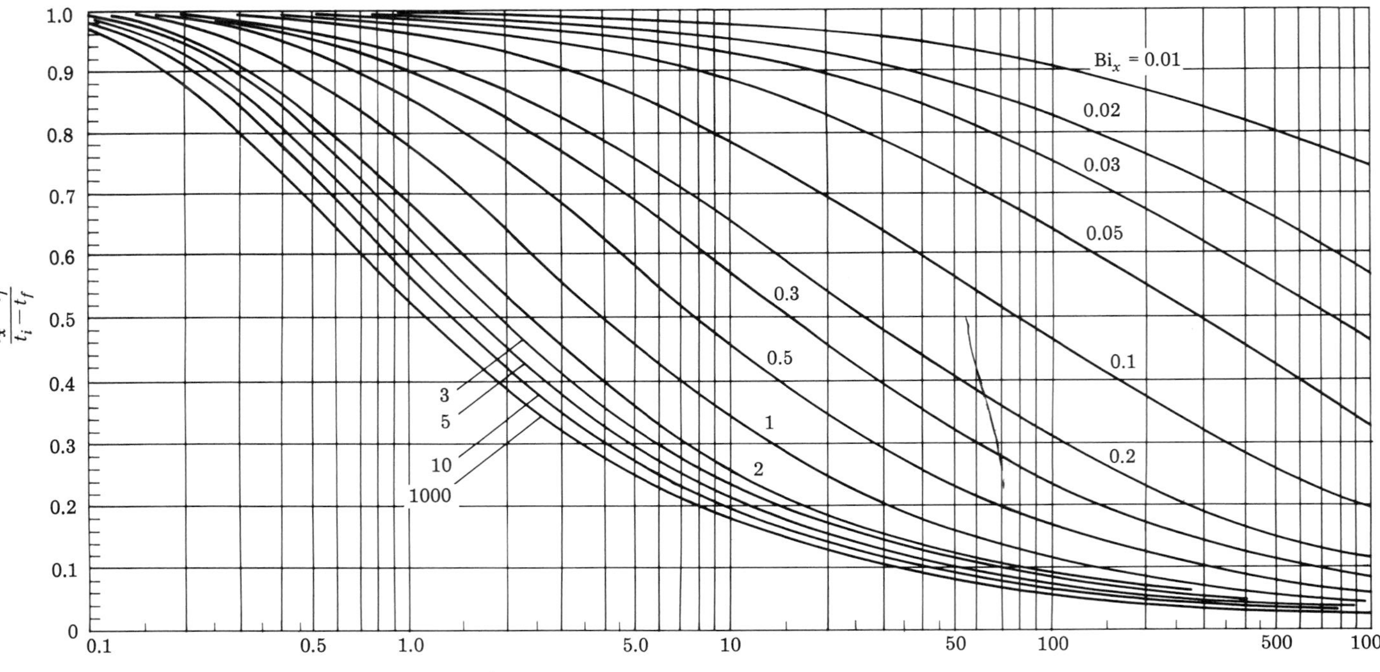

Figure 3.15. Temperature response in a semi-infinite solid, $Bi_x = hx/k$ a parameter.

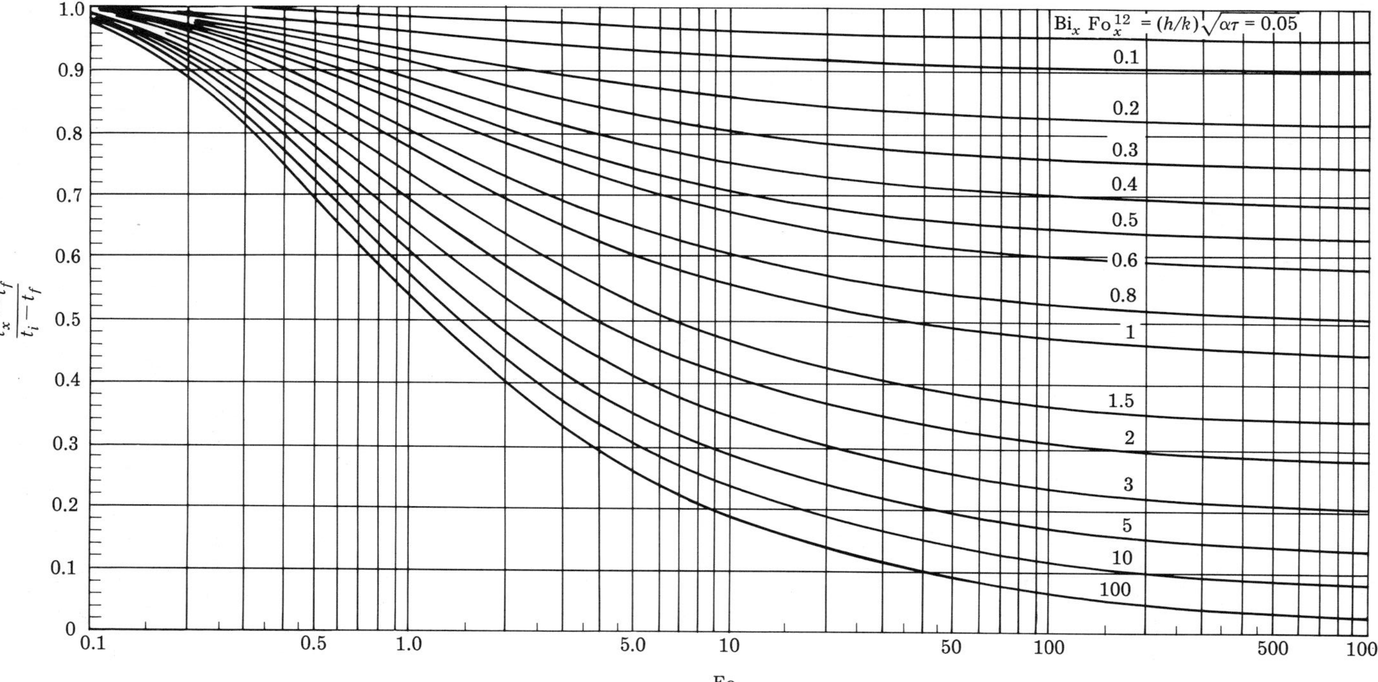

Figure 3.16 Temperature response in a semi-infinite solid, $Bi_xFo_x^{1/2} = (h/k)\sqrt{\alpha\tau}$ a parameter.

so that Fig. 3.15 gives the local Fourier number and allows the time to be calculated:

$$\mathrm{Fo}_x = 1.82 = \frac{\alpha\tau}{x^2} = \frac{0.03\tau}{(6/12)^2}$$
$$\tau = 15.2 \text{ h}.$$

(b) To find the depth at which the temperature reaches 32°F in 6 h, it is more convenient to use Fig. 3.16 since the plotted parameter does not involve x. Thus

$$\mathrm{Bi}_x\mathrm{Fo}_x^{1/2} = \frac{h\sqrt{\alpha\tau}}{k} = \frac{15 \times \sqrt{0.03 \times 6}}{0.85} = 7.49.$$

With this value and the temperature ratio still the same, $(t_x - t_f)/(t_i - t_f) = 0.440$, Fig. 3.16 gives

$$\mathrm{Fo}_x = 2.07 = \frac{\alpha\tau}{x^2} = \frac{0.03 \times 6}{x^2}$$
$$x = 0.30 \text{ ft} = 3.6 \text{ in. } (9.14 \text{ cm}).$$

Thus the pipe should be buried below 3.6 in. to avoid freezing in 6 h.

(c) For a time lapse of 2 h, one has

$$\frac{h\sqrt{\alpha\tau}}{k} = \frac{15 \times \sqrt{0.03 \times 2}}{0.85} = 4.23.$$

At this value the temperature ratio for very large Fo_x (i.e., the surface) is seen from Fig. 3.16 to approach the value

$$\frac{t_x - t_f}{t_i - t_f} \rightarrow 0.13.$$

Thus

$$\frac{t_s - t_f}{t_i - t_f} = \frac{t_s - 10}{60 - 10} = 0.13$$
$$t_s = 16.5°\text{F } (-8.6°\text{C}).$$

3.7 TRANSIENT CONDUCTION IN MORE THAN ONE DIMENSION

The one-dimensional solutions for the slab, cylinder, and semi-infinite solid given in Sec. 3.6 may be used to solve certain transient problems in two or three dimensions. The method involves the use of a product superposition principle, the details of which are beyond the scope of this text but are discussed more fully in Refs. 2, 4, and 9. One of the limitations under which the product superposition principle applies has to do with the nature of the imposed initial condition at time zero. The case of the uniform initial temperature, as used in Sec. 3.6, satisfies this limitation.

The product superposition method may best be illustrated by the consideration of a specific case. Consider, for example, the circular cylinder, radius R, and of finite length, $2L$, shown in Fig. 3.17. If the cylinder is initially at a uniform temperature t_i, and then exposed to a fluid at t_f with a heat transfer coefficient h (t_f and h the same for *all* exposed surfaces), then the dimensionless temperature ratio at any point within the cylinder may be shown to be the *product* of the one-dimensional solutions obtained for the infinite slab and the infinite cylinder whose intersection forms the finite cylinder. The slab and infinite cylinder forming the intersecting finite cylinder are also shown in Fig. 3.17. The dimensionless temperature ratio at a point in the finite cylinder located at a radius r and a distance x from the midplane is the product of the temperature ratio for an infinite cylinder

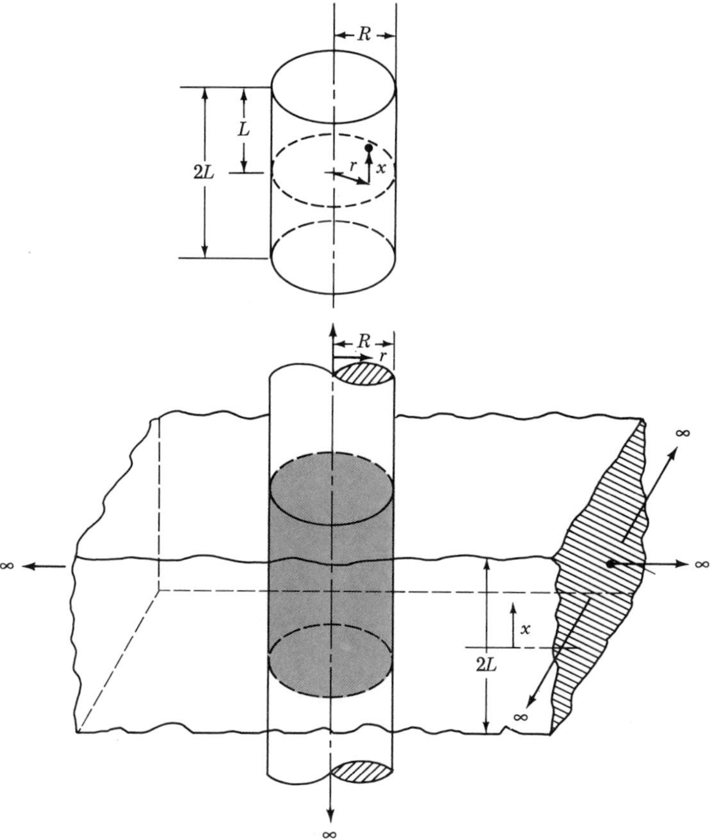

Figure 3.17. Finite cylinder formed by the intersection of an infinite cylinder and an infinite slab.

at radius r (outside radius R) and that for an infinite slab at location x (half-thickness L):

$$\frac{t_{x,r} - t_f}{t_i - t_f} = \left[\frac{t_x - t_f}{t_i - t_f}\right]_{\text{slab},L} \times \left[\frac{t_r - t_f}{t_i - t_f}\right]_{\text{cyln},R} \tag{3.52}$$

The two ratios on the right side of Eq. (3.52) are found by the methods of Sec. 3.6. Obviously, these two ratios must be evaluated at the same time, τ, and since

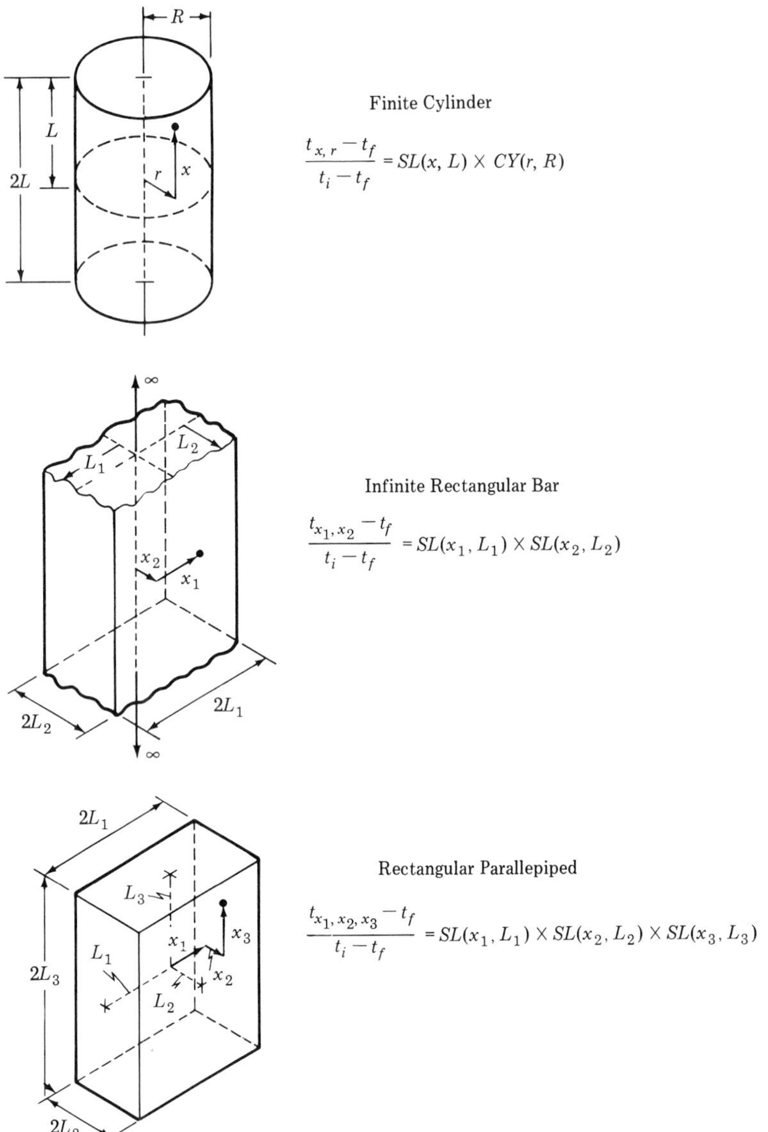

Figure 3.18. Product solutions for two- and three-dimensional conduction.

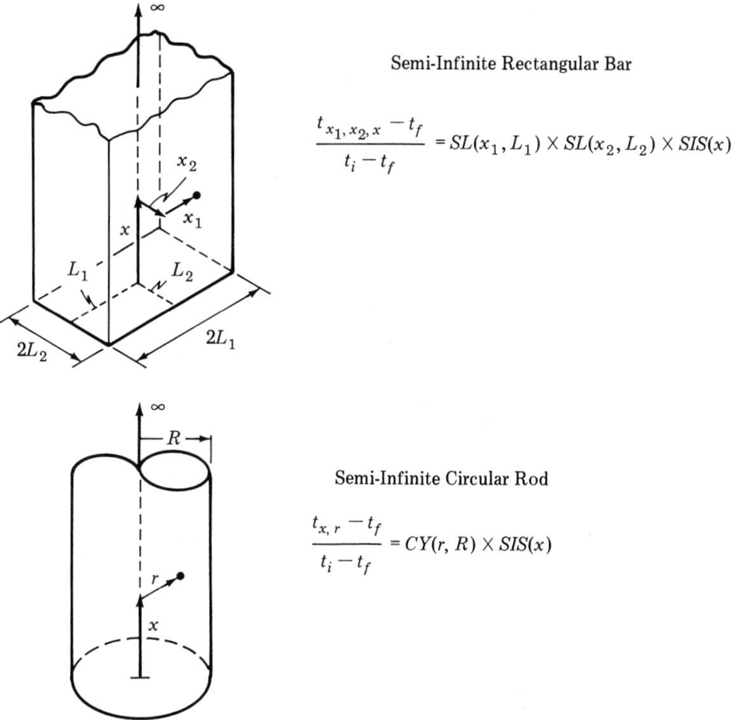

Figure 3.18. (Continued)

the characteristic dimensions (L and R) are generally different, the Fourier numbers in the two instances will probably be different—as will the Biot numbers, for the same reason.

The same multiplicative superposition principle may be applied to determine the temperature–time response in other two- and three-dimensional geometries formed by the intersection of the one-dimensional cases discussed in Sec. 3.6. Several of the possible cases, such as the long rectangular bar or the rectangular parallelepiped, are depicted in Fig. 3.18. The appropriate products are also indicated in Fig. 3.18, and the following notation is used to denote the various one-dimensional solutions involved (the infinite slab, the infinite cylinder, and the semi-infinite solid):

$$\text{Infinite slab: } \frac{t_x - t_f}{t_i - t_f} = \text{SL}(x, L).$$

$$\text{Infinite cylinder: } \frac{t_r - t_f}{t_i - t_f} = \text{CY}(r, R). \tag{3.53}$$

$$\text{Semi-infinite solid: } \frac{t_x - t_f}{t_i - t_f} = \text{SIS}(x).$$

The functions indicated in Eq. (3.53) are those summarized in Eq. (3.42) for the slab, Eq. (3.43) for the cylinder, and Eq. (3.51) for the semi-infinite solid. The approximate equations quoted in the foregoing for the slab and cylinder may also be used for SL(x, L) and CY(r, R) in lieu of the temperature response charts of Figs. 3.8 through 3.13.

The notation indicated for SL(x, L) and CY(r, R) is used to emphasize the characteristic length used in each instance, and again it should be pointed out that the functions used in the products indicated in Fig. 3.18 are to be evaluated at the same real time τ. Also, as posed in Fig. 3.18, the solids involved are all presumed to have a uniform initial temperature t_i. In terms of the functions just defined, the solution quoted in Eq. (3.52) for the finite cylinder, length $2L$ radius R, is

$$\frac{t_{x,r} - t_f}{t_i - t_f} = \mathrm{SL}(x, L) \times \mathrm{CY}(r, R). \tag{3.54}$$

EXAMPLE 3.7

An aluminum cylinder ($k = 215$ W/m-°C, $\alpha = 8.42 \times 10^{-5}$ m^2/s) has a diameter of 5.0 cm and a length of 10 cm. It is initially at a uniform temperature of 200°C and then plunged into a quenching bath at 10°C where a surface heat transfer coefficient of 1000 W/m-°C may be assumed to exist. What is the temperature on the centerline of the cylinder at a distance 1 cm from one end after a time lapse of 1 min?

Solution. The length of the cylinder, compared with its diameter, suggests that it is probably not long enough for the conduction in it to be treated as simply radial. Hence the problem must be treated as two-dimensional, and the product superposition principle just discussed may be applied. In this instance one uses the product of the solution for a slab and that for an infinite cylinder.

For the slab, the half-thickness is half of the cylinder length, $L = 10/2 = 5.0$ cm and the point in question, 1.0 cm from the end (or slab surface), is at a distance $x = 5.0 - 1.0 = 4.0$ cm from the slab midplane. Thus for the given physical properties and time $\tau = 1$ min, the dimensionless parameters required are:

$$\mathrm{Fo} = \frac{\alpha\tau}{L^2} = \frac{8.42 \times 10^{-5} \times 1 \times 60}{(0.05)^2} = 2.02,$$

$$\mathrm{Bi} = \frac{hL}{k} = \frac{1000 \times 0.05}{215} = 0.232,$$

$$\xi = \frac{x}{L} = \frac{4.0}{5.0} = 0.80.$$

Then Figs. 3.8 and 3.9 give the temperature response at the slab centerline and the position correction

$$\frac{t_c - t_f}{t_i - t_f} = 0.67, \qquad \frac{t_x - t_f}{t_c - t_f} = 0.93.$$

Thus the temperature response at the location x is given, in accord with Eq. (3.42), by

$$\begin{aligned} \mathrm{SL}(x, L) = \frac{t_x - t_f}{t_i - t_f} &= \frac{t_c - t_f}{t_i - t_f}\left(\frac{t_x - t_f}{t_c - t_f}\right) \\ &= 0.67 \times 0.93 \\ &= 0.623. \end{aligned}$$

In the above the notation of Eq. (3.53), $\mathrm{SL}(x, L)$, has been used to denote the dimensionless temperature given by the slab solution.

For the infinite cylinder part of the solution, the radius is $R = 5.0/2 = 2.5$ cm, and the point in question lies on the cylinder centerline. Thus at $\tau = 1$ min, the required dimensionless parameters are

$$\mathrm{Fo} = \frac{\alpha\tau}{R^2} = \frac{8.42 \times 10^{-5} \times 1 \times 60}{(0.025)^2} = 8.08,$$

$$\mathrm{Bi} = \frac{hR}{k} = \frac{1000 \times 0.025}{215} = 0.116,$$

$$\rho = \frac{r}{R} = 0.$$

Note that even though the real time, τ, is the same in both the slab and cylinder, the dimensionless times (the Fourier numbers) are different since the characteristic lengths are different in the two geometries. Figure 3.11 gives the temperature response at the cylinder centerline, and the position correction is clearly 1.0:

$$\frac{t_c - t_f}{t_i - t_f} = 0.17, \qquad \frac{t_r - t_f}{t_c - t_f} = 1.0.$$

Thus the temperature response at the location r is

$$\begin{aligned} \mathrm{CY}(r, R) = \frac{t_r - t_f}{t_i - t_f} &= \frac{t_c - t_f}{t_i - t_f}\left(\frac{t_r - t_f}{t_c - t_f}\right) \\ &= 0.17 \times 1.0 \\ &= 0.17. \end{aligned}$$

Again, the notation of Eq. (3.53) has been introduced to denote the dimensionless temperature in the infinite cylinder as $\mathrm{CY}(r, R)$.

The dimensionless temperature in the finite cylinder at x and r is given by the product superposition principle as noted in either Fig. 3.18 or Eq. (3.54):

$$\begin{aligned} \frac{t_{x,r} - t_f}{t_i - t_f} &= \mathrm{SL}(x, L) \times \mathrm{CY}(r, L) \\ &= 0.623 \times 0.17 = 0.106. \end{aligned}$$

Thus the temperature at the desired location is, since $t_i = 200$°C and $t_f = 10$°C,

$$\frac{t_{x,r} - 10}{200 - 10} = 0.106$$

$$t_{x,r} = 30.1°\text{C } (86.2°\text{F}).$$

In this example the center temperature response and the position correction for each of the two one-dimensional geometries used in the product were determined by using the temperature response charts of Figs. 3.8 through 3.13. The approximate equations given by Eqs. (3.44) through (3.49) could have been used instead, with virtually the same results.

It should also be noted that this example was relatively straightforward in that the temperature at the designated location was sought *at a given time*. This meant that the Fourier numbers in the two geometries comprising the product solution were both known. If, on the other hand, one were seeking the time at which the temperature at the point reached a certain value, then, in general, an iterative solution would be required since one does not know, a priori, what fraction of the sought-for dimensionless temperature is contributed by each of the geometries at the to-be-determined time. One needs to assume the time and proceed as in the example—assertaining whether the desired temperature is reached at that time and revising the assumption accordingly. An iterative calculation can be avoided using the closed-form approximate equations by noting that the Fourier numbers in the two geometries are interrelated by

$$\frac{(\text{Fo})_{\text{SL}}}{(\text{Fo})_{\text{CY}}} = \frac{\alpha\tau/L^2}{\alpha\tau/R^2} = \left(\frac{R}{L}\right)^2.$$

EXAMPLE 3.8

Figure 3.19 depicts a very long bar of glass ($k = 0.9$ W/m-°C, $\alpha = 4.0 \times 10^{-7}$ m^2/s), 4 cm square. The bar is initially heated to a uniform temperature of 160°C and then placed to cool in a fluid at 20°C where the heat transfer coefficient is $h = 125$ W/m^2-°C. Find the temperature on one of the corners of the bar, 1 cm from the exposed end, after a lapse of 5 min.

Solution. Since the bar is very long, it may be treated as a semi-infinite bar of the form depicted in Fig. 3.18. The temperature at some location in the bar may be represented as the product of the solutions in two slabs and a semi-infinite solid:

$$\frac{t_{x1,x2,x} - t_f}{t_i - t_f} = \text{SL}(x_1, L_1) \times \text{SL}(x_2, L_2) \times \text{SIS}(x).$$

In the expression above x_1 and x_2 are the coordinates of the point in question from the midplanes of the two slabs (half-thicknesses L_1 and L_2) which comprise the rectangular dimensions of the bar, and x is the distance of the point into the semi-infinite solid, i.e., the distance from the free end of the bar.

In the problem at hand the two slabs each have the same half-thickness, $L_1 = L_2 = 4/2 = 2.0$ cm, and since the temperature is desired at the corner of the bar

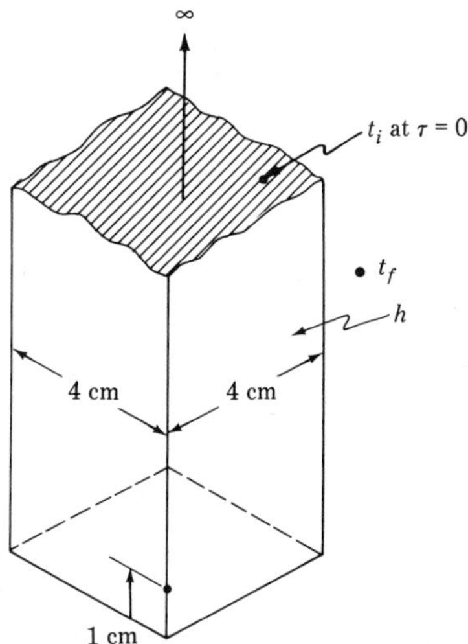

Figure 3.19

the coordinate location in each slab is the same: $x_1 = x_2 = 2.0$ cm. Thus the solution is identical in the two slabs, with the following dimensionless parameters:

$$\text{Fo} = \frac{\alpha\tau}{L^2} = \frac{4 \times 10^{-7} \times 5 \times 60}{(0.02)^2} = 0.300,$$

$$\text{Bi} = \frac{hL}{k} = \frac{125 \times 0.02}{0.9} = 2.778,$$

$$\xi = \frac{x}{L} = 1.0.$$

While the temperature response charts, Figs. 3.8 through 3.10, may be used, in this instance let the approximate equations be employed. Equations (3.44) and (3.45) give

$$\text{Fo}^* = 0.167 - \frac{0.067}{1 + (6/\text{Bi})}$$

$$= 0.167 - \frac{0.067}{1 + (6/2.778)} = 0.1458,$$

$$C = \frac{0.315}{1 + (2.5/\text{Bi})}$$

$$= \frac{0.315}{1 + (2.5/2.778)} = 0.1658,$$

$$\frac{t_c - t_f}{t_i - t_f} = \exp\left[\frac{\text{Fo}^* - \text{Fo}}{0.35 + (1/\text{Bi}) + 0.05 \exp(-4/\text{Bi})}\right]$$

$$= \exp\left[\frac{0.1458 - 0.300}{0.35 + (1/2.778) + 0.05\exp(-4/2.778)}\right] = 0.808,$$

$$\frac{t_x - t_f}{t_c - t_f} = 1 - \frac{(1 + C)\xi^2 - C\xi^3}{1 + (2 - C)/\text{Bi}}$$

$$= 1 - \frac{1}{1 + (2 - 0.1658)/2.778} = 0.398.$$

Thus the temperature response in each of the slabs at the given location is

$$\text{SL}(x_1, L_1) = \text{SL}(x_2, L_2) = \frac{t_x - t_f}{t_i - t_f} = \frac{t_c - t_f}{t_i - t_f}\left(\frac{t_x - t_f}{t_c - t_f}\right)$$

$$= 0.808 \times 0.398 = 0.322.$$

The notation $\text{SL}(x, L)$ has been used to denote the dimensionless temperature solution in the slab in accord with the notation of Eq. (3.53) and Fig. 3.18.

For the solution in the semi-infinite solid, one applies Eqs. (3.50) and (3.51), or Fig. 3.15, to obtain, for the data given,

$$x = 1.0 \text{ cm},$$

$$\text{Fo}_x = \frac{\alpha\tau}{x^2} = \frac{4 \times 10^{-7} \times 5 \times 60}{(0.01)^2} = 1.20,$$

$$\text{Bi}_x = \frac{hx}{k} = \frac{125 \times 0.01}{0.9} = 1.389,$$

$$\text{SIS}(x) = \frac{t_x - t_f}{t_i - t_f} = 0.69 \quad \text{(Fig. 3.15)}.$$

Finally, then, the application of the product superposition principle yields the dimensionless temperature at the location x_1, x_2, x:

$$\frac{t_{x1,x2,x} - t_f}{t_i - t_f} = \text{SL}(x_1, L_1) \times \text{SL}(x_2, L_2) \times \text{SIS}(x),$$

$$= 0.322 \times 0.322 \times 0.69,$$

so that the sought-for temperature is

$$\frac{t_{x1,x2,x} - 20}{160 - 20} = 0.072,$$

$$t_{x1,x2,x} = 30.0°\text{C } (86.0°\text{F}).$$

3.8 THE TRANSIENT RESPONSE OF BODIES WITH NEGLIGIBLE INTERNAL RESISTANCE

As pointed out in Sec. 3.4 when the Biot number was introduced, it could be interpreted as the ratio of the internal resistance of a body, l/k, to the external surface resistance, $1/h$. The case of specified surface temperature discussed in Sec. 3.5 may be taken as limiting cases for $\mathrm{Bi} \to \infty$. The other solutions given in Sec. 3.6 were applicable for the finite Bi. This section will consider cases in which the internal resistance is negligible and $\mathrm{Bi} \to 0$. Such a case may arise if the body has a high-enough thermal conductivity, compared with the surface film coefficient, that the interior temperature of the body may be taken as uniform at any instant of time. Thus, the entire temperature-time history of the body is regulated by the surface resistance. Under these circumstances, a heat balance on the body yields

$$hA_s(t - t_f) = -\rho V c_p \frac{dt}{d\tau}. \tag{3.55}$$

In Eq. (3.55) A_s and V represent the exposed surface area and volume of the body, respectively. The density and specific heat of the body are denoted by ρ and c_p, while h represents the surface heat transfer coefficient between the body and an ambient fluid at temperature t_f.

Equation (3.55) may be rewritten as

$$\frac{d(t - t_f)}{t - t_f} = -\frac{hA_s}{\rho c_p V}\, d\tau.$$

Let it be presumed that the body is initially at a temperature t_i before it is thrust into the fluid at temperature t_f. Integration of the equation above and application of the initial condition that $t = t_i$ at $\tau = 0$ yields the following expression for the subsequent temperature of the body t, as a function of time:

$$\begin{aligned} \frac{t - t_f}{t_i - t_f} &= e^{-(hA_s/\rho c_p V)\tau} \\ &= e^{-\tau/\varphi}. \end{aligned} \tag{3.56}$$

The parameter

$$\varphi = \frac{\rho c_p V}{hA_s} \tag{3.57}$$

is termed the *time constant* of the body—the larger the value of φ, the slower the body is to respond to the change in temperature. For given thermal properties, the time constant φ is proportional to the ratio V/A_s; thus, the smaller the surface area of a body, compared with its volume, the slower it will respond.

The instantaneous rate of heat flow between the body and the ambient fluid is given by

$$q = hA_s(t - t_f),$$

or, in dimensionless form,

$$\frac{q}{hA_s(t_i - t_f)} = e^{-\tau/\varphi}. \tag{3.58}$$

The cumulative heat flow between time $\tau = 0$ and $\tau = \tau$ is

$$Q = \int_0^\tau q \, d\tau,$$

so Eq. (3.58) leads to

$$\frac{Q}{\rho V c_p (t_i - t_f)} = 1 - e^{-\tau/\varphi}. \tag{3.59}$$

Equations (3.56) through (3.59) afford simple means for the rapid determination of the thermal response of a body as long as the basic assumption of negligible internal resistance (i.e., uniform interior temperature) may be made. The question then arises: Under what circumstances may this assumption be made? As noted at the outset of this section, negligible internal resistance implies a vanishingly small Biot number. Examination of the position correction factor for the infinite slab given in Fig. 3.9 ($\mathrm{Bi} = hL/k$) shows that for this geometry the interior temperature is virtually independent of position (within, say, 5%) when Bi is less than about 0.10. The same observation may be made for the infinite cylinder where $\mathrm{Bi} = hR/k$ (see Fig. 3.12).

In the general case discussed in the foregoing, no definite characteristic length exists on which to base the Biot number since the particular body shape was not specified. However, the ratio of the body volume to its surface area, V/A_s appearing in Eq. (3.57) for the definition of the time constant φ, has the dimensions of a length. This length may be used to define a Biot number:

$$L_c = \frac{V}{A_s}, \qquad \mathrm{Bi} = \frac{hL_c}{k}. \tag{3.60}$$

It is readily seen that for the infinite slab $L_c = L$, the half-thickness, while for the long cylinder $L_c = R/2$. As a general rule, then, one may apply the equations of this section and expect errors in the dimensionless temperature response no larger than about 5% if the Biot number noted in Eq. (3.60) is less than about 0.1.

EXAMPLE 3.9

(a) A 1-in.-diameter pure copper sphere is heated to 1200°F and then placed to cool in air at 200°F, where a surface heat transfer coefficient of 12 Btu/h-ft^2-°F may be expected to exist. Find the temperature at the center of the sphere after the lapse of 5 min.

(b) Repeat the problem if the sphere is replaced by a pure copper cube having the same volume.

Solution. Table A.1 gives, for copper, $k = 386$ W/m-°C, $c_p = 0.383$ kJ/kg-°C, $\rho = 8954$ kg/m^3. According to Appendix D, these quantities have the following values in English units:

$$k = \frac{386}{1.7308} = 223 \text{ Btu/h-ft-°F},$$

$$c_p = \frac{0.383}{4.1868} = 0.0915 \text{ Btu/lb}_\text{m}\text{-°F},$$

$$\rho = \frac{8954}{16.018} = 559 \text{ lb}_\text{m}/\text{ft}^3.$$

(a) For the sphere with radius $R = 0.5$ in., the characteristic length, $L_c = V/A_s$, is

$$L_c = \frac{V}{A_s} = \frac{\frac{4}{3}\pi R^3}{4\pi R^2} = \frac{R}{3}$$
$$= 0.1667 \text{ in.}$$

Thus the Biot number is

$$\text{Bi} = \frac{hL_c}{k} = \frac{12 \times (0.1667/12)}{223}$$
$$= 0.00075.$$

The value of Bi is sufficiently small that the internal resistance may be neglected and Eq. (3.56) applied. Thus

$$\varphi = \frac{\rho c_p}{h}\frac{V}{A_s} = \frac{559 \times 0.0915}{12} \times \frac{0.1667}{12} = 0.0592 \text{ h}.$$

Thus Eq. (3.56) gives, at $\tau = 5$ min,

$$\frac{\tau}{\varphi} = \frac{5/60}{0.0592} = 1.408,$$

$$\frac{t - t_f}{t_i - t_f} = e^{-\tau/\varphi}$$

$$\frac{t - 200}{1200 - 200} = e^{-1.408} = 0.2446$$

$$t = 444\text{°F (229°C)}.$$

(b) A cube of side dimension d will have the same volume of the sphere if $d^3 = \frac{4}{3}\pi R^3$, or $d = 0.806$ in. for $R = 0.5$ in. Identical calculations as above give

$$L_c = \frac{V}{A_s} = \frac{d^3}{6d^2} = \frac{d}{6} = 0.1343 \text{ in.},$$

$$\text{Bi} = \frac{hL_c}{k} = \frac{12 \times (0.1343/12)}{223} = 0.0006 < 0.1,$$

$$\varphi = \frac{\rho c_p}{h}\frac{V}{A_s} = \frac{559 \times 0.0915}{12} \times \frac{0.1343}{12} = 0.0477 \text{ h},$$

$$\frac{\tau}{\varphi} = \frac{5/60}{0.0477} = 1.747,$$

$$\frac{t - t_f}{t_i - t_f} = e^{-\tau/\varphi}$$

$$\frac{t - 200}{1200 - 200} = e^{-1.747}$$

$$t = 374°\text{F } (190°\text{C}).$$

The cube cools faster than the sphere since its surface area is larger for the same volume.

REFERENCES

1. Carslaw, H. S., and J. C. Jaeger, *Conduction of Heat in Solids,* New York, Oxford University Press, 1957.
2. Özisik, M. N., *Heat Conduction,* New York, Wiley, 1980.
3. Rohsenow, W. M., and J. P. Hartnet, eds., *Handbook of Heat Transfer,* New York, McGraw-Hill, 1973.
4. Chapman, A. J., *Heat Transfer,* 4th ed., New York, Macmillan, 1984.
5. Hahne, E., and U. Grigull, ''Formfaktor and Formweiderstand der stationaren mehrdimensionalen Warmeleitung,'' *Int. J. Heat Mass Transfer,* Vol. 18, 1975, p. 751.
6. Heisler, M. P., ''Temperature Charts for Induction and Constant-Temperature Heating,'' *Trans. ASME,* Vol. 69, No. 3, 1947, pp. 227–236.
7. Wolf, H., *Heat Transfer,* New York, Harper & Row, 1983.
8. Chen, R. Y., and T. L. Kuo, ''Closed Form Solutions for Constant Temperature Heating of Solids,'' *Mech. Eng. News,* Vol. 16, No. 1, Feb. 1979, p. 20.
9. Schneider, P. J., *Conduction Heat Transfer,* Reading, Mass., Addison-Wesley, 1955.

PROBLEMS

3.1 A rectangular plate has the dimensions of 15 cm × 25 cm and has its edges maintained at 0°C except at one of the 15-cm edges. On this edge the temperature is maintained at

$$t = (100°\text{C}) \sin\left(\frac{\pi x}{15}\right), \qquad 0 \le x \le 15 \text{ cm}.$$

Draw the isothermal lines within the plate for 100°C, 90°C, 80°C, . . . , 10°C.

3.2 For the plate described in Prob. 3.1, calculate the temperature along the midline normal to the 15-cm edges at points 1, 2, 3, 4, 5, 7, 10, 15, and 20 cm from the edge on which the harmonic temperature is specified.

3.3 A rectangular plate, 15 cm × 20 cm, has all its edges except one of the 15-cm sides maintained at 0°C. The 15-cm side is maintained at 200°C. Calculate the temperature at the center of the plate to an accuracy of 0.1°C using Eq. (3.17). How many terms in the series are required?

3.4 A rectangular plate 15 cm × 20 cm has two 20-cm sides maintained at 100°C, one of its 15-cm sides maintained at 300°C, and its other 15-cm side maintained at 500°C. Find the temperature at the center of the plate.

3.5 A long square bar of steel ($k = 50$ W/m-°C) has a central circular hole as pictured in the accompanying figure. The bar dimensions are 10 cm × 10 cm and the hole has a diameter of 4 cm. A fluid in the hole maintains the surface temperature there at 150°C and the outer surface of the bar is at 50°C. Use the conduction shape factor to find the heat flow rate, per unit length, through the bar.

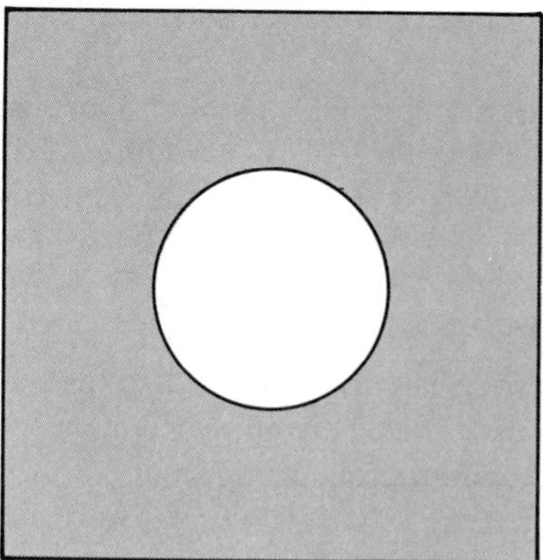

Problems 3.5 and 3.6

3.6 A tall chimney has a cross section of the geometry of the accompanying figure. The chimney is 20 ft square in outside dimensions, contains a circular inner flue 10 ft in diameter, and is made of common brick. The internal flue gases keep the inner surface at 200°F, and the ambient air maintains the outer surface at 90°F. Using the conduction shape factor, find the heat flow rate, per unit length, through the chimney wall.

3.7 For the same geometry of Prob. 3.6, suppose that the hot gases flowing in the circular flue have a temperature of 300°F and that a heat transfer coefficient of 2.5 Btu/h-ft^2-°F exists at the flue surface. If the chimney surface is at 90°F, find the temperature of the flue surface and the heat flow rate, per unit length, through the chimney.

3.8 A square chimney, 20 ft × 20 ft, has a concentric square internal flue 10 ft × 10 ft. The chimney wall material has a thermal conductivity of 0.3 Btu/h-ft-°F. If the inner flue wall is at 300°F and the outer surface is at 100°F, find the heat flow, per unit length, through the chimney wall using the conduction shape factor for a square hole in a square solid.

3.9 Repeat Problem 3.8 by treating the chimney as the combination of four plane walls and four corner sections. Compare the results of the two representations.

3.10 The earth is sometimes used as a heat sink in certain heat pump applications. A hot fluid is injected in a well and heat stored in the ground for later retrieval. Imagine, then, a 15-cm-diameter pipe sink vertically into the ground to a depth of 250 m. If a hot fluid maintains the pipe wall at 200°C and the ground surface temperature is 20°C, find the heat loss from the pipe if the earth has a conductivity of 1.4 W/m-°C.

3.11 Repeat Prob. 3.10 if the pipe is covered with a 3-cm layer of insulation (k = 0.075 W/m-°C), the pipe surface temperature still being maintained at 200°C.

3.12 A cylindrical gasoline storage tank (diameter 1 m, length 2.5 m) is buried horizontally in the ground (k = 1.2 W/m-°C) with its axis 2 m below the surface. If the tank surface is maintained at 70°C and the earth's surface at 20°C, find the rate of heat loss from the tank.

3.13 A pipe carrying chilled water for an air-conditioning system has a diameter of 30 cm and is buried horizontally in the ground so that its center is 90 cm below the surface of the ground. If the pipe wall temperature is 5°C and the earth's surface temperature is 30°C, use the conduction shape factor to find the heat gain to the pipe, per unit length. Take k = 1.4 W/m-°C for the earth.

3.14 Suppose that the pipe of Prob. 3.13 has a layer of insulation (k = 0.06 W/m-°C) 5 cm thick applied to it before it is buried. If the pipe surface temperature is still maintained at 5°C, find the temperature of the insulation surface and the heat gain to the pipe, per unit length.

3.15 Two long pipes, maintained at 500°F and 300°F, are buried very deep in the ground (k = 0.5 Btu/h-ft-°F). If the pipe diameters are 3 in. and 6 in. and if they are spaced 20 in. apart, on centers, use the conduction shape factor to find the rate of heat flow between them.

3.16 Radioactive wastes are sealed in a 1-m-diameter spherical container and buried in the earth with the sphere center at a depth of 15 m. If the waste material generates 1200 W of heat, the soil conductivity is k = 1.2/W/m-°C, and the earth's surface temperature is 15°C, find the temperature of the sphere's surface.

3.17 A slab, 25 cm thick and infinite in extent in the other directions, is initially heated to a uniform temperature of 300°C. It then has its surface temperature suddenly reduced to and maintained at 50°C. If the material has a diffusivity of 5.7×10^{-5} m^2/s, plot the variation of the centerline temperature and the heat removed, per unit area, as functions of time from 0 to 10 min. Take $k = 85$ W/m-°C.

3.18 A slab of material 2 in. thick ($\rho = 50$ lb_m/ft^3, $c_p = 0.2$ Btu/lb_m-°F, $k = 0.5$ Btu/h-ft-°F) is initially at a uniform temperature of 95°F. Its surface temperatures are suddenly raised to 210°F. Determine the temperature at a point $\frac{1}{2}$ in. from the surface and the total heat flow into the slab after 3 min has elapsed.

3.19 A slab of plastic 2.5 cm thick is initially at 70°C and then placed between two steel plates, each at 135°C. The plastic is to be heated just long enough to raise its centerline temperature to 129.8°C. If the plastic has a thermal conductivity of 0.16 W/m-°C and a thermal diffusivity of 7.5×10^{-8} m^2/s, find **(a)** the time required for heating, and **(b)** the temperature at a depth of 0.625 cm from the surface at this time.

3.20 A slab of 1.5% carbon steel, 2.5 cm thick and infinite in extent in the other directions, is heated to 980°C and then quenched in an oil bath at 90°C. If the convective heat transfer coefficient is 570 W/m^2-°C, find the time required to reduce the temperature at the center of the slab to 425°C. What is the temperature at a depth of 0.5 cm from the surface at this time? How much heat, per unit area, has been removed from the slab up to this time? Solve with the use of the temperature response charts and repeat using the approximate equations.

3.21 Plot the time variation of the surface temperature of the slab in Prob. 3.20 from 0 to 200 s. Use either the temperature response charts or the approximate equations.

3.22 A large sheet of material ($\rho = 3200$ kg/m^3, $c_p = 0.837$ kJ/kg-°C, $k = 19$ W/m-°C) is 5.0 cm thick and has an initial uniform temperature of -30°C. The temperature of the surrounding air is suddenly raised to 20°C, with a surface heat transfer coefficient of 8.5 W/m^2-°C. How much time will elapse before the centerline temperature becomes 10°C? What is the average temperature of the sheet at this time? Solve using both the temperature response charts and the approximate equations.

3.23 A concrete wall 5 in. thick and insulated on the inside is initially at equilibrium with the surrounding ambient air at 80°F on the outside. The outside air temperature suddenly drops to 60°F. A heat transfer coefficient of 1.5 Btu/h-ft^2-°F exists at the outside surface. Find the temperature of the inside and outside surfaces after 8 h and 24 h have elapsed. Use either the temperature response charts or the approximate equations.

3.24 A large plate of Alusil aluminum alloy 15 cm thick is initially at a uniform temperature of 25°C. It is placed in an oven having an ambient atmosphere of 525°C and in which the surface heat transfer coefficient is 500 W/m^2-°C.

(a) Find the length of time required to raise the midplane temperature to 425°C.
(b) Find the surface temperature at this time.
(c) Repeat these determinations if the material were stainless steel (18% Cr, 8% Ni).

3.25 Frozen foods are sometimes processed by freezing large slabs of the food which are then subsequently cut into smaller pieces for packaging. Consider, then, a slab of material (ρ = 5 $\text{lb}_\text{m}/\text{ft}^3$, c_p = 0.5 Btu/lb_m-°F, k = 0.5 Btu/h-ft-°F) that is initially at a uniform temperature of 70°F. It is then quick-frozen by thrusting it into a brine bath at −130°F, where the surface heat transfer coefficient is 1.0 Btu/h-ft^2-°F. The centerline temperature is to be reduced to at least −20°F, but the surface temperature must not fall below −60°F to prevent freeze damage to the food. Find, using either the temperature response charts or the approximate equations, the maximum slab thickness that may be processed in this manner. Also find the time required and the average slab temperature at this time.

3.26 Large slabs of steel (ρ = 8000 kg/m^3, k = 17.3 W/m-°C, c_p = 0.419 kJ/kg-°C) are to be annealed by placing them in an oven in which the ambient gas temperature is 980°C and the heat transfer coefficient is 115 W/m^2-°C. The slabs are at a uniform temperature of 150°C before being placed in the oven. It is desired to raise the *average* temperature of the slab to 780°C and the centerline temperature must reach at least 700°C in order for the annealing process to be effective. Determine the maximum slab thickness that may be annealed in this manner. How long does the annealing process take? This problem can be solved using either the temperature response charts or the approximate equations; however, the latter permits the solution to take a closed form while the former requires an iterative calculation.

3.27 Repeat Prob. 3.26, but subject to the restriction that the *average* temperature be raised to 780°C without the surface temperature exceeding 900°C. The use of the temperature response charts will require an iterative solution, while use of the approximate equations will require an iterative procedure or trial-and-error solution of a complex algebraic equation.

3.28 It is desired to cook a 1-in.-thick slab of meat (c_p = 1.0 Btu/lb_m-°F, ρ = 80 $\text{lb}_\text{m}/\text{ft}^3$, k = 0.4 Btu/h-ft-°F), initially at 80°F, by placing it in an oven of uniform temperature and heating it from both sides. A surface heat transfer coefficient of 4 Btu/h-ft^2-°F exists at both surfaces. In order to be certain that undesirable bacteria are killed during cooking, it is necessary to heat the slab of meat until its coolest portion reaches at least 250°F—and then continue the cooking for an additional 20 min. To avoid overcooking it is desired that no portion of the meat exceed 310°F at the end of the total cooking time. Find **(a)** the temperature of the air in the oven, **(b)** the total cooking time, **(c)** the midplane temperature at the end of the cooking time, and **(d)** the average temperature of the slab at the end of cooking. Solution of this problem with the temperature response charts will require an iterative calculation; however, the approximate equations allow a closed-form solution.

3.29 A cylindrical bar of stainless steel (18% Cr, 8% Ni) 10 cm in diameter is removed from an annealing furnace where it has been maintained at 980°C. It is placed in air at 30°C to cool. If the surface heat transfer coefficient is 11 W/m^2-°C, what is its centerline temperature after 2 h has passed? Solve using both the temperature response charts and the approximate equations.

3.30 A cylindrical bar of Duralumin is chilled to −100°C and then heated in an atmosphere at 38°C with a heat transfer coefficient of 140 W/m^2-°C. If the bar is 2.5 cm in diameter, when does the surface temperature reach 15°C? Use either the temperature response charts or the approximate equations.

3.31 A 12-cm-diameter bar (k = 16 W/m-°C, α = 3.35 × 10^{-5}m^2/s) is initially at 650°C. It is placed to cool in air at 40°C through a heat transfer coefficient of 68 W/m^2-°C. How long will it be until the surface reaches 340°C? What will the center temperature be at this time? Solve using both the temperature response charts and the approximate equations.

3.32 A long circular rod 6 cm in diameter is initially at a uniform temperature. It is placed in an oven where the temperature is 480°C and the heat transfer coefficient is 100 W/m^2-°C. After a period of time the surface temperature of the rod is found to be 280°C. If c_p = 0.5 kJ/kg-°C, ρ = 8000 kg/m^3, k = 50 W/m-°C, what is the corresponding temperature of the centerline? What restriction must be placed on this answer? Use either the temperature response charts or the approximate equations.

3.33 Perform the same calculation as posed in Prob. 3.26 for a long circular bar of the same material, instead of a slab, finding the maximum diameter that may be annealed and the length of the annealing process. Discuss the difference in the answers for the slab and cylinder geometries.

3.34 A very thick slab composed of 90% Ni, 10% Cr, is initially at 350°C. It is suddenly exposed to a fluid at 20°C through a heat transfer coefficient of 100 W/m^2-°C. Find the temperature of the surface and that of a point at a depth of 5 cm after 10 min has elapsed.

3.35 A very thick wall of concrete is initially at a temperature of 55°C and then suddenly placed in contact with air at 10°C through a heat transfer coefficient of 25 W/m^2-°C. Calculate the temperature of the surface and a point at a depth of 8 cm after 30 min has elapsed.

3.36 Before the onset of a cold front the ground temperature is 50°F. When the front moves in, the ambient air temperature drops to 0°F and the wind velocity produces a surface heat transfer coefficient of 20 Btu/h-ft^2-°F. Using the soil properties of Example 3.6, find **(a)** the time required for a point 8 in. below the ground surface to reach freezing, and **(b)** the depth at which a freezing temperature will not be reached during a 24-h period.

3.37 A large bronze casting may be approximated as a very thick slab. It is initially at 250°C and then exposed to air at 25°C with a heat transfer coefficient of 150 W/m^2-°C.
(a) Determine how long it takes the temperature at a depth of 5 cm to reach 200°C.
(b) Find the surface temperature at this time.
(c) At the same instant of time, how deep is the point where the temperature is 210°C?

3.38 Plot the temperature-time history for the geometric center of a steel bar ($k = 43$ W/m-°C, $\alpha = 1.172 \times 10^{-5}$ m^2/s) that is 7.5 cm in diameter and 7.5 cm long. The cylinder is initially at 820°C and allowed to cool in air at 40°C with a heat transfer coefficient of 280 W/m^2-°C. Let time range from 0 to 1200 s. Use either the temperature response charts or the approximate equations.

3.39 For the cylinder described in Prob. 3.38, find, after 4 min has passed, the temperature at the geometric center of the cylinder, at the center of the circular ends, and at the midpoint of the lateral cylindrical side.

3.40 A common brick (5.7 cm $\times$ 8.9 cm $\times$ 20 cm) is fired in a kiln at 1425°C. It is allowed to cool in air at 40°C through a heat transfer coefficient of 30 W/m^2-°C. How long does it take for the surface temperature at the center of one of the 8.9 $\times$ 20 sides to reach 65°C? This problem requires an iterative solution if the temperature response charts are used; however, a closed-form solution may be obtained using the approximate equations.

3.41 A 4-in. $\times$ 4-in. wood timber is initially at 75°F. It is suddenly exposed to flames at 1000°F through a heat transfer coefficient of 3.0 Btu/h-ft^2-°F. If the ignition temperature of the wood is 900°F, how much time will elapse before any portion of the timber starts burning? For the wood use $\rho = 50$ lb$_m$/ft^3, $c_p = 0.6$ Btu/lb$_m$-°F, $k = 0.2$ Btu/h-ft-°F. Solve using both the temperature response charts and the approximate equations.

3.42 The accompanying figure depicts a long bar of cast iron 10 cm in diameter, with a flat circular end. The bar is initially at 30°C and then immersed in a fluid at 210°C through a heat transfer coefficient of 250 W/m^2-°C. Compute the temperature **(a)** at the center of the flat end, and **(b)** at a point on the axis 10 cm from the end, after 4 min has elapsed.

3.43 For the rectangular glass bar described in Example 3.8, find the temperature **(a)** at the center of one face for a distance of 2 cm from the end of the bar, and **(b)** at the center of the exposed square end of the bar, after 10 min has elapsed.

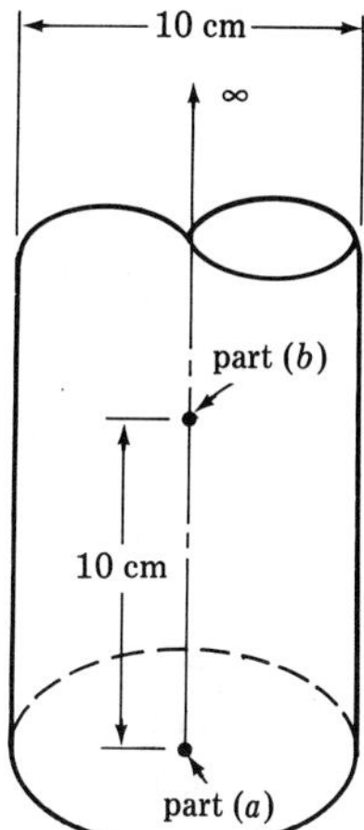

Problem 3.42

3.44 A brass (k = 60 Btu/h-ft-F, α = 1.14 ft^2/h) bar 4 in. in diameter is heated initially to a uniform temperature of 1000°F. It is then placed to cool in air at 100°F, where the surface heat transfer coefficient is 36 Btu/h-ft^2-°F. After 5 min has passed find **(a)** the temperature at the center of the bar and of the surface if the bar is taken as infinitely long, and **(b)** the temperature at the bar axis and on its surface, midway between the ends, if the bar is only 8 in. long.

3.45 The accompanying sketch depicts the corner of an otherwise very large billet of steel (k) = 40 W/m-°C, ρ = 8000 kg/m^3, c_p = 0.50 kJ/kg-°C). The billet, originally at 260°C, is placed in a furnace where the ambient temperature is 1200°C

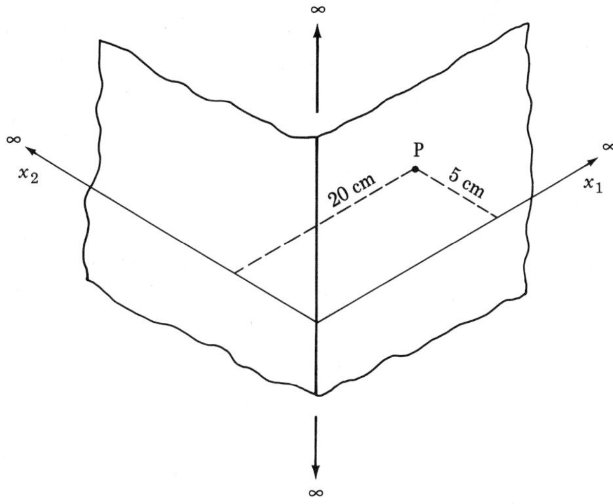

Problem 3.45

and the heat transfer coefficient is 600 W/m^2-°C. Find the temperature at the point P after 25 min has elapsed. What is the temperature of the corner at this time?

3.46 A rolled roast beef may be represented as a cylinder 5 in. in diameter and 10 in. long. It is removed from the refrigerator, where it is at 40°F, and placed in a preheated oven at 325°F, where a heat transfer coefficient of 1.5 Btu/h-ft^2-°F can be expected. How long does it take for the center of the roast to reach 170°F (well done)? For properties of beef use those for water at the mean of 40°F and 325°F. Most cookbooks specify about 35 min/lb$_m$ for a well-done roast. How do these calculations compare?

3.47 A very long copper cylinder, 5 cm in diameter, is heated to 30°C and then plunged into a liquid bath at 0°C. If, after 3 min, the cylinder center temperature is 4°C, determine the average heat transfer coefficient at the surface if the internal thermal resistance is neglected. Then find the Biot number and justify the neglection of the internal resistance.

3.48 A fine wire (0.07 cm diameter) is initially heated to 175°C. It is suddenly exposed to an environment at 40°C through a heat transfer coefficient of 60 W/m^2-°C. After first calculating the Biot number to establish if the internal resistance can be neglected, find the wire temperature and the heat loss (per unit length) after 10 s if the wire is made of **(a)** copper; **(b)** aluminum.

3.49 A plate 1 in. thick is made of 10% nickel steel. It is heated to some initial temperature and then exposed to air at 100°F through a heat transfer coefficient of h = 2 Btu/h-ft^2-°F. After first justifying neglection of internal resistance, find the initial temperature of the plate if its temperature after 10 min is 1000°F.

3.50 The heat transfer coefficient on a sphere is to be measured by heating a 3-cm-diameter pure aluminum sphere to an initial temperature of 175°C and immersing it into a fluid at 25°C. After a time lapse of 42 s, the center of the sphere is measured to be 100°C. Estimate the heat transfer coefficient at the sphere surface.

3.51 A thermocouple is to used to measure the temperature of a flowing gas stream. The thermocouple junction (made of copper constantan, 60% Cu, 40% Ni) may be approximated as a sphere. If the heat transfer coefficient between the junction and the gas stream is 500 W/m^2-°C, calculate the diameter of the junction if the indicated temperature rise is to be 95% of the imposed initial temperature difference in a time of 3 s.

CHAPTER 4

Numerical Methods for Heat Conduction

4.1 INTRODUCTORY REMARKS

The problems of heat conduction treated in Chapters 2 and 3 have stressed the use of mathematical methods. This approach, which treats the conducting body as a continuum, yields much information of a general nature concerning the particular problem being treated. By the successful application of analysis one can ascertain the temperature at *any* point, at *any* time, within the given body. Since the results are given in closed, analytical form, it is possible to deduce much useful information—the effect of various parameters, the effect of altering the body geometry, etc.

Although a great number of problems have been solved analytically, only a limited number of relatively simple geometrical shapes (e.g., cylinders, spheres, infinite slabs, etc.) can be handled. Also, only those boundary conditions which can be easily expressed mathematically may usually be applied. There are many heat conduction problems of considerable practical value for which no analytical solution is feasible. These problems usually involve geometrical shapes of a mathematically inconvenient sort. The absence of an analytical solution does not remove the need of an answer, and some other approach must be sought.

Numerical techniques exist which are able to handle almost any problem of any degree of complexity. The detail and accuracy of the answer obtained depends mainly on the amount of effort one wishes to expend. The various numerical methods all yield *numerical* values for temperatures at selected, *discrete,* points within the body being considered and only at *discrete* time intervals. Thus answers are obtained only for a given set of conditions, a given set of discrete points and discrete time intervals. One must give up the generality of the analytical solution in order to obtain an answer.

Several different techniques of numerical analysis of heat conduction problems exist. The references at the end of the chapter may be consulted in this regard. Some of the techniques developed have been based on the use of hand calculations or desk calculators. The availability of high-speed digital computers has considerably reduced the value of these methods. The formulations presented and the techniques recommended are not the most efficient or concise from the standpoint of hand or desk calculator solution; they are presented in forms most suitable for computer programming.

Both steady-state and transient methods are treated. Formulations of steady state numerical methods are not presented in the most concise fashion, but rather are formulated in a way that carries over readily to the transient methods.

The numerical methods presented here are all based on the representation of the derivatives in the heat conduction equation by finite difference approximations. In recent years a new and particularly powerful technique called the "finite element method" has been developed for the numerical analysis of heat conduction problems. A useful explanation of this technique is too lengthy for this text, and the reader is referred to Ref. 1 for a detailed treatment.

Attention is first directed toward the development of finite difference approximations to the heat conduction equation and the degree of accuracy of these approximations. On the basis of the finite difference formulation, an electrical network analogy becomes apparent. Generalizations to complex problems are then made on the basis of this analogy because of its greater physical appeal.

4.2 FINITE DIFFERENCE APPROXIMATIONS

The basic principle of the numerical approach to a heat conduction problem is the replacement of the differential equation for the continuous temperature distribution in a heat conducting solid by a finite difference equation which must be satisfied at only certain points in the solid. The relation of the finite difference expression to the differential equation can be understood best by deriving the latter from the former, via the use of a Taylor's expansion. Consider, then, a function of two independent variables:

$$f = f(\xi, \eta). \tag{4.1}$$

The general variables ξ and η have been used since the results will be applied to spatial variables (x, y, and z) as well as time (τ).

Let h_1 represent an increment in the variable ξ. A *forward* expansion of the function at $\xi = \xi + h_1$, $\eta = \eta$, in terms of its value at $\xi = \xi$, $\eta = \eta$, is

$$\begin{aligned} f(\xi + h_1, \eta) = f(\xi, \eta) + h_1\left(\frac{\partial f}{\partial \xi}\right)_{\xi,\eta} &+ \frac{h_1^2}{2}\left(\frac{\partial^2 f}{\partial \xi^2}\right)_{\xi,\eta} \\ &+ \frac{h_1^3}{6}\left(\frac{\partial^3 f}{\partial \xi^3}\right)_{\xi,\eta} + \mathcal{O}(h_1^4). \end{aligned} \tag{4.2}$$

The notation $\mathcal{O}(h_1^4)$ is used to indicate that subsequent terms are of the order of h_1^4, and *higher*. If terms of the order of h_1^2, and greater, are neglected, the following *forward* finite difference approximation to the first derivative is obtained:

$$\left[\left(\frac{\partial f}{\partial \xi}\right)_{\xi,\eta}\right]_{\text{fwd}} \approx \frac{1}{h_1}[f(\xi + h_1, \eta) - f(\xi, \eta)]. \tag{4.3}$$

Similarly, for a forward displacement of h_2 in η, the forward finite difference approximation for the first derivative is

$$\left[\left(\frac{\partial f}{\partial \eta}\right)_{\xi,\eta}\right]_{\text{fwd}} \approx \frac{1}{h_2}[f(\xi, \eta + h_2) - f(\xi, \eta)]. \tag{4.4}$$

In like fashion, *backward* finite difference expressions for the first derivatives may be obtained by writing Taylor's expansions for an increment in ξ of $-h_1$ or an increment of η of $-h_2$:

$$\begin{aligned} f(\xi - h_1, \eta) = f(\xi, \eta) - h_1\left(\frac{\partial f}{\partial \xi}\right)_{\xi,\eta} &+ \frac{h_1^2}{2}\left(\frac{\partial^2 f}{\partial \xi^2}\right)_{\xi,\eta} \\ &- \frac{h_1^3}{6}\left(\frac{\partial^3 f}{\partial \xi^3}\right)_{\xi,\eta} + \mathcal{O}(h_1^4). \end{aligned} \tag{4.5}$$

Again neglecting terms of order h_1^2, and greater, one obtains

$$\left[\left(\frac{\partial f}{\partial \xi}\right)_{\xi,\eta}\right]_{\text{bkwd}} \approx \frac{1}{h_1}[f(\xi, \eta) - f(\xi - h_1, \eta)]. \tag{4.6}$$

Also,

$$\left[\left(\frac{\partial f}{\partial \eta}\right)_{\xi,\eta}\right]_{\text{bkwd}} \approx \frac{1}{h_2}[f(\xi, \eta) - f(\xi, \eta - h_2)]. \tag{4.7}$$

The *central* finite difference approximation to the second derivatives may be found by adding Eqs. (4.2) and (4.5):

$$f(\xi + h_1, \eta) + f(\xi - h_1, \eta) = 2f(\xi, \eta) + h_1^2\left(\frac{\partial^2 f}{\partial \xi^2}\right)_{\xi,\eta} + \mathcal{O}(h_1^4). \tag{4.8}$$

When terms of the order of h_1^4, and greater, are neglected,

$$\left[\left(\frac{\partial^2 f}{\partial \xi^2}\right)_{\xi,\eta}\right]_{\text{cent}} \approx \frac{1}{h_1^2}[f(\xi + h_1, \eta) - 2f(\xi, \eta) + f(\xi - h_1, \eta)]. \tag{4.9}$$

Also,

$$\left[\left(\frac{\partial^2 f}{\partial \eta^2}\right)_{\xi,\eta}\right]_{\text{cent}} \approx \frac{1}{h_2^2}[f(\xi, \eta + h_2) - 2f(\xi, \eta) + f(\xi, \eta - h_2)]. \tag{4.10}$$

It is worth noting that the first derivative finite difference approximations neglect terms of the order of h^2, while the second derivative approximations neglect terms of the order of h^4. It should be pointed out that other approximations to the second derivative may be written (e.g., forward, backward, etc.). However, the central difference given here is most commonly used for heat conduction analyses.

Section 4.3 applies these approximations to steady-state heat conduction, and subsequent sections illustrate their solutions. Section 4.7 and subsequent sections apply the difference equations to the case of transient conduction.

4.3 STEADY-STATE NUMERICAL METHODS

From Chapter 2 the steady-state conduction equation is known to be

$$k\left(\frac{\partial^2 t}{\partial x^2} + \frac{\partial^2 t}{\partial y^2} + \frac{\partial^2 t}{\partial z^2}\right) + q^* = 0. \tag{4.11}$$

The thermal conductivity, k, is left as a factor in Eq. (4.11) since this will be a useful form when extensions are made to nonsteady conduction. Also, reference to the derivation of the heat conduction equation in Sec. 2.1 reveals that the expression given on the left side of Eq. (4.11) represents the net rate, *per unit volume,* at which heat is stored at a point in the conducting body, and q^* represents the internal generated heat, *per unit volume.*

Consider, first, the one-dimensional case:

$$k\frac{d^2 t}{dx^2} + q^* = 0. \tag{4.12}$$

Figure 4.1 depicts a slender bar whose lateral faces are insulated. Thus all heat conduction occurs in the x direction and Eq. (4.12) applies. If, however, the second derivative in Eq. (4.12) is replaced by the finite difference approximation in Eq. (4.9) (with the coordinate x replacing ξ, the space increment δx replacing h_1, and

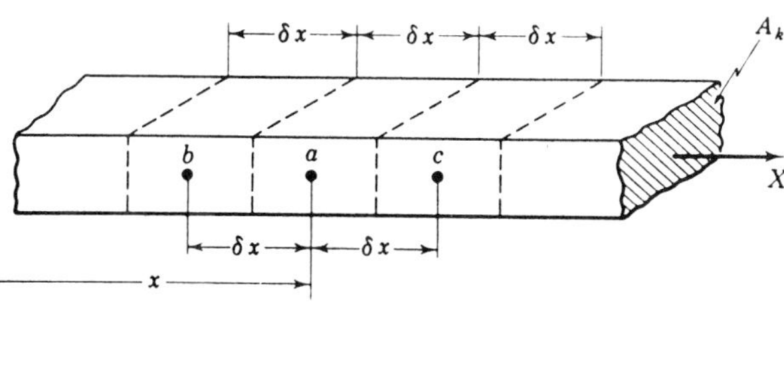

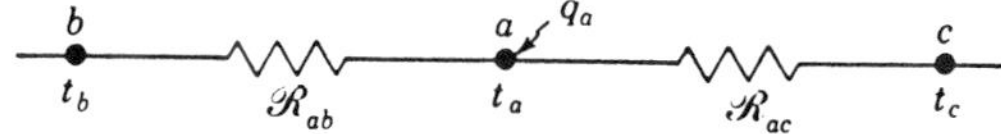

Figure 4.1. Finite difference approximation and corresponding nodal network for one-dimensional conduction.

the temperature t replacing f), the following expression results, which relates the temperatures at x, $x + \delta x$, and $x - \delta x$:

$$\frac{k}{(\delta x)^2}[t(x + \delta x) - 2t(x) + t(x - \delta x)] + q_x^* = 0.$$

In keeping with the central difference approximation used, the internally generated heat is evaluated at the central point x.

If, as suggested in Fig. 4.1, the points at x, $x - \delta x$, and $x + \delta x$ are denoted by a, b, c, respectively, the equation is expressed more simply as

$$\frac{k}{(\delta x)^2}(t_b - 2t_a + t_c) + q_a^* = 0. \tag{4.13}$$

Equation (4.13) is the result of a Taylor's expansion of the heat conduction equation around the point a, and should thus be interpreted as a relation for the temperature t_a in terms of the temperatures at the neighboring points, t_b and t_c. A similar expression may be written for the point b—in terms of t_a and the temperature of the point δx to the left of b. Likewise, an expression may be written for point c, involving t_c, t_a, and the point δx to the right of c. Thus if the body is subdivided into n distinct points, n such equations may be written. This set of n equations may then be solved for the temperatures at the n points presuming the q's are known. Consequently, one obtains the temperatures at these specific points without any consideration of the temperatures between them. The development of Eq. (4.9), upon which Eq. (4.13) is based, showed that the accuracy of the results obtained will increase with decreasing δx (i.e., by increasing the number of points), and, in fact, the error will be of the order of $(\delta x)^4$. Methods of solving the set of equations resulting from the application of Eq. (4.13) at each body point will be discussed in Sec. 4.6.

The representation just developed for the one-dimensional case may be extended to the two-dimensional and three-dimensional cases. Each term in Eq. (4.11) may be approximated in identical form to Eq. (4.13). The two-dimensional case is illustrated in Fig. 4.2. A plate of unit thickness, which is insulated on its plane faces, is subdivided in the x direction by increments of δx and in the y direction by increments of δy. The resulting finite difference approximation for the temperature at point a in terms of the temperatures at the neighboring points b, c, d, and e is

$$\frac{k}{(\delta x)^2}(t_b - 2t_a + t_c) + \frac{k}{(\delta y)^2}(t_d - 2t_a + t_e) + q_a^* = 0. \tag{4.14}$$

As before, such an equation is to be satisfied at each point in the plate—in terms of its four neighboring points. A body divided into an $n \times m$ network will require the solution of $(n \times m)$ such equations. Quite apparently, considerable simplification will result if one chooses $\delta x = \delta y$. This could prove to be valuable if hand calculations were to be employed; however, it is of little consequence when digital computers are used.

The three-dimensional case is not illustrated, but it should be apparent that at each point in a three-dimensional body one must satisfy an equation of the form

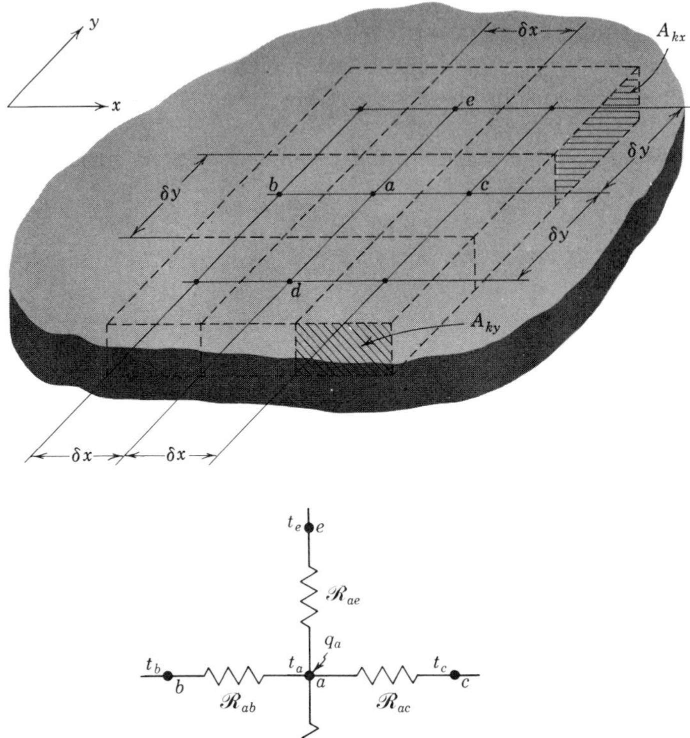

Figure 4.2. Finite difference approximation and corresponding nodal network for two-dimensional conduction.

$$\frac{k}{(\delta x)^2}(t_b - 2t_a + t_c) + \frac{k}{(\delta y)^2}(t_d - 2t_a + t_e) + \frac{k}{(\delta z)^2}(t_f - 2t_a + t_g) + q_a^* = 0. \tag{4.15}$$

The points f and g are the neighboring points in the z direction—at a spacing of δz.

4.4 NODAL NETWORK REPRESENTATIONS FOR THE STEADY STATE

The finite difference equations presented in the preceding section may be interpreted in a physical way that will permit rather convenient generalizations. Consider first the one-dimensional case. It will be recalled from Chapter 2 that for conduction through a slab of thickness Δx, a *unit* thermal resistance could be defined,

$$R = \frac{\Delta x}{k},$$

so the heat *flux* through the slab would be

$$\frac{q}{A} = \frac{t_1 - t_2}{R_{12}}.$$

A *total* thermal resistance, $\mathcal{R}$, may be defined which gives the total heat flow, q, rather than the heat flux:

$$\mathcal{R} = \frac{\Delta x}{kA_k}, \qquad q = \frac{t_1 - t_2}{\mathcal{R}_{12}}. \tag{4.16}$$

The subscript on the area symbol A_k is used to denote the fact that the area considered is the *conduction* area normal to the direction of heat flow.

Now, by reference to Fig. 4.1, imagine the one-dimensional bar to be divided into *lumps* of length δx so located that the nodal points a, b, and c lie at the center of these lumps. The area A_k is the cross-sectional area normal to the one-dimensional coordinate x. Since the left side of Eq. (4.13) represents the heat flow *per unit volume*, multiply it by the volume of the lump surrounding point a, namely $(A_k \times \delta x)$:

$$\frac{kA_k}{\delta x}(t_b - 2t_a + t_c) + q_a = 0,$$

where

$$q_a = q_a^*(A_k \delta x)$$

is the total heat generated, per unit time, in the lump volume around point a. The term $\delta x/kA_k$ is seen to be the thermal resistance between point a and either b or c, so the above expression may be written

$$\frac{t_b - t_a}{\mathcal{R}} + \frac{t_c - t_a}{\mathcal{R}} + q_a = 0, \qquad \mathcal{R} = \frac{\delta x}{kA_k}. \tag{4.17}$$

The latter expression suggests the electrical network analogy identified earlier and illustrated in Fig. 4.1. The lumps into which the conducting body is divided are presumed to be of uniform temperature—equal to the temperature of the node it surrounds. In the network representation, these isothermal lumps are symbolized by the nodal points a, b, and c, and the thermal resistances are symbolized by the electrical resistors shown connecting the nodes. The expression given in Eq. (4.17) is simply a statement of the total heat flow into the node a through the resistances connecting this node to all neighboring nodes. The lump temperatures applied to the nodes are analogous to electrical potentials in the network representation, and the heat balance of Eq. (4.17) is the sum of the currents flowing into a node if q_a is interpreted as a current flow into node a from some external potential source. In the steady state this summation must be zero.

In a similar fashion, the two-dimensional plate, of unit thickness, pictured in Fig. 4.2 may be subdivided into lumps $\delta x \times \delta y \times 1$ in volume, centered on the nodal points $a, b, \ldots, e$. The finite difference expression given in Eq. (4.14) reduces to

$$\frac{t_b - t_a}{\mathcal{R}_x} + \frac{t_c - t_a}{\mathcal{R}_x} + \frac{t_d - t_a}{\mathcal{R}_y} + \frac{t_e - t_a}{\mathcal{R}_y} + q_a = 0,$$

$$\mathcal{R}_x = \frac{\delta x}{kA_{kx}}, \quad A_{kx} = \delta y \times 1, \tag{4.18}$$

$$\mathcal{R}_y = \frac{\delta y}{kA_{ky}}, \quad A_{ky} = \delta x \times 1.$$

The equivalent nodal network is shown in Fig. 4.2. The three-dimensional case can be treated similarly, and the resultant expression should be apparent.

All the results above may be generalized into a single expression. For a nodal network representation of any steady-state problem, if a thermal resistance is provided between a given node and each of its neighbors with which it conducts heat, each node i must satisfy the equations

$$\sum_j \frac{t_j - t_i}{\mathcal{R}_{ij}} + q_i = 0,$$

$$\mathcal{R}_{ij} = \frac{\delta_{ij}}{kA_{k_{ij}}}. \tag{4.19}$$

In Eq. (4.19) j denotes all neighboring nodes connected to node i, δ_{ij} denotes the conduction distance between node i and node j, $A_{k_{ij}}$ is the cross-sectional area for heat conduction normal to δ_{ij}, and q_i is the heat generated in or added to the volume lump at i. This representation permits the inclusion of unequal nodal spacings in any particular coordinate direction. Particular attention must then be paid to the evaluation of the various resistors, since a different conduction distance, δ_{ij}, would be involved in each case. Similarly, the volume of the lump and the conduction area associated with each node might be different. As will be seen later, the volume of the body lump associated with each node will become of considerable significance in the nonsteady case. In general, it is desirable to use equal net spacings, particularly if hand calculations are used, since all resistances become identical and can be eliminated from Eq. (4.19) altogether. In the instance of equal net spacings and zero heat addition, the two-dimensional case given by Eq. (4.18) reduces to

$$t_b + t_c + t_d + t_e - 4t_a = 0. \tag{4.20}$$

With the use of digital computers, the advantage of equal net spacings becomes of less value.

The existence of convective losses from body lumps bounded by a free surface exposed to an ambient fluid may be accounted for by adding to the network another node representing the fluid. The node representing the body lump is connected to the node representing the fluid by a resistor whose value is found from the heat transfer coefficient (a unit conductance) according to

$$\mathcal{R} = \frac{1}{hA_c}. \tag{4.21}$$

Here A_c denotes the surface area of the body lump exposed to the convecting ambient fluid. The expression which must be satisfied at each point then takes the form: At each node i,

$$\sum_j \frac{t_j - t_i}{\mathcal{R}_{ij}} + q_i = 0, \qquad \mathcal{R}_{ij} = \begin{cases} \dfrac{\delta_{ij}}{kA_{k_{ij}}} & \text{for conduction,} \\[2ex] \dfrac{1}{h_{ij}A_{c_{ij}}} & \text{for convection.} \end{cases} \tag{4.22}$$

In Eq. (4.22) q_i represents the heat added at a node by means other than surface convection. In cases in which internal heat generation is present, the q_i's are known. In the event that the node under consideration is one subjected to a boundary condition of specified temperature, t_i is then known and q_i becomes an unknown—namely the heat flux necessary to maintain the node at the desired temperature. In the remaining discussions of this chapter no internal heat generation will be considered, and thus the latter instances of specified boundary temperature will be the only cases in which the q_i term must be included.

The convective resistance was included in order to write the general form of the equation as it is shown in Eq. (4.22). Actually, this convective condition is just one of several boundary conditions that could be encountered. Before illustrating the solution of the set of equations resulting from the application of Eq. (4.22) to actual networks, these boundary conditions must be discussed. A sample nodal network for a complex system is shown in Fig. 4.3 as an illustration of the application of the concepts discussed here.

4.5 BOUNDARY CONDITIONS

The discussion so far has been restricted mainly to points interior to the body in which the heat conduction is taking place. In order to apply the heat balances represented by Eq. (4.19) or (4.22), the imposed boundary conditions must also

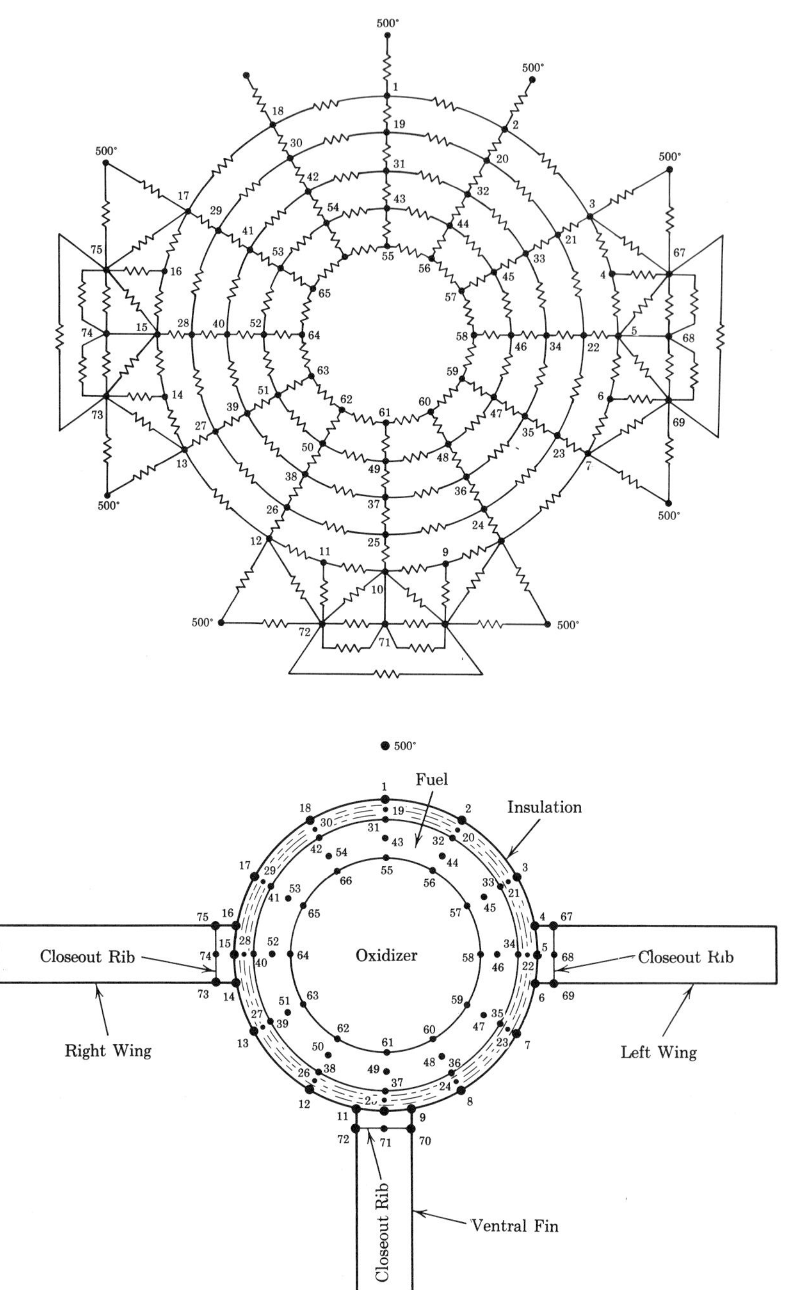

Figure 4.3. Complex nodal representation for the thermal analysis of a missile fuel tank. (From "Temperature Control Systems for Space Vehicles," *ASD Rep. TRD-62-493, Part II,* Air Force Flight Dynamics Laboratory, Wright-Patterson Air Force Base, Ohio, 1963.)

be satisfied. One such boundary condition, that of convection, has already been mentioned. Other typical conditions will be discussed in connection with two-dimensional conduction; however, the special form of these conditions for the one-dimensional case and their generalization to the three-dimensional case are sufficiently apparent to be left as exercises for the reader. Attention will first be directed to the instance in which regular network spacings (i.e., equal spacings in any one direction) are used, and the boundary coincides with a nodal point in this equal spacing.

Regular Spacing, Plane Boundary

Figure 4.4 illustrates a two-dimensional case in which nodal spacings of δx in the x direction and δy in the y direction are used. The conducting solid is presumed to

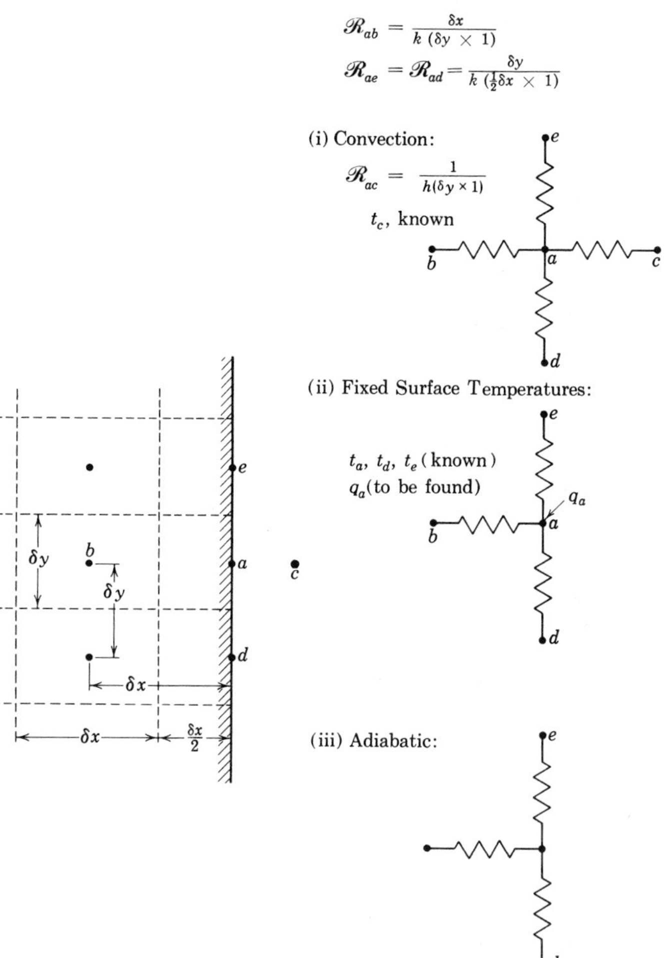

Figure 4.4. Boundary conditions at a plane, two-dimensional surface.

have a unit thickness in the plane of the paper. A plane boundary parallel to the y direction coincides with one column of nodal points. The limits of the volume lumps surrounding each node are shown. It is apparent that the lump volumes associated with the boundary nodes are one-half of the volume of a regular, interior node. Evaluation of the heat balances on node a will be considered.

The value of the conduction resistor connecting a and b is the same as for any other resistor connecting two interior points spaced in the x direction:

$$\mathcal{R}_{ab} = \frac{\delta_{ij}}{kA_{k_{ij}}} = \frac{\delta x}{k(\delta y \times 1)}. \tag{4.23}$$

For resistors $\mathcal{R}_{ad}$ and $\mathcal{R}_{ae}$, the conduction distance is δy, as for an interior resistance in the same direction; however, the conduction area between e and a or d and a is one-half of what it would be otherwise:

$$\mathcal{R}_{ad} = \mathcal{R}_{ae} = \frac{\delta y}{k(\frac{1}{2}\delta x \times 1)}. \tag{4.24}$$

The three resistances just discussed will remain unchanged regardless of the boundary condition to be imposed. The resistance $\mathcal{R}_{ac}$ (if any), the heat input at node a, etc., *will* depend on the imposed condition. Some conditions are listed below.

Convection to a Fluid of Known Temperature. If a fluid of known temperature is in contact with the body surface, and if the heat transfer coefficient, h, for convective heat transfer between the solid surface and the fluid is known, the fluid may be represented by a node—call it c—maintained at the known fluid temperature. Then the resistance connecting a and c is

$$\mathcal{R}_{ac} = \frac{1}{hA_c} = \frac{1}{h(\delta y \times 1)}. \tag{4.25}$$

The application of the heat balance in Eq. (4.22) is, at node a,

$$\frac{t_b - t_a}{\mathcal{R}_{ab}} + \frac{t_d - t_a}{\mathcal{R}_{ad}} + \frac{t_e - t_a}{\mathcal{R}_{ae}} + \frac{t_c - t_a}{\mathcal{R}_{ac}} = 0. \tag{4.26}$$

The $\mathcal{R}$'s are given in Eqs. (4.23) through (4.25). The applicable network is shown in Fig. 4.4. For a square network ($\delta x = \delta y = \delta$) this equation reduces to

$$t_b + \tfrac{1}{2}(t_d + t_e) + \frac{h\delta}{k} t_c - \left(2 + \frac{h\delta}{k}\right) t_a = 0. \tag{4.27}$$

No heat balance is necessary at node c, since its temperature is known.

Specified Surface Temperature. A temperature distribution may be specified and maintained at the boundary. If so, the three temperatures t_a, t_d, and t_e are known. The resistance between these nodes was given above. In order that the temperature at node a remain fixed, an unknown amount of heat must flow between the node and its surroundings. The network representation is shown and Eq. (4.22) yields, at node a:

$$\frac{t_b - t_a}{\mathcal{R}_{ab}} + \frac{t_d - t_a}{\mathcal{R}_{ad}} + \frac{t_e - t_a}{\mathcal{R}_{ae}} + q_a = 0. \tag{4.28}$$

Since t_a is known, the unknown quantities in the equation above are q_a and t_b. For a square network, Eq. (4.28) becomes

$$t_b + \tfrac{1}{2}(t_d + t_e) - 2t_a + \frac{q_a}{k} = 0. \tag{4.29}$$

If it is not desired to find the heat flux at a, this nodal equation can be eliminated completely from consideration.

If the surface is *isothermal*, $t_a = t_d = t_e$, so the nodal balance reduces to

$$\frac{t_b - t_a}{\mathcal{R}_{ab}} + q_a = 0, \tag{4.30}$$

or

$$t_b - t_a + \frac{q_a}{k} = 0 \tag{4.31}$$

for square networks.

An Adiabatic Boundary. If the boundary is insulated, no interaction with the surroundings occurs, and node c is unnecessary. Thus

$$\frac{t_b - t_a}{\mathcal{R}_{ab}} + \frac{t_d - t_a}{\mathcal{R}_{ad}} + \frac{t_e - t_a}{\mathcal{R}_{ae}} = 0, \tag{4.32}$$

where the $\mathcal{R}$'s are given above. For the square network

$$t_b + \tfrac{1}{2}(t_d + t_e) - 2t_a = 0. \tag{4.33}$$

Regular Spacing, Exterior Corner

Figure 4.5 illustrates a two-dimensional case in which the boundary of the solid is an exterior corner, and the regular nodal grid results in a node being located at the corner. In this instance the lump volume associated with nodes b and d is one-half the regular volume and that of node a is one-fourth that of an interior point. Without detailed discussion, the results are

$$\mathcal{R}_{ab} = \frac{\delta x}{k(\frac{1}{2}\delta y \times 1)}, \qquad \mathcal{R}_{ad} = \frac{\delta y}{k(\frac{1}{2}\delta x \times 1)}. \tag{4.34}$$

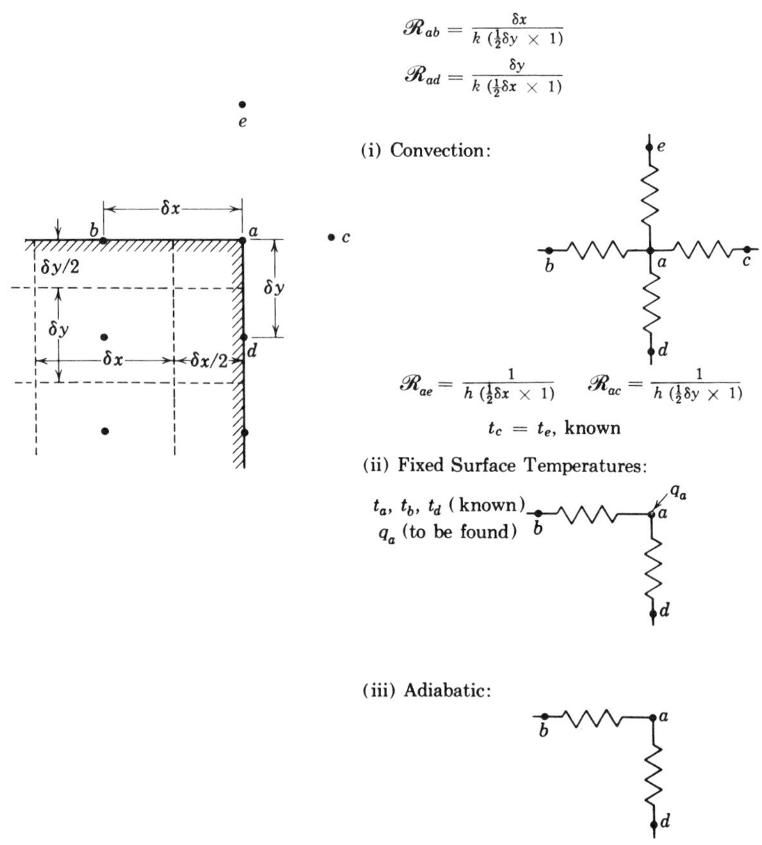

$\mathcal{R}_{ab} = \frac{\delta x}{k\,(\frac{1}{2}\delta y \times 1)}$
$\mathcal{R}_{ad} = \frac{\delta y}{k\,(\frac{1}{2}\delta x \times 1)}$
(i) Convection:
$\mathcal{R}_{ae} = \frac{1}{h\,(\frac{1}{2}\delta x \times 1)}$ $\mathcal{R}_{ac} = \frac{1}{h\,(\frac{1}{2}\delta y \times 1)}$
$t_c = t_e$, known
(ii) Fixed Surface Temperatures:
t_a, t_b, t_d (known)
q_a (to be found)
(iii) Adiabatic:

Figure 4.5. Boundary conditions at a two-dimensional exterior corner.

Convection. Using nodes c and e to represent the fluid (with $t_e = t_c$), one finds that

$$\mathcal{R}_{ae} = \frac{1}{h(\frac{1}{2}\delta x \times 1)}, \qquad \mathcal{R}_{ac} = \frac{1}{h(\frac{1}{2}\delta y \times 1)}. \tag{4.35}$$

Equation (4.26) still applies, with the $\mathcal{R}$'s now given by Eqs. (4.34) and (4.35), and $t_e = t_c$. For the square network,

$$t_b + t_d + 2\frac{h\delta}{k}t_c - 2\left(1 + \frac{h\delta}{k}\right)t_a = 0. \tag{4.36}$$

Specified Surface Temperature. With t_a, t_b, and t_d known, Eq. (4.22) gives

$$\frac{t_b - t_a}{\mathcal{R}_{ab}} + \frac{t_d - t_a}{\mathcal{R}_{ad}} + q_a = 0, \tag{4.37}$$

and for the square network,

$$t_b + t_d - 2t_a + \frac{2q_a}{k} = 0. \tag{4.38}$$

If the surface is isothermal, either Eq. (4.37) or (4.38) yields $q_a = 0$, showing that no heat balance is necessary in this instance.

An Adiabatic Boundary. One has, simply,

$$\frac{t_b - t_a}{\mathcal{R}_{ab}} + \frac{t_d - t_a}{\mathcal{R}_{ad}} = 0 \tag{4.39}$$

which is, for the square network,

$$t_b + t_d - 2t_a = 0. \tag{4.40}$$

Irregular Boundary Points

In many practical applications of the numerical methods discussed here, the boundaries of the solid are often shaped such that it is either inconvenient or impossible to arrange net spacings so that the boundary points coincide with regular points of the net. In such cases special relations must be developed for the resistances connecting the boundary points to interior points. The case of a two-dimensional network is shown in Fig. 4.6. For simplicity of discussion, a square network, of spacing δ, is considered. A curved boundary is shown which passes between nodes in the square network. Points b and e represent boundary points on the network connected to the interior point a. The points b' and e' represent points spaced the regular distance, δ, from point a, and lying outside the boundary of the solid. The symbol δ' will denote the distance from node a to node b, the real boundary point, while δ'' denotes the spacing from a to e. Apparently, both δ' and δ'' are less than δ.

In order to apply the heat balance of Eq. (4.22) to nodes a, b, c, d, and e, the resistances $\mathcal{R}_{ab}$, $\mathcal{R}_{ac}$, $\mathcal{R}_{ad}$, and $\mathcal{R}_{ae}$ must be evaluated. The method described here is developed in full in Ref. 5 wherein it is shown that the proposed representations neglect terms of the order of δ^3, and greater. It will be recalled from the discussion of Secs. 4.2 and 4.3 that the representation given by Eq. (4.22) for interior points neglects terms of the order of δ^4, and greater.

The recommended procedure consists of defining the limits of the solid lump surrounding node a as that rectangle which cuts the network lengths a-b, a-c, a-d, and a-e into equal parts. The results may be observed in Fig. 4.6—the lump volume associated with node a being shaded. The nodal point a is not located at the geometric center of the lump. With this representation, the conduction areas between node a and its neighbors are seen to be

$$A_{k_{ab}} = A_{k_{ac}} = \frac{\delta + \delta''}{2},$$

$$A_{k_{ad}} = A_{k_{ae}} = \frac{\delta + \delta'}{2}.$$

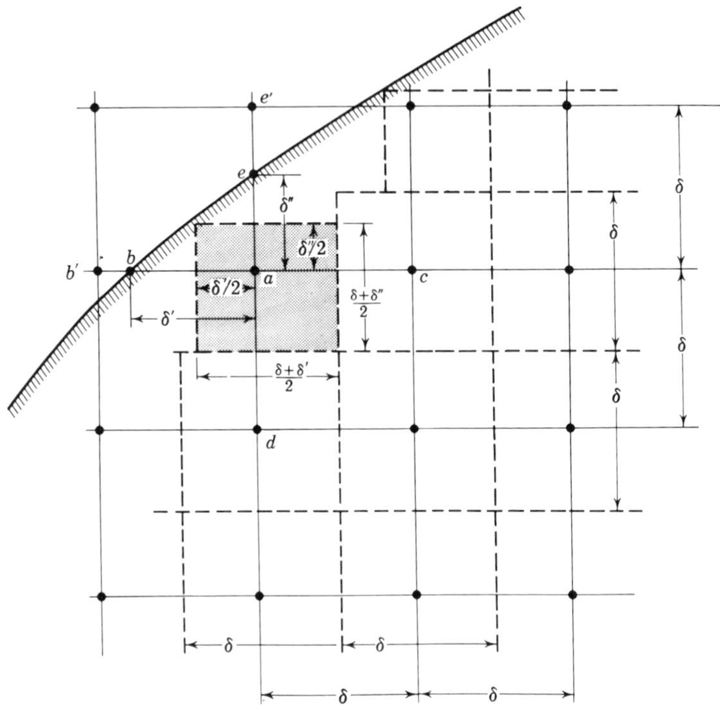

Figure 4.6. Irregular boundary points.

Thus the resistances needed become

$$\begin{aligned}
\mathcal{R}_{ab} &= \frac{\delta'}{k(\delta + \delta'')/2},\\
\mathcal{R}_{ac} &= \frac{\delta}{k(\delta + \delta'')/2},\\
\mathcal{R}_{ad} &= \frac{\delta}{k(\delta + \delta')/2},\\
\mathcal{R}_{ae} &= \frac{\delta''}{k(\delta + \delta')/2}.
\end{aligned} \tag{4.41}$$

Application of these relations is made in the examples of following sections.

4.6 NUMERICAL SOLUTION OF STEADY-STATE NETWORK EQUATIONS

The foregoing sections have been devoted to the development of finite difference approximations for the heat conduction equation, network representations, boundary conditions, etc. The result of applying these concepts to a given conduction situation is that the *differential* equation for heat conduction in a body is replaced by an *algebraic* expression at each nodal subdivision of the body at which the temperature (or the heat flux in the case of an isothermal node) is desired. The expression for the temperature at a node involves the temperatures of the neighboring nodes. Thus for a case in which N nodes of unknown temperature are involved, N algebraic equations are obtained which involve these unknowns. It remains, then, to be shown how such a system of N equations may be solved.

Several methods are available for the solution of a system of simultaneous algebraic equations (e.g., Refs. 2 through 4). Certain methods particularly suited to the form of the equations involved here (usually linear) have been developed to a high degree and may be found discussed in detail in Refs. 4 through 7. Those methods particularly adaptable to digital computer use will be described here. These methods will be briefly discussed and then illustrated by example. The examples will be chosen to illustrate as many of the points developed in the foregoing sections as possible.

Solution by Iteration

The iteration technique is a method of successive approximations and follows a fixed sequence of operations. As such, no opportunity exists for modification of the sequence of the operations at the discretion of the user. Hence the method is particularly adaptable to digital computer use.

Equation (4.22), the nodal heat balance, may be written for each node i:

$$t_i = \frac{q_i + \sum_j \frac{t_j}{\mathcal{R}_{ij}}}{\sum_j 1/\mathcal{R}_{ij}}. \tag{4.42}$$

The symbol j, it is recalled, represents all neighboring nodes with which node i exchanges heat. In general, the j nodes will number from one to, perhaps, six, depending on the dimensionality of the problem and the boundary conditions. The iteration method, properly called the Gauss–Seidel method, consists of the following sequence of operations:

1. An initial set of values of the nodal temperatures, t_i, is assumed.
2. New values of the nodal temperatures are calculated from Eq. (4.42), always using the *most recent* values of the t_j's. That is, as a new value of a given node's temperature is found, it is used in any subsequent calculation in which it appears.
3. This process is repeated, cycling through the N nodes, over and over, until

   $$|[t_i]_{n+1} - [t_i]_n| \leq \epsilon$$

   for all nodes. The subscripts n and $n + 1$ indicate the values at a given node on two successive iterations. The tolerance ϵ is a selected convergence criterion.

An alternative method, the Jacoby method, calculates new temperatures at each node based on all the temperatures of the previous iteration instead of always using the most recent value. However, the Gauss-Seidel method is generally preferred because of its more rapid convergence. The above discussion is directed toward the instances in which the heat fluxes, q_i, are known and the temperatures are unknown. If a node is one of specified temperature, t_i is known and q_i becomes the sought-for unknown.

The process above will converge since Eq. (4.42) represents a weighted averaging process. Somewhere in the network, known, fixed, temperatures (or q's) are specified, and the iteration procedure continually averages these fixed values with the incorrect unknowns. Slowness of convergence should not be confused with accuracy of convergence. The accuracy of the final results depends upon the network spacing parameters and the convergence criterion ϵ. Improved accuracy may be obtained by using finer nets or smaller values of ϵ at the expense of increasing the number of computations and the required computing time, and no definite rule can be formulated for the choice of these parameters.

Quite apparently, the closeness of the initial assumptions for the t_i's to the correct solution has a profound effect on the computing time. In some instances it may be advisable to make an initial calculation using a rather coarse net and then to use the results obtained to interpolate initial values for a much finer net.

The example that follows illustrates the iteration method.

EXAMPLE 4.1

Solve, by numerical means, the problem stated in Example 2.9 for the case of the iron rod. It is desired to find the temperature distribution and the total heat loss from a 1.25-cm-diameter iron rod ($k = 57$ W/m-°C) which is 30 cm long, heated to 120°C on one end, exposed to a convecting fluid at 20°C through a heat transfer coefficient of 9.0 W/m^2-°C, and insulated on its free end.

Solution. Figure 4.7 illustrates the situation at hand. In keeping with the one-dimensional treatment employed in Chapter 2 for extended surfaces, it will be assumed that conduction occurs only in the longitudinal direction, although there is convection to the ambient fluid through the lateral surface of the rod. The length of the rod is subdivided into seven equally spaced nodal points, 5 cm apart, with one node at each end of the rod for application of the boundary conditions at those ends. Node 1 is an isothermal node maintained at $t_1 = 120$°C, and node 7 is a boundary node on an adiabatic surface: $q_7 = 0$. All other nodes, 2 through 6, are regular interior nodes. The choice of nodal spacing results in five equal 5-cm lumps associated with each of the interior nodes and half-size, 2.5-cm, lumps at each end node. Each node exchanges heat by conduction with its neighbors. The ambient fluid is represented as an additional node at a constant temperature $t_f = 20$°C, and each node of the rod exchanges heat by convection with this fluid node. The resultant nodal-resistance network is also shown in Fig. 4.7.

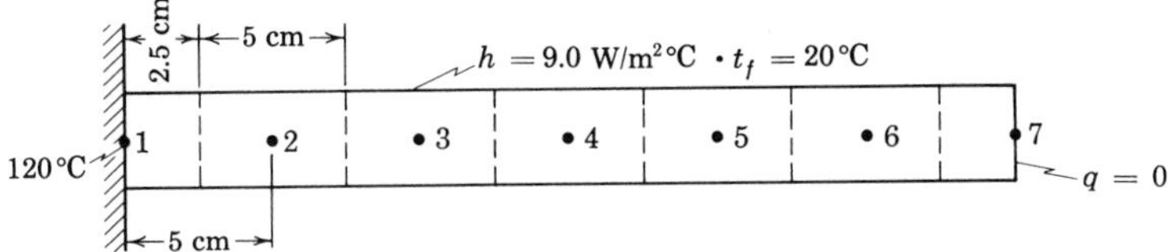

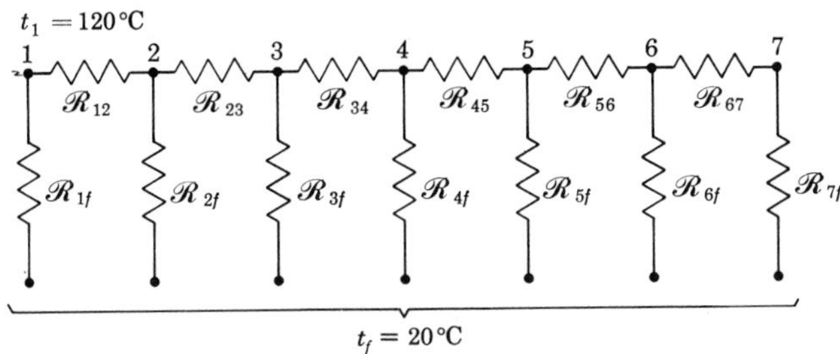

Figure 4.7. Nodal representation of a one-dimensional spine

Using d to denote the rod diameter, the conduction area for the conduction resistors between the nodes is the same for all nodal pairs and is equal to the cross-sectional area of the rod:

$$A_k = \frac{\pi d^2}{4} = \frac{\pi(0.0125)^2}{4} = 1.227 \times 10^{-4}\ \text{m}^2.$$

Since the spacing between nodes is uniform, the interior conduction resistors are all the same, and Eq. (4.22) gives

$$\mathcal{R}_{12} = \mathcal{R}_{23} = \cdots = \mathcal{R}_{67} = \frac{\delta x}{kA_k} = \frac{0.05}{57 \times 1.227 \times 10^{-4}}$$
$$= 7.148°\text{C/W}.$$

The surface area for convection for the five interior nodes is the same and equal to the surface of the associated lumps:

$$A_c = \pi d\ \delta x = \pi \times 0.0125 \times 0.05 = 1.964 \times 10^{-3}\ \text{m}^2.$$

Thus, according to Eq. (4.22), the convection resistors between each of these interior nodes and the fluid node are all the same:

$$\mathcal{R}_{2f} = \mathcal{R}_{3f} = \mathcal{R}_{4f} = \mathcal{R}_{5f} = \mathcal{R}_{6f} = \frac{1}{hA_c} = \frac{1}{9.0 \times 1.964 \times 10^{-3}}$$
$$= 56.59°\text{C/W}.$$

The convection area for the two end nodes, 1 and 7, is half that of the other nodes since the associated lumps are half as long as the others. Thus the convection resistors connecting these nodes to the fluid node are twice the others:

$$\mathcal{R}_{1f} = \mathcal{R}_{7f} = 113.8°\text{C/W}.$$

Note that the conduction resistors are rather small compared with the convection resistors, indicating that the problem is dominated by longitudinal conduction—consistent with the basic premise by which extended surfaces are modeled.

One may now apply the basic heat balance given by Eq. (4.42) at each node:

$$\text{At node } i\text{: } t_i = \frac{q_i + \sum_j \frac{t_j}{\mathcal{R}_{ij}}}{\sum_j 1/\mathcal{R}_{ij}}.$$

For node 1 the temperature is known, $t_1 = 120°C$, and the heat flow required to maintain this temperature, q_1, is unknown. At each of the other nodes the temperature, t_i, is unknown and $q_i = 0$ since there is no internal heat generation. The summations indicated in the expression above are made over the nodes with which node i communicates heat, namely the neighboring nodes for conduction (two in each case except for nodes 1 and 7) and the fluid node for convection (in every instance). No summation is applied to the fluid node since it is of known temperature and the heat flow to the fluid must equal that out of the source, q_1. The results at each of the nodes, knowing that $t_f = 20°C$, are:

$$\begin{aligned}
q_1 &= 17.67 - 0.1399t_2, \\
t_2 &= 0.4703t_3 + 57.62, \\
t_3 &= 0.4703(t_2 + t_4) + 1.188, \\
t_4 &= 0.4703(t_3 + t_5) + 1.188, \\
t_5 &= 0.4703(t_4 + t_6) + 1.188, \\
t_6 &= 0.4703(t_5 + t_7) + 1.188, \\
t_7 &= 0.9406t_6 + 1.188.
\end{aligned}$$

Solution of the foregoing set of equations for the seven unknowns (q_1 and t_2 through t_7) by the Gauss–Seidel iteration technique requires the making of an initial assumption for the temperatures. The results shown in Table 4.1 are based on the assumption of a linear distribution in temperature from the known $t_1 = 120°C$ to $t_7 = 20°C$ (taken as equal to t_f as a first guess). The assumed temperatures result in those listed as "iteration 0" in Table 4.1. Using these values one proceeds down the list of equations above and calculates revised values of $q_1, t_2, t_3, \ldots, t_7$, using the most recently calculated values. For example, in iteration number 1, the value of $t_2 = 98.38°C$ results from the assumed $t_3 = 86.67°C$ while the calculated $t_3 = 80.38°C$ uses $t_2 = 98.38°C$ from the first iteration and $t_4 = 70.00°C$ from the initial assumption. The process is repeated, over and over, until the desired degree of convergence is reached. In Table 4.1 the results are shown using $\epsilon = 0.05°C$, and 18 iterations are necessary to achieve this degree of convergence. The results using $\epsilon = 0.5°C$ (7 iterations) and $\epsilon = 0.005°C$ (30 iterations) are also shown, and the effect of the convergence criterion on accuracy and computation time is apparent. Also shown, for comparison purposes, are the analytical results obtained in Example 2.9.

Table 4.1

Number of Iterations	q_1 W	t_2 °C	t_3 °C	t_4 °C	t_5 °C	t_6 °C	t_7 °C
0	(t_1 = 120)	103.33	86.67	70.00	53.33	36.67	20.00
1	3.214	98.38	80.38	64.07	48.57	33.44	32.64
2	3.907	95.42	76.20	59.86	45.07	37.73	36.68
3	4.321	93.46	73.29	56.85	45.67	39.92	38.73
4	4.596	92.09	71.24	56.17	46.38	41.22	39.96
5	4.787	91.12	70.46	56.14	46.97	42.07	40.76
6	4.922	90.76	70.27	56.33	47.47	42.68	41.33
7	4.973	90.67	70.32	56.58	47.87	43.14	41.77
8	4.985	90.69	70.45	56.84	48.21	43.50	42.11
9	4.982	90.75	70.60	57.06	48.48	43.79	42.38
10	4.974	90.82	70.74	57.26	48.71	44.03	42.60
·	·	·	·	·	·	·	·
·	·	·	·	·	·	·	·
·	·	·	·	·	·	·	·
17	4.925	91.12	71.28	57.97	49.51	44.84	43.36
18 (ϵ = 0.05°C)	4.925	91.14	71.32	58.01	49.56	44.89	43.41
ϵ = 0.005°C, 30 iterations	4.908	91.22	71.46	58.20	49.77	45.10	43.61
ϵ = 0.5°C, 7 iterations	4.973	90.67	70.32	56.58	47.87	43.14	41.77
Analytic solution	4.834	91.095	71.265	57.980	49.541	44.382	43.382

Solution by Matrix Inversion

As was noted in the introductory discussion of this section, the problem at hand is that of solving N algebraic (usually linear) equations in N unknowns, where N is the number of nodal points at which either an unknown temperature or an unknown heat flux must be found. The application of Eq. (4.22) at each node generates a set of N equations of the form

$$\begin{aligned}
a_{11}t_1 + a_{12}t_2 + a_{13}t_3 + \cdots + a_{1N}t_N &= C_1 \\
a_{21}t_1 + a_{22}t_2 + \cdots \qquad\qquad &= C_2 \\
a_{31}t_1 + \cdots \qquad\qquad\qquad &= C_3 \\
\vdots \qquad\qquad\qquad\qquad &\ \ \vdots \\
a_{N1}t_1 + a_{N2}t_2 \qquad + \cdots + a_{NN}t_N &= C_N
\end{aligned} \tag{4.43}$$

The coefficients, $a_{11}, a_{12}, \cdots a_{NN}$ involve the resistances between the nodes and will be different from zero only when the nodes indicated by the subscripts are in thermal communication. The C's are constant terms resulting from specified boundary conditions or heat fluxes.

If one defines the following matrix representations:

$$[A] = \begin{bmatrix} a_{11} & a_{12} & \cdots & a_{1N} \\ a_{21} & a_{22} & \cdots & \\ a_{31} & & & \\ \cdot & & & \\ \cdot & & & \\ \cdot & & & \\ a_{N1} & a_{N2} & \cdots & a_{NN} \end{bmatrix}, \quad [C] = \begin{bmatrix} C_1 \\ C_2 \\ \cdot \\ \cdot \\ \cdot \\ C_N \end{bmatrix}, \quad [t] = \begin{bmatrix} t_1 \\ t_2 \\ \cdot \\ \cdot \\ \cdot \\ t_N \end{bmatrix},$$

the set of equations above may be written

$$[A][t] = [C].$$

Reference 3 may be consulted for information on the rules of matrix algebra. If $[A]^{-1}$ represents the *inverse* of $[A]$, the solution for the temperatures is given by

$$[t] = [A]^{-1}[C]. \tag{4.44}$$

If $[A]^{-1}$ has the elements

$$[A]^{-1} = \begin{bmatrix} b_{11} & b_{12} & \cdots & b_{1N} \\ b_{21} & \cdots & & \\ \cdot & & & \\ \cdot & & & \\ \cdot & & & \\ b_{N1} & b_{N2} & \cdots & b_{NN} \end{bmatrix},$$

the required t's are

$$\begin{aligned} t_1 &= b_{11}C_1 + b_{12}C_2 + b_{13}C_3 + \cdots + b_{1N}C_N, \\ t_2 &= b_{21}C_1 + \cdots \\ &\cdot \\ &\cdot \\ &\cdot \\ t_N &= b_{N1}C_1 + b_{N2}C_2 + \cdots \qquad + b_{NN}C_N. \end{aligned} \tag{4.45}$$

The problem, then, reduces to the inversion of a matrix. The process of matrix inversion (Ref. 3) is usually a laborious one when done by hand, and this method is not recommended in that instance, although the matrix $[A]$ usually contains a large number of zero elements. Most digital computer installations, however, have matrix inversion and matrix multiplication routines available as parts of the system library. In such cases the operations above may be carried out rather simply. Even so, if the number of nodes, and hence the rank of the matrix $[A]$, is large, the inversion process may become lengthy. Generally speaking, the matrix inversion method will usually be speedier for a modest number of nodes, but as the number of nodes increases, the iteration technique may prove to be faster.

EXAMPLE 4.2

Find the temperature distribution for Example 4.1 by matrix inversion.

Solution. Since the heat flow is not desired in this case, the heat balance at node 1 may be eliminated and only the temperatures t_2 through t_7 need to be found. The equations listed in Example 4.1 for nodes 2 through 7 may be rewritten in the form of Eq. (4.43):

$$\begin{aligned}
t_2 - 0.4703t_3 &= 57.62,\\
-0.4703t_2 + t_3 - 0.4703t_4 &= 1.188,\\
-0.4703t_3 + t_4 - 0.4703t_5 &= 1.188,\\
-0.4703t_4 + t_5 - 0.4703t_6 &= 1.188,\\
-0.4703t_5 + t_6 - 0.4703t_7 &= 1.188,\\
-0.9406t_6 + t_7 &= 1.188.
\end{aligned}$$

This set of equations, in matrix form, is

$$[A][t] = [c].$$

Thus the coefficient matrix, $[A]$, and the matrix of nonhomogeneous terms are

$$[A] = \begin{bmatrix}
1 & -0.4703 & 0 & 0 & 0 & 0\\
-0.4703 & 1 & -0.4703 & 0 & 0 & 0\\
0 & -0.4703 & 1 & -0.4703 & 0 & 0\\
0 & 0 & -0.4703 & 1 & -0.4703 & 0\\
0 & 0 & 0 & -0.4703 & 1 & -0.4703\\
0 & 0 & 0 & 0 & -0.9406 & 1
\end{bmatrix}$$

$$[C] = \begin{bmatrix}
57.62\\
1.188\\
1.188\\
1.188\\
1.188\\
1.188
\end{bmatrix}.$$

Inversion of the matrix $[A]$ and multiplication by the matrix $[C]$ gives

$$\begin{aligned}
t_2 &= 91.24°\text{C}, & t_5 &= 49.79°\text{C},\\
t_3 &= 71.48°\text{C}, & t_6 &= 45.13°\text{C},\\
t_4 &= 58.22°\text{C}, & t_7 &= 43.63°\text{C}.
\end{aligned}$$

These results compare favorably with those found by iteration.

EXAMPLE 4.3

A very long shaft has the shape of an 8-cm circular cylinder with a 2-cm × 2-cm square hole located on its axis. The shaft is so long that conduction down its length may be neglected. The surface of the square hole is maintained at 100°C and the external cylindrical surface is maintained at 0°C. The thermal conductivity of the material of which the shaft is composed is 1.0 W/m-°C. Considering two-dimensional conduction only, establish a finite difference nodal network, determine the thermal resistances, and write the nodal heat balance equations needed to determine the distribution of temperature in the shaft cross section.

Solution. The geometry under consideration is shown in Fig. 4.8, and examination of the geometry and boundary conditions shows that, by symmetry, only one-eighth of the cross section need be analyzed. A square, $\delta x = \delta y = 1$ cm, nodal network has been chosen as shown in the figure. The nodal points were selected to fall along the lines of symmetry and on the surface of the square hole. As a result of the curved outer boundary, the regular nodal points 6′, 10′, 11′ and 12′ do not fall on the surface and irregular points 6, 11, and 12 are chosen instead. The symmetry requirement along the radial lines 0–*a* and 0–*b* is handled by including in the network symmetrical points for all nodes which are thermally connected to nodes located on these lines. Specifically, image points for nodes 4 (twice), 7, and 8 are included.

The equivalent electrical resistance is also shown in Fig. 4.8. No resistors are shown for connections along the isothermal surfaces (i.e., points 1–2 and points 6–11–12–13) since no conduction takes place between these nodes as a result of the isothermal condition. For the same reason, no heat balance expressions will be written for these nodes since their temperatures are known and there is no stated desire to find the associated heat fluxes.

Thus heat balance expressions need to be written only for nodes 3, 4, 5, 7, 8, and 9. As may be seen from the resistance network, 17 resistances must be calculated; however, only 13 are independent because of the symmetry conditions already mentioned. Application of the defining relation, Eq. (4.22), for the conduction resistors, $\mathcal{R}_{ij} = \delta_{ij}/kA_{k_{ij}}$, requires the definition of the lumps surrounding each node. Adopting the scheme discussed in the last portion of Section 4.5 in which the lump boundaries are taken as halfway between adjacent nodes, leads to the lumps shown in Fig. 4.8. Nodes 3, 4, 5, and 9 have full lumps (size $\delta x \times \delta y$, $\delta x = \delta y = 1$ cm) around them, while nodes 7 and 8 have irregular lumps whose dimensions may be deduced from the spacing to the irregular points 6, 11, and 12. The latter dimensions may readily be deduced from the geometry of the problem and are shown in Fig. 4.8. The lump associated with node 13 is one-half the size of a regular full lump.

Of the 13 resistances required, those connected to nodes 4 and 5 (i.e., $\mathcal{R}_{14}$, $\mathcal{R}_{34}$, $\mathcal{R}_{45}$, $\mathcal{R}_{48}$, $\mathcal{R}_{25}$, $\mathcal{R}_{59}$) all have conduction distances δ_{ij} equal to the uniform spacing $\delta x = \delta y$ and conduction areas equal to $A_{k_{ij}} = \delta x \times 1 = \delta y \times 1$ (for unit depth). The resistor connecting nodes 9 and 13 also has these same geometrical

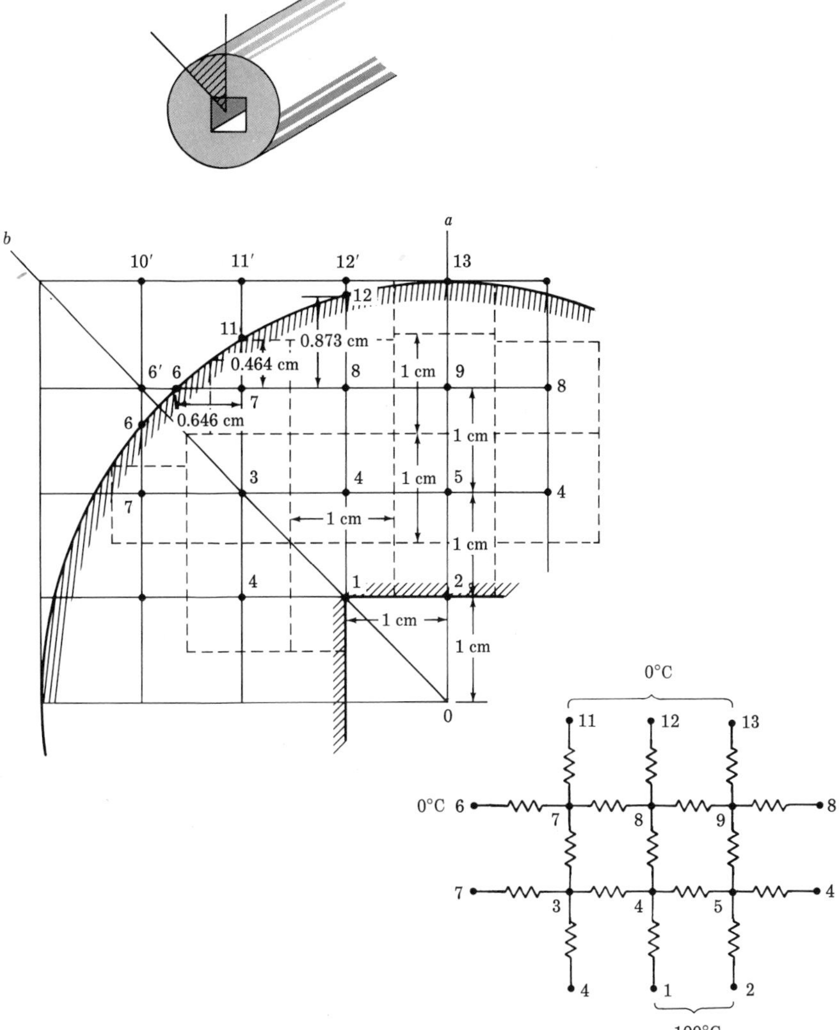

Figure 4.8. Nodal representation of a hollow shaft.

properties. Thus, with $\delta x = \delta y = 1$ cm, $k = 1.0$ W/m-°C, and for a unit depth of 1 m into the paper, Eq. (4.22) gives

$$\mathcal{R}_{14} = \mathcal{R}_{34} = \mathcal{R}_{45} = \mathcal{R}_{48} = \mathcal{R}_{25} = \mathcal{R}_{59} = \mathcal{R}_{9-13} = \frac{\delta_{ij}}{kA_{k_{ij}}}$$

$$= \frac{1.0}{1.0 \times (1.0 \times 1.0)}$$

$$= 1.0\text{°C/W}.$$

For the irregular lumps at nodes 7 and 8 one must apply the representation for δ_{ij} and $A_{k_{ij}}$ developed in the discussion leading to Eqs. (4.41). Comparing the geometry given in Fig. 4.8 for the lump at node 7 to that for the general irregular lump shown in Fig. 4.6, one has

$$\delta = 1 \text{ cm}, \quad \delta' = 0.646 \text{ cm}, \quad \delta'' = 0.464 \text{ cm},$$

so that Eqs. (4.41) give

$$\mathcal{R}_{67} = \frac{\delta'}{k(\delta + \delta'')/2} = \frac{0.646}{1.0 \times (1.0 + 0.464)/2} = 0.8822°\text{C/W},$$

$$\mathcal{R}_{78} = \frac{\delta}{k(\delta + \delta'')/2} = \frac{1.0}{1.0 \times (1.0 + 0.464)/2} = 1.3660°\text{C/W},$$

$$\mathcal{R}_{37} = \frac{\delta}{k(\delta + \delta')/2} = \frac{1.0}{1.0 \times (1.0 + 0.646)/2} = 1.2152°\text{C/W},$$

$$\mathcal{R}_{7-11} = \frac{\delta''}{k(\delta + \delta')/2} = \frac{0.464}{1.0 \times (1.0 + 0.646)/2} = 0.5640°\text{C/W}.$$

Similar reasoning at node 8 yields

$$\mathcal{R}_{8-9} = \frac{1.0}{1.0 \times (1.0 + 0.873)/2} = 1.0678°\text{C/W},$$

$$\mathcal{R}_{8-12} = \frac{0.873}{1.0 \times (1.0 + 1.0)/2} = 0.8730°\text{C/W}.$$

With all the resistances determined, heat balances may be written at each of the nodes 3, 4, 5, 7, 8, and 9. For no internally generated heat, Eq. (4.22) states that

$$\sum_j \frac{t_j - t_i}{\mathcal{R}_{ij}} = \sum_j \frac{t_j}{\mathcal{R}_{ij}} - t_i \sum_j \frac{1}{\mathcal{R}_{ij}} = 0,$$

which for $t_1 = t_2 = 100°\text{C}$, $t_6 = t_{11} = t_{12} = t_{13} = 0°\text{C}$ gives:

Node 3:	$3.6458t_3 - 2.0t_4 - 1.6458t_7 = 0.$
Node 4:	$-1.0t_3 + 4.0t_4 - 1.0t_5 - 1.0t_8 = 100.$
Node 5:	$-2.0t_4 + 4.0t_5 - 1.0t_9 = 100.$
Node 7:	$-0.8229t_3 + 4.4616t_7 - 0.7321t_8 = 0.$
Node 8:	$-1.0t_4 - 0.7321t_7 + 3.8410t_8 - 0.9365t_9 = 0.$
Node 9:	$-1.0t_5 - 1.8730t_8 + 3.8730t_9 = 0.$

These nodal heat balances may be solved for the unknown temperature, by using either the iterative technique (after making an initial assumption for each) or the matrix inversion technique, to yield

$$t_3 = 33.88°\text{C}, \quad t_7 = 9.91°\text{C},$$
$$t_4 = 53.61°\text{C}, \quad t_8 = 22.30°\text{C},$$
$$t_5 = 58.26°\text{C}, \quad t_9 = 25.83°\text{C}.$$

4.7 NONSTEADY NUMERICAL METHODS

The material presented thus far has been limited to the case of steady-state conduction. Of considerable practical importance is the application of numerical techniques to the analysis of nonsteady conduction problems. The finite difference representation of the second derivatives in the space variables, the nodal resistance network representations, the boundary conditions, etc., developed in Secs. 4.3 through 4.5 for the steady state will be equally applicable for nonsteady conduction analysis. The significant difference between the steady and nonsteady cases lies in finite difference representations which may be written for the partial derivative of temperature with respect to time. As will be recalled from Sec. 4.1, two possible finite difference expressions may be written for a first derivative, the forward difference and the backward difference. These representations lead to two different numerical methods for treatment of nonsteady problems.

In either instance, it is important to note that in addition to calculating temperatures at points spaced discrete intervals apart in space, in the nonsteady case the temperatures at these points are calculated at discrete intervals of time. That is, after the temperatures are known at all the spatial points, a finite increment of time is selected and all the temperatures are recalculated at the end of this time. Time is then progressively incremented, the spatial temperatures being calculated at each increment.

The nonsteady methods to be described here may be applied to the solution of steady-state problems. With time-independent boundary conditions, an assumed temperature distribution may be regarded as an initial distribution of a nonsteady problem. As the nonsteady solution is carried forward in time, the desired steady-state solution is eventually approached to any desired degree of accuracy.

Difference Equations

The nonsteady conduction equation is

$$k\left(\frac{\partial^2 t}{\partial x^2} + \frac{\partial^2 t}{\partial y^2} + \frac{\partial t}{\partial z^2}\right) = \rho c_p \frac{\partial t}{\partial \tau}. \tag{4.46}$$

Written in this form, each side of the conduction equation represents the time rate of heat storage, per unit volume, at a point. Finite difference approximations will now be written for this expression, but as noted above, two possible representations are possible. For consciseness of expression, the representations for the one-dimensional case will be written first. In this instance,

$$k \frac{\partial^2 t}{\partial x^2} = \rho c_p \frac{\partial t}{\partial \tau}. \tag{4.47}$$

Explicit Formulation. The explicit formulation is obtained by using the forward difference expression for the first derivative in place of the time derivative on the right side of Eq. (4.47). The central difference for the second derivative is used for the left side. In other words, to expand Eq. (4.47) about $x = x$ and $\tau = \tau$, use Eqs. (4.4) and (4.9) with t replacing f, τ replacing η, x replacing ξ, $\delta\tau$ replacing h_2, and δx replacing h_1. The result is

$$\frac{k}{(\delta x)^2}[t(x + \delta x, \tau) - 2t(x, \tau) + t(x - \delta x, \tau)] = \frac{\rho c_p}{\delta\tau}[t(x, \tau + \delta\tau) - t(x, \tau)]. \tag{4.48}$$

If, as used in the steady-state case and as suggested in Fig. 4.9, the subscripts a, b, and c are used to denote nodal locations at x, $x - \delta x$, and $x + \delta x$, respectively, and if t' is used to denote a temperature at time $\tau + \delta\tau$ while t is used simply to denote a temperature at time τ, Eq. (4.48) is more concisely stated as

$$\frac{k}{(\delta x)^2}(t_b - 2t_a + t_c) = \frac{\rho c_p}{\delta\tau}(t'_a - t_a). \tag{4.49}$$

This expression gives the *future* temperature at node a, t'_a, in terms of the *current* temperatures at node a and its surrounding nodes. This forward difference approximation for the nonsteady conduction equation is also termed the "explicit" formulation.

The equivalent two-dimensional, explicit, formulation is easily shown to be (see Fig. 4.9)

$$\frac{k}{(\delta x)^2}(t_b - 2t_a + t_c) + \frac{k}{(\delta y)^2}(t_d - 2t_a + t_e) = \frac{\rho c_p}{\delta\tau}(t'_a - t_a). \tag{4.50}$$

Implicit Formulation. The implicit representation is obtained by expanding Eq. (4.47) about $x = x$ and $\tau = \tau + \delta\tau$. This is done by use of Eqs. (4.7) and (4.9) with t replacing f, $\tau + \delta\tau$ replacing η, x replacing ξ, $\delta\tau$ replacing h_2, and δx replacing h_1. The result is

$$\frac{k}{(\delta x)^2}[t(x + \delta x, \tau + \delta\tau) - 2t(x, \tau + \delta\tau) + t(x - \delta x, \tau + \delta\tau)] = \frac{\rho c_p}{\delta\tau}[t(x, \tau + \delta\tau) - t(x, \tau)].$$

Using the same abbreviated notation as employed above, one obtains

$$\frac{k}{(\delta x)^2}(t'_b - 2t_a + t'_c) = \frac{\rho c_p}{\delta\tau}(t'_a - t_a). \tag{4.51}$$

This implicit formulation gives the *future* temperature of point a, t'_a, in terms of the current temperature a, t_a, and the *future* temperatures of the neighboring points. The two-dimensional case reduces to

$$\frac{k}{(\delta x)^2}(t'_b - 2t'_a + t'_c) + \frac{k}{(\delta y)^2}(t'_d - 2t'_a + t'_e) = \frac{\rho c_p}{\delta\tau}(t'_a - t_a). \tag{4.52}$$

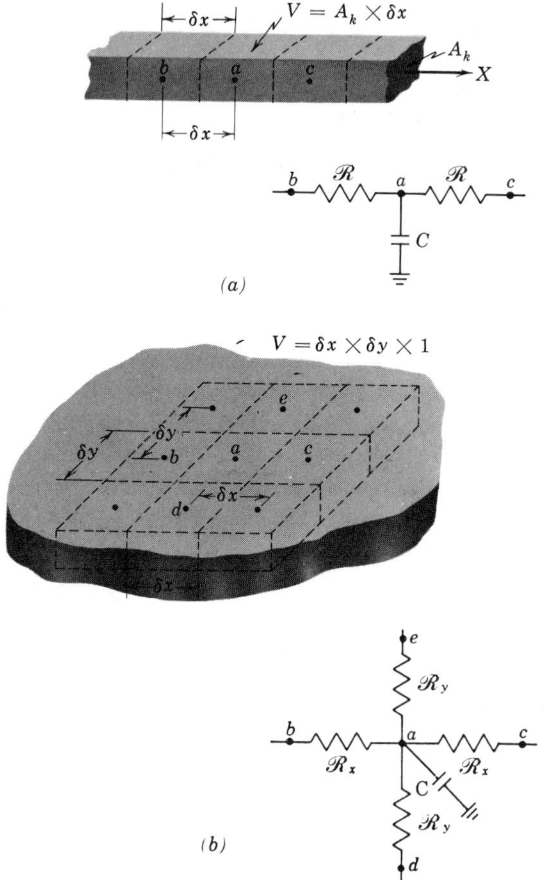

Figure 4.9. Nodal representation of nonsteady conduction: (a) one-dimensional; (b) two-dimensional.

Explicit Versus Implicit Formulation. A rather obvious advantage of the explicit representation over the implicit is the fact the forward difference equation gives the *future* temperature of a single node in terms of *current* temperatures of that node and its neighbors. Thus, if at the end of a certain time period, all the nodal temperatures are known, then each of the nodal temperatures at the end of the next moment, $\delta\tau$, may be explicitly found, node by node. The implicit equation, however, expresses a *future* nodal temperature in terms of its current value and the *future* values of its neighbors' temperatures. Thus, to progress from one time step to the next, a system of equations of the form of Eq. (4.51) or (4.52) must be solved.

It would appear, then, that the explicit representation would be preferred to the implicit representation. However, as will be seen later, in the explicit case a serious restriction must be placed on the magnitude of the time step $\delta\tau$ in relation to the spatial increment δx. Thus, although more direct, the explicit formulation may actually involve more calculation time than the less-direct implicit method.

Nodal Networks in the Nonsteady State

Directing attention for the moment to the one-dimensional case shown in Fig. 4.9, one may divide the conducting solid into lumps centered on the nodal points, as in the steady-state cases discussed earlier. The total heat flow at a node may be had by multiplying Eq. (4.49), or (4.51), by the lump volume, $V_a = A_k\,\delta x$. Thus in the explicit case,

$$\frac{t_b - t_a}{\delta x/kA_k} + \frac{t_c - t_a}{\delta x/kA_k} = \frac{V_a \rho c_p}{\delta \tau}(t_a' - t_a).$$

The resistances $\mathcal{R}_{ab} = \mathcal{R}_{ac} = \delta x/kA_k$ are defined as before. The term

$$C_a = V_a \rho c_p$$

is recognized as the thermal capacity of the lump volume surrounding the node a. Thus the expression above becomes

$$\frac{t_b - t_a}{\mathcal{R}_{ab}} + \frac{t_c - t_a}{\mathcal{R}_{ac}} = C_a \frac{t_a' - t_a}{\delta \tau}. \tag{4.53}$$

Similarly, for the implicit case

$$\frac{t_b' - t_a'}{\mathcal{R}_{ab}} + \frac{t_c' - t_a'}{\mathcal{R}_{ac}} = C_a \frac{t_a' - t_a}{\delta \tau}. \tag{4.54}$$

Either Eq. (4.53) or (4.54) suggests the analogous electrical network illustrated in Fig. 4.9, wherein a capicitance is associated with each node—representing the thermal capacitance of the associated lump.

Generalizing on the concepts just discussed, and drawing on the concepts developed in Sec. 4.4, one may conclude that if a complex geometry (i.e., two or more dimensions) is subdivided into a nodal network of, perhaps unequal spacings, the following expression must be satisfied at each nodal point i:

$$\text{Explicit: } \sum_j \frac{t_j - t_i}{\mathcal{R}_{ij}} + q_i = \frac{C_i}{\delta \tau}(t_i' - t_i). \tag{4.55}$$

$$\text{Implicit: } \sum_j \frac{t_j' - t_i'}{\mathcal{R}_{ij}} + q_i = \frac{C_i}{\delta \tau}(t_i' - t_i). \tag{4.56}$$

The term q_i again denotes the rate of heat addition (external or internal) to the node. Thus the future temperature of node i is determined by

$$\text{Explicit: } t_i' = t_i + \delta\tau\left(\sum_j \frac{t_j - t_i}{\mathcal{R}_{ij} C_i} + \frac{q_i}{C_i}\right). \tag{4.57}$$

$$\text{Implicit: } t_i' = t_i + \delta\tau\left(\sum_j \frac{t_j' - t_i'}{\mathcal{R}_{ij} C_i} + \frac{q_i}{C_i}\right), \tag{4.58}$$

where

$$\mathcal{R}_{ij} = \frac{\delta_{ij}}{kA_{k_{ij}}}, \qquad \text{for conduction,}$$

$$\mathcal{R}_{ij} = \frac{1}{h_{ij}A_{c_{ij}}}, \qquad \text{for convection,} \tag{4.59}$$

$$C_i = V_i \rho c_p.$$

These equations just given express the temperatures at a given node in terms of discrete spatial steps (i.e., δ_{ij}) and discrete time steps (i.e., $\delta\tau$). The method of solution depends on which formulation (explicit or implicit) is used but the important point to be noted is that the continuous solution of the original differential equation is replaced by the numerical solution of these difference equations at particular points in space and at particular intervals of time.

Boundary Conditions in the Nonsteady State

Equations of the form of Eq. (4.57) or (4.58) must be satisfied at each nodal point into which a conducting body is subdivided. As in the steady-state case, certain special precautions must be taken at boundary points. In general, this amounts to calculating the resistances (conduction and convection) connected to the boundary nodes by use of special relations rather than those given in Eq. (4.59). The boundary conditions given in Sec. 4.5 for the steady case are equally applicable to the nonsteady case. Thus no repetition will be made here for those conditions.

The case of time-dependent boundary temperatures enters as a possibility in the nonsteady case, however. In such cases, it is usual practice to treat such a case as that of a fixed temperature boundary condition [see Eqs. (4.28) through (4.30)]. During any time increment, the fixed boundary temperatures are chosen as the mean of the temperatures occurring at the boundary points at the beginning and the end of that time increment. As the calculations proceed in time, these boundary point temperatures must be continually revised according to the prescribed temperature function imposed at the surface. One of the examples given later will illustrate this procedure.

The Stability of the Solution

As has been pointed out previously, the numerical methods presented in this chapter represent approximate solutions to the original differential equations since derivatives are replaced by finite differences. Terms of the order of the fourth power of the spatial step size are neglected and, in the transient case, terms of the order of the square of the time increment are neglected. Errors introduced by these approximations are termed *truncation* errors, and the degree to which the approximate solution approaches the exact solution is termed the *convergence* of the finite difference representation. *Numerical* errors are also introduced into a solution by virtue of the inability, or impracticality, of performing the numerical computations with a sufficient number of significant figures.

In nonsteady numerical problems, the *stability* of the difference equations must also be considered. This matter is related to the way in which numerical and truncation errors introduced at one point in time either damp out or propagate and amplify in succeeding time steps. Detailed analyses of the stability properties of the equations used in this chapter are quite complex, but simple stability criteria may be developed on an elementary, intuitive, basis.

Rewrite Eqs. (4.57) and (4.58) in the following forms—expressing the sought-for future temperature at node i, t'_i, in terms of the other quantities:

$$\text{Explicit: } t'_i = t_i\left(1 - \sum_j \frac{\delta\tau}{\mathcal{R}_{ij}C_i}\right) + \sum_j \frac{t_j\delta\tau}{\mathcal{R}_{ij}C_i}. \tag{4.60}$$

$$\text{Implicit: } t'_i = \frac{t_i + \sum_j t'_j \dfrac{\delta\tau}{\mathcal{R}_{ij}C_i}}{1 + \sum_j \dfrac{\delta\tau}{\mathcal{R}_{ij}C_i}}. \tag{4.61}$$

For simplicity of discussion, any internal heat generation, q_i, has been considered zero.

It may be noted that in the case of the explicit formulation, the coefficient of the t_i term might become negative—particularly if the time increment, $\delta\tau$, is chosen large enough. If the coefficient of t_i *does* become negative, t'_i could conceivably be less than t_i. This implies that for certain values of $\Sigma_j\ \delta\tau/\mathcal{R}_{ij}C_i$, the greater is the temperature t_i at the time τ, the *smaller* it will be at time $\tau + \delta\tau$. This fact does not make sense, thermodynamically, and it can be seen as a trend which will cause the temperature t_i to oscillate wildly from one time period to the next. Although this is not a precise analysis of stability, one would be certain of a stable procedure so long as

$$\sum_j \frac{\delta\tau}{\mathcal{R}_{ij}C_i} \le 1. \tag{4.62}$$

More sophisticated analyses (Refs. 4 and 5) lead to less stringent stability criteria; however, the above limiting relation is generally used in practical cases.

The implication of Eq. (4.62) is that the choice of the time step $\delta\tau$ is intimately connected with the choice of the spatial increment, δ_{ij}, which is involved in the resistance $\mathcal{R}_{ij}$. As smaller spatial increments are chosen to reduce the associated truncation error, smaller time increments must also be used in order to satisfy Eq. (4.62). Thus, increased accuracy in the spatial network is obtained at the cost of smaller time increments—perhaps leading to prohibitive computation times. This is the principal disadvantage of the explicit method, and it will be examined in more detail in Sec. 4.9.

Examination of Eq. (4.61) for the implicit formulation reveals that no stability limitation exists as a result of possible negative coefficients. Thus the time increment is not restricted by the choice of spatial increments. The only limitation on $\delta\tau$ in this instance is that imposed by the minimization of truncation errors in time.

4.8 SOLUTION OF NETWORK EQUATIONS FOR THE IMPLICIT CASE

The nodal heat balance in the implicit case is given by

$$t_i' = \frac{t_i + \sum_j t_j' \dfrac{\delta\tau}{\mathcal{R}_{ij} C_i}}{1 + \sum_j \dfrac{\delta\tau}{\mathcal{R}_{ij} C_i}}. \tag{4.61}$$

If all the resistances of a network are known, and if the capacitances of all the associated lumps are known, then with a specified *initial* temperature distribution, equations of the type of Eq. (4.61) may be written for each nodal point—care being taken to satisfy the established boundary conditions. The net result is that a set of equations are obtained for the *future* temperatures, t_i', of the nodal points—in terms of the *future* temperatures of neighboring points—once a time step, $\delta\tau$, is chosen. There will be as many equations as there are unknown future temperatures. Once this set of equations is solved, the resulting temperatures become the initial temperatures for the next time step calculation.

Thus the implicit technique reduces to the solution of a set of simultaneous algebraic equations at each time increment. Consequently, the methods discussed in Secs. 4.3 through 4.6 for the solution of the steady-state equations are applicable here—iteration or matrix inversion. The implicit formulation is seen, then, to reduce to a series of steady state calculations at each time step. It has the advantage of not having a restricted time step—in fact, the time step may be varied during the progression of the calculation. It has the disadvantage of requiring a *set* of calculations (i.e., iteration or matrix inversion) at each step in time, leading to increased calculation time and storage requirements as the number of nodes becomes large. Since the calculations do not differ, in principle, from those already shown for the steady state, no examples will be given.

4.9 SOLUTION OF NETWORK EQUATIONS FOR THE EXPLICIT CASE

As discussed earlier, the explicit formulation avoids the need of iterative or matrix inversion techniques, since each future nodal temperature can be individually calculated for a time increment $\delta\tau$ from only the current nodal temperatures. Thus from an equation of the form

$$t_i' = t_i\left(1 - \sum_j \frac{\delta\tau}{\mathcal{R}_{ij} C_i}\right) + \sum_j t_j \frac{\delta\tau}{\mathcal{R}_{ij} C_i}, \tag{4.60}$$

new temperatures are successively calculated at each node, starting with the given initial temperature distribution in a network for a given $\delta\tau$. Time is then incremented and the calculations are repeated. No iterations or matrix inversions are required. Only the stability requirement stated in Eq. (4.62) need be satisfied:

$$\sum_j \frac{\delta\tau}{\mathcal{R}_{ij}C_i} \leq 1. \tag{4.62}$$

Special Stability Criteria

Examination of the stability requirement for special instances is appropriate. The relations to be developed emphasize the *upper* limit of the allowable time increment. In practice it is wise to use a time increment smaller than the maximum. The reason for this practice is a combination of the desires to improve accuracy by reducing truncation error in time and to avoid instabilities by staying safely below the upper limit of $\delta\tau$.

Irregular Networks. For the most general instance in which the network contains irregular net spacings, convective resistors as well as conductive resistors, etc., the criterion given in Eq. (4.62) must be evaluated at each nodal point. Since it is necessary at any one time step to carry *each* node forward in time by the same $\delta\tau$, the time increment for the entire calculation is controlled by the node for which stability criterion yields the smallest time increment. That is,

$$\delta\tau \leq \left(\frac{1}{\sum_j \frac{1}{\mathcal{R}_{ij}\text{i}C_i}} \right)_{\min} \tag{4.63}$$

In general, nets that include lumps which are significantly smaller than the others should be avoided. Such small lumps usually have small capacitances and consequently control the time increment for the entire network.

Equally Spaced Networks. In the event that a net is one of equal spacings, special forms result for Eq. (4.63). Figure 4.10 illustrates one- and two-dimensional cases in which regular spacings are used. In the case of the two-dimensional network, regular spacing implies the use of a square grid. In either the one- or two-dimensional cases a half-lump must be provided at the boundary.

Consider first the one-dimensional case. The body dimensions in the two directions, other than the one in which conduction takes place, are taken as l_1 and l_2, respectively. (Actually the body may be of infinite extent in these directions, in which case one would carry out the calculation on a unit flow area basis and simply take $l_1 = l_2 = 1$.) For the case in which there is conduction only (i.e., no convection at the boundary) the equal spacing of δx results in identical resistors—two connected to each interior node and one at the surface node. The resistors all have the value

$$\mathcal{R} = \frac{\delta x}{k(l_1 l_2)}.$$

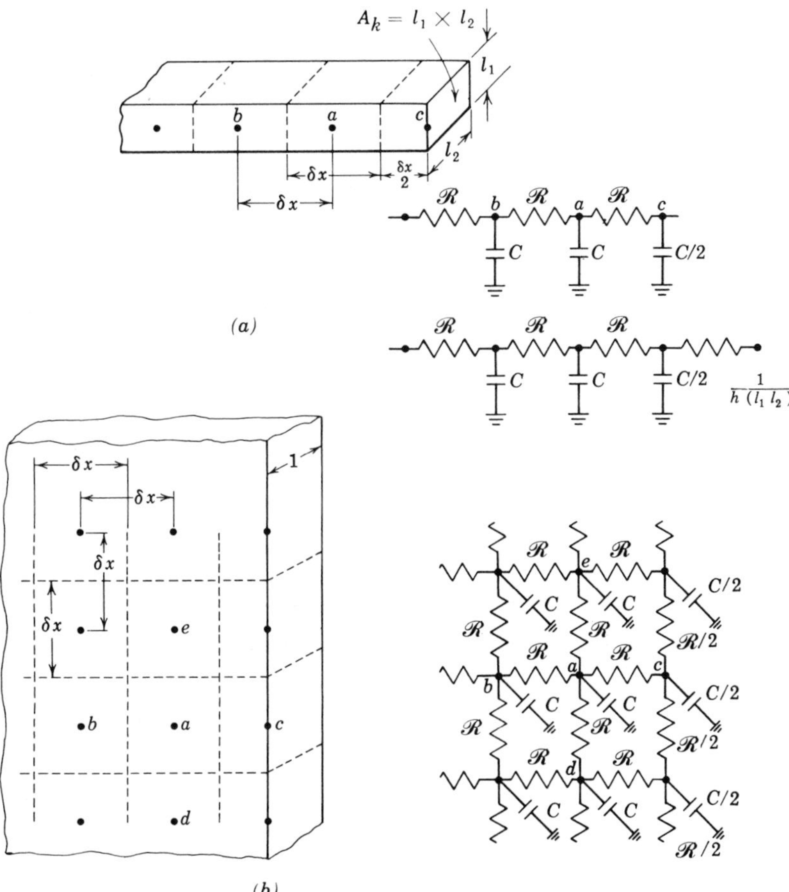

Figure 4.10. Regular boundary points in nonsteady conduction: (a) one-dimensional; (b) two-dimensional.

The capacitors of the interior nodes are all equal to

$$C = \rho V c_p = \rho l_1 l_2 \, \delta x c_p,$$

while the surface node has a capacitance of $C/2$. Since the nodes with capacitance C have two resistors, $\mathcal{R}$, connected to them, and that with capacitance $C/2$ has only one resistor connected to it, the stability criterion of Eq. (4.62) yields the same result for *all* nodes:

$$2 \frac{\delta\tau}{\mathcal{R}C} \leq 1.$$

Rewritten, the expression above becomes

$$\begin{aligned}\delta\tau &\leq \tfrac{1}{2}\mathcal{R}C \\ &\leq \frac{1}{2}\frac{\rho c_p}{k}(\delta x)^2 \\ &\leq \frac{1}{2}\frac{(\delta x)^2}{\alpha}.\end{aligned} \tag{4.64}$$

The latter form shows more clearly the restriction placed on the time increment by the choice of spatial increment.

For the two-dimensional case shown in Fig. 4.10, and for the three-dimensional case with an equally spaced net, arguments similar to that just given yield

$$\text{Two dimensions:}\quad \delta\tau \leq \tfrac{1}{4}\mathcal{R}C \leq \frac{1}{4}\frac{(\delta x)^2}{\alpha}. \tag{4.65}$$

$$\text{Three dimensions:}\quad \delta\tau \leq \tfrac{1}{6}\mathcal{R}C \leq \frac{1}{6}\frac{(\delta x)^2}{\alpha}. \tag{4.66}$$

If the upper limit of the time increment is used, the nodal heat balance, Eq. (4.60), takes on particularly simple forms:

$$\text{One dimension:}\quad t_a' = \frac{t_b + t_c}{2}. \tag{4.67}$$

$$\text{Two dimensions:}\quad t_a' = \frac{t_b + t_c + t_d + t_e}{4}. \tag{4.68}$$

The future temperature at a node is seen to be a simple mean of the surrounding nodes. A similar relation results for the three-dimensional case. These relations prove to be most useful for hand calculations and also form the basis of certain graphical methods (Ref. 5).

In the event that convection occurs at the boundary, the above stability criteria must be altered. In the one-dimensional case, also pictured in Fig. 4.10, the stability criterion yields the same result as given in Eq. (4.64) when applied at the interior nodes. At the surface node, however, there is a conduction resistance and a convection resistance:

$$\mathcal{R}_{\text{cond}} = \frac{\delta x}{k(l_1 l_2)},$$

$$\mathcal{R}_{\text{conv}} = \frac{1}{h(l_1 l_2)},$$

$$C = \tfrac{1}{2}\rho l_1 l_2\, \delta x c_p.$$

Thus Eq. (4.62) yields

$$\frac{\delta\tau}{\frac{1}{2}\rho l_1 l_2\,\delta x c_p}\left[\frac{k l_1 l_2}{\delta x} + h l_1 l_2\right] \le 1$$

or

$$\delta\tau \le \frac{1}{2}\frac{(\delta x)^2}{\alpha}\left[\frac{1}{1 + (h\,\delta x/k)}\right]. \tag{4.69}$$

Since the term in brackets in Eq. (4.69) is always less than 1, it is apparent that the stability criterion applied at the surface node yields a smaller time increment than that resulting from consideration of the interior nodes. Thus the surface node becomes the controlling factor as far as the selection of the time increment is concerned.

A similar analysis when applied to the two-dimensional case shown in Fig. 4.10 and in the three-dimensional case gives

$$\begin{aligned}\text{Two-dimensional:}\quad \delta\tau &\le \frac{1}{4}\frac{(\delta x)^2}{\alpha}\left[\frac{1}{1 + \frac{1}{2}(h\,\delta x/k)}\right].\\ \text{Three-dimensional:}\quad \delta\tau &\le \frac{1}{6}\frac{(\delta x)^2}{\alpha}\left[\frac{1}{1 + \frac{1}{3}(h\,\delta x/k)}\right].\end{aligned} \tag{4.70}$$

Relations similar to those given in Eq. (4.70) may be developed for other special configurations, such as an external corner, etc. These may be found in the references at the end of the chapter. When the surface film coefficient becomes quite large the permissible time increment allowed by the boundary nodes may become considerably less than that allowed by the interior nodes. In such an instance, the computation of the temperature history may become prohibitively long. Certain schemes have been developed whereby such severe limitations may be circumvented.

The implicit and explicit techniques just described are only two, rather direct, of many possible numerical methods available for solution of the heat conduction equation. For certain problems, accelerating techniques are available. These techniques, although more complex to carry out, are hybrid methods which attempt to combine the inherent stability of the implicit approach with the matrix-inversion-free advantages of the explicit approach. Reference 4 may be consulted for information on such techniques.

EXAMPLE 4.4

An infinte slab 30 cm thick is initially at a uniform temperature of 40°C. The temperatures of both surfaces of the slab are suddenly raised to 260°C and maintained at that value. The slab is composed of a material with the following properties: $k = 1.84$ W/m-°C, $\rho = 2300$ kg/m^3, $c_p = 0.8$ kJ/kg-°C, $\alpha = 1.0 \times$

10^{-6} m^2/s. Determine, numerically, the temperature–time history in the slab during the first 1.25 h after the surface temperature is increased, and compare the results with the analytical solution given in Sec. 3.5.

Solution. Because of symmetry considerations only one-half of the slab need be analyzed. Since the slab is infinite in extent except in the direction of its thickness, the conduction may be treated as one-dimensional in that direction. A nodal spacing of 3 cm is chosen, with node 1 at the slab surface and node 6 at the slab centerline, as depicted in Fig. 4.11. Symmetry about the centerline is assured by taking an image point, node 5, on either side of the centerline. Taking the associated lump boundaries as halfway between nodes, the lumps at each interior node are 3 cm thick and that at node 1, on the surface, is a half-lump of 1.5 cm thickness.

The resistance–capacitance network resulting from the chosen nodal network is also shown in Fig. 4.11. Since the nodal spacing is uniform, the conduction distance for all resistors is the same, $\delta_{ij} = 3$ cm. Since the conduction is one-dimensional, the thermal resistances and heat capacities of the lumps may be computed per unit of normal area, $A_{k_{ij}}$. Thus all conduction resistors are the same:

$$\mathcal{R} = \frac{\delta_{ij}}{kA_{k_{ij}}}.$$

The lump volumes are, then, $V = \delta_{ij}A_{k_{ij}}$, so that the heat capacity of each interior lump is

$$C = V\rho c_p = \delta_{ij}A_{k_{ij}}\rho c_p.$$

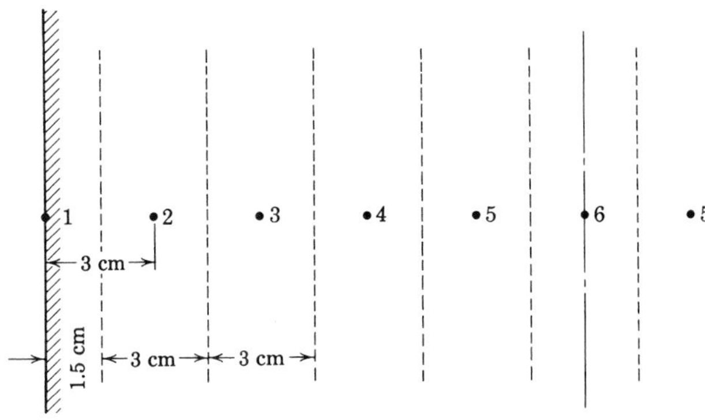

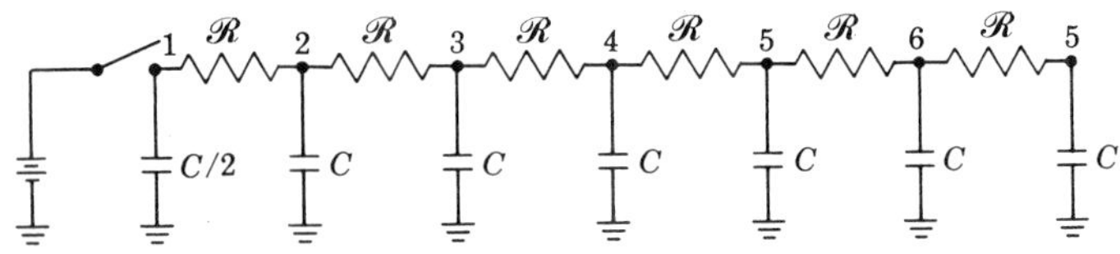

Figure 4.11. Nonsteady circuit for a plane slab.

Then the product $\mathcal{R}C$, needed later, is the same at each interior lump:

$$\mathcal{R}C = \delta_{ij}^2 \frac{\rho c_p}{k} = \frac{\delta_{ij}^2}{\alpha}$$

$$= \frac{(0.03)^2}{1 \times 10^{-6}} = 900 \text{ s.}$$

The surface node, node 1, has a lump with half the volume and half the heat capacity of the other nodes, i.e.,

$$C_1 = \frac{C}{2}.$$

Application of the explicit formulation for the determination of the temperature–time variation at each node [Eqs. (4.60) and (4.61)] requires finding the sum

$$\sum_j \frac{1}{\mathcal{R}_{ij} C_i}$$

at each node. For the interior nodes, 2 through 6, there are two resistances $\mathcal{R}$ connected to each and a heat capacity of C. For the surface node, 1, there is only one resistor, but the heat capacity there is $C/2$. Thus at all nodes, including the surface node, one has

$$\sum_j \frac{1}{\mathcal{R}_{ij} C_i} = \frac{2}{\mathcal{R}C}$$

$$= \frac{2}{900} = \frac{1}{450 \text{ s}}.$$

According to the stability criterion of Eq. (4.63) or (4.64), application of the explicit formulation requires that in order to avoid numerical instability the chosen time step, $\delta\tau$, be limited by

$$\delta\tau < \delta\tau_{\max} = \frac{1}{\displaystyle\sum_j \frac{1}{\mathcal{R}_{ij} C_i}}$$

$$\delta\tau < \delta\tau_{\max} = 450 \text{ s.}$$

With $\Sigma_j\, 1/\mathcal{R}_{ij}C_i = 1/450$ s known at each node and $\mathcal{R}C = 900$ s for each interior node, the temperature–time variation at each interior point may be calculated using the explicit representation of Eq. (4.60):

$$t_i' = t_i\left(1 - \sum_j \frac{\delta\tau}{\mathcal{R}_{ij} C_i}\right) + \sum_j t_j \frac{\delta\tau}{\mathcal{R}_{ij} C_i},$$

$$t_i' = t_i\left(1 - \frac{\delta\tau}{450}\right) + \frac{\delta\tau}{900} \sum_j t_j. \tag{4.71}$$

In Eq. (4.71) $\delta\tau$ is measured in seconds, t_i and t_j are the current temperatures at the nodes, and t_i' is the future temperature at node i. At time $= 0$, the temperature of all interior nodes is known to be 40°C. At node 1, the surface node, the temperature is double-valued: 40°C and 260°C. In keeping with the principle mentioned in association with the boundary conditions, the temperature of node 1 is taken as the mean of these values during the first time step, namely 150°C. Thus, as illustrated in Table 4.2, the temperatures at time $= 0$ for each node are taken as noted in the table and kept constant at these values during the first time step. A time increment $\delta\tau$ is chosen and the new temperatures are calculated for each node from Eq. (4.71). The calculations summarized in Table 4.2 are those that result when $\delta\tau$ is chosen equal to $\delta\tau_{max} = 450$ s. Starting with the initial values at $\tau = 0$, new values are found for each interior node, according to Eq. (4.71), at $\tau = 450$ s knowing that the boundary conditions demand that $t = 260$°C for all $\tau > 0$. Upon completion of each row of the table, time is incremented again by 450 s and the next row constructed in a similar manner. In the instance illustrated in Table 4.2, where $\delta\tau = \delta\tau_{max}$, Eq. (4.71) reduces to $t_i' = \frac{1}{2}\Sigma_j t_j$. That is, the new temperature at a node is the arithmetic average of those of its two neighbors at the preceding time, as in Eq. (4.67). This simplification is not, however, available when other time increments are used as discussed later.

The results of the calculations shown in Table 4.2 are displayed graphically in Fig. 4.12 for three selected times. Also shown for comparison purposes are the analytical predictions of the exact solution given in Sec. 3.5 and Fig. 3.6. The correspondence between the numerical results and the exact solution is apparent. The observed discrepancy is due to the truncation error introduced by the finite difference approximations in both space and time.

Table 4.2. Numerical Calculations for Infinite Slab

Time		Node					
s	h	t_1 °C	t_2 °C	t_3 °C	t_4 °C	t_5 °C	t_6 °C
0	0	150.00	40.00	40.00	40.00	40.00	40.00
450	0.125	260.00	95.00	40.00	40.00	40.00	40.00
900	0.250	260.00	150.00	67.50	40.00	40.00	40.00
1350	0.375	260.00	163.75	95.00	53.75	40.00	40.00
1800	0.500	260.00	177.50	108.75	67.50	46.87	40.00
2250	0.625	260.00	184.38	122.50	77.81	53.75	46.87
2700	0.750	260.00	191.25	131.09	88.13	62.34	53.75
3150	0.875	260.00	195.55	139.69	96.72	70.94	62.34
3600	1.000	260.00	199.84	146.13	105.31	79.53	70.94
4050	1.125	260.00	203.07	152.58	112.83	88.13	79.53
4500	1.250	260.00	206.29	157.95	120.35	96.18	88.13

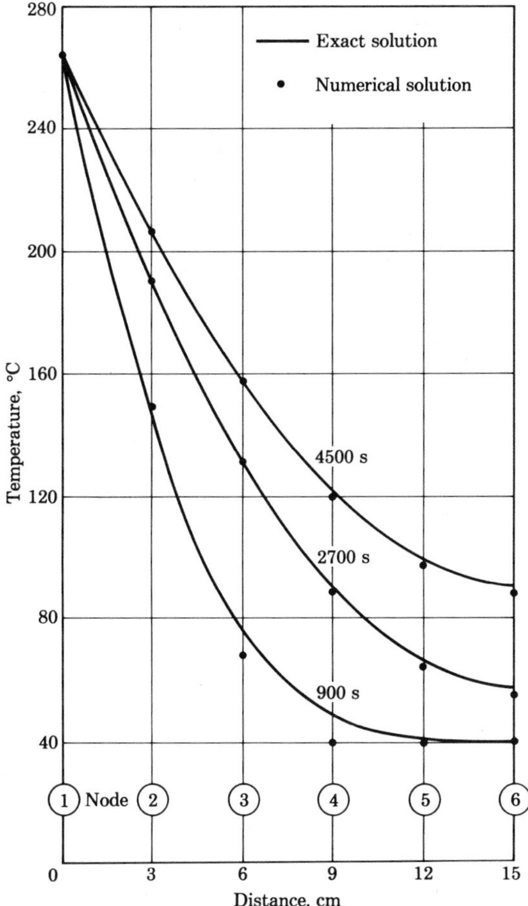

Figure 4.12. Comparison of analytical and numerical results for a plane slab, showing the temperature distribution at various times when using the maximum time step.

The results just discussed and shown in Table 4.2 and Fig. 4.12 were obtained when the time step $\delta\tau$ was taken to be the maximum allowed by the stability criterion. The effect of the choice of the time step on the accuracy of the results is illustrated in Fig. 4.13. In this figure only the temperature–time history for the centerline node, node 6, is shown. The results of the exact analytical solution are compared with those just obtained with $\delta\tau = \delta\tau_{max}$. Also shown are the results that are obtained if a smaller time step, $\delta\tau = \frac{1}{2}\,\delta\tau_{max} = 225$ s is used. The improvement in accuracy is apparent. Figure 4.13 also illustrates the instability of the solution which results if the maximum time step is exceeded. Shown are the calculations resulting when $\delta\tau = 1.2\,\delta\tau_{max} = 540$ s.

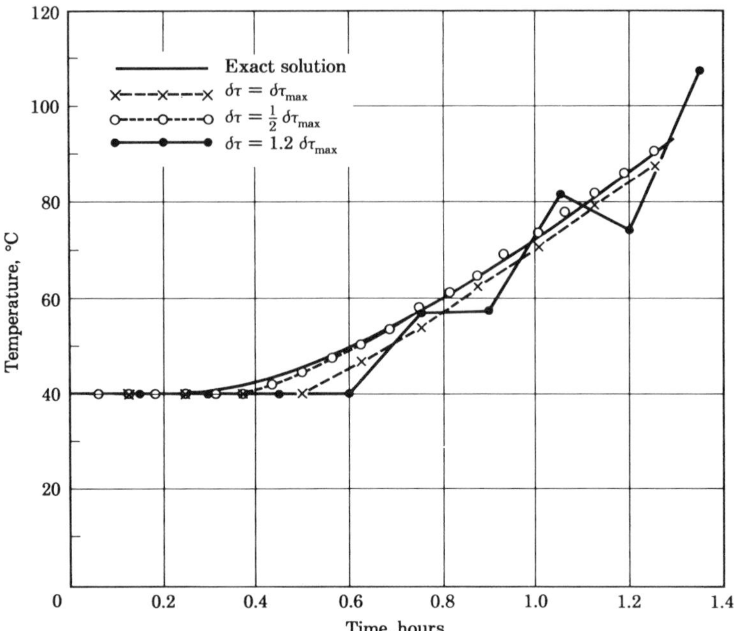

Figure 4.13. Comparison of analytical and numerical results for a plane slab, showing the temperature-time history of the centerline for different time increments.

EXAMPLE 4.5

Determine for the geometry given in Example 4.3, a circular shaft with a concentric square hole, the maximum allowable time step that stability limitations will permit in an analysis of transient conduction in this geometry. Presume that the boundary conditions applied at the exposed surfaces involve only specified temperatures (constant or time-dependent) and do not involve convection. The shaft material has the following properties: $k = 1.0$ W/m-°C, $\rho = 2000$ kg/m^3, $c_p = 0.5$ kJ/kg-°C.

Solution. The maximum permissible time step that may be used is given by Eq. (4.63):

$$\delta\tau \leq \delta\tau_{max} = \left(\frac{1}{\sum_j \dfrac{1}{\mathcal{R}_{ij} C_i}} \right)_{min} .$$

Thus the sum $\Sigma_j\, 1/\mathcal{R}_{ij} C_i$ must be evaluated at each node and the smallest value determined. Only the interior nodes 3, 4, 5, 7, 8, and 9 need be considered since the surface nodes will have known temperatures. The resistances connected to these

Table 4.3. Time Step Calculations for Example 4.5

Node	C_i W-s/°C	$\Sigma_j\, 1/\mathcal{R}_{ij}$ W/°C	$\Sigma_j\, 1/\mathcal{R}_{ij}C_i$ 1/s	$(\Sigma_j\, 1/\mathcal{R}_{ij}C_i)^{-1}$ s
3	100	3.946	0.0365	27.4
4	100	4.0	0.0400	25.0
5	100	4.0	0.0400	25.0
7	59.45	4.462	0.0750	13.3
8	93.0	3.184	0.0410	24.4
9	100	3.873	0.0387	25.8

interior nodes were all evaluated in Example 4.3, per meter of shaft length. Thus one needs, in addition, the heat capacity of each lump:

$$C_i = V_i \rho c_p.$$

The volumes of the various lumps are readily calculated from the dimensions given in Fig. 4.8. The results, per meter of shaft length, are

$$V_3 = V_4 = V_5 = V_9 = 1.0 \times 10^{-4}\ \text{m}^3,$$
$$V_7 = 5.95 \times 10^{-5}\ \text{m}^3, \qquad V_8 = 9.30 \times 10^{-5}\ \text{m}^3.$$

Since $\rho c_p = 2000 \times 0.5 = 10^3\ \text{kJ/m}^{3\circ}\text{C} = 10^6\ \text{W-s/m}^3\text{-}^\circ\text{C}$, the heat capacities of the lumps are

$$C_2 = C_3 = C_4 = C_9 = 100\ \text{W-s/}^\circ\text{C},$$
$$C_7 = 59.45\ \text{W-s/}^\circ\text{C}, \qquad C_8 = 93.0\ \text{W-s/}^\circ\text{C}.$$

The resistances between the various nodes were determined in Example 4.3, so that the sum $\Sigma_j\, 1/\mathcal{R}_{ij}$ may be evaluated at each node. These sums, along with the heat capacities just found, are summarized in Table 4.3, which also shows $\Sigma_j\, 1/\mathcal{R}_{ij}C_i$, and its reciprocal, at each node. The lump at node 7 is seen to have the smallest value of $(\Sigma_j\, 1/\mathcal{R}_{ij}C_i)^{-1}$. This node, then, sets the maximum time step at which all nodes may be carried forward in time:

$$\delta\tau_{\max} = 13.3\ \text{s}.$$

Node 7 dominates the problem because of its small volume (and thus, heat capacity). In general, one should try to avoid choosing a nodal spacing that results in one very small lump whose stability limit unreasonably dominates the increment at which calculations may be carried forward in time.

REFERENCES

1. Desai, C. S., *Elementary Finite Element Methods,* Englewood Cliffs, N.J., Prentice-Hall, 1979.
2. Southwell, R. V., *Relaxation Methods in Theoretical Physics,* New York, Oxford U.P., 1946.

3. Hildebrand, F. B., *Methods of Applied Mathematics,* Englewood Cliffs, N.J., Prentice-Hall, 1952.

4. Patankar, S. V., *Numerical Heat Transfer and Fluid Flow,* New York, McGraw-Hill, 1980.

5. Schneider, P. J., *Conduction Heat Transfer,* Reading, Mass., Addison-Wesley, 1955.

6. Dusinberre, G. M., *Heat Transfer Calculations by Finite Differences,* Scranton, Pa., International Textbook, 1961.

7. Rohsenow, W. M., and J. P. Hartnet, eds., *Handbook of Heat Transfer,* New York, McGraw-Hill, 1973.

PROBLEMS

4.1 The accompanying figure depicts a rectangular plate, 6 cm × 6 cm, with the edges maintained at the temperature shown. For a uniform network spacing $\delta x = \delta y = 2$ cm, find, numerically, the steady-state temperatures at the four interior nodal points if the temperature distribution is assumed to be two-dimensional.

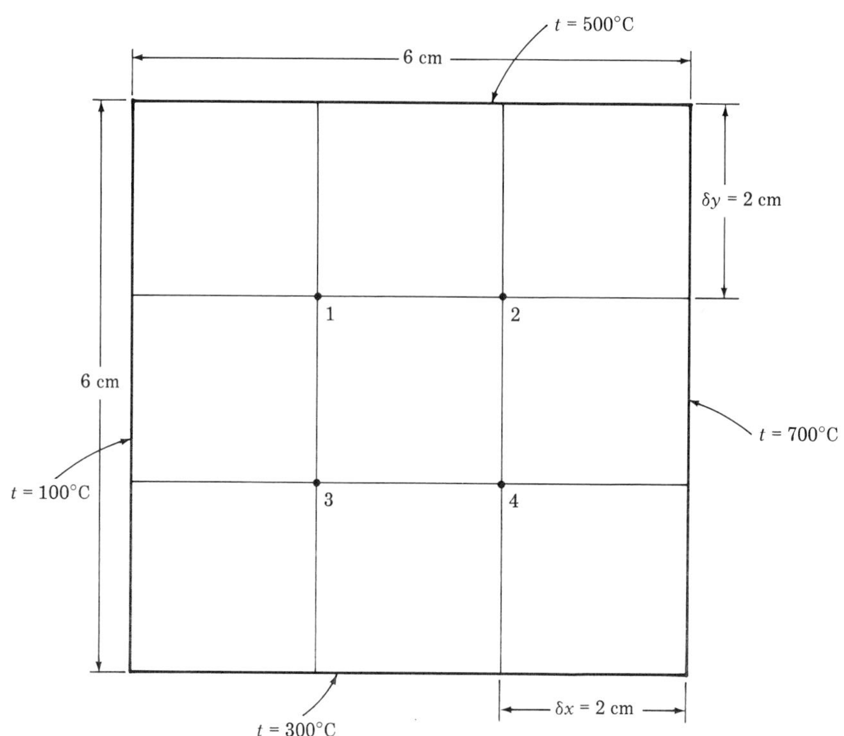

Problem 4.1

4.2 Consider two-dimensional steady-state conduction in the 6 cm × 12 cm rectangular plate shown in the accompanying figure. The thermal conductivity of the plate material is 60 W/m-°C. The left surface of the plate is maintained at 50°C and the right surface at 200°C. The lower surface is insulated and the upper surface is exposed to an ambient fluid at 0°C through a heat transfer coefficient of $h = 40$ W/m^2-°C. Find, numerically, the temperatures at the nine nodal points shown if a square network, $\delta x = \delta y = 3$ cm, is used. Assume that the temperatures of the corners are equal to those specified on the vertical sides.

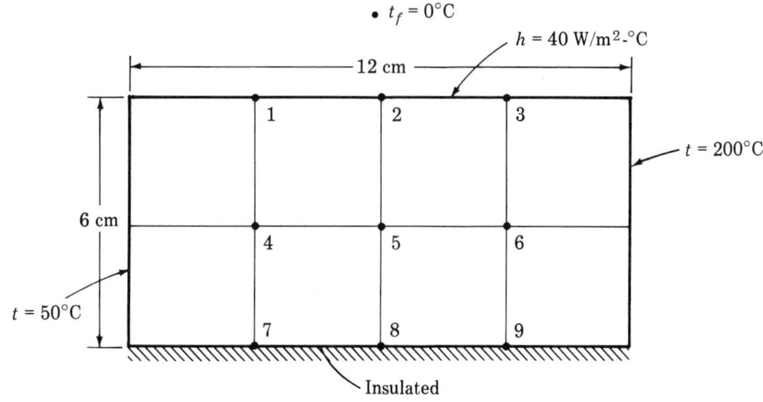

Problem 4.2

4.3 The accompanying figure depicts a straight fin of triangular profile, 8 cm long and 1.6 cm thick at the base. The base of the fin is maintained at 150°C and the lateral fin surfaces are exposed to an ambient fluid at 50°C through a surface heat transfer coefficient of 100 W/m^2-°C. The fin material is stainless steel with $k = 25$ W/m-°C. Presume that the fin width, into the paper, is very large and that the temperature distribution within the fin may be taken as one-dimensional—depending only on the length coordinate x. Using a nodal spacing of $\delta x = 1$ cm, find the steady-state temperature distribution in the fin. Also determine the heat flow from the fin into the fluid, per unit width, and compare the result with that obtained using the fin efficiency introduced in Chapter 2.

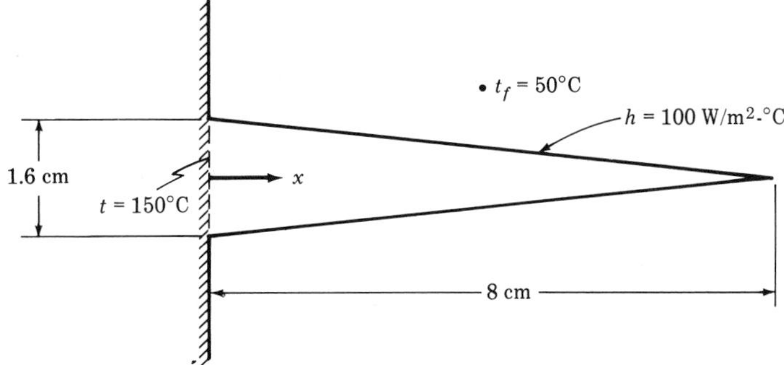

Problem 4.3

4.4 A 0.6-cm-diameter pure copper rod is 48 cm long and is heated to 90°C at *each* end. The rod surface is exposed to an ambient fluid at 25°C through a surface heat transfer coefficient of 25 W/m^2-°C. By numerical means, find the temperature distribution in the rod using a nodal spacing of 4 cm. How much heat is dissipated by the rod? Compare the results with an analytic solution.

4.5 A straight fin of uniform thickness is attached to a surface at 150°C. The fin is 5 cm thick and 15 cm long and is composed of 20% nickel steel. The surface and the end of the fin are exposed to an ambient fluid at 10°C through a heat transfer coefficient of 30 W/m^2-°C. Find, numerically, the temperature distribution in the fin, recognizing that the thickness length ratio is too large to treat the fin as one-dimensional.

4.6 The accompanying figure depicts a section of a chimney made of common brick. The inside surface is maintained at 350°F while the outside surface is maintained at 100°F. Assuming that the chimney is quite tall so that the heat conduction through it may be considered as two-dimensional, find by numerical means the temperature distribution in the chimney. Use a network spacing of 4 in.

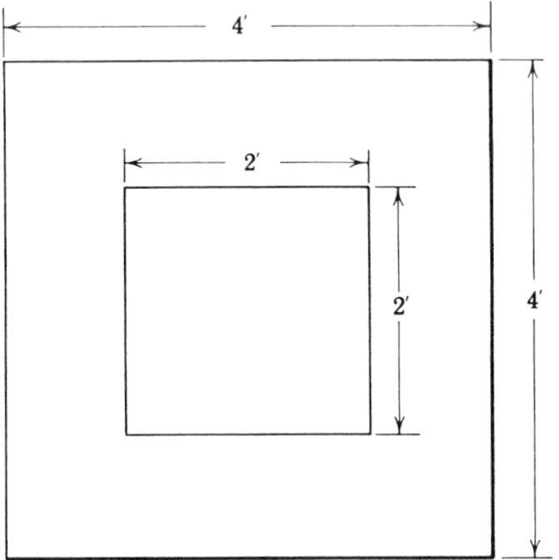

Problem 4.6

4.7 Repeat Prob 4.6 if a gas at 400°F flows inside the chimney with a heat transfer coefficient of 10 Btu/h-ft^2-°F at the surface. Air at 70°F surrounds the outside of the chimney and the heat transfer coefficient there is 2 Btu/h-ft^2-°F. Use a 6-in. net spacing.

4.8 The accompanying figure depicts a concrete block, 30 cm square, with a 15-cm-OD steam pipe buried concentrically inside. The steam pipe is maintained at 150°C. The exterior surfaces of the concrete are maintained at 10°C. Find, numerically, the temperature distribution in the concrete. Presume that the block is long enough to treat the heat flow as two-dimensional.

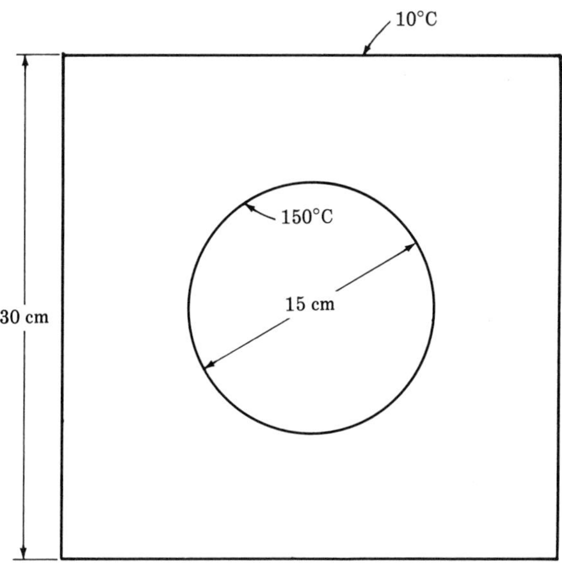

Problem 4.8

4.9 The accompanying figure depicts the cross section of a very long Duralumin I beam. All surfaces, other than the upper and lower faces, are insulated. Determine, numerically, the steady-state rate of heat flow (per foot of beam length) which results when the upper surface is maintained at 40°C and the lower surface at 15°C.

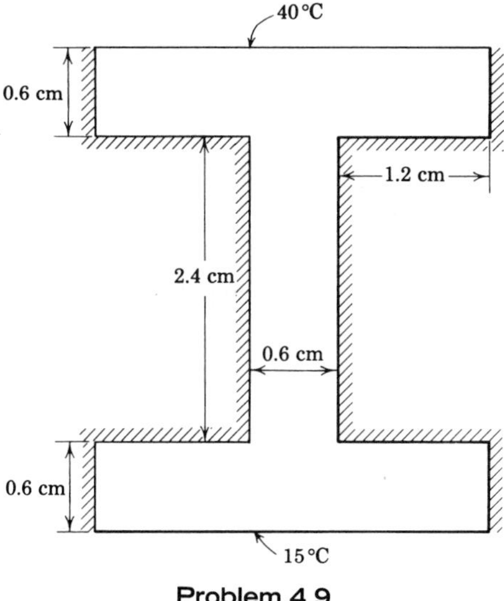

Problem 4.9

4.10 Find, numerically, the steady-state two-dimensional temperature distribution in the geometry shown in the accompanying figure for the boundary conditions as shown. The thermal conductivity of the material is 60 Btu/h-ft-°F.

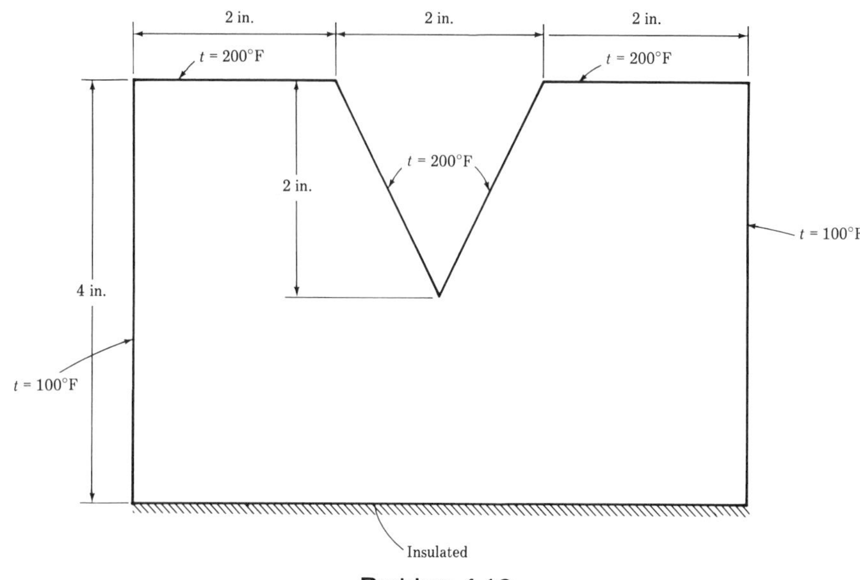

Problem 4.10

4.11 Imagine that the chimney in Prob. 4.6 is initially at 350°F throughout and that the outside surface temperature is subsequently lowered suddenly to 100°F. Determine, numerically, the temperature–time history in the chimney for a time span of 10 h.

4.12 A large plate of pure aluminum is 4 cm thick. It is in contact with convecting fluids on each side. On one side the heat transfer coefficient is 30 W/m^2-°C, and on the other side it is 70 W/m^2-°C. Initially both fluids are at 20°C, and the plate is in equilibrium. Suddenly the fluid temperatures are raised to 90°C. Determine, numerically, how long it takes the center of the plate to reach 80°C.

4.13 For the network chosen in Prob. 4.9, determine the maximum time increment that could be used to numerically analyze the transient response of the I beam to a sudden change in one of the surface temperatures.

4.14 The rod described in Example 4.1 is initially at equilibrium with the ambient fluid at 20°C, the base of the rod being maintained at that temperature. For time greater than zero, the base temperature of the rod is raised, linearly, to 120°C over a time span of 20 min. Find the temperature history for the rod for a total time span of 1 h. Use $\rho = 7840$ kg/m^3, $c_p = 0.46$ kJ/kg-°C.

4.15 A large slab of 1% carbon steel is 1.5 m thick. It is initially at 425°C. Its two surfaces are suddenly exposed to a fluid at 15°C through a heat transfer coefficient of 25 W/m^2-°C. Compute the surface temperature and the center temperature of the slab, 5 h later, using a time increment equal to **(a)** the maximum allowable, and **(b)** two-thirds of the maximum allowable.

4.16 For the geometry specified in Prob. 4.2, presume that the plate is initially uniformly at 200°C before the given boundary conditions are imposed. Determine the maximum allowable time increment for the numerical, explicit calculation of the temperature response within the configuration. The material has a density of 7800 kg/m^3 and a specific heat of 0.45 kJ/kg-°C.

4.17 A steel ($k = 50$ W/m-°C, $\rho = 7800$ kg/m^3, $c_p = 0.47$ kJ/kg-°C) rod 0.3 cm in diameter and 10 cm long is initially at a uniform temperature of 200°C. It is suddenly immersed in a fluid at 40°C, $h = 50$ W/m^2-°C, while one end is maintained at 200°C. Find the temperature distribution in the rod after 100 s has elapsed.

4.18 The accompanying figure depicts a two-dimensional body (1 cm × 2 cm). Initially the material is uniformly at 300°C. The bottom surface is insulated and, suddenly, the top surface is exposed to a fluid at 50°C through a heat transfer coefficient of 200 W/m^2-°C while the left and right surfaces (including the corners) are kept at 300°C. The material involved has $k = 3.0$ W/m-°C, $\rho = 1600$ kg/m^3, $c_p = 0.8$ kJ/kg-°C. Using a square nodal network with $\delta x = \delta y = 0.5$ cm, find the temperatures at the nine nodal points shown, 20 s later.

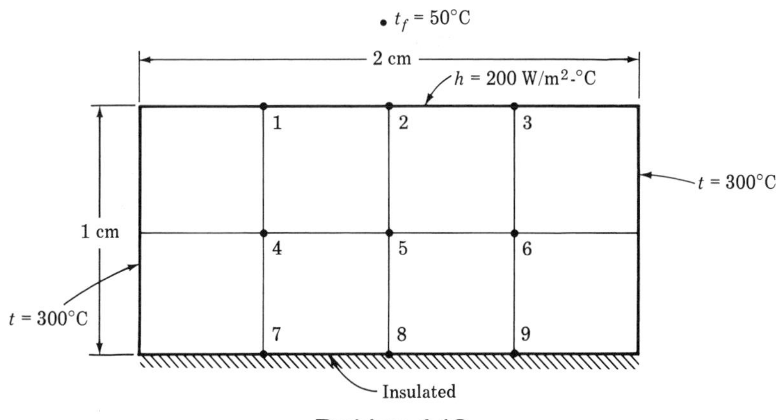

Problem 4.18

4.19 Derive relations analogous to Eqs. (4.34) through (4.40) for the applicable boundary conditions at an *internal* corner.

4.20 Derive the facts given in Eqs. (4.65), (4.66), and (4.70).

CHAPTER 5

The Fluid Mechanical Aspects of Forced Convection and Some Example Solutions

5.1 INTRODUCTORY REMARKS

In the foregoing chapters, primary attention has been directed to the problems of heat conduction in solids. The mode of heat transfer known as convection has been considered only as one type of boundary condition to be applied at the surface of a conducting solid. This boundary condition has been treated in terms of an overall parameter, the surface heat transfer coefficient, h. This coefficient was introduced through Newton's law of cooling [Eq. (1.4)], which relates the heat flux at the surface of a solid body subjected to convection to an ambient fluid by the linear rate equation

$$\frac{q}{A} = h(t_s - t_f), \tag{5.1}$$

where t_s and t_f are the temperatures of the body surface and the fluid, respectively. In these applications h has been presumed known, and the analysis concentrated on the determination of the temperatures in the solid—including the surface temperature t_s. The purpose of the next few chapters is to focus attention on the process of heat convection in a fluid and to develop methods of predicting the value of the heat transfer coefficient that will result under a given set of conditions.

Fluid motion is the distinguishing feature of heat transfer by convection. Although the process of conduction by molecular exchange is still present in a moving fluid, the transport of energy is profoundly influenced by the fluid motion. High-energy (or hot) portions of the fluid are brought into contact with lower-energy (or cooler) portions by virtue of macroscopic displacements in the fluid. When the fluid motion past a surface is caused by the imposition of pressure differences the transfer of heat between the surface and the fluid (if they differ in temperature) is termed *forced convection*. Forced convection may occur internally, as in flow through a pipe, or externally over a submerged body. Even when no external forces are present, fluid motion may occur about a body immersed in a fluid of differing temperature as the result of density differences produced by local heating of the fluid. In this instance the process of heat transfer is termed *free convection*. In complex situations, both forced and free convection may occur simultaneously.

This chapter, and the next, concentrate attention on problems of forced convection, while Chapter 7 examines free and mixed convection.

5.2 FLUID VISCOSITY

In either forced or free convection, an analytical determination of the surface heat transfer coefficient, h, requires determination of the heat flux into the fluid. This heat flux is established by the gradient of the fluid temperature evaluated at the surface, in a direction normal to it, presuming that the fluid thermal conductivity is known.

Thus the determination of h implies the need to find the temperature distribution in a fluid moving past a surface of differing temperature in order that the temperature gradient at the surface can be found. This fluid temperature distribution is found, as will be demonstrated later in this chapter, by applying the energy conservation principle to the fluid, and this, in turn, requires knowing how the fluid velocity and pressure vary near the surface. Consequently, analytic predictions of heat transfer coefficients involve extensive use of the analysis of viscous fluid motion. This chapter presents some of the basic concepts of viscous fluid mechanics and illustrates their application to the solution of forced convection problems. The principal emphasis is placed on the methodology involved. Chapter 6 then presents a collection of useful results applicable to a number of situations having practical engineering interest.

Clearly, then, the fluid property *viscosity* plays an important role in the study of convection, and this property is discussed in some detail next. Also of importance is whether the fluid flow is laminar or turbulent or whether it occurs externally to a surface (as, for instance, airflow over an airfoil or past the exterior of a pipe) or internally (as for flow of a fluid in a tube). These latter influences are treated subsequently.

Dynamic Viscosity

When, as in the processes of convection, a fluid is caused to move past a solid body, velocity gradients are set up within the fluid because of the relative motion of various parts of the fluid. This relative motion is, dominantly, a sliding motion of adjacent fluid elements, producing a shearing resistance. This shearing resistance, combined with the fluid inertia and other applied external forces (pressure, gravity, etc.) determines the movement of the various portions of the fluid. Thus, in order to analyze fluid motion for the purpose of studying convection it is necessary to be able to relate the shearing resistance to the velocity field in the fluid. This is done through the fluid property known as *viscosity*.

The definition of viscosity may be illustrated by considering "parallel," laminar flow such as that past a plane wall as depicted in Fig. 5.1. Some distribution of the fluid velocity, v, exists as illustrated, varying from zero at the wall to some value far away from the wall. Here only laminar motion is being considered—characterized by streamlines that run in a well-ordered manner, with adjacent layers of fluid sliding relative to one another and without any motion taking place normal to the main flow.

Consider a plane, S, in Fig. 5.1, some distance from the surface and parallel to it. The fluid layers on either side of S experience a shearing force, or viscous drag, due to their relative motion. This shearing force arises from the transport of momentum across S due to an exchange of molecules between the fluid layers on either side. The momentum is directed parallel to the main stream of the flow, but it is transported normal to this direction, y in Fig. 5.1, due to the molecular scale motion that results from the thermal agitation of the molecules. The momentum exchange is equivalent to a force, or shear stress, and Newton postulated that the shear stress, τ is directly proportional to the velocity gradient measured normal to S:

$$\tau = \mu\left(\frac{\partial v}{\partial y}\right). \tag{5.2}$$

The proportionality constant, μ, is termed the *coefficient of dynamic viscosity*. This concept of viscosity is based on a molecular scale exchange of momentum. Thus it applies only to laminar flow, and any transport of momentum on a scale larger than the molecular, such as might occur if turbulence were present, is excluded.

The defining relation for dynamic viscosity shows that its dimensions are (force) $\times$ (time)/(length)^2. However, by application of Newton's second law, the dimensions may also be expressed as (mass)/(length) $\times$ (time). Units commonly used are:

SI: N-s/m^2 = kg/m-s.
English: lb$_f$-s/in.^{2}, lb$_f$-h/ft^2, lb$_m$/ft-s, lb$_m$/ft-h.

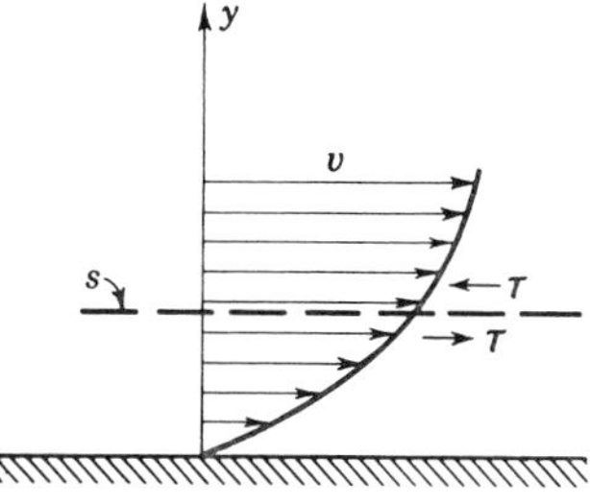

Figure 5.1

As is apparent, there are a great number of viscosity units in use and one must exercise care to select the proper ones for the particular formulation at hand. Appendix D gives conversion factors between the various units involved.

Kinematic Viscosity

Since the forces of viscosity act directly on a fluid, and since the fluid inertia resists these forces, the ratio of the viscous force to the inertia force would be expected to be an important parameter in the analysis of fluid motion. Thus the *kinematic viscosity,* defined as the ratio of the dynamic viscosity to the fluid density (or specific weight), becomes an important fluid property. The kinematic viscosity is denoted by $\boldsymbol{\nu}$:

$$\boldsymbol{\nu} = \frac{\mu}{\rho}. \tag{5.3}$$

Thus the dimensions of kinematic viscosity are $(\text{length})^2/(\text{time})$ with the usual units being:

SI: m^2/s.
English: ft^2/s, ft^2/h

The Viscosity of Liquids and Gases

The molecular theory of the viscosity of gases and liquids predicts that the dynamic viscosity depends primarily on temperature and, to a lesser degree, on pressure. These facts are borne out by experimental observation.

In liquids, as in the case of thermal conductivity and specific heat, the pressure dependence of the dynamic viscosity is found to be quite slight unless extreme pressures are reached. Usually, then, this effect may be ignored and only the temperature dependence need be considered. Likewise, since liquid densities are not very pressure sensitive, the kinematic viscosity may usually be taken as only temperature dependent. Tables A.3, A.4, and A.9 of the Appendix give data for the dynamic and kinematic viscosity of several liquids.

For gases, the picture is somewhat different than that noted above for liquids. Generally speaking, the dynamic viscosity of gases is quite a bit smaller than that for liquids, although the kinematic viscosity may be greater because of lower densities. In most gases the temperature is, again, the most significant factor influencing the dynamic viscosity. The pressure effect may be quite small, as long as the saturation state or critical state is not approached.

Steam, as usual, is an exceptional gas and the dependence of μ on both pressure and temperature may be significant as shown in Table A.5, in both SI and English units. Table A.6 gives values of μ for air at atmospheric pressure (in both SI and English units) and Table A.7 gives μ at atmospheric pressure for several other gases.

The kinematic viscosity of gases is most certainly pressure dependent as well as temperature dependent, owing to the pressure dependence of the density. Thus one normally needs to calculate ν after finding μ and ρ separately. The values of ν given for air in Table A.6 apply only to atmospheric pressure and would need to be recalculated at other pressures after appropriately determining ρ. Likewise, values for the kinematic viscosity of the gases in Table A.7 can be found from the tabulated values of μ after finding ρ from an appropriate equation of state or table of thermodynamic properties.

5.3 SOME BASIC CONCEPTS AND DEFINITIONS

The general discussion of Sec. 5.1 emphasized the importance of viscous hydrodynamics in the study of convective heat transfer. The basic character of viscous fluid motion is profoundly influenced by whether or not the motion is laminar or turbulent and whether the flow takes place over the exterior of an immersed body or through the interior of some passage.

External and Internal Flows

The term "external flow" is used to describe the hydrodynamic situation in which an otherwise unbounded fluid flows past a solid body or surface (or in which a body moves through a stationary unbounded fluid, producing the same relative motion). Figure 5.2 illustrates some external flow situations for both free and forced convection.

In Fig. 5.2(a) a fluid with a known "freestream" velocity, U, and temperature t_f flows past a flat surface maintained at some other temperature, t_s—assumed greater than t_f in this example. At some coordinate location along the surface, x, the fluid velocity parallel to the surface, v_x, varies somewhat as depicted—from zero at the surface (i.e., at $y = 0$, y being the coordinate normal to x) to values approaching the freestream velocity as the distance away from the surface increases. Similarly, the fluid temperature varies from t_s at the surface to t_f, far removed. These variations in velocity and temperature determine the rate of heat flow from the surface into the fluid.

Figure 5.2(b) depicts a somewhat different external flow in which the surface is curved. Here the fluid velocity and temperature vary much like that on the flat surface as long as the coordinate x is taken in the local direction of the surface and y normal to it. The value of the velocity reached at large distances from the surface, $U(x)$, is in this case, variable along the surface as a result of the surface curvature. Figure 5.2(c) suggests the situation that might occur for free convection past a vertical flat surface placed in a fluid of a cooler temperature and with zero velocity far away. Again the fluid temperature varies with the normal distance from the surface much as before. However, the fluid velocity, which results from the buoyant forces produced by local heating, must rise from zero at the surface to some maximum value and then again approach zero as the distance from the surface increases.

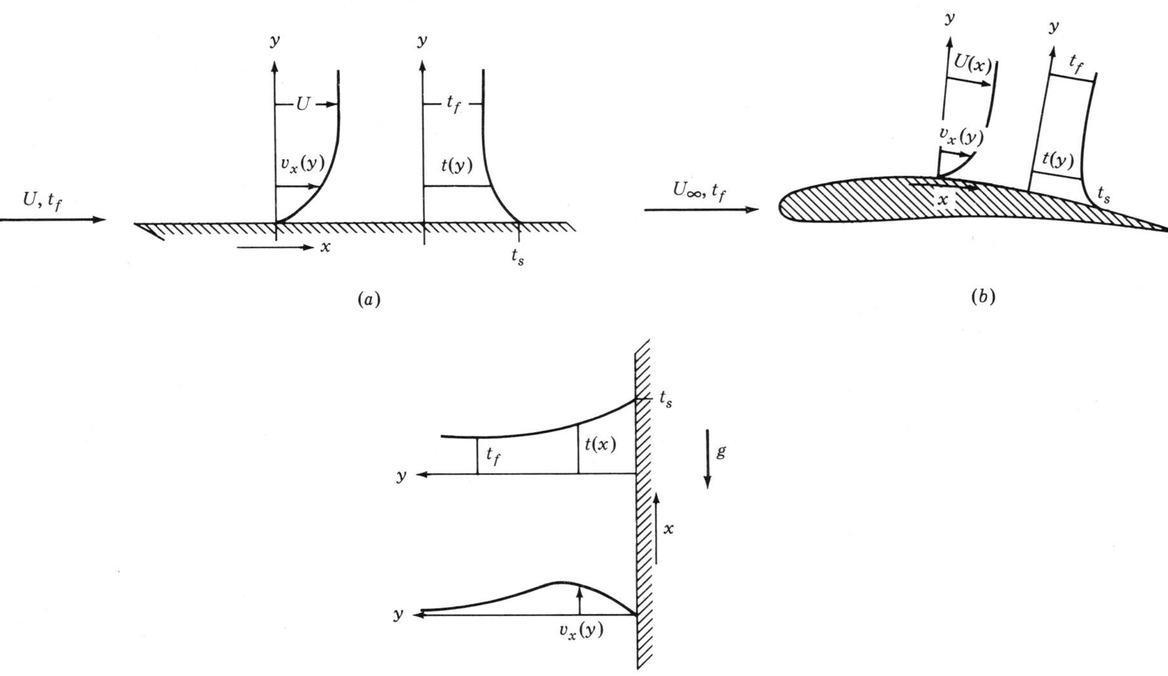

Figure 5.2. External flows: (a) and (b), forced convection; (c) free convection.

In all of these instances of external flow the fluid temperature far removed from the surface, t_f, provides a convenient basis on which to base the definition of the heat tranfer coefficient, h. Accordingly, in such cases h is defined so that the heat flux at the surface ($y = 0$), denoted by $(q/A)_0$, is, from Eq. (5.1),

$$\left(\frac{q}{A}\right)_0 = h(t_s - t_f),$$
$$h = \frac{(q/A)_0}{t_s - t_f}. \tag{5.4}$$

Application of Fourier's law within the fluid gives

$$\left(\frac{q}{A}\right)_0 = -k_f\left(\frac{dt}{dy}\right)_0$$

or (5.5)

$$h = -\frac{k_f(dt/dy)_0}{t_s - t_f}.$$

In Eq. (5.5) k_f represents the thermal conductivity of the fluid and $(dt/dy)_0$ is the fluid temperature gradient evaluated at the surface. The problem of determining the heat transfer coefficient then becomes that of finding $(dt/dy)_0$ for a given hydrodynamic condition of flow past the surface in question. That is, one seeks the distribution of the fluid temperature, as pictured in the sketches of Fig. 5.2, so that the surface gradient can be evaluated.

A distinguishing feature of external flows, just discussed, is the unbounded extent of the fluid domain surrounding the heat-transferring solid surface. Another important hydrodynamic situation that is frequently encountered in convective heat transfer is that of "internal" flow. An internal flow is bounded on all sides by solid surfaces except, possibly, for an inlet and exit. Figure 5.3(a) depicts the most commonly encountered internal flow—that in a pipe or duct. This is an example of forced convection in an internal flow and differs from an external flow in that there is not necessarily a "freestream" velocity and temperature to characterize the hydrodynamic state of the flow or to which the heat transfer coefficient may be referenced. The fluid velocity and temperature vary continuously from their values at one wall (zero in the case of the velocity) to some value at the centerline and then return to their wall values as the opposite surface is approached. The absence of a freestream state necessitates the definition of some other reference fluid velocity and temperature. Internal flows are, then, usually characterized in terms of their mean values, U_m and t_b, obtained by integrating the fluid velocity and temperature over the cross section of the passage, normal to the flow. For constant density flow, the mean velocity is

$$U_m = \frac{\int_A \rho v \, dA}{\int_A \rho \, dA} = \frac{1}{A}\int_A v \, dA.$$

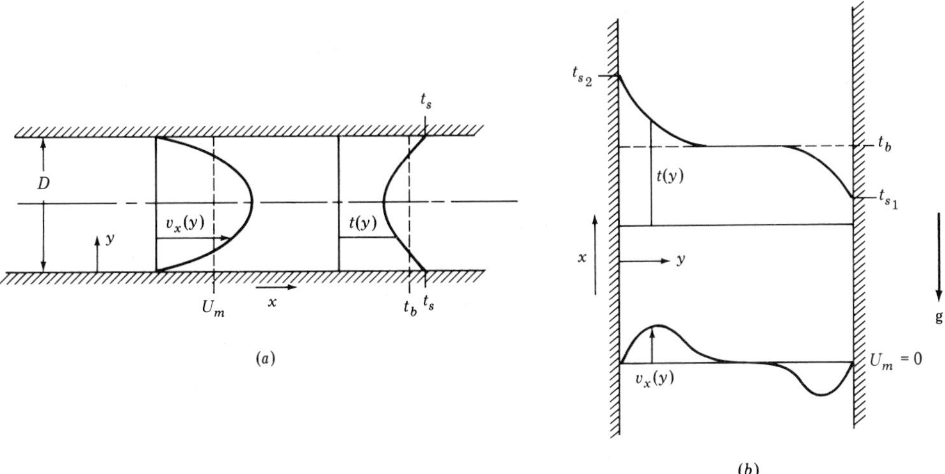

Figure 5.3. Internal flows: (a) forced convection; (b) free convection.

The mean temperature has to be based on a mean energy, which for constant density and specific heat is

$$t_b = \frac{\int_A \rho c_p v t \, dA}{\int_A \rho c_p v \, dA} = \frac{1}{AU_m} \int_A vt \, dA.$$

The exact form of the integrals above depends on the variation of v and t across the passage and the geometric shape of the cross section. These will be evaluated for specific cases later, as appropriate, but the definitions above will suffice for present purposes. (The mean value of the temperature is denoted by t_b since the term "bulk temperature" is commonly applied to this quantity.) The determination of these mean values is an important aspect of the study of forced convection internal flows since U_m will be used to characterize the flow state (i.e., laminar, turbulent, etc.) and t_b is used to define the surface heat transfer coefficient. That is, h is defined in terms of the imposed temperature potential $(t_s - t_b)$ so that the wall heat flux is given by

$$\left(\frac{q}{A}\right)_0 = h(t_s - t_b),$$
$$h = \frac{(q/A)_0}{t_s - t_b}. \tag{5.6}$$

Internal flow may also occur in free convection. Figure 5.3(b) suggests such an instance in which free convection exists between two vertical plates maintained at different temperatures, both of which differ from the bulk, mean, fluid temperature. Rather complex variations of both the temperature and velocity fields may result in such a case as the figure illustrates.

Much of the discussion in the next several sections will emphasize forced convection external flow situations in order to develop the necessary concepts and equations for the analysis of convection. Later in the chapter particular attention will then be directed to forced internal flows. Chapter 7 will be devoted entirely to free convection.

The Boundary Layer Concept—Velocity and Thermal

The foregoing discussion pointed out the importance of viscous hydrodynamics in the study of convection heat transfer. In order to find the heat transfer coefficient one must find the velocity and temperature fields in the fluid as it moves past the heat transferring surface. Exact solutions for the velocity and temperature fields would require the detailed solution of the governing equations of viscous hydrodynamics. These equations consist of a momentum expression which gives a balance between the various forces involved (inertia, pressure, viscous stresses, body forces, etc.) and an energy conservation expression which gives a balance between the energies involved (conducted heat, viscous and pressure work, etc.). There is no need to write these equations here since exact solutions to them are known for only a few cases, most of which have very specialized and impractical boundary conditions. The complicated nature of these equations does not, however, eliminate the *need* for answers to problems of practical importance. For this reason the engineer must be content with the compromise of accepting an approximate solution.

One such approximation could be the simplification which results from neglecting entirely the effects of viscosity. The resulting simplified equations prove to be very valuable in certain fluid mechanical applications—the lifting theory of wing sections, for example. However, such an inviscid approximation is useless for convective heat transfer analyses since it is the detailed viscous interaction near the surface that determines the heat flux there. A compromise between using the full, exact, viscous equations or neglecting viscosity entirely is found in the *boundary layer* concept developed by Ludwig Prandtl in 1904. The concept of the boundary layer has already been discussed in a limited way in Sec. 1.6.

The Velocity Boundary Layer. The Prandtl representation for the boundary layer can be explained best by reference to the case of external flow past a flat surface as depicted in Fig. 5.4. Here is shown an undisturbed stream with a uniform velocity U and uniform temperature t_f flowing past a flat surface. As noted earlier the distribution of the fluid velocity and temperature at some location x from the leading edge of the plate will display variations, as functions of distance away from the surface, much like those shown in the figure.

Considering the velocity variation for the moment, it will be zero at the plate surface, $y = 0$, and rise with increasing y, the normal coordinate, eventually approaching the undisturbed freestream velocity U. The exact nature of the velocity variation depends on a number of parameters, most notably the fluid viscosity. If the fluid has no viscosity, the flow past the surface has a uniform velocity profile

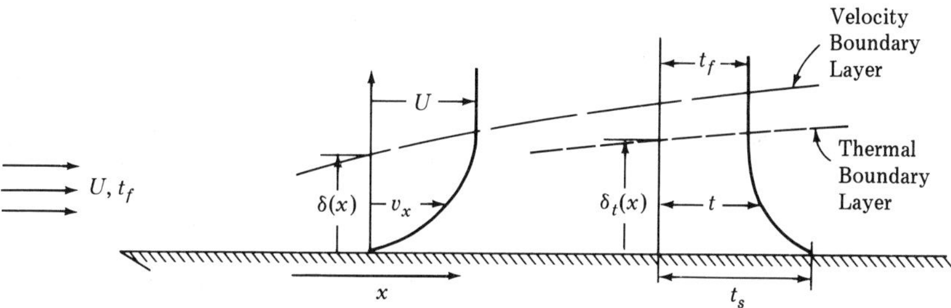

Figure 5.4. The velocity and thermal boundary layers on a flat surface.

equal to U. The fluid viscosity, however, retards the fluid near the surface producing the velocity profile shape shown. (Should the surface be curved rather than flat, even the inviscid velocity would change due to the surface curvature; however, this effect is ignored in the present discussion.) The more viscous the fluid, the thicker will be the region in which a noticeable velocity variation with y exists. Even in fluids of small, but finite viscosity there must always be a region in which the velocity will change from U to zero—perhaps quite thin. In such a region one *must* account for the viscous shear stress due to viscosity.

Prandtl's boundary layer model represents the true flow picture as approximated by two regions. One region, termed the *boundary layer,* is that near the surface in which velocity gradients are large enough so that the influence of viscous shear [as given by the viscosity law of Eq. (5.2)] can not be neglected. The other region, termed the *inviscid region* (or *inviscid core* if internal flow is considered), is defined as the region in which the influence of the presence of the surface has died out enough so that the velocity gradients are sufficiently small to ignore the effects of viscosity. In actual flows the influence of the surface extends to all regions of the fluid, but this influence and the associated changes in velocity decrease rapidly with distance from the surface—particularly if the viscosity is small and the freestream velocity is large (i.e., at a large value of the Reynolds number, a concept to be discussed shortly). Since this decrease is continuous, it is obviously not possible to define a precise location of the limit of the velocity boundary layer. In practice this boundary layer limit, called the *velocity boundary layer thickness* and denoted by δ in Fig. 5.4, is taken to be that distance away from the surface at which the flow velocity has achieved some arbitrary percentage of the freestream velocity—say 90 or 99%. As the flow proceeds along the surface the retarding influence of the viscous action propagates farther into the fluid stream. Hence the velocity boundary layer thickness is a function of the coordinate measured parallel to the surface, $\delta(x)$.

The Thermal Boundary Layer. By reasoning quite analogous to that just presented for the velocity boundary layer, one may also define a *thermal boundary layer* with respect to the temperature variation in the fluid near the surface. Figure 5.4 also illustrates a typical temperature distribution when a fluid of given freestream temperature, t_f, a surface maintained at a different temperature, t_s. In the case illustrated the surface is assumed to be hotter than the fluid, or $t_s > t_f$. Rather than being related to the exchange of momentum and shear forces as in the case of the velocity boundary layer, the thermal boundary layer results from the exchange of energy between the surface and the inviscid core. Energy is transported in the fluid by movement of the fluid and by heat conduction resulting from temperature gradients.

The fluid temperature must change from the surface value, t_s, to the freestream value, t_f, and the thermal boundary layer is taken to be that region (presumably close to the surface) in which the temperature gradient is large enough that conducted heat must be accounted for. Outside the thermal boundary layer temperature gradients and the associated conducted heat are taken to be negligible, and the fluid temperature is assumed to be uniform at t_f.

The thermal boundary is denoted by $\delta_t(x)$ and can be seen to grow in the direction of the flow as more and more of the fluid becomes heated by the surface. While similar in character to the velocity boundary layer, the thermal layer will not necessarily have the same thickness as the velocity layer. In fact, the relative sizes of these two boundary layers will prove to be an important consideration in the analysis of convective heat transfer, and the determination of the ratio δ_t/δ will be an important result of such an analysis.

Laminar Versus Turbulent Flow

As mentioned earlier, one of the most important characteristics of viscous flow is whether it is *laminar* or *turbulent*. The nature of the fluid velocity and temperature variations in the boundary layer is profoundly influenced by which of these flow conditions is present.

Steady laminar motion is characterized by streamlines that run in a well-ordered manner with adjacent layers of fluid sliding relative to one another and without any motion taking place normal to the streamlines. All momentum and heat transport occurring normal to the flow take place due to the *molecular* viscosity and conductivity. Considering flow past a plane surface for example, Fig. 5.5(a) depicts a typical variation in the fluid velocity parallel to the surface. The laminar shear stress experienced by the fluid at some plane S is due solely to molecular level momentum exchange across S as discussed in Sec. 5.2 in connection with the definition of dynamic viscosity.

Under certain conditions, however, a flow may undergo instabilities and change into a flow state known as *turbulent*. In such instances close examination of the flow will reveal that what appears to be steady motion is actually a time-dependent flow oscillating around an *apparent* steady condition. Fluctuations about a mean is observed in the fluid pressure and temperature as well as the velocity; however, for discussion purposes only the variation in velocity is described here. Velocity oscillations are observed to occur in directions which are both parallel and normal to the main flow. Turbulent motion is, then, inherently unsteady in that the velocity

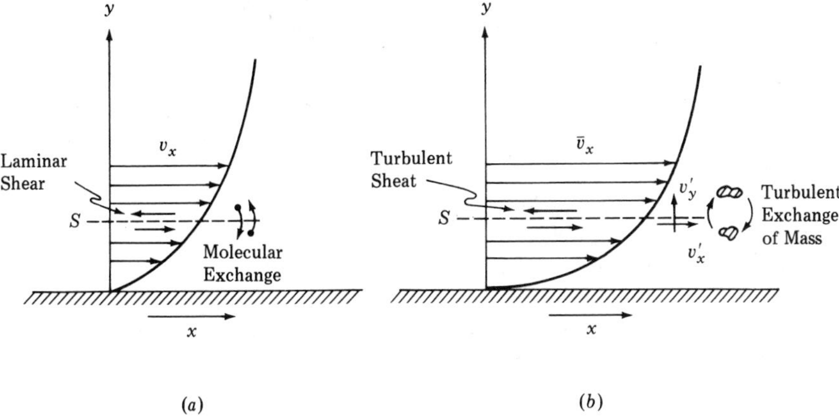

Figure 5.5. Laminar and turbulent flow.

at any point varies in time; however, in "steady" turbulent flow this time variation occurs about a constant mean value.

Considering a two-dimensional motion in the x-y plane, a steady turbulent flow may be represented in the following way.

$$v_x = \bar{v}_x + v_x', \qquad (5.7)$$
$$v_y = \bar{v}_y + v_y',$$

in which v_x and v_y are the local *time-dependent* velocity components, $\bar{v}_x$ and $\bar{v}_y$ are time average values of the components, and v_x' and v_y' are the time-dependent fluctuations about the averages. Figure 5.6 graphically illustrates the meaning of v_x, $\bar{v}_x$, and v_x'. A similar representation applies to the y component of velocity.

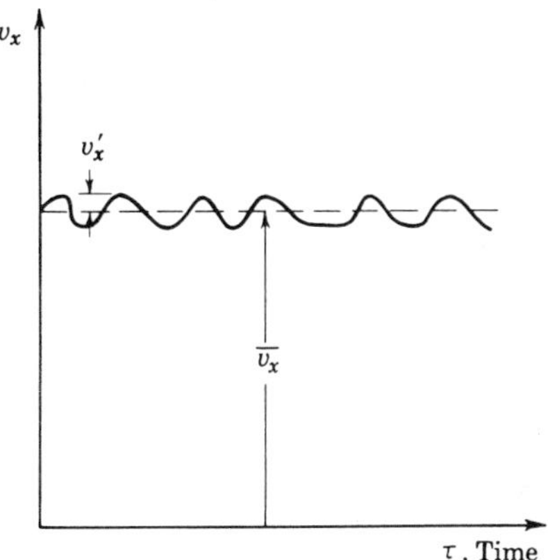

Figure 5.6. Turbulent velocity fluctuations about a time average.

Thus the true velocity at a point is a nonsteady one, and the individual paths of the fluid particles form an interlacing network. This means that a certain amount of mixing occurs on a scale larger than the molecular scale. Returning to the simple case of flow past a plane surface, as depicted in Fig. 5.5(b), the time average velocity in the x direction, $\bar{v}_x$, appears to be steady and parallel to the surface. The time-averaged velocity in the y direction, $\bar{v}_y$ is zero; however, macroscopic mixing occurs in the y direction due to finite fluctuations v'_y even though $\bar{v}_y = 0$. This transverse mixing is a characteristic of turbulent motion that is missing in laminar flow and is of great importance in convective heat transfer since it may significantly alter the rate of heat and momentum exchange in the fluid because of the associated movement of large, macroscopic fluid "lumps." Further details of the analysis of turbulent flow will be postponed until after some experience is gained in understanding laminar flow.

Clearly, then, it is important to know whether a given flow is likely to be in the laminar or turbulent state. This question is answered by examination of the flow parameter known as the *Reynolds number*, discussed extensively in most fluid mechanics texts (Refs. 1 through 3). Important to such considerations is whether one is examining an external or internal flow.

External Flow. Consider, first, the external flow depicted in Fig. 5.7—that of a uniform incident on a flat surface. For such a flow the *local length Reynolds number*, Re_x, is defined as

$$\mathrm{Re}_x = \frac{Ux}{\nu} = \frac{Ux\rho}{\mu}, \tag{5.8}$$

where U is the undisturbed stream velocity, $\nu = \mu/\rho$ is the kinematic viscosity discussed in Sec. 5.2, and x is the local coordinate measured from the leading edge of the surface. Remembering that the dimensions of ν are $(\text{length})^2/(\text{time})$, Re_x is clearly dimensionless, and care should be taken in calculating Re_x to ensure that the units used do, indeed, result in a dimensionless value. As defined, Re_x increases as the flow proceeds along the surface. Experiment shows that the boundary layer

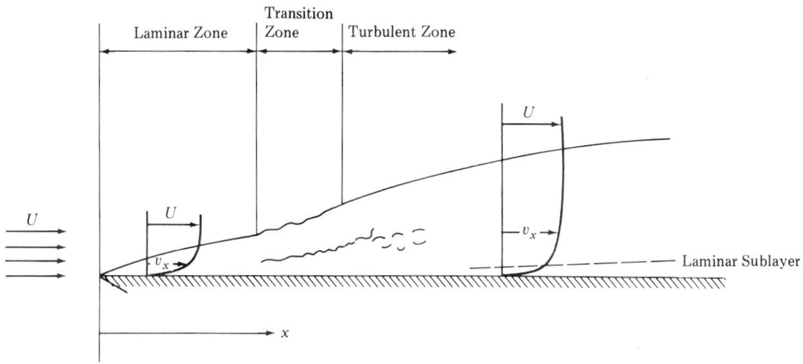

Figure 5.7. Laminar-turbulent transition of a boundary layer on a flat surface.

flow on the surface is laminar for low Re_x and becomes turbulent when Re_x exceeds certain values.

Beginning at the leading edge of the surface where x (and hence Re_x) is small, the laminar viscous shear dominates the inertia forces and a laminar boundary layer grows with an approximately parabolic velocity profile. In this region any transverse disturbances in the flow are damped out. After some distance down the surface (depending on the magnitude of U and ν) the presence of transverse disturbances becomes amplified and a transition zone is encountered in which turbulence begins to develop. Eventually, a region, the turbulent zone, is reached in which the random disturbances become so amplified that the boundary layer may be considered to be fully turbulent steady motion. In the fully turbulent zone the velocity distribution in the boundary layer is observed to be much flatter than in the laminar zone as a result of the enhanced tranverse mixing as depicted in Fig. 5.7. However, near the surface of the turbulent zone the transverse fluctuations must diminish—creating a region that is often characterized as the laminar sublayer.

The transition from the laminar state of flow in the boundary layer to the turbulent state is a complex process which is the subject of extensive investigations. Since it is the result of the amplification of random disturbances one would expect that this transition depends on the freestream fluid velocity, the distance from the leading edge, the roughness of the surface, and the properties of the fluid. For most engineering applications the laminar–turbulent transition may be taken to occur at local length Reynolds numbers between 5×10^5 and 10^6—although under special circumstances laminar layers have been maintained up to value of $Re_x = 3 \times 10^6$, and turbulent layers have been observed for Re_x as low as 80,000. Also, the transition process takes place over a finite distance so that the turbulent condition is not fully established until Re_x reaches some higher value. For simplicity of representation, this text will apply the frequently used criterion that transition will take place at a *critical* Reynolds number of

$$Re_{x,c} \cong 5 \times 10^5. \tag{5.9}$$

However, it must be realized that this criterion is only approximate and that the transition process will occur over a range of Re_x.

Internal Flow. Internal flows undergo a laminar–turbulent transition also. For an internal flow entering, for example, a circular pipe an annular boundary layer forms around the pipe surface, and after a distance (known as the starting length, discussed more fully later) the boundary layer grows together and forms a *fully developed* velocity profile in which the entire pipe is filled with viscous flow. Such a fully developed pipe flow may be laminar or turbulent—again depending on the Reynolds number. In such an instance it is customary to base the Reynolds number on the pipe diameter, D, and the *mean* flow velocity at a cross section, U_m. Thus a *diameter* Reynolds number is defined:

$$Re_D = \frac{U_m D}{\nu} = \frac{U_m D \rho}{\mu}. \tag{5.10}$$

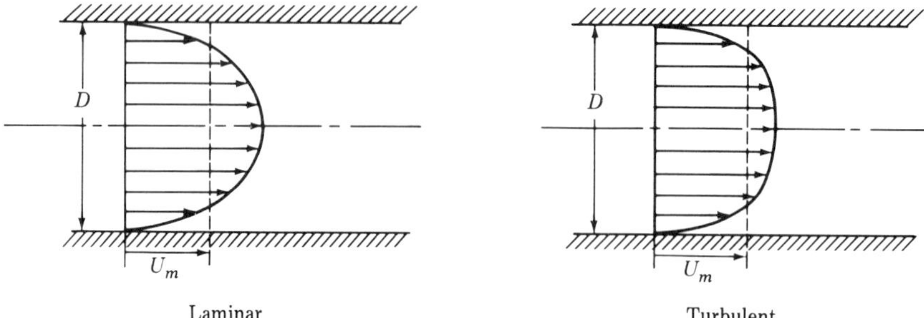

Figure 5.8. Laminar and turbulent fully developed pipe flow.

Two useful alternative forms for Re_D may be found when one introduces either the mass flow rate, $\dot{m}$, or the "mass velocity," the mass flow rate per unit of cross-sectional area:

$$G = \frac{\dot{m}}{A}. \tag{5.11}$$

Since the mass flow rate in a pipe is $\dot{m} = \rho A U_m$ and $\boldsymbol{\nu} = \mu/\rho$, alternative forms for Re_D are:

$$\begin{aligned} \mathrm{Re}_D = \frac{U_m D}{\boldsymbol{\nu}} = \frac{U_m D \rho}{\mu} &= \frac{(\dot{m}/A)D}{\mu}, \\ &= \frac{4\dot{m}}{\pi D \mu} = \frac{GD}{\mu}. \end{aligned} \tag{5.12}$$

Typical fully developed pipe flow velocity distributions are illustrated in Fig. 5.8 for the laminar and turbulent cases. For laminar flow the velocity profile is parabolic while for turbulent flow it is, again, much flatter as a result of radial mixing. As in the case of external flow it is difficult to predict with precision when a fully developed pipe flow will be laminar or turbulent. For engineering calculations it is generally agreed that the transition may be expected to occur for a critical diameter Reynolds number of

$$\mathrm{Re}_{D,c} \cong 2300. \tag{5.13}$$

Again, a range of Reynolds numbers for transition may be expected, depending on the pipe roughness and the smoothness of the entering flow. While the value of $\mathrm{Re}_{D,c} \cong 2300$ is useful for predicting the onset of turbulence, a fully developed turbulent flow may not be reached until Re_D exceeds about 4000.

EXAMPLE 5.1

Atmospheric pressure air at 100°F flows past a flat surface with a velocity of 100 ft/s.

(a) Find the local length Reynolds number and whether the flow is laminar or turbulent at distances 0.5 ft and 1.5 ft from the leading edge of the surface.
(b) Estimate the location from the leading edge at which the laminar–turbulent transition takes place.
(c) Repeat part (b) if the air pressure is 25 psia, all other data remaining the same.

Solution. The local length Reynolds number is, from Eq. (5.8),

$$\mathrm{Re}_x = \frac{Ux}{\nu}.$$

In this instance the flow velocity is $U = 100$ ft/s. For atmospheric pressure air, Table A.6 gives the kinematic viscosity at 100°F to be $\nu = 0.6489$ ft²/h.

(a) For the local distance $x = 0.5$ ft, one has

$$\mathrm{Re}_x = \frac{Ux}{\nu} = \frac{100 \times 3600 \times 0.5}{0.6489} = 2.774 \times 10^5.$$

This value is less than $\mathrm{Re}_{x,c} = 5 \times 10^5$; hence the flow is probably laminar. Note that since U was given in ft/s and ν in ft²/h, the factor 3600 s/h must be included to ensure that Re_x is dimensionless. At a distance $x = 1.5$ ft, one has

$$\mathrm{Re}_x = \frac{Ux}{\nu} = \frac{100 \times 3600 \times 1.5}{0.6489} = 8.322 \times 10^5,$$

turbulent flow.

(b) Since the critical length Reynolds number is, by Eq. (5.9), $\mathrm{Re}_{x,c} = 5 \times 10^5$, the critical distance at which transition is likely to occur is found from

$$\mathrm{Re}_{x,c} = \frac{Ux_c}{\nu}$$

$$5 \times 10^5 = \frac{100 \times 3600 \times x_c}{0.6489}$$

$$x_c = 0.901 \text{ ft } (0.275 \text{ m}).$$

(c) For an air pressure of 25 psia the value of the kinematic viscosity given in Table A.6, and used above, is not applicable since ν is pressure dependent and Table A.6 is based on 1 atm pressure. However, as noted in Sec. 5.2, the dynamic viscosity, μ, of gases is not pressure sensitive. Thus one may use, from Table A.6, $\mu = 4.599 \times 10^{-2}$ $\mathrm{lb_m}$/ft-h at 100°F. The air density may be accurately estimated from the ideal gas law:

$$\rho = \frac{p}{RT} = \frac{25 \times 144}{(1545/28.97)(100 + 460)} = 0.1205 \ \mathrm{lb_m/ft^3},$$

so that the kinematic viscosity becomes

$$\nu = \frac{\mu}{\rho} = \frac{4.599 \times 10^{-2}}{0.1205} = 0.3817 \text{ ft}^2/\text{h}.$$

Repeating now the calculations of part (b),

$$\text{Re}_{x,c} = \frac{Ux_c}{\nu}$$

$$5 \times 10^5 = \frac{100 \times 3600 \times x_c}{0.3817}$$

$$x_c = 0.530 \text{ ft } (0.162 \text{ m}).$$

The distance at which transition takes place is shortened, owing to the decrease of the kinematic viscosity as a result of the higher pressure.

EXAMPLE 5.2

A fluid at a temperature of 80°C flows with a mean velocity of 0.1 m/s through a circular pipe with an inside diameter of 5 cm. Find the diameter Reynolds number and whether the flow is laminar or turbulent if the fluid is (a) water; (b) ethylene glycol.

Solution. For pipe flow the diameter Reynolds number is given by Eq. (5.10):

$$\text{Re}_D = \frac{U_m D}{\nu}.$$

In this instance the mean flow velocity is $U_m = 0.1$ m/s and the diameter is $D =$ 5 cm $= 0.05$ m.

(a) For water at 80°C, Table A.3 gives the kinematic viscosity to be $\nu = 0.3653 \times 10^{-6}$ m²/s, so that

$$\text{Re}_D = \frac{U_m D}{\nu} = \frac{0.1 \times 0.05}{0.3653 \times 10^{-6}} = 13{,}690.$$

Since $\text{Re}_D > \text{Re}_{D,c} = 2300$, as noted in Eq. (5.13) the water flow should be turbulent. Again, note that the units employed in the calculation above were such that Re_D is dimensionless.

(b) For ethylene glycol at 80°C, Table A.4 gives $\nu = 2.98 \times 10^{-6}$ m²/s, so that

$$\text{Re}_D = \frac{U_m D}{\nu} = \frac{0.1 \times 0.05}{2.98 \times 10^{-6}} = 1678.$$

Owing to its greater viscosity, the ethylene glycol flow is probably laminar under the same flow conditions as the water.

Local and Average Heat Transfer Coefficients and Their Relation to Local and Average Surface Friction

The discussion so far in this chapter has emphasized the importance of fluid viscosity in the determination of the heat flow between a surface and an adjacent convecting fluid. Thus it should not be surprising to find that surface heat transfer coefficients are closely related to the viscous, "frictional" drag forces exerted by the fluid on the surface. Before illustrating the analytical processes of estimating h, it is useful to examine some parameters related to viscous drag. The presentation here will be limited to external flows for ease of discussion; corresponding concepts applicable to internal flows will be presented later in the chapter.

Consider, then, two-dimensional external flow of a fluid whose undisturbed temperature is t_f past a surface maintained at a uniform temperature t_s as depicted in Fig. 5.9. The flow is presumed to have a unit depth into the plane of the paper, and x is taken as the coordinate measured parallel to the surface, starting at the leading edge of the surface. The surface may be plane, as suggested in Fig. 5.9(a), or, as in Fig. 5.9(b), the surface may be curved and x measured along the curvature of the surface.

At some location along the surface velocity and thermal boundary layers have been established as shown. Associated with the velocity profile is a shear stress exerted on the surface, τ_0, given by

$$\tau_0 = \mu\left(\frac{\partial v_x}{\partial y}\right)_0. \tag{5.14}$$

Here $(\partial v_x/\partial y)_0$ represents the flow velocity gradient measured at the surface. De-

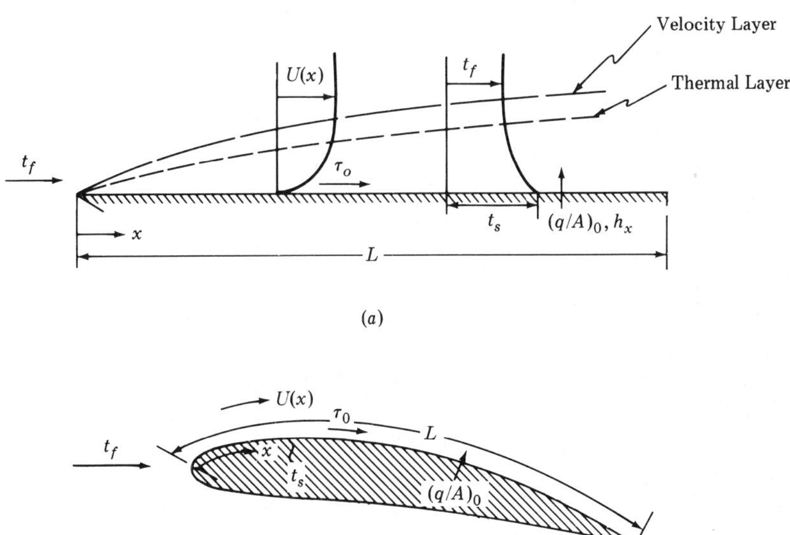

Figure 5.9. Local and total heat transfer and fluid friction.

termination of τ_0 requires the finding of this gradient. In the above, the laminar shear law has been assumed to hold, for the purposes of discussion; should the flow be turbulent an appropriate modification of the exchange coefficient μ will have to be made as illustrated later. Knowledge of the surface shear stress, τ_0, is important to the engineer since it allows one to estimate the fluid drag force on the surface—a quantity related to the energy required to maintain the flow past the surface. In addition, as later analyses will show, τ_0 is directly related to the heat transfer coefficient at the surface. It is important to note at this point that the surface shear, τ_0, is a *local* quantity that is dependent on the location x as measured from the leading edge, and knowledge of the dependence of τ_0 on x will be necessary.

Also at the location x, one may associate with the thermal boundary layer a *local* value of the wall heat flux or the corresponding heat transfer coefficient:

$$\left(\frac{q}{A}\right)_0 = -k\left(\frac{\partial t}{\partial y}\right)_0, \qquad h_x = \frac{(q/A)_0}{t_s - t_f} = \frac{-k(\partial t/\partial y)_0}{t_s - t_f}. \tag{5.15}$$

Again, for illustrative purposes, a laminar law has been used to relate the heat flux to the wall temperature gradient, $(\partial t/\partial y)_0$. As in the case of the wall shear stress, one should note that the surface heat flux, $(q/A)_0$, and the associated heat transfer coefficient, h_x, are *local* quantities, dependent on the surface coordinate x. The notation h_x is used to emphasize this local dependence of the heat transfer coefficient. One seeks, then, to solve the viscous flow field near the surface in order to find $(\partial t/\partial y)_0$.

One should note the similarity of the rate laws representing the local wall shear and local heat flux:

$$\tau_0 = \mu\left(\frac{\partial v_x}{\partial y}\right)_0,$$

$$\left(\frac{q}{A}\right)_0 = -k\left(\frac{\partial t}{\partial y}\right)_0.$$

Both quantities are related to the wall gradients of the associated driving potential (momentum, or velocity, in the case of a shear; temperature in the case of heat flux) through an appropriate transport coefficient (μ or k). A concept known as *Reynolds analogy* will be developed to interrelate these wall gradients and, hence, the wall shear and wall heat flux. (In the instance of turbulent flow the exchange coefficients μ and k will have to be modified to account for momentum and energy transport associated with the transverse turbulent fluctuations.)

The local quantities τ_0 and h_x are often represented in dimensionless forms to make them more general and independent of any system of units. The wall shear stress, having the same dimensions as pressure, is rendered dimensionless by referring it to the freestream dynamic pressure, $\frac{1}{2}\rho U^2$. Thus a local *skin friction coefficient*, C_f, is defined:

$$C_f = \frac{\tau_0}{\frac{1}{2}\rho U^2} = \text{fn}\,(x), \tag{5.16}$$

where emphasis is again made to the dependence on the local coordinate x.

The heat transfer coefficient has the dimensions of a conductance (see Secs. 1.4 and 1.6); hence it may be rendered dimensionless by dividing it by the ratio of the fluid thermal conductivity to a linear dimension. For the local heat transfer coefficient the appropriate linear dimension is the local coordinate x. Thus the commonly used dimensionless form for h_x is

$$\text{Nu}_x = \frac{h_x x}{k} = \text{fn}\,(x). \tag{5.17}$$

This dimensionless form is known as the *Nusselt number,* named for Wilhelm Nusselt, one of the early investigators of convective heat transfer. The Nusselt number just defined is termed the *local* Nusselt number since it is based on the local heat transfer coefficient and the local coordinate. The Nusselt number appears to be of the same form as the Biot number encountered in Chapter 3 in connection with conduction heat transfer (see Sec. 3.4). However, one should note that the Nusselt number is defined in terms of the *fluid* thermal conductivity, while the Biot number is defined in terms of the thermal conductivity of the solid. Note that Eq. (5.17) emphasizes the fact that the local Nusselt number is, in general, dependent on the surface coordinate.

As defined in the foregoing, the local shear stress τ_0, or skin friction coefficient C_f, and the local heat flux $(q/A)_0$, or Nusselt number Nu_x, describe the fluid drag or heat transfer at a particular location along the surface. It is often desirable to know the total drag and heat transfer for a surface of some finite length, L, as illustrated in Fig. 5.9. In such an instance one needs to perform integrations over the surface to determine average values. For the fluid drag it is useful to define an *average* wall shear stress over the length L, $\overline{\tau_0}$, as

$$\overline{\tau_0} = \frac{1}{L}\int_0^L \tau_0\, dx. \tag{5.18}$$

Then, for unit depth, the total drag force on the surface is

$$F_{\text{drag}} = \overline{\tau_0} \times (L \times 1) = \overline{\tau_0} \times A_s, \tag{5.19}$$

where $A_s = L \times 1$ is used to designate the total exposed surface area. The *drag coefficient,* C_D is defined analogously to the skin friction coefficient by nondimensionalizing $\overline{\tau_0}$ with the freestream dynamic pressure:

$$C_D = \frac{\overline{\tau_0}}{\frac{1}{2}\rho U^2} = \frac{F_{\text{drag}}}{(\frac{1}{2}\rho U^2) \times A_s}. \tag{5.20}$$

Introduction of Eq. (5.18) relates the drag coefficient to the local skin friction coefficient:

$$C_D = \frac{1}{\frac{1}{2}\rho U^2} \frac{1}{L} \int_0^L \tau_0 \, dx$$
$$= \frac{1}{L} \int_0^L C_f \, dx. \tag{5.21}$$

The drag coefficient is seen to be the integrated average of the local skin friction coefficient, and once it is determined from the known dependence of C_f on x, Eq. (5.20) permits the calculation of the total drag force, F_{drag}, as a function of the freestream properties and the surface geometry.

The total heat transferred from the surface to the fluid may similarly be determined by integration. Thus, using q to denote this *total* heat flow,

$$q = \int_0^L \left(\frac{q}{A}\right)_0 \times (1 \times dx)$$

$$= (t_s - t_f) \int_0^L h_x \, dx,$$

where a unit depth has again been used. It has also been assumed that both t_s and t_f are independent of x—conditions that will be frequently imposed in the analyses that follow. In analogy to the definition of the drag coefficient as the integrated average of the local skin friction coefficient, one may define an integrated average heat transfer coefficient in terms of the local one. The symbol h with no subscript will be used to denote this average heat transfer coefficient in the remainder of this text:

$$h = \frac{1}{L} \int_0^L h_x \, dx. \tag{5.22}$$

Then the total heat transfer is simply

$$q = hA_s(t_s - t_f), \tag{5.23}$$

where $A_s = L \times 1$ is again the total exposed surface area.

The average heat transfer coefficient h and the total surface length L may be used to define an average, or total surface, Nusselt number:

$$\text{Nu}_L = \frac{hL}{k}, \tag{5.24}$$

where the subscript L on Nu_L is used to emphasize that it is based on the total surface. The quantity Nu_L can be expressed directly as a function of the local Nusselt number, Nu_x, as follows, using Eqs. (5.17) and (5.22):

$$\text{Nu}_L = \frac{hL}{k} = \frac{L}{k}\left[\frac{1}{L} \int_0^L h_x \, dx\right]$$

$$\text{Nu}_L = \frac{L}{k}\left[\frac{1}{L}\int_0^L \text{Nu}_x \frac{k}{x}\,dx\right], \tag{5.25}$$

$$= \int_0^L \frac{1}{x}\,\text{Nu}_x\,dx.$$

Note that $\text{Nu}_L \neq (1/L)\int_0^L \text{Nu}_x\,dx$, as one might be inclined to write!

In all the studies of various convective heat transfer situations that follow in this and subsequent chapters, both local friction and heat transfer will be sought (i.e., τ_0 and h_x) and usually expressed in dimensionless form through C_f and Nu_x. Similarly, average behavior for an entire surface will be sought (i.e., drag and h) in terms of the dimensionless parameters C_D and Nu_L. The quantities related to skin friction (τ_0, $\bar{\tau}_0$, C_f, C_D) are determined by the flow conditions in the velocity boundary layer, and the quantities related to heat transfer (h_x, h, Nu_x, Nu_L) are determined by the temperature distribution within the thermal boundary layer. Thus determination of these two regions, and their influence on one another, will be the central focus of the analyses of convective heat transfer that follow.

EXAMPLE 5.3

The analysis to be presented in Sec. 5.6 for laminar flow of a uniform stream (velocity U, temperature t_f) past a flat surface maintained at a temperature t_s shows that the local wall shear stress and local heat transfer coefficient vary with the distance from the leading edge of the surface in the following ways:

$$\tau_0 = 0.332\mu U\left(\frac{U}{\nu x}\right)^{1/2}, \tag{1}$$

$$h_x = 0.332k\,\text{Pr}^{1/3}\left(\frac{U}{\nu x}\right)^{1/2}. \tag{2}$$

The quantities μ, ν, and k are the dynamic viscosity, kinematic viscosity, and thermal conductivity of the fluid, respectively. The quantity Pr is the "Prandtl number" of the fluid, an important fluid property to be discovered in the analyses that follow. For the current example it will suffice to know that Pr is a dimensionless combination of other fluid properties, $\text{Pr} = \mu c_p/k$, and is tabulated in the appendix along with the others.

Imagine atmospheric pressure air, with a freestream velocity $U = 20$ m/s and temperature $t_f = 40°\text{C}$, flowing past a flat surface maintained at $t_s = 80°\text{C}$. At a distance $x = 0.25$ m from the leading edge of the surface:

(a) Verify that the flow is laminar.
(b) Calculate the local wall shear, heat transfer coefficient, and heat flux.
(c) Calculate the local skin friction coefficient and local Nusselt number.

Solution. The air temperature varies through the boundary layers formed on the surface—from t_s at the surface to t_f in the freestream. Since the properties ρ, μ,

ν, k, and Pr are all temperature dependent, some representative mean values must be selected. For low-speed flow such as involved in this example, it is customary to evaluate the fluid properties at the mean of t_s and t_f. Thus at a mean temperature $(40 + 80)/2 = 60°\text{C}$, Table A.6 gives the following properties of air at 1 atm:

$$\rho = 1.0596 \text{ kg/m}^3, \qquad \mu = 20.03 \times 10^{-6} \text{ kg/m-s},$$
$$\nu = 18.90 \times 10^{-6} \text{ m}^2/\text{s}, \qquad k = 28.52 \times 10^{-3} \text{ W/m-°C},$$
$$\text{Pr} = 0.708.$$

(a) For $x = 0.25$ m, the local length Reynolds number [Eq. (5.8)] is

$$\text{Re}_x = \frac{Ux}{\nu} = \frac{20 \times 0.25}{18.90 \times 10^{-6}} = 2.646 \times 10^5.$$

This is less than the critical value of 5×10^5 suggested by Eq. (5.9), so the flow is laminar as required for the equations for τ_0 and h_x.

(b) The expression for τ_0 given in Eq. (1) yields

$$\tau_0 = 0.332\mu U\left(\frac{U}{\nu x}\right)^{1/2}$$
$$= 0.332 \times 20.03 \times 10^{-6} \times 20 \times \left(\frac{20}{18.90 \times 10^{-6} \times 0.25}\right)^{1/2}$$
$$= 0.274 \text{ kg/m-s}^2 = 0.274 \text{ N/m}^2 \; (3.97 \times 10^{-5} \text{ lb}_\text{f}/\text{in.}^2),$$

while Eq. (2) for h_x gives

$$h_x = 0.332k \text{ Pr}^{1/3}\left(\frac{U}{\nu x}\right)^{1/2}$$
$$= 0.332 \times 28.52 \times 10^{-3} \times (0.708)^{1/3}\left(\frac{20}{18.90 \times 10^{-6} \times 0.25}\right)^{1/2}$$
$$= 17.36 \text{ W/m}^2\text{-°C} \; (3.60 \text{ Btu/h-ft}^2\text{-°F}).$$

Thus the local heat flux, by Eq. (5.15), is

$$\left(\frac{q}{A}\right)_0 = h_x(t_s - t_f),$$
$$= 17.36(80 - 40) = 694.5 \text{ W/m}^2 \; (220.2 \text{ Btu/h-ft}^2).$$

(c) The definition of the local skin friction coefficient in Eq. (5.16) gives

$$C_f = \frac{\tau_0}{\frac{1}{2}\rho U^2} = \frac{0.274}{\frac{1}{2} \times 1.0596 \times (20)^2} = 0.00129.$$

The local Nusselt number is given by Eq. (5.17):

$$\text{Nu}_x = \frac{h_x x}{k} = \frac{17.36 \times 0.25}{28.52 \times 10^{-3}} = 152.2.$$

EXAMPLE 5.4

If the surface described in Example 5.3 has a total length, L, of 0.4 m, find (a) the average wall shear, $\overline{\tau_0}$, and average heat transfer coefficient, h; (b) the total drag force and the total heat flow, for a unit depth; (c) the surface drag coefficient, C_D, and the average surface Nusselt number, Nu_L.

Solution. One should first verify that the flow is still laminar at the end of the surface where $x = L$. Using the properties found in Example 5.3, one finds the total surface length Reynolds number to be

$$\mathrm{Re}_L = \frac{UL}{\nu} = \frac{20 \times 0.4}{18.90 \times 10^{-6}} = 4.233 \times 10^5.$$

Thus the flow is laminar over the entire length of the surface and the equations of Example 5.3 may be used.

(a) The average wall shear is found from its definition in Eq. (5.18) and the dependence of $\overline{\tau_0}$ on x given in Eq. (1) of Example 5.3:

$$\begin{aligned}\overline{\tau_0} &= \frac{1}{L}\int_0^L \tau_0\, dx \\ &= \frac{1}{L}\int_0^L 0.332\mu U\left(\frac{U}{\nu x}\right)^{1/2} dx \\ &= \frac{1}{L}\, 0.332\mu U\left(\frac{U}{\nu}\right)^{1/2}\int_0^L x^{1/2}\, dx \qquad (1) \\ &= \frac{1}{L}\, 0.332\mu U\left(\frac{U}{\nu}\right)^{1/2}[2x^{1/2}]_0^L \\ &= 0.664\mu U\left(\frac{U}{\nu L}\right)^{1/2}.\end{aligned}$$

For the data and properties given in Example 5.3 and with $L = 0.4$ m, one has

$$\begin{aligned}\overline{\tau_0} &= 0.664 \times 20.03 \times 10^{-6} \times 20 \times \left(\frac{20}{18.90 \times 10^{-6} \times 0.4}\right)^{1/2} \\ &= 0.433 \text{ kg/m-s}^2 = 0.433 \text{ N/m}^2\ (6.28 \times 10^{-5} \text{ lb}_f/\text{in.}^2).\end{aligned}$$

Similarly, the average heat transfer coefficient is found from its definition in Eq. (5.22) and the dependence of h_x on x given in Eq. (2) of Example 5.3:

$$\begin{aligned}h &= \frac{1}{L}\int_0^L h_x\, dx \\ &= \frac{1}{L}\int_0^L 0.332k\, \mathrm{Pr}^{1/3}\left(\frac{U}{\nu x}\right)^{1/2} dx\end{aligned}$$

$$h = \frac{1}{L} 0.332k \, \mathrm{Pr}^{1/3}\left(\frac{U}{\nu}\right)^{1/2} \int_0^L x^{-1/2}\, dx \tag{2}$$

$$= \frac{1}{L} 0.332k \, \mathrm{Pr}^{1/3}\left(\frac{U}{\nu}\right)^{1/2} [2x^{1/2}]_0^L$$

$$= 0.664k \, \mathrm{Pr}^{1/3}\left(\frac{U}{\nu L}\right)^{1/2}.$$

Consequently, the average heat transfer coefficient is

$$h = 0.664 \times 28.52 \times 10^{-3} \times (0.708)^{1/3} \times \left(\frac{20}{18.90 \times 10^{-6} \times 0.4}\right)^{1/2}$$
$$= 27.45 \text{ W/m}^2\text{-°C } (4.83 \text{ Btu/h-ft}^2\text{-°F}).$$

(b) The total drag force on the surface for unit depth and $L = 0.4$ m is then, from Eq. (5.19),

$$F_{\text{drag}} = \overline{\tau_0} \times (L \times 1)$$
$$= 0.433 \times 0.4 \times 1 = 0.173 \text{ N } (0.0389 \text{ lb}_f),$$

while the total heat flow is given by Eq. (5.23):

$$q = hA_s(t_s - t_f)$$
$$= h \times L \times 1 \times (t_s - t_f)$$
$$= 27.45 \times 0.4 \times 1 \times (80 - 40) = 439.2 \text{ W } (1499 \text{ Btu/h}).$$

(c) The drag coefficient and the total surface Nusselt number are found from their definitions in Eqs. (5.20) and (5.24):

$$C_D = \frac{\overline{\tau_0}}{\frac{1}{2}\rho U^2} = \frac{0.433}{\frac{1}{2} \times 1.0596 \times (20)^2} = 0.00204,$$

$$\mathrm{Nu}_L = \frac{hL}{k} = \frac{27.45 \times 0.4}{28.52 \times 10^{-3}} = 385.0.$$

EXAMPLE 5.5

Use Eqs. (1) and (2) of Example 5.3 to derive expressions for the skin friction coefficient and local Nusselt number for laminar flow on a plane surface as functions of the local Reynolds number. Then do the same for the drag coefficient and total surface Nusselt number, as functions of the total surface length Reynolds number, using Eqs. (1) and (2) of Example 5.4.

Solution. Equation (1) of Ex. 5.3 gave the dependence of the local wall shear, τ_0, on the position x as

$$\tau_0 = 0.332 \, \mu U \left(\frac{U}{\nu x}\right)^{1/2}.$$

The local skin friction coefficient is defined in Eq. (5.16) as

$$C_f = \frac{\tau_0}{\frac{1}{2}\rho U^2}.$$

Thus

$$C_f = \frac{0.332}{1/2}\frac{\mu}{\rho}\frac{1}{U}\left(\frac{U}{\nu x}\right)^{1/2},$$

and since $\nu = \mu/\rho$, one has

$$\begin{aligned} C_f &= 0.664\left(\frac{\nu}{Ux}\right)^{1/2} \\ &= 0.664\ \mathrm{Re}_x^{-1/2}. \end{aligned} \tag{1}$$

Similarly, since Eq. (2) of Example 5.3 gives the local heat transfer coefficient to be

$$h_x = 0.332k\ \mathrm{Pr}^{1/3}\left(\frac{U}{\nu x}\right)^{1/2},$$

and Eq. (5.17) defines the local Nusselt number to be

$$\mathrm{Nu}_x = \frac{h_x x}{k},$$

one has

$$\begin{aligned} \mathrm{Nu}_x &= 0.332\mathrm{Pr}^{1/3}\left(\frac{Ux}{\nu}\right)^{1/2} \\ &= 0.332\mathrm{Pr}^{1/3}\mathrm{Re}_x^{1/2}. \end{aligned} \tag{2}$$

For $\mathrm{Re}_x = 2.646 \times 10^5$ as found in Example 5.3, Eqs. (1) and (2) just derived yield $C_f = 0.00129$ and $\mathrm{Nu}_x = 152.2$, as found in Example 5.3.

In like fashion, Eq. (1) of Example 5.4 for $\overline{\tau_0}$ and Eq. (5.20) defining the drag coefficient give

$$\begin{aligned} C_D = \frac{\overline{\tau_0}}{\frac{1}{2}U^2} &= \frac{0.664}{1/2}\frac{\mu}{\rho}\frac{1}{U}\left(\frac{U}{\nu L}\right)^{1/2} \\ &= 1.328\left(\frac{\nu}{UL}\right)^{1/2} \\ &= 1.328\mathrm{Re}_L^{-1/2}. \end{aligned} \tag{3}$$

Also, Eq. (2) of Example 5.4 for the average heat transfer coefficient and Eq. (5.24) for the total surface Nusselt number yield

$$\begin{aligned} \mathrm{Nu}_L = \frac{hL}{k} &= 0.664\mathrm{Pr}^{1/3}\left(\frac{UL}{\nu}\right)^{1/2}, \\ &= 0.664\mathrm{Pr}^{1/3}\mathrm{Re}_L^{1/2}. \end{aligned} \tag{4}$$

For the total surface length Reynolds number of Example 5.4, $Re_L = 4.233 \times 10^5$, Eqs. (3) and (4) yield the same results found in Example 5.4.

Equations (3) and (4) can also be obtained from Eqs. (1) and (2) by applying the integral relations of Eqs. (5.21) and (5.25).

5.4 GOVERNING EQUATIONS FOR THE LAMINAR BOUNDARY LAYER

In this section the approximate equations governing the flow of laminar boundary layers in external flow past a surface will be developed. Modifications to these equations that are necessary in the event the boundary layer is turbulent will be discussed in Sec. 5.7, and corresponding equations for internal flows will be presented in Sec. 5.9. The present discussion will also be limited to incompressible, or nearly constant density, flows. The effect of compressibility, which is important in high-speed flow, will be introduced later.

The physical situation for a laminar external boundary layer, discussed earlier in Sec. 5.3 is depicted again in Fig. 5.10(a). In this figure the laminar flow of a viscous fluid past a surface is illustrated, and the flow domain is divided into two regions: the velocity and thermal boundary layers near the surface where viscous and heat conduction effects are important, and the "inviscid" region in which these

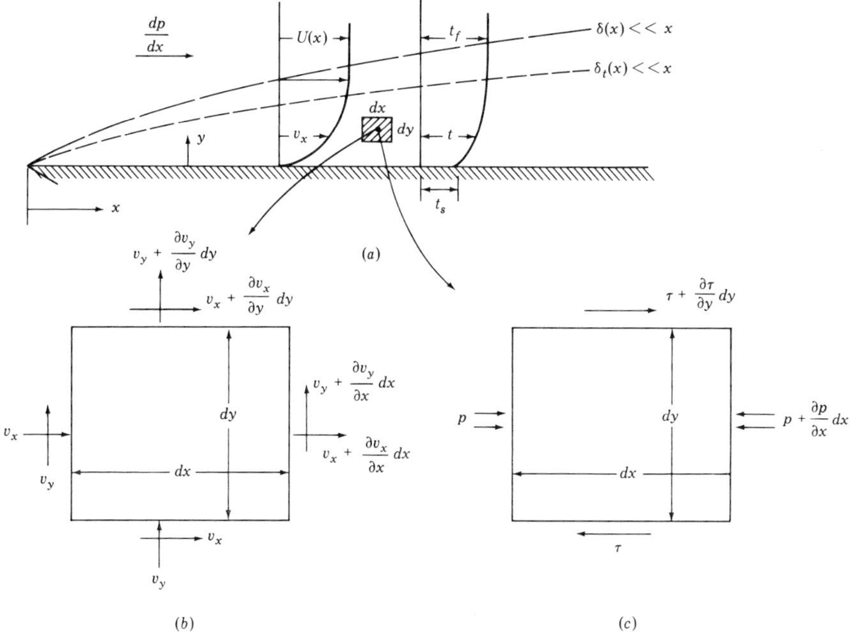

Figure 5.10. Elemental control volume for mass and force balances in a laminar boundary layer.

effects are ignored. Attention is first directed to the velocity boundary layer and the associated balance of momentum and mechanical forces.

In Fig. 5.10(a) the velocity boundary layer is shown as the region, $\delta(x)$, in which there is a strong variation of the velocity parallel to the surface—from 0 at the surface and approaching an inviscid value $U(x)$ as $y \to \delta$. The freestream inviscid velocity $U(x)$ is shown as a possible function of the coordinate parallel to the surface, x, in the event the surface is curved (as suggested earlier in Fig. 5.9). In such an instance x is measured locally parallel to the surface and y normal to it. If the surface is a flat plate, $U(x)$ is constant. The pressure gradient parallel to the surface in the freestream is denoted by dP/dx and is likewise constant if the surface is flat or some function of x if curved. In any event, the quantities $U(x)$ and dP/dx are to be regarded as known quantities—obtained by first solving the hydrodynamic problem of a nonviscous flow past the given surface geometry.

The basic presumption underlying Prandtl's boundary layer concept is that the layer is quite thin compared with the linear dimension of the surface. That is, $\delta(x) << x$. As will be seen, the requirement that the boundary layer be thin is that the local length Reynolds number, $\text{Re}_x = Ux/\nu$, be large (say, greater than about 1000). Under such circumstances the dominant momentum flux is that parallel to the surface—in the x direction. Similarly, the dominant viscous force is that parallel to the surface as determined by the gradient of v_x in the y direction, $\tau = \mu(\partial v_x/\partial y)$, since the greatest velocity change will be that of v_x as it changes from 0 to $U(x)$ over the small distance δ. While there is a flow component normal to x (i.e., v_y) the thinness assumption implies that it is small compared to v_x and its variation with y even smaller. Consequently, there will be a negligible flux of momentum in the y direction and shear stresses in that direction will also be small. As a consequence one may also presume that the pressure variation normal to the surface, $\partial P/\partial y$, is negligible so that in the boundary layer the only pressure force that need be accounted for is that parallel to the surface—dP/dx, already known from the solution of the inviscid freestream flow.

The simplifications above resulting from assuming that the boundary layer is thin enable one to write *approximate* expressions for the balance of mass, momentum, and energy in the boundary layer—resulting in expressions considerably simpler than those that would have been obtained by application of the complete equations of viscous fluid mechanics. These approximate boundary layer equations may be written in both differential and integral forms, depending on the method of solution to be used, and both representations are given below.

The Differential Laminar Boundary Layer Equations

The differential forms of the laminar boundary layer equations may be developed by applying the principles of mass conservation, energy conservation, and the balance of momentum and forces on an elemental control volume within the boundary layer. As illustrated in Fig. 5.10(a), a control volume of dimensions dx and dy (differential displacements in the coordinates parallel and normal to the surface) is chosen. Also shown in parts (b) and (c) of the figure are enlarged views of this elemental control volume showing the quantities necessary to write the conservation of mass and the balance of momentum and forces for the element. The symbols v_x and v_y denote the fluid velocity components in the coordinate directions. The elemental volume of Fig. 5.10(b) depicts the velocities entering the left and bottom faces as v_x and v_y, and these velocities vary with both x and y, leaving the top and right faces with the values shown, to a first-order approximation. If a unit depth into the paper is assumed and if ρ is used to denote the fluid density, constant, then the following mass fluxes in and out of the element result:

$$\begin{aligned}
&\text{Entering left face:} && \rho v_x\, dy. \\
&\text{Entering bottom face:} && \rho v_y\, dx. \\
&\text{Leaving right face:} && \rho\left(v_x + \frac{\partial v_x}{\partial x}\, dx\right) dy. \\
&\text{Leaving top face:} && \rho\left(v_y + \frac{\partial v_y}{\partial y}\, dy\right) dx.
\end{aligned} \tag{5.26}$$

Conservation of mass for the element requires that the sum of the first two lines of Eqs. (5.26) equal the sum of the second two (assuming that the steady state exists). Thus

$$\rho v_x\, dy + \rho v_y\, dx = \rho\left(v_x + \frac{\partial v_x}{\partial x}\, dx\right) dy + \rho\left(v_y + \frac{\partial v_y}{\partial y}\, dy\right) dx.$$

Cancellation of terms and division by the element volume ($dx \times dy \times 1$) yields the so-called *continuity* equation of the boundary layer—mass conservation per unit volume:

$$\frac{\partial v_x}{\partial x} + \frac{\partial v_y}{\partial y} = 0. \tag{5.27}$$

Newton's second law states that in the steady state the net flux of momentum out of a control volume equals the sum of forces on the volume. As a vector expression, this principle may be applied to the fluid element of Fig. 5.10 in each of the coordinate directions. Considering, first, the x direction parallel to the surface (the direction in which the dominant momentum flux takes place in a thin boundary layer) one needs the fluxes of x-directed momentum through the control volume surfaces. These are readily obtained by multiplying the mass fluxes in Eq. (5.26) with the appropriate x-direction velocity component:

$$\begin{aligned}
&\text{Entering left face:} && \rho v_x^2\, dy. \\
&\text{Entering bottom face:} && \rho v_y v_x\, dx. \\
&\text{Leaving right face:} && \rho\left(v_x + \frac{\partial v_x}{\partial x}\, dx\right)^2 dy. \\
&\text{Leaving top face:} && \rho\left(v_y + \frac{\partial v_y}{\partial y}\, dy\right)\left(v_x + \frac{\partial v_x}{\partial y}\, dy\right) dx.
\end{aligned} \tag{5.28}$$

The balance of these momentum fluxes must equal the net force in the x direction. The elemental control volume to the right of Fig. 5.10 shows these forces, and only those in the x direction are shown. On the left and right faces the forces acting are those of pressure, differing because of a possible pressure gradient dP/dx. These pressures act over an area of $dy \times 1$. On the top and bottom faces the x-direction forces are those due to viscous shear stresses and act over an area of $dx \times 1$. Thus the forces are:

On left face: $P\,dy.$

On bottom face: $\tau\,dx.$

On right face: $\left(P + \dfrac{dP}{dx}\,dx\right) dy.$

On top face: $\left(\tau + \dfrac{\partial \tau}{\partial y}\,dy\right) dx.$

The net flux of the momenta given in Eq. (5.28) must equal the sum of the forces just noted:

$$\rho\left(v_x + \frac{\partial v_x}{\partial x}\,dx\right)^2 dy + \rho\left(v_y + \frac{\partial v_y}{\partial y}\,dy\right)\left(v_x + \frac{\partial v_x}{\partial y}\,dy\right) dx - \rho v_x^2\,dy - \rho v_y v_x\,dx$$
$$= P\,dy + \left(\tau + \frac{\partial \tau}{\partial y}\,dy\right) dx - \left(P + \frac{\partial P}{\partial x}\,dx\right) dy - \tau\,dx.$$

This force–momentum balance may be simplified by use of the continuity equation ($\partial v_x/dx = -\partial v_y/\partial y$), the cancellation of some terms, division by the element volume $dx \times dy \times 1$, and neglection of higher-order terms to yield the *momentum* equation of the boundary layer:

$$v_x \frac{\partial v_x}{\partial x} + v_y \frac{dv_x}{\partial y} = -\frac{1}{\rho}\frac{dP}{dx} + \frac{1}{\rho}\frac{\partial \tau}{\partial y}. \tag{5.29}$$

For laminar flow one may relate the shear stress τ to the velocity by

$$\tau = \mu \frac{\partial v_x}{\partial y} = \rho \boldsymbol{\nu} \frac{\partial v_x}{\partial y},$$

so that

$$v_x \frac{\partial v_x}{\partial x} + v_y \frac{\partial v_x}{\partial y} = -\frac{1}{\rho}\frac{dP}{dx} + \frac{\partial}{\partial y}\left(\boldsymbol{\nu}\frac{\partial v_x}{\partial y}\right) \tag{5.30}$$

The form of the equation of motion given in Eq. (5.30) is applicable only to laminar motion since the laminar shear stress law has been used to replace τ in Eq. (5.29). In the event the flow is turbulent, an appropriate modification will have to be made to account for the effects of transverse turbulent fluctuations. This will be done later in the chapter.

As a result of the basic assumption that the boundary layer be thin, the flux of momentum in the y direction is ignored. Thus no momentum expression is written for that direction, and since there are negligible shear forces on the left and right faces of the element, one simply has $\partial P/\partial y = 0$. This fact has automatically been incorporated in Eqs. (5.29) and (5.30) by using $\partial P/\partial x = dP/dx$. The gradient dP/dx in the boundary layer is regarded as a known quantity, obtained by first solving the inviscid flow problem in the freestream external to the boundary layer.

It should be noted that in the foregoing momentum analysis the presence of body forces due to gravity has been ignored. Such forces are important in free convection and an appropriate modification will have to be made when this subject is treated in Chapter 7. For the present, the momentum equation above must be regarded as applicable only to pure forced convection.

Similar to the application of mass conservation and the momentum principle to the elemental control volume in the velocity boundary layer, one may apply the energy conservation principle to obtain an additional equation for the thermal boundary layer. Figure 5.11 depicts an elemental volume of the boundary layer with the appropriate energy terms shown. For such a thermodynamic "open system," the energy conservation principle states that

$$\dot{Q} - \dot{W} = \Sigma\, \dot{m}i,$$

in which $\dot{Q}$ is the rate at which heat is added to the element, $\dot{W}$ is the rate at which work is done by the element, and $\Sigma\, \dot{m}i$ is the net flux of enthalpy out of the element. (In the foregoing statement of energy conservation, the kinetic energy term has been omitted from the summation in the right side with the understanding that the work term, $\dot{W}$, includes only that work done against the dissipative forces

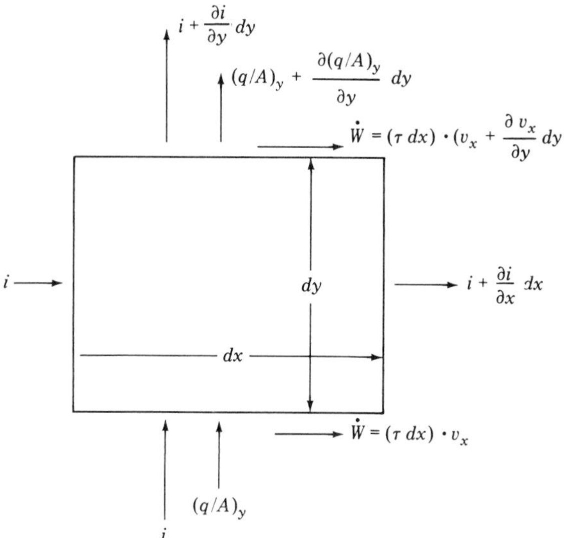

Figure 5.11. Elemental control volume for energy balance in a laminar boundary layer.

of viscosity or work done to deform the fluid, but does not include work done to accelerate the fluid—i.e., increase its kinetic energy. The symbol $\dot{m}$ represents the mass flow rate of the fluid through the element with the symbol i being used to denote enthalpy instead of the usual h to avoid confusion with the heat transfer coefficient.) If i represents the enthalpy of the fluid entering the bottom and left faces of the element, the enthalpies leaving the top and right faces depend on the gradients shown in the figure. Thus the following enthalpy fluxes result when the mass flow rates noted in Eq. (5.26) are used:

Entering left face: $\rho v_x i\, dy.$

Entering bottom face: $\rho v_y i\, dx.$

Leaving right face: $\rho\left(v_x + \dfrac{\partial v_x}{\partial x}\, dx\right)\left(i + \dfrac{\partial i}{\partial x}\, dx\right) dy.$

Leaving top face: $\rho\left(v_y + \dfrac{\partial v_y}{\partial y}\, dy\right)\left(i + \dfrac{\partial i}{\partial y}\, dy\right) dx.$

The heat entering the element, $\dot{Q}$, does so as the result of conduction in the fluid. Consistent with the assumption of a thin boundary layer, one presumes that the dominant temperature gradient is that normal to the surface and that the gradient in the direction of the surface is negligible. Thus the only heat conduction terms accounted for are $(q/A)_y$ at the bottom face and $(q/A)_y + \dfrac{\partial}{\partial y}(q/A)_y\, dy$ at the top, the subscript y being used to denote the heat flux in that direction. These heat fluxes both take place over an area equal to $dx \times 1$, so that the *net* heat conducted into the element is $-\dfrac{\partial}{\partial y}(q/A)_y\, dx\, dy$.

As noted, the work term, $\dot{W}$, in the energy conservation statement includes only work done against dissipative viscous shear forces or work done to deform, or compress, the fluid. Since the fluid is being taken as incompressible, the latter form is absent and the only mechanical work to be accounted for is that done against the viscous shear forces acting at the bottom and top faces of the element. At the bottom face the shear *force* is (per unit depth) $\tau(dx \times 1)$ and acts through a distance equal to v_x, per unit time. Thus the work rate done at the bottom face is $\tau v_x\, dx$. Similarly, the work done at the top face is $\tau\, dx$ multiplied by the velocity there, $v_x + (\partial v_x/\partial y)\, dy$. Thus the *net* rate of work done by the element, per unit depth, is $-\tau(\partial v_x/\partial y)\, dx\, dy$.

When all the foregoing is incorporated into the energy conservation principle, one has

$$\rho\left(v_x + \frac{\partial v_x}{\partial x}\, dx\right)\left(i + \frac{\partial i}{\partial x}\, dx\right) dy + \rho\left(v_y + \frac{\partial v_y}{\partial y}\right)\left(i + \frac{\partial i}{\partial y}\, dy\right) dx$$
$$- \rho v_x i\, dy - \rho v_y i\, dx = \frac{\partial}{\partial y}\left(\frac{q}{A}\right)_y dx\, dy + \tau\left(\frac{\partial v_x}{\partial y}\right) dx\, dy.$$

Once again expanding terms, canceling, using continuity, division by the element volume, and neglecting higher-order terms, one obtains

$$\rho\left(v_x \frac{\partial i}{\partial x} + v_y \frac{\partial i}{\partial y}\right) = -\frac{\partial}{\partial y}\left(\frac{q}{A}\right)_y + \tau\left(\frac{\partial v_x}{\partial y}\right).$$

If it is presumed that the fluid follows Joule's law (i.e., that the enthalpy depends only on temperature, a fact that is true for an ideal gas and closely approximated by many real gases and liquids), the enthalpy changes in the equation above may be rewritten in terms of the specific heat, c_p, and the fluid temperature, t, as $di = c_p\, dt$. Thus the preceding equation becomes

$$\rho c_p\left(v_x \frac{\partial t}{\partial x} + v_y \frac{\partial t}{\partial y}\right) = \tau\left(\frac{\partial v_x}{\partial y}\right) - \frac{\partial}{\partial y}\left(\frac{q}{A}\right)_y. \tag{5.31}$$

This general form of the *energy equation* of the boundary layer may be specialized for laminar flow by introducing the laminar rate laws for the shear stress and heat flux:

$$\tau = \mu\left(\frac{\partial v_x}{\partial y}\right); \qquad \left(\frac{q}{A}\right)_y = -k\left(\frac{\partial t}{\partial y}\right) = \rho c_p \alpha\left(\frac{\partial t}{\partial y}\right),$$

so that

$$v_x \frac{\partial t}{\partial x} + v_y \frac{\partial t}{\partial y} = \frac{\mu}{\rho c_p}\left(\frac{\partial v_x}{\partial y}\right)^2 + \frac{\partial}{\partial y}\left(\alpha \frac{\partial t}{\partial y}\right). \tag{5.32}$$

Here the thermal diffusivity of the fluid, $\alpha = k/\rho c_p$, has been introduced. The term involving $(\partial v_x/\partial y)^2$ represents the rate at which mechanical energy is being dissipated into thermal energy and may under certain circumstances be neglected.

Again, the rewriting of Eq. (5.31) into the form of Eq. (5.32) can be done only for laminar flow, some alteration being necessary if turbulence is present.

In summary, the differential equations of continuity, motion, and energy for the laminar boundary layer (in the absence of body forces) are:

$$\begin{aligned}
\frac{\partial v_x}{\partial x} + \frac{\partial v_y}{\partial y} &= 0,\\
v_x \frac{\partial v_x}{\partial x} + v_y \frac{\partial v_x}{\partial y} &= -\frac{1}{\rho}\frac{dP}{dx} + \frac{1}{\rho}\frac{\partial \tau}{\partial y}\\
&= -\frac{1}{\rho}\frac{dP}{dx} + \frac{\partial}{\partial y}\left(\nu \frac{\partial v_x}{\partial y}\right),\\
v_x \frac{\partial t}{\partial x} + v_y \frac{\partial t}{\partial y} &= \frac{\tau}{\rho c_p}\left(\frac{\partial v_x}{\partial y}\right) - \frac{1}{\rho c_p}\frac{\partial}{\partial y}\left(\frac{q}{A}\right)_y\\
&= \frac{\mu}{\rho c_p}\left(\frac{\partial v_x}{\partial y}\right)^2 + \frac{\partial}{\partial y}\left(\alpha \frac{\partial t}{\partial y}\right).
\end{aligned} \tag{5.33}$$

With the dP/dx regarded as known from the inviscid freestream, the three equations in Eq. (5.33) determine the three independent variables v_x, v_y, and t. Then the local shear stress, τ_0, and local heat transfer coefficient, h_x, may be found from their defining relations in Eqs. (5.14) and (5.15), and expressed dimensionlessly in terms of the skin friction coefficient, c_f, and local Nusselt number, Nu_x.

The Integral Boundary Layer Equations

The differential boundary layer equations just developed will yield, upon successful solution, detailed information as to the variation of the fluid velocity components and temperature throughout the boundary layer and were based on an elemental control volume within the boundary layer. An alternative formulation is one in which the overall behavior of the boundary layer is expressed in terms of a finite control volume encompassing the entire layer—resulting in integral expressions of momentum and energy. While such expressions may be obtained by integrating the differential equations, just obtained, it is instructive to develop them from first principles which emphasize the physical processes taking place.

Consider first the velocity boundary layer as depicted in Fig. 5.12. A control volume, $ABCD$, is chosen, as shown, which extends a distance dx along the surface and a distance H in the normal y direction. The distance H is chosen so that it extends beyond the boundary layer limit, δ, into the inviscid freestream. The mass flow rate into the control volume past face AD is, for unit depth,

$$\int_0^H \rho v_x \, dy, \tag{5.34}$$

and the momentum it carries into the domain is

$$\int_0^H \rho v_x^2 \, dy. \tag{5.35}$$

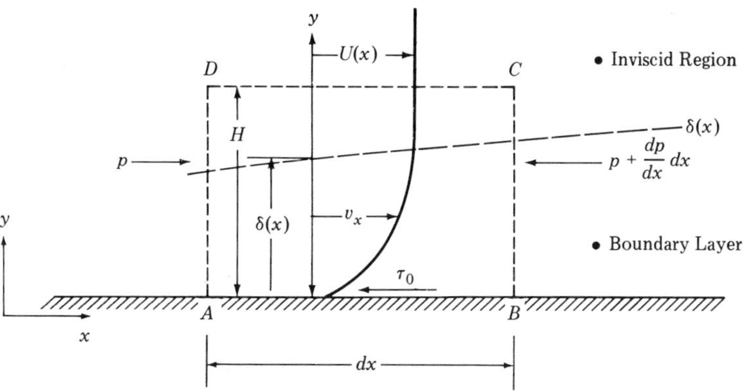

Figure 5.12. Integral analysis of a laminar boundary layer.

The mass flow out of the volume and the momentum taken with it are, to a first-order approximation,

$$\int_0^H \rho v_x \, dy + \frac{d}{dx}\left[\int_0^H \rho v_x \, dy\right] dx \tag{5.36}$$

and

$$\int_0^H \rho v_x^2 \, dy + \frac{d}{dx}\left[\int_0^H \rho v_x^2 \, dy\right] dx. \tag{5.37}$$

The mass flow rate across the top of the domain, DC, must be the difference between Eqs. (5.34) and (5.36) since no mass enters through the surface at AB:

$$\frac{d}{dx}\left[\int_0^H \rho v_x \, dy\right] dx. \tag{5.38}$$

This mass flow crosses the surface DC with an x-directed velocity component equal to the freestream velocity $U(x)$ since H extends into the freestream. Thus the x-directed momentum passing the top, DC, is

$$U(x) \frac{d}{dx}\left[\int_0^H \rho v_x \, dy\right] dx. \tag{5.39}$$

The net momentum flux represented by Eqs. (5.35), (5.37), and (5.39) must be balanced by the forces acting on the boundaries of the domain. The difference between the pressure acting on face DA and face BC is $(dP/dx)\,dx$ and acts over an area equal to $H \times 1$. No x-directed force acts on the surface CD since it is located in the freestream where viscosity is ignored. Let τ_0 represent the wall shear stress acting at the surface where $y = 0$. Since this shear acts over an area of $dx \times 1$, it exerts a force $\tau_0\, dx$ on the domain. When these forces just mentioned are equated to the net momentum flux given by subtracting Eqs. (5.35) and (5.39) from Eq. (5.37), one has

$$\frac{d}{dx}\left[\int_0^H \rho v_x^2 \, dy\right] dx - U(x) \frac{d}{dx}\left[\int_0^H \rho v_x \, dy\right] dx = -\left[\frac{dP}{dx}\, dx\right] H - \tau_0 \, dx.$$

Dividing by the domain length dx and noting that the two integrals on the left are identical for $y > \delta$ (the boundary layer thickness) so that H may be replaced by δ, one obtains finally the integral form of the boundary layer momentum equation:

$$\frac{d}{dx}\int_0^\delta v_x^2 \, dy - U(x) \frac{d}{dx}\int_0^\delta v_x \, dy = -\frac{\delta}{\rho}\frac{dP}{dx} - \frac{\tau_0}{\rho}. \tag{5.40}$$

Use of the momentum integral equation above requires knowledge of the freestream flow for the quantities $U(x)$ and dP/dx. If the velocity v_x is known as a function of y, Eq. (5.40) may be used to determine $\delta(x)$ when one is able to express the wall shear τ_0 in terms of known quantities. For laminar boundary layers one may write τ_0 in terms of the wall value of the velocity gradient: $\tau_0 = \mu(\partial v_x/\partial y)_0$, so that

$$\frac{d}{dx}\int_0^{\delta} v_x^2\,dy - U(x)\frac{d}{dx}\int_0^{\delta} v_x\,dy = -\frac{\delta}{\rho}\frac{dP}{dx} - \nu\left(\frac{\partial v_x}{\partial y}\right)_0. \tag{5.41}$$

The latter form, Eq. (5.41), may not be applied to a turbulent boundary layer; however, Eq. (5.40) may be used if the velocities involved are time average values and an appropriate expression for τ_0 employed.

An integral energy equation for the boundary layer may be derived in an analogous fashion to that just carried out for momentum. In Fig. 5.13 a control domain *ABCD* is again chosen to extend a distance $y = H$ beyond both the thermal and velocity layers. In addition to the velocity distribution, shown varying from 0 to $U(x)$, one must also account for the variation of temperature in the domain—shown as varying from some surface value t_s at $y = 0$ to t_f at $y = H$, the freestream temperature. Following the same approach as for the momentum equation, but with much less discussion, the *excess* rate at which enthalpy is carried out the right face of the domain at *BC* over that carried in the left face at *DA* is

$$\frac{d}{dx}\left[\int_0^H \rho(c_p t)v_x\,dy\right]dx. \tag{5.42}$$

The mass flow rate across the top of the control volume, *DC*, is already known in Eq. (5.38). This flow transports fluid at a temperature equal to t_f since this face of the domain extends into the freestream at that temperature. Thus the enthalpy flux across *DC* is

$$c_p t_f \frac{d}{dx}\left[\int_0^H \rho v_x\,dy\right]dx. \tag{5.43}$$

There is no enthalpy flux at *AB* since there is no flow through the wall; however, there *is* heat conducted into the control volume across *AB* due to the fluid temperature gradient there. Let the heat flux there be denoted by $(q/A)_0$, for the moment,

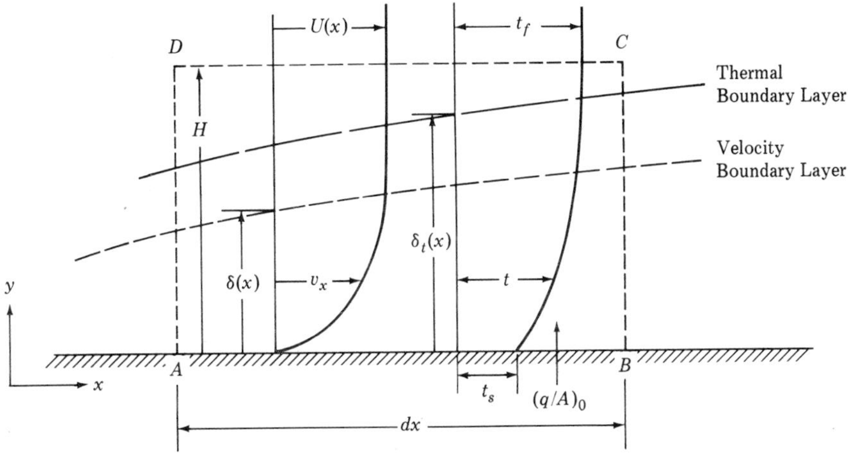

Figure 5.13. Integral analysis of a laminar boundary layer.

so that the total heat flow through the area $dx \times 1$ is $(q/A)_0\,dx$. No heat conduction takes place at the top of the domain, DC, since the temperature gradient in the freestream is taken to be negligible in the boundary layer concept. Thus in the steady state the difference of enthalpy fluxes in Eqs. (5.42) and (5.43) must equal the heat conducted at DC:

$$\frac{d}{dx}\left[\int_0^H \rho(c_p t)v_x\,dy\right]dx - c_p t_f \frac{d}{dx}\left[\int_0^H \rho v_x\,dy\right]dx = \left(\frac{q}{A}\right)_0 dx.$$

In this energy balance, that mechanical energy dissipated into thermal energy through viscosity has been neglected, unlike the differential analysis in which it was included.

If the equation above is divided by the domain length, dx, and the fact that the two integrals are identical for $y > \delta_t$ (the thermal layer thickness), the integral energy equation is obtained:

$$\frac{d}{dx}\int_0^{\delta_t}(t - t_f)v_x\,dy = \frac{(q/A)_0}{\rho c_p}.$$

In the above the fact that ρ, c_p, and t_f are constant have been used. This equation is applicable to turbulent flow as well as laminar if time average velocities and temperatures are used. For laminar flow the wall heat flux is expressible in terms of the temperature gradient at $y = 0$, $(q/A)_0 = -k(\partial t/\partial y)_0 = -\alpha\rho c_p(\partial t/\partial y)_0$, so that

$$\frac{d}{dx}\int_0^{\delta_t}(t - t_f)v_x\,dy = -\alpha\left(\frac{\partial t}{\partial y}\right)_0. \tag{5.44}$$

Here the fluid thermal diffusivity, α, has been introduced.

In summary, the integral equations for the laminar boundary layer, neglecting body forces and viscous dissipation, are:

$$\begin{aligned}\frac{d}{dx}\int_0^{\delta} v_x^2\,dy - U(x)\frac{d}{dx}\int_0^{\delta} v_x\,dy &= -\frac{\delta}{\rho}\frac{dP}{dx} - \frac{\tau_0}{\rho}\\ &= -\frac{\delta}{\rho}\frac{dP}{dx} - \nu\left(\frac{\partial v_x}{\partial y}\right)_0,\\ \frac{d}{dx}\int_0^{\delta_t}(t - t_f)v_x\,dy &= \frac{(q/A)_0}{\rho c_p}\\ &= -\alpha\left(\frac{\partial t}{\partial y}\right)_0.\end{aligned} \tag{5.45}$$

5.5 SIMILARITY PARAMETERS IN CONVECTIVE HEAT TRANSFER

Before proceeding with example solutions of the boundary layer equations for specific physical situations, it will prove useful to examine the general form such solutions might take. In Sec. 3.4 the general form of the solution to one-dimensional nonsteady heat conduction problems was shown to be expressible in terms of three dimensionless similarity parameters: a dimensionless spatial variable; a dimensionless time variable, the Fourier number; and a dimensionless heat transfer coefficient, the Biot number. These similarity parameters were identified by deducing the dimensionless form of the governing one-dimensional heat conduction equation. A similar procedure may be applied in the case of convective heat transfer to identify significant dimensionless similarity parameters by examining the dimensionless form of the governing equations just presented in the foregoing.

The differential formulations of the boundary layer equations provide the most useful form for the present purposes and are summarized in Eqs. (5.33). However, those equations, it will be remembered, were limited to pure forced convection inasmuch as buoyant forces were ignored in the development of the momentum equation. To make the present discussion equally applicable to free convective heat transfer (or even mixed convection), an additional term to account for buoyant forces will be added to the momentum equation. The reader is asked to accept the form of this additional force at this time—a more detailed explanation of it will be given in Chapter 7 when free convection is discussed in detail. In a heated fluid located in a gravity field, g, if t represents the local fluid temperature and t_f the uniform temperature of the bulk of the fluid far removed from the heating surface, it may be shown that the buoyant force, per unit mass of the fluid, is $g\beta(t - t_f)$. Here β represents the coefficient of volume expansion of the fluid, defined as

$$\beta = -\frac{1}{\rho}\left(\frac{\partial \rho}{\partial t}\right)_p.$$

This coefficient of volume expansion is, then, a fluid physical property dependent only on the thermodynamic state of the fluid. Details as to its evaluation for particular fluids will be delayed until Chapter 7, where it will be used extensively in free convection calculations. For the present it is adequate to recognize it as a physical property with the dimensions of reciprocal temperature.

If the buoyant force just defined is added to the momentum equation of the boundary layer as an additional force, the governing equations given in Eqs. (5.33) become, when the properties ν and α are taken to be constant:

$$\frac{\partial v_x}{\partial x} + \frac{\partial v_y}{\partial y} = 0,$$

$$v_x \frac{\partial v_x}{\partial x} + v_y \frac{\partial v_x}{\partial y} = -\frac{1}{\rho}\frac{dP}{dx} + \nu \frac{\partial^2 v_x}{\partial y^2} + g\beta(t - t_f), \tag{5.46}$$

$$v_x \frac{\partial t}{\partial x} + v_y \frac{\partial t}{\partial y} = \frac{\mu}{\rho c_p}\left(\frac{\partial v_x}{\partial y}\right)^2 + \alpha \frac{\partial^2 t}{\partial y^2}.$$

In the above, it has been presumed that the buoyant force acts in the direction of the surface coordinate x, i.e., that the surface in question is vertical. This need not actually be the case and one might need to multiply the buoyant term by the cosine of the angle of inclination of x with g; however, the above will suffice to deduce the *form* of the similarity parameter involving this force.

As noted in the discussion following Eqs. (5.33), what one seeks from these equations is the solution for v_x, v_y and t for a given physical geometry (e.g., external flow past a flat surface, external flow over a cylinder, internal flow in a pipe, etc.) and for given conditions (e.g., the freestream velocity and temperature, the average pipe flow velocity and bulk temperature, the surface temperature, etc.). Once the fluid temperature, t, is known, one finds the heat transfer coefficient:

$$h = \frac{-k(\partial t/\partial y)_s}{(t_s - t_f)}. \tag{5.47}$$

For a given physical geometry and flow conditions such a solution may be generalized by writing the foregoing equations in a dimensionless form. This is done by first selecting certain characteristic quantities to which the various variables may be referenced. For instance, one selects a characteristic length, call it l, which characterizes the size of the body in question (e.g., the length of a flat plate, the diameter of a cylinder or pipe, etc.). Then if dimensionless spatial variables are formed by dividing the true variable (x or y) by l, geometrically similar points are located by specifying similar values of the dimensionless variable. Similarly, one chooses a characteristic velocity, U (usually the freestream velocity in an external flow or an average velocity in an internal flow), to which all velocities are referenced. Since density is constant in this consideration, a characteristic pressure can be defined as the dynamic pressure associated with the characteristic velocity: $\frac{1}{2}\rho U^2$. To avoid difficulties with a temperature scale, one needs to work in terms of a temperature difference, say $(t - t_f)$, where t_f is the freestream or bulk temperature. Then a natural characteristic temperature difference is that evaluated at the specified surface temperature: $\Delta t = t_s - t_f$. Thus, in summary, the characteristic parameters chosen are:

Characteristic length:	l.
Characteristic velocity:	U.
Characteristic pressure:	$\frac{1}{2}\rho U^2$.
Characteristic temperature difference:	$\Delta t = t_s - t_f$.

With these parameters chosen, the independent and dependent variables are nondimensionalized according to the following definitions:

Dimensionless coordinates: $\xi = \dfrac{x}{l}, \; \eta = \dfrac{y}{l}.$

Dimensionless velocities: $v_\xi = \dfrac{v_x}{U}, \; v_\eta = \dfrac{v_y}{U}.$

Dimensionless pressure: $\Pi = \dfrac{p}{\frac{1}{2}\rho U^2}.$

$$\text{Dimensionless temperature:}\quad \phi = \frac{t - t_f}{\Delta t} = \frac{t - t_f}{t_s - t_f}.$$

Then, in two geometrically similar cases, similar spatial points result for equal values of ξ and η. Likewise, dynamic and energetic similarity are attained when v_ξ, v_η, Π, and ϕ are the same in the two cases.

If the dimensionless variables just defined are introduced into the governing equations in Eq. (5.46) and terms collected, the following results are obtained:

$$\frac{\partial v_\xi}{\partial \xi} + \frac{\partial v_\eta}{\partial \eta} = 0,$$

$$v_\xi \frac{\partial v_\xi}{\partial \xi} + v_\eta \frac{\partial v_\xi}{\partial \eta} = -\frac{1}{2}\frac{d\Pi}{d\xi} + \frac{\nu}{Ul}\frac{\partial^2 v_\xi}{\partial \eta^2} + \frac{g\beta l\,\Delta t}{U^2}\phi, \tag{5.48}$$

$$\frac{\rho c_p U l}{k}\left(v_\xi \frac{\partial \phi}{\partial \xi} + v_\eta \frac{\partial \phi}{\partial \eta}\right) = \frac{\mu U^2}{k\,\Delta t}\left(\frac{\partial v_\xi}{\partial \xi}\right)^2 + \frac{\partial^2 \phi}{\partial \eta^2}.$$

In Eqs. (5.48), after the introduction of the dimensionless variables, the momentum equation has been divided by the coefficient of the inertia term and the energy equation by the coefficient of the conduction term. These facts will lead to important conclusions later. Examination of Eqs. (5.48) reveals that identical solutions are obtained for the dimensionless variables when the various coefficients involving the groupings of the characteristic parameters (l, U, Δt) and the fluid properties (ν, β, ρ, c_p, k, μ) are the same. These groupings, then, constitute the relevant similarity parameters that are being sought. Since it is the fluid temperature distribution that is desired in order to find the heat transfer coefficient, imagine, then, that the three dimensionless equations given in Eqs. (5.48) are solved for the dimensionless temperature, ϕ. This solution may be represented by the functional relation

$$\phi = f\left(\xi, \eta, \frac{\nu}{Ul}, \frac{g\beta l\,\Delta t}{U^2}, \frac{\mu U^2}{k\,\Delta t}, \frac{\rho c_p U l}{k}\right). \tag{5.49}$$

The foregoing solution for ϕ is seen to depend on, in addition to the dimensionless spatial variables, four dimensionless groupings of the physical parameters of the problem. These four groupings could be used as similarity parameters as they stand; however, it is customary to rearrange them in the following way:

$$\phi = f\left(\xi, \eta, \frac{\nu}{Ul}, \frac{g l^3\beta\,\Delta t}{\nu^2}\cdot\frac{\nu^2}{U^2 l^2}, \frac{U^2}{c_p\,\Delta t}\cdot\frac{\mu c_p}{k}, \frac{Ul}{\nu}\cdot\frac{\mu c_p}{k}\right). \tag{5.50}$$

In this representation certain groups are seen to appear more than once, and since only a functional dependence is being sought, Eq. (5.50) indicates that the four groups in Eq. (5.49) may be alternatively expressed in terms of the following four groups:

$$\phi = f\left(\xi, \eta, \frac{Ul}{\nu}, \frac{g l^3\beta\,\Delta t}{\nu^2}, \frac{U^2}{c_p\,\Delta t}, \frac{\mu c_p}{k}\right). \tag{5.51}$$

The four dimensionless groups of parameters in Eq. (5.51) are the sought-for similarity parameters for convective heat transfer. The first of these, $Ul/\boldsymbol{\nu}$, is recognized as the Reynolds number already discussed. The other groups also bear special symbols and names:

$$\text{Re} = \frac{Ul}{\boldsymbol{\nu}} = \text{the Reynolds number.}$$

$$\text{Gr} = \frac{gl^3\beta\ \Delta t}{\boldsymbol{\nu}^2} = \text{the Grashof number.} \tag{5.52}$$

$$\text{Ec} = \frac{U^2}{c_p\ \Delta t} = \text{the Eckert number.}$$

$$\text{Pr} = \frac{\mu c_p}{k} = \frac{\boldsymbol{\nu}}{\alpha} = \text{the Prandtl number.}$$

The physical significance of these parameters will be discussed shortly, but it should be emphasized here that each is a dimensionless quantity and when they are calculated numerically, one must carefully note the units involved so that they are, indeed, without dimensions.

Remembering now that the ultimate result desired is the surface heat transfer coefficient, one would obtain it by evaluating the fluid temperature gradient at the surface as given in Eq. (5.47):

$$h = \frac{-k(\partial t/\partial y)_s}{t_s - t_f},$$

where the subscript s denotes the surface at $y = 0$. If the dimensionless temperature $\phi = (t - t_s)/\Delta t$, and the dimensionless coordinate $\eta = y/l$ are introduced, the definition of h above assumes the dimensionless form:

$$\text{Nu} = \frac{hl}{k} = -\left(\frac{\partial\phi}{\partial\eta}\right)_s = \text{the Nusselt number.} \tag{5.53}$$

Thus the Nusselt number discussed in Sec. 5.3 is seen to enter naturally through the nondimensionalization of the definition of h. Combination of Eqs. (5.51), (5.52), and (5.53) implies

$$\text{Nu} = f(\xi, \eta, \text{Re}, \text{Gr}, \text{Ec}, \text{Pr}).$$

Thus one expects the local Nusselt number to depend on position and on the parameters Re, Gr, Ec, and Pr. An average Nusselt number (or average h) would be obtained by integration as described earlier. Thus in either the local or average sense one expects the Nusselt number to have the following functional form:

$$\text{Nu} = f(\text{Re}, \text{Gr}, \text{Ec}, \text{Pr}). \tag{5.54}$$

The implications of Eq. (5.54) are twofold. First, it predicts the form in which an analytic solution to a convective heat transfer will be expressed. Second, in the event that an experimental investigation is conducted, instead of seeking h as a function of U, l, ν, g, β, Δt, c_p, ρ, and k, the above shows that the data should correlate according to Eq. (5.54)—reducing the number of variables to four. This implies a considerable saving of effort in the taking of data and its subsequent reduction.

Some special forms of the results represented by Eq. (5.54) may be encountered in certain circumstances, depending on the importance of various terms included in the governing equations. When the dimensionless form of the momentum equation given in Eq. (5.48) was derived, the coefficient of the viscous term, ν/Ul, was the result of dividing the equation by the coefficient of the inertia term on the left side of the equation. Thus the factor ν/Ul represents the relative size of the viscous forces compared to the inertia force. Ultimately, this coefficient appeared in the final correlation of Eq. (5.54) as its reciprocal, the Reynolds number $\text{Re} = Ul/\nu$. Thus the Reynolds number has the physical interpretation of a measure of the magnitude of the ratio of the inertia forces to the viscous forces. A large Reynolds number implies that inertia dominates the viscous forces, and conversely.

Similar interpretations may be deduced for the other dimensionless parameters given in Eq. (5.54) and defined in Eq. (5.52). The Prandtl number

$$\text{Pr} = \frac{\nu}{\alpha} = \frac{\mu c_p}{k}$$

arises from the rate laws incorporated in the governing equations for shear stress and heat flux:

$$\tau = \mu\left(\frac{\partial v_x}{\partial y}\right) = \rho\nu\left(\frac{\partial v_x}{\partial y}\right),$$

$$\frac{q}{A} = -k\left(\frac{\partial t}{\partial y}\right) = -\rho c_p \alpha\left(\frac{\partial t}{\partial y}\right).$$

Thus Pr is a measure of the ratio of momentum diffusion through the fluid, due to viscosity, to heat diffusion by conduction. Consequently, Pr becomes a measure of the relative sizes of the two boundary layers—the velocity boundary layer and the thermal boundary layer, discussed in some detail in Sec. 5.3. Unlike the other dimensionless variables in Eq. (5.54), the Prandtl number is seen to be a fluid property, even though it is dimensionless. Thus values of Pr may be calculated from values of the other properties already discussed, and values are tabulated for the various fluids given in Tables A.3 through A.7 and A.9. Even though the kinematic viscosity, $\nu = \mu/\rho$, and thermal diffusivity, $\alpha = k/\rho c_p$, are pressure dependent, their ratio, $\text{Pr} = \nu/\alpha = \mu c_p/k$, is virtually pressure independent, except for steam. Thus, usually, only the temperature dependence of Pr must be accounted for and even this is slight for most gases.

The Grashof number, $\text{Gr} = l^3 g\beta\,\Delta t/\nu^2$, in Eq. (5.54) or (5.51) originated from the term $g\beta\,\Delta t/U^2$ in Eq. (5.48). Since

$$\frac{g\beta\,\Delta t}{U^2} = \frac{\text{Gr}}{\text{Re}^2},$$

examination of Eq. (5.48) shows that $g\beta\ \Delta t/U^2$ is a measure of the ratio of buoyant forces to inertia forces (it is sometimes called the Froude number). Thus, from the foregoing interpretation of the Reynolds number, the Grashof number is a measure of the relative size of the buoyant forces compared to the viscous forces. When $\mathrm{Gr}/\mathrm{Re}^2 \ll 1$, Eq. (5.48) shows that the buoyant force is negligible compared to the viscous or inertia forces and the condition is termed *pure forced convection*. When $\mathrm{Gr}/\mathrm{Re}^2 \gg 1$ the converse is true and one has what is termed *pure free convection*. The Eckert number, $\mathrm{Ec} = U^2/c_p\ \Delta t$, originated from the term $\mu U^2/k\ \Delta t$ in Eq. (5.48) since

$$\frac{\mu U^2}{k\ \Delta t} = \mathrm{EcPr}.$$

As the coefficient of the dissipation term in the energy equation EcPr is a measure of the relative importance of viscous dissipation in a particular problem. If $\mathrm{EcPr} \gg 1$, then dissipation is significant; if $\mathrm{EcPr} \ll 1$, it is not.

Consequently, in the results that follow in this and subsequent chapters, one may expect some special forms of the general correlation of Eq. (5.54). For pure forced convection the Grashof number is missing and one expects the form

$$\mathrm{Nu} = f(\mathrm{Re}, \mathrm{Ec}, \mathrm{Pr}) \tag{5.55}$$

when viscous dissipation is important, or the form

$$\mathrm{Nu} = f(\mathrm{Re}, \mathrm{Pr}) \tag{5.56}$$

when dissipation is not important.

For cases of pure free convection the Reynolds number becomes unimportant, with the Grashof number appearing instead. Usually in such cases the Eckert number is sufficiently small compared to $1/\mathrm{Pr}$ that viscous dissipation can be ignored also. Thus pure free convection correlations usually take the form

$$\mathrm{Nu} = f(\mathrm{Gr}, \mathrm{Pr}). \tag{5.57}$$

It should be emphasized that the foregoing sought only to identify the dimensionless parameters that are relevant to convective heat transfer. The general case cited in Eq. (5.54) or the special ones for pure forced or free convection in Eqs. (5.55) through (5.57) indicate only *functional* dependencies. The precise form of the functions will depend on the particular physical geometry involved (e.g., flat surface, cylinder, etc.) and the hydrodynamic and thermal boundary conditions imposed (e.g., forced or free convection, internal or external flow, etc.). These forms will be developed later, in this chapter and the two following chapters, for a number of useful situations.

EXAMPLE 5.6

As illustrated in Fig. 5.14, atmospheric pressure air flows with a freestream velocity U and freestream temperature t_f normal to a horizontal cylinder of diameter D. A

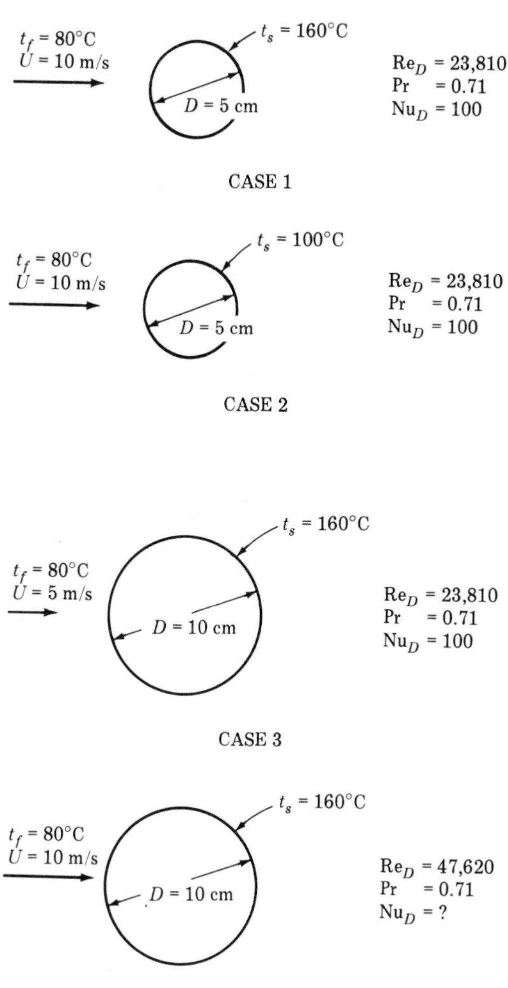

Figure 5.14.

hot fluid flowing inside the cylinder maintains the surface temperature at some value t_s. The effects of varying these parameters on the average heat transfer coefficient and heat flux at the surface are to be observed. In Case 1 the air flows with $U = 10$ m/s, $t_f = 80$°C past a cylinder of diameter $D = 5$ cm and the surface temperature $t_s = 160$°C. Under these conditions the surface heat flux is measured to be 4800 W/m^2.

(a) If, as in Case 2 of Fig. 5.14, the surface temperature is changed to $t_s = 100$°C (by altering the flow of the internal fluid), determine the resultant heat flux.
(b) Determine the heat flux for Case 3, in which the same experiment is performed

for a cylinder with diameter $D = 10$ cm and an air velocity $U = 5$ m/s, all other parameters the same as in Case 1.

(c) Do the same for Case 4, in which the larger cylinder, $D = 10$ cm, is used and the original air velocity is used, $U = 10$ m/s, all other data unchanged.

In each instance presume that the physical properties of the air remain unchanged at the following vlaues (taken, approximately, from Table A.6): $\nu = 21 \times 10^{-6}$ m^2/s, $k = 30 \times 10^{-3}$ W/m-°C, Pr $= 0.71$.

Solution. Assume that the effects of buoyant forces and viscous dissipation are negligible, so that the situation is one of pure forced convection wherein the average heat transfer coefficient is related to the physical flow parameters according to the dimensionless form of Eq. (5.56):

$$\text{Nu} = f(\text{Re}, \text{Pr}).$$

The form of the function in this equation depends on the physical geometry at hand and the hydrodynamic and thermal boundary conditions. In this example all the cases cited are the same—that of the horizontal cylinder of known surface temperature subjected to cross flow of a fluid of known velocity and temperature. While the exact form of the function for such a situation is not yet known, it is known that the *same* functional form will apply in all four cases noted in Fig. 5.14.

For Case 1 the average surface heat transfer coefficient may be found from the given heat flux and temperatures:

$$h = \frac{q/A}{t_s - t_f} = \frac{4800}{160 - 80} = 60 \text{ W/m}^2\text{-°C}.$$

Choosing the cylinder diameter D as the characteristic length (the only dimension available) the diameter Nusselt and Reynolds numbers may be calculated from the data given. The Prandtl number is known from the given air properties. Thus for Case 1, the dimensionless parameters involved in the dimensionless form above are:

$$\text{Nu}_D = \frac{hD}{k} = \frac{60 \times 0.05}{30 \times 10^{-3}} = 100,$$

$$\text{Re}_D = \frac{UD}{\nu} = \frac{10 \times 0.05}{21 \times 10^{-6}} = 23{,}810,$$

$$\text{Pr} = 0.71.$$

Attention to units ensures that Nu_D and Re_D are, indeed, dimensionless.

(a) For Case 2, the only quantity differing from Case 1 is the surface temperature. Since U, D, and ν are unchanged, the Reynolds number is unchanged and Pr is taken as constant in this example. Thus, with identical values of Re_D and Pr, whatever is the form of the function $\text{Nu}_D = f(\text{Re}_D, \text{Pr})$, the same Nusselt number will result: $\text{Nu}_D = 100$. Since D and k are unchanged from Case 1, the heat transfer coefficient must also be the same, $h = 60\ \text{W/m}^2\text{-°C}$. For a surface temperature of $t_s = 100°\text{C}$, the heat flux for Case 2 is then

$$\frac{q}{A} = h(t_s - t_f)$$
$$= 60 \times (100 - 80) = 1200\ \text{W/m}^2.$$

(b) For Case 3 the cylinder diameter is increased to 10 cm and the freestream air velocity is reduced to 5 m/s. Thus the Reynolds number is

$$\text{Re}_D = \frac{UD}{\nu} = \frac{5 \times 0.1}{21 \times 10^{-6}} = 23{,}810,$$

the same as in Case 1. Since the Prandtl number Pr is still unchanged, these identical values of Re_D and Pr again yield the same Nusselt number as in Case 1: however, the heat transfer coefficient is halved since the diameter is doubled:

$$\text{Nu}_D = 100 = \frac{hD}{k} = \frac{h \times 0.1}{30 \times 10^{-3}},$$
$$h = 30\ \text{W/m}^2\text{-°C}.$$

For $t_s = 160°\text{C}$ and $t_f = 80°\text{C}$, the heat flux in Case 3 is

$$\frac{q}{A} = h(t_s - t_f)$$
$$= 30 \times (160 - 80) = 2400\ \text{W/m}^2.$$

(c) For Case 4, the cylinder diameter is $D = 10$ cm and the air velocity is $U = 10$ m/s. Thus the Reynolds number is

$$\text{Re}_D = \frac{UD}{\nu} = \frac{10 \times 0.1}{21 \times 10^{-6}} = 47{,}620,$$

twice that in Case 1. The Prandtl number, Pr, is again unchanged. In this instance, then, the Nusselt number can be deduced only if the particular form of the function $\text{Nu}_D = f(\text{Re}_D, \text{Pr})$ were known. Since it is not, further analysis of this case is not possible. The particular form of the function f for this example of external forced convection across a cylinder is given in the next chapter along with the forms for several other situations having practical application.

This particular example was simplified through the assumption of constant fluid properties (i.e., ν, k, and Pr). In many applications the correlating function for Nu must take into account possible dependencies of physical properties on temperature as might result when the thermal conditions (t_f and t_s in Example 5.6) are altered or when a different fluid is involved. This example also presumed that the effects of buoyancy and viscous dissipation were negligible so that the Grashof and Eckert numbers did not appear in the correlating function for Nu. The justifications for this simplification are given in the next example.

EXAMPLE 5.7

Justify the neglection of buoyancy and viscous dissipation effects in Example 5.6.

Solution. As noted in the foregoing discussion of dimensionless parameters, the significance of buoyant free convection forces is indicated by the magnitude of the Grashof number, defined in Eq. (5.52) as

$$\mathrm{Gr} = \frac{gl^3\beta\,\Delta t}{\nu^2}.$$

In this definition l is a characteristic dimension of the geometry. Since the buoyant forces act vertically, and the significant vertical dimension for a horizontal cylinder, as chosen in Example 5.6, is the diameter, one bases the Grashof number on the cylinder diameter, $l = D$. The temperature difference Δt is that existing between the cylinder surface and the freestream, $\Delta t = t_s - t_f$. The coefficient of volume expansion, β, for air may be approximated by the ideal gas result that $\beta = 1/T_f$, where T_f is the *absolute* temperature of the air. Thus in the situation posed in Example 5.6, $\beta = 1/(80 + 273.15) = 1/353.15°\mathrm{K}$. So, for the conditions specified in Case 1 of Example 5.6 ($D = 5$ cm, $\Delta t = 160 - 80°\mathrm{C}$, $\nu = 21 \times 10^{-6}$ m^2/s), one finds the diameter Grashof number to be

$$\begin{aligned}\mathrm{Gr}_D &= \frac{gD^3\beta\,\Delta t}{\nu^2}\\ &= \frac{9.8 \times (0.05)^2(1/353.15)(160 - 80)}{(21 \times 10^{-6})^2}\\ &= 6.29 \times 10^5.\end{aligned}$$

Note that the calculated value of Gr_D is properly dimensionless. As noted in the discussion of Sec. 5.5, the relative magnitude of the buoyant forces and inertia forces is measured by the ratio $\mathrm{Gr}/\mathrm{Re}^2$. In Example 5.6, the diameter Reynolds number was calculated to be $\mathrm{Re}_D = 23{,}810$, so for this case

$$\begin{aligned}\frac{\mathrm{Gr}_D}{\mathrm{Re}_D^2} &= \frac{6.29 \times 10^5}{(2.381 \times 10^4)^2}\\ &= 0.001.\end{aligned}$$

Since $\mathrm{Gr}_D/\mathrm{Re}_D^2 << 1$, one concludes that the buoyant free convection forces are negligible when compared with the inertia force of the freestream. Hence the neglection of buoyancy and the treatment of the condition as one of pure forced convection is justified.

The importance of the magnitude of the energy dissipated through viscous shear is measured by the magnitude of the Eckert number and how the product EcPr compares with unity. The Eckert number, defined in Eq. (5.52), is

$$\mathrm{Ec} = \frac{U^2}{c_p\,\Delta t}.$$

Table A.6 indicates that the specific heat of air at 80°C is approximately $c_p = 1.01$ kJ/kg-°C, so that for the data of case 1 in Example 5.6, the value of the *dimensionless* Eckert number is

$$\mathrm{Ec} = \frac{U^2}{c_p\,\Delta t} = \frac{10^2}{1.01 \times 10^3 \times (160 - 80)} = 0.00124.$$

Thus the product EcPr is

$$\mathrm{EcPr} = 0.00124 \times 0.71 = 0.0009,$$

a value sufficiently less than 1 to justify the neglect of viscous dissipation effects.

5.6 LAMINAR FORCED CONVECTION PAST A FLAT SURFACE, DISSIPATION NEGLECTED

Perhaps the simplest example of the application of boundary layer analysis to the solution of a convection problem is that of laminar flow of a fluid of uniform velocity U and uniform temperature t_f past a flat surface maintained at a uniform temperature t_s and placed parallel to the incident flow. In spite of its simplicity such a flow condition is often encountered in practical situations; in many cases the results for flow past a flat surface may be used as an approximation to situations in which the flow occurs over a moderately curved surface. Such a situation is depicted in Fig. 5.15, wherein the coordinate x is measured from the leading edge of the surface and parallel to it, and y is the coordinate normal to x. The velocity and thermal boundary layers are shown developing along the surface.

It will be recalled that the application of the boundary layer equations requires prior solution of the external, inviscid stream to obtain the pressure gradient dP/dx and the freestream velocity $U(x)$ as functions of x. The inherent simplicity of the flat surface geometry is that this inviscid solution is readily seen to be: $dP/dx = 0$ and $U =$ constant. Thus, as shown in Fig. 5.15, the velocity variation through the boundary layer is from 0 at the surface to U far away, at all locations x.

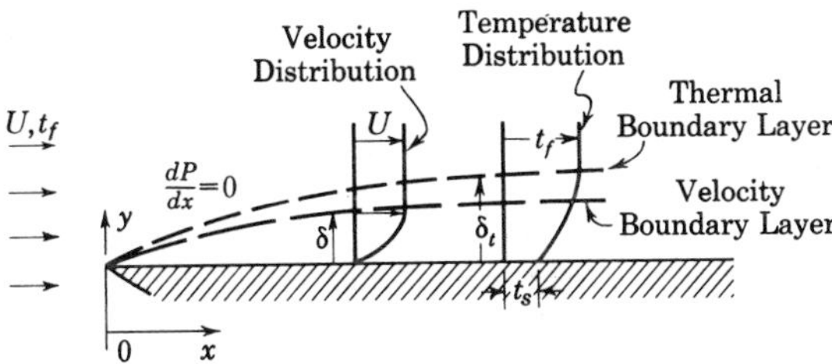

Figure 5.15. Velocity and thermal boundary layers for laminar flow past a flat surface.

Similarly, the temperature distribution through the thermal layer is from the surface value t_s to the uniform freestream value t_f.

The example solutions to be presented in the following will assume viscous dissipation to be negligible. The effects of significant dissipation will be considered later. Also, the present solution will be limited to the laminar state of flow, turbulent effects being introduced later.

To show the basic concepts used in applying either the differential or integral formulations for the boundary layer, the flat surface problem will be solved using both representations.

Differential Solution

To obtain the solution for forced convective heat transfer for laminar flow past a flat surface as posed above and depicted in Fig. 5.15, one may apply the differential form of the boundary layer equations given in Eqs. (5.33) with $dP/dx = 0$ and the dissipation term omitted. If the properties $\boldsymbol{\nu}$ and α are taken as constant, one has

$$\frac{\partial v_x}{\partial x} + \frac{\partial v_y}{\partial y} = 0, \tag{5.58}$$

$$v_x \frac{\partial v_x}{\partial x} + v_y \frac{\partial v_x}{\partial y} = \boldsymbol{\nu}\left(\frac{\partial^2 v_x}{\partial y^2}\right), \tag{5.59}$$

$$v_x \frac{\partial t}{\partial x} + v_y \frac{\partial t}{\partial y} = \alpha\left(\frac{\partial^2 t}{\partial y^2}\right). \tag{5.60}$$

The boundary conditions to be applied are:

$$\begin{aligned} &\text{Velocity:} && \text{at } y = 0,\ v_x = v_y = 0, \\ & && \text{as } y \to \infty,\ v_x \to U. \end{aligned} \tag{5.61}$$

$$\begin{aligned} &\text{Temperature:} && \text{at } y = 0,\ t = t_s, \\ & && \text{as } y \to \infty,\ t \to t_f. \end{aligned} \tag{5.62}$$

Before proceeding with the solution of the set of equations above, one should note the similarity between the forms of the momentum and energy equations. Equation (5.60) is identical in form to Eq. (5.59) if t is replaced by v_x, and the boundary conditions are of similar form if t is replaced by $t - t_s$. The only difference is that the coefficient $\boldsymbol{\nu}$ appears in the momentum equation, whereas α appears in the energy equation. Thus it should not be surprising to know that the velocity and temperature distributions *are* quite similar—with any difference being dependent on the ratio of $\boldsymbol{\nu}$ and α, the Prandtl number $\text{Pr} = \boldsymbol{\nu}/\alpha$. Also, it should be noted that if the fluid properties are taken as independent of temperature, then the solutions of Eqs. (5.58) and (5.59) for v_x and v_y can be carried out separately from that for t in Eq. (5.60). That is, the velocity and temperature problems become "uncoupled."

Considerable insight into the solution of the problem posed by Eqs. (5.58) through (5.62) can be had by first making a rather crude order-of-magnitude analysis. The terms on the left of Eq. (5.59) represent the fluid inertia force, and the first term, $v_x(\partial v_x/\partial x)$, is typical. Since v_x may be as great as U, the freestream velocity, this term may at some downstream location x be as large as $U(U/x)$. Since in the boundary layer y varies from 0 at the surface to δ at the boundary layer limit, the viscous force on the right of Eq. (5.59), $\boldsymbol{\nu}(\partial^2 v_x/\partial y^2)$, can be of the order of $\boldsymbol{\nu}U/\delta^2$. In the boundary layer these two forces are expected to be of the same order of magnitude. Thus one anticipates that

$$\frac{U^2}{x} \sim \frac{\boldsymbol{\nu}U}{\delta^2} \quad \text{or} \quad \delta \sim \sqrt{\frac{\boldsymbol{\nu}x}{U}}. \tag{5.63}$$

Thus the boundary layer may be expected to grow as $\sqrt{x}$ as it proceeds down the surface. More generally, Eq. (5.63) shows that the ratio of the local boundary layer thickness, δ, to the local distance from the leading edge, x, is

$$\frac{\delta}{x} \sim \sqrt{\frac{\boldsymbol{\nu}}{Ux}} = \mathrm{Re}_x^{-1/2}. \tag{5.64}$$

That is, the dimensionless boundary layer thickness, δ/x, is inversely proportional to the square root of the local length Reynolds number, Re_x. This result implies that the "thin" boundary layer assumption means that Re_x must be sufficiently large that δ/x can be considered small, but still less than that which would cause the onset of turbulence.

Based on the result of Eq. (5.63), one may simplify the solution of the governing equations by introducing the *similarity* assumption of Blasius (Refs. 2 and 4). In the absence of any characteristic dimension, Blasius theorized that if at some location x, instead of plotting the local velocity v_x against y, as in Fig. 5.15, one were to plot the dimensionless velocity v_x/U against the dimensionless distance formed by dividing y by the local boundary layer thickness, i.e., y/δ, one should obtain the same distribution at *all* locations. That is, since Eq. (5.63) gives $\delta \sim \sqrt{\boldsymbol{\nu}x/U}$, Blasius assumed that the sought-for function $v_x = \mathrm{fn}(x, y)$ should be expressible in the form

$$\begin{aligned}\frac{v_x}{U} &= \mathrm{fn}\left(\frac{y}{\delta}\right)\\ &= \mathrm{fn}\left(y\sqrt{\frac{U}{\boldsymbol{\nu}x}}\right).\end{aligned}$$

Thus the Blasius similarity assumption is: If a dimensionless spatial variable η is defined

$$\eta = y\sqrt{\frac{U}{\boldsymbol{\nu}x}}, \tag{5.65}$$

the dimensionless velocity distribution v_x/U should be a function of *only* this variable.

$$\frac{v_x}{U} = f'(\eta). \tag{5.66}$$

For simplicity of later algebra, the function given in Eq. (5.66) is taken as the derivative of a yet-to-be-determined function $f(\eta)$.

Substitution of the assumed form of Eq. (5.66) into Eq. (5.58) and subsequent integration shows that continuity is satisfied if v_y/U is of the form

$$\frac{v_y}{U} = \frac{1}{2}\sqrt{\frac{\nu}{Ux}}\,(\eta f' - f). \tag{5.67}$$

Thus the velocity components in the boundary layer, v_x and v_y, can be found if the function $f(\eta)$ can be found.

An analogous similarity assumption may be made for the temperature profile in the boundary layer. Assuming that the thermal layer is of the same order of magnitude as the velocity layer (an assumption the significance of which will be explored shortly), one defines a dimensionless temperature

$$\phi = \frac{t - t_s}{t_f - t_s} \tag{5.68}$$

and assumes that ϕ, likewise, depends only on the parameter η:

$$\phi = \phi(\eta). \tag{5.69}$$

If one then substitutes the assumptions of Eqs. (5.66) and (5.69), along with the result of continuity in Eq. (5.67), into the governing equations (5.59) through (5.62), considerable algebra will give the following equations for determining $f(\eta)$ and $\phi(\eta)$

$$f''' + \tfrac{1}{2}ff'' = 0, \tag{5.70}$$
$$f(0) = f'(0) = 0, \qquad f'(\infty) = 1.$$
$$\phi'' + \tfrac{1}{2}\,\mathrm{Pr}\, f\phi' = 0, \tag{5.71}$$
$$\phi(0) = 0, \qquad \phi(\infty) = 1.$$

In Eq. (5.71), the Prandtl number, Pr, discovered in the similarity analysis of Sec. 5.5 is seen to enter the problem naturally. The absence of any quantities other than f, ϕ, or η in the expressions above justifies the similarity assumptions by which f and ϕ were regarded as functions of η only.

The equation and boundary conditions given in Eq. (5.70) provide a solution for the function $f(\eta)$ from which the velocity components in the boundary layer, v_x and v_y, may be determined from Eqs. (5.66) and (5.67). Once $f(\eta)$ is known, the equation and boundary conditions of Eq. (5.71) permit determination of the function $\phi(\eta)$ so that the temperature distribution is known from Eq. (5.68). Equation (5.70) for $f(\eta)$ is a nonlinear ordinary differential equation, and its solution may be obtained numerically or by series techniques (Ref. 2). The resulting function $f'(\eta) = v_x/U$ is shown in Fig. 5.16 along with some experimental results, and the apparent agreement is good. Equation (5.71) for $\phi(\eta)$ is a first-order linear

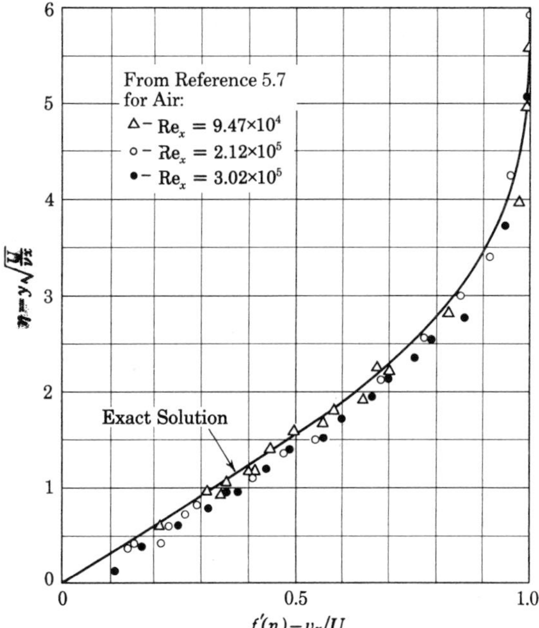

Figure 5.16. Dimensionless results for the velocity distribution in the laminar boundary layer on a flat surface as given by the exact solution of the boundary layer equations. Experimental values are shown for comparison.

equation in ϕ', and its solution may be written in terms of the function $f(\eta)$. The solution for $\phi(\eta)$ is clearly a function of the Prandtl number Pr, and the results for selected values of Pr are given in Fig. 5.17.

The nature of the boundary condition of Eq. (5.70) that $f'(\infty) = 1$ implies that the full freestream velocity U is not achieved anywhere in the fluid domain except in the limit $y \rightarrow \infty$. However, for practical purposes it is desirable to terminate the use of the boundary layer solution at some finite limit, the boundary layer thickness δ, at which the viscous effects are no longer regarded as important. Examination of the function $f'(\eta) = v_x/U$ in Fig. 5.16 shows that v_x/U is about 0.99 when $\eta \approx 5.0$. Thus it is customary to define the boundary layer limit as $y = \delta$ when $\eta = 5.0$. Thus the growth of a laminar boundary may be predicted by

$$\delta = 5.0\sqrt{\frac{\nu x}{U}}, \qquad \frac{\delta}{x} = 5.0\,\mathrm{Re}_x^{-1/2}. \tag{5.72}$$

The result supports the order-of-magnitude determination given in Eqs. (5.63) and (5.64) and which led to the definition of η.

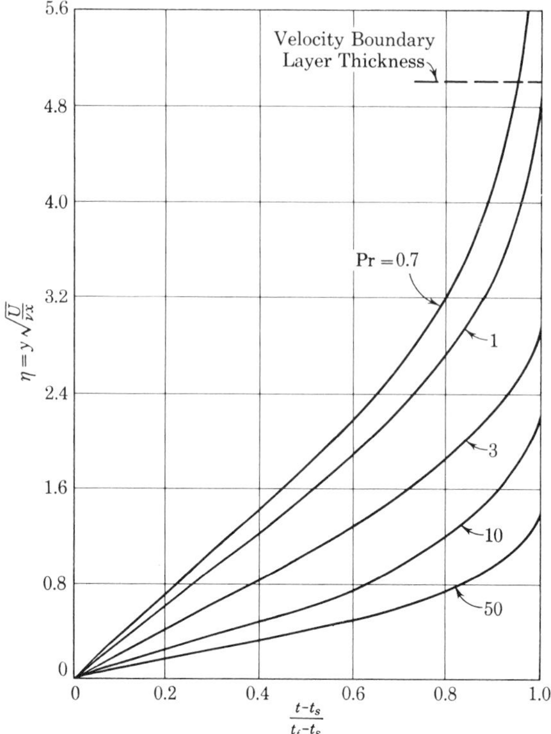

Figure 5.17. Temperature distributions in the laminar boundary layer on a heated flat surface, viscous dissipation neglected.

Examination of the result for the temperature distribution $\phi = (t - t_s)/(t_f - t_s)$ shown in Fig. 5.17 indicates that for a fluid with Pr = 1, the functions ϕ and f' are identical (as examination of the governing equations has already predicted). Further, one may observe that for Pr > 1, the thermal boundary layer will be thinner than the velocity layer, whereas for Pr < 1 it is thicker.

For the purposes of predicting surface shear stresses and surface heat transfer coefficients, one is not so much interested in the full solution for the boundary layer velocity and temperature distribution, as given by the functions $f'(\eta)$ and $\phi(\eta)$ in Figs. 5.16 and 5.17, as in the derivatives of these functions at the surface where $y = \eta = 0$. These derivatives lead to the velocity and temperature *gradients* at $y = 0$, which are related to the shear stress and heat flux there. The solutions that led to Figs. 5.16 and 5.17 also yield these important results (Ref. 2):

$$f''(0) = 0.332, \qquad \phi'(0) = 0.332\text{Pr}^{1/3}. \tag{5.73}$$

The expression for $\phi'(0)$ given in Eq. (5.73) is actually a rather accurate approximation to the more complicated exact result. As would be expected, the slope of the temperature distribution, $\phi'(0)$, is a function of the fluid Prandtl number.

Since Eq. (5.65) defines $\eta = y\sqrt{U/\nu x}$, the definitions of $f' = v_x/U$ and $\phi = (t - t_s)/(t_f - t_s)$ give

$$\left(\frac{\partial v_x}{\partial y}\right)_0 = Uf''(0)\left(\frac{U}{\nu x}\right)^{1/2} = 0.332U\left(\frac{U}{\nu x}\right)^{1/2},$$

$$\left(\frac{\partial t}{\partial y}\right)_0 = (t_f - t_s)\phi'(0)\left(\frac{U}{\nu x}\right)^{1/2} = 0.332\text{Pr}^{1/3}(t_f - t_s)\left(\frac{U}{\nu x}\right)^{1/2},$$

so that the definition of the wall shear stress, τ_0, the local heat transfer coefficient, h_x, and Eqs. (5.14) and (5.15) yield

$$\tau_0 = \mu\left(\frac{\partial v_x}{\partial y}\right)_0 = 0.332\mu U\left(\frac{U}{\nu x}\right)^{1/2},$$
$$h_x = \frac{-k(\partial t/\partial y)_0}{t_s - t_f} = 0.332\text{Pr}^{1/3}k\left(\frac{U}{\nu x}\right)^{1/2}. \tag{5.74}$$

Incidentally, these are the relations used for illustrative purposes in Examples 5.3 through 5.5. As shown in those examples, the dimensionless forms of the foregoing expressions give the local skin friction coefficient and the local Nusselt number:

$$C_f = \frac{\tau_0}{\frac{1}{2}\rho U^2} = \frac{0.332}{\frac{1}{2}}\frac{\nu}{U}\left(\frac{U}{\nu x}\right)^{1/2}$$
$$= 0.664\text{Re}_x^{-1/2},$$
$$\text{Nu}_x = \frac{h_x x}{k} = 0.332\text{Pr}^{1/3}x\left(\frac{U}{\nu x}\right)^{1/2} \tag{5.75}$$
$$= 0.332\text{Pr}^{1/3}\text{Re}_x^{1/2}.$$

Here the local Reynolds number Re_x has been introduced. Also as shown in those examples, for a surface of total length L, the average wall shear stress, $\bar{\tau}_0$, the average heat transfer coefficient, h, are found by integration over the surface from $x = 0$ to $x = L$:

$$\bar{\tau}_0 = \frac{1}{L}\int_0^L \tau_0\,dx = 0.664\mu U\left(\frac{U}{\nu L}\right)^{1/2},$$

$$h = \frac{1}{L}\int_0^L h_x\,dx = 0.664\text{Pr}^{1/3}k\left(\frac{U}{\nu L}\right)^{1/2},$$

so that the drag coefficient and the total surface Nusselt number are expressed in terms of the total surface Reynolds number $\text{Re}_L = UL/\nu$:

$$C_D = \frac{\bar{\tau}_0}{\frac{1}{2}\rho U^2} = 1.328\text{Re}_L^{-1/2},$$
$$\text{Nu}_L = \frac{hL}{k} = 0.664\text{Pr}^{1/3}\text{Re}_L^{1/2}. \tag{5.76}$$

Equations (5.75) for the local case and Eqs. (5.76) for the total surface case then constitute the solution for laminar forced convection past a flat surface. The forms obtained for the Nusselt number as functions of the Prandtl and Reynolds numbers agrees with the prediction found by nondimensionalizing the governing equations: that $\mathrm{Nu} = f(\mathrm{Pr}, \mathrm{Re})$.

One might also note the great similarity between the solution for the surface friction (τ_0 or C_f) and that for surface heat transfer (h_x or Nu_x). This is the result from the already mentioned similarity between the momentum and energy equations of the boundary layer. The above shows that, for the flat surface at least,

$$\mathrm{Nu}_x = \frac{C_f}{2}\,\mathrm{Pr}^{1/3}\mathrm{Re}_x,$$

interrelating the two important quantities Nu_x and C_f. This interrelation is a very important aspect of convective heat transfer and will be expanded later to other geometries and, in modified form, even to turbulent flow.

Integral Solution

Before discussing examples of the use of the foregoing results or comparing them with experiment, it is instructive to examine the method of solving the same problem using the integral formulation of the boundary layer. This is done only to show the method and the close correspondence of the results of the two approaches. In practice, for other geometries or hydrodynamic boundary conditions, one would apply only that method which provides the easiest solution.

For the case at hand, laminar flow past a flat surface, the integral equations (5.45) become the following when one applies $U(x) = U$, a constant, and $dP/dx = 0$:

$$\begin{aligned} \frac{d}{dx}\int_0^{\delta} v_x(v_s - U)\,dy &= -\nu\left(\frac{\partial v_x}{\partial y}\right)_0, \\ \frac{d}{dx}\int_0^{\delta_t} v_x(t - t_f)\,dy &= -\alpha\left(\frac{\partial t}{\partial y}\right)_0. \end{aligned} \tag{5.77}$$

Once again one should note the marked similarity between the momentum and energy equations.

The momentum equation is solved by assuming a reasonable polynomial form for the velocity distribution which satisfies as many boundary conditions as possible. Then, even though this distribution is approximate, the integral in the momentum equation may not differ too much from that resulting from the exact solution. Once this integration is accomplished, the momentum equation yields a differential equation for the boundary layer thickness which can then be solved for $\delta(x)$. For the case to be illustrated here, let the velocity distribution through the boundary layer be approximated by a cubic polynomial,

$$v_x = a + by + cy^2 + dy^3. \tag{5.78}$$

Four boundary conditions are required to find the constants in Eq. (5.78). The facts that v_x must be zero at the surface and U at the boundary layer limit δ, along with the reasonable desire to have the velocity reach a uniform value at $y = \delta$, readily yield

$$v_x = 0 \qquad \text{for } y = 0,$$
$$v_x = U \qquad \text{for } y = \delta,$$
$$\frac{\partial v_x}{\partial y} = 0 \qquad \text{for } y = \delta.$$

An additional condition may be found from Eq. (5.59) by noting that at the surface where $v_x = v_y = 0$, one must have $(\partial^2 v_x/\partial y^2) = 0$. Thus a fourth condition is

$$\frac{\partial^2 v_x}{\partial y^2} = 0 \qquad \text{for } y = 0.$$

When the four conditions just noted are applied to Eq. (5.78), one finds the assumed velocity distribution to be of the form

$$\frac{v_x}{U} = \frac{3}{2}\left(\frac{y}{\delta}\right) - \frac{1}{2}\left(\frac{y}{\delta}\right)^3. \tag{5.79}$$

This velocity distribution is, of course, approximate, but in comparison with the exact solution of the differential formulation given in the foregoing will show a close correspondence. Figure 5.18 plots the approximate representation of Eq. (5.79) with that obtained from the analytic solution in Fig. 5.16 and the close

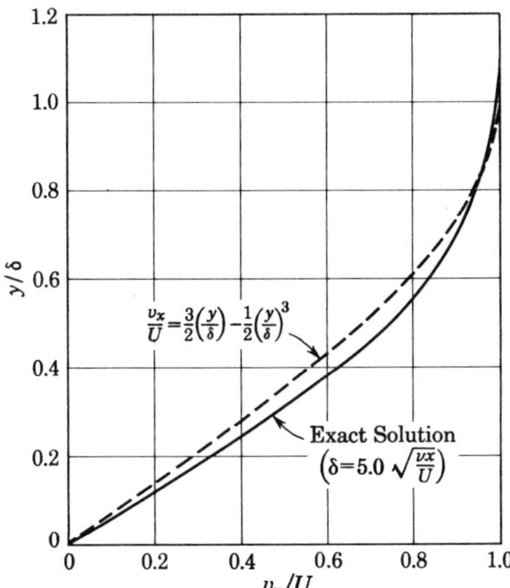

Figure 5.18. Comparison of the velocity distribution for a laminar boundary layer on a flat surface as given by the exact solution and the momentum integral method.

correspondence is apparent. Substitution of this assumed distribution into the integral momentum equation (5.77) gives a differential equation for δ:

$$\frac{d}{dx}\left\{U^2 \int_0^\delta \left[\frac{3}{2}\left(\frac{y}{\delta}\right) - \frac{1}{2}\left(\frac{y}{\delta}\right)^3\right]\left[\frac{3}{2}\left(\frac{y}{\delta}\right) - \frac{1}{2}\left(\frac{y}{\delta}\right)^3 - 1\right] dy\right\} = -\boldsymbol{\nu}\left(\frac{3}{2}\frac{U}{\delta}\right)$$

or

$$\frac{d}{dx}\left(U^2 \frac{39}{280}\delta\right) = \frac{3}{2}\frac{U\boldsymbol{\nu}}{\delta}$$

$$\delta\frac{d\delta}{dx} = \frac{140}{13}\frac{\boldsymbol{\nu}}{U}.$$

This simple differential equation for δ may be readily integrated, with the fact that $\delta = 0$ at $x = 0$, to give

$$\delta^2 = \frac{280}{13}\frac{\boldsymbol{\nu}x}{U},$$

$$\delta = \sqrt{\frac{280}{13}}\sqrt{\frac{\boldsymbol{\nu}x}{U}}, \tag{5.80}$$

$$\frac{\delta}{x} = 4.64\mathrm{Re}_x^{-1/2}.$$

Comparison with Eq. (5.72) shows that while not the same as the "exact" solution, the result obtained by this simpler method for the boundary layer growth is quite acceptable.

The wall shear stress and skin friction coefficient are easily obtained from the velocity distribution of Eq. (5.79):

$$\tau_0 = \mu\left(\frac{\partial v_x}{\partial y}\right)_0 = \mu\left(\frac{3}{2}\frac{U}{\delta}\right),$$

$$\begin{aligned} C_f = \frac{\tau_0}{\frac{1}{2}\rho U^2} &= 3\frac{\boldsymbol{\nu}}{U\delta} \\ &= 3\frac{\boldsymbol{\nu}}{U}\sqrt{\frac{13}{280}}\sqrt{\frac{U}{\boldsymbol{\nu}x}} \\ &= 0.646\mathrm{Re}_x^{-1/2}. \end{aligned} \tag{5.81}$$

This result is seen to be reasonably close to that given in Eq. (5.75) by the differential solution.

The thermal boundary layer and the integral energy equation may be analyzed by an analogous approach. An *assumed* polynomial form is taken for the temperature:

$$t = a + by + cy^2 + dy^3.$$

By the same reasoning used for the velocity distribution the boundary conditions to be satisfied by t are

$$t = t_s \quad \text{and} \quad \frac{\partial^2 t}{\partial y^2} = 0 \quad \text{for } y = 0,$$

$$t = t_f \quad \text{and} \quad \frac{\partial t}{\partial y} = 0 \quad \text{for } y = \delta_t,$$

where $\delta_t(x)$ is the sought for thermal layer thickness. These conditions reduce the assumed form for t to

$$\frac{t - t_s}{t_f - t_s} = \frac{3}{2}\left(\frac{y}{\delta_t}\right) - \frac{1}{2}\left(\frac{y}{\delta_t}\right)^3. \tag{5.82}$$

This result, along with the form already assumed for v_x in Eq. (5.79), may be introduced into the integral energy equation (5.77):

$$\frac{d}{dx}\left\{U(t_f - t_s)\int_0^{\delta_t}\left[\frac{3}{2}\left(\frac{y}{\delta}\right) - \frac{1}{2}\left(\frac{y}{\delta}\right)^3\right]\left[\frac{3}{2}\left(\frac{y}{\delta_t}\right) - \frac{1}{2}\left(\frac{y}{\delta_t}\right)^3 - 1\right]dy = -\alpha(t_f - t_s)\frac{3}{2}\frac{1}{\delta_t}.\right.$$

Considerable algebraic manipulation gives

$$\frac{d}{dx}\left\{\frac{\delta_t^2}{\delta}\left[\frac{3}{20} - \frac{3}{280}\left(\frac{\delta_t}{\delta}\right)^2\right]\right\} = \frac{3}{2}\frac{\alpha}{U\,\delta_t}. \tag{5.83}$$

This result is simplified by assuming that while not equal, the thermal and velocity boundary layers are probably of the same order of magnitude. Thus if $(\delta_t/\delta) \approx 1$, the term 3/280 may be neglected when compared to 3/20, so that

$$\frac{d}{dx}\left[\frac{3}{20}\frac{\delta_t^2}{\delta}\right] \approx \frac{3}{2}\frac{\alpha}{U\,\delta_t}$$

$$\left(\frac{\delta_t}{\delta}\right)\frac{d}{dx}\left[\delta\left(\frac{\delta_t}{\delta}\right)^2\right] = \frac{10\alpha}{U\,\delta}.$$

The implication of assuming $(\delta_t/\delta) \approx 1$ will be examined further shortly.

Since δ is an already known function of x as given in Eq. (5.80), the latter expression may be considered as an equation for δ_t. This solution for δ_t is better done in terms of the ratio δ_t/δ. If Eq. (5.80) for δ is introduced, further algebraic manipulation gives

$$\left(\frac{\delta_t}{\delta}\right)^3 + \frac{4}{3}x\frac{d}{dx}\left[\left(\frac{\delta_t}{\delta}\right)^3\right] = \frac{13}{14}\frac{\alpha}{\nu}. \tag{5.84}$$

The latter equation is a first-order linear equation in $(\delta_t/\delta)^3$ which has the solution

$$\left(\frac{\delta_t}{\delta}\right)^3 = \frac{13}{14}\frac{\alpha}{\nu} + Cx^{-3/4}. \tag{5.85}$$

The constant C in this equation must be zero to avoid an indeterminate solution at $x = 0$, so

$$\frac{\delta_t}{\delta} = \left(\frac{13}{14}\frac{\alpha}{\nu}\right)^{1/3}.$$

Thus the important result is obtained that the ratio of the thermal and velocity boundary layer thicknesses is a function of only the Prandtl number to the one-third power:

$$\frac{\delta_t}{\delta} = \left(\frac{13}{14}\frac{1}{\mathrm{Pr}}\right)^{1/3} \approx \mathrm{Pr}^{-1/3}. \tag{5.86}$$

This result was already predicted by the differential solution as indicated in Eqs. (5.73). The simplification above, wherein 3/280 was neglected with respect to 3/20, is thus seen to be applicable only when Pr is near 1.

Since the velocity layer thickness, δ, is already a known function of Re_x as given in Eq. (5.80), one now knows that the thermal layer varies as

$$\frac{\delta_t}{x} = \sqrt{\frac{280}{13}}\sqrt[3]{\frac{13}{14}}\,\mathrm{Re}_x^{-1/2}\mathrm{Pr}^{-1/3}.$$

Thus the local heat transfer coefficient is readily found from the assumed temperature polynomial in Eq. (5.82):

$$h_x = \frac{-k\left(\dfrac{\partial t}{\partial y}\right)_0}{t_s - t_f} = \frac{3}{2}k\frac{1}{\delta_t} = \frac{3}{2}\sqrt{\frac{13}{280}}\sqrt[3]{\frac{14}{13}}\frac{k}{x}\mathrm{Re}_x^{1/2}\mathrm{Pr}^{1/3}, \tag{5.87}$$

$$\mathrm{Nu}_x = \frac{h_x x}{k} = \frac{3}{2}\sqrt{\frac{13}{280}}\sqrt[3]{\frac{14}{13}}\,\mathrm{Pr}^{1/3}\mathrm{Re}_x^{1/2} = 0.331\mathrm{Pr}^{1/3}\mathrm{Re}_x^{1/2}.$$

As in the case of the local skin friction coefficient, this result for the local Nusselt number is seen to be virtually identical to that obtained by the solution of the differential formulation of the problem. Quite clearly, then, the corresponding relations for the average values on a flat surface of total length L are

$$C_D = 1.292\mathrm{Re}_L^{-1/2}, \tag{5.88}$$

$$\mathrm{Nu}_L = \frac{hL}{k} = 0.662\mathrm{Pr}^{1/3}\mathrm{Re}_L^{1/2}.$$

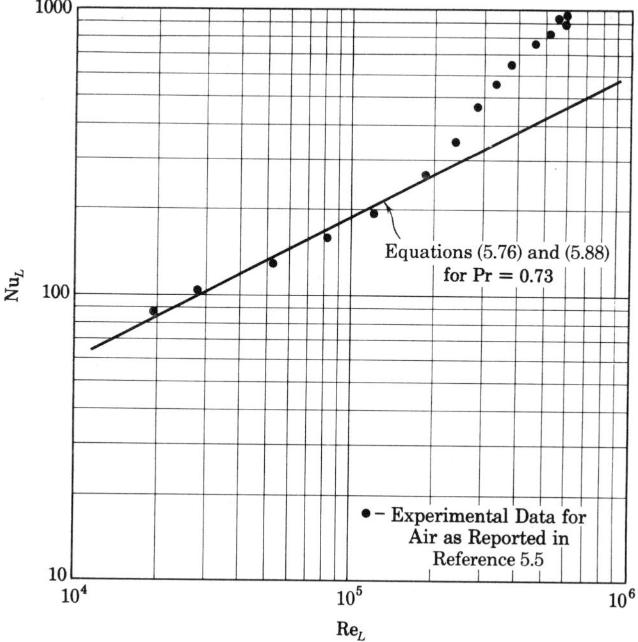

Figure 5.19. Comparison of theory and experimental for heat transfer in laminar flow of air past a flat surface.

Figure 5.19 compares the results given by either the differential solution [Eq. (5.76)] or the integral solution [Eq. (5.88)] with experimentally measured results for air. The close agreement, at least in the range of Re_x for which laminar flow can be expected, is rather good.

As noted earlier, the main purpose of showing the solution to the same problem by the two methods, differential and integral, was to show the basic methodology of each and to show that similar results can be expected. Obviously, there is no need to solve a given problem by both methods—only that which is most readily adapted to a given set of geometrical and hydrodynamic conditions. The numerous correlations given in the following chapters, for a number of cases of practical interest, have been obtained by one or the other of these techniques (or pure empirical experiment) with only the results noted.

The solutions and results just obtained were for the special geometry of the flat surface. Also involved were two other important assumptions: that the flow was laminar and that the effects of viscous dissipation could be neglected. The inclusion of the latter effect, viscous dissipation, complicates the nature of the differential equation to be solved. Such results are quoted in the next chapter, where, as expected, the Eckert number is seen to become significant. The appearance of turbulence produces such a profound effect in the basic nature of the hydrodynamic processes taking place that it is appropriate to examine this phenomenon in more detail, as in Sec. 5.7.

EXAMPLE 5.8

Water flows with a freestream velocity of $U = 1.5$ m/s and a freestream temperature of $t_f = 80°C$ past a flat surface of total length $L = 10$ cm and maintained at a temperature $t_s = 40°C$. Using the results obtained from the solution of the differential formulation for the boundary layer:

(a) Find the thickness of the velocity boundary layer and the thermal boundary layer at a location $x = 5$ cm from the leading edge.
(b) At the same location find the fluid velocity and temperature at a distance $y = 0.0226$ cm away from the surface.
(c) Find, per meter of depth into the paper, the total drag force and heat flow on one side of the surface. Use the properties of water evaluated at the mean of the surface and freestream temperatures.

Solution: For water at a mean temperature of $(80 + 40)/2 = 60°C$, Table A.3 gives the necessary properties:

$$\rho = 983.1 \text{ kg/m}^3, \qquad \nu = 0.4748 \times 10^{-6} \text{ m}^2/\text{s},$$
$$k = 0.6507 \text{ W/m-°C}, \qquad \text{Pr} = 3.00.$$

(a) At a location $x = 5$ cm from the leading edge, the local length Reynolds number is

$$\text{Re}_x = \frac{Ux}{\nu} = \frac{1.5 \times 0.05}{0.4748 \times 10^{-6}} = 1.580 \times 10^5,$$

well within the laminar range. Equation (5.72) gives the local velocity layer thickness:

$$\frac{\delta}{x} = 5.0\text{Re}_x^{-1/2} = 0.01258,$$

so that for $x = 5$ cm,

$$\delta = 0.0629 \text{ cm } (0.0248 \text{ in.}).$$

Equation (5.86) showed that for either the differential or integral formulations the ratio of the thermal and velocity layer thicknesses is

$$\frac{\delta_t}{\delta} = \text{Pr}^{-1/3} = (3)^{-1/3} = 0.6934,$$

so the thermal layer thickness at the same location is

$$\delta_t = 0.6934 \times 0.0629 = 0.0436 \text{ cm } (0.0172 \text{ in.}).$$

(b) At $x = 5$ cm and $y = 0.0226$ cm, the parameter η defined in Eq. (5.65) is

$$\eta = y\sqrt{\frac{U}{\nu x}} = \frac{0.0226}{100}\left(\frac{1.5}{0.4748 \times 10^{-6} \times 0.05}\right)^{1/2}$$
$$= 1.80.$$

For this value of η, Figs. 5.16 and 5.17 give the dimensionless velocity and temperature to be, approximately,

$$f' = \frac{v_x}{U} = 0.58, \qquad \phi = \frac{t - t_s}{t_f - t_s} = 0.79,$$

from which the fluid velocity and temperature are found to be

$$\frac{v_x}{1.5} = 0.58, \qquad v_x = 0.87 \text{ m/s (2.85 ft/s)},$$

$$\frac{t - 40}{80 - 40} = 0.79, \qquad t = 71.6°\text{C (160.9°F)}.$$

(c) For a total plate length $L = 10$ cm, the surface length Reynolds number is

$$\text{Re}_L = \frac{UL}{\nu} = \frac{1.5 \times 0.1}{0.4748 \times 10^{-6}} = 3.159 \times 10^5,$$

still within the laminar range. Equations (5.76) then give the drag coefficient and average Nusselt number to be

$$C_D = 1.328\text{Re}_L^{-1/2} = 1.328(3.159 \times 10^5)^{-1/2} = 0.00236,$$
$$\text{Nu}_L = 0.664\text{Pr}^{1/3}\text{Re}_L^{1/2} = 0.664(3)^{1/3}(3.159 \times 10^5)^{1/2} = 538.$$

Thus, from Eq. (5.20), the drag force on the plate for a total surface area, per meter of depth, of $A_s = 0.1 \times 1 = 0.1$ m^2:

$$C_D = \frac{\text{drag}}{\frac{1}{2}\rho U^2 A_s}$$

$$\text{drag} = 0.00236 \times \tfrac{1}{2} \times 983.1 \times (1.5)^2 \times 0.1$$
$$= 0.26 \text{ N } (0.058 \text{ lb}_f).$$

The average heat transfer coefficient is

$$\text{Nu}_L = \frac{hL}{k}$$

$$h = 538 \times \frac{0.6507}{0.1} = 3500 \text{ W/m}^2\text{-°C (616 Btu/h-ft}^2\text{-°F)},$$

so that the heat flow is

$$q = hA_s(t_s - t_f)$$
$$= 3500 \times 0.1 \times (80 - 40) = 14{,}000 \text{ W (47,800 Btu/h)}.$$

EXAMPLE 5.9

Repeat Example 5.8, but use the results of the integral analysis of the boundary layer.

Solution. The fluid properties found in Example 5.8 still apply.

(a) The local length Reynolds number at $x = 5$ cm found in Example 5.8 applies in this instance:

$$\text{Re}_x = 1.580 \times 10^5.$$

Thus Eq. (5.80) gives the velocity layer thickness to be, at $x = 5$ cm,

$$\frac{\delta}{x} = 4.64\text{Re}_x^{-1/2} = 0.01167$$

$$\delta = 0.0583 \text{ cm } (0.0230 \text{ in.}),$$

a value slightly less than that predicted by Eq. (5.72) in Example 5.8. The same relation [Eq. (5.86)] applies, so that the thermal layer thickness is

$$\frac{\delta_t}{\delta} = \text{Pr}^{-1/3} = (3)^{-1/3} = 0.6934$$

$$\delta_t = 0.0405 \text{ cm } (0.0159 \text{ in.}).$$

(b) The fluid velocity and temperature are found from the polynomial expressions assumed for the dimensionless distributions. For the velocity one needs the parameter y/δ; so at $y = 0.0226$ cm, with $\delta = 0.0583$ cm just found,

$$\frac{y}{\delta} = 0.3877.$$

Equation (5.79) gives the dimensionless velocity to be

$$\frac{v_x}{U} = \frac{3}{2}\left(\frac{y}{\delta}\right) - \frac{1}{2}\left(\frac{y}{\delta}\right)^3 = 0.552,$$

so that

$$v_x = 0.552 \times 1.5 = 0.83 \text{ m/s } (2.72 \text{ ft/s}).$$

This velocity is somewhat less than that predicted by the more accurate differential solution. The dimensionless temperature is given by Eq. (5.82) in terms of the parameter y/δ_t. Using the value of δ_t found in part (a), one has

$$\frac{y}{\delta_t} = \frac{0.0226}{0.0405} = 0.558,$$

$$\frac{t - t_s}{t_f - t_s} = \frac{3}{2}\left(\frac{y}{\delta_t}\right) - \frac{1}{2}\left(\frac{y}{\delta_t}\right)^3 = 0.750$$

$$\frac{t - 40}{80 - 40} = 0.750$$

$$t = 70.0°\text{C } (158.0°\text{F}).$$

(c) For a total plate Reynolds number of $\text{Re}_L = 3.159 \times 10^5$ for a plate with

$L = 10$ cm as found in Example 5.8, Eqs. (5.88) give the drag coefficient and average Nusselt number to be

$$C_D = 1.292\mathrm{Re}_L^{-1/2} = 0.0230,$$
$$\mathrm{Nu}_L = 0.662\mathrm{Pr}^{1/3}\mathrm{Re}_L^{1/2} = 537.$$

Calculations identical to those outlined in part (c) of Example 5.8, yield, finally,

$$\begin{aligned} \text{drag} &= 0.25 \text{ N } (0.056 \text{ lb}_f), \\ h &= 3490 \text{ W/m}^2\text{-°C } (615 \text{ Btu/h-ft}^2\text{-°F}), \\ q &= 13{,}960 \text{ W } (47{,}600 \text{ Btu/h}). \end{aligned}$$

The values for the drag force and heat flow are virtually the same as those found in Example 5.8 by the differential formulation.

5.7 TURBULENT BOUNDARY LAYERS

Most of the discussion so far in this chapter, and the governing equations for the boundary layer in particular, have been limited to the *laminar* state of flow. As noted earlier, in Sec. 5.3, under certain circumstances the fluid motion in a boundary layer may become *turbulent*—a flow state in which the various flow properties (velocity, temperature, etc.) appear to be "steady" on a time-averaged basis but which, on close examination, are actually time-dependent quantities fluctuating about steady, average values.

Following the concept introduced in Sec. 5.3 and Eq. (5.7), a *steady turbulent* flow may be characterized as one for which the local flow properties may be represented as

$$\begin{aligned} v_x &= \bar{v}_x + v_x', \\ v_y &= \bar{v}_y + v_y', \\ t &= \bar{t} + t', \\ P &= \bar{P} + P'. \end{aligned} \tag{5.89}$$

Here the flow properties (v_x, v_y, etc.) are time dependent but fluctuate about constant average values ($\bar{v}_x$, $\bar{v}_y$, etc.) by amounts (v_x', v_y', etc.) that have vanishing time averages. In a more formal sense, steady turbulent flows are those represented as in Eq. (5.89), in which, over some selected time interval $\Delta\tau$,

$$\bar{v}_x = \frac{1}{\Delta\tau}\int_{\tau}^{\tau+\Delta\tau} v_x \, d\tau = \text{constant},$$

$$\bar{v}_x' = \frac{1}{\Delta\tau}\int_{\tau}^{\tau+\Delta\tau} v_x' \, d\tau = 0;$$

and similarly for v_y, v_y', t, t', P, and P'. In the above the bar over a quantity represents its time average, and τ is being used to denote time.

Clearly, the foregoing representation is dependent on the time interval, $\Delta\tau$, over which the time average is taken. This interval must be large enough that the fluctuations can be observed but small enough that the time averages are constant. Thus what constitutes a steady turbulent motion is dependent on the physical situation being examined. What constitutes steady turbulence for an engineer studying water flow in a pipe may not satisfy the needs of a meteorologist studying the motion of the earth's atmosphere.

It must also be pointed out that while the time average of the various individual fluctuations are zero, the time average of their products need not be. That is, while $\bar{v}_x'$, $\bar{v}_y'$, and $\bar{t}'$ are all zero, the following may not:

$$\overline{v_x' v_y'} \neq 0, \text{ necessarily,}$$
$$\overline{v_x' t'} \neq 0, \text{ necessarily.}$$

In fact, as will be seen presently, the latter two quantities have very special significance in turbulent boundary layers.

The effect of the turbulent fluctuations on the fluid motion in a boundary layer can be seen by reexamining the governing differential equations for the laminar case [Eqs. (5.33)], which are repeated here for convenience:

$$\begin{aligned}
\frac{\partial v_x}{\partial x} + \frac{\partial v_y}{\partial y} &= 0, \\
v_x \frac{\partial v_x}{\partial x} + v_y \frac{\partial v_x}{\partial y} &= -\frac{1}{\rho}\frac{dP}{dx} + \frac{1}{\rho}\frac{\partial \tau}{\partial y} \\
&= -\frac{1}{\rho}\frac{dP}{dx} + \frac{\partial}{\partial y}\left(\nu \frac{\partial v_x}{\partial y}\right), \\
v_x \frac{\partial t}{\partial x} + v_y \frac{\partial t}{\partial y} &= \frac{\tau}{\rho c_p}\left(\frac{\partial v_x}{\partial y}\right) - \frac{1}{\rho c_p}\frac{\partial}{\partial y}\left(\frac{q}{A}\right) \\
&= \frac{\nu}{c_p}\left(\frac{\partial v_x}{\partial y}\right)^2 + \frac{\partial}{\partial y}\left(\alpha \frac{\partial t}{\partial y}\right).
\end{aligned} \tag{5.33}$$

The second forms of each of the momentum equation and the energy equation were obtained from the first forms by application of the laminar rate laws:

$$\begin{aligned}
\tau_l &= \mu \frac{\partial v_x}{\partial y} = \rho\nu \frac{\partial v_x}{\partial y}, \\
\left(\frac{q}{A}\right)_l &= -k \frac{\partial t}{\partial y} = -\rho c_p \alpha \frac{\partial t}{\partial y}.
\end{aligned} \tag{5.90}$$

In the latter expressions the subscript l has been added to emphasize that τ_l and $(q/A)_l$ are the *laminar* shear stress and heat flux resulting from the *molecular* scale exchange of momentum and energy and involve μ and k (or ν and α), which are fluid physical properties.

When one examines the motion in a turbulent boundary layer as shown, for example, for flow past a surface in Fig. 5.20, account must also be made for the additional exchange of momentum and energy due to the turbulent fluctuations v_x', v_y', t'.

Consider first the exchange of momentum across some plane, say S, in Fig. 5.20. As in the pure laminar case a molecular level exchange of x-directed momentum takes place across this plane leading to a laminar shear stress that is proportional to the gradient in the time-averaged velocity $\bar{v}_x$. Thus the laminar shear is the same as in Eq. (5.90) except with $\bar{v}_x$ replacing v_x. In addition, however, the turbulent fluctuations v_x' and v_y' give rise to an additional exchange of momentum on a macroscopic scale much larger than the molecular scale. Fluid masses that move upward (i.e., positive v_y') will be associated with, mainly, a negative v_x' since they are moving from a region where, on the whole, a smaller mean velocity prevails. The momentum change, per unit of area of S, experienced by such masses will then be $(\rho v_y')(-v_x')$. By similar reasoning, a negative v_y' will most often be associated with a positive v_x', with an x-directed momentum change of $(-\rho v_y')(v_x')$. Thus the time-averaged value of the flux of x-directed momentum through S will be $-\rho\overline{v_x'v_y'}$, an amount different from zero. This momentum exchange, on a macroscopic scale, gives rise to an additional "turbulent" shear stress

$$\tau_t = -\rho\overline{v_x'v_y'},$$

which should be added to the laminar stress already defined.

Similar reasoning may be applied to deduce that the turbulent fluctuations v_y' and t' contribute to an additional energy transport across S, over the molecular scale conduction, which may be interpreted as a turbulent heat flux:

$$\left(\frac{q}{A}\right)_t = \rho c_p \overline{v_y't'}.$$

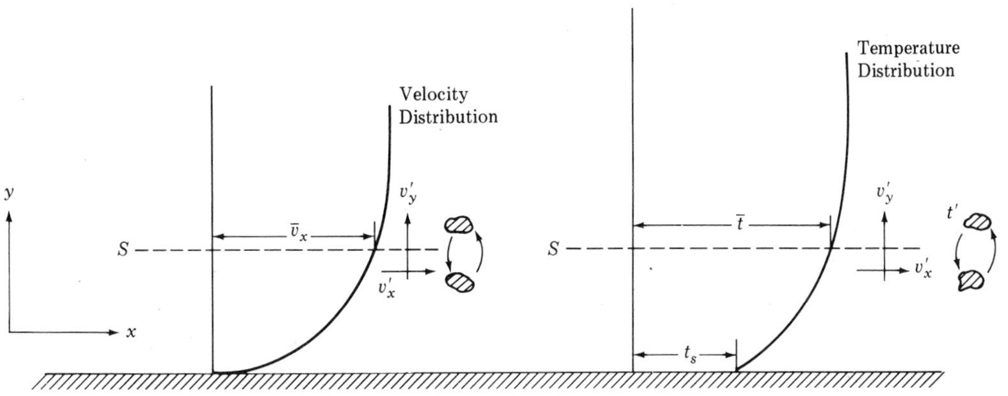

Figure 5.20. Turbulent momentum and energy exchange in a boundary layer.

The foregoing representations for the turbulent shear stress and turbulent heat flux are sound physically, but contribute little to the actual solution of a problem since the quantities $\overline{v_x'v_y'}$ and $\overline{v_y't'}$ are, as yet, unknown. An alternative representation that is often used is that in which two new coefficients, ϵ and ϵ_H, are defined such that

$$\begin{aligned} \tau_t &= -\rho\overline{v_x'v_y'} = \rho\epsilon\frac{\partial \bar{v}_x}{\partial y}, \\ \left(\frac{q}{A}\right)_t &= \rho c_p\overline{v_x't'} = -\rho c_p\epsilon_H\frac{\partial \bar{t}}{\partial y}. \end{aligned} \tag{5.91}$$

This is done so that for turbulent flow the total apparent shear stress and heat flux, the sum of their laminar and turbulent parts, are conveniently expressed as

$$\begin{aligned} \tau &= \tau_l + \tau_t \\ &= \rho(\nu + \epsilon)\frac{\partial \bar{v}_x}{\partial y}, \\ \frac{q}{A} &= \left(\frac{q}{A}\right)_l + \left(\frac{q}{A}\right)_t, \\ &= -\rho c_p(\alpha + \epsilon_H)\frac{\partial \bar{t}}{\partial y}. \end{aligned}$$

In this formulation, the effect of turbulence is represented as additional contributions to the shear stress and heat flux through the augmentation of the laminar kinematic viscosity ν, and diffusivity, α, by the amounts ϵ and ϵ_H, respectively. For this reason ϵ is called the *eddy viscosity* and ϵ_H the *eddy diffusivity*.

The introduction of the eddy viscosity and diffusivity is made to make the formal representation of the turbulent boundary layer as similar as possible to that for the laminar layer. Dropping now the overbar notation with the understanding that v_x, v_y, t, and P represent time-averaged values if a flow is turbulent, the following notation is used to represent the shear stress and heat flux in either the laminar or turbulent case:

$$\begin{aligned} \tau &= \rho(\nu + \epsilon)\frac{\partial v_x}{\partial y}, \\ \frac{q}{A} &= -\rho c_p(\alpha + \epsilon_H)\frac{\partial t}{\partial y}, \end{aligned} \tag{5.92}$$

with $\epsilon = \epsilon_H = 0$ for laminar flow. Then the general governing equations for the boundary layer are the same as derived in Sec. 5.4 for laminar flow, with $\nu + \epsilon$ replacing ν and $\alpha + \epsilon_H$ replacing α. The differential forms given in Eqs. (5.33) become

$$\frac{\partial v_x}{\partial x} + \frac{\partial v_y}{\partial y} = 0,$$

$$
\begin{aligned}
v_x \frac{\partial v_x}{\partial x} + v_y \frac{\partial v_x}{\partial y} &= -\frac{1}{\rho}\frac{dP}{dx} + \frac{1}{\rho}\frac{\partial \tau}{\partial y} \\
&= -\frac{1}{\rho}\frac{dP}{dx} + \frac{\partial}{\partial y}\left[(\nu + \epsilon)\frac{\partial v_x}{\partial y}\right], \\
v_x \frac{\partial t}{\partial x} + v_y \frac{\partial t}{\partial y} &= \frac{\tau}{\rho c_p}\left(\frac{\partial v_x}{\partial y}\right) - \frac{1}{\rho c_p}\frac{\partial}{\partial y}\left(\frac{q}{A}\right) \\
&= \frac{\nu + \epsilon}{c_p}\left(\frac{\partial v_x}{\partial y}\right)^2 + \frac{\partial}{\partial y}\left[(\alpha + \epsilon_H)\frac{\partial t}{\partial y}\right].
\end{aligned}
\tag{5.93}
$$

In like fashion, the integral governing equations in Eq. (5.45) become

$$
\begin{aligned}
\frac{d}{dx}\int_0^{\delta} v_x^2\, dy - U(x)\frac{d}{dx}\int_0^{\delta} v_x\, dy &= -\frac{\delta}{\rho}\frac{dP}{dx} - \frac{\tau_0}{\rho} \\
&= -\frac{\delta}{\rho}\frac{dP}{dx} - (\nu + \epsilon)\left(\frac{\partial v_x}{\partial y}\right)_0, \\
\frac{d}{dx}\int_0^{\delta_t} (t - t_f)v_x\, dy &= \frac{(q/A)_0}{c_p} \\
&= -(\alpha + \epsilon_H)\left(\frac{\partial t}{\partial y}\right)_0.
\end{aligned}
\tag{5.94}
$$

While the use of the concepts of the eddy viscosity, ϵ, and eddy diffusivity, ϵ_H, are convenient in that the basic governing equations are formally quite similar for the turbulent and laminar cases, it must be pointed out that this simplicity is rather deceptive. The laminar viscosity and diffusivity, ν and α, are known physical properties of the fluid; the eddy counterparts, ϵ and ϵ_H, are *not*. These eddy coefficients depend on the nature of the turbulent fluctuating quantities v_x', v_y', t'—as can be seen in the definitions given in Eq. (5.91). Thus ϵ and ϵ_H are quantities that depend on the state of turbulence of the flow. Consequently, some knowledge of how the fluctuating components or of how the eddy coefficients vary through the flow field is still required before a solution can be obtained. This constitutes one of the major points of inquiry in the study of turbulence. Here, certain simple empirical representations will be employed.

The general characteristics of turbulent boundary layers are similar to those of laminar layers—thin regions exist near the surface of a body where the time-averaged velocity and temperature vary rapidly from uniform values in the free-stream to a fixed value at the surface (zero in the case of velocity, t_s in the case of temperature). As a result of the transverse fluctuations, there exists a more uniform velocity or temperature distribution near the outer edge of the boundary layer than in the corresponding laminar case. Figure 5.21 illustrates, in a pictorial way, the typical distribution for the velocity layer; an analogous result would result for the thermal layer.

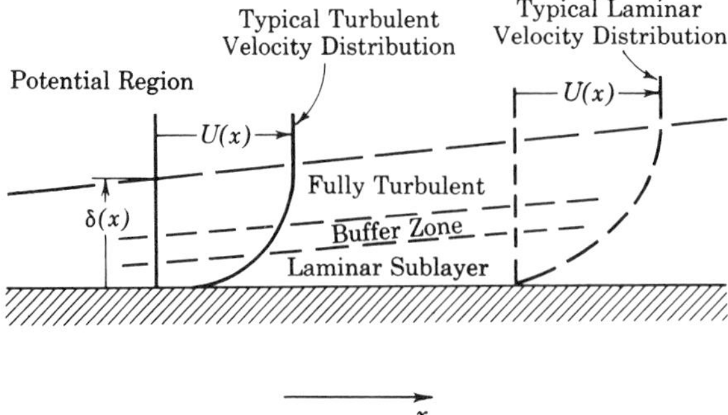

Figure 5.21. Turbulent boundary layer on a flat surface.

Often, the turbulent layer is further divided, as suggested in Fig. 5.21. Near the surface one may imagine a region in which the transverse fluctuations must be suppressed by the presence of the surface. In this region one would expect that the eddy viscosity and diffusivity are dominated by the laminar ones, or $\epsilon << \nu$ and $\epsilon_H << \alpha$. Thus the apparent shear stress and heat flux in Eqs. (5.92) follow the laminar laws and this region is sometimes called the *laminar sublayer*. Near the outer edge of the turbulent layer a region is observed to exist in which the opposite is true: $\epsilon >> \nu$, $\epsilon_H >> \alpha$. Such a region, in which the eddy quantities dominate and Eqs. (5.92) reduce to (5.91), is referred to as the *fully turbulent zone*. There may be a region between the laminar sublayer and the fully turbulent zone in which both effects must be accounted for ($\epsilon \approx \nu$, $\epsilon_H \approx \alpha$), and it is called the *buffer zone*. These three regions (laminar sublayer, buffer zone, and fully turbulent zone) are not distinctly identifiable, of course, and there must actually be some continuous variation from one to another.

The laminar, or molecular, Prandtl number has been defined as

$$\mathrm{Pr} = \frac{\nu}{\alpha}$$

and is clearly a fluid property. It has already been noted that the relative size of the laminar velocity and thermal layers is dependent on this quantity and that certain simplifications result for fluids with $\mathrm{Pr} \approx 1$. For turbulent flow one may also define a *turbulent Prandtl* number

$$\mathrm{Pr}_t = \frac{\epsilon}{\epsilon_H}.$$

Quite obviously the turbulent Prandtl number is *not* a fluid property but varies through the flow field as ϵ and ϵ_H vary. Because of the similarity of the turbulent mixing mechanisms for momentum and energy exchange, it is often possible to hypothesize that in the fully turbulent zone $\mathrm{Pr}_t \approx 1$ even though the molecular $\mathrm{Pr} \neq 1$.

As the foregoing discussion indicated, knowledge of how the eddy coefficients, ϵ and ϵ_H, vary in the flow field (or, equivalently, how v_x', v_y', t' vary) is necessary to use the governing equations to deduce such things as the velocity distribution, temperature distribution, surface shear stress, and surface heat transfer coefficients for turbulent flow. Much literature exists which explores this avenue of analysis. However, here a reverse approach will be used. Empirical representations for the surface skin friction and velocity profile will be taken as the starting point—from which deductions may be made as to the corresponding representations for ϵ and ϵ_H.

The Turbulent Skin Friction Coefficient

On the basis of extensive measurements in smooth tubes, Blasius (Ref. 6) deduced the following relation for the skin friction coefficient in turbulent flow on a smooth flat surface:

$$C_f = \frac{\tau_0}{\frac{1}{2}\rho U^2} = 0.045\left(\frac{\nu}{U\delta}\right)^{1/4}, \qquad 5 \times 10^5 < \mathrm{Re}_x < 10^7. \tag{5.95}$$

In Eq. (5.95), δ is the local boundary layer thickness at the location x from the leading edge of the surface. Thus the term $U\delta/\nu$ is seen to be a Reynolds number based on δ and is sometimes called the *thickness Reynolds number*.

Power Law Velocity Distribution

Velocity profiles in turbulent boundary layers have been obtained by a number of investigators. As would be expected, turbulent layers behave much in a laminar fashion near the surface (i.e., in the laminar sublayer) with an eventual transition to the fully turbulent condition farther away. A detailed description of this variation is complex, and one such representation is described in some detail in Sec. 5.8. However, for certain applications, the overall behavior of a turbulent boundary layer may be represented adequately by the following power law:

$$\frac{v_x}{U} = \left(\frac{y}{\delta}\right)^{1/7}. \tag{5.96}$$

The local layer thickness, δ, must be known in order to apply this law for v_x, and the method of deducing δ as a function of x will be noted shortly.

Figure 5.22 compares the above-noted one-seventh power law with data measured for air, and the good agreement is apparent. Clearly, the local length Reynolds number must exceed that for the laminar–turbulent transition (5×10^5), and experiment indicates that Eq. (5.96) is an adequate representation for the range $5 \times 10^5 < \mathrm{Re}_x < 10^7$. However, Eq. (5.96) can be used only to describe the overall behavior near the wall since a quick examination shows that $\partial v_x/\partial y \to \infty$ as $y \to 0$. Thus near the surface the one-seventh power law leads to a physically unacceptable shear stress there, and some other representation, such as that noted in the next section, must be applied.

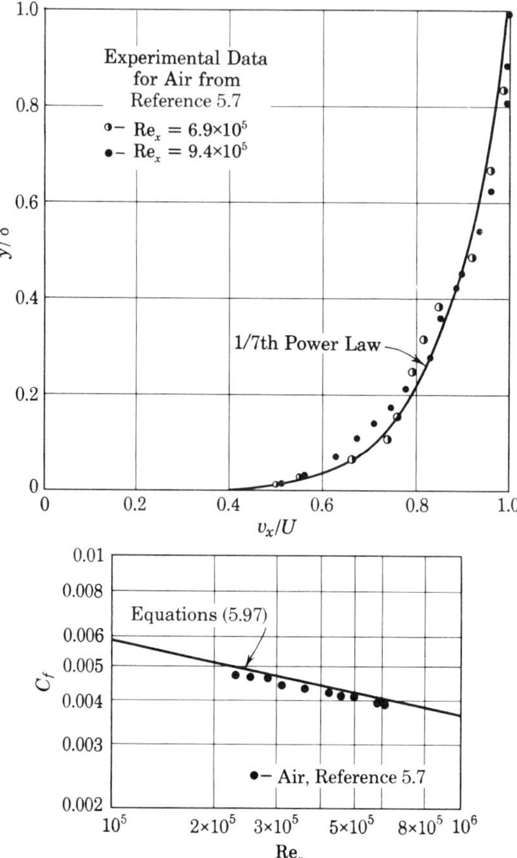

Figure 5.22. Velocity distribution and skin friction coefficient for turbulent flow of air past a flat surface.

As noted earlier in Eq. (5.94), the integral momentum equation may be applied to a turbulent boundary layer if the wall shear stress, τ_0, is adequately represented. For flow on a flat surface [so that $dP/dx = 0$, $U(x) = U =$ constant], Eq. (5.94) is (remembering that v_x is now the time-averaged velocity)

$$\frac{d}{dx}\int_0^{\delta} v_x(v_x - U)\,dy = -\frac{\tau_0}{\rho}.$$

If v_x is assumed to follow the one-seventh power law of Eq. (5.96) and the Blasius law for the skin friction coefficient [Eq. (5.95)] is used to represent τ_0, the above equation gives

$$U^2\frac{d}{dx}\int_0^{\delta}\left(\frac{y}{\delta}\right)^{1/7}\left[\left(\frac{y}{\delta}\right)^{1/7} - 1\right]dy = -\frac{1}{\rho}\left(\frac{1}{2}\,\rho U^2\right)0.045\left(\frac{\nu}{U\delta}\right)^{1/4}$$

or

$$\delta^{1/4}\, d\,\delta = \frac{0.045}{2}\,\frac{72}{7}\left(\frac{\nu}{U}\right)^{1/4} dx.$$

Integration under the condition that $\delta = 0$ at $x = 0$ (i.e., that x is measured from the beginning of the *turbulent* layer, an assumption to be discussed more fully later) yields

$$\begin{aligned}\frac{\delta}{x} &= \left(\frac{0.045}{2}\,\frac{72}{7}\,\frac{5}{4}\right)^{4/5}\left(\frac{\nu}{Ux}\right)^{1/5},\\ &= 0.371\mathrm{Re}_x^{-1/5}.\end{aligned}$$

The foregoing result predicts that a turbulent layer grows in thickness at a greater rate than does a laminar one, for which Eq. (5.72) showed δ/x growing as $\mathrm{Re}_x^{-1/2}$. The latter expression for δ/x may be combined with the Blasius law in Eq. (5.95) to express the skin friction coefficient in terms of the more convenient length Reynolds number rather than the thickness Reynolds number:

$$\begin{aligned}C_f &= 0.045\left(\frac{\nu}{Ux}\,\frac{x}{\delta}\right)^{1/4}\\ &= 0.045\left(\frac{\nu}{Ux}\,\frac{\mathrm{Re}_x^{1/5}}{0.371}\right)^{1/4}\\ &= 0.0577\mathrm{Re}_x^{-1/5}.\end{aligned}$$

The latter expression for C_f and the preceding one for δ/x may be used as they stand. However, Schlicting (Ref. 2) suggests that inaccuracies in the one-seventh power law be accounted for by empirically adjusting these expressions to the following form:

$$\left.\begin{aligned}\frac{\delta}{x} &= 0.381\mathrm{Re}_x^{-1/5},\\ C_f &= 0.0592\mathrm{Re}_x^{-1/5}.\end{aligned}\right\}\qquad 5 \times 10^5 < \mathrm{Re}_x < 10^7, \tag{5.97}$$

Figure 5.22 also compares experiment with Eq. (5.97) for the skin friction coefficient, and the agreement is good. For applications where the length Reynolds number exceeds 10^7 as noted in Eq. (5.97), Schlicting further recommends the following expirical relation:

$$C_f = 0.37(\log_{10}\mathrm{Re}_x)^{-2.584}, \qquad 10^7 < \mathrm{Re}_x. \tag{5.98}$$

The relations of Eqs. (5.97) and (5.98), then, will be used in what follows to represent the boundary layer thickness and local skin friction coefficient for turbulent flow on a flat surface. These relations are to be used in lieu of selecting a representation for ϵ, the eddy viscosity, and solving the governing boundary layer equations.

Universal Velocity Distribution

As noted in the foregoing, the one-seventh power law velocity distribution does not represent well the detailed velocity profile in a turbulent boundary layer, particularly near the surface. A more detailed representation is provided by the so-called "universal velocity distribution" (or "law of the wall"). This representation is based on a concept introduced by Prandtl known as the "mixing length." The details of the mathematics underlying this concept will not be presented here, only the results. The reader may wish to consult Ref. 2 for a more complete presentation.

In analogy to the concept used in the molecular exchange of momentum of a molecular mean free path, Prandtl hypothesized that the turbulent eddies which arise from the fluctuations v_x' and v_y' must travel some distance, the "mixing length," before losing their identities—and momentum. By further reasoning that this mixing length must be zero at the laminar sublayer, that it increase linearly with distance into the turbulent boundary layer, and that the shear stress is constant through the layer at the wall value, τ_0, Prandtl was able to predict that the time-averaged velocity in the boundary layer followed a logarithmic law,

$$v_x = C_1\sqrt{\frac{\tau_0}{\rho}}\ln y - C_2, \tag{5.99}$$

in which C_1 and C_2 are empirical constants to be determined. Also, this logarithmic law presumes prior knowledge of the wall shear stress, so that rather than being deduced from the velocity distribution, τ_0 must be obtained from some other source—namely the Blasius skin friction law already given in Eqs. (5.97) and (5.98).

The logarithmic velocity distribution of Eq. (5.99) is usually written in a dimensionless form by introducing a characteristic "shear velocity," v_x^*, defined as

$$v_x^* = \sqrt{\frac{\tau_0}{\rho}}. \tag{5.100}$$

Then a dimensionless velocity and y coordinate are defined:

$$v_x^+ = \frac{v_x}{v_x^*}, \tag{5.101}$$

$$y^+ = y\frac{v_x^*}{\nu}. \tag{5.102}$$

With these definitions, Eq. (5.99) becomes

$$v_x^+ = C_1 \ln y^+ + C_3, \tag{5.103}$$

with C_3 being a new constant.

This logarithmic velocity law clearly does not apply in the laminar sublayer but only in the buffer and fully turbulent zones, where values of C_1 and C_3 must yet be determined. Before quoting the generally accepted values of these constants, it is useful to note that the form of the velocity distribution given in Eq. (5.103) is tantamount to assuming the nature of the behavior of the eddy viscosity, ϵ, in the turbulent boundary layer. Since the total shear stress is given by

$$\tau = \rho(\nu + \epsilon)\frac{\partial v_x}{\partial y},$$

the assumption that $\tau = \tau_0$ throughout the boundary layer, as used in Prandtl's mixing length concept, leads to

$$\frac{dv_x^+}{dy^+} = \frac{1}{1 + (\epsilon/\nu)}. \tag{5.104}$$

Thus a knowledge of the dimensionless velocity distribution, v_x^+, as a function of y^+ amounts to knowing how the eddy viscosity ϵ varies in the boundary layer.

The universal velocity distribution results from representing the turbulent boundary layer as separable into three sublayers as suggested earlier: a *laminar* sublayer in which the eddy viscosity is negligible ($\epsilon << \nu$); a buffer zone where the eddy viscosity is of the same order as the kinematic viscosity ($\epsilon \approx \nu$); and a fully turbulent zone in which the eddy viscosity dominates ($\epsilon >> \nu$). As already noted, the logarithmic distribution is not applicable in the laminar sublayer, but if one puts $\epsilon/\nu \approx 0$ in Eq. (5.104), a linear velocity distribution, $v_x^+ = y^+$, results. For the buffer and fully turbulent zones the logarithmic distribution *is* used, with appropriately determined values of the constants C_1 and C_3. Equation (5.104) then gives the variation of ϵ in these zones as

$$\begin{aligned}\frac{\epsilon}{\nu} &= \frac{1}{dv_x^+/dy^+} - 1\\ &= \frac{y^+}{C_1} - 1.\end{aligned}$$

Figure 5.23 illustrates experimental data on turbulent boundary layers, and on the basis of data like these, the generally accepted representation for the universal velocity distribution is taken to be:

$$\begin{aligned}&\text{Laminar sublayer: } 0 < y^+ < 5: && v_x^+ = y^+\\ & && \frac{\epsilon}{\nu} = 0.\\ &\text{Buffer zone: } 5 < y^+ < 30: && v_x^+ = 5.0 \ln y^+ - 3.05,\\ & && \frac{\epsilon}{\nu} = \frac{y^+}{5} - 1, \quad 0 < \frac{\epsilon}{\nu} < 5.\\ &\text{Fully turbulent zone: } y^+ > 30: && v_x^+ = 2.5 \ln y^+ + 5.5,\\ & && \frac{\epsilon}{\nu} = \frac{y}{2.5} - 1, \quad \frac{\epsilon}{\nu} > 11.\end{aligned} \tag{5.105}$$

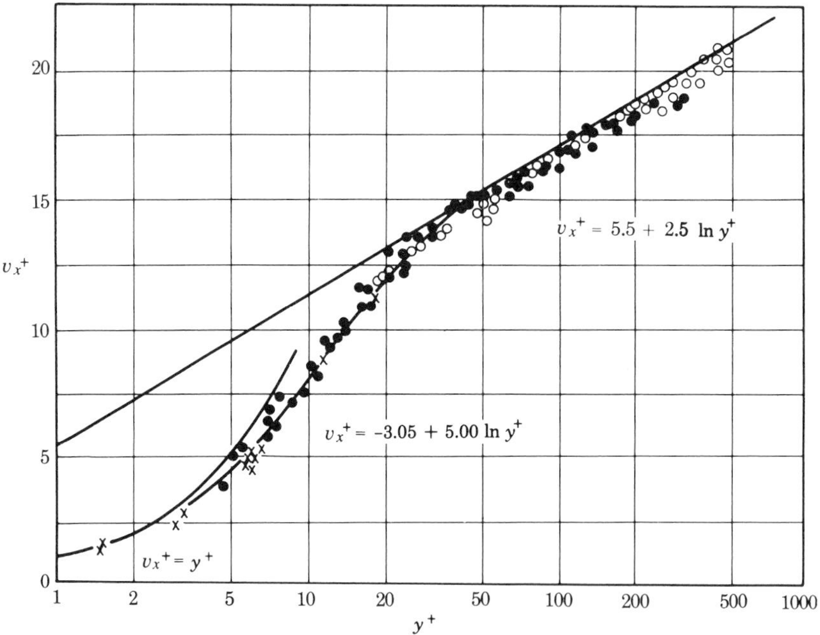

Figure 5.23. Universal velocity distribution. (From H. Reichardt, "Heat Transfer through Turbulent Boundary Layers," *NCAA Tech. Memo. 1047,* Washington, D.C., 1943.)

It must be remembered that in the use of the universal distribution of Eq. (5.105), one must first find the wall shear τ_0 from Eqs. (5.97) and (5.98) in order that the shear velocity v_x^* involved in the definitions of v_x^+ and y^+ can be found. For computational purposes this is perhaps better expressed in terms of the freestream velocity and the skin friction coefficient:

$$v_x^* = \sqrt{\frac{\tau_0}{\rho}} = \sqrt{U^2 \frac{C_f}{2}} = U\sqrt{\frac{C_f}{2}}. \tag{5.106}$$

EXAMPLE 5.10

Atmospheric pressure air at 90°F flows with a velocity of 50 ft/s past a flat surface maintained at 70°F. At a distance of 2 ft from the leading edge, find the thickness of the boundary layer, the laminar sublayer, and the buffer zone. At this location find the ratio ϵ/ν at a distance of 0.01 in. from the surface, 0.1 in. from the surface, and at the edge of the boundary layer. Evaluate the air properties at the mean between the freestream and surface temperatures.

Solution. At a mean temperature of $(90 + 70)/2 = 80°\text{F}$, Table A.6 gives $\rho =$

0.07350 lb_m/ft^3 and ν = 0.6086 ft^2/h. Thus at the location x = 2 ft, the local length Reynolds number is

$$\mathrm{Re}_x = \frac{Ux}{\nu} = \frac{50 \times 3600 \times 2.0}{0.6086} = 5.915 \times 10^5,$$

a value within the turbulent range. Consequently, Eq. (5.97) gives the dimensionless boundary layer thickness and the local skin friction coefficient to be

$$\frac{\delta}{x} = 0.381\mathrm{Re}_x^{-1/5} = 0.381 \times (5.915 \times 10^5)^{-1/5} = 0.0267,$$

$$C_f = 0.0592\mathrm{Re}_x^{-1/5} = 0.0592 \times (5.915 \times 10^5)^{-1/5} = 0.00415,$$

from which the boundary layer thickness at x = 2 ft is

$$\delta = 0.0267 \times 2 = 0.0534 \text{ ft} = 0.641 \text{ in. } (1.63 \text{ cm}).$$

The shear velocity v_x^* is needed for the universal velocity distribution and is found from its definition in Eq. (5.106):

$$v_x^* = \sqrt{\frac{\tau_0}{\rho}} = \sqrt{\frac{C_f}{2}U^2} = U\sqrt{\frac{C_f}{2}}$$

$$= 50\sqrt{\frac{0.00415}{2}} = 2.278 \text{ ft/s}.$$

Then from Eq. (5.102), the dimensionless coordinate y^+ is

$$y^+ = y\frac{v_x^*}{\nu} = y \times \frac{2.278 \times 3600}{0.6086} = y \times 1.347 \times 10^4, \quad y \text{ in ft}.$$

The universal distribution of Eq. (5.105) gives the limit of the laminar sublayer to be y^+ = 5 and the limit of the buffer zone to be y^+ = 30; thus these two limits are:

Limit of laminar sublayer: $y^+ = 5,$

$$y = \frac{5}{1.347 \times 10^4}$$

$$= 0.00031 \text{ ft} = 0.00445 \text{ in. } (0.0113 \text{ cm}).$$

Limit of buffer zone: $y^+ = 30,$

$$y = \frac{30}{1.347 \times 10^4}$$

$$= 0.002226 \text{ ft} = 0.0267 \text{ in. } (0.0678 \text{ cm}).$$

Compared to the full layer thickness, which was found above to be δ = 0.641 in., the laminar sublayer and the buffer zone are seen to be quite small. Thus the dominant portion of the turbulent layer is the fully turbulent zone. This fact that the laminar sublayer and the buffer zone are such a small part of the whole layer will be used later to simplify some problems.

In the laminar sublayer the eddy viscosity is neglected, so that $\epsilon/\nu \approx 0$ there. At a location 0.01 in. away from the surface, one is in the buffer zone since

$$y^+ = y \times 1.347 \times 10^4 = \frac{0.01}{12} \times 1.347 \times 10^4$$
$$= 11.23 < 30.$$

So at this location the ratio ϵ/ν, which is given by Eq. (5.105), has grown to

$$\frac{\epsilon}{\nu} = \frac{y^+}{5} - 1 = 1.246.$$

Here it is seen ϵ and ν are of the same order of magnitude. However, at $y = 0.1$ in., one moves into the fully turbulent zone and Eq. (5.105) in this region shows that ϵ/ν has grown considerably:

$$y^+ = y \times 1.347 \times 10^4 = \frac{0.1}{12} \times 1.347 \times 10^4,$$
$$= 112.3,$$
$$\frac{\epsilon}{\nu} = \frac{y^+}{2.5} - 1 = 43.9.$$

Here the eddy viscosity is seen to be more than an order of magnitude greater than ν, and at the outer limit of the layer, $y = \delta = 0.641$ in., it is several orders of magnitude greater:

$$y^+ = y \times 1.347 \times 10^4 = \frac{0.641}{12} \times 1.347 \times 10^4$$
$$= 719.8,$$
$$\frac{\epsilon}{\nu} = \frac{y^+}{2.5} - 1 = 287.$$

5.8 LOCAL HEAT TRANSFER IN TURBULENT BOUNDARY LAYERS ON A FLAT SURFACE

Section 5.7 introduced the concepts of the eddy viscosity and eddy diffusivity to represent the additional contributions to the transfer of momentum and energy in a turbulent boundary layer which result from the unsteady fluctuations in velocity and temperature that distinguish turbulent flow from laminar. Empirical laws for the skin friction coefficient and the velocity distribution were introduced and amounted to empirical representations for the eddy viscosity, ϵ. However, no mention was made of the temperature distribution in a turbulent boundary layer or, correspondingly, of the variation of the eddy diffusivity. While one might introduce the corresponding concept of a "universal temperature distribution," the analysis of a turbulent thermal boundary layer will be done by using the similarity between heat and momentum transfer already alluded to in earlier portions of this chapter.

The Similarity Between Heat and Momentum Transfer

The similarity between heat and momentum transfer may best be seen by first examining the laminar case already discussed in Sec. 5.6. For laminar flow Eq. (5.70) for the dimensionless velocity distribution $f' = v_x/U$ may be rewritten

$$\frac{d^2}{d\eta^2}\left(\frac{v_x}{U}\right) + \frac{1}{2} f \frac{d}{d\eta}\left(\frac{v_x}{U}\right) = 0, \qquad \frac{v_x(0)}{U} = 0, \quad \frac{v_x(\infty)}{U} = 1. \tag{5.107}$$

Equation (5.71) for the dimensionless temperature $\phi = (t - t_s)/(t_f - t_s)$ may be written

$$\frac{d^2\phi}{d\eta^2} + \frac{1}{2} f \,\mathrm{Pr}\, \frac{d\phi}{d\eta} = 0, \qquad \phi(0) = 0, \quad \phi(\infty) = 1. \tag{5.108}$$

Since laminar motion is being considered, Pr in Eq. (5.108) is the molecular Prandtl number. If Pr = 1 (a condition approached by many, but certainly not all, gases and some liquids), examination of Eqs. (5.107) and (5.108) shows that

$$\phi = \frac{v_x}{U}, \qquad \frac{t - t_s}{t_f - t_s} = \frac{v_s}{U}. \tag{5.109}$$

Thus, for the special case of laminar flow when Pr = 1, the dimensionless temperature and velocity profiles are identical.

More important, however, is the fact that Eq. (5.109) implies that

$$\frac{1}{t_f - t_s}\frac{\partial t}{\partial y} = \frac{1}{U}\frac{\partial v_s}{\partial y}, \tag{5.110}$$

so that for laminar flow in which the shear stress and heat flux are given by

$$\tau = \rho \boldsymbol{\nu} \frac{\partial v_x}{\partial y},$$

$$\frac{q}{A} = -\rho c_p \alpha \frac{\partial t}{\partial y},$$

one may deduce that

$$\frac{q/A}{\tau} = -c_p \frac{t_f - t_s}{U}. \tag{5.111}$$

The implication of Eq. (5.111) is that for Pr = 1, the ratio of the heat flux to the shear stress in a laminar boundary layer is *constant!* This latter fact will be extended to turbulent layers in the next section in what is called the *Reynolds analogy*, but before doing so it is instructive to make another deduction for the laminar case first.

Equation (5.111) for laminar flow may be rewritten

$$\frac{q/A}{t_s - t_f} = \frac{\tau c_p}{U}.$$

When evaluated at the surface where the heat flux is $(q/A)_0$ and the shear stress is τ_0, the left side of the equation above is recognized as the surface heat transfer coefficient, h. Thus

$$h = \frac{\tau_0 c_p}{U}. \tag{5.112}$$

Equation (5.112) may be interpreted as a statement of Reynolds analogy for laminar flow. It states that for fluids with Pr = 1, the heat transfer coefficient and the surface shear stress are uniquely related. The implication of this similarity between heat and momentum transfer is quite important. High heat transfer rates are invariably accompanied by high viscous drag forces; in some circumstances, viscous drag measurements may be used to deduce heat transfer coefficients.

Equation (5.112) may be written in a dimensionless form as follows, recognizing that Pr = 1 implies $\boldsymbol{\nu} = \alpha$ or $\mu c_p = k$:

$$\begin{aligned} \frac{hx}{k} &= \frac{\tau_0 c_p x}{kU}, \\ &= \frac{\tau_0 x}{\mu U}, \\ &= \frac{C_f}{2}\frac{\rho U x}{\mu}, \\ \text{Nu}_x &= \frac{C_f}{2}\text{Re}_x. \end{aligned} \tag{5.113}$$

Examination of Eq. (5.75) when Pr = 1 bears out the statement given in Eq. (5.113).

The similarity of heat and momentum transfer implied in Eqs. (5.111), (5.112), and (5.113) for laminar flow may be extended to the turbulent case when Pr $\neq$ 1, providing considerable simplifications, as is illustrated next.

The Reynolds Analogy for Turbulent Heat Transfer

In the case of turbulent flow past a flat surface, the total apparent shear stress and heat flux are given by

$$\tau = \rho(\nu + \epsilon)\frac{\partial v_x}{\partial y},$$
$$\frac{q}{A} = -\rho c_p(\alpha + \epsilon_H)\frac{\partial t}{\partial y}. \tag{5.114}$$

Reynolds assumed, as a simple model, that the turbulent boundary layer consisted of only the fully turbulent zone—that is, he presumed that the laminar sublayer and the buffer zone are negligible. In fact, Example 5.10 just showed that the latter two zones are often quite small when compared with the fully turbulent zone. Thus, in the boundary layer $\epsilon >> \nu$ and $\epsilon_H >> \alpha$, so that ν and α may be taken as negligible and the ratio of the two expressions in Eq. (5.114) yields

$$\frac{q/A}{\tau} = -c_p\frac{\epsilon_H}{\epsilon}\frac{\partial t}{\partial v_x} = -\frac{c_p}{\mathrm{Pr}_t}\frac{\partial t}{\partial v_x},$$

in which $\mathrm{Pr}_t = \epsilon/\epsilon_H$ is the *turbulent* Prandtl number mentioned in Sec. 5.7. Reynolds further assumed that since the eddy viscosity and eddy diffusivity arise from the same mechanism of transverse fluctuation, then $\mathrm{Pr}_t \approx 1$, i.e., $\epsilon \approx \epsilon_H$, so that

$$\frac{q/A}{\tau} = -c_p\frac{\partial t}{\partial v_x}. \tag{5.115}$$

The similarity between heat and momentum transfer discussed for laminar flow in Sec. 5.7 noted that the ratio $(q/A)/\tau$ was constant. Reynolds assumed that this same similarity exists for the fully turbulent boundary layer and integrated Eq. (5.115) across the boundary layer from $t = t_s$, $v_x = 0$ to $t = t_f$, $v_x = U$, while taking $(q/A)/\tau$ as constant and equal to the wall values, to obtain

$$\frac{(q/A)_0}{\tau_0} = -c_p\frac{t_f - t_s}{U}$$

or

$$h = \frac{\tau_0 c_p}{U}.$$

While the latter expression appears to be the same as that in Eq. (5.112) for laminar flow, one should note that although it has been presumed that the turbulent Prandtl number is unity, $\mathrm{Pr}_t \approx 1$, it has *not* been assumed that the molecular Prandtl number $\mathrm{Pr} = \nu/\alpha$ is 1. Thus the statement above may also be written

$$h = \frac{\tau_0 k\mathrm{Pr}}{\mu U},$$
$$\mathrm{Nu}_x = \frac{C_f}{2}\mathrm{Re}_x\mathrm{Pr}. \tag{5.116}$$

Equation (5.116) is known as the *Reynolds analogy* for turbulent heat transfer on a flat surface. When Eq. (5.116) is combined with Eqs. (5.97) and (5.98), Reynolds analogy for turbulent heat transfer on a flat surface is

$$\begin{aligned} \text{Nu}_x &= 0.0296\text{Re}_x^{0.8}\text{Pr}, \qquad 5 \times 10^5 < \text{Re}_x < 10^7, \\ \text{Nu}_x &= \frac{0.185\text{Re}_x\text{Pr}}{(\log_{10} \text{Re}_x)^{2.584}}, \qquad 10^7 < \text{Re}_x. \end{aligned} \tag{5.117}$$

Equations (5.117) may be used to predict turbulent heat transfer coefficients with reasonable accuracy; however, comparison with experiments will be deferred until after other improvements have been discussed. Also, integrated forms of Eqs. (5.117) for the average plate length Nusselt number will be obtained later.

The Colburn Analogy

On the basis of experimental evidence Colburn (Ref. 8) empirically modified the Reynolds analogy as

$$\text{Nu}_x = \frac{C_f}{2} \text{Re}_x\text{Pr}^{1/3}, \tag{5.118}$$

or for the flat surface:

$$\begin{aligned} \text{Nu}_x &= 0.0296\text{Re}_x^{0.8}\text{Pr}^{1/3}, \qquad 5 \times 10^5 < \text{Re}_x < 10^7, \\ \text{Nu}_x &= \frac{0.185\text{Re}_x\text{Pr}^{1/3}}{(\log_{10} \text{Re}_x)^{2.584}}, \qquad 10^7 < \text{Re}_x. \end{aligned} \tag{5.119}$$

The Von Kármán Analogy

Von Kármán (Ref. 9) improved on Reynolds' analysis by including the laminar sublayer and the buffer zone given in the universal velocity distribution of Eq. (5.105). Applying Eq. (5.115) for the fully turbulent zone between $y^+ = 30$ and the freestream, again with the ratio $(q/A)_0/\tau_0$ constant, gives

$$\frac{(q/A)_0}{\tau_0} = -c_p \frac{t_f - t_{30}}{U - v_{x30}}. \tag{5.120}$$

The temperature at the limit of the buffer zone, t_{30}, is found by integrating

$$\begin{aligned} \left(\frac{q}{A}\right)_0 &= -\rho c_p(\alpha + \epsilon_H) \frac{\partial t}{\partial y} \\ &= -\rho c_p\left(\frac{1}{\text{Pr}} + \frac{\epsilon_H}{\nu}\right)\sqrt{\frac{\tau_0}{\rho}} \frac{\partial t}{\partial y^+} \end{aligned}$$

between $y^+ = 0$, $t = t_s$ and $y^+ = 30$, $t = t_{30}$. In this integration one uses $\epsilon_H/\nu = 0$ in the interval $0 < y^+ < 5$ (the laminar sublayer) and for $5 < y^+ < 30$ one uses

$$\frac{\epsilon_H}{\nu} = \frac{\epsilon}{\nu} = \frac{y^+}{5} - 1$$

as given in Eq. (5.105) for the buffer zone. Considerable algebra gives

$$t_s - t_{30} = \sqrt{\frac{\rho}{\tau_0}}\frac{(q/A)_0}{\rho c_p}\left[5\text{Pr} + 5\ln\frac{(1/\text{Pr} - 1) + 6}{(1/\text{Pr} - 1) + 1}\right]. \tag{5.121}$$

The velocity v_{x30} required in Eq. (5.120) is readily found from Eq. (5.105):

$$v_{x30} = \sqrt{\frac{\tau_0}{\rho}}\,(5\ln 30 - 3.05)$$

$$= \sqrt{\frac{\tau_0}{\rho}}\,(5\ln 6 + 5).$$

If the latter expression and Eq. (5.121) are introduced into Eq. (5.120) more algebraic manipulation yields finally the von Kármán analogy:

$$h = \frac{(q/A)_0}{t_s - t_f} = \frac{\tau_0 c_p}{U + \sqrt{\tau_0/\rho}\,\{5(\text{Pr} - 1) + 5\ln[1 + \frac{5}{6}(\text{Pr} - 1)]\}},$$

$$\text{Nu}_x = \frac{\frac{1}{2}C_f\text{Re}_x\text{Pr}}{1 + 5\sqrt{C_f/2}\,\{(\text{Pr} - 1) + \ln[1 + \frac{5}{6}(\text{Pr} - 1)]\}}. \tag{5.122}$$

With Eq. (5.97), one has

$$\text{Nu}_x = \frac{0.0296\text{Re}_x^{0.8}\text{Pr}}{1 + 0.860\text{Re}_x^{-0.1}\{(\text{Pr} - 1) + \ln[1 + \frac{5}{6}(\text{Pr} - 1)]\}} \tag{5.123}$$

for $5 \times 10^5 < \text{Re}_x < 10^7$. Von Kármán's analogy [Eq. (5.123)] is found to give reasonable predictions for turbulent heat transfer on a flat surface. Comparison with experiment is deferred until later, although the reader may wish to refer to Fig. 5.31, in which the counterpart of Eq. (5.123) for pipe flow is compared with experiment.

As was mentioned earlier, the main purpose of this chapter was to illustrate the *method* of analysis in the prediction of heat transfer coefficients. The examples noted are just that, *examples*. More elaborate theories exist for the prediction of heat transfer and skin friction; however, the basic concepts involved are not different, just the amount of algebra. Chapter 6 will present a summary of the foregoing analyses, and others, that are recommended for the solution of practical engineering problems.

EXAMPLE 5.11

For the conditions stated in Example 5.10, atmospheric air with a velocity U = 50 ft/s and temperature t_f = 90°F flowing past a flat surface at t_s = 70°F, find the local heat transfer coefficient and heat flux at a location 2 ft from the leading edge. Use the predictions given by (a) the Reynolds analogy; (b) the Colburn analogy; (c) the von Kármán analogy. Compare the results of these three representations.

Solution. At a mean temperature of $(90 + 70)/2 = 80°\text{F}$, Table A.6 gives

$$\nu = 0.6086 \text{ ft}^2/\text{h}, \qquad k = 0.01511 \text{ Btu/h-ft-°F}, \qquad \text{Pr} = 0.711.$$

As already found in Example 5.10, the local Reynolds number and skin friction coefficient are, at $x = 2$ ft,

$$\text{Re}_x = \frac{Ux}{\nu} = 5.915 \times 10^5,$$

$$C_f = 0.0592\text{Re}_x^{-1/2} = 0.00415.$$

(a) Reynolds analogy [Eq. (5.116) or (5.117)] gives the local Nusselt number:

$$\text{Nu}_x = \frac{C_f}{2}\text{Re}_x\text{Pr} = 0.0296\text{Re}_x^{0.8}\text{Pr}$$

$$= \frac{0.00415}{2} \times 5.915 \times 10^5 \times 0.711 = 872.4.$$

Thus the local heat transfer coefficient and heat flux are

$$\text{Nu}_x = \frac{h_x x}{k}$$

$$872.4 = \frac{h_x \times 2.0}{0.01511}$$

$$h_x = 6.59 \text{ Btu/h-ft}^2\text{-°F } (37.4 \text{ W/m}^2\text{-°C}).$$

$$\left(\frac{q}{A}\right)_x = h_x(t_s - t_f)$$

$$= 6.59(70 - 90)$$

$$= -131.8 \text{ Btu/h-ft}^2 \ (-415.8 \text{ W/m}^2)$$

(into the surface).

(b) For the Colburn analogy, Eq. (5.119) gives

$$\text{Nu}_x = 0.0296\text{Re}_x^{0.8}\text{Pr}^{1/3}$$

$$= 0.0296(5.915 \times 10^5)^{0.8}(0.711)^{1/3} = 1095.$$

$$\text{Nu}_x = \frac{h_x x}{k}, \qquad h_x = 8.27 \text{ Btu/h-ft}^2\text{-°F } (47.0 \text{ W/m}^2\text{-°C}).$$

$$\left(\frac{q}{A}\right)_x = h_x(t_s - t_f) = -165.4 \text{ Btu/h-ft}^2 \ (521.8 \text{ W/m}^2).$$

(c) For the von Kármán analogy, Eq. (5.123) yields

$$\mathrm{Nu}_x = \frac{0.0296\mathrm{Re}_x^{0.8}\mathrm{Pr}}{1 + 0.860\mathrm{Re}_x^{-0.1}\{(\mathrm{Pr} - 1) + \ln[1 + \frac{5}{6}(\mathrm{Pr} - 1)]\}}$$

$$= 1001.$$

$$\mathrm{Nu}_x = \frac{h_x x}{k}, \qquad h_x = 7.56 \text{ Btu/h-ft}^2\text{-°F } (42.9 \text{ W/m}^2\text{-°C}).$$

$$\left(\frac{q}{A}\right)_x = h_x(t_s - t_f) = -151.2 \text{ Btu/h-ft}^2 \ (478.0 \text{ W/m}^2).$$

Von Kármán's analogy is probably the most accurate of those used above. Colburn's analogy gives values within 10% of those given by von Kármán but is certainly simpler to use. The Reynolds analogy results differ considerably from the others and should be regarded as only a reasonable estimate. It should be noted that because of the complexity of the physical processes taking place in convective heat transfer agreement within ±10% of various analytical formulations or empirical predictions is generally regarded as quite acceptable.

5.9 VISCOUS FLOW IN PIPES OR TUBES—FULLY DEVELOPED FLOW

Of equal practical importance as the cases of flow past a flat surface just considered is that of flow through circular pipes or tubes. The remaining sections of this chapter will be devoted to the analysis of heat transfer for this important geometry.

As a viscous fluid, initially of uniform velocity, enters a pipe or tube a boundary layer builds up along the surface of the pipe. This is illustrated in Fig. 5.24. Near the entrance of the pipe this growth of the boundary layer is much like the flow along the flat surface—particularly if the curvature of the pipe is not large. However, the flow cannot be the same as in the flat surface case due to the presence of the opposite wall—where a boundary layer is also developing. As the flow proceeds down the pipe, the boundary layer growing along the pipe wall gets

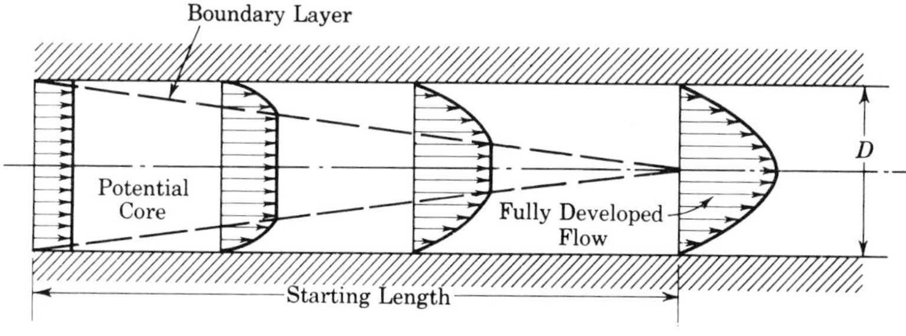

Figure 5.24. Starting length in pipe flow.

thicker, eventually growing together from opposite sides and filling the pipe with "boundary layer" flow. The state of flow cannot actually be considered as boundary layer flow since the fundamental assumption that the viscous flow region is thin compared to the geometric dimensions of the pipe is no longer satisfied. However, experiment shows that many of the empirical laws quoted in the foregoing sections for flow along a flat surface may also be used in the case of pipe flow when appropriate definitions are made for the undisturbed stream velocity, boundary layer thickness, etc.

Pipe flow is termed *fully developed flow* once the viscous layers have grown together and filled the pipe cross section. The pipe length required to establish this fully developed flow is called the *starting* or *entrance* length. In the starting length the velocity distribution across a diameter consists of an inviscid core near the center of the pipe (the velocity being uniform) which joins the boundary layer region at each surface where the velocity varies from the inviscid core value to zero at the wall. As one moves along the pipe in the starting region, the viscous or boundary layer portion of the velocity distribution curve increases while the inviscid core portion decreases. For constancy of mass flow rate, the mean velocity across the cross section (mass flow rate divided by density and cross-sectional area) must remain constant—for incompressible flow. Hence the velocity of the inviscid core region must increase as the flow proceeds down the starting length. This increase is also shown in Fig. 5.24.

The transition from the laminar to turbulent state of flow may occur either in the boundary layer in the starting length or in the fully developed flow—or not at all—depending on the conditions of the flow (i.e., velocity, viscosity, etc.). Thus it is possible to have a laminar or turbulent fully developed region. The transition in the starting region can be predicted reasonably well by use of the criterion based on the local length Reynolds number, Re_x, as noted in Sec. 5.3.

For fully developed pipe flow the length Reynolds number loses its significance, and it is customary to employ the *diameter* Reynolds number, Re_D, introduced earlier in Sec. 5.3 and Eq. (5.10):

$$\mathrm{Re}_D = \frac{U_m D}{\nu},$$

U_m being the mean velocity of the flow in the pipe. Recall that it was noted earlier that the laminar–turbulent transition in pipe flow is generally regarded as occurring at a value of $\mathrm{Re}_D \approx 2300$.

A formula, developed by Langhaar (Ref. 10), which is useful for predicting the length of the starting section of laminar pipe flow is

$$\frac{L_s}{D} = 0.0575\mathrm{Re}_D. \tag{5.124}$$

This formula must be treated as approximate since the fully developed condition is reached asymptotically, and hence the starting length, L_s, is difficult to define. If the laminar–turbulent transition takes place in the starting region, the starting length decreases from the value indicated by Eq. (5.124). No adequate theory exists which will predict the starting length in turbulent flow since the nature of the pipe entrance, the pipe roughness, etc., will have a serious effect on the growth and transition of the boundary layer in the pipe. Starting lengths of 25 to 40 pipe diameters are typical for turbulent flow.

In fully developed flow it is customary to represent the surface shear stress in a manner similar to that used for the skin friction coefficient in external flow. If τ_R is used to denote the fluid shear stress at the pipe wall (i.e., at the pipe radius R) the *Darcy friction factor,* f, is defined

$$f = \frac{4\tau_R}{\frac{1}{2}\rho U_m^2}. \tag{5.125}$$

Here U_m again represents the mean flow velocity over the pipe cross section. Some workers prefer a friction factor, the Fanning factor, which is one-fourth of that just defined, in analogy to the skin friction coefficient; however, the Darcy factor just defined is more often used.

Wall friction in a pipe must be balanced by a gradient in pressure along the pipe length. Thus if dP is the pressure difference between two cross sections dx apart,

$$-\pi R^2\, dP = 2\pi R \tau_R\, dx,$$

so that the pressure gradient along the pipe is

$$\begin{aligned} -\frac{dP}{dx} &= \frac{2\tau_R}{R} = \frac{4\tau_R}{D}, \\ &= \frac{1}{2}\,\rho U_m^2\,\frac{f}{D}. \end{aligned}$$

For fully developed flow in which the fluid properties remain constant U_m and f are constant along the pipe, and the pressure drop due to friction over a pipe length L is, then,

$$-\Delta P = f\,\frac{L}{D}\,\frac{1}{2}\,\rho U_m^2. \tag{5.126}$$

Thus the friction factor is related to the pressure gradient in the pipe and is used to calculate pressure losses due to friction. As will be seen later, application of the various analogies between heat and momentum transfer will relate f to the heat transfer coefficient in the pipe.

Laminar Flow

For fully developed laminar flow, the velocity distribution across a pipe cross section is given by the well-known parabolic form,

$$\frac{v}{U_m} = 2\left[1 - \left(\frac{r}{R}\right)^2\right] \tag{5.127}$$

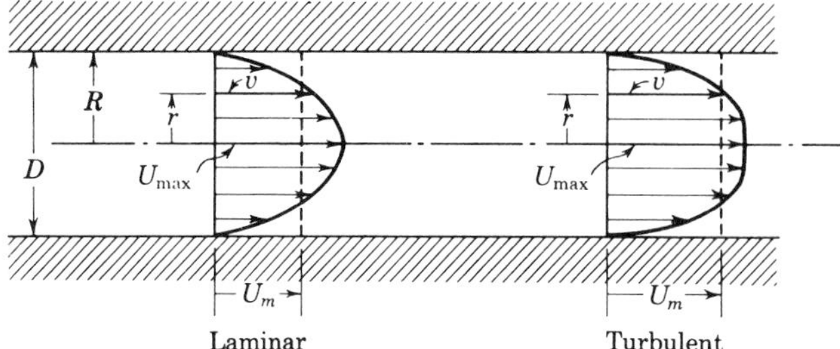

Figure 5.25. Fully developed profiles for laminar and turbulent pipe flow.

as illustrated in Fig. 5.25. In Eq. (5.127), v is the local velocity at the radial location r, and R is the pipe radius. Since the surface shear stress is

$$\tau_R = -\mu\left(\frac{\partial v}{\partial r}\right)_{r=R},$$

the negative sign arising from the fact that r is measured positive away from the centerline, one has

$$\tau_R = 4\mu\,\frac{U_m}{R},$$

or by Eq. (5.125),

$$f = 64\,\frac{\mu}{U_m D\rho} = \frac{64}{\mathrm{Re}_D}. \tag{5.128}$$

Thus the friction factor is easily found for fully developed laminar flow from the pipe diameter Reynolds number.

Turbulent Flow

For fully developed turbulent flow, as in the case of flow past a flat surface, the velocity profile is much flatter than the parabolic laminar profile. Experiments have shown that while pipe flow is not a flow with a zero pressure gradient, as in the flat surface case, many of the relations for flat surfaces may be applied with reasonable accuracy to fully developed flow in a pipe with the boundary layer thickness, δ, replaced by R and the free stream velocity U replaced by U_{max}. The symbol U_{max} is used to denote the maximum tube axial velocity along the centerline.

Power Law Velocity Distribution. The power law velocity distribution given in Eq. (5.96) becomes, for pipe flow,

$$\frac{v}{U_{max}} = \left(\frac{R - r}{R}\right)^{1/7}. \tag{5.129}$$

As in the case of flat surface flow, this relation is only approximate and cannot be expected to describe the fully developed velocity profile near the pipe wall. It does give an adequate overall representation and yields the flat profile suggested in Fig. 5.25. Since it is customary to relate pipe flow parameters to the mean velocity, U_m, Eq. (5.129) may be integrated to yield

$$\begin{aligned} U_m &= \frac{1}{\pi R^2}\int_0^R v(2\pi r)\,dr \\ &= \frac{98}{120}U_{\max} = 0.817U_{\max} \approx 0.8U_{\max}. \end{aligned} \tag{5.130}$$

Schlicting (Ref. 2) suggests the latter adjustment that $U_m \approx 0.8U_{\max}$ as being more physically realistic.

A friction factor for fully developed pipe flow based on the one-seventh power law may now be deduced. The Blasius relation for flat plate shear stress in terms of the boundary layer thickness may be translated into that for pipe flow by replacing in Eq. (5.95) τ_0 with τ_R, U with $U_{\max}$, and δ with $R/2$:

$$\frac{\tau_R}{\frac{1}{2}\rho U_{\max}^2} = 0.045\left(\frac{\nu}{U_{\max}R}\right)^{1/4}.$$

With $U_{\max} = U_m/0.8$, one finds

$$\begin{aligned} f &= \frac{4\tau_R}{\frac{1}{2}\rho U_m^2} = \left[4 \times 0.045\,\frac{(2 \times 0.8)^{1/4}}{(0.8)^2}\right]\left(\frac{\nu}{U_m D}\right)^{1/4}, \\ f &= 0.316\mathrm{Re}_D^{-1/4}, \quad 10^4 < \mathrm{Re}_D < 5 \times 10^4. \end{aligned} \tag{5.131}$$

For higher Re_D, one may use the following empirical relation (Ref. 11):

$$f = 0.184\mathrm{Re}_D^{-1/5}, \quad 3 \times 10^4 < \mathrm{Re}_D < 10^6. \tag{5.132}$$

Both Eqs. (5.131) and (5.132) are for smooth tubes and result principally from the one-seventh power law. Equivalent relations resulting from the universal distribution and some including the effects of surface roughness are noted next.

Universal Velocity Distribution. For fully developed pipe flow, the universal velocity distribution given in Eq. (5.105) may be used as long as the dimensionless variables are defined as

$$v^* = \sqrt{\frac{\tau_R}{\rho}},$$

$$y^+ = (R - r)\frac{v^*}{\nu},$$

$$v^+ = \frac{v}{v^*}.$$

Since

$$f = \frac{4\tau_R}{\frac{1}{2}\rho U_m^2}$$

$$= \frac{8}{(U_m^+)^2}$$

the friction factor can be predicted from the mean value of the dimensionless velocity distribution. Equations (5.105) may be integrated (Ref. 12) to show

$$U_m^+ = 2.5 \ln \left(\frac{R}{\nu} \sqrt{\frac{\tau_R}{\rho}} \right) + 1.75,$$

from which

$$\frac{1}{\sqrt{f}} = 0.88 \ln (\mathrm{Re}_D \sqrt{f}) - 0.91.$$

Reference 2 suggests that the above equation fits experimental data better if empirically adjusted to

$$\frac{1}{\sqrt{f}} = 0.87 \ln (\mathrm{Re}_D \sqrt{f}) - 0.8. \qquad (5.133)$$

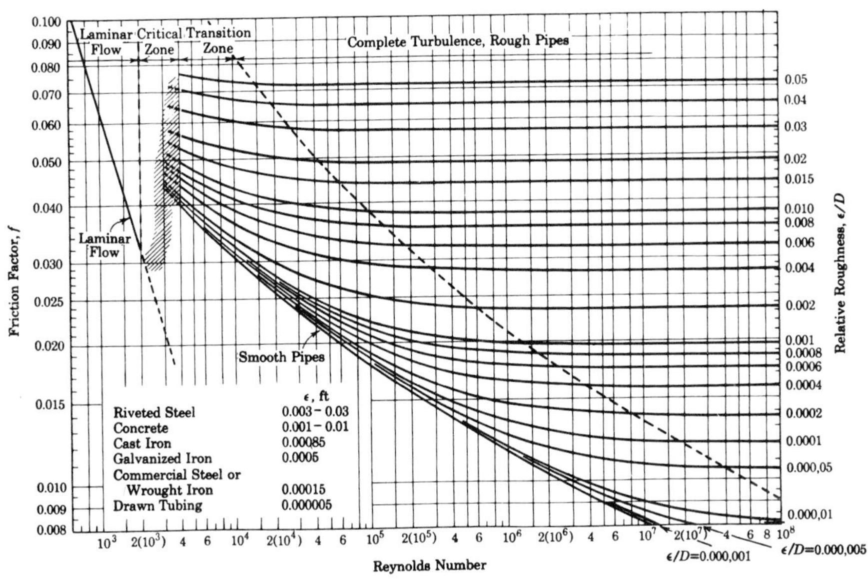

Figure 5.26. Moody diagram for pipe friction. (From L. F. Moody, "Friction Factors for Pipe Flow," *Trans. ASME,* Vol. 66, 1944, p. 671. Used by permission.)

Equation (5.133) is probably the most reliable relation for smooth pipes; however, Eqs. (5.131) and (5.133) are easier to use and give comparable results.

All of Eqs. (5.131) through (5.133) are applicable only to "smooth" tubes—tubes in which the surface irregularities lie well within the laminar sublayer. More detailed analyses (Refs. 13 through 15) include the effects of surface roughness as measured by the relative roughness ϵ/D (ϵ being the absolute pipe surface roughness). Moody (Ref. 15) combined these studies into the diagram shown in Fig. 5.26. This diagram can be used to determine f with a fair degree of accuracy. The curve for $\epsilon/D = 0$ is equivalent to Eq. (5.131), (5.132), or (5.133).

5.10 HEAT TRANSFER IN FULLY DEVELOPED PIPE FLOW

In the case of flow in a pipe in which heat transfer is taking place, there is a difference between the pipe wall temperature and the fluid temperature. This difference produces a temperature profile in the fluid with some sort of variation such as suggested in Fig. 5.27. Here the case in which the surface is hotter than the fluid is shown, and some variation of the fluid temperature occurs between the value at the surface to that at the centerline. The exact form of this variation depends on the thermal boundary conditions imposed at the surface, whether the flow is laminar or turbulent, and possible entry length effects. Since there is no easily identified characteristic temperature of the flow, such as the free stream temperature in the case of external flow, some definition must be made to characterize the fluid temperature and the heat transfer coefficient. As discussed earlier in Sec. 5.3 and Eq. (5.6), this is done by defining the *bulk* fluid temperature (or "mixing cup" temperature) t_b as the energy-averaged temperature across the cross section,

$$t_b = \frac{\int_0^R v\rho c_p \, 2\pi r t \, dr}{\int_0^R v\rho c_p \, 2\pi r \, dr}. \tag{5.134}$$

Then the local heat transfer coefficient is defined as

$$h = \frac{(q/A)_s}{t_s - t_b}. \tag{5.135}$$

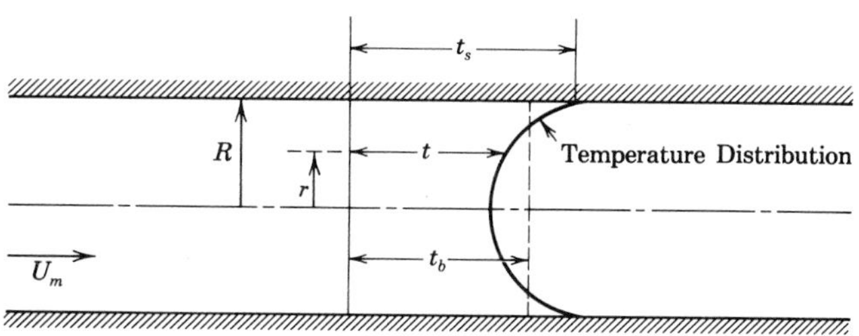

Figure 5.27. Temperature distribution in pipe flow.

Clearly, the bulk temperature is an integrated average for the cross section and is useful in expressing the heat transferred to the fluid between two different cross sections (where the bulk temperature is t_{b_1} and t_{b_2}) by writing an energy balance on the fluid:

$$q = \dot{m}c_p(t_{b_2} - t_{b_1}),$$

$\dot{m}$ being the mass flow rate. In some differential length of pipe, dx, the two statements above enable one to write

$$dq = \dot{m}c_p\,dt_b = h2\pi R\,dx(t_s - t_b). \tag{5.136}$$

As in the case of the development of the velocity profile discussed in Sec. 5.9, there are also starting length considerations with respect to the development of the temperature profile in pipe flow. Figures 5.28 and 5.29 depict a fluid of initially uniform temperature entering a pipe, the wall of which is hotter than the fluid. As the flow proceeds down the pipe, a thermal boundary layer develops at the surface in which the fluid temperature is altered by heat transfer with the surface; however, a central core, in which the inlet temperature is retained, exists. Eventually, the thermal layer completely fills the pipe at the thermal starting length L_{s_t} and the temperature profile becomes fully developed. In this condition there is a continuous variation in the fluid temperature from the surface value to some minimum value at the pipe centerline and no central core of uniform temperature exists. After the

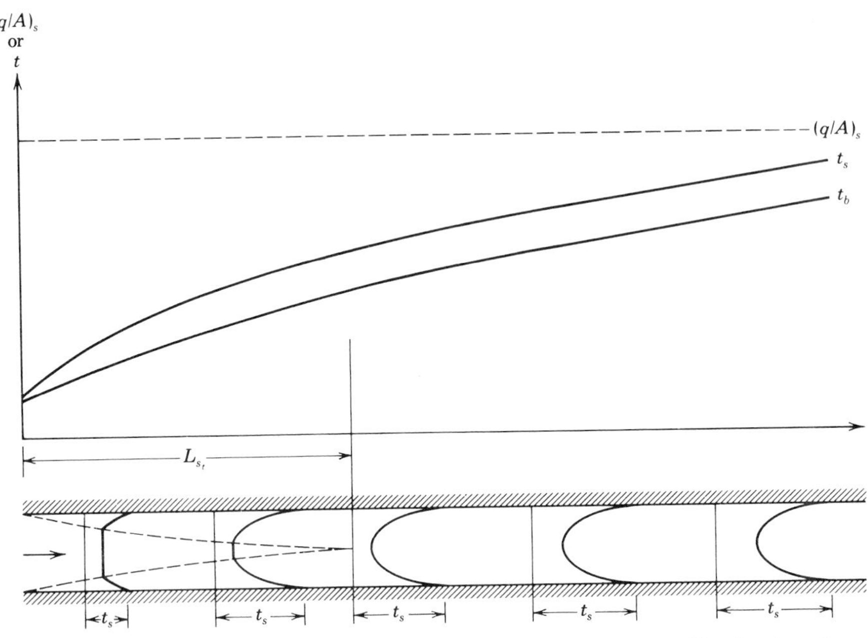

Figure 5.28. Developing and fully developed temperature profiles in pipe flow, constant wall heat flux.

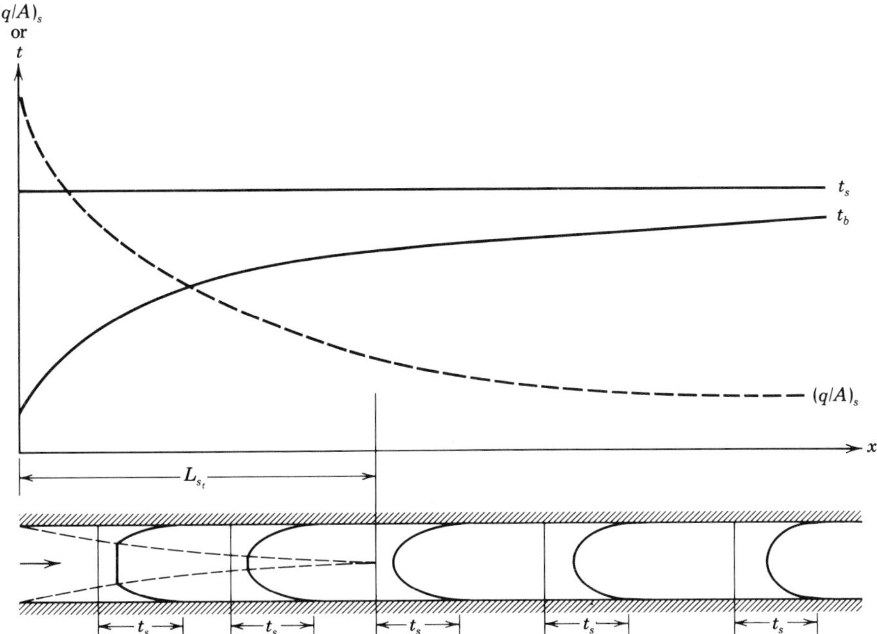

Figure 5.29. Developing and fully developed temperature profiles in pipe flow, constant wall temperature.

fully developed temperature profile has been established, the profile may or may not remain unchanged, depending on the nature of the imposed boundary conditions.

Figure 5.28 depicts the case in which the heat flux at the surface, $(q/A)_s$ is held uniform. Since heat is being transferred to the fluid, Eq. (5.136) requires that the bulk temperature increase but that the difference between the surface temperature t_s, and the bulk temperature, t_b, remain fixed. Thus, both t_s and t_b increase with x, while the dimensionless profile $(t_s - t)/(t_s - t_b)$ remains unchanged.

Figure 5.29 depicts a different situation in which the surface temperature t_s is held fixed. The bulk temperature must still increase, and in the fully developed region the temperature profile appears to flatten as t_b increases. However, the dimensionless profile $(t_s - t)/(t_s - t_b)$ remains fixed and the rate of heat transfer, $(q/A)_s$, decreases in accord with Eq. (5.136).

Thus, a fully developed thermal profile is established when

$$\frac{\partial}{\partial x}\frac{t_s - t}{t_s - t_b} = 0.$$

Clearly, the dimensionless temperature gradient at the surface is also constant, with length, in thermally fully developed flow. Hence, for fluids with constant properties, the heat transfer coefficient and Nusselt number are constant in fully developed pipe flow.

The comments above have been made without reference to whether the flow involved is, hydrodynamically, laminar or turbulent since the thermal starting length concept is observed in both instances. The exact nature of the shape of the developing and fully developed temperature profiles is, of course, very dependent on whether the flow is laminar or turbulent.

In all subsequent discussions in this chapter, and those which follow, whenever the temperature of a fluid flowing in a pipe is referred to it is presumed that it is the *bulk* temperature unless otherwise specified.

Laminar Flow

In the instance of laminar pipe flow, the problem of the determination of the developing temperature profile and the fully developed profile has been subjected to analysis by a number of investigators. Such analyses are not necessarily so complex as to be beyond the scope of this text; however, as will be seen shortly, the magnitudes of typical entry lengths for fluids viscous enough to maintain fully laminar pipe flow are quite large. As a result many applications of practical interest involve pipes which never achieve fully developed flow. Also, while the analyses almost always assume constant fluid properties, real fluids of high viscosity usually exhibit strong temperature dependencies and considerable deviation from theoretical predictions result. For these reasons, most engineering calculations for laminar pipe flow are done using either empirical relations or empirical adjustments to the theoretical results; hence only the results of the analytical predictions are presented here. Their empirical counterparts will be presented in Chapter 6.

Actually, there are two basic types of thermal entry problems. The simpler is that which assumes the thermal conditions develop in the presence of a fully developed velocity profile—such as would be the case if the point at which heat transfer begins is preceded by an unheated hydrodynamic starting length. Highly viscous oils with large Prandtl numbers provide a reasonable approximation to this assumption. The more complex case is that called the *combined* entry length problem in which both the velocity and temperature profiles develop simultaneously. The results quoted here refer only to the first case.

The case in which one considers fully developed laminar flow, both hydrodynamic and thermal, can be rather simply analyzed (Ref. 16) for the condition of constant wall heat flux (so that $t_s - t_b$ is constant) to show that the diameter Nusselt number is constant and equal to

$$\mathrm{Nu}_D = \frac{hD}{k} = \frac{48}{11} = 4.36. \tag{5.137}$$

Nusselt (Ref. 7), in one of the earliest heat transfer analyses, showed that the corresponding solution for the case of a fixed wall temperature is given by

$$\mathrm{Nu}_D = 3.66. \tag{5.138}$$

The deductions above, then, are the asymptotic values approached, in either

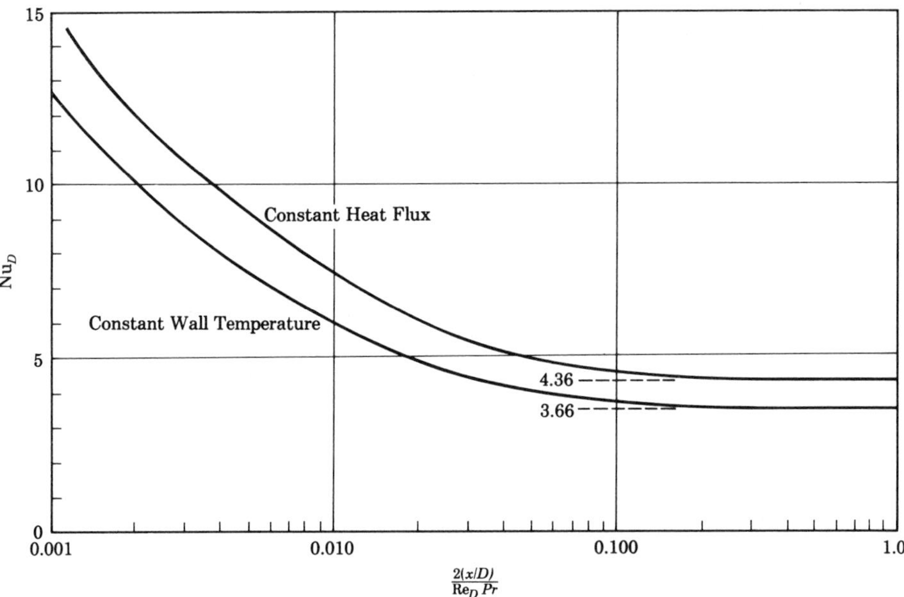

Figure 5.30. Local Nusselt number in the thermal starting region for laminar flow in a pipe.

case, in the thermal entry region. Analyses due to Nusselt as well as Sellars et al. (Ref. 18), summarized in Ref. 11, determine the local diameter Nusselt number in the developing region for the constant wall temperature and constant wall heat flux cases as functions of distance along the pipe. The results of these analyses are shown in Fig. 5.30. The approach to the fully developed values of Eqs. (5.137) and (5.138) is apparent. On the basis of the results shown in Fig. 5.30, Kays (Ref. 11) recommends that the thermal starting length for laminar pipe flow be defined:

$$\frac{2(x/D)}{\mathrm{Re}_D\ \mathrm{Pr}} \approx 0.1$$

or

$$\frac{L_{S_t}}{D} = 0.05\mathrm{Re}_D\mathrm{Pr}. \tag{5.139}$$

When compared with the hydrodynamic starting length given by Eq. (5.124), the equation above shows that the thermal starting length differs by a factor of Pr. Thus for cold water with $\mathrm{Pr} \approx 10$, the thermal length is about 10 times the hydrodynamic length, and for diameter Reynolds numbers near the transition value, L_{S_t} can be as great as 1000 pipe diameters. For viscous oils (see Table A.4) $\mathrm{Pr} \approx 10^4$ and it is apparent that fully developed thermal flow will rarely ever be achieved in practice.

Turbulent Flow

The thermal starting length in turbulent flow is a very difficult quantity to determine. Kays (Ref. 11) summarizes many calculations, and these results show that L_{s_t}/D is a complex function of Re and Pr that cannot be expressed in a simple way. In general, turbulent thermal starting lengths are shorter than laminar ones; the higher the value of Pr, the shorter is the starting length. Except for the liquid metals, for which Pr is quite small, one can expect thermal starting lengths of the order of 20 to 30 pipe diameters. Because of the complexity of determining turbulent thermal starting lengths, such effects are usually treated by empirical adjustments to the fully developed results. Hence only the fully developed analyses will be presented here.

The treatment of fully developed turbulent heat transfer in pipes and tubes will be treated here by use of heat and momentum analogies discussed in Sec. 5.8 for flow past a flat surface. While it is possible to develop these analogies anew for the fully developed pipe flow case, it is more simply done by noting certain corresponding facts between the two cases. For flow on the flat surface, the local surface shear stress was related to the skin friction coefficient by

$$\tau_0 = \frac{C_f}{2} \rho U^2$$

and the surface heat transfer coefficient was defined:

$$h = \frac{(q/A)_0}{t_s - t_f}.$$

For fully developed pipe flow the wall shear and heating transfer coefficient are given by

$$\tau_R = \frac{f}{8} \rho U_m^2,$$

$$h = \frac{(q/A)_s}{t_s - t_b}.$$

In the flat surface flow the analogies were based on the similarity of the exchange of heat and momentum between the viscous layers and the reservoir of heat and momentum represented by the free stream at U and t_f. In the case of fully developed pipe flow, corresonding similarities may be drawn for exchanges between the viscous layers and the reservoir of heat and momentum represented by the bulk flow at U_m and t_b. Thus the analogies developed in Sec. 5.8 may be translated into correspnding analogies for pipe flow by simply replacing $C_f/2$ by $f/8$, U by U_m, and t_f by t_b. The corresponding results are noted below.

The relations obtained below all give *local* Nusselt numbers (and hence local heat transfer coefficients) in a pipe. The method of calculating results which account for the changing bulk temperature as the fluid flows down the pipe will be treated in Chapter 6.

The Reynolds Analogy. For fully developed pipe flow, Reynolds analogy in Eq. (5.116) becomes

$$\text{Nu}_D = \frac{f}{8}\,\text{Re}_D\text{Pr}. \tag{5.140}$$

Values for f may be found from Eqs. (5.131) through (5.133), or Fig. 5.26, as appropriate. However, this will not be done here as the Colburn analogy, given next, is found to provide better results.

The Colburn Analogy. For fully developed pipe flow the Colburn analogy of Eq. (5.118) becomes

$$\text{Nu}_D = \frac{f}{8}\,\text{Re}_D\text{Pr}^{1/3}. \tag{5.141}$$

Again, f may be found from Fig. 5.26 or Eq. (5.133), as appropriate; however, for smooth pipes Eq (5.132) or (5.133) gives expressions which are easily evaluated for computational purposes:

$$\begin{aligned} \text{Nu}_D &= 0.0395\text{Re}_D^{0.75}\text{Pr}^{1/3}, \quad 10^4 < \text{Re}_D < 5 \times 10^4, \\ \text{Nu}_D &= 0.023\text{Re}_D^{0.8}\text{Pr}^{1/3}, \quad 3 \times 10^4 < \text{Re}_D < 10^6. \end{aligned} \tag{5.142}$$

Figure 5.31 compares the prediction of Eq. (5.142) with some measured values for air. Considering the influence of such factors as the variation of property values

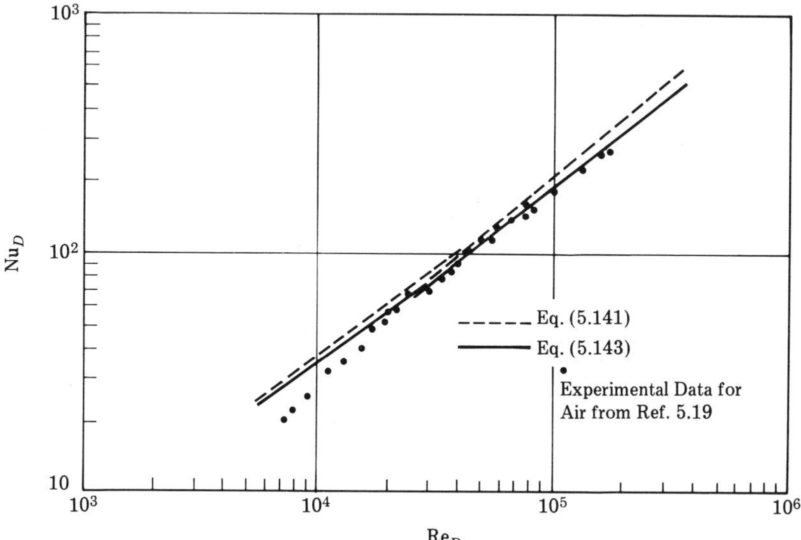

Figure 5.31. Turbulent heat transfer in pipes. Comparison between experimental data for air and the predictions of the Colburn and von Kármán analogies.

with temperature, etc., the agreement is not bad—typical of results obtained in convective heat transfer.

The Von Kármán Analogy. The von Kármán analogy of Eq. (5.122) similarly has a counterpart for fully developed pipe flow. For smooth pipes, one has

$$\mathrm{Nu}_D = \frac{(f/8)\,\mathrm{Re}_D\mathrm{Pr}}{1 + 5\sqrt{f/8}\{(\mathrm{Pr} - 1) + \ln[1 + \frac{5}{6}(\mathrm{Pr} - 1)]\}}, \qquad (5.143)$$
$$f = 0.316\mathrm{Re}_D^{-1/4}, \qquad 10^4 < \mathrm{Re}_D < 5 \times 10^4,$$
$$f = 0.184\mathrm{Re}_D^{-1/5}, \qquad 3 \times 10^4 < \mathrm{Re}_D < 10^6.$$

The predictions of Eq. (5.143) are also shown in Fig. 5.31.

5.11 CLOSURE

As mentioned at the outset of this chapter, its main purpose was to illustrate the methods employed to apply the principles of fluid dynamics and boundary layer theory to the *study* of forced convection problems. Even these analyses had to use certain empirical information, particularly for turbulent flow. The examples considered were not all-inclusive, but rather, were chosen to illustrate method. Many, much more complex analyses are available in the published literature; however, the basic concepts remain the same. Chapter 6 presents a summary of some of the results of this chapter, and some other results, that may be used for typical engineering design calculations.

Since the results just presented in the foregoing section for fluid friction and heat transfer in pipe flow will be amplified considerably in Chapter 6 with additional analytical and empirical results, no illustrative examples for this flow condition will be given here. Chapter 6 will present several examples for this important engineering application.

REFERENCES

1. White, F. M., *Fluid Mechanics,* New York, McGraw-Hill, 1979.
2. Schlicting, H., *Boundary Layer Theory,* 7th ed., New York, McGraw-Hill, 1979.
3. White, F. M., *Viscous Fluid Mechanics,* New York, McGraw-Hill, 1974.
4. Basius, H., "Grenzschichten in Flüssigkeiten mit kleiner Reibung," *Z. Math. Phys.,* Vol. 56, 1908, p. 1. English translation in *NACA Tech. Memo. 1256,* Washington, D.C., 1950.
5. Parmelee, G. V., and R. G. Huebscher, "Heat Transfer by Forced Convection Along a Smooth Flat Surface," *Heat. Piping Air Cond.,* Vol. 19, No. 8, 1947, p. 115.
6. Blasius, H., "Das Ahnlichkeitsgesetz bei Reibungvorgängen in Flüssigkeiten," *Forsch. Geb. Ingenieurwes.,* No. 131, 1913.
7. Dhawan, S., "Direct Measurements of Skin Friction," *NACA Tech. Note 2567,* Washington, D.C., 1952.

8. Colburn, A. P., " A Method of Correlating Forced Convection Heat Transfer Data and a Comparison with Liquid Friction," *Trans. AIChE,* Vol. 29, 1933, p. 174.
9. von Kármán, T., "The Analogy Between Fluid Friction and Heat Transfer," *Trans. ASME,* Vol. 61, 1939, p. 705.
10. Langhaar, H. L., "Steady Flow in the Transition Length of a Straight Tube," *Trans. ASME,* Vol. 64, 1942, p. A-55.
11. Kays, W. M., and M. E. Crawford, *Convection Heat and Mass Transfer,* 2nd ed., New York, McGraw-Hill, 1980.
12. Özisik, M. N., *Basic Heat Transfer,* New York, McGraw-Hill, 1977.
13. Nikaradse, J., "Strömungsgetze in rauhen Rohren," *Forsch. Geb. Ingenieurwes.,* No. 361, 1933.
14. von Kármán, T., "On Laminar and Turbulent Friction," *NACA Tech. Memo. 1092,* Washington, D.C., 1946.
15. Moody, L. F., "Friction Factors for Pipe Flow," *Trans. ASME,* Vol. 66, 1944, p. 671.
16. Holman, J. P., *Heat Transfer,* 5th ed., New York, McGraw-Hill, 1981.
17. Nusselt, W., "Die Abhängigkeit der Wärmeübergangszahl von der Rohrlänge," *Z.V.D.I.,* Vol. 54, 1910, p. 1154.
18. Sellars, J. R., M. Tribus, and J. S. Klein, "Heat Transfer to Laminar Flow in a Round Tube or Flat Conduit—The Graetz Problem Extended," *Trans. ASME,* Vol. 78, 1956, p. 441.
19. Deissler, R. G., and C. S. Eian, "Analytical and Experimental Investigation of Fully Developed Turbulent Flow of Air in a Smooth Tube with Heat Transfer with Variable Fluid Properties," *NACA Tech. Note 2629,* Washington, D.C., 1952.

PROBLEMS

5.1 A fluid at 1 atm pressure and a temperature of 40°C flows past a flat surface 30 cm long. Find the maximum freestream velocity of the fluid for which laminar flow at the trailing edge is probable if the fluid is **(a)** air, **(b)** ammonia, **(c)** water, or **(d)** ethylene glycol.

5.2 Repeat Prob. 5.1 if all data are the same except that the fluid pressure is raised to 5 atm.

5.3 A fluid with a temperature of 60°C flows past a flat surface with a freestream velocity of 50 m/s. Determine whether the flow is probably laminar or turbulent at a location 50 cm from the leading edge if the fluid is **(a)** saturated water, **(b)** the lubricating oil described in Table A.4, or **(c)** air at atmospheric pressure.

5.4 If the mean flow velocity in a circular tube is 6 ft/s, what must the tube size be to ensure that fully developed turbulent flow probably exists when the fluid (at 1 atm pressure, 70°F) is **(a)** air or **(b)** water.

5.5 Find the maximum mass velocity (i.e., mass flow rate per unit of cross-sectional area, kg/s-m^2) for which one may expect fully developed laminar flow in a 3-cm-diameter tube if the fluid (at 1 atm pressure, 20°C) is **(a)** water, **(b)** ethylene glycol, **(c)** air, or **(d)** helium.

5.6 Steam at a pressure of 200 psia and a temperature of 600°F flows with a velocity of 2 ft/s through a 2-in. schedule 40 pipe. Determine whether the flow is probably laminar or turbulent.

5.7 Repeat Example 5.3 if the fluid is water with a freestream velocity of 1.5 m/s and a freestream temperature of 80°C flowing past a flat surface at 40°C. Calculate the desired quantities at a distance of 5 cm from the leading edge.

5.8 If the surface described in Prob. 5.7 has a total length of $L = 10$ cm, find the same quantities determined in Example 5.4.

5.9 Repeat Example 5.3 if the fluid is atmospheric pressure air at 60°F flowing with a freestream velocity of 30 ft/s past a flat surface maintained at 115°F. Find the desired quantities at a distance of 5 in. from the leading edge.

5.10 If the surface described in Prob. 5.9 has a total length of $L = 10$ in., find the same quantities determined in Example 5.4.

5.11 Find the indicated dimensionless parameters for the specified fluids and conditions.

(a) Find Re_x for: $U = 20$ m/s
$x = 0.6$ m
air at 25°C, 1 atm.

(b) Find Re_D for: $\dot{m} = 0.5$ kg/s
$D = 10$ cm
water at 40°C.

(c) Find Nu_L for: $h = 10$ W/m^2-°C
$L = 1$ m
CO_2 at 50°C, 1 atm.

(d) Find Nu_x for: $h = 110$ Btu/h-ft-^{2}°F
$x = 3$ ft
steam at 100 psia, 400°F.

(e) Find Pr for: $\mu = 2.04 \times 10^{-5}$ kg/m-s
$c_p = 0.958$ kJ/kg-°C
$k = 0.02631$ W/m-°C.

5.12 Find the indicated dimensionless parameters for the specified fluids and conditions.

(a) Find Re_x for: $U = 1.5$ m/s
$x = 2$ in.
water at 50°C.

(b) Find Nu_L for: $h = 2$ Btu/h-ft^2-°F
$L = 1$ ft
air at 60°F, 1 atm.

(c) Find Gr for: $l = 0.5$ m
$\Delta t = 20$°C
water at 30°C.

(d) Find Ec for: $U = 185$ m/s
$\Delta t = 15$°C
air at 15°C, 1 atm.

(e) Find Pr for: $\rho = 925$ kg/m^3
$\mu = 0.0001$ kg/m-s
$\alpha = 0.5 \times 10^{-7}$ m^2/s.

5.13 Consider the convective heat transfer case discussed in Example 5.6 and illustrated in Fig. 5.14: cross flow of air past a horizontal cylinder. Two experiments are performed in which the air temperature is $t_f = 100$°C, the cylinder surface temperature is $t_s = 40$°C, and the cylinder diameter is $D = 0.3$ m. The data of the two experiments give the following average heat transfer coefficients between the surface and the air:

$$h = 45 \text{ W/m}^2\text{-°C when } U = 20 \text{ m/s},$$
$$h = 35 \text{ W/m}^2\text{-°C when } U = 15 \text{ m/s}.$$

When viscous dissipation and buoyance effects are neglected, the functional form for the Nusselt number is known to be $\text{Nu}_D = C\text{Re}_D^m\text{Pr}^n$, in which C, m, and n are empirical constants. Estimate the surface heat transfer coefficients that would be expected if for identical specified temperatures, a cylinder with $D = 0.6$ m is placed in an airstream with a freestream velocity U of **(a)** 15 m/s or **(b)** 30 m/s.

5.14 Justify the neglection of viscous dissipation and buoyancy effects for the situations specified in Prob. 5.13.

5.15 A body of some shape has a characteristic length of $l = 0.5$ m. When placed in an airstream at $t_f = 100$°C, $U = 25$ m/s, the heat flux from the surface is found to be 2000 W/m^2 when the surface temperature is 140°C. If a second body of the same shape but five times as large is placed in an airstream with $U = 5$ m/s, find the resultant surface heat transfer coefficient for air and surface temperatures that are the same as in the original case.

5.16 Water with a bulk temperature of 100°F flows in a circular tube the surface of which is maintained at 80°F. Two experiments are performed when the tube diameter is $D = 2$ in. These two experiments are performed at different mean flow velocities, U_m, and the following surface heat transfer coefficients are measured:

$$h = 570 \text{ Btu/h-ft}^2\text{-°F when } U_m = 3 \text{ ft/s},$$
$$h = 410 \text{ Btu/h-ft}^2\text{-°F when } U_m = 2 \text{ ft/s}.$$

The functional form for the Nusselt number is known to be $\text{Nu}_D = C\text{Re}_D^m\text{Pr}^n$ in which C, m, and n are empirical constants. Estimate the surface heat transfer coefficients that would be expected for water flowing at the same temperature in a tube with $D = 4$ in. maintained at the same temperature if the mean flow velocity is **(a)** 2 ft/s or **(b)** 4 ft/s.

5.17 Water at 90°F flows over the outside of a circular tube, normal to it, maintained with a surface temperature of 40°F. Two experiments are performed for a tube diameter of 2 in. at different freestream velocities of the water. The following average heat transfer coefficients are measured:

$$h = 940 \text{ Btu/h-ft}^2\text{-°F when } U = 5 \text{ ft/s},$$
$$h = 1420 \text{ Btu/h-ft}^2\text{-°F when } U = 10 \text{ ft/s}.$$

When viscous dissipation effects and buoyancy effects are neglected, the functional form for the Nusselt number is known to be $\text{Nu}_D = C\text{Re}_D^m\text{Pr}^n$, in which C, m and n are constants. Estimate the average heat transfer coefficients that would be expected if for identical temperatures, a tube of diameter 1.5 in. is placed in a water stream at **(a)** 10 ft/s, or **(b)** 20 ft/s.

5.18 Steam at a pressure of 4000 kN/m^2 and a bulk temperature of 500°C flows in a pipe ($D = 19$ cm) maintained with a surface temperature of 450°C. Two experiments are performed at different flow velocities, and the following surface heat fluxes are measured:

$$\text{at } U_m = 3.6 \text{ m/s}, \; q/A = 9400 \text{ W/m}^2,$$
$$\text{at } U_m = 5.0 \text{ m/s}, \; q/A = 12{,}200 \text{ W/m}^2.$$

The functional form for the Nusselt number is known to be $\text{Nu}_D = C\text{Re}_D^m\text{Pr}^n$, where C, m, and n are constants. Estimate the surface heat flux, for identical temperatures, when **(a)** $U_m = 6.0$ m/s, $D = 12$ cm; or **(b)** $U_m = 2.5$ m/s, $D = 25$ cm.

5.19 The importance of the magnitude of the energy dissipated through viscous shear was shown to be measured by how the product EcPr compares with unity. For $U = 10$ m/s and $\Delta t = 20$°C, calculate the magnitude of this parameter for a fluid at 40°C if the fluid is **(a)** atmospheric air, **(b)** water, or **(c)** the lubricating oil given in Table A.4.

5.20 Compare the product EcPr for air ($c_p \approx 1.0$ kJ/kg-°C, Pr ≈ 0.71) flowing at a typical automobile speed of $U = 25$ m/s or at the speed of a high speed aircraft, $U = 300$ m/s, if $\Delta t = 10$°C in both cases.

5.21 Verify the algebra leading from Eq. (5.46) to Eq. (5.48). Then show that the form of Eq. (5.49) and the definitions of Eqs. (5.52) and (5.53) yield Eq. (5.54).

5.22 Verify the algebra leading from Eqs. (5.58) through (5.62) to Eqs. (5.70) and (5.71).

5.23 Verify the algebra leading from Eqs. (5.77) and (5.78) to Eqs. (5.80) and (5.81).

5.24 Verify the algebra leading from Eq. (5.82) to Eq. (5.87).

5.25 Air at atmospheric pressure and 5°C flows past a flat surface at 45 m/s. Plot the laminar boundary layer thickness as a function of the distance from the leading edge up to the distance corresponding to a local length Reynolds number of 5×10^5.

5.26 Saturated water at 80°C flows with a freestream velocity of 1.5 m/s past a flat surface maintained at 40°C. Using the results of the differential solution given in Figs. 5.16 and 5.17, plot the velocity and temperature profiles in the boundary layer at stations 2.5, 5.0, and 7.5 cm from the leading edge. For ease in reading Figs. 5.16 and 5.17, use even values of the parameter η and calculate the corresponding values of y, at each x. Evaluate the fluid properties at the mean of the freestream and surface temperatures.

5.27 Repeat Prob. 5.26 using the results of the integral method as given by Eqs. (5.79), (5.80), (5.82), and (5.86).

5.28 Atmospheric air at 15°C flows with a freestream velocity of 7.5 m/s over a flat surface 24 cm long and maintained at 45°C. Using air properties at the mean between the freestream and surface temperatures, find **(a)** the local heat transfer coefficient at 6.0, 12.0, 18.0, and 24.0 cm from the leading edge; and **(b)** the average heat transfer coefficient for the entire surface.

5.29 For the data of Prob. 5.26, find **(a)** the local heat transfer coefficient at 2.5, 5.0, and 7.5 cm from the leading edge; and **(b)** the average heat transfer coefficient for the entire surface if it is 7.5 cm long.

5.30 Glycerine at a freestream temperature of 40°C flows with a velocity of 3 m/s past a flat surface maintained at 10°C.

(a) Find the local boundary layer thickness, the local skin friction coefficient and the local heat transfer coefficient at points (1) 0.1 m, (2) 0.2 m, and (3) 0.3 m from the leading edge.
(b) If the surface is 0.3 m long and 0.3 m wide, find the total heat transferred and the total drag force on one side.

5.31 Atmospheric pressure air with a freestream of $U = 80$ ft/s, $t_f = 70$°F flows past a flat surface maintained at 175°F. The surface is 6 in. long and 12 in wide.

(a) Determine the local boundary layer thickness, heat transfer coefficient, and skin friction coefficient at points (1) 2 in., (2) 4 in., and (3) 6 in. from the leading edge.
(b) Find the total heat transferred and the total drag force on one side.

5.32 If a first-degree polynomial is used to represent the velocity and temperature distribution in a laminar boundary layer on a flat surface, apply the integral equations of the boundary layer to obtain expressions for the dependence of the local boundary layer thickness, the local skin friction coefficient, and the local Nusselt number on the local Reynolds number.

5.33 Using a fourth-degree polynomial, apply the integral momentum equation of the laminar boundary layer to find how δ/x varies with Re_x on a flat plate. Use the additional condition that $\partial^2 v_x/\partial y^2 = 0$ at $y = \delta$.

5.34 Verify the algebra leading from Eq. (5.96) to the expression, just preceding Eq. (5.97), that $C_f = 0.0577 Re_x^{-1/5}$.

5.35 Water at 40°C flows past a flat surface with a freestream velocity of 0.75 m/s. Using the universal velocity distribution, find, for a distance 0.75 m from the leading edge, **(a)** the skin friction coefficient, and **(b)** the velocity and eddy diffusivity at a distance (1) 0.5 cm away from the surface, and (2) 0.04 cm away from the surface.

5.36 Atmospheric-pressure air flows past a flat surface at a velocity of 12 m/s and at a temperature of 90°C. Using the universal velocity distribution, find, for a distance 1.0 m from the leading edge, **(a)** the skin friction coefficient, and **(b)** the velocity and eddy diffusivity at a distance (1) 0.06 cm away from the surface, and (2) 0.6 cm away from the surface.

5.37 Verify the algebra leading from Eq. (5.120) to Eq. (5.122).

5.38 Find the local heat transfer coefficient for the conditions specified in Prob. 5.35, using **(a)** the Reynolds analogy, **(b)** the Colburn analogy, and **(c)** the von Kármán analogy.

5.39 Find the local heat transfer coefficient for the conditions specified in Prob. 5.36 using **(a)** the Reynolds analogy, **(b)** the Colburn analogy, and **(c)** the von Kármán analogy.

5.40 Water at 60°C flows at the rate of 10 kg/s through a pipe with a 7.5-cm ID.

(a) Find the heat transfer coefficient using (1) the Colburn analogy, and (2) the von Kármán analogy.

(b) Find the eddy viscosity and eddy diffusivity at a point 0.05 cm from the pipe wall.

5.41 Steam at 800°F, 1000 psia, flows in a 6-in. schedule 40 pipe with an average velocity of 20 ft/s. Find the local heat transfer coefficient using **(a)** the Colburn analogy, and **(b)** the von Kármán analogy.

CHAPTER 6

Working Formulas for Forced Convection

6.1 INTRODUCTORY REMARKS

The analyses in Chapter 5 and their applications of boundary layer theory to the solution of forced heat convection problems were meant, as stated, to illustrate the methods of approach and to show typical results. A detailed exposition of the derivation and development of all the relationships having engineering importance for the prediction of forced convection heat transfer coefficients is beyond the scope of this book. In fact, some of the most useful engineering formulas for forced convection coefficients have no real theoretical basis. They are the results of extensive experimental analyses.

The purpose of this chapter is to present a collection of the most useful of the existing relations for the more frequently encountered cases of forced convection. Some of these will be relations having theoretical bases, and some will be empirical dimensionless correlations of experimental data. In some instances more than one relation will be given for a particular case. The relations considered to be the preferred—or "recommended"—ones will be marked by a ★. This presentation of working relations for forced convection will, of course, be limited. For sources of relations for cases not covered here, the reader is advised to consult Refs. 1 through 5.

Since the energy required to move a fluid past a surface in external flow, or through a passage in internal flow, is an important consideration in most engineering applications in which heat transfer is taking place, relations for the skin friction coefficient, drag coefficient, and friction factor will be presented along with those for the heat transfer coefficient or Nusselt number.

Unless otherwise noted, all the relations to be quoted here will be limited (like those derived in Chapter 5) to the imposed condition of a constant surface temperature. That is, the thermal boundary condition imposed at the solid surface (either analytically or experimentally) is that of a uniform temperature. This is in contrast to other possible conditions—such as uniform heat flux, constant temperature difference, etc. Again, the reader is advised to consult the quoted references for relations based on those other conditions.

The theoretical analyses of Chapter 5 assumed that the physical properties (μ, c_p, k, and ρ) were constants. This is, of course, not true in reality. In almost every case there is a significant dependence of these properties on temperature, and in the case of ρ (and ν) there is also a pressure effect for gases. The temperature dependence of the properties means that there may be a significant variation in the quantities through the thermal boundary layer. The accuracy of the results of these theoretical relations and of the dimensionless experimental correlations depends on the temperature chosen for the evaluation of the properties. In most cases one of two mean temperatures is used. One is the bulk fluid temperature, or mixing cup temperature, as defined in Eq. (5.134). This is usually applied in the case of forced convection inside a closed duct or pipe and is the mean fluid temperature at a cross section—the temperature that would result if the fluid flowing through a cross section were to be thoroughly mixed. The symbol t_b will be used to denote the *bulk temperature*. The second mean temperature that is often used is the *mean film temperature*. This is the arithmetic mean of the surface temperature, t_s, and (in the case of flow external to a body) the free stream or undisturbed fluid temperature, t_f. Using t_m to denote the mean film temperature, one has

$$t_m = \frac{t_s + t_f}{2}.$$

Sometimes in the case of internal flow a mean film temperature which is the average of the surface temperature and the bulk temperature will be used in place of the bulk temperature.

In the working formulas to be presented here, it will be stated in each case what temperature should be used for evaluation of the fluid properties. Likewise, the range of the various dimensionless parameters (e.g., Re, Pr, etc.) over which a particular formula is valid will also be stated.

As noted in Chapter 5, pure forced convection heat transfer results (i.e., buoyancy neglected) may be expected to be of the form

$$\text{Nu} = f(\text{Re}, \text{Pr}, \text{Ec}).$$

The Eckert number, Ec, arises from the inclusion of the effects of the dissipation of mechanical energy through viscous friction. When this dissipation is neglected, the expected heat transfer correlation is of the form

$$\text{Nu} = f(\text{Re}, \text{Pr}).$$

The solutions of Chapter 5 were of the foregoing forms. In instances in which empirical methods are used, the resulting formulas are invariably presented in the foregoing forms. However, certain effects such as the temperature dependence of properties, the influences of the starting length, etc., may necessitate the inclusion of additional parameters as some of the following examples will show.

Finally, it should be pointed out that uncertainties in the many fluid properties involved, experimental errors, unaccounted for effects, geometrical deviations, surface roughness, etc., may cause, at times, considerable deviation between theory and experiment or scatter in the correlation of experimental data. As a result such deviations or scatter of the order of magnitude of $\pm 10\%$ in predicted Nusselt numbers are not uncommon in forced convection—and even greater deviations may be encountered in some complex cases.

6.2 FORCED CONVECTION PAST PLANE SURFACES, DISSIPATION NEGLECTED

The case of forced convection of a fluid of known velocity and temperature past a flat surface of known temperature was discussed extensively in Chapter 5. Being the case of the simplest geometry and boundary conditions, the analytic expressions found in Chapter 5 for the flat surface are found to hold rather well with certain modifications as noted in the following.

Laminar Flow

For laminar flow past a flat surface, the analytic results of Sec. 5.6 have been verified experimentally and are recommended for use as noted below.

The local skin friction coefficient and total surface length drag coefficient are predicted well by Eqs. (5.75) and (5.76):

$$\begin{aligned} C_f &= 0.664\mathrm{Re}_x^{-1/2}, \\ C_D &= 1.328\mathrm{Re}_L^{-1/2}, \\ &\begin{bmatrix} \mathrm{Re} < \mathrm{Re}_{x,c} \approx 5 \times 10^5 \\ \text{properties at } t_m \end{bmatrix}. \end{aligned} \tag{6.1}★$$

Clearly, the length Reynolds number, based on the freestream velocity, must be less than that for turbulent transition, usually taken as 5×10^5. The heat transfer for this flow condition is well represented by the analytic results:

$$\begin{aligned} \mathrm{Nu}_x &= 0.332\mathrm{Re}_x^{1/2}\mathrm{Pr}^{1/3}, \\ \mathrm{Nu}_L &= 0.664\mathrm{Re}_L^{1/2}\mathrm{Pr}^{1/3}, \\ &\begin{bmatrix} 0.6 \le \mathrm{Pr} \le 50 \\ \mathrm{Re} < \mathrm{Re}_{x,c} \approx 5 \times 10^5 \\ \text{properties at } t_m \end{bmatrix}. \end{aligned} \tag{6.2}★$$

The lower limit of 0.6 is established for the Prandtl number since that is the lowest possible for gases. The next lower values normally encountered for Pr are those for liquid metals (Table A.9), which are of the order of 0.01. In such instances the thermal boundary layer is very much thicker than the velocity layer and the basic assumption involved in the derivation of Eq. (6.2) that these layers are of the same order is violated. For fluids of such low Pr, Kays and Crawford (Ref. 1) derived the following:

$$\begin{aligned} \mathrm{Nu}_x &= 0.565(\mathrm{Re}_x\mathrm{Pr})^{1/2}, \\ \mathrm{Nu}_L &= 1.130(\mathrm{Re}_L\mathrm{Pr})^{1/2}, \end{aligned} \tag{6.3}\bigstar$$

$$\begin{bmatrix} \mathrm{Pr} < 0.05 \\ \mathrm{Re} < \mathrm{Re}_{x,c} \approx 5 \times 10^5 \\ \text{properties at } t_m \end{bmatrix}.$$

For the entire range of Prandtl numbers, Churchill and Ozoe (Ref. 6) recommend the following semiempirical relations:

$$\begin{aligned} \mathrm{Nu}_x &= \frac{0.3387\mathrm{Re}_x^{1/2}\mathrm{Pr}^{1/3}}{[1 + (0.0468/\mathrm{Pr})^{2/3}]^{1/4}}, \\ \mathrm{Nu}_L &= \frac{0.6674\mathrm{Re}_L^{1/2}\mathrm{Pr}^{1/3}}{[1 + (0.0468/\mathrm{Pr})^{2/3}]^{1/4}}, \end{aligned} \tag{6.4}\bigstar$$

$$\begin{bmatrix} \mathrm{RePr} > 100 \\ \text{properties at } t_m \end{bmatrix}.$$

Turbulent Flow—Local Heat Transfer

For turbulent flow past flat surfaces, only correlations for the local Nusselt number will be given since average values must account for an initial laminar section. Cases for such average, mixed flow are treated in the next section.

The local skin friction is given by the relation presented in Sec. 5.7, empirically adjusted as noted in that section [see Eqs. (5.97) and (5.98)]:

$$\begin{aligned} C_f &= 0.0592\mathrm{Re}_x^{-1/5}, \qquad 5 \times 10^5 < \mathrm{Re}_x < 10^7, \\ C_f &= 0.37\,(\log_{10}\mathrm{Re}_x)^{-2.584}, \qquad \mathrm{Re}_x > 10^7, \\ &\quad [\text{properties at } t_m]. \end{aligned} \tag{6.5}\bigstar$$

For local heat transfer, the analytic expression of the von Kármán analogy in Eq. (5.122) may be used with the foregoing relations for C_f as long as the Prandtl number is not too different from unity and the fluid properties are evaluated at t_m:

$$\mathrm{Nu}_x = \frac{\frac{1}{2}C_f\mathrm{Re}_x\mathrm{Pr}}{1 + 5\sqrt{C_f/2}\,\{(\mathrm{Pr} - 1) + \ln[1 + \frac{5}{6}(\mathrm{Pr} - 1)]\}},$$

However, the equation above is awkward to use and difficult to integrate for average values. Thus, the Colburn analogy [Eq. (5.119)] is recommended:

$$\begin{aligned} \mathrm{Nu}_x &= 0.0296\mathrm{Re}_x^{0.8}\mathrm{Pr}^{1/3}, \qquad 5 \times 10^5 < \mathrm{Re}_x < 10^7, \\ \mathrm{Nu}_x &= 0.185\mathrm{Re}_x(\log_{10}\mathrm{Re}_x)^{-2.584}\mathrm{Pr}^{1/3}, \qquad \mathrm{Re}_x > 10^7, \end{aligned} \tag{6.6}\bigstar$$

$$\begin{bmatrix} 0.6 \le \mathrm{Pr} \le 60 \\ \text{properties at } t_m \end{bmatrix}.$$

Mixed Flow—Average Friction and Heat Transfer

The skin friction and heat transfer results given above for turbulent flow were limited to local effects only. When turbulent flow exists on a flat surface it must have been preceded by a laminar section, and this laminar portion must be accounted for in the determination of average skin friction and heat transfer.

As the discussion of Sec. 5.3 and the pictorial representation of Fig. 5.7 would indicate, the transition from the laminar to turbulent state for external flow is a complex process that takes place over some finite distance. However, a reasonable model may be constructed wherein the transition distance is neglected and it is presumed that transition takes place instantaneously at a distance from the leading edge which corresponds to the critical length Reynolds number. If x_c is used to denote this distance, it is determined as

$$\mathrm{Re}_{x,c} = \frac{Ux_c}{\nu},$$

where $\mathrm{Re}_{x,c}$ is the critical Reynolds number, usually taken as 5×10^5. Under this assumption the drag coefficient for a surface of length $L > x_c$ is found by integrating the laminar local skin friction coefficient [Eq. (6.1)] from 0 to x_c and the turbulent one from x_c to L [Eq. (6.5)]:

$$\begin{aligned} C_D &= \frac{1}{L}\int_0^L C_f\,dx \\ &= \frac{1}{L}\left[\int_0^{x_c} 0.664\left(\frac{\nu}{Ux}\right)^{1/2} dx + \int_{x_c}^L 0.0592\left(\frac{\nu}{Ux}\right)^{1/5} dx\right] \\ &= 1.328\left(\frac{\nu}{Ux_c}\right)^{1/2}\frac{x_c}{L} + 0.074\left(\frac{\nu}{UL}\right)^{1/5}\left[1 - \left(\frac{x_c}{L}\right)^{4/5}\right]. \end{aligned}$$

If the total surface length Reynolds number $\mathrm{Re}_L = UL/\nu$ is introduced and one notes that $x_c/L = \mathrm{Re}_{x,c}/\mathrm{Re}_L$, the foregoing expression for C_D may be expressed

$$C_D = 0.074\mathrm{Re}_L^{-1/5} - A\mathrm{Re}_L^{-1}, \qquad \mathrm{Re}_{x,c} < \mathrm{Re}_L < 10^7, \tag{6.7}$$

where the constant A depends on the value chosen for the critical Reynolds number:

$$A = 0.074\mathrm{Re}_{x,c}^{4/5} - 1.328\mathrm{Re}_{x,c}^{1/2}. \tag{6.8}$$

For the customarily used value of $\mathrm{Re}_{x,c} = 5 \times 10^5$, one finds $A = 1743$. The limit imposed on Eq. (6.7) results form the range of validity of the turbulent skin friction law in Eq. (6.5). For higher Reynolds numbers the second form of Eq. (6.5) must be applied, and Schlicting (Ref. 7), approximates the result as

$$C_D = 0.455(\log_{10}\mathrm{Re}_L)^{-2.584} - A\mathrm{Re}_L^{-1}, \qquad 10^7 < \mathrm{Re}_L < 10^9. \tag{6.9}$$

In a similar fashion one may deduce an average plate length Nusselt number in accord with the result of Eq. (5.25):

$$\mathrm{Nu}_L = \int_0^L \frac{1}{x}\,\mathrm{Nu}_x\,dx = \int_0^{x_c} \frac{1}{x}\,\mathrm{Nu}_x\,dx + \int_{x_c}^L \frac{1}{x}\,\mathrm{Nu}_x\,dx.$$

For the laminar portion, $0 \le x \le x_c$, one uses the expression for the local Nusselt number for laminar flow as in Eq. (6.2). However, in the turbulent portion, $x_c \le x \le L$, the result depends on which of the analogies of Sec. 5.8 is used. Clearly, the von Kármán analogy [Eq. (6.5)] presents difficulties in carrying out the desired integration. For this reason, either the Reynolds analogy or the Colburn analogy is usually used. The Colburn analogy is preferred since it agrees better with experimental data and since it involves the factor $\mathrm{Pr}^{1/3}$, as does the laminar law.

Thus, using the laminar law of Eq. (6.2) and the turbulent law of Eq. (6.6), one finds the average Nusselt number to be

$$\begin{aligned}\mathrm{Nu}_L &= \int_0^{x_c} 0.332\mathrm{Re}_x^{1/2}\mathrm{Pr}^{1/3}\,\frac{1}{x}\,dx + \int_{x_c}^L 0.0296\mathrm{Re}_x^{0.8}\mathrm{Pr}^{1/3}\,\frac{1}{x}\,dx \\ &= [0.664\mathrm{Re}_{x,c}^{1/2} + 0.037(\mathrm{Re}_L^{0.8} - \mathrm{Re}_{x,c}^{0.8})]\mathrm{Pr}^{1/3} \qquad (6.10) \\ &= [0.037\mathrm{Re}_L^{0.8} - \tfrac{1}{2}A]\mathrm{Pr}^{1/3}, \qquad \mathrm{Re}_{x,c} < \mathrm{Re}_L < 10^7.\end{aligned}$$

In Eq. (6.10) the constant A is that defined in terms of the critical Reynolds number in Eq. (6.8). Comparison of Eqs. (6.7) for C_D and (6.10) for Nu_L suggests that the Colburn analogy applies for the total surface length as well as locally. That is,

$$\mathrm{Nu}_L = \frac{C_D}{2}\,\mathrm{Re}_L\mathrm{Pr}^{1/3}.$$

Thus for Reynolds numbers in excess of that specified for Eq. (6.10), one may use Eq. (6.9) to show that

$$\mathrm{Nu}_L = [0.228\mathrm{Re}_L(\log_{10}\mathrm{Re}_L)^{-2.584} - \tfrac{1}{2}A]\mathrm{Pr}^{1/3}, \qquad \mathrm{Re}_{x,c} < \mathrm{Re}_L < 10^9. \qquad (6.11)$$

The integrated relations just derived are found to give reliable results when the fluid properties are evaluated at the mean film temperature t_m. Thus, using the customarily applied transition Reynolds number $\mathrm{Re}_{x,c} = 5 \times 10^5$ so that $A = 1743$ from Eq. (6.8), the following recommendation is made for average skin friction and heat transfer in the case of mixed laminar–turbulent flow on a flat surface:

$$\begin{aligned}C_D &= 0.074\mathrm{Re}_L^{-1/5} - 1743\mathrm{Re}_L^{-1}, \quad 5 \times 10^5 < \mathrm{Re}_L < 10^7, \\ C_D &= 0.445(\log_{10}\mathrm{Re}_L)^{-2.584} - 1743\mathrm{Re}_L^{-1}, \quad 10^7 < \mathrm{Re}_L < 10^9,\end{aligned} \qquad (6.12)\bigstar$$

$$\begin{aligned}\mathrm{Nu}_L &= [0.037\mathrm{Re}_L^{0.8} - 872]\mathrm{Pr}^{1/3}, \quad 5 \times 10^5 < \mathrm{Re}_L < 10^7, \\ \mathrm{Nu}_L &= [0.228\mathrm{Re}_L(\log_{10}\mathrm{Re}_L)^{-2.584} - 872]\mathrm{Pr}^{1/3}, \quad 10^7 < \mathrm{Re}_L < 10^9,\end{aligned} \qquad (6.13)\bigstar$$

$$\begin{bmatrix} 0.6 \le \mathrm{Pr} \le 60 \\ \mathrm{Re}_{x,c} = 5 \times 10^5 \\ \text{properties at } t_m \end{bmatrix}.$$

For critical Reynolds number other than 5×10^5, the constants 1743 in Eq. (6.12) and 872 in Eq. (6.13) must be replaced by A and $A/2$, respectively as calculated from the definition in Eq. (6.8).

EXAMPLE 6.1

Atmospheric-pressure air at 40°C flows with a velocity of 40 m/s past a flat plate 0.75 m long and 0.5 m wide. The surface is maintained at 260°C. Treat the air as incompressible and neglect viscous dissipation.

(a) Find the total heat transferred and the total drag force on one side of the plate.
(b) Find the fractions of the heat transfer and drag force for the laminar portion of the flow.
(c) What error in part (a) results if the boundary layer is assumed to be turbulent from the leading edge?
(d) Repeat parts (a) and (b) if the velocity is doubled.

Solution. The mean film temperature is $t_m = (40 + 260)/2 = 150°\text{C}$. At this temperature Table A.6 gives for air

$$\rho = 0.8342\ \text{kg/m}^3, \qquad \nu = 28.58 \times 10^{-6}\ \text{m}^2/\text{s},$$
$$k = 34.59 \times 10^{-3}\ \text{W/m-°C}, \qquad \text{Pr} = 0.701.$$

(a) The plate length Reynolds number is

$$\text{Re}_L = \frac{UL}{\nu} = \frac{40 \times 0.75}{28.58 \times 10^{-6}} = 1.05 \times 10^6.$$

For transition at 5×10^5, Eqs. (6.12) and (6.13) give

$$\begin{aligned}
\text{Nu}_L &= [0.037\text{Re}_L^{0.8} - 872]\text{Pr}^{1/3} \\
&= [0.037(1.05 \times 10^6)^{0.8} - 872](0.701)^{1/3} = 1381, \\
C_D &= 0.074\text{Re}_L^{-1/5} - 1743\text{Re}_L^{-1} \\
&= 0.074(1.05 \times 10^6)^{-0.2} - 1743(1.05 \times 10^{-6})^{-1} = 0.00296.
\end{aligned}$$

Thus the average heat transfer coefficient is

$$h = \text{Nu}_L \frac{k}{L} = 1381 \times \frac{0.03459}{0.75} = 63.7\ \text{W/m}^2\text{-°C}.$$

The total heat transfer and drag force are then

$$\begin{aligned}
q = Ah(t_s - t_f) &= 0.75 \times 0.5 \times 63.7 \times (260 - 40) \\
&= 5255\ \text{W}\ (17{,}930\ \text{Btu/h}), \\
D = \tfrac{1}{2}\rho U^2 A\, C_D &= \tfrac{1}{2} \times 0.8342 \times (40)^2 \times 0.75 \times 0.5 \times 0.00296 \\
&= 0.741\ \text{N}\ (0.167\ \text{lb}_f).
\end{aligned}$$

(b) For the laminar portion it is necessary to locate the point of transition. For a critical Reynolds number of 5×10^5,

$$\mathrm{Re}_{x,c} = 5 \times 10^5 = \frac{Ux_c}{\nu}$$

$$x_c = 5 \times 10^5 \frac{28.58 \times 10^{-6}}{40} = 0.357 \text{ m}.$$

For the laminar portion the average Nusselt number and drag coefficient are, from Eqs. (6.1) and (6.2),

$$\begin{aligned} \mathrm{Nu} &= 0.664\mathrm{Re}^{1/2}\mathrm{Pr}^{1/3} \\ &= 0.664 \times (5 \times 10^5)^{1/2}(0.701)^{1/3} = 404.5, \\ C_D &= 1.328\mathrm{Re}^{-1/2} \\ &= 1.328(5 \times 10^5)^{-1/2} = 0.00188. \end{aligned}$$

Thus the average heat transfer coefficient for the laminar portion is

$$\begin{aligned} h_{\text{lam}} &= \mathrm{Nu}\,\frac{k}{x_c} \\ &= 404.5 \times \frac{0.03459}{0.357} = 39.2 \text{ W/m}^2\text{-°C}, \end{aligned}$$

and the heat transfer and drag force are

$$\begin{aligned} q_{\text{lam}} &= Ah(t_s - t_f) \\ &= 0.357 \times 0.5 \times 39.2 \times (260 - 40) \\ &= 1539 \text{ W (5251 Btu/h), 30\% of the total.} \\ D &= \tfrac{1}{2}\rho U^2 A C_D \\ D_{\text{lam}} &= \tfrac{1}{2} \times 0.8342 \times (40)^2 \times 0.357 \times 0.5 \times 0.00188 \\ &= 0.224 \text{ N } (0.050 \text{ lb}_f)\text{, 30\% of the total.} \end{aligned}$$

(c) If the assumption is made that the boundary layer is turbulent from the leading edge, the constants 1743 and 872 in Eqs. (6.12) and (6.13) become 0, so that

$$\begin{aligned} \mathrm{Nu}_L &= 0.037\mathrm{Re}_L^{0.8}\mathrm{Pr}^{1/3} \\ &= 0.037(1.05 \times 10^6)^{0.8}(0.701)^{1/3} = 2156, \\ C_D &= 0.074\mathrm{Re}_L^{-1/5} \\ &= 0.074(1.05 \times 10^6)^{-0.2} = 0.00462, \end{aligned}$$

for which

$$\begin{aligned} h &= 99.4 \text{ W/m}^2\text{-°C}, \\ q &= 8202 \text{ W (27,990 Btu/h)}, \\ D &= 1.157 \text{ N } (0.260 \text{ lb}_f). \end{aligned}$$

Thus an error of more than 50% is incurred by neglecting the initial laminar portion.

(d) Doubling the flow velocity changes only the plate length Reynolds number and the position of the transition point. Performing identical calculations gives

$$\begin{aligned}
\text{Re}_L &= 2.10 \times 10^6, \\
\text{Nu}_L &= 2979, \\
C_D &= 0.00320, \\
h &= 137.4 \text{ W/m}^2\text{-°C}, \\
q &= 11{,}338 \text{ W (38,690 Btu/h)}, \\
D &= 3.199 \text{ N } (0.719 \text{ lb}_f), \\
x_c &= 0.179 \text{ m}, \\
h_{\text{lam}} &= 78.3 \text{ W/m}^2\text{-°C}, \\
q_{\text{lam}} &= 1539 \text{ W (5251 Btu/h), 14\% of the total}, \\
D_{\text{lam}} &= 0.449 \text{ N } (0.050 \text{ lb}_f), \text{ 14\% of the total.}
\end{aligned}$$

6.3 FORCED CONVECTION PAST PLANE SURFACES, DISSIPATION INCLUDED

The preceding correlations, based on the analyses of Chapter 5, apply to cases in which the effects of the dissipation of mechanical energy in the boundary layer through viscous friction were neglected. Practical applications exist, particularly in connection with high-speed aerodynamics, in which these dissipative effects are significant. The discussion of Sec. 5.5 showed that when dissipative effects are accounted for an additional dimensionless parameter, the Eckert number, Ec, enters the correlation for Nu along with the Reynolds and Prandtl numbers. The Eckert number was defined, for external flow, as

$$\text{Ec} = \frac{U^2}{c_p(t_s - t_f)}. \tag{6.14}$$

It was further demonstrated that dissipation could be neglected only when EcPr $<< 1$, so that at sufficiently high flow velocities when EcPr ≈ 1 some modification of the results of Sec. 6.2 must be made.

There are applications involving highly viscous fluids, such as lubricating oils, for which Pr is large enough that EcPr may approach unity at relatively modest velocities. However, such instances are rarely encountered in normal practice, so the following discussion is limited to those cases of high-speed aerodynamics in which EcPr $\rightarrow 1$ as a result of large Ec and not large Pr. One would expect, then, that the assumption of constant density, as used in all the analyses of Chapter 5, would have to be modified. However, even though the freestream velocity, U, is large, velocities in the boundary layer near the surface are not, and since heat transfer and skin friction are basically surface phenomena it should not be surprising to find that compressibility (i.e., variable ρ) has little effect on these quantities.

Thus the principal difference as far as skin friction and heat transfer for high-speed flow are concerned is not variable density but rather, the inclusion of the dissipation term in the energy equation. The reader should refer now to the dissipativeless solution of Sec. 5.6 which is posed in Eqs. (5.58) through (5.62). When dissipation is included, the major impact on the solution is the necessity to include the term $\mu(\partial v_x/\partial y)^2$ which was omitted in the energy equation (5.60). The effect of including this term is to alter the resultant temperature distribution, in particular its gradient at the surface, from that which results when this term is omitted. For very high shear rates the amount of mechanical energy dissipated into thermal energy may be of such a magnitude as to cause so much heating within the boundary layer as to cause even a reversal of the heat flow for cases when $t_s > t_f$.

This heating within the boundary layer due to viscous dissipation is termed *aerodynamic heating* and is customarily described in terms of the *recovery temperature*. The recovery temperature, t_r, is defined as the temperature a surface would achieve if rather than being maintained at a specified temperature, it is allowed to come to thermal equilibrium with the viscous, shearing stream flowing past it. That is, one imagines that the surface is allowed to be heated to t_r by the aerodynamic heating effect of the boundary layer such that the heat flow at the surface would be zero. The available kinetic energy in the freestream is measured by $U^2/2$, but only a fraction of this is converted into dissipated thermal energy, as evidenced by the temperature rise $t_r - t_f$. This fraction is represented by the *recovery factor, r*:

$$r = \frac{t_r - t_f}{U^2/2c_p}. \tag{6.15}$$

Knowledge of the recovery factor is, then, required to determine the recovery temperature. The complete analysis of the boundary layer equations which includes the dissipation terms (see Refs. 7 through 10) shows that the recovery factor may be determined simply as a function of the Prandtl number:

$$\begin{aligned} r &= \mathrm{Pr}^{1/2} \quad \text{for laminar flow,} \\ r &= \mathrm{Pr}^{1/3} \quad \text{for turbulent flow.} \end{aligned} \tag{6.16}$$

To illustrate the magnitude of the aerodynamic heating effect, Table 6.1 shows the rise of t_r over t_f for laminar flow of air ($\mathrm{Pr} \approx 0.71$, $c_p \approx 1.02$ kJ/kg-°C) for various freestream velocities as calculated by Eqs. (6.15) and (6.16). The dissipative effect of viscosity at high velocities is readily apparent.

As noted above, the recovery temperature is that which the surface would achieve if allowed to come to thermal equilibrium with the dissipating boundary layer flowing past it. If, however, the surface is maintained at some other fixed temperature, t_s, there will be heat flow between the surface and the stream. Thus t_r, while calculable, is a fictitious quantity. Nonetheless, t_r remains an important parameter since the same analyses that yielded Eqs. (6.16) for the recovery factor (and hence, t_r) show also the important result that: *The same correlations for the heat transfer coefficient found for dissipativeless flow apply in the case with dissipation if the surface heat flux is based on the temperature potential $t_s - t_r$ rather than $t_s - t_f$.*

Table 6.1 Recovery Temperature in Air Due to Viscous Dissipation

Air Velocity m/s	$t_r - t_f$ °C
10	0.04
25	0.3
50	1.0
100	4.1
150	9.2
200	16.4
300	36.8
500	102.3

That is, if one redefines h such that the surface heat flux is

$$\left(\frac{q}{A}\right)_s = h(t_s - t_r), \tag{6.17}$$

the correlations quoted in Sec. 6.2 may be applied equally well to cases in which viscous dissipation is important. Since these correlations express the Nusselt number in terms of the Reynolds and Prandtl numbers, Nu $= f(\text{Re}, \text{Pr})$, one might think that the Eckert number is not involved as the dimensional analysis has predicted. However, the Eckert number *is* involved since the recovery temperature, t_r, is required. The definition of the recovery factor in Eq. (6.15), a known function of Pr, may be combined with the definition of Ec in Eq. (6.14) to show that

$$\frac{t_r - t_f}{t_s - t_f} = r\,\frac{\text{Ec}}{2}. \tag{6.18}$$

The implication of expressing the surface heat flux in terms of the potential $t_s - t_r$, as in Eq. (6.17), is that a surface in a dissipative flow "sees" t_r as its thermal sink rather than t_f. This is further illustrated by combining Eqs. (6.17) and (6.18) to show that

$$\left(\frac{q}{A}\right)_s = h(t_s - t_f)\left(1 - r\,\frac{\text{Ec}}{2}\right). \tag{6.19}$$

Equation (6.19) shows how the heat flux in the case of dissipative flow is modified when expressed in terms of the traditional potential $t_s - t_f$. Since Ec $= U^2/c_p(t_s - t_f)$, then when the surface is cooler than the freestream, $t_s < t_f$, Ec is negative and the heat flux is amplified by dissipation. When the surface is hotter than the freestream, Ec > 0, the heat flow is decreased for Ec $< 2/r$ and even *reversed* when Ec $> 2/r$. That is, for Ec $> 2/r$, heat flows *into* the surface even though the surface is *hotter* than the freestream. Equation (6.19) further shows that for very small Ec (i.e., Ec $<< 2/r$) the results approach that of dissipativeless flow ($t_r \to t_f$), as would be expected. Since r is either $\mathrm{Pr}^{1/2}$ or $\mathrm{Pr}^{1/3}$ [Eq. (6.16)], the statement that $\mathrm{Ec} << 2/r$ is equivalent to the earlier statement that EcPr $<< 1$ for dissipation to be neglected.

The foregoing discussion concentrated mainly on the effects of the inclusion of the dissipative term, $\mu(\partial v_x/\partial y)^2$, in the energy equation and not the effects of variable density that might also be important at flow velocities great enough to make dissipation important. As noted earlier, the compressibility effects for high-speed gas flow are minimized by the fact that velocities in the boundary layer near the surface may still be rather modest. The main effect observed in such cases is the large temperature differences that occur in compressible boundary layers, which may cause significant variation in the fluid properties. Studies reported in Refs. 11 and 12 which account for variable fluid properties indicate that the use of the mean film temperature, t_m, for the evaluation of fluid properties is inappropriate for accurate results. A relatively simple technique recommended by Eckert (Ref. 11) to account for fluid property dependence on temperature is that the properties be evaluated at a reference temperature, t^*, given by

$$t^* = 0.5(t_s + t_f) + 0.22(t_r - t_f), \tag{6.20}$$

where t_r is the recovery temperature and r the recovery factor defined earlier. For low-speed flow $t_r \to t_f$ and t^* becomes t_m, the mean film temperature.

As a result of all the foregoing discussion, the following recommendation is made for heat transfer and fluid friction in the case of dissipative flow past a flat surface: Use

$$\begin{aligned} \left(\frac{q}{A}\right)_s &= h(t_s - t_r), \\ \frac{t_r - t_f}{t_s - t_f} &= r\,\frac{\mathrm{Ec}}{2}, \\ \mathrm{Ec} &= \frac{U^2}{c_p^*(t_s - t_f)}, \\ t^* &= 0.5(t_s + t_f) + 0.22(t_r - t_f), \end{aligned} \tag{6.21}\bigstar$$

with:

Laminar flow:

$$\begin{aligned} r &= \mathrm{Pr}^{*1/2}, \\ \mathrm{Nu}_x^* &= 0.332\mathrm{Re}_x^{*1/2}\mathrm{Pr}^{*1/3}, \\ \mathrm{Nu}_L^* &= 0.664\mathrm{Re}_L^{*1/2}\mathrm{Pr}^{*1/3}, \\ C_f^* &= 0.664\mathrm{Re}_x^{*-1/2}, \\ C_D^* &= 1.328\mathrm{Re}_L^{*-1/2}, \end{aligned} \tag{6.22}\bigstar$$

$$\left[\begin{array}{l} 0.6 \le \mathrm{Pr}^* \le 50 \\ \mathrm{Re}^* < \mathrm{Re}_{x,c} \approx 5 \times 10^5 \\ \text{properties at } t^* \end{array}\right]$$

Turbulent flow:

$$\begin{aligned}
r &= \mathrm{Pr}^{*1/3} \\
\mathrm{Nu}_x^* &= 0.0296\mathrm{Re}_x^{*0.8}\mathrm{Pr}^{*1/3}, \qquad 5\times10^5 < \mathrm{Re}_x^* < 10^7, \\
\mathrm{Nu}_x^* &= 0.185\mathrm{Re}_x^*\,(\log_{10}\mathrm{Re}_x^*)^{-2.584}\mathrm{Pr}^{*1/3}, \qquad \mathrm{Re}_x^* > 10^7, \\
C_f^* &= 0.0592\mathrm{Re}_x^{*-1/5}, \qquad 5\times10^5 < \mathrm{Re}_x^* < 10^7, \\
C_f^* &= 0.37(\log_{10}\mathrm{Re}_x^*)^{-2.584}, \qquad \mathrm{Re}_x^* > 10^7, \\
&\left[\begin{array}{l} 0.6 \le \mathrm{Pr}^* \le 60 \\ \text{properties at } t^* \end{array}\right].
\end{aligned} \tag{6.23}★$$

An iterative calculation is usually necessary in applying Eqs. (6.21) through (6.23) since the fluid properties (and hence, r) depend on t^*, which itself depends on r and t_r. This is illustrated in the next example. No correlation is shown for the average Nu_L and C_D for mixed flow (when $\mathrm{Re}_L^* > \mathrm{Re}_{x,c}$). Integration of Eqs. (8.22) and (8.23) over the entire plate length is pointless since the recovery temperature, t_r, and the reference temperature, t^*, will generally be different in the laminar and turbulent portions. Thus each section has to be treated separately, as the next example illustrates.

EXAMPLE 6.2

In a wind tunnel test, air at -40°C flows with a velocity of 900 m/s past a flat plate 0.6 m long and 0.3 m wide. The air pressure is 3.45 kN/m^2 (approximately 0.5 psia). Find the heat transferred to or from the plate if its surface is maintained at 40°C.

Solution. The data given suggest that part of the plate will be in laminar flow and part in turbulent flow. If this is so, each part must be treated separately since t_r and t^* will be different in each.

Treating the laminar portion first, assume as tentative values $c_p = 1.009$ kJ/kg-°C and Pr $= 0.706$. These assumptions may have to be adjusted later after t^* is found. The Eckert number and the recovery factor are then

$$\mathrm{Ec} = \frac{U^2}{c_p(t_s - t_f)} = \frac{(900)^2}{1.009\times[40-(-40)]\times1000} = 10.03,$$

$$r = \mathrm{Pr}^{1/2} = 0.840.$$

Thus Eq. (6.21) gives the recovery temperature:

$$\frac{t_r - t_f}{t_s - t_f} = r\frac{\mathrm{Ec}}{2}$$

$$\frac{t_r - (-40)}{40 - (-40)} = 0.840 \times \frac{10.03}{2}; \qquad t_r = 297.3°\mathrm{C}.$$

Then the reference temperature, Eq. (6.21), is

$$t^* = 0.5(t_s + t_f) + 0.22(t_r - t_f)$$
$$= 0.5[40 + (-40)] + 0.22[297.3 - (-40)] = 74.2°\text{C}.$$

At this temperature, Table A.6 shows $c_p^* = 1.009$ kJ/kg-°C, $\text{Pr}^* = 0.707$. These values are sufficiently close to the tentative values chosen above. Had they been significantly different, new values would have to be assumed and the above calculations repeated until a satisfactory check is obtained. The remaining fluid properties may then be found from Table A.6, noting that $\boldsymbol{\nu}^*$ and ρ^* may not be read from that table since the pressure is not atmospheric. Thus μ^* must be used and the air density is found from the ideal gas law. Thus

Laminar section:

$$\text{Ec} = 10.03, \quad r = 0.840,$$
$$t_r = 297.3°\text{C}, \quad t^* = 74.20°\text{C},$$
$$c_p^* = 1.009 \text{ kJ/kg-°C}, \quad \mu^* = 20.66 \times 10^{-6} \text{ kg/m-s},$$
$$k^* = 29.57 \times 10^{-3} \text{ W/m-°C}, \quad \text{Pr}^* = 0.707,$$
$$\rho^* = \frac{P}{RT^*} = \frac{3.45}{(8.314/28.97)(74.2 + 273.15)}$$
$$= 0.0346 \text{ kg/m}^3.$$

Corresponding calculations must be carried out for the turbulent portion of the plate, the only difference being the use of $r = \text{Pr}^{1/3}$. Again, one must assume trial values of c_p and Pr which are then verified, or adjusted, and the final results are:

Turbulent section:

$$\text{Ec} = 10.03, \quad r = 0.890,$$
$$t_r = 317.4°\text{C}, \quad t^* = 78.6°\text{C},$$
$$c_p^* = 1.009 \text{ kJ/kg-°C}, \quad \mu^* = 20.86 \times 10^{-6} \text{ kg/m-s},$$
$$k^* = 29.81 \times 10^{-3} \text{ W/m-°C}, \quad \text{Pr}^* = 0.706,$$
$$\rho^* = 0.0342 \text{ kg/m}^3.$$

Note that the recovery temperature is 20°C higher in the turbulent portion than in the laminar portion.

The transition point may now be located using the properties of the laminar portion:

$$\text{Re}_{x,c} = 5 \times 10^5 = \frac{Ux_c\rho^*}{\mu^*} = \frac{900 \times x_c \times 0.0346}{20.66 \times 10^{-6}}$$
$$x_c = 0.332 \text{ m}.$$

Thus more than half the plate is in laminar flow.

The average heat transfer coefficient in the laminar section is found from Eq. (6.22):

$$\frac{hx_c}{k^*} = 0.664\ \mathrm{Re}^{*\,1/2}_{x,c}\mathrm{Pr}^{*1/3}$$

$$= 0.664(5 \times 10^5)^{1/2}(0.707)^{1/3} = 418.3,$$

$$h = \frac{hx_c}{k} \times \frac{k}{x_c}$$

$$= 418 \times \frac{0.02951}{0.332} = 37.2\ \mathrm{W/m^2\text{-}^\circ C},$$

so that the heat transferred in that section is

$$\begin{aligned} q_{\mathrm{lam}} &= hA(t_s - t_r) \\ &= 37.2 \times (0.332 \times 0.3)(40 - 297.3) \\ &= -952.6\ \mathrm{W}\ (3250\ \mathrm{Btu/h}). \end{aligned}$$

The average heat transfer coefficient in the turbulent section has to be found by integration of Eq. (6.23) from x_c to L and division by $L - x_c$:

$$h = k^* \frac{0.0296}{0.8} \frac{1}{L - x_c}\left(\frac{U\rho^*}{\mu^*}\right)^{0.8} (L^{0.8} - x_c^{0.8})\mathrm{Pr}^{*1/3}$$

$$= 0.02981 \times \frac{0.0296}{0.8} \frac{1}{0.6 - 0.332}\left(\frac{900 \times 0.0342}{20.86 \times 10^{-6}}\right)^{0.8}$$

$$\times (0.6^{0.8} - 0.332^{0.8})(0.706)^{1/3}$$

$$= 79.1\ \mathrm{W/m^2\text{-}^\circ C}.$$

The turbulent heat transfer is then

$$\begin{aligned} q_{\mathrm{turb}} &= hA(t_s - t_r) \\ &= 79.1(0.6 - 0.332) \times 0.3 \times (40 - 317.4) \\ &= -1764.2\ \mathrm{W}\ (6020\ \mathrm{Btu/h}). \end{aligned}$$

Thus the total heat transfer is

$$\begin{aligned} q &= q_{\mathrm{lam}} + q_{\mathrm{turb}} \\ &= -952.6 - 1764.2 \\ &= -2716.8\ \mathrm{W}\ (9270\ \mathrm{Btu/h}). \end{aligned}$$

Note that even though $t_s > t_f$, the high rate of dissipation in the boundary layer requires that the surface be cooled to maintain its temperature at 40°C.

6.4 FORCED CONVECTION INSIDE CYLINDRICAL PIPES AND TUBES

The engineering importance of forced flow through pipes and tubes for the purpose of transport, heating, cooling, etc., is apparent. Most heat exchangers involve the heating or cooling of fluids in tubes. Most ducts or pipes used to transport gases or liquids are invariably at temperatures different from the surroundings, producing a flow of heat to or from the fluids. Both laminar and turbulent conditions are encountered.

In the following sections both analytical and empirical correlations recommended for engineering use in this important geometry will be summarized. These correlations will yield methods of predicting the average value of the heat transfer coefficient on the inner tube surface applicable over the finite length of the tube. The manner in which this average value of h can be used to calculate the heat transferred to the fluid over a given length as the bulk temperature changes (or, equivalently, how the change in the bulk temperature can be calculated) will be discussed after the correlations are described.

The power required to move a fluid through a pipe or tube is an important consideration in almost every application in which one needs to know the heat transfer coefficient for this geometry. Hence relations will also be summarized for predicting the friction factor, from which the fluid pressure drop can be calculated from Eq. (5.126).

Laminar Flow

As discussed in Sec. 5.10, the analytical prediction of heat transfer for forced laminar convection in a tube is quite complex, due mainly to the effects of the starting length (both thermal and hydrodynamic) and the temperature dependence of the fluid properties. As given in Eq. (5.138), Nusselt showed that for fully developed laminar flow with a constant tube surface temperature, the Nusselt number was given by

$$\mathrm{Nu}_D = \frac{hD}{k} = 3.66. \tag{6.24}$$

For tubes that are so long that the starting length may be ignored, the average Nusselt number over the tube length is also given by Eq. (6.24)—with the fluid properties evaluated at the mean of the inlet and outlet bulk temperatures. What constitutes a tube long enough to use this relation will be noted shortly.

Hausen (Ref. 13) developed the following semiempirical relation for thermally fully developed flow:

$$\mathrm{Nu}_D = 3.66 + \frac{0.0668(D/L)\mathrm{Re}_D\mathrm{Pr}}{1 + 0.4[(D/L)\mathrm{Re}_D\mathrm{Pr}]^{2/3}}, \tag{6.25}$$

in which the effect of the tube length is included. Clearly, as $L \to \infty$, the Nusselt value is approached. Equation (6.25), again, yields an average h over the tube length and the properties are to be evaluated at the mean bulk temperature. However, Eq. (6.25) assumes the flow is hydrodynamically fully developed before entering the pipe and is, hence, generally not applicable.

Thus for "short" tubes in which the starting lengths are important, it would appear that one must resort to the analytical results noted in Sec. 5.10 (see Fig. 5.30 and Ref. 11 of Chapter 5). However, these results are awkward to use, and it is generally preferred to use the following empirical relation of Sieder and Tate (Ref. 14):

$$\mathrm{Nu}_D = 1.86\left[\left(\frac{D}{L}\right)\mathrm{Re}_D\mathrm{Pr}\right]^{1/3}\left(\frac{\mu}{\mu_s}\right)^{0.14}, \tag{6.26}\bigstar$$

$$\begin{bmatrix} 0.48 < \mathrm{Pr} < 16{,}700 \\ (D/L)\mathrm{Re}_D\mathrm{Pr} > 10 \\ \text{properties, except } \mu_s, \text{ at mean } t_b \\ \mu_s \text{ at } t_s \end{bmatrix}.$$

Equation (6.26) gives the average h to be applied over the pipe length L. The quoted limit of $(D/L)\mathrm{Re}_D\mathrm{Pr} > 10$ establishes what is meant by a short tube for the use of Eq. (6.26). For values below this limit, Eq. (6.24) may be used.

The pipe friction factor for laminar flow is similarly dependent on whether or not the tube or pipe is very long with respect to the hydrodynamic starting length. For fully developed laminar flow or for pipes in which the starting length is ignored, the friction factor is given by the classical results quoted in Eq. (5.128):

$$f = \frac{64}{\mathrm{Re}_D}. \tag{6.27}$$

However, the effect of the starting length may be accounted for by using the following expression for an apparent friction factor derived by Shah (Ref. 15) as a curve fit to the analytic results noted in Sec. 5.10:

$$f = \frac{1}{\mathrm{Re}_D}\,\frac{64 + 1.25z + 0.00289z^{5/2}}{1 + 0.00021z^2}, \tag{6.28}\bigstar$$

$$\begin{bmatrix} z = (D/L)\mathrm{Re}_D \\ \text{properties at mean } t_b \end{bmatrix}.$$

For small values of z (i.e., long tubes), say $z < 1$, the results above approach those of Eq. (6.27).

Turbulent Flow

For fully developed turbulent flow the friction factor relations for smooth pipes developed in Sec. 5.9 and noted in EQs. (5.131) and (5.132) are, generally, applicable. These relations can then be used, via the various heat transfer-momentum transfer analogies to yield relations for the Nusselt number. The relation of von

Kármán given in Eq. (5.143) may be used with generally acceptable results as long as the fluid properties are evaluated at the bulk temperature and as long as the Prandtl number is not too different from 1:

$$\mathrm{Nu}_D = \frac{(f/8)\,\mathrm{Re}_D\mathrm{Pr}}{1 + 5\sqrt{f/8}\{(\mathrm{Pr} - 1) + \ln[1 + \frac{5}{6}(\mathrm{Pr} - 1)]\}}, \tag{6.29}$$

However, this equation is awkward to use, and that of Colburn [Eq. (5.142)] is much simpler:

$$\begin{aligned} \mathrm{Nu}_D &= 0.0395\mathrm{Re}_D^{0.75}\mathrm{Pr}^{1/3}, && 10^4 < \mathrm{Re}_D < 5 \times 10^4, \\ \mathrm{Nu}_D &= 0.023\mathrm{Re}_D^{0.8}\mathrm{Pr}^{1/3}, && 3 \times 10^4 < \mathrm{Re}_D < 10^6. \end{aligned} \tag{6.30}$$

A variation on Eq. (6.30) which partially accounts for the effects of variable fluid properties is the widely used relation known as the Dittus–Boelter (Ref. 16) equation:

$$\begin{aligned} \mathrm{Nu}_D &= 0.023\mathrm{Re}_D^{0.8}\mathrm{Pr}^n, \\ f &= 0.316\mathrm{Re}_D^{-1/4}, && 10^4 < \mathrm{Re}_D < 5 \times 10^4, \\ f &= 0.184\mathrm{Re}_D^{-1/5}, && 3 \times 10^4 < \mathrm{Re}_D < 10^6. \end{aligned} \tag{6.31}\bigstar$$

$$\left[\begin{aligned} n &= 0.4 \text{ for } t_s > t_b \\ n &= 0.3 \text{ for } t_s < t_b \\ &0.7 < \mathrm{Pr} < 160 \\ &10^4 < \mathrm{Re}_D < 10^6 \\ &|t_s - t_b| < 6^\circ\mathrm{C} \text{ for liquids} \\ &|t_s - t_b| < 60^\circ\mathrm{C} \text{ for gases} \\ &\text{properties at } t_b \end{aligned}\right].$$

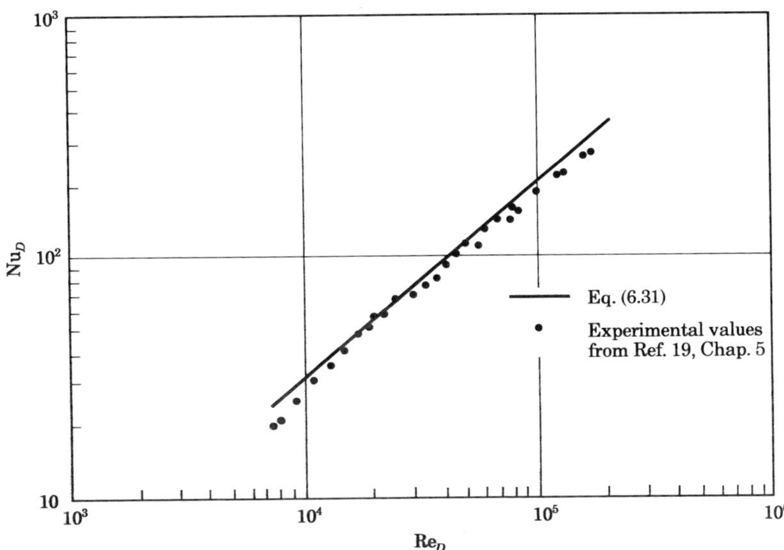

Figure 6.1. Turbulent heat transfer in pipes. Comparison of the Dittus-Boelter equation with experimental data.

The Dittus–Boelter relation (6.31) is compared in Fig. 6.1 with the same experimental data shown in Fig. 5.31 with the Colburn and von Kármán analogies and the agreement is as good. One advantage for the use of the Dittus–Boelter equation is that knowledge of the pipe surface temperature, t_s, is not required—only whether or not the fluid in the pipe is being heated, $t_s > t_b$, or cooled, $t_s < t_b$. This fact will be an advantage later in Chapter 10 when certain iterative calculations in which the surface temperature is unknown are carried out. However, errors of as much as $\pm 20\%$ are frequently encountered with Eq. (6.31) and there are limitations placed on the magnitude of the temperature difference, $t_s - t_b$, particularly for liquids.

For differences between t_s and t_b greater than those specified for the Dittus–Boelter relation (6.31), the following empirical equation due to Sieder and Tate (Ref. 14) will give better results, although knowledge of the surface temperature is required:

$$\begin{aligned} \mathrm{Nu}_D &= 0.027\mathrm{Re}_D^{0.8}\mathrm{Pr}^{1/3}\left(\frac{\mu}{\mu_s}\right)^{0.14}, \\ f &= 0.316\mathrm{Re}_D^{-1/4}, \qquad 10^4 < \mathrm{Re}_D < 5 \times 10^4, \\ f &= 0.184\mathrm{Re}_D^{-1/5}, \qquad 3 \times 10^4 < \mathrm{Re}_D < 10^6. \end{aligned} \tag{6.32}★$$

$$\left[\begin{array}{l} 0.7 < \mathrm{Pr} < 160 \\ 10^4 < \mathrm{Re}_D < 10^6 \\ \text{properties, except } \mu_s, \text{ at } t_b \\ \mu_s \text{ at } t_s \end{array}\right].$$

Petukhov (Ref. 17) has recently produced a much more accurate correlation which agrees well with the data of many workers. However, it does require the use of a different relation for the friction factor:

$$\begin{aligned} \mathrm{Nu}_D &= \frac{(f/8)\mathrm{Re}_D\mathrm{Pr}}{1.07 + 12.7\sqrt{f/8}(\mathrm{Pr}^{2/3} - 1)}\left(\frac{\mu}{\mu_s}\right)^n, \\ f &= (1.82 \log_{10} \mathrm{Re}_D - 1.64)^{-2}, \end{aligned} \tag{6.33}★$$

$$\left[\begin{array}{l} n = 0.11 \text{ for liquids, } t_s > t_b \\ n = 0.25 \text{ for liquids, } t_s < t_b \\ n = 0 \text{ for gases} \\ 0.5 < \mathrm{Pr} < 200 \text{ for } 6\% \text{ accuracy} \\ 200 < \mathrm{Pr} < 2000 \text{ for } 10\% \text{ accuracy} \\ 10^4 < \mathrm{Re}_D < 5 \times 10^6 \\ 0 < \mu/\mu_s < 40 \\ \text{properties, except } \mu_s, \text{ at } t_b \\ \mu_s \text{ at } t_s. \end{array}\right].$$

The relation for the friction factor, f, given in Eq. (6.33) is for smooth pipes. For rough pipes f may be found from the Moody diagram in Fig. 5.26. Since the exponent n in Eq. (6.33) is zero for gases, Petukhov's correlation may be used for gases without prior knowledge of the pipe surface temperature.

Of the three recommended correlations for fully developed turbulent pipe flow [Eqs. (6.31) through (6.33)], that of Petukhov [Eq. (6.33)] is generally regarded as the most accurate and should probably be used if the associated computational difficulties are not excessive.

Equations (6.31) through (6.33) presume the existence of fully developed flow and that the effect of the starting length is negligible. They should be applied only when $L/D \gtrsim 60$. As written, then, these equations give *local* values of h since they are based on the local bulk temperature of the fluid. The bulk temperature changes with distance as the fluid flows along the pipe, and the determination of this change is described later; however, for an average h over a given pipe length, it is generally suitable to use the above correlations based on the average bulk temperature between inlet and outlet. This is shown later in Example 6.4.

For short tubes it may be necessary to account for starting length effects. As noted in Chapter 5, not much information is available on turbulent starting lengths other than the general statement that they are usually much shorter than in laminar flow. Nusselt (Ref. 18) found the following expression to be applicable for shorter tubes:

$$\mathrm{Nu}_D = 0.036\mathrm{Re}_D^{0.8}\mathrm{Pr}^{1/3}\left(\frac{D}{L}\right)^{1/18}, \quad \begin{bmatrix} 10 < L/D < 400 \\ \text{properties at } t_b \end{bmatrix}. \tag{6.34}$$

EXAMPLE 6.3

Water flows with a bulk temperature of 30°C at a velocity of 1 m/s in a tube with an inside diameter of 2.5 cm. The tube surface is maintained at 50°C. Find the heat transfer coefficient using (a) Eq. (6.33); (b) Eq. (6.32); (c) Eq. (6.31).

Solution. At $t_b = 30°\mathrm{C}$ and $t_s = 50°\mathrm{C}$, Table A.3 gives

$$\mu = 0.7978 \times 10^{-3}\ \mathrm{kg/m\text{-}s}, \qquad \nu = 0.8012 \times 10^{-6}\ \mathrm{m^2/s},$$
$$k = 0.6150\ \mathrm{W/m\text{-}°C}, \qquad \mathrm{Pr} = 5.42,$$
$$\mu_s = 0.5471 \times 10^{-3}\ \mathrm{kg/m\text{-}s},$$

Thus the Reynolds number is

$$\mathrm{Re}_D = \frac{U_m D}{\nu} = \frac{1 \times 0.025}{0.8012 \times 10^{-6}} = 3.12 \times 10^4, \text{ turbulent.}$$

(a) Equation (6.33) yields

$$\begin{aligned} f &= (1.82 \log_{10} \mathrm{Re}_D - 1.64)^{-2} \\ &= [1.82 \times \log_{10} (3.12 \times 10^4) - 1.64]^{-2} \\ &= 0.02338, \end{aligned}$$

$$\begin{aligned}
\mathrm{Nu}_D &= \frac{(f/8)\mathrm{Re}_D\mathrm{Pr}}{1.07 + 12.7\sqrt{f/8}(\mathrm{Pr}^{2/3} - 1)}\left(\frac{\mu}{\mu_s}\right)^{0.11} \\
&= \frac{(0.02338/8) \times 3.12 \times 10^4 \times 5.42}{1.07 + 12.7\sqrt{0.02338/8}(5.42^{2/3} - 1)}\left(\frac{0.7978}{0.5471}\right)^{0.11} \\
&= 206.0, \\
h &= \frac{\mathrm{Nu}_D k}{D} = 206.0 \times \frac{0.6150}{0.025} \\
&\qquad = 5066\ \mathrm{W/m^2\text{-}^\circ C}\ (892.2\ \mathrm{Btu/h\text{-}ft^2\text{-}^\circ F}).
\end{aligned}$$

(b) Equation (6.32) yields

$$\begin{aligned}
\mathrm{Nu}_D &= 0.027\mathrm{Re}_D^{0.8}\mathrm{Pr}^{1/3}\left(\frac{\mu}{\mu_s}\right)^{0.14} \\
&= 0.027 \times (3.12 \times 10^4)^{0.8} \times 5.42^{1/3} \times \left(\frac{0.7978}{0.5471}\right)^{0.14} \\
&= 196.9, \\
h &= \frac{\mathrm{Nu}_D k}{D} = 196.9 \times \frac{0.6150}{0.025} \\
&\qquad = 4844\ \mathrm{W/m^2\text{-}^\circ C}\ (853.1\ \mathrm{Btu/h\text{-}ft^2\text{-}^\circ F}).
\end{aligned}$$

This value is within 4% of that found in part (a).

(c) Equation (6.31) yields, with $n = 0.4$,

$$\begin{aligned}
\mathrm{Nu}_D &= 0.023\mathrm{Re}_D^{0.8}\mathrm{Pr}^{0.4} \\
&= 0.023 \times (3.12 \times 10^4)^{0.8} \times 5.42^{0.4} \\
&= 178.1, \\
h &= \frac{\mathrm{Nu}_D k}{D} = 178.1 \times \frac{0.6150}{0.025} \\
&\qquad = 4381\ \mathrm{W/m^2\text{-}^\circ C}\ (771.5\ \mathrm{Btu/h\text{-}ft^2\text{-}^\circ F}).
\end{aligned}$$

This value differs by 14% from that predicted by the more accurate relation of Petukhov [Eq. (6.33)]; however, Eq. (6.31), the Dittus–Boelter equation, has been applied beyond its specified limits since $t_s - t_b = 20°\mathrm{C} > 6°\mathrm{C}$.

EXAMPLE 6.4

Steam flows through a 3-in. schedule 40 pipe with an average velocity of 20 ft/s. It enters at 600 psia, 800°F, and leaves at a temperature of 600°F. The pipe wall is maintained at a uniform temperature of 500°F. Find the heat transfer coefficient at the inner pipe surface (a) at the entrance conditions; (b) at the exit conditions; (c) at the mean of the entrance and exit conditions. Compare the results of part (c) with the mean of the results of parts (a) and (b). Assume that the pressure drop along the pipe is negligible so that the pressure at the pipe exit is the same as at the entrance, 600 psia. Example 6.5 will examine this assumption.

Solution. Table B.1 gives the inside diameter of a 3-in. schedule 40 pipe to be $D = 3.068$ in. The difference between the pipe surface temperature and the steam temperature, $t_s - t_b$, exceeds the limit of 60°C imposed for gases in the application of Eq. (6.31). Thus either Eq. (6.32) or (6.33) must be used, and Eq. (6.33) is chosen here because of its greater reliability.

(a) At the pipe entrance the steam pressure is 600 psia and $t_b = 800$°F. Thus Table A.5 yields the following properties:

$$\rho = 0.8409 \text{ lb}_m/\text{ft}^3, \qquad \mu = 0.06176 \text{ lb}_m/\text{ft-h}$$
$$k = 0.0355 \text{ Btu/h-ft-°F}, \qquad \text{Pr} = 0.96.$$

The value of μ_s at the pipe surface temperature is not needed since $n = 0$ in Eq. (6.33) for gases. Thus Eq. (6.33) gives

$$\text{Re}_D = \frac{U_m D \rho}{\mu} = \frac{20 \times 3600 \times (3.068/12) \times 0.8409}{0.06176} = 2.506 \times 10^5,$$

$$\frac{f}{8} = \tfrac{1}{8}(1.82 \log_{10}\text{Re}_D - 1.64)^{-2}$$

$$= \tfrac{1}{8}[1.82 \times \log_{10}(2.506 \times 10^5) - 1.64]^{-2} = 0.00187,$$

$$\text{Nu}_D = \frac{(f/8)\text{Re}_D\text{Pr}}{1.07 + 12.7\sqrt{f/8}(\text{Pr}^{2/3} - 1)}\left(\frac{\mu}{\mu_s}\right)^0$$

$$= \frac{0.00187 \times 2.506 \times 10^5 \times 0.96}{1.07 + 12.7\sqrt{0.00187}(\text{Pr}^{2.3} - 1)}$$

$$= 425.3,$$

$$h = \text{Nu}_D \frac{k}{D}$$

$$= 425.3 \times \frac{0.0355}{3.068/12} = 59.1 \text{ Btu/h-ft}^2\text{-°F } (335.6 \text{ W/m}^2\text{-°C}).$$

(b) At the pipe exit $t_b = 600$°F, and since the pressure drop is being neglected (an assumption to be examined in Example 6.5) the pressure is still 600 psia there. Thus Table A.5 gives

$$\rho = 1.0575 \text{ lb}_m/\text{ft}^3, \qquad \mu = 0.04991 \text{ lb}_m/\text{ft-h},$$
$$k = 0.0296 \text{ Btu/h-ft-°F}, \qquad \text{Pr} = 1.11.$$

Repeating the calculations of part (a), the abbreviated results are

$$\mathrm{Re}_D = \frac{U_m D \rho}{\mu} = 3.900 \times 10^5,$$

$$\frac{f}{8} = \tfrac{1}{8}(1.82 \log_{10} \mathrm{Re}_D - 1.64)^{-2} = 0.00172,$$

$$\mathrm{Nu}_D = \frac{(f/8)\mathrm{Re}_D \mathrm{Pr}}{1.07 + 12.7\sqrt{f/8}(\mathrm{Pr}^{2/3} - 1)}\left(\frac{\mu}{\mu_s}\right)^0 = 670.4,$$

$$h = \mathrm{Nu}_D \frac{k}{D} = 77.6 \text{ Btu/h-ft}^2\text{-°F } (440.6 \text{ W/m}^2\text{-°C}).$$

(c) If, now, one evaluates the heat transfer coefficient at the mean of the conditions between the entrance and the exit of the pipe, one needs the properties of steam at 600 psia and $t_b = (800 + 600)/2 = 700$°F. From Table A.5,

$$\rho = 0.9323 \text{ lb}_\text{m}/\text{ft}^3, \qquad \mu = 0.05592 \text{ lb}_\text{m}/\text{ft-h},$$
$$k = 0.03212 \text{ Btu/h-ft-°F}, \qquad \mathrm{Pr} = 1.01.$$

The same calculations, using Eq. (6.33), give

$$\mathrm{Re}_D = \frac{U_m D \rho}{\mu} = 3.069 \times 10^5,$$

$$\frac{f}{8} = \tfrac{1}{8}(1.82 \log_{10} \mathrm{Re}_D - 1.64)^{-2} = 0.00179,$$

$$\mathrm{Nu}_D = \frac{(f/8)\mathrm{Re}_D \mathrm{Pr}}{1.07 + 12.7\sqrt{f/8}(\mathrm{Pr}^{2/3} - 1)}\left(\frac{\mu}{\mu_s}\right)^0 = 518.1,$$

$$h = \mathrm{Nu}_D \frac{k}{D} = 65.1 \text{ Btu/h-ft}^2\text{-°F } (369.7 \text{ W/m}^2\text{-°C}).$$

The mean of the values of h found in parts (a) and (b) at the pipe entrance and exit is $h_{av} = (59.1 + 77.6)/2 = 68.4$, a value within 5% of that evaluated at the mean conditions. Except for fluids with extremely temperature dependent properties, this is the generally observed result: A suitable mean value for the heat transfer coefficient in a pipe is the one calculated for fluid properties at the average of the inlet and exit conditions of the fluid.

Had the preceding calculations been performed using the Sieder and Tate formula (6.32) rather than the Petukhov formula (6.33), the specific values of h would differ somewhat from those found above, but the same result concerning the mean would be observed.

Heat Transfer to Liquid Metals in Pipes or Tubes

Other than the correlation given in Eq. (6.3) for laminar flow past a flat surface, the correlations given so far in this chapter have been limited to fluids with Prandtl numbers greater than about 0.6. This limitation includes all gases and most liquids. However, liquid metals are a notable exception. Owing to their very high thermal conductivity, liquid metals exhibit Prandtl numbers several orders of magnitude lower than other liquids, usually of the order of 0.01, as may be observed in Table A.9. As noted earlier, the principal influence of the low value of Pr is that the thermal boundary layer is much thicker than the velocity layer. Thus, the fluid velocity is nearly uniform at its undisturbed value through much of the thermal boundary layer, altering considerably the solution of the energy equation.

The Martinelli analogy (Ref. 19) presents a detailed analytical expression for fully developed turbulent flow in tubes which applies to a wide range of Pr, including the liquid metals. The resulting expression, however, is difficult to use and is not always reliable. For this reason, most current engineering calculations for liquid metal heat transfer in tubes are carried out using any of a number of empirical or semiempirical relations.

For fluids of low Pr, the significant heat transfer parameter is the Péclet number, defined as

$$\text{Péclet number} = \text{Pe} = \text{PrRe} = \frac{Ul}{\alpha}. \tag{6.35}$$

Seban and Shimazaki (Ref. 20) applied the Martinelli analogy to the case of fixed surface temperature, and determined that for very low Prandtl numbers, the results are well approximated by the formula

$$\text{Nu}_D = 5.0 + 0.025\text{Pe}_D^{0.8}, \quad \begin{bmatrix} \text{Pe}_D > 100 \\ L/D > 60 \\ \text{properties at } t_b \end{bmatrix}. \tag{6.36}\bigstar$$

Other correlations for constant surface temperature include that due to Azer and Chao (Ref. 21):

$$\text{Nu}_D = 5.0 + 0.05\text{Pe}_D^{0.77}\text{Pr}^{0.25}, \quad \begin{bmatrix} \text{Pe}_D > 15{,}000 \\ \text{Pr} < 0.1 \\ \text{properties at } t_b \end{bmatrix}, \tag{6.37}\bigstar$$

and that due to Sliecher et al. (Ref. 22):

$$\text{Nu}_D = 4.8 + 0.0156\text{Pe}_D^{0.85}\text{Pr}^{0.08}, \quad \begin{bmatrix} 0.004 < \text{Pr} < 0.1 \\ \text{Re}_D < 5 \times 10^5 \\ \text{properties at } t_b \end{bmatrix}. \tag{6.38}\bigstar$$

Heat Transfer in Tubes of Finite Length

The discussion so far in this section has been limited to methods of predicting surface heat transfer coefficients for flows inside circular tubes. Some of the recommended relations (such as in laminar flow) give average values of h to be used over a given tube length, while others (such as in the fully developed turbulent cases) give local values of h which may be used to find applicable averages as in Example 6.4. All these relations depend on knowing the bulk fluid temperature, or more usually, the average of the bulk temperature as the fluid moves along the pipe. In order to apply these relations meaningfully and in order to determine the total heat transferred over a given pipe length (or, conversely, to determine the required length needed to transfer a given amount of heat) one needs to know how the bulk temperature of the fluid depends on the axial location in the pipe.

The question above is answered relatively easily by considering the case of fixed pipe wall temperature depicted in Fig. 6.2. Here a fluid, flowing at the mass rate $\dot{m}$, enters the tube with a bulk temperature t_{b_i}. The tube has a constant wall temperature t_s, taken as greater than t_{b_i} for purposes of discussion. As the fluid flows along the pipe its bulk temperature increases due to heat transfer to the fluid from the wall. At the end of the pipe, i.e., at L, the temperature has become t_{b_o}.

The way in which t_b varies with x, the distance along the pipe, can be deduced by writing a heat balance on an element of the tube dx long, as noted in Fig. 6.2. If h is the surface heat transfer coefficient and C is the circumference of the pipe, a heat balance on dx yields

$$dq = h(C\,dx)(t_s - t_b) = \dot{m}c_p\,dt_b. \tag{6.39}$$

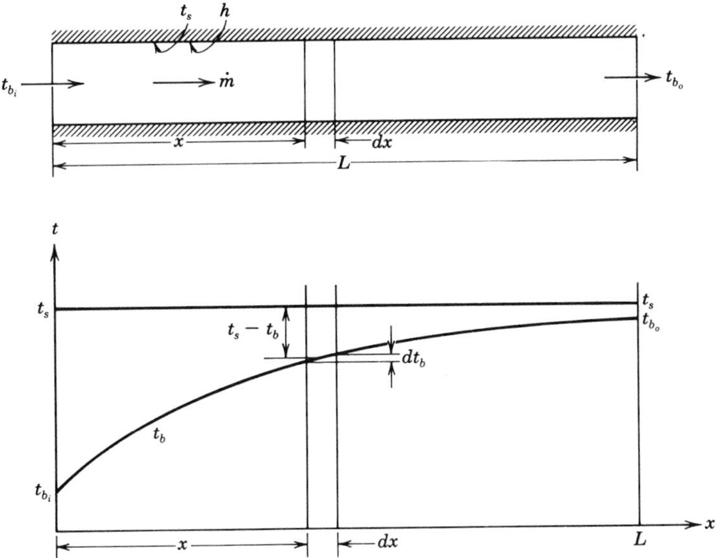

Figure 6.2. Temperature variation in a pipe with constant wall temperature.

where dt_b is the increase of the fluid bulk temperature over the distance dx. The above may be rewritten in the following form since t_s is constant:

$$\frac{d(t_b - t_s)}{t_b - t_s} = -\frac{hC}{\dot{m}c_p}\,dx. \tag{6.40}$$

When integrated from $x = 0$, $t_b = t_{b_i}$ to $x = x$, $t_b = t_b$, Eq. (6.40) gives, for h taken constant,

$$\ln\left(\frac{t_b - t_s}{t_{b_i} - t_s}\right) = -\frac{hC}{\dot{m}c_p}x,$$
$$\frac{t_b - t_s}{t_{b_i} - t_s} = \exp\left(-\frac{hC}{\dot{m}c_p}x\right). \tag{6.41}$$

Equation (6.41) shows that the bulk temperature varies exponentially with distance along the pipe.

To find the total heat transferred over a pipe length L, Eq. (6.39) may be integrated to yield

$$q = hCL(t_s - t_b)_m = \dot{m}c_p(t_{b_o} - t_{b_i}), \tag{6.42}$$

where $(t_s - t_b)_m$ is the mean of the temperature difference $(t_s - t_b)$ over the length L. This mean temperature difference is found by evaluating Eq. (6.41) for $x = L$, $t_b = t_{b_o}$:

$$\ln\left(\frac{t_{b_o} - t_s}{t_{b_i} - t_s}\right) = -\frac{hCL}{\dot{m}c_p}.$$

The factor $hCL/\dot{m}c_p$ is found from Eq. (6.42) to be $(t_{b_o} - t_{b_i})/(t_s - t_b)_m$, so that the above expression becomes

$$(t_s - t_b)_m = \frac{(t_s - t_{b_i}) - (t_s - t_{b_o})}{\ln\left(\dfrac{t_s - t_{b_i}}{t_s - t_{b_o}}\right)}. \tag{6.43}$$

Thus the appropriate mean value to use for $(t_s - t_b)$ as it varies from $(t_s - t_{b_i})$ to $(t_s - t_{b_o})$ is the *log mean* as given in Eq. (6.43). Then the total heat transferred over the pipe length L is given by Eq. (6.42).

The log mean appearing here will be encountered again later in Chapter 10 when dealing with heat exchangers.

In the discussion above, it should be remembered that it has been assumed that the surface heat transfer coefficient h is a constant over the pipe length. Thus some suitable average must have been computed previously. The formulation given in Eqs. (6.42) and (6.43) may be used to find the required length of pipe to raise the bulk temperature to a desired value or to find the outlet bulk temperature for a given pipe length as discussed in Example 6.6.

EXAMPLE 6.5

In Example 6.4 the flow of steam at a velocity of 20 ft/s through a 3-in. pipe, maintained with a surface temperature of 500°F, was considered. The steam enters the pipe at 600 psia, 800°F, and leaves at 600°F. Using the results of part (c) of Example 6.4, in which the flow properties were evaluated at the average bulk temperature of 700°F:

(a) Find the length of pipe necessary for the steam temperature to drop from 800°F to 600°F.
(b) Verify the neglection of the pressure drop along the pipe.

Solution. At the mean bulk temperature of 700°F the following steam properties, Reynolds number, Nusselt number, and heat transfer coefficient were found in Example 6.4(c):

$$\begin{aligned} \rho &= 0.9323 \text{ lb}_\text{m}/\text{ft}^3, \qquad \mu = 0.05592 \text{ lb}_\text{m}/\text{ft-h}, \\ k &= 0.03212 \text{ Btu/h-ft-°F}, \qquad c_p = 0.5821 \text{ Btu/lb}_\text{m}\text{-°F}, \\ \text{Pr} &= 1.01, \\ \text{Re}_D &= 3.069 \times 10^5, \qquad \text{Nu}_D = 518.1, \\ h &= 65.1 \text{ Btu/h-ft}^2\text{-°F}. \end{aligned}$$

(a) Since the inlet and outlet steam temperatures are known, the length of the pipe can be found from the heat balance of Eq. (6.42),

$$q = hCL(t_s - t_b)_m = \dot{m}c_p(t_{b_o} - t_{b_i})$$

if the mean temperature difference $(t_s - t_b)_m$ can be found. In terms of the known pipe surface temperature, $t_s = 500°\text{F}$, and the specified inlet and outlet bulk temperatures, $t_{b_i} = 800°\text{F}$, $t_{b_o} = 600°\text{F}$, Eq. (6.43) gives this mean temperature difference to be

$$\begin{aligned} (t_s - t_b)_m &= \frac{(t_s - t_{b_i}) - (t_s - t_{b_o})}{\ln\left(\dfrac{t_s - t_{b_i}}{t_s - t_{b_o}}\right)} \\ &= \frac{(500 - 800) - (500 - 600)}{\ln\left(\dfrac{500 - 800}{500 - 600}\right)} = -182.0°\text{F}. \end{aligned}$$

(Note that the *arithmetic* mean of the inlet and outlet temperature differences would be $-200°\text{F}$, considerably different from the logarithmic mean.)

The mass flow rate of steam through the pipe may be found from

$$\dot{m} = \frac{\pi}{4} D^2 \rho U_m.$$

With the pipe diameter $D = 3.068$ in. as used in Example 6.4, $U_m = 20$ ft/s, and the mean density as noted above,

$$\dot{m} = \frac{\pi}{4}\left(\frac{3.068}{12}\right)^2 \times 0.9323 \times 20 \times 3600 = 3446 \text{ lb}_\text{m}/\text{h}.$$

Thus the heat lost by the steam during its traverse through the pipe is given by the right half of Eq. (6.42):

$$\begin{aligned} q &= \dot{m}c_p(t_{b_o} - t_{b_i}) \\ &= 3446 \times 0.5821(600 - 800) = -4.012 \times 10^5 \text{ Btu/h}. \end{aligned}$$

This heat is transferred to the pipe wall by the logarithmic mean temperature difference as given by the left half of Eq. (6.42):

$$q = hCL(t_s - t_b)_m,$$

which, since $C = \pi D$, yields the necessary pipe length:

$$\begin{aligned} -4.012 \times 10^5 &= 65.1 \times \pi \times \left(\frac{3.068}{12}\right) \times L \times (-182.0) \\ L &= 42.2 \text{ ft } (12.9 \text{ m}). \end{aligned}$$

(b) The verification of the neglection of the pressure drop along the pipe in Example 6.4 requires the finding of the friction factor, f. One may use the expression for f given in Eq. (6.29), the Moody diagram of Fig. 5.26, or Eq. (6.33). It is most logical to use the latter, Eq. (6.33), since this was also the basis of finding the heat transfer coefficient in Example 6.4 from Petukhov's formula. Thus, for $\text{Re}_D = 3.069 \times 10^5$, as noted above and found in Example 6.4,

$$\begin{aligned} f &= (1.82 \log_{10} \text{Re}_D - 1.64)^{-2} \\ &= [1.82 \log_{10} (3.069 \times 10^5) - 1.64]^{-2} = 0.01436. \end{aligned}$$

[Note: Had Eq. (6.29) been used, one would find $f = 0.01470$, and if Fig. 5.26 had been employed for smooth pipes, $\epsilon/D = 0$, one would obtain $f = 0.014$. These values are quite close to that found using Petukhov's formula.] The definition of the friction factor given in Eq. (5.126) gives the pressure drop of the steam between the pipe entrance and exit to be, using the pipe length L just found,

$$\begin{aligned} -\Delta P &= f\frac{L}{D}\frac{1}{2}\rho U_m^2 \\ &= 0.01436\left(\frac{42.2}{3.068/12}\right) \times \frac{1}{2} \times 0.9323 \times \frac{20^2}{32.2} \times \frac{1}{144} \\ &= 0.10 \text{ lb}_\text{f}/\text{in.}^2 \ (0.66 \text{ kN/m}^2). \end{aligned}$$

Compared to the inlet pressure of 600 psia, this pressure drop is, indeed, negligible. Thus the treatment of the flow as constant pressure, as done in Example 6.4, is justified.

EXAMPLE 6.6

Water flows at the rate of $\dot{m} = 3.0$ kg/s through a pipe with an inside diameter $D = 6$ cm and a length $L = 5$ m. The pipe wall temperature is maintained at 90°C. If the water enters at a bulk temperature of $t_{b_i} = 20$°C, what is its exit temperature?

Solution. In Example 6.5, the unknown pipe length was found from the heat balance of Eq. (6.42) since the inlet and outlet bulk fluid temperatures were known. In this instance the same heat balance may be used to find the unknown outlet temperature since the pipe length is known. However, a representative, average heat transfer coefficient for the entire pipe length must be known to apply Eq. (6.42), and only the conditions at the pipe inlet are known. Thus one must employ an iterative procedure in which an outlet temperature is *assumed,* an average h determined based on this assumption, the outlet temperature *calculated* from Eq. (6.42), and the result compared with the assumption. Specifically, one must proceed as follows:

1. Assume a bulk fluid outlet temperature, t_{b_o}.
2. At a mean bulk temperature $t_{b_m} = (t_{b_i} + t_{b_o})/2$, find an average h, as in Examples 6.4 and 6.5.
3. Based on the given t_{b_i} and surface temperature t_s, find the mean temperature difference $(t_s - t_b)_m$ from Eq. (6.43) using the assumed t_{b_o}.
4. Apply the heat balance of Eq. (6.42) using the heat transfer coefficient found in step 2 and $(t_s - t_b)_m$ found in step 3 to calculate the outlet bulk temperature, t_{b_o}.
5. Compare the calculated t_{b_o} of step 4 with the assumed value in step 1. If the agreement of these two values is satisfactory, the problem is terminated; otherwise, the assumed t_{b_o} must be revised and the procedure repeated. Usually, the accuracy of given data and the heat transfer correlations used do not justify carrying this iterative procedure to too great a refinement—an agreement of 2 to 3°C between the assumed and calculated values is often adequate.

For the particular data specified in the problem posed here, follow the foregoing procedure:

1. Assume an outlet water temperature of $t_{b_o} = 35$°C. For the specified inlet temperature of $t_{b_i} = 20$°C, the mean bulk temperature is

$$t_{b_m} = \frac{t_{b_i} + t_{b_o}}{2} = \frac{20 + 35}{2} = 27.5\text{°C}.$$

2. At $t_{b_m} = 27.5$°C, Table A.3 gives the following properties of water:

$c_p = 4.180$ kJ/kg-°C, $\rho = 997.4$ kg/m^3, $\mu = 0.8442 \times 10^{-3}$ kg/m-s,
$\nu = 0.8473 \times 10^{-6}$ m^2/s, $k = 0.6113$ W/m-°C, Pr $= 5.78$.

For the surface temperature $t_s = 90°\text{C}$, Table A.3 gives

$$\mu_s = 0.3150 \times 10^{-3} \text{ kg/m-s}.$$

The mean fluid velocity, needed to find the Reynolds number, is

$$U_m = \frac{\dot{m}}{\rho A},$$

which, for the pipe diameter of $D = 6$ cm $= 0.06$ m gives

$$U_m = \frac{3.0}{997.4 \times \pi/4 \times (0.06)^2} = 1.064 \text{ m/s}.$$

Thus the pipe Reynolds number is

$$\begin{aligned} \text{Re}_D &= \frac{U_m D}{\nu} \\ &= \frac{1.064 \times 0.06}{0.8473 \times 10^{-6}} = 7.535 \times 10^4. \end{aligned}$$

This indicates turbulent flow so that Eq. (6.31), (6.32), or (6.33) might be used to find the heat transfer coefficient. Equation (6.31) is not applicable since the temperature-difference limit for liquids is exceeded for the data given. Equation (6.33) is chosen over (6.32) because of its greater accuracy. Thus

$$\begin{aligned} \frac{f}{8} &= \tfrac{1}{8}(1.82 \log_{10} \text{Re}_D - 1.64)^{-2} \\ &= \tfrac{1}{8}[1.82 \log_{10} (7.535 \times 10^4) - 1.64]^{-2} = 0.00239. \end{aligned}$$

Then the Nusselt number is

$$\begin{aligned} \text{Nu}_D &= \frac{(f/8)\text{Re}_D\text{Pr}}{1.07 + 12.7\sqrt{f/8}(\text{Pr}^{2/3} - 1)}\left(\frac{\mu}{\mu_s}\right)^{0.11} \\ &= \frac{0.00239 \times 7.535 \times 10^4 \times 5.78}{1.07 + 12.7\sqrt{0.00239}(5.78^{2/3} - 1)}\left(\frac{0.8443}{0.3150}\right)^{0.11} \\ &= 473.3, \end{aligned}$$

so that the average heat transfer coefficient at the inside pipe surface is

$$h = \text{Nu}_D \frac{k}{D} = 473.3 \times \frac{0.6113}{0.06} = 4822 \text{ W/m}^2\text{-°C}.$$

3. The surface temperature $t_s = 90°\text{C}$ and the inlet water temperature $t_{b_i} = 20°\text{C}$ are given. Thus the assumed outlet temperature, $t_{b_o} = 35°\text{C}$, allows a mean temperature difference, $(t_s - t_b)_m$, to be found from Eq. (6.43):

$$(t_s - t_b)_m = \frac{(t_s - t_{b_i}) - (t_s - t_{b_o})}{\ln\left(\dfrac{t_s - t_{b_i}}{t_s - t_{b_o}}\right)}$$

$$= \frac{(90 - 20) - (90 - 35)}{\ln\left(\dfrac{90 - 20}{90 - 35}\right)} = 62.20°\text{C}.$$

4. The heat balance of Eq. (6.42) permits finding the outlet bulk temperature, remembering that the pipe length, $L = 5$ m, is given:

$$h\pi DL(t_s - t_b)_m = \dot{m}c_p(t_{b_o} - t_{b_i})$$

$$4822 \times \pi \times 0.06 \times 5 \times 62.20 = 3.0 \times 4.180 \times 10^3 \times (t_{b_o} - 20)$$

$$t_{b_o} = 42.55°\text{C}.$$

5. This calculated $t_{b_o} = 42.55°$C is at considerable variance with the value of 35°C assumed in step 1. Hence the calculations must be revised.

Without discussion, a final iteration with an acceptable agreement is summarized below:

1. Assume that $t_{b_o} = 42°$C. Thus $t_{b_m} = 31°$C.
2. At $t_{b_m} = 31°$C,

$c_p = 4.180$ kJ/kg-°C, $\rho = 995.4$ kg/m^3, $\mu = 0.7822 \times 10^{-3}$ kg/m
$\nu = 0.7859 \times 10^{-6}$ m^2/s, $k = 0.6164$ W/m-°C, Pr $= 5.30$.

Thus

$$U_m = \frac{\dot{m}}{\rho A} = 1.066 \text{ m/s},$$

$$\text{Re}_D = \frac{U_m D}{\nu} = 8.138 \times 10^4,$$

$$\frac{f}{8} = \tfrac{1}{8}(1.82 \log_{10} \text{Re}_D - 1.64)^{-2} = 0.00235,$$

$$\text{Nu}_D = \frac{(f/8)\text{Re}_D\text{Pr}}{1.07 + 12.7\sqrt{f/8}(\text{Pr}^{2/3} - 1)}\left(\frac{\mu}{\mu_s}\right)^{0.11} = 481.3,$$

$$h = 4945 \text{ W/m}^2\text{-°C}.$$

3. For the assumed $t_{b_o} = 42°$C, the mean temperature difference becomes

$$(t_s - t_b)_m = \frac{(t_s - t_{b_i}) - (t_s - t_{b_o})}{\ln\left(\dfrac{t_s - t_{b_i}}{t_s - t_{b_o}}\right)} = 58.3°\text{C}.$$

4. For $h = 4945\ \text{W/m}^2\text{-°C}$ and $(t_s - t_b)_m = 58.3$, Eq. (6.42) gives a revised t_{b_o}:

$$h\pi D_L(t_s - t_b)_m = \dot{m}c_p(t_{b_o} - t_{b_i})$$
$$t_{b_o} = 41.7°\text{C}.$$

5. This calculated value of 41.7°C is deemed sufficiently close to the assumed value of 42°C to terminate the calculations, with the result that the outlet water temperature is probably

$$t_{b_o} = 42°\text{C (108°F)}.$$

6.5 FORCED CONVECTION IN NONCIRCULAR SECTIONS

The correlations of Sec. 6.4 were all limited to flows in circular pipes and tubes. Many engineering applications occur in which noncircular cross sections are involved. Both analytical and empirical information on forced convection in noncircular sections are sparse. In some instances the correlations for circular passages may be used if the Nusselt and Reynolds numbers are based on the *hydraulic,* or *effective,* diameter defined as four times the cross-sectional area divided by the wetted perimeter:

$$D_h = \frac{4A}{P}. \tag{6.44}$$

Laminar Flow

For the case of laminar flow in noncircular sections, even the use of the hydraulic diameter does not yield very accurate data, particularly if the section involves sharp corners. Thus, little information is available. However, the limiting case of fully developed flow in noncircular sections (i.e., corresponding to $\text{Nu}_D = 3.66$ for the circular section) has been solved numerically for a number of cases by several workers. Kays (Ref. 1) summarizes these findings for rectangular ducts with constant surface temperature as given in Table 6.2. In Table 6.2 the limiting, constant Nusselt number is that based on the hydraulic diameter which, for a rectangular duct of dimension $a \times b$, is

$$\text{Nu}_{D_h} = \frac{hD_h}{k},$$

$$D_h = \frac{2ab}{a + b},$$

in accord with Eq. (6.44).

For other cross sections and for the starting length solution in a few cases, the reader is referred to Refs. 1 and 5.

Table 6.2 Nusselt Numbers of Fully Developed Laminar Flow in Rectangular Ducts with Constant Wall Temperature

b/a	Nu_{D_h}
1.0	3.61
1.43	3.73
2.0	4.12
3.0	4.79
4.0	5.33
8.0	6.49
∞	8.24

Turbulent Flow

For turbulent flow, which may be taken to occur for $Re_{D_h} \gtrsim 2300$, the correlations quoted in Sec. 6.4 for circular tubes may be applied for reasonable results if Nu_D and Re_D are replaced with Nu_{D_h} and Re_{D_h}, respectively. In view of the uncertainties involved, there is probably no reason to employ relations any more sophisticated than the Dittus–Boelter formula given in Eq. (6.31). In particular, this latter expression has been found to give useful predictions for the physically significant case of fully developed turbulent flow in the annular space between two pipes. In this instance the hydraulic diameter is easily shown to be

$$D_h = D_o - D_i,$$

where D_o and D_i are the outside and inside diameters, respectively. The h thus found from Eq. (6.31) is applicable at either surface of the annulus.

6.6 FORCED CONVECTION FOR EXTERNAL FLOW NORMAL TO TUBES, TUBE BANKS, AND NONCIRCULAR SECTIONS

The last four sections have dealt with forced convective heat transfer for either external flow past flat surfaces or internal flow in pipes and ducts. Of equal engineering importance are the cases resulting from the flow of a fluid external to a nonflat surface. The flow of fluids normal to single cylinders, other noncircular shapes, and over banks of tubes are cases occurring frequently in engineering practice. The flow of fluids around such bluff bodies introduces special complications.

Flow Normal to a Single Circular Cylinder

Figure 6.3(a) depicts a circular cylinder placed normal to a fluid stream, the properties of which at points well upstream of the cylinder are denoted by U_∞, t_∞, and ρ_∞. Classical ideal, inviscid flow theory predicts a pressure distribution around the surface of the cylinder as suggested in Fig. 6.3(b). The pressure at the forward stagnation point equals the sum of the free stream static pressure and the free stream dynamic pressure. Then, the pressure first drops as the fluid speeds up around the forward part of the cylinder. Since the ideal fluid is not retarded by surface friction, it slows around the rear half of the cylinder exactly in reverse of the acceleration around the forward half. Consequently, the pressure rises around the rear half of the cylinder.

The behavior of a boundary layer in a region of increasing pressure, such as on the rear of a cylinder in cross flow, is quite different than those discussed previously for flat surfaces. As a viscous boundary layer moves into a region of rising pressure, its velocity profile becomes successively more and more flattened. The retarded layer, deficient in momentum, has increasing difficulty overcoming the rising pressure. Such a process is depicted in Fig. 6.4. Eventually, a point is reached at which

$$\left(\frac{\partial v_x}{\partial y}\right)_{y=0} = 0.$$

Downstream of this point reverse flow sets in, and the boundary layer pulls away, or "separates," from the surface. This separation causes a profound change in the pressure distribution along the surface.

Figure 6.3(b) indicates typical pressure distributions in real fluid flows around a circular cylinder compared with the ideal fluid distribution. Significant differences are seen to exist between the laminar and turbulent cases. A laminar layer will separate at a polar coordinate of about 80° from the forward stagnation point; a turbulent layer, having, on the average, more kinetic energy, will remain attached up to about 110°. The separation phenomenon is accompanied by the creation and shedding of eddies in the wake of the cylinder, as suggested in Fig. 6.3(a).

The peculiar flow processes just described, particularly the separation of the

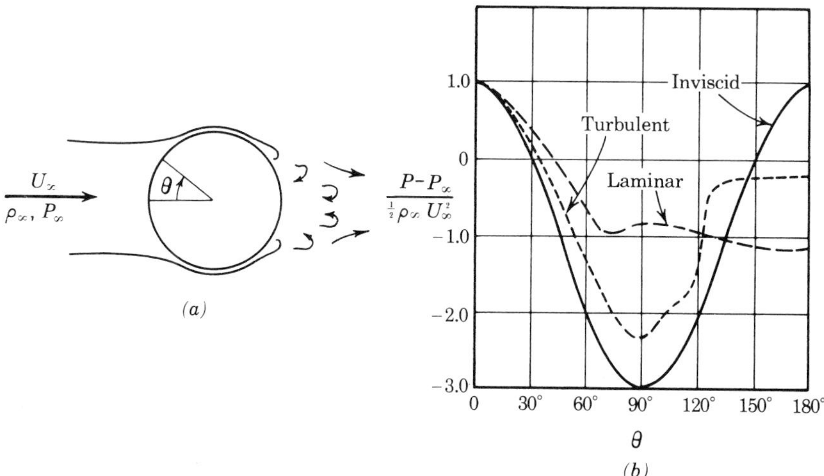

Figure 6.3. Flow normal to a circular cylinder.

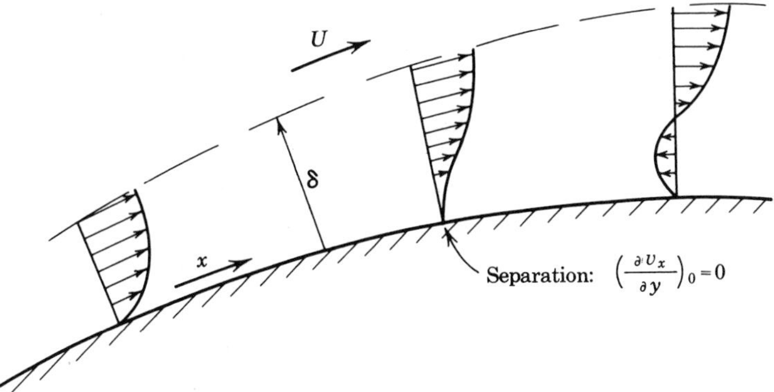

Figure 6.4. Boundary layer flow in an adverse pressure gradient.

boundary layer, have significant influence on both the viscous drag and the surface heat transfer coefficient. As noted, one of the most important influences is whether the boundary layer separation has been posponed by the onset of turbulence. The latter fact is usually described in terms of the Reynolds number based on the undisturbed freestream and the cylinder diameter:

$$\mathrm{Re}_D = \frac{U_\infty D}{\nu}. \tag{6.45}$$

Experiments indicate that the existence of a turbulent boundary layer and an associated delay in the separation phenomenon occurs at about

$$\mathrm{Re}_D \approx 3 \times 10^5. \tag{6.46}$$

The drag coefficient for a cylinder in cross flow is defined in terms of the dynamic pressure of the freestream, $\frac{1}{2}\rho_\infty U_\infty^2$, and the projected frontal area of the cylinder, $D \times L$, where L is the cylinder length. No single relation is available to describe the observed drag coefficient over all ranges of Re_D; however, White (Ref. 23) suggests that up to the transition point the following relation will give reasonable results:

$$C_D = \frac{\text{drag}}{(\frac{1}{2}\rho_\infty U_\infty^2)DL},$$

$$C_D = 1 + 10.0\,\mathrm{Re}_D^{-2/3}, \qquad (6.47)\bigstar$$

$$\begin{bmatrix} 1 \le \mathrm{Re}_D \le 2 \times 10^5 \\ \text{properties at } t_m \end{bmatrix}.$$

The surface heat transfer is equally complex. For a cylinder whose surface is maintained at a temperature different from the freestream, the local heat transfer coefficient varies quite peculiarly with location on the cylinder surface. This problem was investigated extensively by Giedt (Ref. 24) for air flowing past heated cylinders. Figure 6.5 shows typical results. At the lower Reynolds numbers the boundary layer is laminar and a minimum value of the local Nusselt number is seen to occur at the separation point. Subsequent to separation, a rise in Nu is seen to occur as transverse mixing is enhanced by the creation of the eddies mentioned above. At higher Reynolds numbers a first minimum in Nu is observed when the laminar-turbulent transition takes place and a second when separation occurs.

The foregoing processes are sufficiently complex that an analytical description is virtually impossible. Also, it is usually the average heat transfer coefficient,

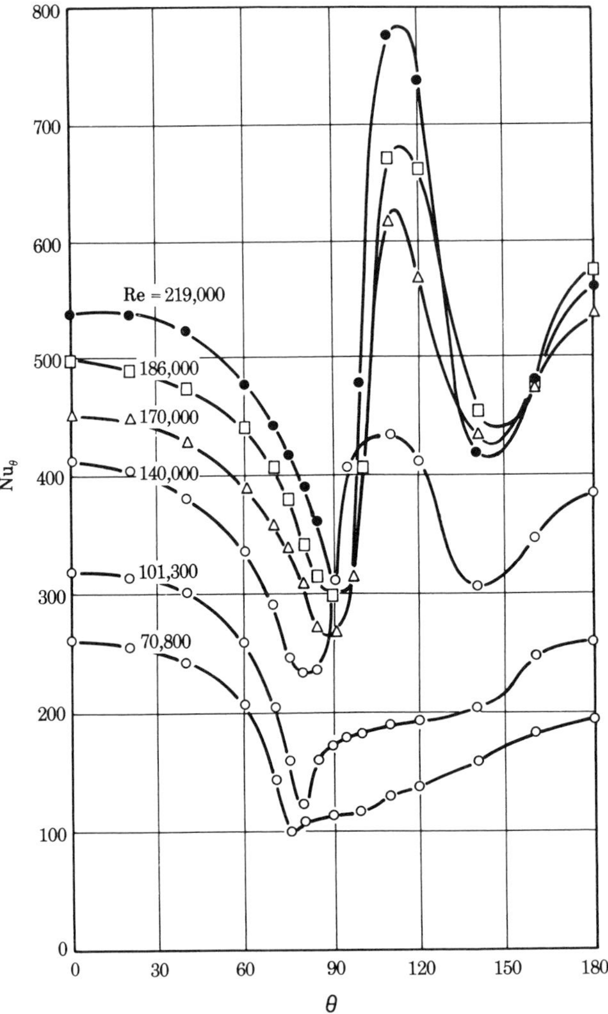

Figure 6.5. Circumferential variation of the local heat transfer coefficient for cross flow of air over a cylinder. (From W. H. Geidt, "Variation of Point Unit Heat Transfer Coefficient Around a Cylinder Normal to an Air Stream," *Trans. ASME,* Vol. 71, 1949, p. 375. Used by permission.)

Table 6.3 Values of *C* and *m* for Zhukauskas' Equation for Flow Normal to Singular Circular Cylinders*

Range of Re_D	C	m
1–40	0.75	0.4
40–100	0.51	0.5
10^3–2×10^5	0.26	0.6
2×10^5–10^6	0.076	0.7

*From Ref. 25.

rather than the local, which is desired for engineering applications. Hence empirical correlations are normally used.

Zhukauskas (Ref 25) has performed the most recent and most comprehensive study of flow normal to single tubes and banks of tubes. For single tubes the following correlation yields rather accurate results when used in conjunction with Table 6.3 and when the mean film temperature is defined as $t_m = (t_\infty + t_s)/2$:

$$\mathrm{Nu}_D = C\mathrm{Re}_D^m \mathrm{Pr}^n \left(\frac{\mathrm{Pr}_\infty}{\mathrm{Pr}_s}\right)^{1/4},$$

$$\begin{bmatrix} 0.7 < \mathrm{Pr} < 500 \\ 1 < \mathrm{Re}_D < 10^6 \\ C \text{ and } m \text{ from Table 6.3} \\ n = 0.37,\ \mathrm{Pr} \le 10 \\ n = 0.36,\ \mathrm{Pr} > 10 \\ \text{properties at } t_m, \text{ except } \mathrm{Pr}_\infty, \mathrm{Pr}_s \\ \mathrm{Pr}_\infty \text{ at } t_\infty,\ \mathrm{Pr}_s \text{ at } t_s \end{bmatrix}. \tag{6.48}★$$

Churchill and Bernstein (Ref. 26) have proposed, based on an analysis of many data covering wide ranges of Pr and Re_D, the following correlation:

$$\mathrm{Nu}_D = 0.3 + \frac{0.62\mathrm{Re}_D^{1/2}\mathrm{Pr}^{1/3}}{[1 + (0.4/\mathrm{Pr})^{2/3}]^{1/4}}\left[1 + \left(\frac{\mathrm{Re}_D}{2.82 \times 10^5}\right)^{5/8}\right]^{4/5},$$

$$\begin{bmatrix} \text{for all } \mathrm{Re}_D\mathrm{Pr} > 0.2 \\ \text{properties at } t_m \end{bmatrix}. \tag{6.49}★$$

Flow Normal to Noncircular Cylinders

The results for flow normal to noncircular cylinders is less complete than that above for circular cylinders. For gases, Jakob (Ref. 2) recommends the use of the following correlation for the shapes shown in Table 6.4:

$$\mathrm{Nu}_D = C\mathrm{Re}_D^m\mathrm{Pr}^{1/3},$$

$$\begin{bmatrix} C \text{ and } m \text{ from Table 6.4} \\ \text{properties at } t_m \end{bmatrix}. \tag{6.50}★$$

Table 6.4 Values of C and m for Forced Convection of Gases to Noncircular Cylinders*

Geometry	Re_D	C	m
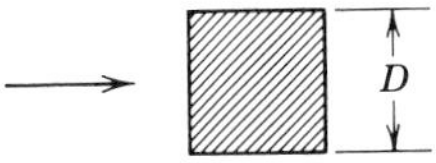 Square, flow normal to face	$5 \times 10^3 - 10^5$	0.102	0.675
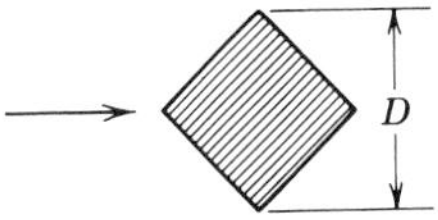 Square, flow toward corner	$5 \times 10^3 - 10^5$	0.246	0.588
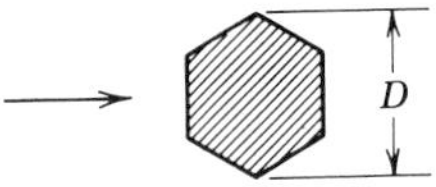 Hexagon, flow toward corner	$5 \times 10^3 - 1.95 \times 10^4$ $1.95 \times 10^4 - 10^5$	0.160 0.0385	0.638 0.782
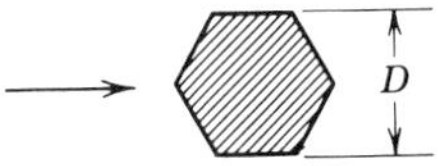 Hexagon, flow normal to face	$5 \times 10^3 - 10^5$	0.153	0.638
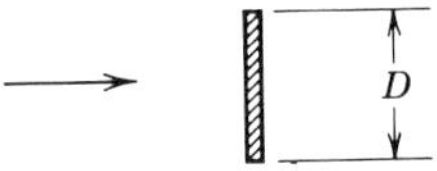 Vertical plate	$4 \times 10^3 - 1.5 \times 10^4$	0.228	0.731

*From Ref. 2.

EXAMPLE 6.7

Atmospheric-pressure air flows with a freestream velocity and temperature of $U_\infty = 10$ m/s, $t_\infty = 25°C$ normal to a circular cylinder of diameter $D = 1.25$ cm, and maintained with a surface temperature of $t_s = 125°C$.

(a) Estimate the average surface heat transfer coefficient using Zhukauskas' correlation of Eq. (6.48).
(b) Repeat the calculation of h using Churchill and Bernstein's formula of Eq. (6.49) and compare the results.
(c) Calculate the drag force, per unit of cylinder length, exerted by the air on the cylinder.

Solution. For a mean film temperature $t_m = (t_\infty + t_s)/2 = (25 + 125)/2 = 75°C$, and at these temperatures, Table A.6 gives

$$\nu = 20.41 \times 10^{-6} \text{ m}^2/\text{s}, \qquad k = 29.57 \times 10^{-3} \text{ W/m-°C}, \qquad \text{Pr} = 0.707.$$

Equation (6.48) will also require Pr evaluated at $t_\infty = 25°\text{C}$ and $t_s = 125°\text{C}$, and at these temperatures, Table A.6 gives

$$\text{Pr}_\infty = 0.713, \qquad \text{Pr}_s = 0.703.$$

The Reynolds number, based on the freestream velocity, is

$$\text{Re}_D = \frac{U_m D}{\nu} = \frac{10 \times 0.0125}{20.41 \times 10^{-6}} = 6124.$$

(a) For the application of Zhukauskas' equation, Eq. (6.48) indicates that $n = 0.37$ and Table 6.3 gives $C = 0.26$, $m = 0.6$. Thus

$$\begin{aligned}
\text{Nu}_D &= C\text{Re}_D^m \text{Pr}^n \left(\frac{\text{Pr}_\infty}{\text{Pr}_s}\right)^{1/4} \\
&= 0.26\text{Re}_D^{0.6}\text{Pr}^{0.37}\left(\frac{\text{Pr}_\infty}{\text{Pr}_s}\right)^{1/4} \\
&= 0.26(6124)^{0.6}(0.707)^{0.37}(0.713/0.703)^{1/4} \\
&= 43.0.
\end{aligned}$$

Thus the heat transfer coefficient is

$$\begin{aligned}
h &= \text{Nu}_D \frac{k}{D} \\
&= 43.0 \times \frac{0.02957}{0.0125} \\
&= 102 \text{ W/m}^2\text{-°C } (17.9 \text{ Btu/h-ft}^2\text{-°F}).
\end{aligned}$$

(b) The Churchill–Bernstein formula (6.49) gives

$$\begin{aligned}
\text{Nu}_D &= 0.3 + \frac{0.62\text{Re}_D^{1/2}\text{Pr}^{1/3}}{[1 + (0.4/\text{Pr})^{2/3}]^{1/4}}\left[1 + \left(\frac{\text{Re}_D}{2.82 \times 10^5}\right)^{5/8}\right]^{4/5} \\
&= 0.3 + \frac{0.62(6124)^{1/2}(0.707)^{1/3}}{[1 + (0.4/0.707)^{2/3}]^{1/4}}\left[1 + \left(\frac{6124}{2.82 \times 10^5}\right)^{5/8}\right]^{4/5} \\
&= 41.0,
\end{aligned}$$

$$\begin{aligned}
h = \text{Nu}_D \frac{k}{D} &= 41.0 \times \frac{0.02957}{0.0125} \\
&= 97.0 \text{ W/m}^2\text{-°C } (17.1 \text{ Btu/h-ft}^2\text{-°F}).
\end{aligned}$$

This value is within 5% of that found in part (a), and such agreement is regarded as quite good.

(c) The drag force on the cylinder is found from the drag coefficient. The Reynolds number of $\text{Re}_D = 6124$ is within the range of validity of Eq. (6.47), so

$$\begin{aligned}
C_D &= 1 + 10.0\text{Re}_D^{-2/3} \\
&= 1 + 10.0(6124)^{-2/3} = 1.003.
\end{aligned}$$

Thus, by Eq. (6.47), the drag force per meter of cylinder length is

$$\begin{aligned}\frac{\text{drag}}{L} &= C_D \tfrac{1}{2}\rho_\infty U_\infty^2 D \\ &= 1.003 \times \tfrac{1}{2} \times 1.1843 \times 10^2 \times 0.0125 \\ &= 0.74\ \text{N/m}\ (0.55\ \text{lb}_f/\text{ft}).\end{aligned}$$

In the calculation above, the density of the freestream at $t_\infty = 25°\text{C}$ ($\rho_\infty = 1.1843\ \text{kg/m}^3$) has been obtained from Table A.6.

Flow over Banks of Tubes

Many heat exchangers, air conditioning cooling or heating coils, etc., involve a bank (or bundle) of circular tubes over which a fluid may flow. Typical geometric arrangements are shown in Fig. 6.6. Figure 6.6(a) depicts the *aligned* arrangement of a tube bank in which successive rows of tubes are aligned with one another, as viewed with respect to the velocity of the incident flow. Also employed is the *staggered* arrangement shown in Fig. 6.6(b) wherein successive rows are displaced transversely with respect to the oncoming flow. In either arrangement, the significant bank dimensions are the tube diameter, D, the longitudinal spacing, S_L, and the transverse tube spacing, S_T.

The flow pattern through a tube bank is generally quite complex. The heat transfer to or from a particular tube is not only influenced by the boundary layer separation phenomenon discussed for a single tube but also influenced by effects such as the "shading" of one tube by another, the impingement of the wake of one tube on another, etc. The heat transfer for any particular tube is then not only determined by the incident fluid conditions, U_∞ and t_∞, but also by D, S_L, S_T and the tube position in the bank. The heat transfer coefficient for the first row of tubes is much like that for a single cylinder in cross flow, while that for tubes in the inner rows is generally larger because of the wake generated by the previous tubes.

For heat transfer correlations, in tube banks, the Reynolds number used is defined as

$$\text{Re}_{D_m} = \frac{U_{\max} D}{\nu}, \tag{6.51}$$

where $U_{\max}$ is the maximum average fluid velocity occurring at the minimum free area of the bank. For the aligned tube arrangement, $U_{\max}$ is clearly

$$U_{\max} = U_\infty \frac{S_T}{S_T - D}, \tag{6.52}$$

while for the staggered arrangement, $U_{\max}$ is given by Eq. (6.52) when the diagonal spacing, $S_D = [S_L^2 + (S_T/2)^2]^{1/2}$ is such that $2(S_D - D) > S_T - D$. For $2(S_D - D) < S_T - D$, $U_{\max}$ is then

$$U_{\max} = U_\infty \frac{S_T}{2(S_D - D)}. \tag{6.53}$$

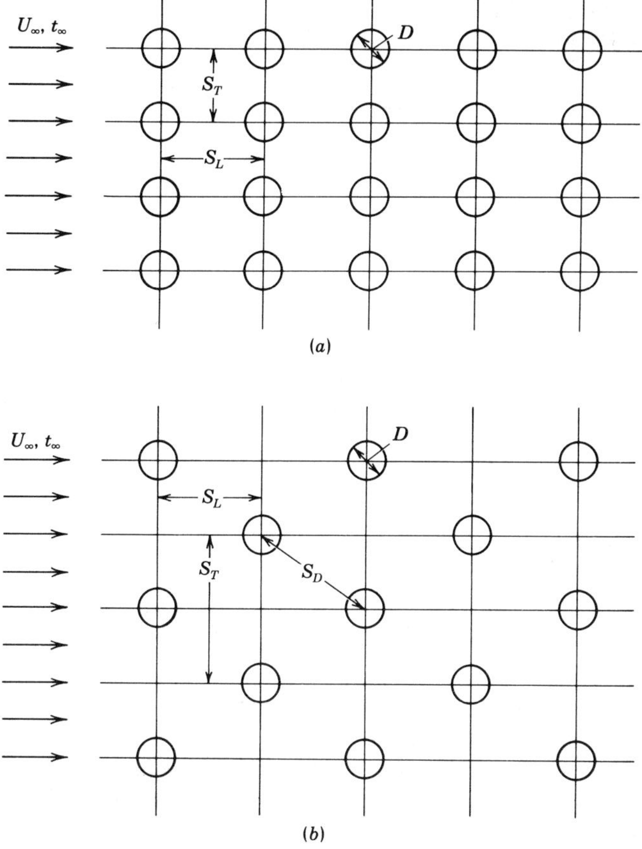

Figure 6.6. Tube banks: (a) aligned; (b) staggered.

Generally, one wishes to know an *average* heat transfer coefficient for the entire tube bank. For many years the correlation of Grimson (Ref. 27) has been used and may still be expected to give useful results. However, the more recent and extensive work of Zhukauskas summarized in Ref. 25 is recommended here. For *many* rows of tubes, say 20 or more, this correlation for the average heat transfer coefficient is

$$\mathrm{Nu}_D = C\mathrm{Re}_{Dm}^n \mathrm{Pr}^{0.36}\left(\frac{\mathrm{Pr}_\infty}{\mathrm{Pr}_s}\right)^{1/4}, \quad \begin{bmatrix} 0.7 < \mathrm{Pr} < 500 \\ 10 < \mathrm{Re}_{D_m} < 10^6 \\ \text{C and } n \text{ from Table 6.5} \\ \text{properties at } t_m \text{ except } \mathrm{Pr}_\infty, \mathrm{Pr}_s \\ \mathrm{Pr}_\infty \text{ at } t_\infty, \mathrm{Pr}_s \text{ at } t_s \end{bmatrix}. \qquad (6.54)\bigstar$$

As noted earlier, the correlation above is for the inner rows of a bank, or for banks of many rows. Using 20 rows as a normalizing case, Zhukauskas recommends the

Table 6.5 Values of C and n for Zhukauskas' Equation for Flow over Banks of Tubes*

Re_{D_m} Range	Aligned	C	n	Staggered	C	n
10–100		0.8	0.4		0.9	0.4
100–1000		See note 1			See note 1	
1000–2 × 10^5	$S_T/S_L < 0.7$:	See note 2		$S_T/S_L < 2$:	$0.35(S_T/S_L)^{1/5}$	0.6
	$S_T/S_L > 0.7$:	0.27	0.63	$S_T/S_L > 2$:	0.40	0.6
2 × 10^5–10^6		0.021	0.84		0.022	0.84

Notes:
1. In this range of Re_D use the correlation for a single tube, Eq. (6.48).
2. In this range of Re_D for aligned patterns, the heat transfer is poor, and designs of this geometry are not recommended.
*From Ref. 25.

use of Fig. 6.7 wherein the correction factor

$$\frac{(Nu_D)_{N\ \text{rows}}}{(Nu_D)_{20\ \text{rows}}}$$

is plotted. Thus Eq. (6.54) is used to determine Nu_D for 20 rows and Fig. 6.7 to correct the result for cases with less than 20 rows.

In the same study that led to Eq. (6.54) for the heat transfer in a bank of tubes, Zhukauskas also presents graphical representations for calculating the pressure drop of the fluid as it flows through the bank. However, the studies reported by Jakob (Ref. 28) presents an empirical equation that is simpler to use. Jakob defines an apparent friction factor, f, so that the pressure drop may be found from (where N

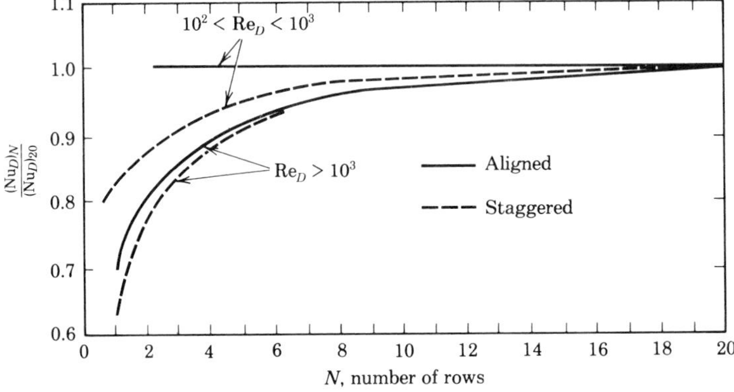

Figure 6.7. Correction factor for the number of rows in tube banks. (From A. Zhukauskas," Heat Transfer for tubes in Crossflow," *Advances in Heat Transfer,* New York, Academic Press, 1972, p. 93. Used by permission.)

is the number of tube rows)

$$-\Delta P = fN\left(\frac{1}{2}\,\rho_\infty U_{\max}^2\right)\left(\frac{\mu_s}{\mu_\infty}\right)^{0.14},$$

$$f = \left[C_1 + \frac{C_2}{(S_T/D - 1)^n}\right]\mathrm{Re}_{D_m}^{-m}, \tag{6.55}★$$

$$\left[\begin{array}{l} \text{properties at } t_m \text{ except } \rho_\infty, \mu_\infty, \mu_s \\ \rho_\infty \text{ and } \mu_\infty \text{ at } t_\infty \\ \mu_s \text{ at } t_s \\ \text{staggered banks: } C_1 = 1.0,\ C_2 = 0.470 \\ \qquad n = 1.08,\ m = 0.16 \\ \text{aligned banks: } C_1 = 0.176,\ C_2 = 0.34S_L/D \\ \qquad n = 0.43 + (1.13D/S_T) \\ \qquad m = 0.15 \end{array}\right].$$

EXAMPLE 6.8

Air-conditioning systems often use "cooling coils" to reduce the temperature of air for human comfort. A "coil" consists of a bank of tubes placed in an air-conditioning duct such that the air flows normal to the tubes while a cooling medium (chilled water or a refrigerant) flows through the tubes. Figure 6.8 depicts such a

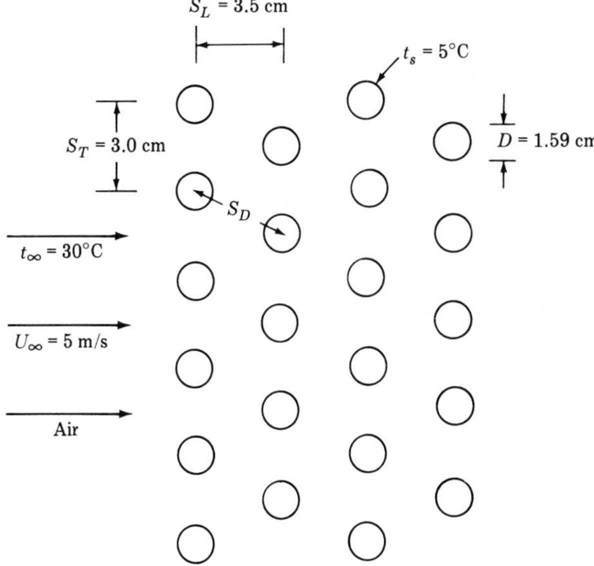

Figure 6.8. The tube bank of Example 6.8.

tube bank, consisting of a staggered array of tubes with a diameter $D = 1.59$ cm. There are four rows of tubes with a longitudinal spacing $S_L = 3.5$ cm and a transverse spacing $S_T = 3.0$ cm. Chilled water flowing through the tubes maintains the tube surface temperature at $t_s = 5°C$. Atmospheric-pressure air is incident on the bank with freestream conditions: $U_\infty = 5$ m/s, $t_\infty = 30°C$.

(a) Estimate the surface heat transfer coefficient.
(b) Find the pressure drop of the air across the tube bundle.

Solution. Application of the heat transfer and friction factor relations of Eqs. (6.54) and (6.55) require some air properties at the tube surface temperture, $t_s = 5°C$, some at the freestream temperature, $t_\infty = 30°C$, and others at the mean temperature $t_m = (t_s + t_\infty)/2 = 17.5°C$. Table A.6 yields:

At $t_s = 5°C$: $\mu_s = 17.45 \times 10^{-6}$ kg/m-s, $\Pr_s = 0.717$.

At $t_\infty = 30°C$: $\rho_\infty = 1.1644$ kg/m^3, $\mu_\infty = 18.65 \times 10^{-6}$ kg/m-s, $\Pr_\infty = 0.712$.

At $t_m = 17.5°C$: $\nu = 14.87 \times 10^{-6}$ m^2/s, $k = 25.45 \times 10^{-3}$ W/m-°C, $\Pr = 0.714$.

The diagonal spring in the staggered tube bank is readily found:

$$S_D = \left[S_L^2 + \left(\frac{S_T}{2}\right)^2\right]^{1/2} = \left[3.5^2 + \left(\frac{3.0}{2}\right)^2\right]^{1/2} = 3.808 \text{ cm}.$$

The frontal flow area between tubes is $S_T - D = 3.0 - 1.59 = 1.41$ cm, while that for the divided stream through the diagonal openings is $2(S_D - D) = 2(3.808 - 1.59) = 4.436$ cm. Since the frontal area is less than the diagonal, $(S_T - D) < 2(S_D - D)$, the maximum flow velocity occurs in the space $S_T - D$ and is given by Eq. (6.52):

$$U_{\max} = U_\infty \frac{S_T}{S_T - D}$$

$$= 5 \times \frac{3.0}{3.0 - 1.59} = 10.638 \text{ m/s}.$$

Thus the Reynolds number based on this velocity and the tube diameter is, by Eq. (6.51),

$$\text{Re}_{D_m} = \frac{U_{\max} D}{\nu}$$

$$= \frac{10.638 \times 0.0159}{14.87 \times 10^{-6}} = 1.138 \times 10^4.$$

(a) For this Reynolds number, Table 6.5 gives the following constants for use in Eq. (6.54) since $S_T/S_L = 3.0/3.5 = 0.857$:

$$C = 0.35(S_T/S_L)^{1/5} = 0.35(0.857)^{1/5} = 0.339,$$
$$n = 0.6.$$

So Eq. (6.54) gives the Nusselt number to be

$$\begin{aligned}\text{Nu}_D &= C\text{Re}_{D_m}^n \text{Pr}^{0.36}\left(\frac{\text{Pr}_\infty}{\text{Pr}_s}\right)^{1/4} \\ &= 0.339(1.138 \times 10^4)^{0.6}(0.714)^{0.36}\left(\frac{0.712}{0.717}\right)^{1/4} \\ &= 81.44 = (\text{Nu}_D)_{20}.\end{aligned}$$

As noted, the calculation above yields the Nusselt number for a large number of tube rows, 20, and the correction for a lesser number is gotten from Fig. 6.7. For $N = 4$, Fig. 6.7 gives

$$\frac{(\text{Nu}_D)_4}{(\text{Nu}_D)_{20}} \cong 0.88,$$

so that

$$\text{Nu}_D = (\text{Nu}_D)_4 = 0.88 \times 81.44 = 71.66$$

and

$$\begin{aligned}h = \text{Nu}_D \frac{k}{D} &= 71.66 \times \frac{0.02545}{0.0159} \\ &= 114.7 \text{ W/m}^2\text{-°C } (20.2 \text{ Btu/h-ft}^2\text{-°F}).\end{aligned}$$

(b) For the pressure drop, Eq. (6.55) gives the apparent friction factor to be

$$\begin{aligned}f &= \left[1.0 + \frac{0.472}{(S_T/D - 1)^{1.08}}\right] \text{Re}_{D_m}^{-0.16} \\ &= \left[1.0 + \frac{0.472}{(3.0/1.59 - 1)^{1.08}}\right](1.138 \times 10^4)^{-0.16} = 0.345.\end{aligned}$$

The pressure drop is, then,

$$\begin{aligned}-\Delta P &= fN\left(\frac{1}{2}\rho_\infty U_{\max}^2\right)\left(\frac{\mu_s}{\mu_\infty}\right)^{0.14} \\ &= 0.345 \times 4 \times \left(\frac{1}{2} \times 1.1644 \times 10.638^2\right)\left(\frac{17.45}{18.65}\right)^{0.14} \\ &= 90.1 \text{ N/m}^2 \ (0.013 \text{ lb}_f/\text{in}^2).\end{aligned}$$

REFERENCES

1. Kays, W. M., and M. E. Crawford, *Convective Heat and Mass Transfer*, 2nd ed., New York, McGraw-Hill, 1980.

2. Jakob, M., *Heat Transfer,* Vols. 1 and 2, New York, Wiley, 1949, 1957.
3. Knudsen, J. G., and D. L. Katz, *Fluid Dynamics and Heat Transfer,* New York, McGraw-Hill, 1958.
4. Eckert, E. R. G., and R. M. Drake, *Analysis of Heat and Mass Transfer,* New York, McGraw-Hill, 1972.
5. Rohsenow, W. M., and J. P. Hartnet, eds., *Handbook of Heat Transfer,* New York, McGraw-Hill, 1973.
6. Churchill, S. W., and H. Ozoe, "Correlations for Laminar Forced Convection in Flow over an Isothermal Flat Plate and in Developing and Fully Developed Flows in an Isothermal Tube," *J. Heat Transfer, Trans. ASME,* Vol. 95, 1973, p. 416.
7. Schlicting, H., *Boundary Layer Theory,* 7th ed., New York, McGraw-Hill, 1979.
8. Chapman, A. J., *Heat Transfer,* 4th ed., New York, Macmillan, 1984.
9. Eckert, E. R. G., "Engineering Relations for Friction and Heat Transfer to Surfaces in High Velocity Flow," *J. Aerosp. Sci.,* Vol. 22, 1955, p. 585.
10. Wimbrow, W. R., "Experimental Investigations of Temperature Recovery Factors on Bodies of Revolution at Supersonic Speeds," *NACA Tech. Note 1975,* Washington, D.C., 1949.
11. Eckert, E. R. G., "Survey on Heat Transfer at High Speeds," *Aeronaut. Res. Lab. Rep. 189,* Office of Aerospace Research, Wright-Patterson Air Force Base, Ohio, 1961.
12. Lin, C. C., ed., *Turbulent Flows and Heat Transfer, High Speed Aerodynamics and Jet Propulsion,* Vol. V, Princeton, N.J., Princeton University Press, 1959.
13. Hausen, H., "Darstellung des Wärmeüberganges in Rohren durch vergallgemeinerte Potenzbeziehungen," *Z.A.V.D.I. Beihefte Verfahrenstech.,* No. 4, 1943, p. 91.
14. Sieder, E. N., and E. G. Tate, "Heat Transfer and Pressure Drop of Liquids in Tubes," *Ind. Eng. Chem.,* Vol. 28, 1936, p. 1429.
15. Shah, R. K., "A Correlation for Laminar Hydrodynamic Entry Length Solutions for Circular and Non-circular Ducts," *J. Fluids Eng. Trans. ASME,* Vol. 100, 1978, p. 177.
16. Dittus, F. W., and L. M. K. Boelter, *Univ. Calif., Berkeley, Publ. Eng.,* Vol. 2., 1930, p. 443.
17. Petukhov, B. S., "Heat Transfer and Friction in Turbulent Pipe Flow with Variable Physical Properties," in *Advances in Heat Transfer,* Vol. 6, New York, Academic Press, 1970, p. 504.
18. Nusselt, W., "Der Wärmeaustausch zwischen Wand und Wasser im Roher," *Forsch. Geb. Ingenieurwes.,* Vol. 2, 1931, p. 309.
19. Martinelli, R. C., "Heat Transfer to Molten Metals," *Trans. ASME,* Vol. 69, No. 8, 1947, p. 447.
20. Seban, R. A., and T. T. Shimazaki, "Heat Transfer to a Fluid Flowing Turbulently in a Smooth Pipe with Walls at Constant Temperature." *Trans. ASME,* Vol. 73, 1951, p. 803.
21. Azer, N. Z., and B. T. Chao, "Turbulent Heat Transfer in Liquid Metals—Fully Developed Pipe Flow with Constant Wall Temperature," *Int. J. Heat Mass Transfer,* Vol. 3, 1961, p. 77.

22. Sleicher, C. A., A. S. Awad, and R. H. Notter, "Temperature and Eddy Diffusivity Properties in N_aK," *Int. J. Heat Mass Transfer,* Vol. 16, 1973, p. 1565.

23. White, F. M., *Viscous Fluid Flow,* New York, McGraw-Hill, 1974.

24. Giedt, W. H., "Investigation of the Variation of Point Unit Heat Transfer Coefficient Around a Cylinder Normal to an Air Stream," *Trans ASME,* Vol. 71, 1949, p. 375.

25. Zhukauskas, A., "Heat Transfer from Tubes in Crossflow," in *Advances in Heat Transfer,* Vol. 8, New York, Academic Press, 1972, p. 93.

26. Churchill, S. W., and M. Bernstein, "A Correlating Equation for Forced Convection from Gases and Liquids to a Circular Cylinder in Crossflow," *J. Heat Transfer, Trans. ASME,* Vol. 94, 1977, p. 300.

27. Grimson, E. D., "Correlation and Utilization of New Data on Flow Resistance and Heat Transfer for Cross Flow of Gases over Tube Banks," *Trans. ASME,* Vol. 59, 1937, p. 583.

28. Jakob, M., "Heat Transfer and Flow Resistance in Cross Flow of Gases over Tube Banks," *Trans. ASME,* Vol. 60, 1938, p. 384.

PROBLEMS

6.1 Glycerine at 40°C flows at 4 m/s past a flat surface maintained at 10°C. If the plate is 0.3 m long and 0.5 m wide, find the heat transferred from one side of the surface using both Eqs. (6.2) and (6.4). What is the drag force for one side? Neglect the effects of viscous dissipation.

6.2 Water at 90°C flows with a velocity of 1 m/s past a flat surface maintained at 50°C. Calculate the local heat transfer coefficient and the local skin friction coefficient at 0.01 and 0.05, 0.1, and 0.15 m from the leading edge. Find the average heat transfer coefficient and the drag coefficient for a total surface length of 0.15 m. Neglect the effects of viscous dissipation.

6.3 Atmospheric-pressure air at 20°C flows with a velocity of 25 m/s past a flat surface maintained at 80°C. The surface is 15 cm long. Using both Eqs. (6.2) and (6.4), find the average heat transfer coefficient and the drag coefficient for the surface if the effects of viscous dissipation are neglected.

6.4 A flat plate, 1.2 m × 0.5 m, is towed, parallel to its long side, at a velocity of 2 m/s through the lubricating oil given in Table A.4. The oil is at 80°C and the plate is electrically heated so that it is maintained at 100°C. Estimate the power required **(a)** to tow the plate, and **(b)** to heat the plate. Neglect the effects of viscous dissipation.

6.5 Atmospheric air at 0°F flows past a flat surface maintained at 100°F. The surface is 2 ft long and 1 ft wide. What air velocity is required to remove 500 Btu/h from one side of the surface?

6.6 Atmospheric air at 15°C flows with a velocity of 25 m/s past a flat surface, 50 cm long, maintained at 75°C. Find **(a)** the local heat transfer coefficient and local skin friction coefficient at locations 12.5, 25, 37.5, and 50 cm from the leading edge, and **(b)** the average heat transfer coefficient and drag coefficient for one side of the surface. Neglect the effects of viscous dissipation.

6.7 A flat surface has the dimensions of 3 in. × 18 in. It is maintained at 190°F and is placed in an airstream at 1 atm, 50°F, flowing with a velocity of 90 ft/s. Find the total heat transferred and the total drag force on one side of the surface, neglecting the effects of viscous dissipation, if the leading edge is **(a)** the 18-in. side, and **(b)** the 3-in. side.

6.8 Nitrogen at atmospheric pressure and 50°C flows with a velocity of 20 m/s past a flat surface held at 0°C. The surface is 0.8 m long and 0.2 m wide. Find **(a)** the total heat transferred and the total drag force on one side of the surface, and **(b)** the fraction of the heat transferred and drag force in the laminar portion of the boundary layer. Neglect the effects of viscous dissipation.

6.9 Justify the neglection of the effects of viscous dissipation in Probs. 6.5 through 6.8.

6.10 Justify the neglection of the effects of viscous dissipation in Probs. 6.1 through 6.4.

6.11 Atmospheric-pressure air at a temperature of 50°C flows over a flat surface heated to 100°C. The surface is 0.2 m long and 0.1 m wide. The value of Re_L is 40,000. **(a)** What is the total rate of heat transfer from one side of the surface? **(b)** If the air velocity is doubled and the air pressure increased to 10 atm, what is the heat transfer rate?

6.12 An isothermal flat surface has the dimensions of $L \times 2L$. Based on a transition Reynolds number of 5×10^5, determine the conditions of a fluid stream flowing past the surface for which the total heat transfer rate is the same when either the L-side or $2L$-side is the leading edge.

6.13 Air at atmospheric pressure flows with a freestream velocity and temperature of 185 m/s, 15°C, past a flat surface. Accounting for the effects of viscous dissipation, find the heat flux and the skin friction coefficient at a location 2.5 cm from the leading edge if the plate is maintained at **(a)** 20°C, **(b)** 30°C, and **(c)** 40°C.

6.14 Repeat Prob. 6.13 if the effects of viscous dissipation are neglected and compare the results.

6.15 Compute the recovery temperature, the local heat transfer and skin friction coefficients, and the local heat flux that may be expected to occur at a point 0.3 m behind the leading edge of the fin of a supersonic plane traveling at a speed of 660 m/s at **(a)** sea level (1 atm, 15°C), and **(b)** 20,000-ft altitude (0.46 atm, −25°C). The fin surface is to be maintained at 100°C in both cases.

6.16 Air at an altitude of 1500 m (84 kN/m^2, 5°C) flows past a flat surface maintained at 10°C with a velocity of 200 m/s. Find the local heat flux at a point 2.5 cm from the leading edge **(a)** if viscous dissipation is neglected, and **(b)** if viscous dissipation is not neglected.

6.17 A flat surface 2 ft long and 1 ft wide is placed in an airstream at 0.50 psia, −40°F, flowing with a velocity of 3020 ft/s. How much cooling must be provided to maintain the surface at 100°F if viscous dissipation is accounted for?

6.18 Air at a pressure of 35 kN/m^2 and a temperature of −40°C flows past a flat surface 10 cm long and 30 cm wide with a freestream Mach number of 3.0. The surface is maintained at 150°C. Estimate the total heat transferred to or from the surface.

6.19 Air at 3 psia, 0°F, flows past a flat surface maintained at 200°F with a velocity of 4200 ft/s. If the surface is 18 in. long, find the cooling required, for one side, to maintain the given surface temperature.

6.20 A flat surface of an aircraft is located in an airstream with the conditions: P = 1.0 psia, t_f = −30°F, U = 2500 ft/s. The surface is 1 ft long and 0.5 ft wide, and is maintained at a surface temperature t_s = 110°F.

(a) Perform some sort of calculation to ascertain whether or not viscous dissipation should be accounted for.
(b) Based on the conclusions of part (a), find how much cooling or heating must be supplied to maintain the given surface temperature.

6.21 The lubricating oil listed in Table A.4 flows with an average velocity of 0.3 m/s and an average bulk temperature of 160°C in a tube 2.3 cm in diameter, the surface of which is maintained at 140°C. Find the average surface heat transfer coefficient and the pressure drop **(a)** if the tube is 2 m long, and **(b)** if the tube is 6 m long.

6.22 Atmospheric-pressure air flows with a mean bulk temperature of 50°C through a tube 0.6 cm in diameter at a mass rate of 0.5 kg/h. If the tube surface is maintained at 90°C, and it is 0.3 m long, find the average surface heat transfer coefficient and the pressure drop.

6.23 The cooling oil used in a transformer has properties closely resembling those of glycerine. Such an oil flows at a mean velocity of 0.5 m/s through a circular tube 5 cm in diameter and 2 m long. If the oil bulk temperature is 50°C and the tube wall is at 30°C, find the rate of heat transfer from the oil and the pressure drop through the tube.

6.24 The lubricating oil listed in Table A.4 flows with a mean velocity and bulk temperature of 0.04 m/s, 100°C in a circular tube, the surface of which is maintained at 40°C. The tube has a diameter of 2.5 cm and is 10 m long. Determine the heat transfer coefficient at the tube surface and the pressure drop through the tube.

6.25 Air at atmospheric pressure flows at the volume rate of 1 m^3/s with an average bulk temperature of 15°C through an air conditioning duct 20 cm in diameter with a surface temperature of 30°C. Determine the average surface heat transfer coefficient using Eqs. (6.31) through (6.33). What is the friction factor for the duct?

6.26 Water at 80°F enters a 2-in. schedule 40 pipe and leaves at 120°F. It flows at the rate of 50,000 lb_m/h. The pipe surface is maintained at 140°F. Find the heat transfer coefficient at the pipe surface for **(a)** the entrance conditions, **(b)** the exit conditions, and **(c)** the average of the entrance and exit conditions.

6.27 Superheated steam at 4000 kN/m^2 pressure and 500°C flows at the rate of 1.25 kg/s through an 8-in., schedule 80 pipe with a surface temperature of 450°C. Find the heat transfer coefficient at the pipe surface and the pipe friction factor.

6.28 Water in a circulating chilled water air conditioning system enters the chilling unit at 15°C and leaves at 5°C. The tubes in the chilling unit are 3.8 cm in diameter, the surface temperature is 2°C, and the average water velocity is 0.6 m/s. Estimate the heat transfer coefficient and friction factor at the tube surface.

6.29 Air at 40 atm pressure flows at a velocity of 6 m/s through a tube 2.5 cm in diameter at a mean bulk temperature of 250°C. The tube surface is a 80°C. Calculate the expected heat transfer coefficient and friction factor at the tube surface.

6.30 For fully developed flow in a circular tube with a surface temperature of 150°C, find the surface heat transfer coefficient and friction factor in each case if the mean flow velocity is 6 m/s in each case: **(a)** air, 90°C, 1 atm pressure, tube diameter = 2.5 cm; **(b)** air, 90°C, 1 atm pressure, tube diameter = 1.25 cm; **(c)** air, 90°C, 10 atm pressure, tube diameter = 2.5 cm; **(d)** water, 90°C, 10 atm pressure, tube diameter = 2.5 cm; and **(e)** lubricating oil, 90°C, tube diameter = 2.5 cm.

6.31 The cooling water in the tube of a steam condenser flows at velocity of 2 m/s and a bulk temperature of 60°C. The condenser tube has an inside diameter of 5 cm, and the steam condensing on it maintains the tube surface at 100°C. Determine the heat transfer coefficient at the inner tube surface using Eqs. (6.31) through (6.33) and compare the various results.

6.32 A fluid at a mean bulk temperature of 40°C flows with a velocity of 10 m/s inside a nominal-2-in. schedule 80 pipe, the surface of which is maintained at 50°C. Find the heat transfer coefficient if the fluid is **(a)** helium at 1 atm pressure, **(b)** air at 1 atm pressure, **(c)** water, or **(d)** glycerine, if the pipe is 5 m long.

6.33 Water enters a pipe (ID = 5 cm) at 20°C and leaves at 50°C. The water flows with a mean velocity of 3.5 m/s, and the pipe surface is maintained at 60°C. Find the heat transfer coefficient at the pipe surface for **(a)** the entrance conditions, **(b)** the exit conditions, and **(c)** the average of the entrance and exit conditions.

6.34 Liquid bismuth at 500°C flows with a velocity of 1 m/s in a 1.25-cm-diameter tube which is maintained at 600°C. Determine the heat transfer coefficient at the inner tube surface.

6.35 Liquid sodium at 538°C flows at the rate of 6 m/s through a pipe 2.5 cm in diameter. Estimate the convective heat transfer coefficient using Eqs. (6.36) through (6.38).

6.36 Repeat Prob. 6.22 for mercury at 316°C.

6.37 A 44% potassium, 56% sodium, liquid metal mixture is used as the heat transfer medium in a nuclear reactor. What heat transfer coefficient can be expected at the inner surface of a pipe of 2.5 cm ID if the medium is at 371°C and flows with a velocity of 3 m/s?

6.38 Natural gas at a pressure of 100 psia (approximated as methane) flows at 90°F through a $\frac{3}{4}$-in.-diameter pipe at the rate of 10 ft/s. If the pipe wall is at 80°F, estimate the surface heat transfer coefficient.

6.39 Water flowing at the rate of 2.6 kg/s is heated from 7°C to 20°C as it flows through a smooth 2-in. schedule 40 pipe. The inside pipe wall temperature is 90°C. What is the required length of pipe? What is the pressure drop of the water as it flows through the pipe?

6.40 Water flowing at the mass rate of 0.75 kg/s is to be cooled from 50°C to 20°C. Which of the following pipes offers the least pressure loss?

(a) The water flows through a 1.25-cm-ID pipe maintained at 10°C.
(b) The water flows through a 2.5-cm-ID pipe maintained at 15°C.

6.41 Superheated steam at 6000 kN/m^2 pressure, 400°C, enters a pipe 2.5 cm in diameter with a velocity of 12 m/s. The pipe wall is maintained at 500°C. How long must the pipe be to heat the steam to 430°C? What is the pressure drop of the steam as it flows through the pipe?

6.42 Water at 10°C and flowing at the mass rate of 2.5 kg/s enters a pipe 5 cm in diameter. The pipe wall is maintained at 80°C. If the pipe is 5 m long, what is the outlet temperature of the water? What is the pressure drop through the pipe?

6.43 Superheated steam flows at a mass rate of 2 kg/s through a pipe 8 cm in diameter. The steam enters at 4000 kN/m^2 pressure and 450°C. The pipe is 10 m long and has its surface maintained at 350°C. Find the outlet temperature of the steam and the pressure drop through the pipe.

6.44 Water, flowing at the rate of 1.2 kg/s, is to be heated from 10°C to 30°C by passing it through a 4-cm-diameter pipe 3.5 m long. Estimate the required surface temperature of the pipe.

6.45 Water at 60°F enters a 1-in.-diameter tube 10 ft long. If the water flow rate is 2 lb_m/s and the tube wall is maintained at 200°F, determine the exit temperature of the water and the pressure drop.

6.46 Liquid sodium is to be used as the transport fluid in a nuclear power plant. It enters a 2.5-cm-diameter tube of a heat exchanger at 180°C and flows at the rate of 3 kg/s. If the tube wall is maintained at 240°C, how long should the tube be to raise the sodium to a temperature of 230°C?

6.47 Ethylene glycol is to be used as the heating medium for a constant-temperature water bath. The glycol enters a 0.3-cm-diameter thin-walled tube at 85°C and a flow rate of 0.01 kg/s. The tube forms a coil within a well-mixed water bath at 25°C. What is the required tube length if the glycol outlet temperature is 35°C?

6.48 Superheated steam flows at a mass flow rate of 7 kg/s through a pipe with an inside diameter of 4 in. The steam enters the pipe at a pressure of 400 psia and a bulk temperature 800°F. The pipe is 30 ft long and has its surface temperature maintained at 700°F. Neglecting the pressure drop in the pipe, find **(a)** the outlet temperature of the steam, and **(b)** the temperature of the steam at a point halfway along the pipe—i.e., at 15 ft from the entrance. Then **(c)** justify the neglect of the pressure drop along the length of the pipe.

6.49 Air at 1 atm and 20°C enters a 2-m-long rectangular duct with a cross section of 8 cm × 16 cm. The duct surface temperature is 125°C, and the mass flow rate of the air is 0.10 kg/s. Find the heat transferred to the air and the outlet temperature of the air.

6.50 Water flows in a rectangular duct having a cross section of 0.5 cm × 1.0 cm with a bulk temperature of 20°C. If fully developed laminar flow exists and if the duct wall temperature is 60°C, calculate the heat transferred to the water, per meter of duct length.

6.51 Air at atmosphere pressure and 30°C flows with a velocity of 5 m/s through a square air conditioning duct, 20 cm × 20 cm, the surface of which is at 20°C. Estimate the surface heat transfer coefficient.

6.52 Superheated steam flows through the annulus formed between a 4-in. and a 6-in. schedule 40 steel pipe. The steam is at 200 psia, 400°F, and its velocity is 5 ft/s.

Water at 120°F is flowing at 2 ft/s through the inner pipe. Estimate the heat transferred, per foot of pipe length, to the water.

6.53 A "hot wire annemometer" is a device used to measure the velocity of a fluid by determining the heat loss from a fine wire, heated to a known temperature, placed in the fluid stream. Consider, then, such a device made of a wire 0.025 cm in diameter, placed normal to an airstream at 1 atm, 15°C. When the wire is electrically heated so that its surface temperature is 100°C, the heat loss from the wire is measured to be 40 W per meter of wire length. Determine the velocity of the airstream.

6.54 Air at 190°F and 1 atm flows past a heated wire $\frac{1}{16}$ in. in diameter. If the air velocity is 20 ft/s, find the heat loss, per unit length, from the wire if its surface temperature is 300°F.

6.55 Air at 80°F flows with a velocity of 90 ft/s normal to a single cylinder with a 1-in. outside diameter. If the cylinder surface temperature is 250°F, find the heat transfer rate and drag force on the cylinder, per foot of length, when the air pressure is **(a)** 1 atm, **(b)** 2 atm, and **(c)** 4 atm.

6.56 A fluid at 20°C flows with a velocity of 10 m/s across a 5-cm-OD tube whose surface is maintained at 100°C. Determine the heat transfer rate and drag force, per meter of length, if the fluid is **(a)** atmospheric air, **(b)** water, or **(c)** ethylene glycol.

6.57 Hot combustion gases, having properties close to those of nitrogen, flow at a pressure of 175 kN/m^2 and a temperature of 60°C normal to a 5-cm-OD cylinder at a velocity of 15 m/s. What is the surface heat transfer coefficient if the pipe is held at 120°C? What is the drag force, per unit length, on the pipe?

6.58 Repeat Prob. 6.57 for the flow of the same gases at the same conditions past a square pipe, 5 cm on the side, placed with one side normal to the flow. The pipe is, again, at 120°C.

6.59 A fluid at 40°C flows with a velocity of 10 m/s normal to a cylinder heated to 120°C and having a diameter of 5 mm. Find the heat transfer coefficient at the cylinder surface and the drag force, per unit length, if the fluid is (a) helium at 1 atm, (b) air at 1 atm, and (c) water.

6.60 A tube bank uses an aligned arrangement with $S_T = S_L = 2.0$ cm and tubes 1.0 cm in diameter. There are eight rows of tubes. The tube surfaces are at 80°C, and atmospheric air at 20°C approaches the bank, normal to the tubes, with an undisturbed velocity of 5 m/s. Estimate the average surface heat transfer coefficient for the bank of tubes. How much heat, per square meter of frontal area of the bank normal to the incident flow, is transferred to the air? What is the pressure drop in the air as it flows over the bank?

6.61 Repeat Prob. 6.60 if a staggered arrangement is used, maintaining the same S_T and S_L.

6.62 A tube bank uses an aligned arrangement with $S_T = 2.5$ cm, $S_L = 2.0$ cm, and tubes 1.5 cm in diameter. There are four rows of tubes. The tube surfaces are at 5°C, and atmospheric air at 25°C approaches the bank, normal to the tubes, with an undisturbed velocity of 6 m/s. Estimate the average heat transfer coefficient for the bank and the pressure drop across the bank.

6.63 Repeat Prob. 6.62 if a staggered arrangement is used, maintaining the same S_T and S_L.

6.64 Air at atmospheric pressure approaches a tube bank with $U_\infty = 1$ m/s, $t_\infty = 30$°C. The tube bank consists of 1-cm-diameter tubes, 10 rows deep with surface temperatures of 80°C. The longitudinal and transverse tube spacings are $S_L = S_T = 1.25$ cm. Find the heat transfer coefficient and pressure drop across the tube bank if the tubes have **(a)** an aligned arrangement, or **(b)** a staggered arrangement.

6.65 Air at atmospheric pressure approaches a tube bank with $U_\infty = 3.5$ ft/s, $t_\infty = 90$°F. The tube consists of 0.5-in.-diameter tubes, eight rows deep. The tube surface temperatures are 175°F. The longitudinal and transverse tube spacings are $S_L = S_T = 0.625$ in. Find the heat transfer coefficient and the pressure drop for the tube bank if the tubes have **(a)** an aligned arrangement, or **(b)** a staggered arrangement.

6.66 Water at 95°C flows in a 1-in. 16-gage copper tube with a velocity of 0.6 m/s. The outside surface is exposed to air at 20°C through a surface heat transfer coefficient of 8.5 W/m^2-°C. Determine the heat loss from a 1-m length of the tube.

6.67 Superheated steam at 400 psia, 600°F, flows with a velocity of 20 ft/s through a 6-in. steel schedule 40 pipe. The outer surface is exposed to a stream of air (1 atm, 70°C) flowing normal to the pipe at 40 ft/s. Estimate the rate of heat loss, per foot of length, from the pipe.

CHAPTER 7

Heat Transfer by Free Convection

7.1 INTRODUCTORY REMARKS

The examples of convective heat transfer considered up to now have all been examples of *forced* convection between a fluid and a solid body wherein the fluid motion relative to the solid surfaces was caused by an external input of work (i.e., by means of a pump, fan, propulsive motion of an aircraft, etc.) This chapter will be concerned with *free* or *natural* convection in which the fluid velocity at points remote from the surface will be essentially zero. Near the surface there will be some fluid motion if the surface is at a temperature different from that of the free fluid. If such is the case, there will be a density difference between the fluid near the surface and that far removed from the surface. This density difference will produce a positive or negative buoyant force (depending on whether the surface is hotter or colder than the fluid) in the fluid near the surface. The buoyant force results in a fluid motion, substantially in the vertical direction, past the surface—with consequent convective heat transfer taking place. The force of gravity is, then, the driving force which produces the fluid motion and maintains the convective process.

The heating of rooms and buildings by the use of "radiators" is a familiar example of heat transfer by free convection. Heat losses from hot pipes, ovens, etc., surrounded by cooler air are due to free convection, at least in part.

7.2 SIMILARITY CONSIDERATIONS IN FREE CONVECTION

Free convection may take place as an external flow about a body submerged in a fluid of large extent and at a temperature different from that of the body. Free convection may also occur as an internal flow, and the exchange of heat in an air space between two vertical walls maintained at different temperatures is a commonly encountered example of such an instance. Positive and negative buoyant forces produced by local heating or cooling in the space between the walls will induce fluid motion in the space between them, profoundly influencing the exchange of heat.

Pure free convection is that condition which prevails if the ambient fluid velocity far removed from the surface is virtually zero, so that the only fluid motion is due to the locally induced buoyancy. Mixed free and forced convection may occur if a forced fluid motion also exists as the result of an imposed pressure difference. As noted in Sec. 5.5, in which the various similarity parameters for convective heat transfer were developed, the significance of the inertia forces in forced convection were described in terms of the Reynolds number,

$$\mathrm{Re} = \frac{Ul}{\nu},$$

where U and l are a characteristic velocity and a characteristic length. The significance of the buoyant free convective forces were described in terms of the Grashof number:

$$\mathrm{Gr} = \frac{l^3 g\beta\,\Delta t}{\nu^2}. \tag{7.1}$$

The similarity analysis of Sec. 5.5 further showed that the ratio $\mathrm{Gr}/\mathrm{Re}^2$ determines when a flow may be characterized as either pure forced or pure free convection:

$$\begin{aligned} \mathrm{Gr}/\mathrm{Re}^2 &<< 1: \quad \text{pure forced convection,} \\ \mathrm{Gr}/\mathrm{Re}^2 &>> 1: \quad \text{pure free convection.} \end{aligned} \tag{7.2}$$

The analyses of Chapter 5 and the correlations of Chapter 6 were all restricted to cases of pure forced convection, $\mathrm{Gr}/\mathrm{Re}^2 << 1$, and in such instances the Grashof number lost its significance since buoyant forces were being neglected. Thus the heat transfer correlations were all of the form

$$\mathrm{Nu} = f(\mathrm{Re}, \mathrm{Pr}),$$

with the possible inclusion of the Eckert number, Ec, when dissipation was important.

The main emphasis of this chapter will be directed toward cases of pure free convection, $\mathrm{Gr}/\mathrm{Re}^2 >> 1$. In such cases there is no identifiable freestream velocity to which a Reynolds number may be referenced. Thus the Reynolds number loses its significance and becomes replaced by the dominating Grashof number, so that heat transfer correlations take the form

$$\mathrm{Nu} = f(\mathrm{Gr}, \mathrm{Pr}).$$

Dissipation is almost always negligible in free convection problems. Thus the Eckert number usually does not appear.

The last part of this chapter will take some note of "mixed" convection, a combination of forced and free. Such cases have to account for significant effects of both the Reynolds and Grashof numbers.

The functional forms for forced and free convection just noted in the foregoing indicates that the Grashof number will perform for free convection much the same function as the Reynolds number does for forced convection. This is, indeed, the case; however, another parameter, the Rayleigh number, is more often used for this purpose. The Rayleigh number, Ra, is defined as

$$\begin{aligned} \mathrm{Ra} &= \mathrm{GrPr}, \\ &= \frac{l^3 g\beta\, \Delta t\, \mathrm{Pr}}{\nu^2}, \end{aligned} \tag{7.3}$$

simply the product of the previously defined Grashof and Prandtl numbers. In terms of the Rayleigh number, correlations for free convection would be expected to take the form

$$\mathrm{Nu} = f(\mathrm{Ra}, \mathrm{Pr}), \tag{7.4}$$

and the latter form, Eq. (7.4), will be used in most of the relations presented in this chapter.

Buoyancy-induced free convection flow may be laminar or turbulent. Free convection results from a fundamentally unstable situation. Considering the case of a heated surface placed in a cooler fluid as an example, the locally heated fluid must rise due to buoyancy and the cooler fluid must descend. Disturbances result from the interaction between these flows and may be amplified in the resultant boundary layer so that turbulence is established. The laminar–turbulent transition depends on the relative magnitude of the buoyant forces and the resisting viscous forces—as expressed through the Grashof or Rayleigh numbers. A detailed analysis of the stability of free convection boundary layers is rather complex (Ref. 1); however, as a general rule one may expect that transition will occur for a critical Rayleigh number of

$$\mathrm{Ra}_{x,c} \cong 10^9, \tag{7.5}$$

where Ra_x represents the local length Rayleigh number in which x, the distance traveled by the boundary layer, is used as the characteristic length:

$$\mathrm{Ra}_x = \frac{x^3 g\beta\, \Delta t\, \mathrm{Pr}}{\nu^2}. \tag{7.6}$$

In the definitions of the Grashof number [Eq. (7.1)] or the Rayleigh number [Eqs. (7.3) and (7.6)], Δt symbolizes the driving temperature potential between the surface at t_s and the ambient fluid at t_f: $\Delta t = t_s - t_f$. The symbol β is used to denote the coefficient of thermal expansion of the fluid and is the origin of the buoyant forces characteristic of free convection. The coefficient of thermal expansion is defined as the negative of the fractional change of the fluid density with respect to temperature:

$$\beta = -\frac{1}{\rho}\left(\frac{\partial \rho}{\partial t}\right)_p. \tag{7.7}$$

Thus β is determined by the governing $P - \rho - t$ equation of state for the fluid involved. Many gases (e.g., air) may be adequately represented by the ideal gas equation, $\rho = P/RT$ (T being the *absolute* temperature), so that β is readily calculated from

$$\beta = \frac{1}{T}. \tag{7.8}$$

For work in this text, Eq. (7.8) may be used for most of the gases tabulated in Appendix A (hence values of β are not given), except for steam. For the rare cases involving steam, β would have to be calculated, approximately, from the tabulated values of density.

The case of liquids is more complex because of the lack of equations of state to permit evaluation of β from Eq. (7.7). As long as extreme pressures are avoided, the coefficient of expansion of liquids can be taken as dependent on temperature only. Table A.3 gives an extensive tabulation of β for water and Table A.4 gives limited data for other liquids.

7.3 THE GOVERNING EQUATIONS OF FREE CONVECTION

This section will present the basic governing equations for pure free convection in the case of external flow. These equations are presented to illustrate the methodology of the analytical prediction of free convection heat transfer. Following some brief examples of such solutions, the chapter will close by quoting a collection of useful results (some analytic, some empirical) for use in the solution of practical problems, including internal flows and mixed free and forced convection.

For purposes of discussion consider the surface to be a plane vertical wall maintained at some temperature, t_s, placed in a fluid, the temperature of which is t_f, far removed from the surface. For the case of pure free convection, the fluid velocity far from the wall will be zero. If one considers the case in which the surface is hotter than the fluid, $t_s > t_f$, the distribution of the fluid velocity and temperature in a direction normal to the surface will be something like that shown in Fig. 7.1. The temperature decreases from the value t_s imposed at the wall asymptotically to the value of the fluid far away. The fluid velocity, on the other hand, must increase from zero at the wall to a maximum value in the immediate neighborhood of the wall and then decrease to a zero value far away. The motion of the fluid is restricted mainly to a region close to the surface. For fluids of low viscosity this region is relatively thin, and it has been found that the same simplifications leading to the boundary layer equations of momentum and energy derived in Chapter 5 are applicable. However, certain modifications to these equations must be made to account for the buoyant forces that may be present.

It will be remembered that the forced convection equations [Eqs. (5.33) and (5.45) for laminar flow or Eqs. (5.93) and (5.94) for turbulent flow] were written with the coordinate direction x taken parallel to the surface and that no body forces in that direction were included in the force balance of the momentum equation. If for the free convection case, the x axis is taken again parallel to the surface, as suggested in Fig. 7.1, then the coordinate x is in the direction of the buoyant

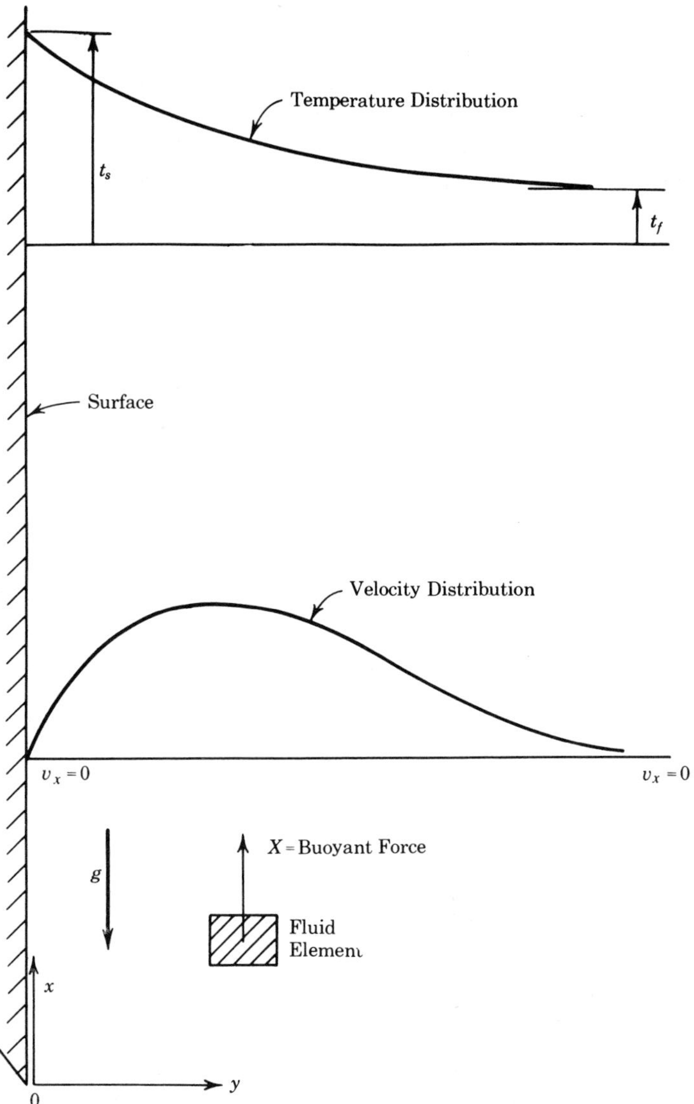

Figure 7.1. Temperature and velocity distributions near a heated vertical surface in free convection flow.

force—i.e., the vertical direction. Thus, when the momentum equation is written in this direction, the additional buoyant force must be included. If one adopts the convention that the pressure term in the momentum equation represents the difference between the local fluid pressure and the pressure in the undisturbed fluid, at the same value of x, far removed from the surface, then the buoyant body force, per unit of mass, will simply be

$$\frac{1}{\rho} g(\rho_f - \rho),$$

in which ρ is the fluid density in the boundary layer, ρ_f the density of the undisturbed fluid, and g is the acceleration of gravity. The foregoing force is taken as positive when measured in the direction opposite to that of the gravity vector. For small density differences, $\rho_f - \rho$, the coefficient of expansion defined in Eq. (7.7) may be taken as nearly constant, so that the following approximation may be used to relate the density difference $\rho_f - \rho$ to the corresponding temperature difference between the undisturbed fluid at t_f and the boundary layer at t:

$$\beta = -\frac{1}{\rho}\left(\frac{\partial \rho}{\partial t}\right)_p \approx -\frac{1}{\rho}\frac{\rho_f - \rho}{t_f - t}.$$

Then the buoyant force noted above becomes

$$\frac{1}{\rho} g(\rho_f - \rho) = g\beta(t - t_f).$$

When the differential formulation for the boundary layer is used, the foregoing term must be added to the momentum equation. The energy equation remains unchanged except for the readily justified neglection of the viscous dissipation term involving $(\partial v_x/\partial y)^2$. Thus if the possible existence of turbulence is provided for by using the form of the equations given in Eq. (5.93), one has the following set of governing differential equations for free convection:

$$\begin{aligned}
\frac{\partial v_x}{\partial x} + \frac{\partial v_y}{\partial y} &= 0, \\
v_x \frac{\partial v_x}{\partial x} + v_y \frac{\partial v_x}{\partial y} &= -\frac{1}{\rho}\frac{dP}{dx} + \frac{\partial}{\partial y}\left[(\nu + \epsilon)\frac{\partial v_x}{\partial y}\right] + g\beta(t - t_f), \qquad (7.9) \\
v_x \frac{\partial t}{\partial x} + v_y \frac{\partial t}{\partial y} &= \frac{\partial}{\partial y}\left[(\alpha + \epsilon_H)\frac{\partial t}{\partial y}\right].
\end{aligned}$$

It must be remembered that the pressure in the gradient term, dP/dx, represents the difference between the pressure in the boundary layer and that in the undisturbed fluid at the same vertical location x. That is, the hydrostatic pressure in the undisturbed fluid has been subtracted. Thus in some cases, such as flow past a vertical plane surface, the gradient dP/dx may be zero even when there is a vertical gradient in the undisturbed hydrostatic pressure.

In like fashion, integral expressions may be written for the governing equations of free convection. In the development of the integral equations for forced convection [Eqs. (5.45) for laminar flow or (5.94) for turbulent flow], the buoyant force was again omitted. Accounting for this effect in the summation of forces represented by the integral momentum equation leads to the inclusion of the additional term

$$\int_0^{\delta} g\beta(t - t_f)\, dy.$$

Thus the integral equations for free convection may be obtained from Eqs. (5.94) by adding the foregoing term to the right side of the momentum equation. The energy equation remains unchanged since dissipation was already neglected. When this is done and one uses the fact that the external velocity is $U(x) = 0$, the following are obtained:

$$\begin{aligned}\frac{d}{dx}\int_0^{\delta} v_x^2\, dy &= -\frac{\delta}{\rho}\frac{dP}{dx} - \frac{\tau_0}{\rho} + \int_0^{\delta} g\beta(t - t_f)\, dy \\ &= -\frac{\delta}{\rho}\frac{dP}{dx} - (\nu + \epsilon)\left(\frac{\partial v_x}{\partial y}\right)_0 + \int_0^{\delta} g\beta(t - t_f)\, dy, \\ \frac{d}{dx}\int_0^{\delta_t} (t - t_f) v_x\, dy &= \frac{(q/A)_0}{c_p} \\ &= -(\alpha + \epsilon_H)\left(\frac{\partial t}{\partial y}\right)_0.\end{aligned} \tag{7.10}$$

Solutions to free convection problems have been obtained using both differential formulation of Eqs. (7.9) and the integral formulation of Eqs. (7.10), and some of the results are summarized in Sec. 7.4. Such solutions turn out to be much more complex than for forced convection, even for the simple geometry of the plane vertical surface. The additional complexity involved in the solution of free convection is that the inclusion of the buoyancy term in the momentum equation introduces the fluid temperature into that equation. This means that one cannot solve the velocity problem and the temperature problem separately as was done in the case of forced convection. Thus a simultaneous solution of both the momentum and energy equations must be performed.

7.4 ANALYTICAL SOLUTIONS OF FREE CONVECTION PAST A VERTICAL SURFACE

This section will present some sample solutions of the free convection equations for the simple geometry of a vertical plane surface. These are presented to illustrate the techniques involved, using both the differential and integral formulations and treating both laminar and turbulent flows. Although more complex than the solutions presented in Chapter 5 for forced convection, due to the coupling of the momentum and energy equations, the basic concepts involved are the same. Hence only the highlights of the solutions will be given, with much of the algebraic details omitted.

Solution of the Laminar Case by the Differential Formulation

Schmidt and Beckmann (Ref. 2) analyzed the problem of laminar free convection past a vertical plane surface by application of the differential governing equations. The geometry of the problem is that illustated in Fig. 7.1 with the x coordinate measured upward along the surface from its leading edge and y normal to it. For a flat surface $dP/dx = 0$. When this fact is used, and the limitation of laminar flow is applied (i.e., $\epsilon = \epsilon_H = 0$), the governing equations [Eqs. (7.9)] become

$$\frac{\partial v_x}{\partial x} + \frac{\partial v_y}{\partial y} = 0,$$

$$v_x \frac{\partial v_x}{\partial x} + v_y \frac{\partial v_x}{\partial y} = \boldsymbol{\nu} \frac{\partial^2 v_x}{\partial y^2} + g\beta(t - t_f), \tag{7.11}$$

$$v_x \frac{\partial t}{\partial x} + v_y \frac{\partial t}{\partial y} = \alpha \frac{\partial^2 t}{\partial y^2}.$$

For a fixed surface temperature t_s and an ambient fluid condition far from the surface of t_f and zero velocity, the boundary conditions for Eqs. (7.11) are:

$$\begin{aligned} &\text{At } y = 0: \quad v_x = v_y = 0, \quad t = t_s. \\ &\text{As } y \to \infty: \quad v_x = 0, \quad t = t_f. \end{aligned} \tag{7.12}$$

The basic aim is to find the temperature distribution, $t = t(x, y)$, so that the surface temperature gradient can be found to evaluate the heat transfer coefficient. The set of equations in Eqs. (7.11) may be rendered ordinary, by the choice of a similarity parameter as done in the Blasius solution for laminar forced convection. The reasoning leading to the choice of a similarity parameter is rather involved but not unlike that used in the Blasius solution. The results are summarized in the following.

Define a dimensionless similarity parameter η as follows:

$$\eta = \left(\frac{g\beta\,\Delta t}{4\boldsymbol{\nu}^2}\right)^{1/4} \frac{y}{x^{1/4}}, \tag{7.13}$$

where the notation $\Delta t = t_s - t_f$ has been used. Next, it is assumed that the x velocity component parallel to the surface is of the form

$$v_x = 2(g\beta\,\Delta t x)^{1/2} F'(\eta), \tag{7.14}$$

in which $F'(\eta)$ is the derivative of an as-yet-undetermined function of η *only*, $F(\eta)$. Determination of the function $F(\eta)$ is, then, tantamount to finding the fluid velocity

v_x. One further assumes that the dimensionless temperature, φ, is also a function of η *only:*

$$\varphi(\eta) = \frac{t - t_f}{t_s - t_f} = \frac{t - t_f}{\Delta t}. \tag{7.15}$$

The form assumed for v_x in Eq. (7.14) may be used with the continuity equation, the first equation of Eqs. (7.11), to find v_y. When the resulting expression for v_y is substituted along with Eq. (7.14) for v_x and Eq. (7.15) for t into the differential equations (7.11), considerable algebra yields

$$\begin{aligned} &F''' + 3FF'' - 2(F')^2 + \varphi = 0, \\ &\varphi'' + 3\mathrm{Pr}F\varphi' = 0, \\ &F(0) = 0,\ F'(0) = 0, \qquad \varphi(0) = 1, \\ &F'(\infty) = 0, \qquad \varphi(\infty) = 0. \end{aligned} \tag{7.16}$$

The boundary conditions in Eqs. (7.16) result from applying the conditions given in Eq. (7.12).

The fact that the set of equations above involves only the parameter η [i.e., $F(\eta)$, $\varphi(\eta)$, and their derivatives] confirms the choice of that quantity as a similarity parameter and the assumed functional form for v_x in Eq. (7.14). In fact, the implication of Eq. (7.14) is that the quantity $(g\beta\ \Delta t x)^{1/2}$ represents a characteristic velocity with respect to which the fluid velocity, v_x, may be nondimensionalized. The absence of a specified freestream velocity for such nondimensionalization is one of the complicating aspects of free convection; however, the results just obtained in Eqs. (7.16) have provided an alternative: $(g\beta\ \Delta t x)^{1/2}$.

Equations (7.16) still represent a coupled set of differential equations, so that a simultaneous solution for $F(\eta)$ and $\varphi(\eta)$ is still required. Clearly, such a solution depends on the fluid Prandtl number, Pr. Ostrach (Ref. 3) obtained a numerical solution for a wide range of Pr, and the results are shown in Figs. 7.2 and 7.3. Figure 7.2 presents the function $F'(\eta)$, which, as just noted, amounts to a dimensionless velocity profile according to Eq. (7.14). The curve for $\mathrm{Pr} = 0.72$ is representative of air, and one may note from Fig. 7.2 that for Prandtl numbers greater than this, the limit of the free convection boundary layer is reached for $\eta \approx 5$. Thus the definition of η in Eq. (7.13) would indicate that the boundary layer limit, $y = \delta$, is given by

$$5 \approx \left(\frac{g\beta\ \Delta t}{4\nu^2}\right)^{1/4} \frac{\delta}{x^{1/4}}$$

or

$$\frac{\delta}{x} = 5\sqrt{2}\left(\frac{\nu^2}{x^3 g\beta\ \Delta t}\right)^{1/4} \approx 7.0\mathrm{Gr}_x^{-1/4}, \tag{7.17}$$

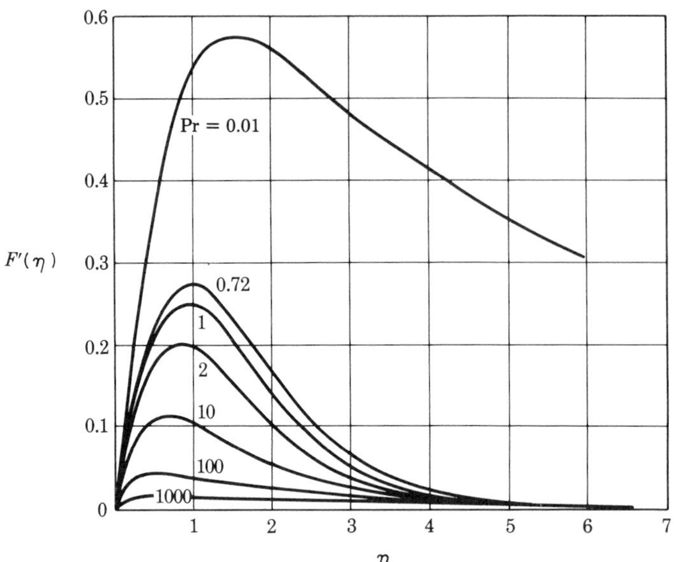

Figure 7.2. Velocity profiles for laminar free convection on a flat surface. (From S. Ostrach, "An Analysis of Laminar Free Convection Flow and Heat Transfer About a Flat Plate Parallel to the Direction of the Generating Body Force," *NACA Tech. Note 2635,* Washington, D.C., 1952.)

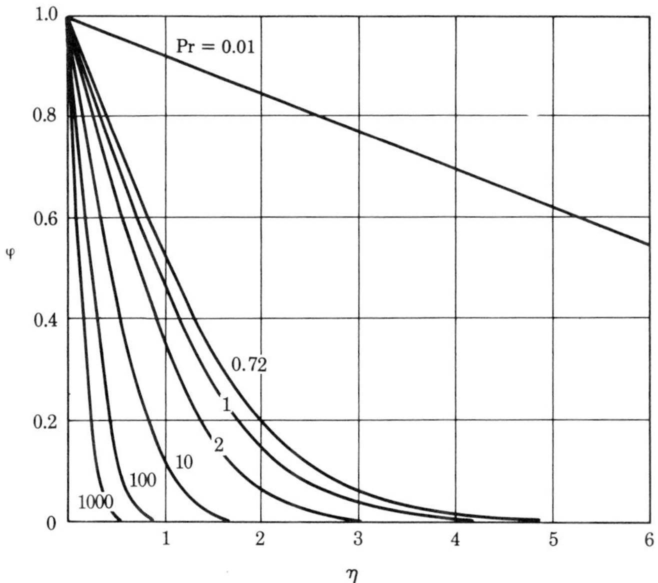

Figure 7.3. Temperature profiles for laminar free convection on a flat surface. (From S. Ostrach, "An Analysis of Laminar Free Convection Flow and Heat Transfer About a Flat Plate Parallel to the Direction of the Generating Body Force," *NACA Tech. Note 2635,* Washington D.C., 1952.)

where

$$\mathrm{Gr}_x = \frac{x^3 g\beta\ \delta t}{\nu^2} \tag{7.18}$$

is recognized as the local Grashof number.

The function $\varphi(\eta)$ in Fig. 7.3 is the dimensionless temperature profile given by Eq. (7.15). This figure, when compared with Fig. 7.2, indicates, as found for forced convection, that the thermal layer is thinner than the velocity layer when $\mathrm{Pr} > 1$, while for $\mathrm{Pr} < 1$ the reverse is true.

For determination of the surface heat transfer coefficient it is necessary to know the temperature gradient at the surface (i.e., $y = \eta = 0$). Ostrach's numerical solution also gives values for $\varphi'(0)$ (i.e., the slope of the curves in Fig. 7.3 at $\eta = 0$) as a function of Pr. Ede (Ref. 4), however, suggests the following approximate expressions for these results:

$$\varphi'(0) = -0.676\mathrm{Pr}^{1/2}(0.861 + \mathrm{Pr})^{-1/4}. \tag{7.19}$$

Thus the local surface heat transfer coefficient can be found:

$$\begin{aligned} h_x &= \frac{-k\left(\dfrac{\partial t}{\partial y}\right)_0}{t_s - t_f} \\ &= -k\varphi'(0)\frac{d\eta}{dy} \\ &= -k\varphi'(0)\left(\frac{g\beta\ \Delta t}{4\nu^2}\right)^{1/4}\frac{1}{x^{1/4}}. \end{aligned}$$

Introduction of Eq. (7.19) and the definition of the local Nusselt number yield

$$\mathrm{Nu}_x = \frac{h_x x}{k} = 0.478\left(\frac{g\beta\ \Delta t}{\nu^2}\right)^{1/4} x^{3/4}\mathrm{Pr}^{1/2}(0.861 + \mathrm{Pr})^{-1/4}.$$

The local Grashof number, Gr_x, of Eq. (7.18) is seen to enter naturally:

$$\mathrm{Nu}_x = 0.478\mathrm{Gr}_x^{1/4}\mathrm{Pr}^{1/2}(0.861 + \mathrm{Pr})^{-1/4}. \tag{7.20}$$

Integration over a finite plate length L gives the average heat transfer coefficient and surface length Nusselt number in terms of

$$\begin{aligned} \mathrm{Nu}_L &= \int_0^L \frac{1}{x}\mathrm{Nu}_x\,dx \\ &= \frac{4}{3}0.478\left(\frac{g\beta\ \Delta t}{\nu^2}\right)^{1/4} L^{3/4}\mathrm{Pr}^{1/2}(0.861 + \mathrm{Pr})^{-1/4} \qquad (7.21) \\ &= 0.637\mathrm{Gr}_L^{1/4}\mathrm{Pr}^{1/2}(0.861 + \mathrm{Pr})^{-1/4}, \end{aligned}$$

where

$$\mathrm{Gr}_L = \frac{L^3 g\beta\,\Delta t}{\nu^2}$$

is the total surface length Grashof number. Equations (7.20) and (7.21) are often written in terms of the Rayleigh number, Re = GrPr, introduced earlier in Eq. (7.3):

$$\begin{aligned} \mathrm{Nu}_x &= 0.478\mathrm{Ra}_x^{1/4}\left(1 + \frac{0.861}{\mathrm{Pr}}\right)^{-1/4}, \\ \mathrm{Nu}_L &= 0.637\mathrm{Ra}_L^{1/4}\left(1 + \frac{0.861}{\mathrm{Pr}}\right)^{-1/4}. \end{aligned} \tag{7.22}$$

The discussion of this section presumed that the surface was hotter than the ambient fluid and the coordinate x was measured upward in the direction of the buoyant force. The results apply equally well to the case in which the surface is cooler than the fluid as long as the coordinate direction is reversed.

Solution of the Laminar Case by the Integral Formulation

As in the forced convection problem, the case of free convection past a vertical plane surface may be solved by application of the integral boundary layer equations. For laminar flow with the pressure gradient again neglected. Eqs. (7.10) become

$$\begin{aligned} \frac{d}{dx}\int_0^{\delta} v_x^2\,dy &= -\nu\left(\frac{\partial v_x}{\partial y}\right)_0 + \int_0^{\delta} g\beta(t - t_f)\,dy, \\ \frac{d}{dx}\int_0^{\delta_t} (t - t_f)v_x\,dy &= -\alpha\left(\frac{\partial t}{\partial y}\right)_0. \end{aligned} \tag{7.23}$$

Eckert (Ref. 5) obtained a solution to these equations by assuming polynomial expressions for the velocity and temperature profiles of the form

$$\begin{aligned} \frac{v_x}{v_x^*} &= \frac{y}{\delta}\left(1 - \frac{y}{\delta}\right)^2, \\ \frac{t - t_f}{t_s - t_f} &= \left(1 - \frac{y}{\delta}\right)^2. \end{aligned} \tag{7.24}$$

The forms assumed for the dimensionless velocity and temperature profiles clearly imply the additional assumption that Pr is not too different from unity, so that one may take the velocity and thermal boundary layers as approximately equal.

As noted in the discussion of the solution to the differential formulation, there is no readily apparent characteristic velocity in free convection for nondimensionalizing v_x. The quantity v_x^* in Eq. (7.24) represents some such characteristic velocity which must be determined in terms of the physical parameters of the problem. The differential solution indicated that v_x^* will involve $(g\beta\,\Delta t x)^{1/2}$ in some way.

The details of Eckert's solution to this problem will not be given here. Basically, the method consisted of substituting Eqs. (7.24) into Eqs. (7.23), with $\delta \approx \delta_t$ and determining the two unknown quantities: $\delta(x)$ and $v_x^*(x)$. These then permitted the calculation of the local and average heat transfer coefficients, with the results

$$\begin{aligned} \mathrm{Nu}_x &= 0.508\mathrm{Ra}_x^{1/4}\left(1 + \frac{0.952}{\mathrm{Pr}}\right)^{-1/4}, \\ \mathrm{Nu}_L &= 0.678\mathrm{Ra}_L^{1/4}\left(1 + \frac{0.952}{\mathrm{Pr}}\right)^{-1/4}. \end{aligned} \tag{7.25}$$

These results compare favorably with those in Eq. (7.22) obtained from the differential analysis.

Solution of the Turbulent Case by the Integral Formulation

For Rayleigh numbers in excess of about 10^9, free convection boundary layers are observed to be in a turbulent state of flow. If this is the case the integral equations must be expressed in terms of the surface shear stress, τ_0, and heat flux $(q/A)_0$. Thus Eqs. (7.10) become

$$\begin{aligned} \frac{d}{dx}\int_0^{\delta} v_x^2\,dy &= -\frac{\tau_0}{\rho} + \int_0^{\delta} g\beta(t - t_f)\,dy, \\ \frac{d}{dx}\int_0^{\delta_t} (t - t_f)v_x\,dy &= \frac{(q/A)_0}{c_p}. \end{aligned} \tag{7.26}$$

Eckert and Jackson (Ref. 6) obtained a solution to these equations in much the same fashion as just described for the integral solution of the laminar case, except that some appropriate empirical law must be used to represent τ_0 and $(q/A)_0$. The problem, again, involves the discovery of an appropriate characteristic velocity for nondimensionalizing v_x.

The solution of Eckert and Jackson involved the following assumptions:

1. The thermal and velocity layer thicknesses are the same.
2. The surface is entirely in turbulent flow.
3. The dimensionless velocity and temperature profiles are analogous to those used for laminar flow with the exception that the one-seventh power law is introduced:

$$\frac{v_x}{v_x^*} = \left(\frac{y}{\delta}\right)^{1/7}\left(1 - \frac{y}{\delta}\right)^4,$$

$$\frac{t - t_f}{t_s - t_f} = 1 - \left(\frac{y}{\delta}\right)^{1/7}.$$

4. The surface shear stress follows the Blasius law for forced flow [Eq. (5.95)]:

$$C_f = \frac{\tau_0}{\frac{1}{2}\rho v_x^{*2}} = 0.045\left(\frac{\nu}{v_x^*\delta}\right)^{1/4}.$$

5. The Colburn analogy [Eq. (5.118)] may be applied to relate the surface shear stress and heat transfer coefficient:

$$\mathrm{Nu}_x = \frac{C_f}{2}\mathrm{Re}_x\mathrm{Pr}^{1/3}.$$

With these assumptions the following solution was obtained from Eqs. (7.26):

$$\mathrm{Nu}_x = 0.0295\mathrm{Ra}_x^{2/5}\mathrm{Pr}^{1/15}(1 + 0.494\mathrm{Pr}^{2/3})^{-2/5},$$
$$\mathrm{Nu}_L = 0.0246\mathrm{Ra}_L^{2/5}\mathrm{Pr}^{1/15}(1 + 0.494\mathrm{Pr}^{2/3})^{-2/5}. \qquad (7.27)$$

These relations agree well with experimental data for air and water, but they have not been checked extensively for other fluids.

EXAMPLE 7.1

The vertical door of a hot oven is 0.5 m high and maintained at 200°C. It is exposed to atmospheric air at 20°C. Find (a) the local heat transfer coefficient halfway up the door, and (b) the average heat transfer coefficient for the entire door using both Eqs. (7.22) and (7.25). (c) Estimate the thickness of the free convection boundary layer at the top of the door. As will be noted in later correlations, it is customary to evaluate the air properties at the mean film temperature except for β, which should be evaluated at the ambient air temperature.

Solution. At the mean film temperature $t_m = (t_s + t_f)/2 = (200 + 20)/2 =$ 110°C, Table A.6 gives

$$\nu = 24.10 \times 10^{-6}\ \mathrm{m^2/s}, \qquad k = 31.94 \times 10^{-3}\ \mathrm{W/m\text{-}^\circ C}, \qquad \mathrm{Pr} = 0.704.$$

At the ambient air temperature $t_f = 20$°C, $T_f = 293.15$°K, the coefficient of expansion for air is given by the relation of Eq. (7.8):

$$\beta = \frac{1}{T_f} = \frac{1}{293.15^\circ\mathrm{K}}.$$

The characteristic temperature difference is

$$\Delta t = t_s - t_f = 200 - 20 = 180^\circ\mathrm{C}.$$

(a) Halfway up the door the local coordinate is $x = 0.25$ m. Thus the local Grashof and Rayleigh numbers are, with $g = 9.8\ \mathrm{m/s^2}$,

$$\mathrm{Gr}_x = \frac{x^3 g\beta\,\Delta t}{\nu^2} = \frac{(0.25)^3 \times 9.8 \times (1/293.15) \times 180}{(24.10 \times 10^{-6})^2} = 1.619 \times 10^8,$$

$$\mathrm{Ra}_x = \mathrm{Gr}_x\mathrm{Pr} = 1.619 \times 10^8 \times 0.704 = 1.140 \times 10^8.$$

Since $\mathrm{Ra}_x < 10^9$, the flow is laminar and Eqs. (7.22) or Eqs. (7.25) apply. Equation (7.22) predicts the local heat transfer coefficient to be

$$\mathrm{Nu}_x = 0.478\mathrm{Ra}_x^{1/4}\left(1 + \frac{0.861}{\mathrm{Pr}}\right)^{-1/4}$$

$$= 0.478(1.140 \times 10^8)^{1/4}\left(1 + \frac{0.861}{0.704}\right)^{-1/4} = 40.45,$$

$$h_x = \mathrm{Nu}_x \frac{k}{x} = 40.45 \times \frac{0.03194}{0.25} = 5.17\ \mathrm{W/m^2\text{-}^\circ C}\ (0.91\ \mathrm{Btu/h\text{-}ft^2\text{-}^\circ F}).$$

Equation (7.25), on the other hand, predicts that

$$\mathrm{Nu}_x = 0.508\mathrm{Ra}_x^{1/4}\left(1 + \frac{0.952}{\mathrm{Pr}}\right)^{-1/4} = 42.38,$$

$$h_x = \mathrm{Nu}_x \frac{k}{x} = 5.41\ \mathrm{W/m^2\text{-}^\circ C}\ (0.95\ \mathrm{Btu/h\text{-}ft^2\text{-}^\circ F}).$$

Thus the two predictions give comparable results.

(b) For the whole surface $L = 0.5$ m, so the surface length Grashof and Rayleigh numbers are

$$\mathrm{Gr}_L = \frac{L^3 g\beta\, \Delta t}{\nu^2} = \frac{(0.5)^3 \times (1/293.15) \times 180}{(24.10 \times 10^{-6})^2} = 1.295 \times 10^9,$$

$$\mathrm{Ra}_L = \mathrm{Gr}_L \mathrm{Pr} = 1.295 \times 10^9 \times 0.704 = 9.117 \times 10^8.$$

Thus the boundary layer is still laminar at the top of the door. The average heat transfer coefficient for the whole door is, by Eq. (7.22),

$$\mathrm{Nu}_L = 0.637\mathrm{Ra}_L^{1/4}\left(1 + \frac{0.861}{\mathrm{Pr}}\right)^{-1/4}$$

$$= 0.637(9.11 \times 10^8)^{1/4}\left(1 + \frac{0.861}{0.704}\right)^{-1/4} = 90.67,$$

$$h = \mathrm{Nu}_L \frac{k}{L} = 90.67 \times \frac{0.03194}{0.5} = 5.79\ \mathrm{W/m^2\text{-}^\circ C}\ (1.02\ \mathrm{Btu/h\text{-}ft^2\text{-}^\circ F}),$$

whereas Eq. (7.25) yields

$$\mathrm{Nu}_L = 0.678\mathrm{Ra}_x^{1/4}\left(1 + \frac{0.952}{\mathrm{Pr}}\right)^{-1/4} = 95.13,$$

$$h = \mathrm{Nu}_L \frac{k}{L} = 6.07\ \mathrm{W/m^2\text{-}^\circ C}\ (1.08\ \mathrm{Btu/h\text{-}ft^2\text{-}^\circ F}).$$

Again the results are comparable.

(c) The boundary layer thickness at the top of the door may be estimated by Eq. (7.17):

$$\frac{\delta}{L} = 7.0\mathrm{Gr}_L^{-1/4}.$$

In part (b), Gr_L was found to be 1.295×10^9, so the layer thickness is

$$\frac{\delta}{L} = 7.0(1.295 \times 10^9)^{-1/4} = 0.037,$$

$$\delta = 0.037 \times L = 0.037 \times 0.5 = 0.018 \text{ m}$$
$$= 1.8 \text{ cm } (0.73 \text{ in.}).$$

7.5 WORKING CORRELATIONS FOR FREE CONVECTION

The analyses just shown illustrate the techniques employed to predict heat transfer coefficients in free convection. Similar analyses have been applied to numerous other geometric configurations, and many experimental investigations have been carried out for a vast number of situations. This section will be devoted to a presentation, with a minimum of discussion, of the results of some of these studies that may be used for practical engineering calculations. As in the case of forced convection, the applicable range of the parameters involved will be noted as will the recommendation for the evaluation of the fluid properties. All correlations will be given in the forms suggested in Sec. 7.2:

$$\text{Nu} = f(\text{Gr}, \text{Pr}),$$
$$\text{Nu} = f(\text{Ra}, \text{Pr}).$$

No relations for skin friction will be quoted since free convection motions occur due to natural forces. Thus there is generally no need to estimate energy requirements to overcome pressure or friction losses.

Free Convection Past Vertical Plane Surface

From an analysis of the data of many experimenters, McAdams (Ref. 7) proposed a correlation for the plate length Nusselt number on vertical plane surfaces of the form

$$\text{Nu}_L = C\text{Ra}_L^m, \tag{7.28}$$

in which the coefficient C and the exponent m are functions of Ra_L, the Rayleigh number based on the total plate length L. Values of these quantities are not reported here since a more accurate correlation will be quited shortly. However, it is interesting to note that in the laminar range ($10^4 < \text{Ra}_L < 10^9$) McAdams recommended that $m = \frac{1}{4}$ in accord with the forms of Eqs. (7.22) and (7.25). Also, for the turbulent range ($10^9 < \text{Ra}_L$) McAdams suggested $m = \frac{1}{3}$. This latter fact is interesting in that since the Rayleigh number involves L^3 and the Nusselt number simply L, then Eq. (7.28) shows that h is independent of L in this range. Such behavior is not uncommon in other free convection correlations in the turbulent regime.

Recently, Churchill and Chu (Ref. 8) have performed an extensive correlation of a great number of workers and recommend the following relation:

$$\mathrm{Nu}_L = 0.68 + 0.670\mathrm{Ra}_L^{1/4}\left[1 + \left(\frac{0.492}{\mathrm{Pr}}\right)^{9/16}\right]^{-4/9}, \quad 0 < \mathrm{Ra}_L < 10^9,$$

$$\mathrm{Nu}_L = \left\{0.825 + 0.387\mathrm{Ra}_L^{1/6}\left[1 + \left(\frac{0.492}{\mathrm{Pr}}\right)^{9/16}\right]^{-8/27}\right\}^2, \quad 10^9 < \mathrm{Ra}_L,$$

(7.29)★

$$\begin{bmatrix} 0 < \mathrm{Pr} < \infty \\ \text{properties, except } \beta, \text{ at } t_m \\ \beta \text{ at } t_m \text{ for liquids, at } t_f \text{ for gases} \end{bmatrix}.$$

Actually, the second of the two equations above may be applied over the entire range of Ra; however, better accuracy is obtained from the use of the first equation. In the application of the above correlations it is useful to remember that for gases, the coefficient of volume expansion, β, is equal to the reciprocal of the absolute temperature, T_f, if the ideal gas approximation is used.

The form of Eqs. (7.29) is not unlike that of the analytical predictions in Sec. 7.4 or the form suggested by McAdams in Eq. (7.28); however, the additive constant 0.68 is not present in these other forms. The presence of this constant results from the need at very low Ra, where the boundary layer approximations do not hold, to approach a nonzero value of Nu as the heat transfer becomes predominately that of conduction in the ambient fluid.

Free Convection past Vertical Cylinders

If the surface in question is a vertical cylinder, one would expect very little difference between the heat transfer coefficient in this instance and that for the vertical plane just considered as long as the circumferential curvature of the cylinder is not great. Indeed, this is the case, and Gebhart (Ref. 1) shows that the correlation for a vertical plate may be applied to the case of the vertical cylinder as long as

$$\frac{D}{L} > \frac{35}{\mathrm{Gr}_L^{1/4}}. \tag{7.30}$$

EXAMPLE 7.2

Repeat part (b) of Example 7.1 (i.e., find the average heat transfer coefficient for the oven door) by applying the correlation of Eq. (7.29) and compare the results with those found in Example 7.1.

Solution. For the data specified in Example 7.1 the following fluid properties and Rayleigh number were found:

$$k = 0.03194 \text{ W/m-°C}, \qquad \mathrm{Pr} = 0.704, \qquad \mathrm{Ra}_L = 9.117 \times 10^8.$$

Thus Eq. (7.29) predicts that

$$\mathrm{Nu}_L = 0.68 + 0.670\mathrm{Ra}_L^{1/4}\left[1 + \left(\frac{0.492}{\mathrm{Pr}}\right)^{9/16}\right]^{-4/9}$$

$$= 0.68 + 0.670(9.117 \times 10^8)^{1/4}\left[1 + \left(\frac{0.492}{0.704}\right)^{9/16}\right]^{-4/9} = 89.95,$$

$$h = \mathrm{Nu}_L \frac{k}{L} = 89.95 \times \frac{0.03194}{0.5} = 5.75 \text{ W/m}^2\text{-°C (1.01 Btu/h ft}^2\text{-°F)}.$$

This result agrees very well with the analytical predictions of 5.79 W/m^2-°C from Eq. (7.22) or 6.07 W/m^2-°C from Eq. (7.25).

EXAMPLE 7.3

Find the average heat transfer coefficient for the oven door in Examples 7.1 and 7.2 if all the data are the same with the exception that the air pressure is 2 atm instead of 1 atm. Use the correlation of Eq. (7.29) and compare the result with that obtained from the appropriate analytical result of Sec. 7.4.

Solution. The thermal conductivity of the air is not strongly affected by pressure; hence the value used in Example 7.1 still applies: $k = 0.03194$ W/m-°C. While the thermal diffusivity, α, and kinematic viscosity, $\boldsymbol{\nu}$, *are* pressure dependent (since density is), the Prandtl number

$$\mathrm{Pr} = \frac{\boldsymbol{\nu}}{\alpha} = \frac{\mu c_p}{k}$$

is virtually pressure independent inasmuch as none of μ, k, or c_p are sensitive to pressure. Thus Pr $= 0.704$, as in Example 7.1. However, the Rayleigh number *is* altered by changing the air pressure. Since

$$\mathrm{Ra}_L = \frac{L^3 g\beta\,\Delta t}{\boldsymbol{\nu}^2}\,\mathrm{Pr},$$

$$\boldsymbol{\nu} = \frac{\mu}{\rho},$$

$$\rho = \frac{P}{RT},$$

Ra is proportional to P^2 if all other quantities remain unchanged. Thus, since $\mathrm{Ra}_L = 9.117 \times 10^8$ for $P = 1$ atm, then at $P = 2$ atm

$$\mathrm{Ra}_L = 9.117 \times 10^8 \times (2)^2 = 3.647 \times 10^9.$$

Since this exceeds the transition value, the free convective boundary layer has become turbulent, at least at the top of the oven door.

For the range of Ra_L just found, the second form of Eq. (7.29) is applicable. Thus

$$\text{Nu}_L = \left\{0.825 + 0.387\text{Ra}_L^{1/6}\left[1 + \left(\frac{0.492}{\text{Pr}}\right)^{9/16}\right]^{-8/27}\right\}^2$$

$$= \left\{0.825 + 0.387(3.647 \times 10^9)^{1/6}\left[1 + \left(\frac{0.492}{0.704}\right)^{9/16}\right]^{-8/27}\right\}^2 = 183.5,$$

$$h = \text{Nu}_L\,\frac{k}{L} = 183.5 \times \frac{0.03194}{0.5} = 11.72\ \text{W/m}^2\text{-°C}\ (2.06\ \text{Btu/h-ft}^2\text{-°F}).$$

Since the flow is turbulent, the only available analytical prediction is that due to Eckert and Jackson as given by Eq. (7.27):

$$\text{Nu}_L = 0.0246\text{Ra}_L^{2/5}\text{Pr}^{1/15}(1 + 0.494\text{Pr}^{2/3})^{-2/5}$$

$$= 0.0246(3.647 \times 10^9)^{2/5}(0.704)^{1/15}[1 + 0.494(0.704)^{2/3}]^{-2/5},$$

$$= 140.7,$$

$$h = \text{Nu}_L\,\frac{k}{L} = 140.7 \times \frac{0.03194}{0.5} = 8.99\ \text{W/m}^2\text{-°C}\ (1.58\ \text{Btu/h-ft}^2\text{-°F}).$$

This value is nearly 25% less than that found from Eq. (7.29). One should probably place the greatest confidence in the result given by the correlation of Eq. (7.29) because of the extensive experimentation underlying it. Also, the analysis leading to Eq. (7.27) involved a number of simplifying assumptions that have not been verified.

It is worth noting that for gases there is generally a significant pressure effect on free convection coefficients, while for liquids this is not likely to be true because of the relative insensitivity of density to pressure variations.

Free Convection Around Horizontal Plates

The cases considered so far had the principal body dimension in line with the gravity vector, i.e., vertical. The flow produced by the free convection was, then, parallel to the surface regardless of whether the surface was hotter or colder than the ambient fluid. However, in the case of free convection about heated, or cooled, horizontal plates, the principal body dimension is normal to the direction of the buoyant forces and considerable differences occur in the flow patterns. Consider the situations depicted in Fig. 7.4, that of a heated surface facing down or facing up.

In the case of a heated plate facing down, as depicted in Fig. 7.4(a), the less dense, heated, fluid forms at the bottom surface of the plate and, seeking a way to rise, flows laterally to the edge and then up. A flow pattern, somewhat as sketched, is then established. However, if the heated surface faces up, as in Fig. 7.4(b), the less dense fluid forms on the top surface, and its tendency to rise is inhibited by the denser, cooler, fluid above it that tends to move down. Thus, in

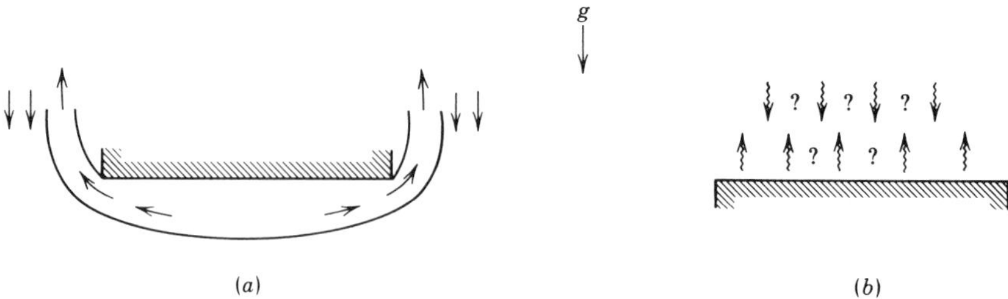

Figure 7.4. Heated plates facing (a) down; (b) up.

this case an unstable situation is created, and the exact nature of the flow pattern is strongly influenced by other effects—such as possible small inclinations of the plate surface, the influence of fluid motion on the other side of the plate, etc. This same instability is observed in other free convective situations in which there is a heated surface with a large dimension placed at a significant angle to the gravity vector (90° in this case, but perhaps inclined at large angles in other cases). Of particular interest are the rectangular enclosures considered later.

Because of the foregoing effects, reliable correlations for horizontal plates are difficult to find. Based on the work of McAdams (Ref. 7) Goldstein et al. (Ref 9), and Lloyd and Moran (Ref. 10), the following recommendations are made:

Characteristic length:

$$L_c = \frac{\text{plate area}}{\text{plate perimeter}}.$$

Heated plate facing up, cooled plate facing down:

$$\begin{aligned} \mathrm{Nu}_{L_c} &= 0.54\mathrm{Ra}_{L_c}^{1/4}, \qquad 2.6 \times 10^4 < \mathrm{Ra}_{L_c} < 10^7, \\ \mathrm{Nu}_{L_c} &= 0.15\mathrm{Ra}_{L_c}^{1/3}, \qquad 10^7 < \mathrm{Ra}_{L_c} < 3 \times 10^{10}. \end{aligned} \tag{7.31}$$ ★

Heated plate facing down, cooled plate facing up:

$$\mathrm{Nu}_{L_c} = 0.27\mathrm{Ra}_{L_c}^{1/4}, \qquad 3 \times 10^5 < \mathrm{Ra}_{L_c} < 3 \times 10^{10},$$

$$\left[\begin{array}{l} \text{properties, except } \beta \text{, at } t_m \\ \beta \text{ at } t_m \text{ for liquids, } t_f \text{ for gases} \end{array}\right].$$

Free Convection Past Inclined Plates

The availability of correlations for inclined plates is quite limited. For surfaces inclined at angles of 60°, *or less,* with the vertical, Churchill and Chu (Ref. 8) recommend that Eqs (7.29) be used with g replaced with $g \cos\theta$ (θ = angle with vertical) for $\mathrm{Ra}_L < 10^9$. For $\mathrm{Ra}_L > 10^9$, this modification is not used.

For inclinations with the vertical exceeding about 60°, the situation becomes quite complex and the reader is referred to the works reported in Refs. 11 and 12.

EXAMPLE 7.4

A large vat containing ethylene glycol is to be heated by an electrical heating element that consists of a square, horizontal, plate, 15 cm by 15 cm in size. The glycol is at 0°C and quiescent far removed from the heater. Find the power input to the heating element required to maintain its surface (top and bottom) at 40°C.

Solution. The free convection correlations of Eq. (7.31) apply for the two surfaces of the heating element. At a mean film temperature $t_m = (t_s + t_f)/2 = (0 + 40)/2 = 20°\text{C}$, Table A.4 gives

$$\nu = 19.18 \times 10^{-6}\ \text{m}^2/\text{s}, \quad k = 0.249\ \text{W/m-°C}, \quad \text{Pr} = 204, \quad \beta = 0.65 \times 10^{-3}\ \text{°K}^{-1}.$$

Since glycol is a liquid, the coefficient of expansion is evaluated at t_m (rather than t_f, as for gases), as required by Eq. (7.31).

The correlations of Eq. (7.31) are based on the characteristic length:

$$L_c = \frac{\text{plate area}}{\text{plate perimeter}} = \frac{15 \times 15}{4 \times 15} = 3.75\ \text{cm}.$$

Thus the Grashof number based on this length and $\Delta t = t_s - t_f = 40 - 0 = 40°\text{C}$ is

$$\text{Gr}_{Lc} = \frac{L_c^3 g\beta\, \Delta t}{\nu^2} = \frac{(0.0375)^3 \times 9.8 \times 0.65 \times 10^{-3} \times 40}{(19.18 \times 10^{-6})^2} = 3.653 \times 10^4,$$

so that the Rayleigh number is

$$\text{Ra}_{Lc} = \text{Gr}_{Lc} \times \text{Pr} = 3.653 \times 10^4 \times 204 = 7.451 \times 10^6.$$

The top surface of the heating element is a heated horizontal plate facing up, so that the free convective coefficient there (call it h_T) is given by Eq. (7.31):

$$\text{Nu}_{Lc} = 0.54\text{Ra}_{Lc}^{1/4} = 0.54(7.451 \times 10^6)^{1/4} = 28.21,$$

$$h_T = \text{Nu}_{Lc}\frac{k}{L_c} = 28.21 \times \frac{0.249}{0.0375} = 187.3\ \text{W/m}^2\text{-°C}.$$

The bottom surface of the heater is a heated horizontal plate facing down so that Eq. (7.31) gives the heat transfer coefficient there, h_B, to be

$$\begin{aligned} \text{Nu}_{L_c} &= 0.27\text{Ra}_{L_c}^{1/4} \\ &= 0.27(7.451 \times 10^6)^{1/4} = 14.11, \\ h_B &= \text{Nu}_{L_c}\frac{k}{L_c} = 14.11 \times \frac{0.249}{0.0375} = 93.7\ \text{W/m}^2\text{-°C}. \end{aligned}$$

Consequently, the power input required for the heating element is the total heat transferred from the top and bottom surfaces of the plate to the glycol:

$$\begin{aligned} q &= (h_T + h_B) \times A \times (t_s - t_f) \\ &= (187.3 + 93.7)(0.15 \times 0.15)(40 - 0) \\ &= 253\ \text{W (863 Btu/h)}. \end{aligned}$$

Free Convection Around Long Horizontal Cylinders

For horizontal cylinders sufficiently long that end effects may be neglected, McAdams (Ref. 7) again suggested a correlation of the form

$$\text{Nu}_D = C\text{Ra}_D^m,$$

where the Nusselt and Rayleigh numbers are based on the cylinder diameter. However, Churchill and Chu (Ref. 13), in another extensive analysis, recommend the following relation which is valid for a wide range of Ra_D:

$$\text{Nu}_D = \left\{0.60 + 0.387\text{Ra}_D^{1/6}\left[1 + \left(\frac{0.559}{\text{Pr}}\right)^{9/16}\right]^{-8/27}\right\}^2, \tag{7.32★}$$

$$\left[\begin{array}{l} 0 < \text{Pr} < \infty \\ 10^{-5} < \text{Ra}_D < 10^{12} \\ \text{properties, except } \beta, \text{ at } t_m \\ \beta \text{ at } t_m \text{ for liquids, } t_f \text{ for gases} \end{array}\right].$$

Free Convection Around Spheres

For free convection about a sphere, Yuge (Ref. 14) determined the correlation:

$$\text{Nu}_D = 2 + 0.43\text{Ra}_D^{1/4}, \tag{7.33★}$$

$$\left[\begin{array}{l} \text{Pr} \approx 1 \\ 1 < \text{Ra}_D < 10^5 \\ \text{properties, except } \beta, \text{ at } t_m \\ \beta \text{ at } t_m \text{ for liquids, } t_f \text{ for gases} \end{array}\right].$$

The lead constant of 2 in Eq. (7.33) results from the fact that for no buoyant forces ($\text{Ra}_D \to 0$), all heat loss from the sphere is by conduction in the ambient fluid.

EXAMPLE 7.5

A horizontal 4-in. schedule 40 pipe, 3 ft long, has a surface temperature of 200°F. It is placed in a quiescent fluid at 40°F. Find the heat loss from the pipe if the ambient fluid is (a) atmospheric air; (b) water.

Solution. Table B.1 gives the outside diameter of a 4-in. schedule 40 pipe to be $D = 4.500$ in.

(a) For atmospheric air at $t_m = (t_s + t_f)/2 = (200 + 40)/2 = 120°\text{F}$, Table A.6 gives

$$\nu = 0.6902 \text{ ft}^2/\text{h}, \qquad k = 0.01602 \text{ Btu/h-ft-°F}, \qquad \text{Pr} = 0.709.$$

The coefficient of expansion in evaluated at the ambient fluid temperature, and by the ideal gas approximation

$$\beta = \frac{1}{T_f} = \frac{1}{40 + 460} = \frac{1}{500 \text{ °R}}.$$

Thus the diameter Grashof and Rayleigh numbers are

$$\begin{aligned} \text{Gr}_D &= \frac{D^3 g\beta\, \Delta t}{\nu^2} \\ &= \frac{(4.5/12)^3 \times 32.2 \times (3600)^2 \times (1/500) \times (200 - 40)}{(0.6902)^2} \\ &= 1.478 \times 10^7, \\ \text{Ra}_D &= \text{Gr}_D\text{Pr} = 1.478 \times 10^7 \times 0.709 = 1.048 \times 10^7. \end{aligned}$$

The correlation of Eq. (7.32) for a horizontal cylinder gives the surface heat transfer coefficient:

$$\begin{aligned} \text{Nu}_D &= \left\{0.60 + 0.387\text{Ra}_D^{1/6}\left[1 + \left(\frac{0.559}{\text{Pr}}\right)^{9/16}\right]^{-8/27}\right\}^2 \\ &= \left\{0.60 + 0.387(1.048 \times 10^7)^{1/6}\left[1 + \left(\frac{0.559}{0.709}\right)^{9/16}\right]^{-8/27}\right\}^2 \\ &= 28.65, \end{aligned}$$

$$h = \text{Nu}_D \frac{k}{D} = 28.65 \times \frac{0.01602}{4.5/12} = 1.22 \text{ Btu/h-ft}^2\text{-°F}.$$

Thus the heat loss from the pipe is

$$\begin{aligned} q &= hA(t_s - t_f) \\ &= h\pi DL(t_s - t_f) \\ &= 1.22 \times \pi \times \frac{4.5}{12} \times 3 \times (200 - 40) \\ &= 690 \text{ Btu/h (202 W)}. \end{aligned}$$

(b) For water all the fluid properties, including β, are evaluated at $t_m = 120°\text{F}$, so Table A.3 yields

$$\boldsymbol{\nu} = 0.02185 \text{ ft}^2/\text{h}, \qquad k = 0.3693 \text{ Btu/h-ft-°F},$$
$$\text{Pr} = 3.65, \qquad \beta = 0.253 \times 10^{-3} \text{ °R}^{-1}.$$

Similar calculations to those just shown in part (a) give

$$\text{Gr}_D = \frac{D^3 g\beta\, \Delta t}{\boldsymbol{\nu}^2} = 1.866 \times 10^9,$$

$$\text{Ra}_D = \text{Gr}_D\text{Pr} = 6.811 \times 10^9,$$

$$\text{Nu}_D = \left\{0.60 + 0.387\text{Ra}_D^{1/6}\left[1 + \left(\frac{0.559}{\text{Pr}}\right)^{9/16}\right]^{-8/27}\right\}^2 = 256.7,$$

$$h = \text{Nu}_D \frac{k}{D} = 253 \text{ Btu/h-ft}^2\text{-°F},$$

$$q = h\pi DL(t_s - t_f) = 1.431 \times 10^5 \text{ Btu/h } (4.193 \times 10^4 \text{ W}).$$

Note that the free convective heat transfer coefficient is significantly greater when the pipe is immersed in water than when it is placed in air. Typically, free convective coefficients in liquids are greater than in gases due to the facts that liquids generally have smaller kinematic viscosities and larger Prandtl numbers and thermal conductivities. The smaller values of $\boldsymbol{\nu}$ lead to greater values of Ra which, together with the greater values of Pr, lead to larger Nusselt numbers. The heat transfer coefficients are then greater since h is proportional to the product of Nu and k.

EXAMPLE 7.6

Repeat Example 7.5(a) (i.e., the case in which the ambient fluid is air) if the pipe is placed in the vertical position instead of the horizontal position.

Solution. The same fluid properties found in Example 7.5(a) apply. However, since the pipe is now vertical instead of horizontal, the characteristic dimension is the length L instead of the diameter D. Thus the Grashof number is

$$\text{Gr}_L = \frac{L^3 g\beta\, \Delta t}{\boldsymbol{\nu}^2} = \left(\frac{L}{D}\right)^3 \text{Gr}_D$$
$$= \left(\frac{3.0}{4.5/12}\right)^3 \times 1.478 \times 10^7 = 7.567 \times 10^9.$$

The Rayleigh number is then

$$\text{Ra}_L = \text{Gr}_L\text{Pr} = 7.567 \times 10^9 \times 0.709 = 5.365 \times 10^9.$$

To apply the equation for a vertical plane to this case of a vertical cylinder, one must first examine whether or not the condition of Eq. (7.30) is met:

$$\frac{D}{L} > \frac{35}{\mathrm{Gr}_L^{1/4}}.$$

From the data above,

$$\frac{D}{L} = \frac{4.5/12}{3} = 0.125,$$

$$\frac{35}{\mathrm{Gr}_L^{1/4}} = \frac{35}{(7.567 \times 10^9)^{1/4}} = 0.119.$$

Thus $D/L > 35/\mathrm{Gr}_L^{1/4}$ and Eq. (7.29) may be applied to find the heat transfer coefficient:

$$\mathrm{Nu}_L = \left\{0.825 + 0.387\mathrm{Ra}_L^{1/6}\left[1 + \left(\frac{0.492}{\mathrm{Pr}}\right)^{9/16}\right]^{-8.27}\right\}^2$$

$$= \left\{0.825 + 0.387(5.365 \times 10^9)^{1/6}\left[1 + \left(\frac{0.492}{0.709}\right)^{9/16}\right]^{-8/27}\right\}^2 = 207.3,$$

$$h = \mathrm{Nu}_L \frac{k}{L} = 207.3 \times \frac{0.01602}{3.0} = 1.11 \text{ Btu/h-ft}^2\text{-°F}.$$

So the total heat transferred is

$$q = h\pi DL(t_s - t_f)$$

$$= 1.11 \times \pi \times (4.5/12) \times 3 \times (200 - 40)$$

$$= 628 \text{ Btu/h (184 W)}.$$

The heat transfer coefficient and heat flow are seen to be somewhat less than for the horizontal position [Example 7.5(a)] due, principally, to the fact that the boundary layer must travel a longer distance and, hence, becomes thicker—offering greater thermal resistance. Similar results would be obtained for the case in which water is the ambient fluid. The heat transfer in the vertical position would be less than in the horizontal position—but still much greater than in the case of air.

Free Convection in Inclined Rectangular Enclosures

The configuration formed by a rectangular enclosure in which two of the opposite surfaces are maintained at different temperatures and the other two are adiabatic is one of considerable interest for heat exchange by free convection between the two surfaces of given temperature. The air space between the walls of a building or in the space between the layers of glass in double glazed windows are cases of this geometry with the heated surfaces in a vertical position and are of significant engineering interest. The space enclosed between the absorbing surface of a solar collector and its glass, or plastic, cover plate is an example of this geometry in

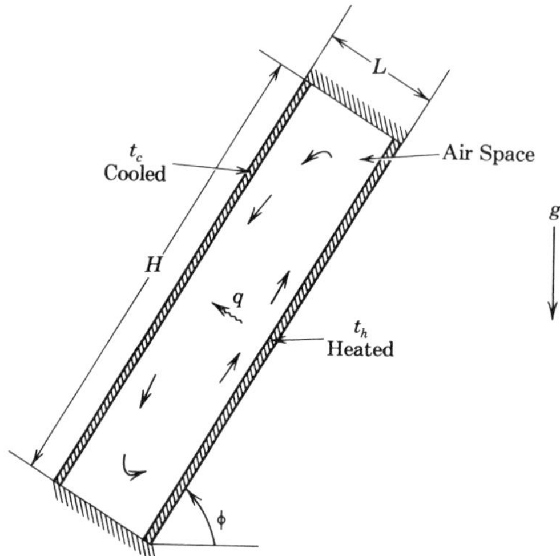

Figure 7.5. Free convection in an inclined rectangular enclosure.

which the isothermal surfaces are inclined. Current wide interest in solar energy collection has led to recent extensive studies of free convective exchange in such enclosures.

Figure 7.5 illustrates the basic geometry and nomenclature to be considered. The hotter of the two surfaces is denoted by t_h and the cooler by t_c. These two surfaces are parallel and spaced a distance L apart. Their dimension normal to L is given by H and the dimension normal to the plane of the paper is presumed to be large when compared with the gap spacing L. The other two surfaces forming the rectangular enclosure are taken to be adiabatic. An important geometric parameter of the enclosure is the *aspect ratio,* Ar:

$$\mathrm{Ar} = \frac{H}{L}. \tag{7.34}$$

The gravity vector is taken as vertical and the inclination of the *heated surface* with the horizontal is denoted by the angle φ. Thus, for $\varphi = 90°$, the air gap is vertical; for $\varphi = 0°$ it is horizontal with the heated surface on the bottom—an unstable situation, as noted earlier. For $\varphi = 180°$ the layer is also horizontal, but is in the stable configuration with the cooler surface on the bottom.

The effective heat transfer coefficient for the enclosure is defined as the h for which the heat flow, per unit of area of the isothermal surfaces, q/A, is given by

$$\frac{q}{A} = h(t_h - t_c). \tag{7.35}$$

The Rayleigh and Nusselt numbers for this configuration are defined as

$$\mathrm{Ra}_L = \frac{L^3 g\beta(t_h - t_c)\mathrm{Pr}}{\nu^2}, \tag{7.36}$$

$$\mathrm{Nu}_L = \frac{hL}{k}. \tag{7.37}$$

For given values of the geometric parameters and the surface temperatures, the value of h, or Nu, is strongly dependent on the inclination angle φ, which determines the form of the free convective flow pattern set up between the two surfaces. Based mainly on the studies of Hollands and his coworkers reported in Refs. 15 and 16 and on the work of Arnold et al., in Ref. 17, the following recommendations are made for various ranges of the inclination φ. In all instances the fluid properties are evaluated at the average temperature:

$$t_{\mathrm{av}} = \frac{t_h + t_c}{2}, \tag{7.38}$$

$\varphi = 90°$: Nu_{90} = largest of: Nu_1, Nu_2, Nu_3

$$\mathrm{Nu}_1 = 0.0605\mathrm{Ra}_L^{1/3},$$

$$\mathrm{Nu}_2 = \left\{1 + \left[\frac{0.104\mathrm{Ra}_L^{0.293}}{1 + (6310/\mathrm{Ra}_L)^{1.36}}\right]^3\right\}^{1/3}, \tag{7.39}\bigstar$$

$$\mathrm{Nu}_3 = 0.242\left(\frac{\mathrm{Ra}_L}{\mathrm{Ar}}\right)^{0.272},$$

$$\begin{bmatrix} 5 < \mathrm{Ar} < 110 \\ 10^2 < \mathrm{Ra}_L < 2 \times 10^7 \end{bmatrix}.$$

$\varphi = 60°$: Nu_{60} = largest of: Nu_1, Nu_2

$$\mathrm{Nu}_1 = \left[1 + \left(\frac{0.0936\mathrm{Ra}_L^{0.314}}{1 + G}\right)^7\right]^{1/7},$$

$$G = \frac{0.5}{[1 + (\mathrm{Ra}_L/3160)^{20.6}]^{0.1}}, \tag{7.40}\bigstar$$

$$\mathrm{Nu}_2 = \left(0.104 + \frac{0.175}{\mathrm{Ar}}\right)\mathrm{Ra}_L^{0.283},$$

$$\begin{bmatrix} 5 < \mathrm{Ar} < 110 \\ 10^2 < \mathrm{Ra}_L < 2 \times 10^7 \end{bmatrix}.$$

$90° > \varphi > 60°$: Linear interpolation between Nu_{90} and Nu_{60}. (7.41)★

$$60° > \varphi \geq 0°: \mathrm{Nu} = 1 + 1.44[1 - G]^{\bullet}[1 - G(\sin 1.8\varphi)^{1/6}] + [0.664G^{-1/3} - 1]^{\bullet},$$

$$G = \frac{1708}{\mathrm{Ra}_L \cos\varphi}, \tag{7.42}\bigstar$$

Notation $[\ ]^\bullet$ means that the bracketed term should be made zero if negative,

$$\begin{bmatrix} \text{Ar} > 12 \\ 0 < \text{Ra}_L < 10^5 \end{bmatrix}.$$

$$90^\circ < \varphi < 180^\circ: \text{Nu} = 1 + [\text{Nu}_{90} - 1] \sin \varphi, \tag{7.43}★$$

$$\begin{bmatrix} \text{all Ar} \\ 10^3 < \text{Ra}_L < 10^6 \end{bmatrix}.$$

EXAMPLE 7.7

A flat-plate solar collector has the geometry depicted in Fig. 7.5. The lower absorber plate is maintained at $t_h = 70°\text{C}$. The outer surface is a glass cover plate $t_c = 10°\text{C}$. The spacing between the plates is $L = 8$ cm, and the plates have a length $H = 1$ m. The array is tilted toward the sun so that the lower absorbing (hot) surface makes an angle of $\varphi = 40°$ with the horizontal. If the width of the panel (into the plane of the paper) is 2 m, find the free convective heat exchange between the absorber plate and the glass cover if the space contains air at atmospheric pressure.

Solution. For the average temperature of the two plates, $t_{av} = (t_h + t_c)/2 = (70 + 10)/2 = 40°\text{C}$, Table A.6 gives

$$\boldsymbol{\nu} = 16.96 \times 10^{-6}\ \text{m}^2/\text{s}, \qquad k = 27.10 \times 10^{-3}\ \text{W/m-°C}, \qquad \text{Pr} = 0.710.$$

At the same temperature the coefficient of expansion is

$$\beta = \frac{1}{T_{av}} = \frac{1}{40 + 273.15} = \frac{1}{313.15\text{°K}}.$$

Thus the Rayleigh number based on the plate spacing $L = 8$ cm is, from Eq. (7.36),

$$\begin{aligned} \text{Gr}_L &= \frac{L^3 g\beta(t_h - t_c)\text{Pr}}{\boldsymbol{\nu}^2} \\ &= \frac{(0.08)^3 \times 9.8 \times (1/313.15) \times (70 - 10) \times 0.710}{(16.96 \times 10^{-6})^2} \\ &= 2.373 \times 10^6. \end{aligned}$$

The aspect ratio [Eq. (7.43)] for the given geometry is

$$\text{Ar} = \frac{H}{L} = \frac{1.0}{0.08} = 12.5.$$

The inclination, $\varphi = 40°$, lies in the range $60° > \varphi < 0°$, which indicates that Eq. (7.42) may apply. The aspect ratio satisfies the condition $\text{Ar} > 12$; however,

Ra_L exceeds the limit 10^5 noted for Eq. (7.42). This limit was noted for Ra_L since that was the range over which the experimental data were taken to obtain the correlation of Eq. (7.42). In the absence of any other relation, Eq. (7.42) will still be applied here since it is probable that it should be valid for Ra_L up to, perhaps, 10^7. Thus Eq. (7.42) gives

$$G = \frac{1708}{\mathrm{Ra}_L \cos \varphi} = \frac{1708}{2.373 \times 10^6 \cos 40^\circ} = 0.00094,$$

$$\begin{aligned}\mathrm{Nu} &= 1 + 1.44[1 - G]^\bullet[1 - G(\sin 1.8\varphi)^{1/6}] + [0.664G^{-1/3} - 1]^\bullet \\ &= 1 + 1.44[1 - 0.00094]^\bullet[1 - 0.00094(\sin 72^\circ)^{1/6}] \\ &\qquad + [0.664(0.00094)^{-1/3} - 1]^\bullet \\ &= 1 + 1.44 \times 0.99906 \times 0.99907 + 5.778 = 8.22.\end{aligned}$$

Then from the definition of Nu in Eq. (7.37) and the heat flow in Eq. (7.35), one finds

$$h = \mathrm{Nu}\frac{k}{L} = 8.22\frac{0.02710}{0.08} = 2.78\ \mathrm{W/m^2\text{-}^\circ C},$$

$$\begin{aligned}q &= hA(t_h - t_c) \\ &= 2.78 \times (1 \times 2)(70 - 10) = 334\ \mathrm{W}\ (1140\ \mathrm{Btu/h}).\end{aligned}$$

This convective exchange just found between the absorber plate and the cover of a solar collector represents an energy loss for such a device. Of the approximately 2000 W of solar energy falling on such an absorber, this amounts to about a 15% loss (there are other losses to be discussed in some detail in Chapter 11) and could be reduced by either increasing the spacing between the cover and the absorber plate or by evacuating the space between them. These effects are illustrated in the problems at the end of the chapter.

7.6 MIXED FREE AND FORCED CONVECTION

The analyses and correlations presented so far in this chapter were limited to pure free convection. If there is simultaneously forced flow past the surface in question, one may have to account for possible simultaneous heat transfer due to forced convection. As noted in Sec. 7.2, the relations of pure free convection apply only when the $\mathrm{Gr}/\mathrm{Re}^2$ is much greater than unity. Conversely, the pure forced convection relations of Chapters 5 and 6 apply only when this ratio is very small when compared to unity [see Eq. (7.2)]. Here the Reynolds number, Re, is based on the externally imposed freestream velocity. Thus, when this ratio is near 1,

$$\frac{\mathrm{Gr}}{\mathrm{Re}^2} \approx 1,$$

the situation is described as *mixed* convection.

As might be expected, analytical or empirical correlations for the mixed regime when $Gr/Re^2 \approx 1$ are sparse. This section presents a very brief discussion of some of the results.

Mixed Convection past Vertical Plane Surfaces

Convective heat transfer in the mixed regime for forced flow past a flat surface for which buoyant forces act in the same direction as the freestream velocity, as suggested in Fig. 7.6, has been examined in a numerical analysis by Lloyd and Sparrow (Ref. 18). Their results are compared in Fig. 7.6, for a wide range of Pr, with the

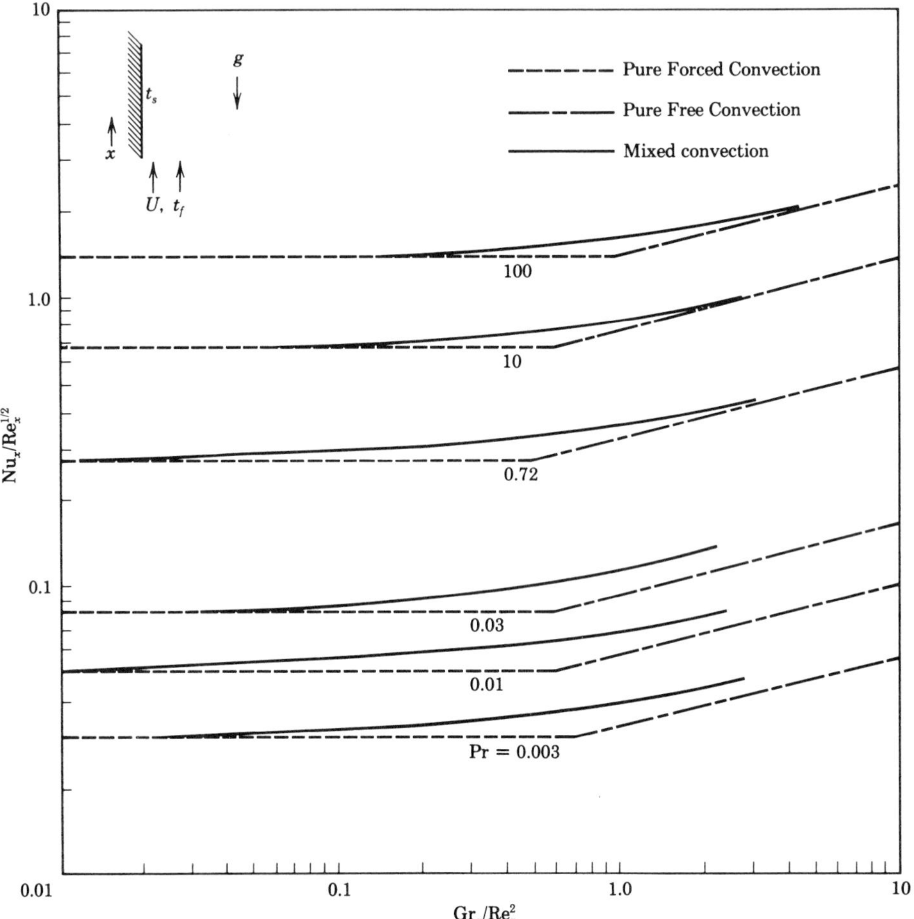

Figure 7.6. Combined free and forced convection past a vertical flat surface. (From J. R. Lloyd and E. M. Sparrow, *Int. J. Heat Mass Transfer*, Vol. 13, 1970, p. 434. Used by permission.)

Table 7.1 Maximum Values of Gr_x/Re_x^2 for Neglection of Buoyancy on Forced Convection past a Vertical Plate with Less Than 5% Error

Pr	100	10	0.72	0.03–0.003
$\mathbf{Gr_x/Re_x^2}$	0.24	0.13	0.08	0.056–0.05

solutions for pure free and forced convection. As would be expected, the mixed flow results approach those for forced convection as Gr_x/Re_x^2 becomes small and those for free convection as Gr_x/Re_x^2 becomes large. The effect of buoyancy on the deviation of mixed convection from pure forced convection is more pronounced for fluids of low Prandtl numbers. The tabulation of Table 7.1 shows the value of Gr_x/Re_x^2 below which the use of the pure forced convection result of Eq. (6.2) may be used with less than 5% deviation. One may use Fig. 7.6 to similarly estimate the value of Gr_x/Re_x^2 above which pure free convection relations may be used.

Mixed Convection in Horizontal Pipes

Forced flow in horizontal pipes of large diameter may experience buoyancy effects when $Gr_D/Re_x^2 \approx 1$. Metais and Eckert (Ref. 19) examined this case and delineated the regions in which free convection contributions are significant. Their study indicated that laminar flow may be expected if $Re_D < 2000$ when $Ra_D(D/L) < 2 \times 10^4$ or if $Re_D < 800$ when $Ra_D(D/L) > 2 \times 10^4$. With these criteria for the presence of laminar or turbulent flow, the following recommendations are made for mixed convection in horizontal pipes (Refs. 19 and 20):

Laminar:

$$Nu_D = 1.75\left\{\left[\left(\frac{D}{L}\right)Re_D Pr\right] + 0.012\left[\left(\frac{D}{L}\right)Re_D Pr Gr_D^{1/3}\right]^{4/3}\right\}^{1/3}\left(\frac{\mu}{\mu_s}\right)^{0.14}, \tag{7.44}★$$

$$\left[\begin{array}{l} Re_D < 2000 \text{ when } Ra_D(D/L) < 2 \times 10^4 \\ Re_D < 800 \text{ when } Ra_D(D/L) > 2 \times 10^4 \\ \text{properties, except } \mu_s, \text{ at mean } t_b \\ \mu_s \text{ at } t_s \end{array}\right].$$

Turbulent:

$$Nu_D = 4.69 Re_D^{0.27} Pr^{0.21} Gr_D^{0.07}\left(\frac{D}{L}\right)^{0.36}, \tag{7.45}★$$

$$\left[\begin{array}{l} Re_D > 2000 \text{ when } Ra_D(D/L) < 2 \times 10^4 \\ Re_D > 800 \text{ when } Ra_D(D/L) > 2 \times 10^4 \\ \text{properties at mean } t_b \end{array}\right].$$

EXAMPLE 7.8

If the oven door described in Example 7.1 is subjected to an upward forced flow of air, find the minimum freestream air velocity for which free convection effects may be neglected.

Solution. In Example 7.1 the following fluid properties and door geometry were determined:

$$\nu = 24.10 \times 10^{-6}\ \text{m}^2/\text{s}, \qquad \text{Pr} = 0.704, \qquad L = 0.5\ \text{m}.$$

The surface length Grashof number was evaluated to be

$$\text{Gr}_L = 1.295 \times 10^9.$$

At Pr = 0.704, Table 7.1 shows that less than 5% error will result in the heat transfer coefficient by treating the flow as pure forced convection if

$$\frac{\text{Gr}_L}{\text{Re}_L^2} < 0.08$$

or

$$\text{Re}_L > \left(\frac{1.295 \times 10^9}{0.08}\right)^{1/2}$$
$$> 1.27 \times 10^5.$$

The minimum freestream velocity needed to achieve this surface length Reynolds number is

$$\text{Re}_L = \frac{UL}{\nu} = \frac{U \times 0.5}{24.10 \times 10^{-6}} > 1.27 \times 10^5$$
$$U > 6.1\ \text{m/s}\ (2.01\ \text{ft/s}).$$

REFERENCES

1. Gebhart, B., *Heat Transfer*, 2nd ed., New York, McGraw-Hill, 1971.
2. Schmidt, E., and W. Beckmann, "Das Temperatur und Geschwindigkeitsfeld vor einer Wärmer abgebenden senkrechten Platte bei näturicher Konvection," *Tech. Mech. Thermodynamick,* Vol. 1, 1930, p. 341.
3. Ostrach, S., "An Analysis of Laminar Free Convection Flow and Heat Transfer About a Flat Plate Parallel to the Generating Body Force," *NACA Rep. 1111,* Washington, D.C., 1953.
4. Ede, A. J., "Advances in Free Convection," *Advances in Heat Transfer,* Vol. 4, New York, Academic Press, 1967.
5. Eckert, E. R. G., *Introduction to Heat and Mass Transfer,* New York, McGraw-Hill, 1963.

6. Eckert, E. R. G., and T. W. Jackson, "An Analysis of Turbulent Free Convection Boundary Layer on a Flat Plate," *NACA Tech. Note 2207,* Washington, D.C., 1950.

7. McAdams, W. H., *Heat Transmission,* 3rd ed., New York, McGraw-Hill, 1954.

8. Churchill, S. W., and H. H. S. Chu, "Correlating Equations for Laminar and Turbulent Free Convection from a Vertical Plate," *Int. J. Heat Mass Transfer,* Vol. 18, 1975, p. 1323.

9. Goldstein, R. J., E. M. Sparrow, and D. C. Jones, "Natural Convection Mass Transfer Adjacent to the Horizontal Plates," *Int. J. Heat Mass Transfer,* Vol. 16, 1973, p. 1025.

10. Lloyd, J. R., and W. R. Moran, "Natural Convection Adjacent to Horizontal Surfaces of Various Planforms." *ASME Paper 74-WA/HT-66,* 1974.

11. Vliet, G. C., "Natural Convection Local Heat Transfer on Constant-Heat-Flux Inclined Surfaces." *J. Heat Transfer, Trans. ASME,* Vol. 91, 1969, p. 511.

12. Fujii, T., and H. Imura, "Natural Convection Heat Transfer from a Plate with Arbitrary Inclination," *Int. J. Heat Mass Transfer,* Vol. 15, 1972, p. 755.

13. Churchill, S. W., and H. H. S. Chu, "Correlating Equations for Laminar and Turbulent Free Convection from a Horizontal Cylinder," *Int. J. Heat Mass Transfer,* Vol. 18, 1975, p. 1049.

14. Yuge, T., "Experiments on Heat Transfer from Spheres Including Combined Natural and Forced Convection," *J. Heat Transfer, Trans. ASME,* Vol. 82, 1960, p. 214.

15. ElSherbing, S. M., G. D. Raithby, and K. G. T. Hollands, "Heat Transfer by Natural Convection Across Vertical and Inclined Air Layers," *J. Heat Transfer, Trans, ASME,* Vol. 104, 1982, p. 96.

16. Hollands, K. G. T., T. E. Unny, G. D. Raithby, and L. Konicek, "Free Convection Heat Transfer Across Inclined Air Layers," *J Heat Transfer, Trans. ASME,* Vol. 98, 1976, p. 189.

17. Arnold, J. N., I. Catton, and D. K. Edwards, "Experimental Investigation of Natural Convection in Inclined Rectangular Regions of Differing Aspect Ratios," *J. Heat Transfer, Trans. ASME,* Vol. 98, 1976, p. 67.

18. Lloyd, J. R., and E. M. Sparrow, "Combined Forced and Free Convection Flow on Vertical Surfaces," *Int. J. Heat Mass Transfer,* Vol. 13, 1970, p. 434.

19. Metais, B., and E. R. G. Eckert, "Forced, Mixed and Free Convection Regimes," *J. Heat Transfer, Trans. ASME,* Vol. 86, 1964, p. 295.

20. Brown, C. K., and W. H. Gauvin, "Combined Free and Forced Convection, Parts I and II." *Can. J. Chem. Eng.,* Vol. 43, 1965, pp. 306, 313.

PROBLEMS

7.1 Verify the algebra leading from Eqs. (7.11) and (7.12) to Eq. (7.16).

7.2 A vertical plate 20 cm long and 10 cm wide is placed in still atmospheric air at 40°C. The plate surface is maintained at 90°C. Find the average heat transfer coefficient on the plate surface by use of **(a)** Eq. (7.22), **(b)** Eq. (7.25), and **(c)** Eq. (7.29).

7.3 Repeat Prob. 7.2 if the fluid is helium.

7.4 If the plate of Prob. 7.2 is placed vertically in water at 40°C, find the average heat transfer coefficient by use of **(a)** Eq. (7.27), and **(b)** Eq. (7.29).

7.5 For the plate of Prob. 7.2, find the local heat transfer coefficient and the boundary layer thickness at a location **(a)** 10 cm from the leading edge, and **(b)** 20 cm from the leading edge.

7.6 A plate 0.3 m high and 0.2 m wide is maintained at 150°C and placed vertically in still air at 40°C. Find the total heat loss from one side of the plate if the air pressure is **(a)** 0.1 atm, **(b)** 1.0 atm, **(c)** 5.0 atm, and **(d)** 10.0 atm.

7.7 Estimate the heat loss rate from one side of a vertical plate (0.6 m high, 0.3 m wide) maintained at 120°C in still nitrogen at −15°C, 70 atm.

7.8 The analysis of Eckert that yields the results given in Eq. (7.25) from the assumed profiles of Eq. (7.24) shows that the nondimensionalizing velocity, v_x^*, is given by

$$v_x^* = \frac{20}{\sqrt{15}}\left(\frac{20}{21} + \text{PR}\right)^{-1/2} (g\beta\,\Delta t x)^{1/2}.$$

Show, from Eq. (7.24), that the maximum velocity in the boundary layer for laminar free convection on a vertical flat surface is given by

$$v_{\text{max}} = \frac{27}{4}\, v_x^*.$$

Then, using these results, calculate the maximum velocity in the laminar boundary layer for the conditions of Prob. 7.2 at a location **(a)** 10 cm from the leading edge, and **(b)** 20 cm from the leading edge.

7.9 A steel plate, 40 cm by 40 cm, is heated to 90°C in a furnace. It is removed and hung vertically in air atmospheric at 30°C. Estimate the initial rate of cooling of the plate.

7.10 Repeat Prob. 7.9 if the plate is placed in water.

7.11 An electrically heated 20-in.-square plate hangs vertically in a still fluid at 1 atm, 70°F. Estimate the required surface temperature if the plate is to dissipate (both sides) 990 Btu/h of heat when the fluid is **(a)** air, and **(b)** water.

7.12 An oil bath at 60°F, using the lubricating oil in Table A.4, is to be heated by immersing a vertical 8-in.-square thin electrically heated plate. What is the maximum power input allowed if the plate temperature is not to exceed 250°F?

7.13 A cylinder 7.5 cm in diameter and 2.0 m long is placed in atmospheric air at 40°C. The surface of the cylinder is maintained at 90°C. Find the total heat loss from the cylinder by free convection if the cylinder is **(a)** horizontal, and **(b)** vertical.

7.14 A circular cylinder 6 ft long and 6 in. in diameter is maintained with a surface temperature of 450°F. It is placed in atmospheric air at 75°F. Estimate the total heat loss from the cylinder if it is **(a)** horizontal, and **(b)** vertical.

7.15 A circular horizontal electrical heater plate is 10 cm in diameter. Its upper and lower surfaces are maintained at 200°C in a tank containing lubricating oil at 20°C. Calculate the total required heat input to the plate.

7.16 Repeat Prob. 7.15 if the plate is placed in atmospheric air at 20°C.

7.17 A cube (6 in. on each side) is suspended in still atmospheric air at 70°F with one face parallel to the ground. All surfaces of the cube are maintained at a uniform temperature of 200°F. Estimate the total heat loss from the cube.

7.18 A heated circular grill, positioned horizontally, has a diameter of 25 cm and a surface temperature of 130°C. It loses heat by free convection to atmospheric air at 25°C simultaneously with radiant exchange to a large room, also at 25°C. If the surface emissivity is 0.75, estimate the total heat loss from the top of the grill.

7.19 A plate 25 cm long and 12 cm wide is maintained at 90°C in atmospheric air at 40°C. Find the total heat transferred from both sides of the plate if it is placed so that the 25-cm dimension **(a)** is vertical, or **(b)** makes an angle of 25° with the vertical.

7.20 A plate 25 cm long and 12 cm wide is maintained at 90°C in water at 40°C. Find the total heat transferred from both sides of the plate if it is placed so that the 25-cm dimension **(a)** is vertical, or **(b)** makes an angle of 25° with the vertical.

7.21 Find the average coefficient of free convective heat transfer between a horizontal 14-cm diameter cylinder with a surface temperature of 90°C and atmospheric air at 30°C.

7.22 Find the average coefficient of free convective heat transfer between a vertical plate, the length of which equals the half-circumference of the cylinder in Prob. 7.21. The surface and air temperature are the same as in Prob. 7.21. Discuss possible reasons for the closeness between the answers of this problem and Prob. 7.21.

7.23 A horizontal cylinder with a surface temperature of 200°C is placed in atmospheric air at 40°C. Find the free convective heat transfer coefficient for diameters of **(a)** 0.025 cm, **(b)** 0.25 cm, **(c)** 2.5 cm, and **(d)** 25 cm.

7.24 A horizontal 12-in. schedule 40 steam pipe passes through a room in which the air is at 100°F. The pipe surface is at 600°F. What is the heat loss by free convection per foot of pipe length?

7.25 Find the coefficient of free convective heat transfer between a horizontal 10-cm pipe with a surface temperature of 40°C and **(a)** methane at −20°C, 350 kN/m^2, and **(b)** water at 90°C.

7.26 An electrical heating element consists of a wire 0.125 cm in diameter. It is immersed horizontally in a bath of lubricating oil at 40°C and the wire surface is maintained at 150°C. Find the free convection heat transfer coefficient.

7.27 A long uninsulated air-conditioning duct consisting of a horizontal circular pipe 8 in. in diameter carries cold air at 50°F. If it is exposed to ambient air at 90°F, estimate the heat loss, per foot of length.

7.28 Compare the free convection heat transfer rates from a horizontal 5-cm-OD 1.5-m-long cylinder maintained at 80°C when exposed to a still environment at 20°C consisting of **(a)** atmospheric pressure air, **(b)** air at 2 atm pressure, **(c)** carbon dioxide at 1 atm, and **(d)** water.

7.29 A horizontal, uninsulated, steam pipe passes through a large room where the ambient still air and the walls are at 25°C. The nominal-5-in. pipe has a surface temperature of 125°C and a surface emissivity of 0.8. Estimate the heat loss, per unit length, by both free convection and radiation.

7.30 A fine wire, diameter = 0.025 mm, is heated electrically in the horizontal position while exposed to still helium at 3 atm, 10°C. If the surface temperature is not to exceed 240°C, determine the power to be supplied, per unit length.

7.31 A 40-W fluorescent light tube is 3.8 cm in diameter and 1.2 m long. Of the input energy to the tube, 60% is converted into light and the remainder is absorbed by the tube wall. Neglecting radiation, estimate the tube operating surface temperature when it is exposed to atmospheric air at 20°C in **(a)** the horizontal position, or **(b)** the vertical position.

7.32 A 550-W electrical resistance heater is designed to operate in a still-water bath at 20°C. The heater is a horizontal circular rod, 1 cm in diameter, 30 cm long.
(a) Estimate its operating surface temperature in water.
(b) If the heater is inadvertently turned on while in atmospheric air at 20°C, estimate its operating temperature.

7.33 A horizontal electrical wire 0.475 cm in diameter dissipates 13 W of heat per meter of length to surrounding atmospheric air at 25°C. Estimate the surface temperature of the wire.

7.34 A horizontal cylinder of unknown diameter is maintained with a surface temperature of 45°C in atmospheric air at 30°C. The heat loss, per unit surface of area, is found to be 158 W/m^2. What will be the heat loss if the surface temperature is raised to 100°C?

7.35 The heat loss by free convection from a horizontal, electrically heated wire to ambient atmospheric air at 50°F is known to be 350 Btu/h-ft^2 when the wire temperature is 150°F. What is the rate of heat loss, per unit surface area, when the wire temperature is raised to 300°F?

7.36 A straight horizontal fin consists of a circular bar 0.6 cm in diameter and 10 cm long. The fin base is at 150°C, and the fin extends into atmospheric air at 50°C. Applying the results of this chapter and Sec. 2.9, estimate the heat loss from the fin to the air if the bar is made of 0.5% carbon steel.

7.37 A cast iron rod 0.75 in. in diameter and 10 in. long extends horizontally into atmospheric air at 90°F from a heat source at 300°F. Estimate the rate of heat exchange between the rod and the air.

7.38 Repeat Prob. 7.36 if the rod is oriented vertically.

7.39 Repeat Prob. 7.37 if the rod is oriented vertically.

7.40 A 1.5-cm-diameter sphere is electrically heated to a surface temperature of 100°C while suspended in still atmospheric air at 20°C. Calculate the energy required to heat the sphere.

7.41 A 5-cm-diameter sphere is supplied 640 W of energy by electrical heating. If it is submerged in a still body of water at 30°C, estimate its surface temperature.

7.42 A sphere of 2.5 cm diameter has an internal heat source which maintains its surface at 90°C. If the sphere exchanges heat by free convection with a fluid at 20°C, find the rate of heat flow, in watts, if the fluid is **(a)** atmospheric air, **(b)** water, and **(c)** ethylene glycol. Assume that Eq. (7.33) holds.

7.43 A double glazed window is 1.25 m high and 0.8 m wide. The two layers of glass are 6 cm apart. If the two glass layers are at 20°C and −10°C, what is the heat loss by free convection through the air space?

7.44 Repeat Prob. 7.43 if the air space thickness is changed to **(a)** 3 cm, and **(b)** 8 cm.

7.45 A solar collector panel, 2 m × 1 m, consists of a lower absorber plate at 70°C with a glass cover plate at 30°C. The air space between the two surfaces is 5 cm thick. If the air in this space is at atmospheric pressure, find the heat loss between the two surfaces by free convection if the 2-m dimension of the hotter surface is inclined at **(a)** 60° or **(b)** 45° with the horizontal.

7.46 Repeat Prob. 7.45 if the air space between the two surfaces is evacuated to a pressure of 0.25 atm.

7.47 A building window pane is 3.5 ft high and 2.5 ft wide. It is separated from the outdoor air by a storm window of the same size, spaced 2 in. away. If the windows are at 70°F and 15°F, determine the rate of heat flow through the air gap.

7.48 A solar collector panel, 6 ft × 3 ft, consists of a lower absorber plate at 150°F with a glass cover plate at 90°F. The space between the plates is 1.5 in. thick and contains air at atmospheric pressure. Find the heat loss by free convection between the two plates if the 6-ft dimension of the hotter surface is inclined at an angle of **(a)** 30° or **(b)** 60° with the horizontal.

7.49 Repeat Prob. 7.48 if the air space between the surfaces is evacuated to a pressure of 0.5 atm.

7.50 A rectangular space of the geometry shown in Fig. 7.5 contains air at 1 atm pressure. The following conditions exist: $H = 1$ m, $L = 5$ cm, $t_h = 80°C$, $t_c = 40°C$. Calculate the heat transfer coefficient due to free convection for values of φ ranging from 0 to 180°, by increments of 20°.

7.51 Atmospheric air at 20°C flows upward along a flat surface 0.4 m high maintained at 50°C. What is the minimum flow velocity for which the effect of buoyancy will be less than 5%?

7.52 Ethylene glycol at 40°C flows past a vertical flat surface heated to 80°C. If the surface is 0.5 m long, what is the minimum upward flow velocity of the glycol for which the buoyancy effect will be less than 5%?

7.53 Atmospheric-pressure air at 80°F flows upward with a velocity of 6 ft/s past a 1.5 ft high vertical flat surface. Determine the maximum surface temperature for which the effects of free convection will be less than 5%.

CHAPTER 8

Heat Transfer in Condensing and Boiling

8.1 INTRODUCTORY REMARKS

The convective processes discussed in Chapters 5 through 7 have all been limited to cases in which the fluid medium remained in a single phase—liquid or gas. A great number of practical engineering applications exists in which a phase change occurs simultaneously with the heat transfer process. The importance of the processes of condensing and boiling to the production of power from vapor cycles, the production of refrigeration, the design of petrochemical processes, etc., is obvious.

While the solid–gas phase change process is of current importance in applications such as the thermal protection of spacecraft and missiles, and the solid–liquid phase change process is important for thermal storage considerations, these applications will not be treated here and only the vapor–liquid phase change processes of condensing and boiling will be discussed. Since fluid motion is still involved, condensing and boiling are classified as convective mechanisms; however, there are profound differences between these mechanisms and single phase convective heat transfer. There are significant differences between the various fluid properties in the two phases (such as density, viscosity, conductivity, specific heat) that influence the fluid mechanical processes markedly. At the same time there is a release or consumption of latent heat that also has profound influence on the observed heat transfer rates during phase change, so that heat transfer coefficients that are orders of magnitude greater than those in single phase convection are commonly observed.

First, the process of the condensation of a vapor will be discussed. Subsequently, the process of boiling will be considered. Because of the great complexity of these processes, much of these considerations will be descriptive in nature.

8.2 GENERAL COMMENTS REGARDING CONDENSATION

The process of condensation of a vapor is usually accomplished by allowing it to come in contact with a surface the temperature of which is maintained at a value lower than the saturation temperature of the vapor for the pressure at which it exists. The removal of thermal energy from the vapor causes it to release its latent heat of vaporization and, hence, to condense onto the surface.

The appearance of the liquid phase on the cooling surface, either in the form of individual drops or in the form of a continuous film, offers a greater thermal resistance to the further removal of heat from the remaining vapor. In most cases of condensation, the condensate is removed by the action of gravity. Naturally, then, the rate of removal of condensate, and with it the rate of heat removal from the vapor, is greater for vertically placed surfaces than for horizontal surfaces. In order to avoid the accumulation of large amounts of liquid at the lower end of a condensing surface it should be short; hence horizontal cylinders are particularly applicable for such purposes. Thus most condensing equipment consists of assemblies of horizontal tubes around which the vapor to be condensed is allowed to flow. The cool temperature of the outer tube surface is maintained by circulating a colder medium, usually water, through the inside of the tube.

Two general types of condensation have been observed to occur in practice. The first, called *film condensation,* usually occurs when a vapor, relatively free of impurities, is allowed to condense on a clean surface. Under these conditions, it is found that the condensate will appear as a continuous film all over the surface and it will flow off the surface under the action of gravity.

The second type of condensation, less frequently observed, is known as *dropwise condensation,* and has been observed to occur either on highly polished surfaces or on surfaces contaminated with certain fatty acids. In this case the condensate is found to appear in the form of individual drops. These drops increase in size and combine with one another until their size is great enough that their weight causes them to run off the surface, leaving the surface exposed to the formation of a new drop.

Since there will be much less of a liquid barrier between the condensing surface and the vapor, it should not be surprising to know that heat transfer rates are five to ten times greater for dropwise condensation than for film condensation. Hence it would appear that dropwise condensation would be preferred for commercial applications. Some of the details of dropwise condensation are discussed in a later section, but it is sufficient to state here that it is a difficult mode to maintain in practice so that conservative design calculations are usually made on the basis of presuming that film condensation will exist. This presumption has the advantage that film condensation is much easier to subject to analysis, as described next.

8.3 ANALYSIS OF LAMINAR FILM CONDENSATION ON VERTICAL SURFACES

Figure 8.1 depicts the physical situation for film condensation on a vertical flat surface. The surface is maintained at some temperature, t_s, which is lower than the saturation temperature of the neighboring vapor. Heat removal from the vapor causes condensation on the surface and a film of condensate forms and flows down the surface under the influence of gravity. The film is presumed to originate at the top of the surface and increases in thickness as it flows down due to the accumulation of condensate. The film thickness, $\delta(x)$, is thus a function of the distance along the surface, x. (The dimension y is measured normal to the surface.) If one can determine the shape of the velocity profile and how δ varies with x, then the mass flow rate of condensate may be found as a function of position. This flow rate is, in turn, related to the amount of vapor condensed, so that one may eventually find the heat transfer coefficient as a function of x.

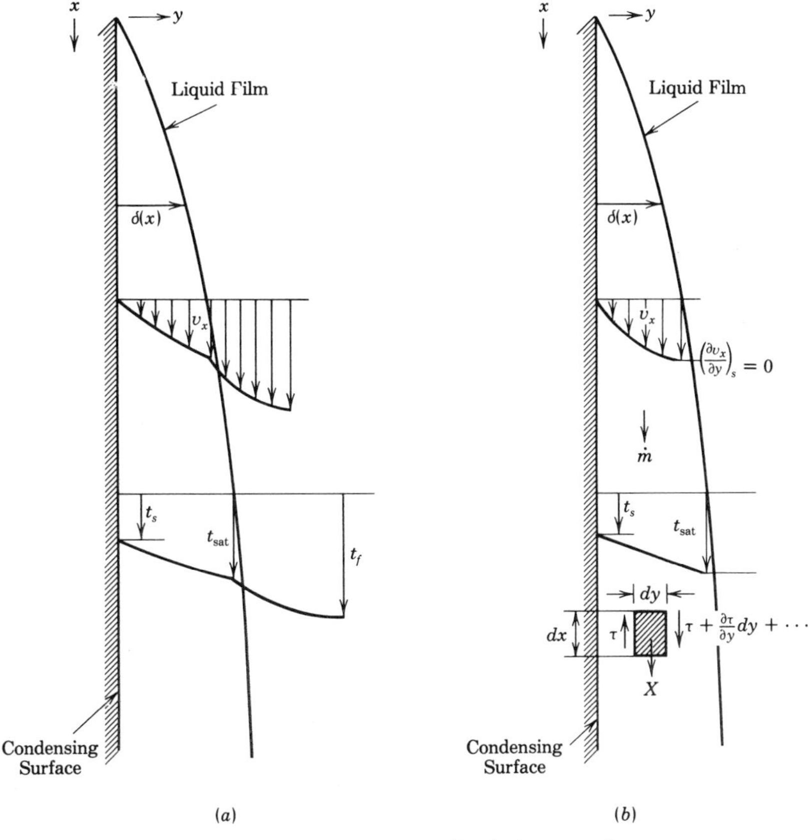

Figure 8.1. Film condensation on a vertical plane surface.

Figure 8.1(a) depicts the temperature and velocity profile in the film for a general situation in which the ambient vapor may be superheated. Thus a temperature gradient may exist in the vapor near the liquid–vapor interface. The interface is at the saturation temperature, t_{sat}, and then subcooling of the liquid film occurs to the surface value t_s. Likewise, if the ambient vapor is moving relative to the liquid at the interface, some interfacial shear may exist as suggested by the velocity gradient shown at $y = \delta$.

Nusselt (Ref. 1) simplified the problem by invoking a number of assumptions. These assumptions were: (1) the film flow is laminar; (2) the vapor is not superheated; (3) the film grows in thickness so slowly that the transfer of heat through it is purely by conduction, giving a linear temperature distribution; (4) there is no interfacial shear, so that $\partial v_x/\partial y = 0$ at $y = \delta$; and (5) the film moves slowly enough to neglect inertia forces within it.

Figure 8.1(b) illustrates some of the assumptions. The neglection of inertia forces means that a force balance on an element of the film, dx by dy, gives (for a unit depth into the plane of the paper)

$$-\left(\frac{\partial \tau}{\partial y}\,dy\right)dx = X(dx\,dy).$$

In the above, τ is the shear stress shown acting on the dx sides of the element and X is the body force, per unit volume, acting on the element. If the flow in the film is laminar then $\tau = \mu(\partial v_x/\partial y)$, v_x being the flow velocity parallel to the surface. The body force is that of gravity acting on the element, less the buoyancy resulting from the displacement of the vapor. Thus

$$X = g(\rho_l - \rho_v),$$

so that the force balance above becomes

$$\frac{\partial^2 v_x}{\partial y^2} = \frac{(\rho_l - \rho_v)g}{\mu_l}. \tag{8.1}$$

In Eq. (8.1), and all subsequent equations in this chapter, the subscript l refers to properties of the liquid component and the subscript v to those of the vapor component.

Integration of Eq. (8.1) twice and application of the conditions that $v_x = 0$ at $y = 0$ and $\partial v_x/\partial y = 0$ at $y = \delta$ yield

$$v_x = \frac{g(\rho_l - \rho_v)\delta^2}{\mu_l}\left[\frac{y}{\delta} - \frac{1}{2}\left(\frac{y}{\delta}\right)^2\right]. \tag{8.2}$$

The mass flow rate in the condensate film is readily found, per unit of width into the paper, by integration of the velocity. If the symbol $\dot{m}'$ is used to denote this flow rate per unit width, then

$$\dot{m}'(x) = \int_0^\delta \rho_l v_x \, dy$$
$$= \frac{g\rho_l(\rho_l - \rho_v)\delta^3}{3\mu_l}. \tag{8.3}$$

Equation (8.3) interrelates the film thickness, δ, and the mass flow rate, $\dot{m}'$, but both are still unknown functions of x. These may be related by noting that the heat flux into the wall may be written in terms of each. By use of Fourier's law and the assumption of a linear temperature distribution in the film, one has for the heat flux:

$$\frac{q}{A} = \frac{k_l(t_s - t_{sat})}{\delta}$$
$$= k_l \frac{\Delta t}{\delta}, \tag{8.4}$$

where $\Delta t = t_{sat} - t_s$ has been introduced. Also, noting that the heat flux into the wall over a distance dx must result in the condensation of an amount of vapor equal to $d\dot{m}'$, one may also state, for unit depth, that

$$\frac{q}{A} = h_{fgg} \frac{d\dot{m}'}{dx} \tag{8.5}$$

in which h_{fg} represents the latent heat of vaporization in accord with usual thermodynamics notation.

Equating Eqs. (8.4) and (8.5) and using Eq. (8.3) gives

$$k_l \frac{\Delta t}{\delta} = h_{fg} \frac{d\dot{m}'}{dx} = h_{fg} \frac{g\rho_l(\rho_l - \rho_v)\delta^2}{\mu_l} \frac{d\delta}{dx},$$

$$\delta^3 \frac{d\delta}{dx} = \frac{\mu_l k_l \, \Delta t}{g\rho_l(\rho_l - \rho_v)h_{fg}}.$$

Integration of the latter equation with the condition that $\delta = 0$ at $x = 0$, gives the following for the dependence of the film thickness on the distance along the surface:

$$\delta = \left[\frac{4\mu_l k_l \, \Delta t \, x}{g\rho_l(\rho_l - \rho_v)h_{fg}}\right]^{1/4}. \tag{8.6}$$

This result may be substituted into Eq. (8.4) to find the heat flux, q/A; however, it is customary to define the local heat transfer coefficient so that

$$\frac{q}{A} = h_x(t_{sat} - t_s)$$
$$= h_x \, \Delta t. \tag{8.7}$$

Combined with Eq. (8.4) this gives

$$h_x = \frac{k_l}{\delta}, \tag{8.8}$$

so that when Eq. (8.6) for δ is introduced, one finds the local heat transfer coefficient:

$$h_x = \left[\frac{g\rho_l(\rho_l - \rho_v)h_{fg}k_l^3}{4\mu_l x\,\Delta t}\right]^{1/4}.$$

In terms of the usual local Nusselt number, the above is

$$\mathrm{Nu}_x = \frac{h_x x}{k_l} = \frac{1}{\sqrt{2}}\left[\frac{g\rho_l(\rho_l - \rho_v)h_{fg}x^3}{\mu_l k_l\,\Delta t}\right]^{1/4}. \quad (8.9)$$

Integrating for an average h and an average Nusselt number on a plate of total length L gives

$$h = \frac{4}{3}\left[\frac{g\rho_l(\rho_l - \rho_v)h_{fg}k_l^3}{4\mu_l L\,\Delta t}\right]^{1/4},$$
$$\mathrm{Nu}_L = \frac{hL}{k_l} = 0.943\left[\frac{g\rho_l(\rho_l - \rho_v)h_{fg}L^3}{\mu_l k_l\,\Delta t}\right]^{1/4}. \quad (8.10)$$

In spite of the several simplifying assumptions made, Nusselt's analysis resulting in Eqs. (8.9) and (8.10) yields rather good results when compared with experiments. More refined analyses of laminar film condensation are summarized in Ref. 2, wherein the effects of some of these assumptions are investigated. The most significant refinements take into account the energy removed in subcooling the film temperature profiles. Sparrow and Gregg (Ref. 3), in a detailed analysis of laminar film condensation on a vertical surface, show that Nusselt's analysis is quite accurate for fluids with Prandtl numbers greater than those for liquid metals as long as the Jakob number is small. The Jakob number, a measure of the relative magnitude of the film subcooling, is defined as

$$\text{Jakob number} = \mathrm{Ja} = \frac{c_{pl}\,\Delta t}{h_{fg}}. \quad (8.11)$$

The numerator in the definition of the Jakob number, $c_{pl}\,\Delta t$, represents the heat removed in subcooling the liquid film from the saturation temperature to the temperature of the surface. For low values of Ja this subcooling is modest and Nusselt's equations give satisfactory results for many practical applications.

Adjustments to Nusselt's equations to account for the effects of values of Ja which are significant are summarized, along with other effects not discussed here, in the next section under the heading "working correlations." That section will also present recommended relations for geometries other than the vertical plane surface.

While Nusselt's results will be modified in the next section for use in practical situations, it is instructive at this point to apply them to a simple example in order to illustrate the order of magnitude of some of the quantities involved.

EXAMPLE 8.1

Saturated steam at 1 atm condenses on a vertical flat surface maintained at t_s = 80°C. Determine the local heat transfer coefficient and the condensate film thickness at a location 0.5 m from the top of the surface.

Solution. It will be assumed that the condensate film formed on the surface is in laminar flow so that the analysis of Nusselt given in the foregoing may be applied. The criterion by which the existence of laminar flow may be determined is given in the next section.

At a pressure of 1 atm, the Steam Tables give the following:

$$t_{sat} = 100°\text{C}, \qquad \Delta t = 100 - 80 = 20°\text{C},$$
$$h_{fg} = 2257.0 \text{ kJ/kg},$$
$$\rho_v = 5.98 \times 10^{-4} \text{ kg/m}^3.$$

The properties of the liquid film will be evaluated at the mean temperature in the film, $t_m = (t_s + t_{sat})/2 = (80 + 100)/2 = 90°\text{C}$. Table A.3 yields

$$\rho_l = 965.1 \text{ kg/m}^3, \qquad \mu_l = 0.3150 \times 10^{-3} \text{ kg/m-s},$$
$$k_l = 0.6727 \text{ W/m-°C}.$$

Equation (8.9) determines the local Nusselt number at the specified location $x = 0.5$ m:

$$\text{Nu}_x = \frac{1}{\sqrt{2}}\left[\frac{g\rho_l(\rho_l - \rho_v)h_{fg}x^3}{\mu_l k_l\, \Delta t}\right]^{1/4}$$
$$= \frac{1}{\sqrt{2}}\left[\frac{9.8 \times 965.1(965.1 - 5.98 \times 10^{-4})2257.0 \times 10^3 \times (0.5)^3}{0.3150 \times 10^{-3} \times 0.6727 \times 20}\right]^{1/4}$$
$$= 3511.$$

The factor 10^3 is included in the numerator of the calculation above to render the Nusselt number approximately dimensionless—a fact worth remembering in subsequent calculations in this chapter. The local heat transfer coefficient is now readily found:

$$h_x = \text{Nu}_x \frac{k}{x}$$
$$= 3511 \times \frac{0.6727}{0.5} = 4723 \text{ W/m}^2\text{-°C } (831.8 \text{ Btu/h-ft}^2\text{-°F}).$$

Equation (8.6) could be used to find the liquid film thickness; however, use of Eq. (8.8) gives the result more directly:

$$\delta = \frac{k_l}{h_x} = \frac{0.6727}{4723}$$

$$= 1.42 \times 10^{-4} \text{ m} = 0.142 \text{ mm } (5.6 \times 10^{-3} \text{ in.}).$$

Several points regarding some orders of magnitude should be noted. First, the local heat transfer coefficient is seen to be much larger than those encountered in the forced and free convection examples discussed in Chapters 6 and 7 for fluids not undergoing a phase change. Such large coefficients are typical of condensing heat transfer and as will be seen later in this chapter, the same is true for cases involving boiling. In addition, it will be noted that the liquid film produced is very thin, 0.142 mm in this case. Even though the heat transfer rate is high, the magnitude of the latent heat produces relatively little liquid. For a surface location twice that just used, say $x = 1.0$ m, the liquid film is still quite small: $\delta = 0.169$ mm.

Finally, one should note that the density change occurring during the condensation process is very large, so that the factor $(\rho_l - \rho_v) \approx \rho_l$. Thus the term $\rho_l(\rho_l - \rho_v)$ in Nusselt's equation for Nu_x, Eq. (8.9) may be quite accurately replaced by $\rho_l(\rho_l - \rho_v) \approx \rho_l^2$. This situation exists for most condensing fluids encountered in practice and the simplification just noted may be applied—simplifying somewhat the calculation of Nu_x and eliminating the need to evaluate the vapor density, ρ_v.

8.4 WORKING CORRELATIONS FOR FILM CONDENSATION

As noted in the foregoing, the analysis of Nusselt for film condensation on a vertical flat surface involved a number of simplifying assumptions. More refined analyses have been performed to account for some of these simplifications. Similarly, a number of studies have been performed to predict condensation on surfaces other than vertical flat ones. This section presents a summary of some of the analytical and empirical relations developed in such studies that are recommended for use in engineering calculations.

In these correlations, the mean film temperature is defined as $t_m = (t_{sat} + t_s)/2$, and usually all liquid properties are evaluated at this temperature except h_{fg}, which should be evaluated at t_{sat}.

As might be expected, one rather significant effect that must be accounted for is whether the liquid condensate film is in the laminar or turbulent state of flow. The criterion by which this may be established is discussed first.

The Laminar–Turbulent Transition

The analysis presented in Sec. 8.3 presumed the existence of laminar flow in the liquid film. This assumption is generally true for short surfaces for which an excessively thick layer of liquid does not build up. However, in some applications, the surface may be long enough that the condensate film becomes so thick that a transition to a turbulent state of flow occurs.

To characterize the transition to turbulence, a Reynolds number must be defined. In this instance of flow with a free surface, a local Reynolds number must be defined in terms of the hydraulic diameter defined in Eq. (6.44) as four times the flow area divided by the wetted perimeter:

$$D_h = \frac{4A}{P}.$$

Consider, as depicted in Fig. 8.2, the flow in a condensate film on a surface of width b. If at some location down the surface the local film thickness is δ, the flow area is $A = \delta b$ and the wetted perimeter is simply the plate width b. (In the event condensation is occurring on a vertical cylinder, the width b is the circumference πD.) Then the hydraulic diameter is

$$D_h = \frac{4\ \delta b}{b} = 4\delta.$$

Then, if v_m represents the *mean* velocity in the film at the point in question, a local Reynolds number may be defined:

$$\text{Re} = \frac{v_m D_h \rho_l}{\mu_l},$$

which becomes

$$\text{Re} = \frac{4 v_m \rho_l \delta}{\mu_l}.$$

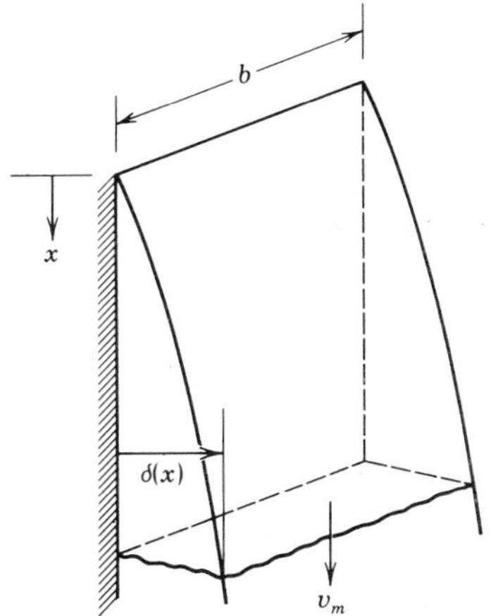

Figure 8.2

The product $v_m \rho_l \delta$ is recognized as the mass flow rate of condensate, per unit width—the quantity $\dot{m}'$ introduced in Sec. 8.3. In condensation work, however, this is given the special symbol Γ:

$$\Gamma = \dot{m}' = v_m \rho_l \delta,$$

so that the local Reynolds number, based on this flow rate, is

$$\text{Re}_\Gamma = \frac{4\Gamma}{\mu_l}. \tag{8.12}$$

The quantity Γ, and hence Re_Γ, is obviously a function of position down the surface, in the direction of the flow, and will have its maximum value at the lower end of the surface (i.e., $x = L$) where *all* the mass condensed must pass. The total mass rate condensed over the whole surface of length L, call it $\dot{m}_T$, is found from the total heat transferred, which in turn, is related to the average heat transfer coefficient:

$$\dot{m}_T = \frac{q}{h_{fg}} = \frac{hA\ \Delta t}{h_{fg}} = \frac{hbL\ \Delta t}{h_{fg}}. \tag{8.13}$$

Thus at the lower edge where all of $\dot{m}_T$ must flow, the maximum value of Γ is

$$\Gamma_{max} = \frac{\dot{m}_T}{b} = \frac{hL\ \Delta t}{h_{fg}}, \tag{8.14}$$

so that the Reynolds number there is

$$\begin{aligned} \text{Re}_{\Gamma_{max}} &= \frac{4\Gamma_{max}}{\mu_l} \\ &= \frac{4hL\ \Delta t}{h_{fg}\mu_l}. \end{aligned} \tag{8.15}$$

Both forms of Eq. (8.15) for $\text{Re}_{\Gamma_{max}}$ will prove to be useful.

Experiment indicates that the critical value of Re_Γ for film condensation is about 1800, laminar films occurring below this value and turbulent ones above it.

Film Condensation on Vertical or Inclined Plates and on Vertical Cylinders

For laminar film condensation on vertical plates, McAdams (Ref. 4) recommends that Nusselt equations, Eq. (8.10), be employed with the constant 0.943 empirically adjusted to 1.13. Further, Rohsenow (Ref. 5), suggests that subcooling of the film be accounted for by replacing the latent heat of vaporization with

$$h'_{fg} = h_{fg}(1 + 0.68\text{Ja}),$$

where Ja is the Jakob number defined in Eq. (8.11). Finally, it is found that the Nusselt relation holds for film condensation on the upper side of a plate inclined at an angle φ with the horizontal if g is replaced with $g \sin \varphi$, as long as φ is not less than about 30°. Thus, the recommended relation for laminar film condensation on vertical or inclined plates is

$$\mathrm{Nu}_L = 1.13\left[\frac{g'\rho_l(\rho_l - \rho_v)h'_{fg}L^3}{\mu_l k_l\,\Delta t}\right]^{1/4}, \qquad (8.16)\bigstar$$

$$\left[\begin{array}{l} \mathrm{Pr} > 0.5 \\ \mathrm{Re}_{\Gamma\max} < 1800 \\ \mathrm{Ja} < 1.0 \\ g' = g \sin\varphi,\ \varphi \geq 30° \\ h'_{fg} = h_{fg}(1 + 0.68\mathrm{Ja}) \\ \text{liquid properties at } t_m \\ h_{fg} \text{ at } t_{sat} \end{array}\right].$$

When turbulence occurs in the film, McAdams further suggests the use of the following empirical relation:

$$\mathrm{Nu}_L = 0.0077\left[\frac{g'\rho_l(\rho_l - \rho_v)L^3}{\mu_l^2}\right]^{1/3}\mathrm{Re}_{\Gamma\max}^{0.4}, \qquad (8.17)\bigstar$$

$$\left[\begin{array}{l} \mathrm{Pr} > 0.5 \\ \mathrm{Re}_{\Gamma\max} > 1800 \\ g' = g \sin\varphi,\ \varphi \geq 30° \\ \text{liquid properties at } t_m \end{array}\right].$$

As noted in Example 8.1, for most cases of practical interest, the relative magnitudes of the liquid and vapor densities enables the use of the approximation

$$\rho_l(\rho_l - \rho_v) \approx \rho_l^2 \qquad (8.18)$$

in Eqs. (8.16) and (8.17) without much error.

Equations (8.16) and (8.17) must be used with caution for plates of small inclination since an obviously incorrect result occurs for $\varphi \to 0$. Horizontal plates are treated later.

Equations (8.16) and (8.17) may be used for *vertical* cylinders as long as the film thickness is quite small compared with the cylinder radius, $\delta << R$. These relations *may not* be used for inclined cylinders. Horizontal cylinders are treated later.

EXAMPLE 8.2

Saturated steam at 70 kN/m^2 (10 psia) condenses on a vertical plane surface 0.2 m wide and maintained at 40°C. Find the mass rate at which steam is condensed on one side of the surface if it is (a) 1 m high; (b) 3 m high.

Solution. At a pressure of 70 kN/m^2, the Steam Tables give $t_{sat} = 89.95$°C, $h_{fg} = 2283.3$ kJ/kg. Thus, from the data given,

$$\Delta t = (t_{sat} - t_s) = (89.95 - 40) = 49.95°\text{C},$$

$$t_m = \frac{t_{sat} + t_s}{2} = \frac{89.95 + 40}{2} = 65°\text{C}.$$

Table A.3 gives the condensate properties at t_m to be

$$\rho_l = 980.5 \text{ kg/m}^3, \qquad \mu_l = 0.4338 \times 10^{-3} \text{ kg/m-s},$$

$$k_l = 0.6553 \text{ W/m-°C}, \qquad c_{pl} = 4.187 \text{ kJ/kg-°C}.$$

(a) For a plate length $L = 1$ m, assume that the condensate film is laminar and verify this assumption later. Using $\rho_l(\rho_l - \rho_v) \approx \rho_l^2$, as suggested in Eq. (8.18), Eq. (8.16) gives the average heat transfer coefficient to be (with $\varphi = 90°$):

$$\text{Ja} = \frac{c_{pl}\,\Delta t}{h_{fg}} = \frac{4.187 \times 49.95}{2283.3} = 0.0916,$$

$$h'_{fg} = h_{fg}(1 + 0.68\text{Ja})$$

$$= 2283.3(1 + 0.68 \times 0.0916) = 2425.5 \text{ kJ/kg},$$

$$\text{Nu}_L = 1.13\left[\frac{g\rho_l^2 h'_{fg} L^3}{\mu_l k_l\,\Delta t}\right]^{1/4}$$

$$= 1.13\left[\frac{9.8 \times (980.5)^2 \times 2425.5 \times 1000 \times 1^3}{0.4338 \times 10^{-3} \times 0.6553 \times 49.95}\right]^{1/4} = 7175.2,$$

$$h = \text{Nu}_L \frac{k_l}{L} = 7175.2\,\frac{0.6553}{1} = 4690 \text{ W/m}^2\text{-°C}.$$

Thus Eq. (8.13) gives the total steam condensed to be

$$\dot{m}_T = \frac{hbL\,\Delta t}{h_{fg}} = \frac{4690 \times 0.2 \times 1 \times 49.95}{2283.3 \times 1000} = 0.0205 \text{ kg/s } (163 \text{ lb}_m\text{/h}).$$

To verify that the flow is laminar over the entire surface the value of Γ_{max} is found from Eq. (8.14),

$$\Gamma_{max} = \frac{\dot{m}_T}{b} = \frac{0.0205}{0.2} = 0.0103 \text{ kg/m-s},$$

so that the Reynolds number at the bottom of the plate is, by Eq. (8.15),

$$\text{Re}_{\Gamma_{max}} = \frac{4\Gamma_{max}}{\mu_l} = \frac{4 \times 0.0103}{0.4338 \times 10^{-3}} = 950.$$

Thus the flow is laminar, as assumed, since this is less than the critical value of 1800, and the desired condensate rate is the value of $\dot{m}_T$ just found above.

(b) For a surface length of $L = 3$ m, the flow at the bottom may well be turbulent, so that Eq. (8.17) must be used to find h. Since this equation involves the unknown $\text{Re}_{\Gamma_{max}}$, an iterative calculation may appear to be necessary. However, the second definition of $\mathrm{Re}_{\Gamma_{max}}$ given in Eq. (8.15) may be substituted into Eq. (8.17) to avoid iteration:

$$\mathrm{Nu}_L = 0.0077\left[\frac{g'\rho_l(\rho_l - \rho_v)L^3}{\mu_l^2}\right]^{1/3} \mathrm{Re}_{\Gamma_{max}}^{0.4},$$

$$\frac{hL}{k_l} = 0.0077\left[\frac{g\rho_l^2 L^3}{\mu_l^2}\right]^{1/3}\left[\frac{4hL\,\Delta t}{h_{fg}\mu_l}\right]^{0.4},$$

where the simplification of Eq. (8.18) has been used as well as the fact that the surface is vertical. All quantities in the foregoing expression are known except the heat transfer coefficient, h. Thus h may be found:

$$\frac{h \times 3}{0.6553} = 0.0077\left[\frac{9.8 \times (980.5)^2 \times 3^3}{(0.4338 \times 10^{-3})^2}\right]^{1/3} \times \left[\frac{4 \times h \times 3 \times 49.95}{2283.3 \times 1000 \times 0.4338 \times 10^{-3}}\right]^{0.4}$$

$$h = 4335 \text{ W/m}^2\text{-°C}.$$

Then, just as in part (a), the total mass rate condensed and the resultant values of Γ and Reynolds numbers at the bottom of the surface are

$$\dot{m}_T = \frac{hbL\,\Delta t}{h_{fg}} = \frac{4335 \times 0.2 \times 3 \times 49.95}{2283.3 \times 1000} = 0.0569 \text{ kg/s } (452 \text{ lb}_m\text{/h}),$$

$$\Gamma_{max} = \frac{\dot{m}_T}{b} = \frac{0.0569}{0.2} = 0.2845 \text{ kg/s-m},$$

$$\mathrm{Re}_{\Gamma_{max}} = \frac{4\Gamma_{max}}{\mu_l} = \frac{4 \times 0.2845}{0.4338 \times 10^{-3}} = 2623.$$

Since $\mathrm{Re}_{\Gamma_{max}} > 1800$, the flow is turbulent, as assumed, and the condensation rate is the value of $\dot{m}_T$ just found.

Laminar Film Condensation on the Top of a Horizontal Plate

As noted above, when a plate is in a horizontal position, the analysis of Nusselt fails since the driving force $g \sin\varphi \to 0$ as $\varphi \to 0$. In such an instance one must recast the analysis to account for the fact that the only driving force is that due to hydrostatic pressure difference created by the film being thicker in the center of the plate than at the edge. This problem was analyzed by Clifton and Chapman (Ref. 6), with the following result:

$$\mathrm{Nu}_L = 2.43\left(\frac{g\rho_l^2 h_{fg} L^3}{\mu_l k_l\,\Delta t}\right)^{1/5} F\left(\frac{\mathrm{Ja}}{\mathrm{Pr}}\right), \qquad (8.19)\bigstar$$

$$\begin{bmatrix} \text{liquid properties at } t_m \\ h_{fg} \text{ at } t_{sat} \\ F(\text{Ja}/\text{Pr}) \text{ from Table 8.1} \end{bmatrix}.$$

The function $F(\text{Ja}/\text{Pr}) = F(k_l\ \Delta t/\mu_l h_{fg})$ is a complex function tabulated in Table 8.1.

Film Condensation on the Outside of Horizontal Cylinders

Analyses similar to that of Nusselt for vertical surfaces may be carried out for other shapes of interest. One very important case is that of condensation on the outside of a horizontal cylinder as illustrated in Fig. 8.3(a). The condensate film forms at the top of the cylinder and flows around it, thickening as it goes. Nusselt analyzed this problem by representing the film at any polar location θ as that along an inclined flat surface making an angle φ with the horizontal. Thus, Nusselt used Eq. (8.8) with g replaced with $g \sin \varphi$. Since $\varphi = \pi/2 - \theta$ and $\theta = x/(D/2)$, Eq. (8.8) may be written as a function of x and then integrated around the cylinder to obtain the average h and Nu. Performing this integration numerically, Nusselt obtained the result noted below. This result agrees well when compared with experiment. The question of transition to turbulence rarely arises in practical applications involving horizontal cylinders.

$$\text{Nu}_D = \frac{hD}{k} = 0.728\left[\frac{g\rho_l(\rho_l - \rho_v)h'_{fg}D^3}{\mu_l k_l\ \Delta t}\right]^{1/4}, \tag{8.20}\bigstar$$

$$\begin{bmatrix} \text{Pr} > 0.5 \\ \text{Ja} < 1.0 \\ h'_{fg} = h_{fg}(1 + 0.68\text{Ja}) \\ \text{liquid properties at } t_m \\ h_{fg} \text{ at } t_{sat} \end{bmatrix}.$$

If one uses the vertical plate relation for a vertical cylinder, one may note that the ratio of the heat transfer coefficient for a horizontal cylinder to what it would have in a vertical position is

Table 8.1 Function F for Use in Eq. (8.14)

$\frac{\text{Ja}}{\text{Pr}} = \frac{k_l\ \Delta t}{\mu_l h_{fg}}$	F
0.176	0.329
0.381	0.322
0.698	0.320
2.27	0.285
4.08	0.264

$$\frac{h_{\text{horiz}}}{h_{\text{vert}}}\frac{D}{L} = \frac{0.728}{1.13}\left(\frac{D^3}{L^3}\right)^{1/4},$$
$$\frac{h_{\text{horiz}}}{h_{\text{vert}}} = 0.644\left(\frac{L}{D}\right)^{1/4}. \tag{8.21}$$

In commercial condensers the ratio L/D may be in the range of 50 to 100, or even more. Thus as much as twice the condensing capacity may be expected for the horizontal case than for the vertical case—a result of the fact that a much thinner film forms in the horizontal position.

The correlation of Eq. (8.20) applies only to a single cylinder. Most commercial condensers involve banks of tubes. If the tubes are arranged in a vertical pattern as suggested in Fig. 8.3(b), the condensate from one tube drips onto that of the next, and so on. Thus, the film thickness is greater for the lower tubes, increasing the resistance to heat transfer. For a vertical stack of N tubes, Nusselt showed that the average coefficient is reduced according to

$$\frac{(\text{Nu}_D)_N}{(\text{Nu}_D)_1} = \left(\frac{1}{N}\right)^{1/4}, \tag{8.22}$$

where the subscripts N and 1 refer to N tubes and a single tube. Thus improved performance in a condenser can be achieved by using a staggered arrangement as suggested in Fig. 8.3(c). For the analyses of complex tube bank arrangements the reader is referred to Ref. 7.

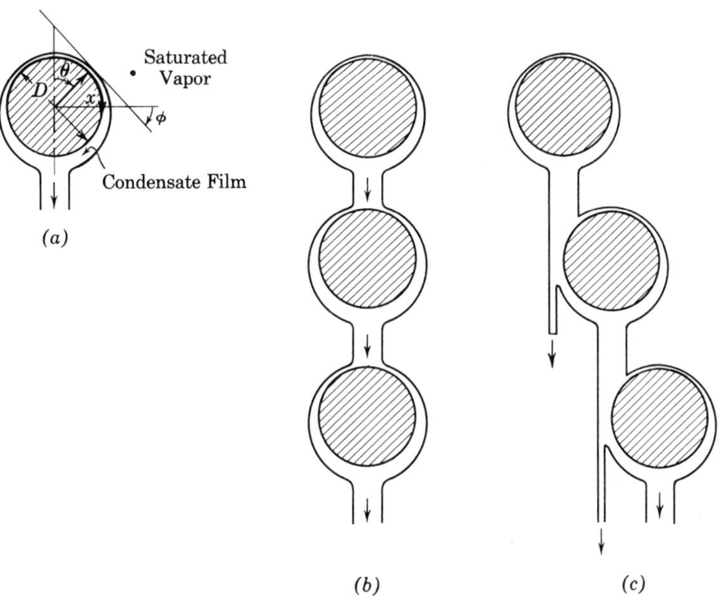

Figure 8.3. Film condensation on horizontal cylinders.

EXAMPLE 8.3

A steam condenser uses horizontal tubes with an outside diameter $D = 1$ in. to condense saturated steam at a pressure of 1 psia. Cooling water flowing through the tubes maintains the surface temperature at 95°F. Find the amount of steam condensed, per foot of tube length, for a single tube—i.e., neglecting the interaction between tubes in the bank.

Solutions. At 1 psia the Steam Tables give $t_{sat} = 101.7°F$, $h_{fg} = 1036.0$ Btu/lb$_m$. Thus

$$\Delta t = t_{sat} - t_s = 101.7 - 95 = 6.7°F,$$

$$t_m = (t_{sat} + t_s)/2 = (101.7 + 95)/2 = 98°F.$$

At t_m Table A.3 gives the following properties of the condensate:

$$\rho_l = 62.02 \text{ lb}_m/\text{ft}^3, \qquad \mu_l = 1.687 \text{ lb}_m/\text{ft-h},$$

$$k_l = 0.3607 \text{ Btu/h-ft-°F}, \qquad c_{pl} = 0.998 \text{ Btu/lb}_m\text{-°F}.$$

For a single horizontal tube Eq. (8.20) gives the heat transfer coefficient to be, using the approximation of Eq. (8.18):

$$\text{Ja} = \frac{c_{pl}\,\Delta t}{h_{fg}} = \frac{0.998 \times 6.7}{1036.0} = 0.0065,$$

$$h'_{fg} = h_{fg}(1 + 0.68\text{Ja}) = 1036.0(1 + 0.68 \times 0.0065)$$

$$= 1040.5 \text{ Btu/lb}_m,$$

$$\text{Nu}_D = 0.728\left[\frac{g\rho_l^2 h'_{fg} D^3}{\mu_l k_l\,\Delta t}\right]^{1/4}$$

$$= 0.728\left[\frac{32.2 \times (3600)^2 \times (62.02)^2 \times 1040.5 \times (1/12)^3}{1.687 \times 0.3607 \times 6.7}\right]^{1/4} = 508,$$

$$h = \text{Nu}_D \frac{k}{D} = 508 \times \frac{0.3607}{1/12} = 2199 \text{ Btu/h-ft}^2\text{-°F}.$$

The steam condensed, per unit length of tube, is then

$$\frac{\dot{m}}{L} = \frac{q/L}{h_{fg}} = \frac{\pi D h\,\Delta t}{h_{fg}}$$

$$= \frac{\pi(1/12)2199 \times 6.7}{1036.0} = 3.72 \text{ lb}_m/\text{h-ft } (1.54 \times 10^{-3} \text{ kg/s-m}).$$

EXAMPLE 8.4

A refrigeration condenser uses a horizontal 3-cm-diameter tube to condense Freon-12 at a saturation temperature of $t_{\text{sat}} = 50°\text{C}$. Cooling water inside the tube maintains the surface temperature at 40°C. Find the condensing heat transfer coefficient.

Solution. For $t_{\text{sat}} = 50°\text{C}$ and $t_s = 40°\text{C}$, $t_m = (t_{\text{sat}} + t_s)/2 = 45°\text{C}$. At t_m Table A.4 gives for Freon-12:

$$\rho_l = 1236.55 \text{ kg/m}^3, \qquad c_{pl} = 1.01175 \text{ kJ/kg-°C}, \quad k_l = 0.068 \text{ W/m}^2\text{-°C}$$
$$\nu_l = 0.191 \times 10^{-6} \text{ m}^2\text{/s}, \quad \mu_l = \rho_l \nu_l = 2.362 \times 10^{-4} \text{ kg/m-s}.$$

For Freon-12 Ref. 8 gives, at $t_{\text{sat}} = 50°\text{C}$,

$$h_{\text{fg}} = 121.43 \text{ kJ/kg}.$$

For a single horizontal tube Eq. (8.20) applies, so that the heat transfer coefficient is, with $\Delta t = t_{\text{sat}} - t_s = 10°\text{C}$,

$$\text{Ja} = \frac{c_{pl}\,\Delta t}{h_{\text{fg}}} = \frac{1.01175 \times 10}{121.43} = 0.0833,$$

$$h'_{\text{fg}} = h_{\text{fg}}(1 + 0.68\text{Ja}) = 121.43(1 + 0.68 \times 0.0833) = 128.31 \text{ kJ/kg},$$

$$\text{Nu}_D = 0.728\left[\frac{g\rho_l^2 h'_{\text{fg}} D^3}{\mu_l k_l\,\Delta t}\right]^{1/4}$$

$$= 0.728\left[\frac{9.8 \times (1236.55)^2 \times 128.31 \times 1000 \times (0.03)^3}{2.362 \times 10^{-4} \times 0.068 \times 10}\right]^{1/4} = 548.9,$$

$$h = \text{Nu}_D \frac{k}{D} = 548.9 \times \frac{0.068}{0.03} = 1244 \text{ W/m}^2\text{-°C (219 Btu/h-ft}^2\text{-°F)}.$$

Condensation Inside Horizontal Tubes

The analyses and results presented so far for condensation ignored the possible existence of a shear force at the liquid–vapor interface. That is, it has been presumed that there is a negligible relative velocity between the vapor and the surface of the liquid film. The process of condensation of a vapor flowing inside a horizontal tube is one of considerable importance in the air-cooled condensers of refrigeration machines. It is also of interest in a number of processes in the chemical and petrochemical industries. In such instances it is quite possible that there is a significant velocity in the vapor core as it flows through the tube with the simultaneous formation of a liquid film at the tube surface. In these cases, the relative velocity of the vapor and the liquid can become so significant that the resultant interfacial shear alters considerably the hydrodynamic and heat transfer processes taking place.

The detailed analyses of such problems in which interfacial shear is significant is beyond the scope of this text, and the interested reader is referred to Ref. 2 for further information.

However, for low vapor velocities in horizontal tubes the results of Chato (Ref. 9), noted next may be applied. The meaning of "low" vapor velocity is expressed in terms of a mean vapor Reynolds number defined by a mean vapor velocity, $U_{m,v}$:

$$\text{Re}_v = \frac{\rho_v U_{m,v} D}{\mu_v}. \tag{8.23}$$

This is better expressed in terms of the mass velocity, $G = \dot{m}/A$, defined in Eq. (5.12), so that if $G_v = \dot{m}_v/A$ represents the mass flow rate of vapor, per unit cross-sectional tube area, the vapor Reynolds number is

$$\text{Re}_v = \frac{DG_v}{\mu_v} = \frac{4\dot{m}_v}{\pi D \mu_v}. \tag{8.24}$$

Then Chato's recommendation for condensation inside a horizontal tube is

$$\text{Nu}_D = 0.555\left[\frac{g\rho_l(\rho_l - \rho_v)h'_{fg}D^3}{\mu_l k_l\, \Delta t}\right]^{1/4}, \tag{8.25}\bigstar$$

$$\left[\begin{array}{l} \text{Re}_v < 3.5 \times 10^4, \text{ at inlet} \\ h'_{fg} = h_{fg}(1 + 0.375\, h_{fg}) \\ \text{liquid properties at } t_m \\ h_{fg} \text{ at } t_{sat} \end{array}\right].$$

For higher values of Re_v Ref. 2 should be consulted.

8.5 EFFECTS OF NONCONDENSABLE GASES ON CONDENSATION

The analyses and correlations presented in Secs. 8.3 and 8.4 presume that the vapor to be condensed was the only component present. Thus the liquid–vapor interfacial temperature was taken to be t_{sat} at the pressure of the bulk vapor. When a noncondensable gas, such as air, is also present, the saturation pressure in the bulk gas–vapor medium is that corresponding to the partial pressure of the vapor in this mixture. In addition, the vapor, in moving toward the condensing surface, must diffuse through the noncondensable component. Thus, a decreasing gradient in the partial pressure of the vapor component must exist as the interface is approached. Consequently, the saturation temperature at the interface will be lower than that in the bulk mixture. The diffusion process and the lowered saturation temperature reduces the condensation rate and, hence, the observed heat transfer coefficient. Rather complex theories exist to predict these effects (Ref. 2) but are beyond the scope of this text; however, the effect of noncondensables may be significant.

Typical calculations show that the condensing heat transfer coefficient for steam may be reduced by as much as 50% when air in amounts as low as 1%, by mass, is present. Consequently, it is customary practice in steam condenser design to provide means of removing as much of the noncondensable gases as possible through venting or the use of ejectors.

8.6 DROPWISE CONDENSATION

As noted at the outset of Sec. 8.2, two basic types of condensation are observed to occur, film and dropwise. All the subsequent discussion was limited to film condensation. It has been observed that when traces of oil, or other similar substances, are present on the condensing surface so that the liquid does not readily wet it, the condensate breaks into droplets—forming what is known as "dropwise condensation." The drops thus formed grow as condensation progresses and either coalesce with neighboring drops or run off the surface after their weight overcomes the restraining forces of surface tension. As the drops run off, a greater fraction of the condensing surface is left free of liquid, thus enhancing the heat transfer process.

Since the entire surface is not covered with a continuous liquid film, heat transfer in the dropwise condensation mode is observed to be five to ten times as great as in the film mode. Thus, if dropwise condensation could be maintained in industrial condensing equipment, significant savings in the size of such devices could be realized. Consequently, a great deal of time and effort has been devoted to the study of this mechanism and in seeking ways to promote its occurrence. The subject is too complex and the results too tentative to be treated adequately here, but the reader may wish to consult Chapter 12 of Ref. 2 for a summary of the subject. The results of studies to date indicate that dropwise condensation is promoted by the use of highly polished surfaces or by the injection of various fatty acids such as oleic or stearic acids. However, the effectiveness of additives lasts for only a few hundred hours, due to the cleansing effects of steam. Consequently, the use of dropwise promoters may require continual injection which would become expensive or incur undesired effects in the rest of the plant, or cycle, using the condenser.

As a result of all the above, most industrial condensing equipment is still designed on the presumption that film condensation will be present.

8.7 HEAT TRANSFER DURING THE BOILING OF A LIQUID

The process of the purposeful conversion of a liquid into a vapor is one of obvious engineering importance. The production of large quantities of steam for electrical power generation through the use of vapor cycles is an example that most readily comes to mind. Many processes involved in the refining of petroleum and the manufacture of chemicals include the vaporization of a liquid.

Careful experiments have shown that if a liquid, such as water, is distilled and completely degasified under vacuum (i.e., if it contains *no* impurities, not even dissolved gases) it will undergo the liquid–vapor phase change without the appearance of bubbles when it is heated in a clean, smooth vessel. Under normally encountered engineering conditions, however, the presence of impurities, dissolved gases, and surface irregularities causes the appearance of vapor bubbles on the heating surface when the rate of heat input is great enough. The creation of these bubbles, their subsequent growth and detachment, etc., profoundly influence the heat transfer process taking place.

Boiling may occur under various conditions. When the heating surface is submerged in an otherwise quiescent pool of liquid, and heat is transferred to the liquid by free convection and bubble agitation, the process is termed *pool boiling*. This term is used in contrast to boiling which may occur simultaneously with fluid motion induced by externally imposed pressure differences and which is referred to as *forced convection boiling*. Most of the discussion to follow concentrates on pool boiling.

As noted in the foregoing, most boiling encountered in normal engineering applications eventually results in the formation, growth, collapse, etc., of vapor bubbles. In order to understand some of the peculiarities of pool boiling it is advantageous to first note some facts concerning the mechanics and thermodynamics of bubbles. Consider the existence of a spherical bubble of radius R in a liquid as depicted in Fig. 8.4. The pressure of the vapor inside the bubble, P_v, must exceed that in the surrounding liquid, P_l, because of the surface tension acting on the liquid–vapor interface. If σ represents the surface tension (force per unit length) then a force balance on an equatorial plane gives

$$\pi R^2(P_v - P_l) = 2\pi R\sigma \qquad (8.26)$$

$$P_v - P_l = \frac{2\sigma}{R}.$$

Several aspects may be noted from Eq. (8.26). To create a bubble of small radius, it would be necessary to develop very large pressures in the vapor. Hence, bubbles are usually found to form at pits existing in surface irregularities where a

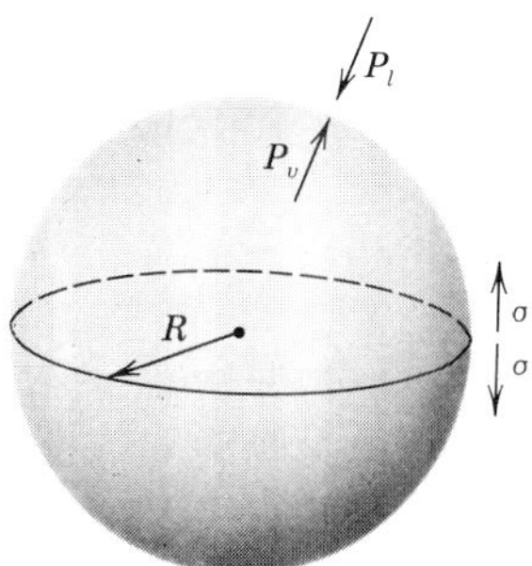

Figure 8.4. Spherical vapor bubble in a liquid.

bubble of finite initial radius may form, or, more likely, at points on the surface where gases dissolved in the surface of the liquid come out of solution. Also, consider the situation when the bubble is in thermal equilibrium—i.e., when the liquid and vapor have the same temperature and the bubble is not growing. The temperature of the vapor in the bubble must be the saturation temperature at the pressure P_v. Since the liquid is at the same temperature it is then at a temperature in excess of the saturation temperature at P_l since $P_l < P_v$. Thus the liquid surrounding the bubble is in a metastable state. That is, the liquid is *superheated* at a temperature in excess of t_{sat} for the liquid, typically taken as the boiling point at the given ambient pressure. Thus, near the heating surface where bubbles are formed, liquid superheat temperatures are found to exist. To cause a bubble to grow, the liquid superheat will exceed that predicted by Eq. (8.26), and liquid superheats of the order of 0.5 to 8°C above the ambient saturation temperature are commonly encountered in the boiling of water. If the bubble detaches from the surface and moves up into cooler liquid, it may recondense and collapse if the superheat is insufficient to maintain it. The foregoing discussion assumed the existence of spherical bubbles. In real cases, the bubbles created on a boiling surface will be nonspherical, depending on local surface irregularities and the contact angle the bubble makes with the surface—a quantity dependent on the way in which the liquid "wets" the surface.

Pool Boiling

With the foregoing comments in mind, it is now possible to describe the various processes that may occur during the pool boiling of a liquid. For a given ambient pressure over a pool of liquid, heated from below by a horizontal surface, the boiling is termed *subcooled pool boiling* if the temperature of the bulk of the liquid is below that of saturation at the given pressure. In this instance, bubbles formed at the surface rise and are eventually recondensed in the liquid. All vapor produced is simply that by evaporation at the free surface. Continued heating will raise the temperature of the bulk of the pool, until one reaches the condition in which the bulk liquid is at, or slightly exceeds, the saturation temperature. This condition is known as *saturated pool boiling*.

Saturated pool boiling has been studied extensively. The reader may wish to consult Refs. 10 and 11 for summaries of these works. In general, the heat flux through the heating surface during saturated pool boiling is a complex function of the temperature excess, $t_s - t_{sat}$, by which the surface temperature exceeds the ambient pressure saturation temperature. The necessity for the existence of a surface temperature greater than t_{sat} has already been discussed. Figure 8.5 shows the typical dependence of the heat flux, a/A, on $\Delta t = t_s - t_{sat}$. The data of Fig. 8.5 are those typically observed for water at atmospheric pressure and were obtained using an electrically heated wire which serves the dual purpose of a heating surface and a resistance thermometer for the measurement of t_s. This curve is similar to those observed for other liquids and other pressures.

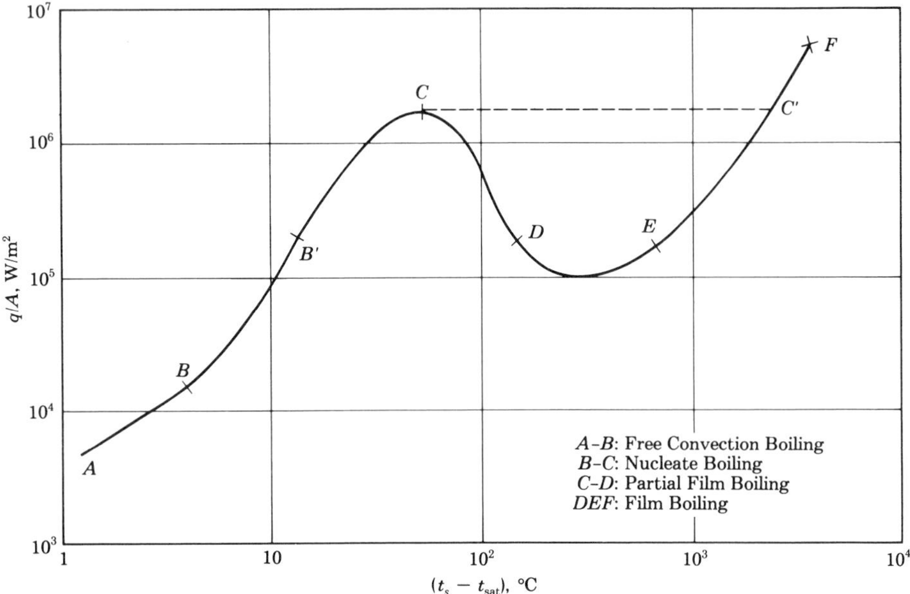

Figure 8.5. Typical boiling curve for water.

Figure 8.5 displays several regimes of pool boiling in which the physical mechanisms taking place are quite different. If one defines the boiling heat transfer coefficient as based on the surface-saturation temperature excess, so that the heat flux is

$$\frac{q}{A} = h(t_s - t_{sat}) = h\,\Delta t, \tag{8.27}$$

then q/A depends not only on Δt directly, but also in accord with how h depends on Δt. In the range of low Δt (approximately 5°C), region A-B in Fig. 8.5, no bubbles are observed even though the liquid is slightly superheated. Heat is transported from the heating surface to the bulk of the liquid mainly by free convection effects. All vapor produced is by evaporation at the liquid surface. In this regime, called *free convection boiling*, h is proportional to $(\Delta t)^{1/4}$ in accord with the results of Chapter 7, so that q/A varies as $(\Delta t)^{5/4}$. Note, however, that the magnitude of the heat flux shown in Fig. 8.5 is much larger than those normally encountered for free convection without phase change.

Continued increase in Δt causes the process to enter the regime of *nucleate boiling* denoted by B-C in Fig. 8.5. In this region (up to Δt of the order of 50°C) bubbles are found to appear on the heating surface. The bubbles first appear at certain favorable sites, or nuclei. As Δt increases, more and more sites are activated. In the lower end of the region B-C, the bubbles leave the surface and may recondense in the bulk of the liquid. As Δt is increased more, the bubbles rise all the way to the liquid surface to release vapor. This bubble activity causes the liquid near the heating surface to become highly agitated and a marked increase in the heat flux occurs, increasing to Δt raised to the third or fourth power. Since the liquid phase is such a better conductor than the vapor, most of the heat transfer in nucleate boiling is from the surface into the liquid and then from the liquid into the vapor of the bubble rather than from the surface directly into the vapor. Eventually, as Δt increases further, so many bubble nucleation sites become activated that they begin to interfere with one another, causing parts of the surface to become covered with vapor and inhibiting liquid motion near the surface. Thus, the increase of q/A is slowed, as suggested by point B' in Fig. 8.5, until a maximum point is reached. This upper limit of the nucleate boiling regime, point C, is referred to as the *peak heat flux* (also called the *critical heat flux* or the *boiling crisis*.)

Additional increase of Δt beyond that at the peak heat flux causes the process to enter the regime of *transition boiling,* or *partial film boiling,* denoted by C-D in Fig. 8.5. At any location on the surface, an unstable condition exists in which the process oscillates between nucleate boiling and *film boiling* in which the surface is blanketed with a film of vapor. Eventually, the regime of *stable film boiling,* D-E-F in Fig. 8.5, is reached in which the surface is completely covered with a blanket of vapor. Heat transfer takes place by conduction through the vapor and then into the liquid. Since the vapor is a poorer conductor than the liquid, the heat flux in the film boiling region, D-E, is measurably less than that at the peak heat flux. If Δt is further increased, above, say, 150°C, q/A again increases as heat exchange by radiation comes into play. Eventually, a point is reached at which the surface melts—point F in Fig. 8.5.

Figure 8.6 presents high-speed photographs obtained by Westwater and Sangangelo (Ref. 12) which illustrate some of the boiling regimes just discussed. These photographs were obtained for the boiling of methanol on a horizontal tube.

The preceding discussion was based on the presumption that the temperature of the heating surface could be maintained at a particular value—that is, that Δt was the independent variable. With Δt thus under control, the boiling curve in Fig. 8.5 could be traversed as described, and it would appear that the desired operating point for a commercial vapor-producing system would be at the peak heat flux of point C. However, in many applications it is the heat flux that is the independent variable (as for an electrically heated wire or a boiler with a given rate of heat production from fuel burning) and the surface temperature assumes a value compatible with q/A. However, the boiling curve is multivalued when viewed with q/A as the independent variable, so that more than one possible surface temperature may correspond to a given heat flux. Also, the partial film boiling region is unstable, so that for a given q/A, the only equilibrium points possible are along A-B-C or D-E-F in Fig. 8.5. If one were to operate too close to the peak heat flux at point C, a slight increase in q/A may make the system choose to operate in the film boiling regime—say at point C'. This action would result in a dramatic increase in the surface temperature, producing important mechanical design problems. It is even possible that the point C' may be so close to the surface melting point, that failure of the system could result.

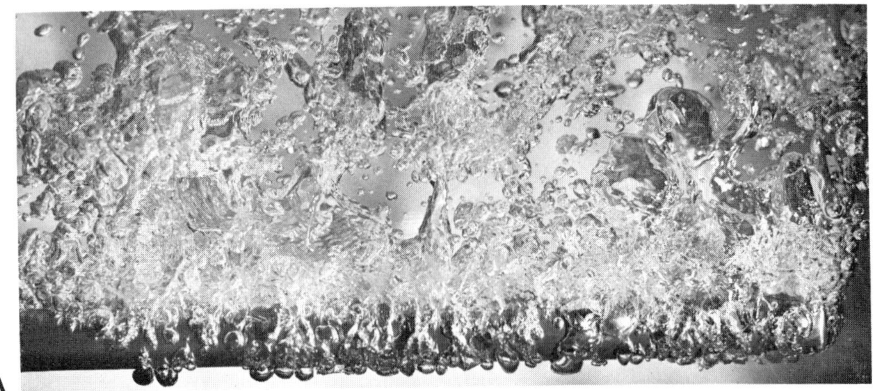
A

B

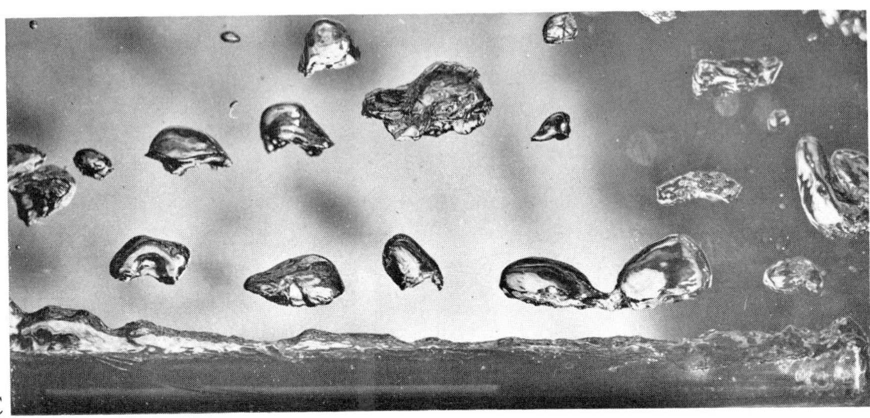
C

Figure 8.6. Photographs of methanol pool boiling from Ref. 12: (a) nucleate boiling; (b) transition boiling; (c) film boiling. Photographs courtesy of Prof. J. W. Westwater, University of Illinois at Champaign-Urbana.

8.8 WORKING CORRELATIONS FOR POOL BOILING

All the discussion in Sec. 8.7 should indicate that boiling heat transfer is a very complex subject. Both the analytical and empirical investigations of boiling are quite involved, and much remains to be done to thoroughly understand it. While this matter is still the object of much current research, this section presents the most reliable information available at this time.

Nucleate Pool Boiling

A wide divergence of opinion exists as to the most reliable way of predicting heat transfer in the nucleate pool boiling regime. While subject to much criticism, the relation most widely used is that due to Rohsenow (Ref. 13) given next. However, one should apply this relation with caution, realizing that large variations from it may be observed in practice. Rohsenow's equation is

$$\frac{q}{A} = \mu_l h_{fg}\left[\frac{g(\rho_l - \rho_v)}{g_c \sigma}\right]^{1/2}\left(\frac{c_{pl}\,\Delta t}{C_{sf} h_{fg} \mathrm{Pr}_l^s}\right)^3, \quad \left[\begin{array}{l} s = 1.0 \text{ for water} \\ s = 1.7 \text{ for other liquids} \\ \text{properties at } t_{sat} \end{array}\right]. \tag{8.28}\bigstar$$

In Eq. (8.28) the subscript l refers to the physical properties of the *saturated liquid* and v to those of the *saturated vapor*. The latent heat of vaporization is given by h_{fg}, σ is the surface tension at the liquid–vapor interface, g is the acceleration of gravity, and g_c is the dimensional constant discussed in Sec. 1.8 and is included in the event an inconsistent system of units is employed. The constant C_{sf} is an empirically determined quantity that is dependent on the composition and roughness of the heating surface and the way in which the liquid wets the surface. Some values of C_{sf} are given in Table 8.2 for water as compiled by Holman (Ref. 14) from a number of sources. Reference 14 may be consulted for values of C_{sf} for other surface-liquid combinations.

Table 8.2 Values of C_{sf} for Water and Various Surfaces

Surface	C_{sf}
Brass	0.0060
Copper—polished	0.0130
Copper—scored	0.0068
Platinum	0.0130
Stainless steel	
Chemically etched	0.0133
Ground and polished	0.0080
Mechanically polished	0.0132
Teflon pitted	0.0058

For water, the surface tension in Eq. (8.28) may be calculated from (Ref. 15):

$$\sigma,\ \mathrm{N/m} = 0.2358\left(1 - \frac{T}{647.15}\right)^{1.256} \times \left[1 - 0.625\left(1 - \frac{T}{647.15}\right)\right], \qquad (8.29)\bigstar$$

in which T is the absolute temperature in degrees Kelvin. For other substances the reader may seek values of σ from various handbooks such as Refs. 16 and 17.

Peak Heat Flux

The correlation of Eq. (8.28) provides an estimate of the heat flux to be expected in the nucleate boiling regime. However, it does not provide information as to the *peak heat flux,* point C in Fig. 8.5. As the discussion of Sec. 8.7 indicated, the peak heat flux is an important parameter in the design of boiling apparatus as well as in connection with the burnout phenomenon.

The peak heat flux depends on many of the same quantities involved in Eq. (8.28), but it is also a strong function of the geometric shape of the heating surface. For very large horizontal flat surfaces Zuber (Ref. 18) predicts that the peak heat flux is given by

$$\left(\frac{q}{A}\right)_{\max} = \frac{\pi}{24} h_{fg}\rho_v^{1/2}\left[1 + \frac{\rho_v}{\rho_l}\right]^{1/2} [\sigma g g_c(\rho_l - \rho_v)]^{1/4}. \qquad (8.30)$$

For typical cases in which the vapor density is much smaller than the liquid density, the equation above is well approximated by

$$\left(\frac{q}{A}\right)_{\max} = 0.131 h_{fg}\rho_v^{1/2}[\sigma g g_c(\rho_l - \rho_v)]^{1/4}. \qquad (8.31)$$

In an extensive and comprehensive series of works, Lienhard and coworkers (Ref. 19) ascertained that Eq. (8.31) underpredicts the peak heat flux by about 14% for large flat surfaces. They also determined additional modifications for other surface geometries. Lienhard expresses the peak heat flux in the same form as Eq. (8.31) with the lead constant, $\pi/24 = 0.131$, approximately modified to account for the heater geometry. The geometric effects of the heating surface is expressed in terms of a dimensionless length, L', defined as

$$L' = \frac{L_c}{\left[\dfrac{g_c\sigma}{g(\rho_l - \rho_v)}\right]^{1/2}}, \qquad (8.32)$$

in which L_c is a characteristic dimension of the heater. Table 8.3 presents a partial tabulation of Lienhard's factors which are to be applied to Eq. (8.31). In summary, then, Lienhard's correlation for the peak heat flux is

Table 8.3 Correction Factor $F(L')$ for Use in Lienhard's Equation

Heater Geometry	$F(L')$	Range of L'	Characteristic Length, L_c
Large horizontal flat surface	1.14	$L' \geq 2.7$	Heater width or diameter
Horizontal cylinder	$0.89 + 2.27 \exp(-3.44\sqrt{L'})$	$L' \geq 0.15$	Cylinder radius
Large sphere	0.84	$L' \geq 4.26$	Sphere radius
Small sphere	$1.734/\sqrt{L'}$	$4.26 \geq L' \geq 0.15$	Sphere radius

$$\left(\frac{q}{A}\right)_{\max} = F(L') \times 0.131 h_{fg}\rho_v^{1/2}[\sigma g g_c(\rho_l - \rho_v)]^{1/4}, \tag{8.33}★$$

$$\begin{bmatrix} \text{properties at } t_{sat} \\ L' \text{ from Eq. (8.32)} \\ F(L') \text{ and } L_c \text{ from Table 8.3} \end{bmatrix}.$$

The first entry in Table 8.3 accounts for Lienhard's observation that Zuber's operation, Eq. (8.31), underpredicts by 14%. Reference 19 may be consulted for values of $F(L')$ applicable for additional geometries.

EXAMPLE 8.5

Nucleate boiling of water at atmospheric pressure occurs on the bottom of a horizontal polished copper pan, 15 cm in diameter. The surface is maintained at 118°C by the external heating source on which the pan sits. Find (a) the boiling heat flux; (b) the peak heat flux.

Solution. At atmospheric ambient pressure the Steam Tables give

$$t_{sat} = 100°\text{C}, \qquad h_{fg} = 2257.0 \text{ kJ/kg},$$
$$\rho_l = 958.31 \text{ kg/m}^3, \qquad \rho_v = 0.5978 \text{ kg/m}^3.$$

For the liquid phase Table A.3 gives at $t_{sat} = 100°\text{C}$,

$$c_{pl} = 4.215 \text{ kJ/kg-°C}, \qquad \mu_l = 0.2822 \times 10^{-3} \text{ kg/m-s}, \qquad \text{Pr}_l = 1.76.$$

The surface tension at the liquid–vapor interface is given by Eq. (8.29) for water at $T = 100 + 273.15 = 373.15°\text{K}$:

$$\sigma = 0.2358\left(1 - \frac{T}{647.15}\right)^{1.256}\left[1 - 0.625\left(1 - \frac{T}{647.15}\right)\right]$$

$$= 0.2358\left(1 - \frac{373.15}{647.15}\right)^{1.256}\left[1 - 0.625\left(1 - \frac{373.15}{647.15}\right)\right]$$

$$= 0.05892 \text{ N/m}.$$

(a) For nucleate pool boiling the heat flux for a specified surface temperature is given by Eq. (8.28):

$$\frac{q}{A} = \mu_l h_{fg}\left[\frac{g(\rho_l - \rho_v)}{g_c\sigma}\right]^{1/2}\left(\frac{c_{pl}\,\Delta t}{C_{s,f}h_{fg}\text{Pr}_l^s}\right)^3.$$

In this instance $s = 1.0$ since the fluid is water. Table 8.2 indicates that for water on polished copper, $C_{s,f} = 0.0130$. The driving temperature potential difference is $\Delta t = t_s - t_{sat} = 118 - 100 = 18°\text{C}$. Thus the heat flux is, with $g_c = 1$ since the consistent SI system of units is involved,

$$\frac{q}{A} = 0.2822 \times 10^{-3} \times 2257.0\left[\frac{9.8 \times (958.31 - 0.5978)}{1 \times 0.05892}\right]^{1/2}$$

$$\times \left[\frac{4.215 \times 18}{0.013 \times 2257.0 \times 1.76}\right]^3$$

$$= 806.2 \text{ kW/m}^2 \ (2.556 \times 10^5 \text{ Btu/h-ft}^2).$$

(b) The peak heat flux is found from Eq. (8.33). To apply this relation one must first determine the dimensionless parameter L' given in Eq. (8.32). For a horizontal heating surface the characteristic dimension required is the diameter $L_c = 15$ cm $= 0.15$ m. Thus, with the physical properties found in part (a), one has

$$L' = \frac{L_c}{\left[\dfrac{g_c\sigma}{g(\rho_l - \rho_v)}\right]^{1/2}}$$

$$= \frac{0.15}{\left[\dfrac{0.05892}{9.8(958.31 - 0.5978)}\right]^{1/2}} = 60.0.$$

This value of L' is greater than 2.7, as required by Table 8.3 for a horizontal plane heater. Thus $F(L') = 1.14$, and the peak heat flux is, from Eq. (8.33),

$$\left(\frac{q}{A}\right)_{max} = F(L') \times 0.131 h_{fg}\rho_v^{1/2}[\sigma g g_c(\rho_l - \rho_v)]^{1/4}$$

$$= 1.14 \times 0.131 \times 2257.0(0.5978)^{1/2}$$

$$\times [0.05892 \times 9.8(958.31 - 0.5978)]^{1/4}$$

$$= 1263.7 \text{ kW/m}^2 \ (4.006 \times 10^5 \text{ Btu/h-ft}^2).$$

Note that this peak heat flux is well in excess of the heat flux determined in part (a).

Film Pool Boiling

When the state of full film boiling is established, the heating surface is completely covered by a blanket of vapor. The temperature of the surface becomes rather large since all heat transfer must be done through the poorly conducting vapor layer.

For film boiling on a submerged horizontal cylinder, the convective picture is that of a layer of vapor flowing upward past a cylinder and bounded by a large liquid domain. Bromley (Ref. 20) likened this situation to that of condensation on a horizontal cylinder—a liquid layer flowing downward past a cylinder and bounded by a large vapor domain. Following the same analysis as Nusselt did for the condensation problem, Bromley obtained the following for film boiling on a horizontal cylinder:

$$\text{Nu}_D = \frac{h_b D}{k_v} = 0.62\left[\frac{g\rho_v(\rho_l - \rho_v)(h_{fg} + 0.4c_{pv}\,\Delta t)D^3}{\mu_v k_v\,\Delta t}\right]^{1/4}, \qquad (8.34)\bigstar$$

$$\left[\begin{array}{l} h_{fg} \text{ and } \rho_l \text{ at } t_{sat} \\ \text{other properties at } (t_s + t_{sat})/2 \end{array}\right].$$

In Eq. (8.34), the symbol h_b is used to denote the *boiling* heat transfer coefficient in order to distinguish it from the radiation coefficient to be defined next.

When the surface temperature is high enough to produce film boiling, it is usually hot enough to cause an appreciable heat transfer by radiation. While radiative heat transfer is not discussed until the next chapter, it is possible to account for its effect here by quoting some relations to be developed in Chapter 9. The radiative heat transfer in this case may be expressed as

$$\left(\frac{q}{A}\right)_r = h_r(t_s - t_{sat}),$$

where h_r is a radiation coefficient defined as

$$h_r = \frac{\sigma\epsilon(T_s^4 - T_{sat}^4)}{T_s - T_{sat}}. \qquad (8.35)$$

In Eq. (8.35) the capital T's are absolute temperature, ϵ is the surface emissivity, and σ is the Stefan-Boltzmann radiation constant (not surface tension!):

$$\sigma = 5.67 \times 10^{-8}\ \text{W/m}^2\text{-°K}^4. \qquad (8.36)$$

Bromley suggested that the combined heat transfer by boiling and radiation in the case of film boiling on a horizontal surface could be calculated by defining a total heat transfer coefficient given by the empirical relation

$$h = h_r + h_b\left(\frac{h_b}{h}\right)^{1/3}. \qquad (8.37)$$

In Eq. (8.37), h_r and h_b are found from Eqs. (8.34) and (8.35). Then the total heat flux from the cylinder is

$$\frac{q}{A} = h(t_s - t_{sat}). \qquad (8.38)$$

Quite obviously, the use of Eq. (8.37) requires an iterative calculation.

EXAMPLE 8.6

Film boiling occurs on the surface of a horizontal 1-cm-diameter cylinder with a surface temperature of $t_s = 300°\text{C}$ when it is submerged in a water bath at 1 atm, $t_{sat} = 100°\text{C}$. If the surface emissivity is $\epsilon = 0.8$, estimate the surface heat flux.

Solution. At the liquid saturation temperature $t_{sat} = 100°\text{C}$, the Steam Tables give

$$\rho_l = 598.31 \text{ kg/m}^3, \qquad h_{fg} = 2257.0 \text{ kJ/kg}.$$

The mean film temperature is $t_m = (t_{sat} + t_s)/2 = (100 + 300/2 = 200°\text{C}$, and the vapor properties are to be evaluated at this temperature and the ambient pressure of 1 atm (101.33 kN/m). The entries in Table A.5 for steam at a pressure of 100 kN/m and a temperature of 200°C are sufficiently accurate for these purposes, so

$$\rho_v = 0.4603 \text{ kg/m}^3, \qquad c_{pv} = 1.979 \text{ kJ/kg-°C}.$$

$$\mu_v = 16.18 \times 10^{-6} \text{ kg/m-s}, \qquad k_v' = 33.37 \times 10^{-3} \text{ W/m-°C}.$$

Equation (8.34) may then be applied to find the film boiling heat transfer coefficient, h_b:

$$\frac{h_b D}{k_v} = 0.62\left[\frac{g\rho_v(\rho_l - \rho_v)(h_{fg} + 0.4c_{pv}\,\Delta t)D^3}{\mu_v k_v\,\Delta t}\right]^{1/4},$$

which gives, since $\Delta t = (t_s - t_{sat}) = (300 - 100) = 200°\text{C}$,

$$\frac{h_b D}{k_v} = 0.62\left[\frac{\begin{array}{c}9.8 \times 0.4603(598.31 - 0.4603)\\ \times [2257.0 + 0.4 \times 1.979 \times 200] \times 10^3 \times (0.01)^3\end{array}}{16.18 \times 10^{-6} \times 33.37 \times 10^{-3} \times 200}\right]^{1/4}$$

$$= 54.64,$$

$$h_b = 54.64\frac{k_v}{D} = 54.64 \times \frac{33.37 \times 10^{-3}}{0.01} = 182.3 \text{ W/m}^2\text{-°C}.$$

The heat exchanged from the cylinder radiation is determined by the radiation coefficient given in Eq. (8.35):

$$h_r = \frac{\sigma\epsilon(T_s^4 - T_{sat}^4)}{T_s - T_{sat}}.$$

The surface and saturation temperatures, T_s and T_{sat}, are the *absolute* values of the specified quantities: $T_s = 300 + 273.15 = 573.15°\text{K}$, $T_{sat} = 100 + 273.15 = 373.15°\text{K}$. Thus, since the emissivity of the cylinder surface is given as $\epsilon = 0.8$ and σ is given by Eq. (8.36), the relation above yields the following for the radiation coefficient:

$$h_r = \frac{5.67 \times 10^{-8} \times 0.8 \times (573.15^4 - 373.15^4)}{573.15 - 373.15}$$

$$= 20.1 \text{ W/m}^2\text{-°C}.$$

According to Bromley's recommendation, the combined coefficient for boiling *and* radiation is given by Eq. (8.37):

$$h = h_r + h_b\left(\frac{h_b}{h}\right)^{1/3}$$

$$= 20.1 + 182.3\left(\frac{182.3}{h}\right)^{1/3}.$$

Iterative solution of the above yields the combined coefficient

$$h = 198 \text{ W/m}^2\text{-°C},$$

so that the total surface heat flux is, from Eq. (8.38),

$$\begin{aligned} \frac{q}{A} &= h(t_s - t_{sat}) \\ &= 198(300 - 100) \\ &= 39.6 \text{ kW/m}^2 \; (12.6 \times 10^3 \text{ Btu/h-ft}^2). \end{aligned}$$

8.9 FORCED CONVECTION BOILING

Forced convection boiling is most often encountered in the case of forced flow of a liquid through a tube with boiling occurring at the inner surface. Bubble growth and the influence of bubble dynamics on the heat transfer mechanism are influenced strongly by the forced velocity of the flow and thus differ significantly from the situation observed in pool boiling. Consequently, the process is sufficiently complicated to preclude the presentation here of any generalized theories governing such mechanisms.

Figure 8.7 depicts in a pictorial way the various regimes that are observed to develop as a subcooled liquid is forced upward through a vertical tube, the surface of which is maintained at temperatures in excess of t_{sat}. The vertical dimension is greatly compressed. Also shown, pictorially, is the manner in which the heat transfer coefficient typically varies along the tube. Near the tube inlet the heat transfer process is that for forced convection of a single-phase liquid as discussed in Chapter 6. However, boiling is soon initiated, and bubbles begin to appear on the surface as the fluid moves along the tube. The bubbles grow, break away from the surface and are taken into the mainstream of the liquid. This is known as the *bubbly-flow* regime and is associated with a sharply increased heat transfer coefficient. As the flow continues along the tube, the fraction of vapor becomes so large that slugs of vapor are formed, separated by slugs of liquid. This *slug-flow* regime is followed by the *annular-flow* regime, in which so much vapor exists that it flows in the central core of the pipe with the liquid in an annular film at the surface. The heat transfer rate continues to rise through the bubbly-, slug-, and annular-flow regimes. However, dry spots begin to occur on the surface, and the heat transfer coefficient drops as the flow passes into the *mist-flow* regime. In the latter regime, the flow is mainly vapor with the liquid phase suspended in it in a mist of liquid droplets. Eventually, all the mist evaporates, and the vapor begins to superheat.

A complete summary of the state of knowledge of forced convection boiling and two-phase flow is given in Ref. 2. Needless to say, the foregoing description of the forced boiling process is sufficiently complicated to preclude the presentation

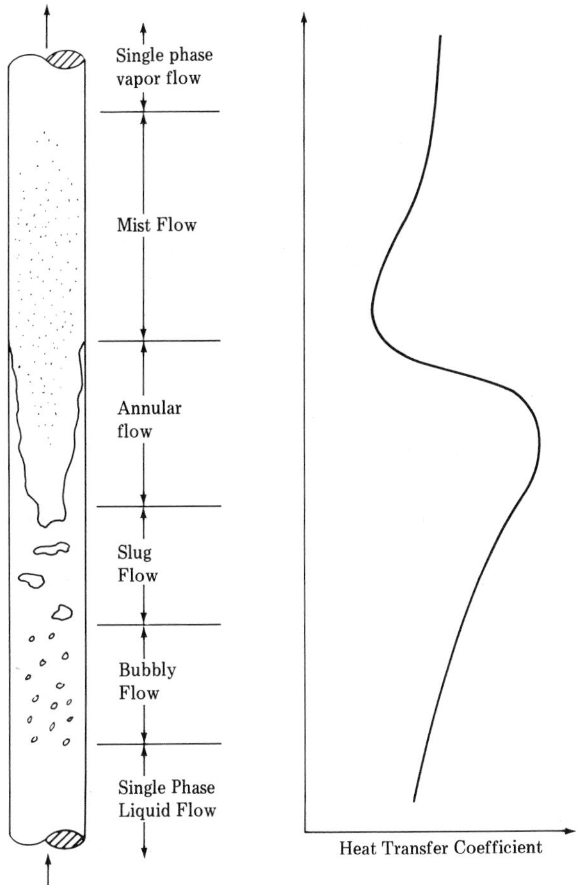

Figure 8.7. Flow regimes for forced convection boiling in a vertical tube.

of the many correlations which have been developed for the various flow regimes.

As a general simple rule, Rohsenow and Griffith (Ref. 21) suggest that forced convection boiling to be treated by superimposing the two effects of forced convection and nucleate boiling, at least up to the slug-flow regime. That is, the total heat transfer is to be found from

$$\left(\frac{q}{A}\right)_{\text{total}} = \left(\frac{q}{A}\right)_{\text{boiling}} + \left(\frac{q}{A}\right)_{\text{conv.}} \tag{8.39}$$

In Eq. (8.39), the boiling contribution from Eq. (8.28) and the convective contribution is found by use of the Dittus-Boelter equation in Eq. (6.31) with the constant 0.023 changed to 0.019.

REFERENCES

1. Nusselt, W., "Die Oberflächenkondensation des Wasserdampfes," *Z.V.D.I.*, Vol. 60, 1916, p. 569.
2. Rohsenow, W. M., and J. P. Hartnett, eds., *Handbook of Heat Transfer*, New York, McGraw-Hill, 1973.
3. Sparrow, E. M., and J. L. Gregg, "A Boundary-Layer Treatment of Laminar Film Condensation, *J. Heat Transfer, Trans. ASME*, Vol. 81, 1959, p. 13.
4. McAdams, W. H., *Heat Transmission*, 3rd ed., New York, McGraw-Hill, 1954.
5. Rohsenow, W. M., "Heat Transfer and Temperature Distribution in Laminar Film Condensation," *Trans ASME*, Vol. 78, 1956, p. 1645.
6. Clifton, J. V., and A. J. Chapman, "Condensation of a Pure Vapor on a Finite Size Horizontal Plate," *ASME Paper 67-WA/HT-18*, New York, 1967.
7. Devore, A., "How to Design Multitube Condensers," *Pet. Refiner*, Vol. 38, No. 6, June 1959, p. 205.
8. Van Wylen G. J., and R. E. Sonntag, *Fundamentals of Classical Thermodynamics*, 2nd ed., New York, Wiley, 1978.
9. Chato, J. C., "Laminar Condensation Inside Horizontal and Inclined Tubes," *J. Am. Soc. Heating, Refrig., Air-Cond. Eng.*, Vol. 4, 1962, p. 52.
10. Leppert, G., and C. C. Pitts, "Boiling," in *Advances in Heat Transfer*, Vol. 1, New York, Academic Press, 1964.
11. Rohsenow, W. M., ed., *Developments in Heat Transfer*, Cambridge, Mass., MIT Press, 1964.
12. Westwater, J. W., and J. G. Santangelo, "Photographic Study of Boiling," *Ind. Eng. Chem.*, Vol. 47, Aug. 1955, p. 1605.
13. Rohsenow, W. M., "A Method of Correlating Heat Transfer Data for Surface Boiling of Liquids," *Trans. ASME*, Vol. 74, 1952, p. 969.
14. Holman, J. P., *Heat Transfer*, 5th ed., New York, McGraw-Hill, 1981.
15. *ASME Steam Tables*, 3rd ed., New York, American Society of Mechanical Engineers, 1977.
16. Bolz, R. E., and G. L. Tuve, eds., *CRC Handbook of Tables for Applied Engineering Sciences*, 2nd ed., Boca Raton, Fla., CRC Press, 1979.
17. *ASHRAE Handbook of Fundamentals*, Atlanta, American Society of Heating, Refrigeration, and Air-Conditioning Engineers, 1981.
18. Zuber, N., "On the Stability of Boiling Heat Transfer," *Trans. ASME*, Vol. 80, 1958, p. 711.
19. Lienhard, J. H., *A Heat Transfer Textbook*, Englewood Cliffs, N.J., Prentice-Hall, 1981.
20. Bromley, L. A., "Heat Transfer in Stable Film Boiling," *Chem. Engr. Prog.*, Vol. 46, 1950, p. 221.
21. Rohsenow, W. M., and P. Griffith, "Correlation of Maximum Heat Flux Data for Boiling of Saturated Liquids," *AIChE-ASME Heat Transfer Symp.*, Louisville, Ky., 1955.

PROBLEMS

8.1 A vertical plate 20 cm wide and 1 m high is maintained at 65°C and is exposed to saturated steam at 1 atm pressure. Find the total heat transferred and the amount of steam condensed, per hour, from both sides of the plate.

8.2 A vertical $\frac{3}{4}$-in. tube is 15 ft long and has its surface maintained at 85°F. Steam at 90°F is condensing on its surface. How much steam is condensed in 1 h?

8.3 For the data of Prob. 8.1, find the thickness of the condensate film and the maximum film velocity at the bottom edge of the plate and at a location halfway down the plate.

8.4 A vertical plate (1 m high, 0.5 m wide) is maintained at 80°C and exposed to saturated atmospheric pressure steam. Estimate the local condensing heat transfer coefficient at the middle of the plate and at the bottom edge of the plate.

8.5 A vertical plate is maintained at 54°C while condensing saturated steam at 1 atm pressure. Determine the average condensing heat transfer coefficient if the plate is **(a)** 1 m high, and **(b)** 2.5 m high.

8.6 Saturated ammonia at 658 kN/m^2 pressure is condensing on a vertical surface 0.6 m high maintained at 8°C. Find the average condensing heat transfer coefficient.

8.7 Saturated Freon-12 at 25°C condenses on the outer surface of a vertical 10-cm-diameter pipe of 1 m length. The pipe has a uniform surface temperature of 15°C. Determine the total mass rate of condensation.

8.8 Saturated steam at 80°C condenses on the outer surface of a 10-cm-diameter vertical tube 1.2 m long. Cooling water flowing through the tube maintains its surface temperature at 40°C. Calculate **(a)** the average heat transfer coefficient over the length of the tube, **(b)** the rate of condensation of steam, and **(c)** the thickness of the tube condensate film at the bottom of the tube.

8.9 A vertical plate 50 cm high and 20 cm wide is to be used to condense saturated steam at 1 atm. At what surface temperature should the plate be maintained to condense the steam at the rate of 25 kg/h from one side?

8.10 Saturated Freon-12 at 25°F condenses on the outer surface of a tube (0.5 in. in diameter, 5 ft long) maintained at 5°F. Find the rate of condensation if the tube is **(a)** vertical, or **(b)** horizontal.

8.11 Saturated steam at 70 kN/m^2 condenses on the outer surface of a vertical tube of OD = 2.5 cm, maintained at 30°C. Determine the total rate of condensation if the tube length is **(a)** 2 m, and **(b)** 4 m.

8.12 Saturated steam at 1 atm condenses on a flat surface, maintained at 125°F, that is 6 in. wide and 3 ft long. Find the rate of condensation if the 3-ft-long side is inclined with the horizontal at an angle of **(a)** 90°, **(b)** 60°, **(c)** 45°, and **(d)** 0°.

8.13 A circular tube 1.6 cm in diameter and 1.8 m long has a surface temperature of 45°C. Saturated steam at 55°C is condensing on its surface. Find the mass of steam condensed, per hour, if the tube is **(a)** horizontal, and **(b)** vertical.

8.14 Saturated steam condenses on the outer surface of a horizontal 1-in.-diameter tube with a surface temperature of 60°F. Find the condensing heat transfer coefficient if the steam pressure is **(a)** 14.7 psia, **(b)** 50 psia, and **(c)** 100 psia.

8.15 Saturated Freon-12 at 40°C condenses on a horizontal 2.5-cm tube with a surface temperature of 32°C. Find the rate at which the Freon is being condensed, per meter of tube length.

8.16 Saturated ammonia at 90°F condenses on the outer surface of a horizontal 1-in.-OD tube maintained at 80°F. Find the condensing heat transfer coefficient.

8.17 Saturated steam at a pressure of 15 kN/m^2 condenses on the outer surface of a horizontal $\frac{5}{8}$-in. tube with a surface temperature of 45°C. Estimate the rate at which steam condenses, per meter of tube length.

8.18 A horizontal tube 1 m long with a surface temperature of 70°C is used to condense saturated steam at 1 atm. What tube diameter is necessary to condense steam at the rate of 125 kg/h?

8.19 Freon-12 condenses at a pressure of 1 atm on the outside of a horizontal tube 10 cm in diameter and 1 m long. What surface temperature is necessary to condense the Freon at the rate of 180 kg/h?

8.20 Saturated methyl chloride, used as a refrigerant, condenses at a pressure of 6 atm (t_{sat} = 27°C, h_{fg} = 375 kJ/kg) on the outer surface of a 2.5-cm-OD horizontal tube with a surface temperature of 15°C. Determine the condensing heat transfer coefficient.

8.21 A steam condenser consists of a 25 × 25 square array of 1.25-cm-OD horizontal tubes, 3 m long. Saturated steam at 50°C condenses on the surface of the tubes which are maintained at 30°C by cooling water flowing inside them. Determine **(a)** the average heat transfer coefficient for the array, and **(b)** the rate at which condensate is produced.

8.22 Saturated steam at atmospheric pressure condenses on a 10 × 10 square array of horizontal tubes, 0.8 cm in diameter. If the tube surfaces are at 27°C, find the mass rate of condensation, per unit length of the tubes.

8.23 A horizontal pipe carrying cold water has a surface temperature of 3°C as it passes through a room where the air is at 35°C with a relative humidity of 75%. The pipe has a diameter of 6 cm and is 10 m long. Assuming that the moisture in the air condenses as a saturated vapor at its partial pressure in the air, estimate the mass rate of condensation.

8.24 Steam at 1 psia condenses on a horizontal bank of $\frac{5}{8}$-in. tubes. The bank is eight tubes high and the tubes are not staggered. The tube surfaces are at 96°F. Find the average heat transfer coefficient for the bank of tubes. What percentage of the steam condensed is condensed by the top row of tubes?

8.25 Calculate the heat flux and the heat transfer coefficient for nucleate pool boiling of water at atmospheric pressure on a mechanically polished stainless steel surface when the surface temperature is **(a)** 110°C, and **(b)** 120°C.

8.26 Repeat Prob. 8.25 if the surface is brass.

8.27 The bottom of a polished copper pan is maintained at 117°C and is used to boil water at 1 atm pressure. If the pan is 0.2 m in diameter, what is the power required?

8.28 It is desired to boil 2 kg/h of water at atmospheric pressure in a polished copper pan 20 cm in diameter. What must the surface temperature be?

8.29 Calculate the heat flux and the heat transfer coefficient for water boiling at 14.7 psia pressure if the surface is ground and polished stainless steel maintained at **(a)** 220°F, or **(b)** 230°F.

8.30 A platinum wire is submerged in water at a pressure of 500 kN/m^2. If the wire surface is 10°C hotter than the water saturation temperature, find the heat flux from the wire.

8.31 Compare the heat flux in nucleate pool boiling of water on mechanically polished stainless steel with a temperature excess of $\Delta t = 10$°C when the pressure is **(a)** 1 atm, **(b)** 2 atm, and **(c)** 5 atm.

8.32 Saturated water at 1 atm is boiled on a heating surface 10.4°C hotter than the water. If the heat flux is measured to be 7×10^5 W/m^2, what is the surface coefficient $C_{s,f}$?

8.33 Water at 1 atm is boiled with a polished copper heating surface. If the heat flux is 1.27×10^5 Btu/h-ft^2, determine the surface temperature.

8.34 What is the peak heat flux for water boiling on a large horizontal flat surface at **(a)** 1 atm, and **(b)** 10 atm. What is the minimum dimension of the heater in each instance?

8.35 Water at 1 atm is to be boiled using an electrically heated polished copper surface maintained at 115°C. Determine (1) the heat flux, (2) the rate of evaporation, and (3) the peak heat flux if the heating element is **(a)** a sphere of 10 cm diameter, or **(b)** a horizontal cylinder 5 cm diameter and 20 cm long.

8.36 Polished copper tubes 2.5 cm in diameter and 0.75 m long are to be used to boil water at 1 atm on the outside of the tubes. If the tubes are to be operated at 75% of the critical heat flux, how many tubes are needed to evaporate 750 kg/h? What is the surface temperature of the tubes?

8.37 Water at 1 atm boils in the film boiling regime at a temperature difference of Δt = 775°F on an electrically heated horizontal wire (D = 0.075 in.). If the surface emissivity is 0.9, find the power required, per foot of length, to heat the wire.

8.38 An electrical heating element consists of a horizontal rod 0.6 cm in diameter. It is maintained at a surface temperature of 260°C in a pool of atmospheric pressure water. Assuming that the element has a surface emissivity of 0.9 and that stable film boiling exists, estimate the power required to operate the heater, per meter of length.

8.39 A steel bar 2.5 cm in diameter has a surface emissivity of 1.0. It is heated to a temperature of 450°C and then suddenly thrust into a water bath at 1 atm pressure. Estimate the initial rate of heat transfer from the bar, per unit length.

8.40 Water at atmospheric pressure flows with a bulk temperature of 95°C and an average velocity of 1.5 m/s through a brass tube 1.5 cm in diameter. If the tube surface is at 110°C, find the combined rate of heat transfer due to forced convection and boiling, per unit of length

CHAPTER 9

Heat Transfer by Radiation

9.1 INTRODUCTORY REMARKS

The fundamental physical phenomenon which forms the basis of all heat transfer studies is the observed fact that the temperature of a body, or a portion of a body, which is hotter than its surroundings tends to decrease with time. This temperature decrease indicates a flow of energy from the body. The entire discussion of the foregoing eight chapters has been limited to cases in which some physical medium was necessary for the transport of the energy from the high temperature source to the low temperature sink, leading to the mechanisms of conduction and convection. Generally speaking, the rate of flow of the thermal energy in these instances was proportional to the difference in temperature between the source and the sink.

If, however, a heated body is physically isolated from its cooler surroundings (i.e., by a vacuum), its temperature is still observed to decrease in time, again showing a loss of energy. In this case an entirely different energy transfer mechanism is taking place, and it is called *thermal radiation*.

A body need not be heated to exhibit the loss of energy by radiation. The "thermal" radiation just referred to is one aspect of a more general phenomenon which might be termed *radiant energy*. The emission of other forms of radiant energy may be caused when a body is excited by such means as an oscillating electrical current, electronic or neutronic bombardment, chemical reaction, etc. Also, when radiant energy strikes a body and is absorbed, it may manifest itself in the form of thermal internal energy, a chemical reaction, an electromotive force, etc., depending on the nature of the incident radiation and the substance of which the body is composed. In either case, emission or absorption, this book will be concerned only with thermal radiation—i.e., radiation produced by or which produces thermal excitation of a body.

Several theories have been proposed to explain the transport of energy by radiation. One theory holds that the body emits discrete *packets,* or *quanta,* of energy and has been successful in explaining the experimental facts observed in such cases as the photoelectric emission of electrons, thermal radiation emission, etc. Another theory assets that radiation may be represented as an electromagnetic wave motion and has been useful in explaining such phenomena as interference of light, polarization of light, etc. At the present time a dual theory is generally accepted, giving radiant energy the characteristics of a wave motion as well as discontinuous emission.

Whichever theory is used, it is convenient to classify all electromagnetic radiant energy emissions in terms of the wavelength when considered as wave motions propagating at the velocity of light—3×10^8 m/s. In this manner, while thermal radiation is found to consist of electromagnetic radiation covering the entire spectrum of wavelengths, most of it is concentrated in the band of wavelengths between 1×10^{-1} and 1×10^2 μm.† Other well-known electromagnetic waves are listed below with their approximate wavelength bands indicated. The subdivision of the electromagnetic spectrum into such bands is somewhat arbitrary.

Cosmic rays:	up to 4×10^{-7} μm
Gamma rays:	4×10^{-7} to 1.4×10^{-4} μm
X-rays:	1×10^{-5} to 2×10^{-2} μm
Ultraviolet rays:	5×10^{-3} to 3.9×10^{-1} μm
Visible light:	3.9×10^{-1} to 7.8×10^{-1} μm
Solar radiation:	1×10^{-1} to 3.0 μm
Infrared radiation:	7.8×10^{-1} to 1×10^3 μm
Thermal radiation:	1×10^{-1} to 1×10^2 μm
Hertzian waves:	1×10^2 to 5×10^{10} μm
Radio waves:	1×10^7 to 5×10^{10} μm

Before proceeding further, it will be necessary to define a number of terms and properties which are used to characterize radiant energy. The following discussion applies mainly to thermal radiation, although some of the concepts are applicable to other forms of radiant emission.

9.2 BASIC DEFINITIONS

General definitions will be made first. Further discussion will show a need for refinement of these definitions to account for a multitude of special considerations, such as the dependence of a quantity on direction or on wavelength.

As noted in the foregoing section, thermal radiation is defined as that electromagnetic radiation between wavelengths of about 1×10^{-1} and 1×10^2 μm. If

†Most electromagnetic radiation wavelengths are expressed in terms of the micrometer, μm, or 1×10^{-6} m. This unit is also called the *micron*.

the thermal radiation emitted by a surface were to be decomposed into its spectrum over the wavelength band, it would be found that the radiation is not equally distributed over all wavelengths. Likewise, radiation incident on a surface, reflected by a surface, absorbed by a surface, etc., may be wavelength dependent. This wavelength dependency will be generally different from case to case, surface to surface, etc. The wavelength dependency of any radiative quantity or surface property will be referred to as a *spectral* dependency. The adjective *monochromatic* will be used to qualify a radiative quantity as one applicable at a single wavelength (e.g., monochromatic emission) while the adjective *total* will indicate that the phenomenon concerned has been evaluated over the entire thermal radiation spectrum (e.g., total emission).

Similarly, radiative quantities and surface properties may exhibit *directional* dependencies. For example, a given surface may, because of roughness, emit radiation preferentially in certain directions. In radiation terminology, the adjective *directional* is used to qualify a quantity applicable to a single direction, and the adjective *hemispherical* applies to a quantity when the effect in question has been summed over *all* directions above the surface involved. Except for the quantity *intensity,* to be defined shortly, the presentation in this text will be limited to hemispherical phenomena only and directional effects will not be considered. Generally, the adjective hemispherical will not be applied, but will be understood.

Emissive Power, Radiosity, and Irradiation

Emissive Power. The term *emissive power* is used to denote the *emitted* thermal radiation leaving a surface, per unit time, per unit area of surface. The *total hemispherical* emissive power of a surface is all the emitted energy, summed over all directions and all wavelengths, and is usually denoted by the symbol E. The total emissive power is found to be dependent upon the temperature of the emitting surface, the substance of which the surface is composed, and the nature of the surface structure (i.e., roughness, etc.).

In general, the emissive power of a given surface element may be both spectrally and directionally dependent. As noted earlier, this text will not consider directional effects, and the emissive power will always be understood to be hemispherical—summed over all directions in the hemisphere above the surface. However, it will be necessary to consider the way in which the emission from a surface is distributed among the wavelengths in the thermal band. The *monochromatic* emissive power, symbolized by E_λ, is then defined as the rate, per unit of area, at which a surface emits thermal radiation at a particular wavelength, λ. Thus the total and monochromatic hemispherical emissive powers are related by

$$E = \int_0^\infty E_\lambda \, d\lambda \tag{9.1}$$

and the functional dependence of E_λ on λ must be known to evaluate E.

It should be noted that the emissive power, total or monochromatic, consists only of *original* emission leaving a surface. It does not include any energy leaving a surface that is the result of the reflection of any incident radiation. The quantity *radiosity* is used to account for such reflections, as noted next.

Radiosity. *Radiosity* is the term used to indicate all the radiation *leaving* a surface, per unit time and unit area. The symbols J_λ and J are used to denote the monochromatic and total radiosities, the monochromatic radiosity being that at a single wavelength and the total being that summed over all wavelengths:

$$J = \int_0^\infty J_\lambda \, d\lambda. \tag{9.2}$$

The radiosity differs from the emissive power in that it includes reflected energy as well as original emission. Again, the work presented here presumes that the radiosity has been summed over all directions over a surface.

Irradiation. *Irradiation* is the term used to denote the rate, per unit area, at which thermal radiation is incident on a surface (from all directions). The irradiation incident on a surface is the result of emissions and reflections from other surfaces, and may thus be spectrally dependent. The symbols G_λ and G are used to denote the monochromatic and total irradiation, respectively, and

$$G = \int_0^\infty G_\lambda \, d\lambda. \tag{9.3}$$

Clearly, the emissive power, radiosity, and irradiation of a surface are interrelated by the reflective, absorptive, and transmissive properties of a surface, as discussed next.

Absorptivity, Reflectivity, and Transmissivity

When radiation is incident on a surface, part of it may be reflected away from the surface, part of it may be absorbed by the surface, and part may be transmitted through the surface. These fractions of reflected, absorbed, and transmitted energy are interpreted as surface properties called *reflectivity, absorptivity,* and *transmissivity,* respectively. Depending on whether or not one is concerned with the surface behavior with respect to incident energy at a specific wavelength or summed over all wavelengths, both *monochromatic* and *total* values of these properties may be defined. The following symbols are used:

ρ_λ = monochromatic reflectivity = fraction reflected at wavelength λ,

ρ = total reflectivity = fraction reflected at all wavelengths,

α_λ = monochromatic absorptivity = fraction absorbed at wavelength λ,

α = total absorptivity = fraction absorbed at all wavelengths,

τ_λ = monochromatic transmissivity = fraction transmitted at wavelength λ,

τ = total transmissivity = fraction transmitted at all wavelengths.

Interrelations between the monochromatic and total properties will be deduced in a later section.

Using the definitions above, energy conservation gives

$$\rho_\lambda + \alpha_\lambda + \tau_\lambda = 1, \qquad \rho + \alpha + \tau = 1. \tag{9.4}$$

In general both the monochromatic and total surface properties are dependent on the surface composition, its roughness, etc., and on its temperature. The monochromatic properties are dependent on the wavelength of the incident radiation, and the total properties are dependent on the spectral distribution of the incident energy.

In the case of gases, these properties are also dependent on the geometrical size and shape of the gas bulk through which the radiation passes. Most gases have high values of τ and low values of α and ρ. For instance, air at atmospheric pressure is virtually transparent to thermal radiation, so that one may take $\alpha \approx \rho \approx 0$ and $\tau \approx 1$. Other gases, notably water vapor and carbon dioxide, may be highly absorptive to thermal radiation—at least at certain wavelengths.

Most solids, except for glass, encountered in engineering practice are opaque to thermal radiation, so that $\tau \approx 0$. The initial parts of this chapter will be devoted to the study of the radiant exchange between opaque solid surfaces separated by thermally transparent media. The questions which arise when the surfaces are separated by absorbing gases will be considered at the end of the chapter.

For thermally opaque solid surfaces, one has

$$\rho + \alpha = 1. \tag{9.5}$$

From the definitions of emissive power, radiosity, and irradiation, one has the useful relations

$$\begin{aligned} J &= E + \rho G, \\ &= E + (1 - \alpha)G. \end{aligned} \tag{9.6}$$

Monochromatic versions of Eqs. (9.5) and (9.6) may also be written.

The particular behavior of real surfaces with respect to ρ, α, and τ will be discussed following further definitions and the discussion of ideal blackbody radiation.

Solid Angle

The concept of radiation intensity, to be discussed next, is intimately tied to the spatial concept of *solid angle*. Figure 9.1 illustrates the ideas involved in the definition of solid angle. If, as in Fig. 9.1(a) an arbitrary surface, A, is located in space, the solid angle subtended by A at some point, say P, is the magnitude of the cone (not necessarily a right circular cone) formed by connecting all points on the boundary of A to the point P. The measure of the solid angle ω contained in this cone is found by generating a spherical surface, radius r and centered at P,

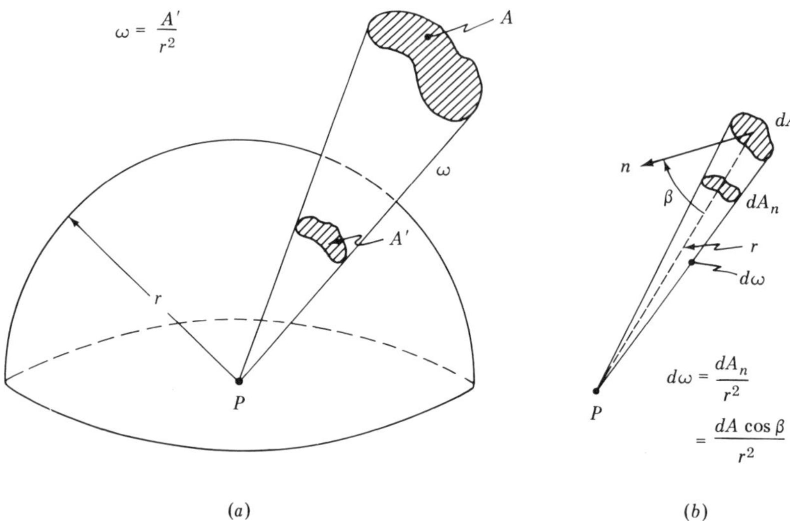

Figure 9.1. Solid angles.

and finding the area of the spherical surface A' in Fig. 9.1(a), that is intercepted by the cone. Then the measure of the solid angle is defined as

$$\omega = \frac{A'}{r^2}. \tag{9.7}$$

The solid angle so defined is dimensionless, as is the plane angle, and the dimensionless unit, the steradian (sr) is used to denote it. Clearly, there are 4π steradians around a point in space.

According to its definition, the determination of the solid angle subtended by an area requires the finding of the projection of that area on a spherical surface. For a finite area this may be a complex problem. However, if one considers a differential area, such as suggested in Fig. 9.1(b), the matter is considerably simplified. If, as in Fig. 9.1(b), it is desired to find the differential solid angle, $d\omega$, subtended at point P by a differential area, dA, located a distance r from P, then one simply has

$$d\omega = \frac{dA_n}{r^2} = \frac{dA \cos \beta}{r^2}, \tag{9.8}$$

where dA_n is the projection of the given dA, normal to r, and β is the angle between r and the normal drawn to dA.

EXAMPLE 9.1

Find the solid angles subtended at the point P by each of the two differential areas, dA_1 and dA_2, oriented as noted in Fig. 9.2.

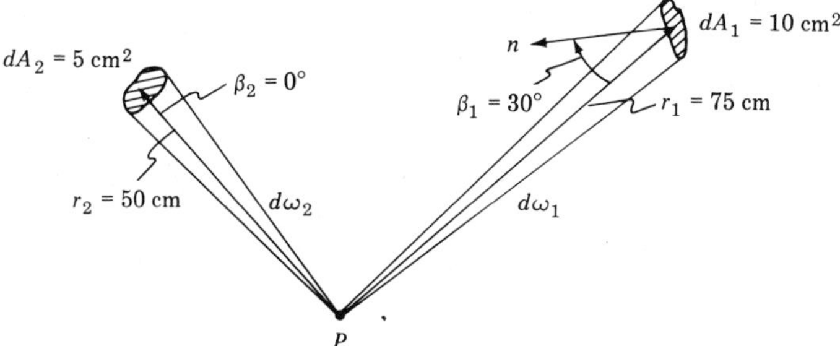

Figure 9.2. Example 9.1.

Solution. The differential solid angle is given by Eq. (9.8):

$$d\omega = \frac{dA \cos \beta}{r^2}.$$

For the surface $dA_1 = 10 \text{ cm}^2$, the radial distance and normal angle are $r_1 = 75$ cm, $\beta_1 = 30°$. Thus the solid angle $d\omega_1$ is

$$d\omega_1 = \frac{dA_1 \cos \beta_1}{r_1^2}$$

$$= \frac{10 \times \cos 30°}{(75)^2} = 1.54 \times 10^{-3} \text{ sr}.$$

For the second surface $dA_2 = 5 \text{ cm}^2$, $r_2 = 50$ cm, and $\beta_2 = 0°$. Thus the projected area is the area dA_2, itself, and the solid angle is

$$d\omega_2 = \frac{dA_2 \cos \beta_2}{r_2^2} = \frac{dA_2}{r_2^2}$$

$$= \frac{5}{(50)^2} = 2.0 \times 10^{-3} \text{ sr}.$$

Intensity of Radiation

In the consideration of the exchange of radiant energy between surfaces it is necessary to define a quantity, known as the *intensity* of radiation. The intensity describes the directional distribution of radiant energy leaving a surface (either by emission or reflection) or the directional distribution of the energy incident upon a surface. A rigorous definition of radiation intensity may be made in terms of the radiation passing a point in space in a particular direction. However, for ease in understanding, a definition of intensity will be made in terms of the energy leaving a "point source" on a surface and how it distributes directionally in the hemisphere above the surface. Whether the energy originates from the surface by original emission or by reflection will be left unspecified for the moment. Also, the definition given in the following may be made for a given wavelength, resulting in a monochromatic intensity, or as a quantity summed over all wavelengths, resulting in a total intensity.

Consider, as illustrated in Fig. 9.3, a small emitting surface of area ΔA, the center of which is denoted as the point Q. Let Δq represent the *rate* at which radiant energy leaves ΔA. It will be useful to think in terms of a radiant energy *flux*. The average flux leaving ΔA is defined as

$$f_{\text{av}} = \frac{\Delta q}{\Delta A}.$$

The radiant flux from the *point source* at Q is defined as

$$f_Q = \lim_{\Delta A \to 0} \frac{\Delta q}{\Delta A}. \tag{9.9}$$

It is important to remember that the flux just defined is based on the energy *leaving* a surface (reflected or emitted) and is calculated per unit area of the surface from which it leaves.

Now the radiant flux from ΔA, or Q, fills the half-space above ΔA. The *intensity* of the radiation at a point in space due to the emission from this point source is defined as the radiant energy passing the spatial point per unit time, per unit solid angle subtended at Q, per unit area of radiating surface projected normal to the direction which the point in space makes with the emitting point. In terms of the drawing in Fig. 9.3, the radiation intensity at some point in space, say P, due to the radiation leaving ΔA may be defined in terms of the radiation falling on an element of the spherical surface (center at Q, radius r) which passes through P. If δa represents an element of the spherical surface surrounding the point P, only a

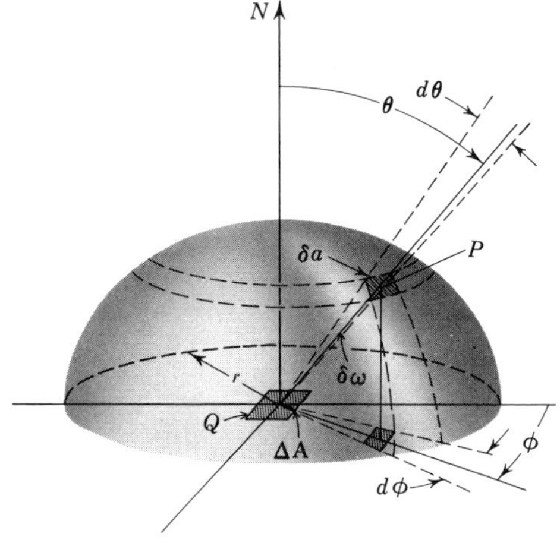

Figure 9.3

fraction of Δq strikes δa. Let $\delta(\Delta q)$ represent this fraction of the energy leaving ΔA which falls on δa. In reference to Fig. 9.3, N is the normal to the plane containing ΔA, the radiating surface, θ is the angle between N and the line connecting P and Q. The solid angle subtended by the receiving area δa at the radiating point Q is $\delta\omega = \delta a/r^2$ since δa is normal to the radial line connecting Q and P.

By definition, the intensity at the element δa due to the radiation from ΔA, call it $I_{\delta a/\Delta A}$, is

$$I_{\delta a/\Delta A} = \frac{\delta(\Delta q)}{\delta\omega(\Delta A \cos\theta)}.$$

The intensity at δa due to the radiation coming from the *point source* Q is, by use of Eq. (9.9),

$$\begin{aligned} I_{\delta a/Q} &= \frac{\delta f}{\delta\omega}\frac{1}{\cos\theta} \\ &= \frac{\delta f}{\delta a}\frac{r^2}{\cos\theta}. \end{aligned} \tag{9.10}$$

In Eq. (9.10), the flux f is that flux measured at the *radiating* source, Q, not at the receiving area δa.

The intensity at the *point* P, in the direction θ, due to the radiation originating from Q is then defined as the limit of Eq. (9.10) as $\delta a \to 0$:

$$\begin{aligned} I &= \frac{df}{d\omega}\frac{1}{\cos\theta} \\ &= \frac{df}{da}\frac{r^2}{\cos\theta}. \end{aligned} \tag{9.11}$$

The quantity $df/d\omega$ represents the radiated flux, per unit spatial solid angle, while df/da is the flux at P of the flux leaving Q—that is, df/da is the areal density at P of the flux emanating from Q.

Equations (9.11) may be rewritten in the following forms for calculation of the fraction of the flux leaving the surface that is contained in the solid angle $d\omega$ or intercepted by the receiving element da:

$$\begin{aligned} df &= I\cos\theta\, d\omega \\ &= I\cos\theta\,\frac{da}{r^2}. \end{aligned} \tag{9.12}$$

Since all of the flux leaving the surface at Q must pass into the hemisphere above it, as depicted in Fig. 9.3, then

$$f = \int_h I\cos\theta\, d\omega, \tag{9.13}$$

in which the notation $\int_h$ is used to denote an integration taken over all the solid angle of the hemisphere. In terms of the polar angle θ and the azimuthal angle φ

(i.e., longitude) shown in Fig. 9.3,

$$d\omega = \frac{da}{r^2} = \frac{r\,d\theta(r \sin\theta\,d\varphi)}{r^2}$$
$$= \sin\theta\,d\varphi\,d\theta.$$

Thus Eq. (9.13) is also

$$f = \int_0^{2\pi}\int_0^{\pi/2} I \sin\theta\cos\theta\,d\theta\,d\varphi. \tag{9.14}$$

Intensity, it is seen, is an inherently directional quantity and in general I may depend on the two angles θ and φ. Thus knowledge of $I = I(\theta, \varphi)$ is necessary in order to evaluate f from Eq. (9.13) or (9.14). Radiation is termed *diffuse* if I is uniform—i.e., independent of θ and φ. Thus, for diffuse radiation

$$\begin{aligned} f &= I\int_0^{2\pi}\int_0^{\pi/2} \sin\theta\cos\theta\,d\theta\,d\varphi \\ &= I\pi, \end{aligned} \tag{9.15}$$

and the surface flux and the intensity are related in a very simple way. Also, Eq. (9.11) shows that if I is constant, the flux intercepted by an area element varies as $\cos\theta$ and inversely as r^2—the familiar *Lambert's law* of diffuse radiation.

The preceding definition and discussion of intensity were based on the assumption that the radiant flux, f, was leaving the surface and flowing *outward* in the direction given by θ and φ and contained in the solid angle $d\omega$. In the event one is interested only in the original emission leaving a surface, the flux is the hemispherical emissive power, E, and the associated intensity will be symbolized by $I_e(\theta, \varphi)$. Thus

$$E = \int_h I_e(\theta, \varphi)\cos\theta\,d\omega = \int_0^{2\pi}\int_0^{\pi/2} I_e(\theta, \varphi)\sin\theta\cos\theta\,d\theta\,d\varphi. \tag{9.16}$$

A surface is termed a *diffuse emitter* if $I_e(\theta, \varphi)$ is uniform. Then

$$E = \pi I_e. \tag{9.17}$$

If one is interested in all the energy leaving a surface, the flux is the hemispherical radiosity, J, and the associated intensity is denoted by $I_{e+r}(\theta, \varphi)$ to indicate that both reflected energy and original emission are involved. Then

$$J = \int_h I_{e+r}(\theta, \varphi)\cos\theta\,d\omega = \int_0^{2\pi}\int_0^{\pi/2} I_{e+r}(\theta, \varphi)\sin\theta\cos\theta\,d\theta\,d\varphi. \tag{9.18}$$

If a surface is a diffuse emitter *and* a *diffuse reflector*, then

$$J = \pi I_{e+r}. \tag{9.19}$$

The geometric variables associated with these concepts are illustrated in Fig. 9.4(a).

In addition to the radiant energy flowing outward from a surface as just dis-

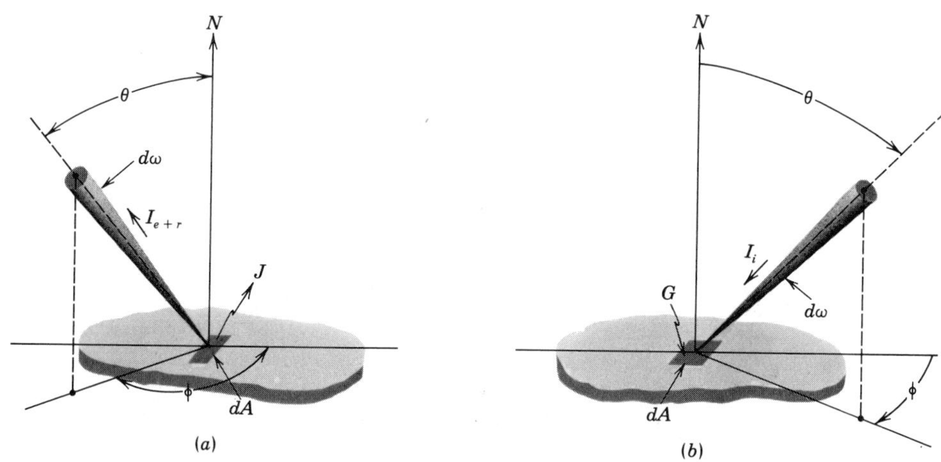

Figure 9.4. Angular relations for radiation intensity.

cussed, one may also associate with any direction (θ, φ) an inwardly directed intensity that would result from radiation originating (by emission or reflection) from other surfaces. If $I_i(\theta, \varphi)$ represents the rate at which energy is incident upon a surface, per unit area of intercepting surface normal to the direction (θ, φ), per unit area of intercepting surface normal to the direction (θ, φ), per unit solid angle, then the same relations given in Eqs. (9.13) and (9.14) apply with I_i replacing I. The flux intercepted by the surface is the irradiation defined earlier. Thus, as suggested in Fig. 9.4(b),

$$G = \int_h I_i(\theta, \varphi) \cos\theta \, d\omega = \int_0^{2\pi} \int_0^{\pi/2} I_i(\theta, \varphi) \sin\theta \cos\theta \, d\theta \, d\varphi. \tag{9.20}$$

If the incident energy is diffuse so that I_i is uniform, then

$$G = \pi I_i. \tag{9.21}$$

As mentioned earlier, all the foregoing definitions for intensity, its relation to emissive power, radiosity, irradiation, etc., may be made in the monochromatic sense as well as the total sense. That is, all the energy fluxes are defined for a given wavelength, and the associated intensities are monochromatic ones (i.e., I_λ, $I_{\lambda,e}$, $I_{\lambda,e+r}$, $I_{\lambda,i}$). The integrations given in Eqs. (9.16) through (9.21) then yield the monochromatic emissive power, radiosity, and irradiation. Equations (9.1) through (9.3) relate the latter quantities to their total counterparts.

EXAMPLE 9.2

Figure 9.5 depicts the situation in which the radiation from a surface, A_1, is intercepted by several receiving surfaces of various locations and orientations. For simplicity, the radial lines drawn from a point Q on A_1 to the various intercepting

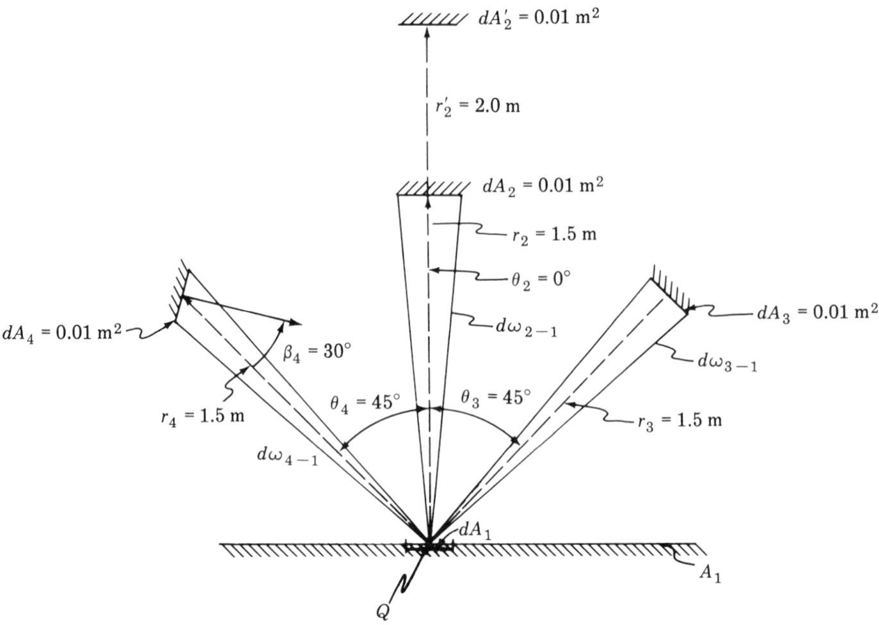

Figure 9.5. Example 9.2.

surfaces all lie in the same plane—that of the paper. The intercepting surfaces are all of the same area ($dA_2 = dA_3 = dA_4 = 0.01\ \text{m}^2$) and are assumed to be small enough compared to the radial distances ($r_2 = r_3 = r_4 = 1.5$ m) that they may be treated as infinitesimal. For a differential portion of the originating surface around the point Q, $dA_1 = 0.01\ \text{m}^2$, an amount of radiant energy $dq_1 = 10$ W leaves the surface and radiates into the half-space above it. Presume that this radiation is distributed *diffusely*.

(a) Find the intensity of the radiation in the space above A_1 due to the emission from dA_1.

(b) Find the fraction of the *flux* from dA_1 and the fraction of the heat flow dq_1 that is intercepted by each of dA_2, dA_3, and dA_4.

(c) Find the same quantities found in part (b) for dA_2 if it is moved to a farther distance, $r_2' = 2.0$ m, away from dA_1, other data remaining the same.

Solution

(a) The flux of energy leaving the originating surface at Q is

$$f = \frac{dq_1}{dA_1} = \frac{10}{0.01} = 1000\ \text{W/m}^2.$$

Since the radiant emission is stated as being diffusely distributed in the half-space above A_1, the radiation intensity is given by Eq. (9.15):

$$I = \frac{f}{\pi} = \frac{1000}{\pi} = 318.3 \text{ W/m}^2\text{-sr}.$$

(b) The fraction of the flux from A_1 that is intercepted by a surface in the space above it is dependent on the solid angle subtended by the intercepting surface and the polar angle made by the intercepting surface with the normal to the emitting surface as given by Eq. (9.12):

$$df = I \cos\theta \, d\omega.$$

Thus it is necessary to determine the solid angle subtended at dA_1 by each of the receivers, dA_2, dA_3, dA_4. Using the notation $d\omega_{2-1}$, $d\omega_{3-1}$, $d\omega_{4-1}$ to denote these solid angles, as noted in Fig. 9.5, application of the methods shown in Example 9.1 and Eq. (9.8) yields the following (noting that dA_2 and dA_3 are oriented normally so that $\cos\beta_2 = \cos\beta_3 = 1$):

$$d\omega_{2-1} = \frac{dA_2 \cos\beta_2}{r_2^2} = \frac{0.01 \times 1}{(1.5)^2} = 4.44 \times 10^{-3} \text{ sr},$$

$$d\omega_{3-1} = \frac{dA_3 \cos\beta_3}{r_3^2} = \frac{0.01 \times 1}{(1.5)^2} = 4.44 \times 10^{-3} \text{ sr},$$

$$d\omega_{4-1} = \frac{dA_4 \cos\beta_4}{r_4^2} = \frac{0.01 \times \cos 30^\circ}{(1.5)^2} = 3.85 \times 10^{-3} \text{ sr}.$$

Note that while each of these areas is of the same size and is located the same distance from dA_1, the surface dA_4 subtends a smaller solid angle due to its orientation and will, hence, intercept a smaller fraction of the flux from dA_1. The fraction of the flux leaving dA_1 that is intercepted by each of these areas may now be found from Eq. (9.12), using the polar angles noted in Fig. 9.5:

$$\begin{aligned} df_{2-1} &= I \cos\theta_2 \, d\omega_{2-1} \\ &= 318.3 \times \cos 0^\circ \times 4.44 \times 10^{-3} = 1.413 \text{ W/m}^2, \\ df_{3-1} &= I \cos\theta_3 \, d\omega_{3-1} \\ &= 318.3 \times \cos 45^\circ \times 4.44 \times 10^{-3} = 0.999 \text{ W/m}^2, \\ df_{4-1} &= I \cos\theta_4 \, d\omega_{4-1} \\ &= 318.3 \times \cos 45^\circ \times 3.85 \times 10^{-3} = 0.867 \text{ W/m}^2. \end{aligned}$$

The fractions just found are fractions of the flux from the *originating* surface, dA_1, so that the heat flow intercepted by each is found by multiplying the above by dA_1:

$$dq_2 = df_{2-1}\, dA_1 = 1.413 \times 0.01 = 0.014 \text{ W},$$

$$dq_3 = df_{3-1}\, dA_1 = 0.999 \times 0.01 = 0.010 \text{ W},$$

$$dq_4 = df_{4-1}\, dA_1 = 0.867 \times 0.01 = 0.009 \text{ W}.$$

First, it should be noted that the energy flows intercepted (dq_2, etc.) are small fractions of the energy, $dq_1 = 10$ W, which left dA_1. This is the result of the fact that these surfaces subtend such a small fraction of the total 2π steradians in the half-space above dA_1. Note also that while dA_2 and dA_3 are the same size and subtend the same solid angle the energy intercepted by dA_3 is smaller than that intercepted by dA_2 because of the polar angle made by dA_3 compared with that by dA_2 (45° vs. 0°). The $\cos\theta$ term in Eq. (9.12) results from the fact that the intercepted flux depends on the area of the emitting surface as *seen* by the receiver. Thus, while the intensity is uniform in the space, the energy flow is not. The energy intercepted by dA_4 is smaller even than that by dA_3 since, although the polar angles and distances are the same, dA_4 subtends a smaller solid angle as the result of its inclination to the line connecting dA_1 and dA_4.

(c) The intercepting areas analyzed in part (b) were all located the same distance from the originating surface. Hence the effects just noted were all the results of only the angular location or orientation of the surfaces. The inverse-square effect of radiant energy flow can be illustrated by comparing the results just obtained for dA_2 if it is moved farther away (to the position denoted in Fig. 9.5 by dA_2' and $r_2' = 20$ m) leaving the polar angle unchanged at $\theta_2' = \theta_2 = 0°$ and keeping dA_2' normal to the line connecting it to dA_1. The intensity of the emission from dA_1, I, is still the same since it depends only on the emitted energy (under the diffuse assumption). All that changes from the previous case is the solid angle subtended by the receiver. Thus, performing the same calculations for the new location, one has for the solid angle

$$d\omega_{2'-1} = \frac{dA_2' \cos\beta_2'}{r_2'^2} = \frac{0.01 \times 1}{(2.0)^2} = 2.5 \times 10^{-3} \text{ sr},$$

considerably smaller than $d\omega_{2-1}$, owing to the inverse-square effect. Then the intercepted flux and energy flow are correspondingly reduced:

$$df_{2'-1} = I \cos\theta_2'\, d\omega_{2'-1}$$

$$= 318.3 \times 1 \times 2.5 \times 10^{-3} = 0.796 \text{ W/m}^2,$$

$$dq_{2'} = df_{2'-1}\, dA_1 = 0.796 \times 0.01 = 0.008 \text{ W}.$$

Compared with $df_{2-1} = 1.413$ W/m^2 and $dq_2 = 0.014$ W, the effect of distance is obvious.

9.3 BLACKBODY RADIATION

In order to describe the radiation characteristics and properties of real surfaces, it is useful to define an ideal surface for purposes of comparison. Such an indeal surface is the *perfect blackbody*. A perfect blackbody is defined as one which absorbs *all* incident radiation regardless of the spectral distribution or directional character of the incident radiation—i.e., $\alpha_\lambda = \alpha = 1$ or $\rho_\lambda = \rho = 0$. The term "black" is used since dark surfaces normally show high values of absorptivity. Since a blackbody absorbs all incident radiation, the only radiation leaving a blackbody surface is original emission. The emissive power of a blackbody, to be denoted by E_b, will be seen to depend on the surface temperature only. This dependence on temperature and the spatial distribution of blackbody radiation as given by the intensity function $I_b(\theta, \varphi)$ will be developed in the following discussion. Also the spectral distribution of the emitted energy as given by the dependence of the monochromatic emissive power, $E_{b\lambda}$, on wavelength, is of great importance and will be discussed in some detail.

While a perfect blackbody is an idealization, it is possible to produce radiation in the laboratory that is very nearly the same as that which would originate from such a surface. Imagine a hollow cavity, such as depicted in Fig. 9.6(a), the walls of which are maintained at a uniform temperature. If a small hole (small compared to the size of the cavity) is provided in the wall, then any ray of incident radiant energy entering the hole will undergo a large number of internal reflections—with a portion of the radiation being absorbed at each reflection. Very little of the incident beam ever finds its way out of the small hole, and thus the plane of the hole appears to be a perfect absorber. Thus, the radiation found to be emanating from the hole will appear to be that coming from a blackbody.

This concept of blackbody radiation may also be explained by consideration of the situation depicted in Fig. 9.6(b). Here one imagines a cavity, the walls of which have been heated to a *uniform* temperature. The surface of the cavity is not necessarily black, and thus the space is filled with a radiation field which is the result of original emission and reflections from the walls. Since the walls are maintained at a uniform temperature, the resultant radiant field is in thermal equilibrium with the walls. Now imagine a very small wafer of material to be placed inside the cavity—small enough that its presence only negligibly alters the existing radiation field. Further, presume that the material of which the wafer is composed is a perfect blackbody and that it is allowed to come to thermal equilibrium with the walls—that is, the wafer temperature is the same as that of the walls. If $E_b(T)$ is the emissive power of a blackbody at temperature T, then $E_b(T)$ is the flux of energy away from the surface of the wafer. Since thermal equilibrium exists, the energy flux incident on the wafer, i.e., the irradiation G, must be the same since it is all absorbed. Changing the orientation or position of the wafer does not change this result, and one may say that $G = E_b(T)$ throughout the cavity. Thus one concludes that the irradiation field produced in a uniformly heated cavity, regardless of the cavity surface properties, is uniform and equal to the emissive power of an ideal

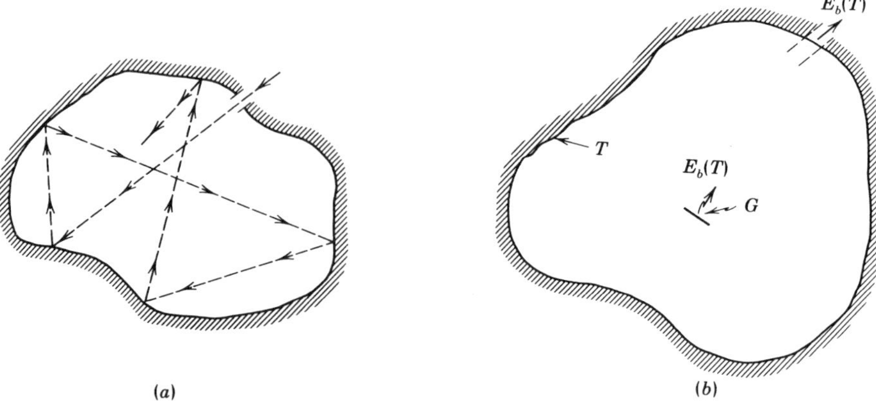

Figure 9.6

blackbody at the temperature of the walls. If a very small hole were made in the wall of the cavity, the radiation escaping from it would be that which would be emitted by a blackbody.

The Stefan–Boltzmann Law for Blackbody Emissive Power

In the preceding discussion showing that the radiation field in an isothermal enclosure is the emissive power of a blackbody at the same temperature, the only surface property involved was the temperature. Thus one concludes that the emissive power of an ideal blackbody is a function of temperature only. Based on experiments, Josef Stefan in 1879 suggested that the total emissive power of a blackbody is proportional to the fourth power of the absolute temperature. Later, Ludwig Boltzmann applied the principles of classical thermodynamics and analytically derived the same fact. Thus the following dependence of E_b on T is called the *Stefan–Boltzmann law:*

$$E_b = \sigma T^4, \tag{9.22}$$

in which σ is the Stefan–Boltzmann constant with the value

$$\begin{aligned}\sigma &= 5.670 \times 10^{-8}\ \text{W/m}^2\text{-°K}^4\\ &= 0.1714 \times 10^{-8}\ \text{Btu/h-ft}^2\text{-°R}^4.\end{aligned}$$

The temperature of the black surface is the *absolute* temperature. For computational purposes it is convenient to note that

$$\begin{aligned}E_b = \sigma T^4 &= 5.670 \times \left(\frac{T}{100}\right)^4\ \text{W/m}^2\ (T\text{ in °K})\\ &= 0.1714\left(\frac{T}{100}\right)^4\ \text{Btu/h-ft}^2\ (T\text{ in °R}).\end{aligned}$$

The Stefan–Boltzmann law for the *total* emissive power gives the total energy emitted by a blackbody, summed over all wavelengths. It is also derivable from Planck's law for the monochromatic emissive power as discussed later.

Spatial Characteristics of Blackbody Radiation

The earlier discussion not only showed that the irradiation field in an isothermal cavity is equal to E_b but also that the irradiation is the same for *all* planes of *any* orientation within the cavity. Thus the integral expression of Eq. (9.20) must yield the same result for all locations and orientations. It may then be shown that the intensity of blackbody radiation, I_b, is uniform. Thus blackbody radiation is diffuse and

$$E_b = \pi I_b. \tag{9.23}$$

Spectral Distribution of Blackbody Radiation

The Stefan–Boltzmann law gives the *total* emissive power of an ideal blackbody. The distribution of the emission from a blackbody over the complete spectrum of wavelengths as given by the monochromatic emissive power $E_{b\lambda} = E_{b\lambda}(\lambda, T)$ is also of great interest. A typical distribution of $E_{b\lambda}$ is shown in Fig. 9.7. The monochromatic emission (emitted energy per unit time and area at *a* wavelength) is seen to rise from very small values for short wavelengths, reach a peak value, and then fall again to small values as the wavelength becomes large. Most of the emission at temperatures encountered in engineering applications lies in the wavelength band between 1×10^{-1} and 1×10^{2} μm, as noted earlier. The area under the $E_{b\lambda}$ versus λ curve is the total emissive power given in Eqs. (9.1) and (9.22):

$$E_b = \int_0^\infty E_{b\lambda}(\lambda, T)\, d\lambda = \sigma T^4. \tag{9.24}$$

Since E_b is a function of only the surface temperature, it follows that the monochromatic blackbody emissive power depends only on temperature *and* wavelength. The analytic prediction of the observed dependence of $E_{b\lambda}$ on T and λ was one of the central problems of interest to physicists of the late nineteenth century. Its eventual discovery by Max Planck in 1901 marked the first achievement of his quantum theory of radiation. Planck's law is given as

$$E_{b\lambda}(\lambda, T) = \frac{C_1}{\lambda^5(e^{C_2/\lambda T} - 1)}, \tag{9.25}$$

where the accepted values of the two constants are

$$C_1 = 3.7413 \times 10^8 \text{ W-}\mu\text{m}^4/\text{m}^2,$$

$$C_2 = 1.4388 \times 10^4\ \mu\text{m-°K}.$$

The units of C_1 and C_2 indicate that λ in Eq. (9.25) is to be expressed in μm.

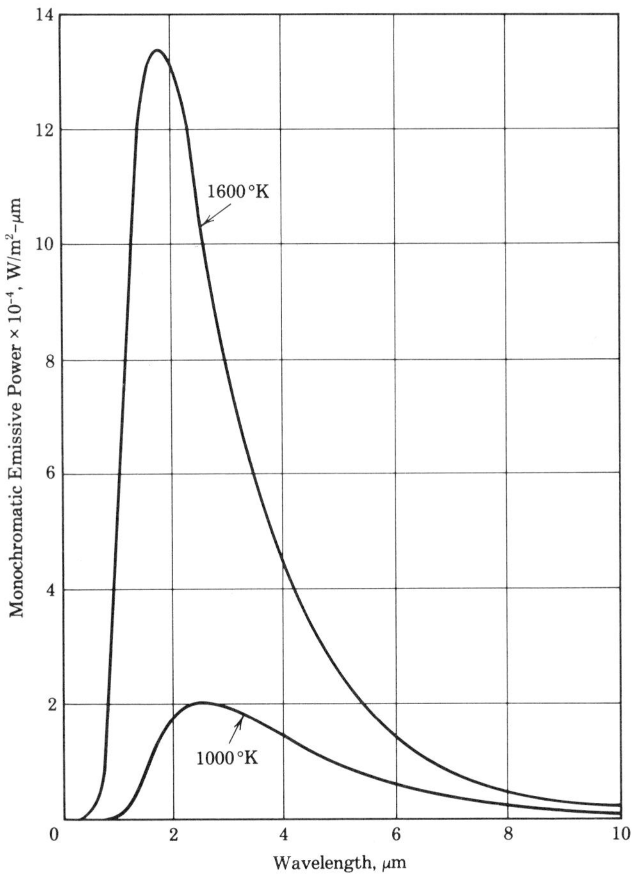

Figure 9.7. Monochromatic emissive power of an ideal blackbody at different temperatures.

Integration of Eq. (9.25) according to Eq. (9.24) will show that the Stefan–Boltzmann constant is related to C_1 and C_2:

$$\sigma = \left(\frac{\pi}{C_2}\right)^4 \frac{C_1}{15}. \tag{9.26}$$

Figure 9.7 shows a plot of Planck's equation for two different temperatures. The area under these curves is proportional to T^4 in accord with the Stefan–Boltzmann law. As may be deduced from Eq. (9.25), the monochromatic blackbody emissive power exhibits a peak value that increases with T and which occurs at shorter wavelengths as T increases. *Wien's law* may be deduced from Planck's, and states that the maximum value of $E_{b\lambda}$ occurs at the wavelength given by

$$\begin{aligned} \lambda_{\max} T &= C_3 \\ &= 2897.8 \ \mu\text{m-°K}. \end{aligned} \tag{9.27}$$

For use in performing certain calculations to be described later it is useful to have tabulated values of the monochromatic blackbody emissive power. This is most efficiently done by noting that the ratio $E_{b\lambda}/\sigma T^5$ is a function of the product λT only:

$$\frac{E_{b\lambda}}{\sigma T^5} = \frac{C_1}{\sigma} \frac{1}{(\lambda T)^5 (e^{C_2/\lambda T} - 1)}. \tag{9.28}$$

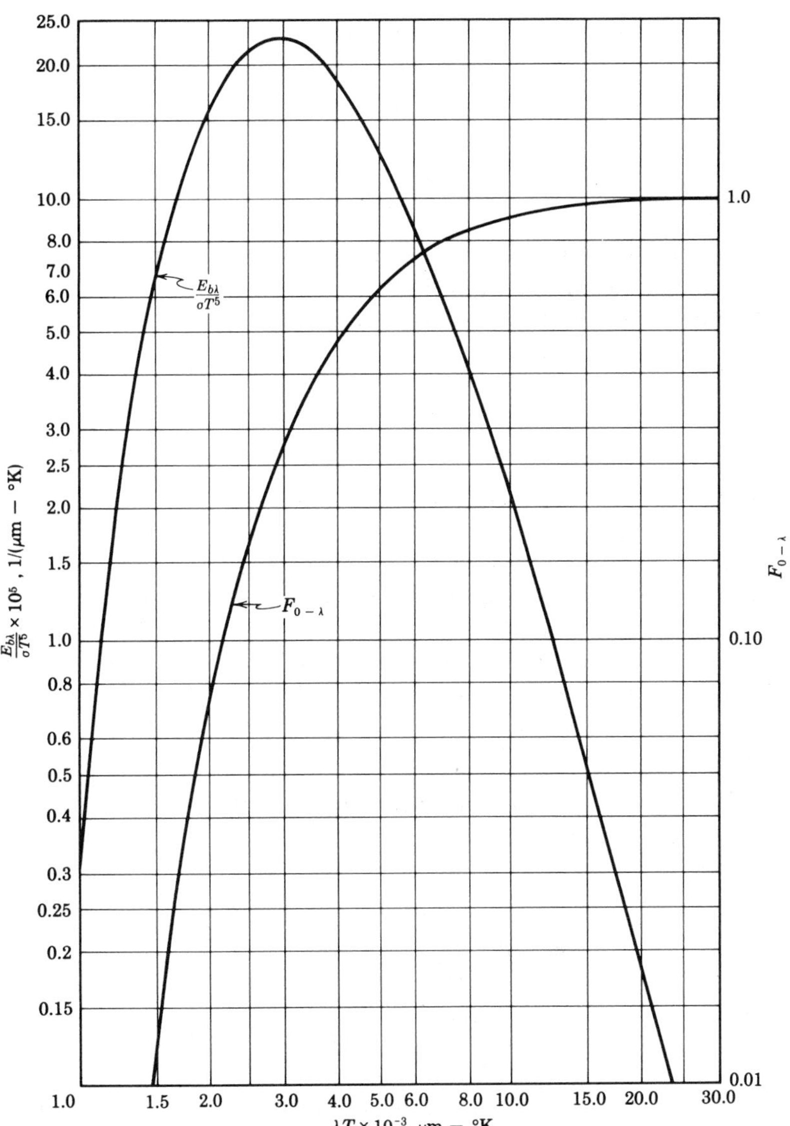

Figure 9.8. Planck blackbody radiation functions.

Equation (9.28) is displayed graphically in Fig. 9.8 and is tabulated in Table A.12. For given λ and T, $E_{b\lambda}$ may be readily evaluated.

Also of great subsequent use is the fraction of the total emission (i.e., the area under the curves in Fig. 9.7) that occurs between specified values of the wavelength. This fraction is readily determined from the partial integral of $E_{b\lambda}$ between $\lambda = 0$ and any value λ. This partial integral, when divided by the total emission is defined as the quantity $F_{0-\lambda}(T)$:

$$F_{0-\lambda}(T) = \frac{\int_0^\lambda E_{b\lambda}\, d\lambda}{\int_0^\infty E_{b\lambda}\, d\lambda} = \frac{\int_0^\lambda E_{b\lambda}\, d\lambda}{\sigma T^4} = \int_0^{\lambda T} \frac{C_1 d(\lambda T)}{\sigma(\lambda T)^5(e^{C_2/\lambda T} - 1)}. \tag{9.29}$$

The latter function is also seen to be a function of only the product λT and is also displayed in Fig. 9.8 and tabulated in Table A.12. The fraction of energy emitted between two wavelengths, λ_1 and λ_2, by a blackbody at temperature T is thus

$$F_{\lambda_1-\lambda_2}(T) = F_{0-\lambda_2}(T) - F_{0-\lambda_1}(T). \tag{9.30}$$

EXAMPLE 9.3

A blackbody has a surface temperature of 600°K. Find (a) the total emissive power; (b) the wavelength at which the maximum monochromatic emissive power occurs; (c) the magnitude of the monochromatic emissive power at this wavelength just found and at wavelengths of 2.0 and 6.0 μm; (d) the fraction of the total emission that occurs between the wavelengths of 2.0 and 6.0 μm.

Solution

(a) At $T = 600$°K, the Stefan–Boltzmann law of Eq. (9.22) gives the total emissive power of a blackbody to be

$$\begin{aligned} E_b &= \sigma T^4 \\ &= 5.670\left(\frac{T}{100}\right)^4 \\ &= 5.670\left(\frac{600}{100}\right)^4 = 7348\ \text{W/m}^2\ (2329\ \text{Btu/h-ft}^2). \end{aligned}$$

(b) The wavelength at which the maximum monochromatic emissive power occurs for a blackbody at temperature T is given by Wien's law in Eq. (9.27):

$$\lambda_{max} T = C_3 = 2897.8 \ \mu\text{m-°K}$$

$$\lambda_{max} = \frac{2897.8}{600} = 4.83 \ \mu\text{m}.$$

(c) The monochromatic emissive power which occurs at a given wavelength λ from a blackbody maintained at temperature T is given by Eq. (9.28):

$$\frac{E_{b\lambda}}{\sigma T^5} = \frac{C_1}{\sigma} \frac{1}{(\lambda T)^5 (e^{C_2/\lambda T} - 1)}.$$

This quantity is given in Table A.12 for various values of the product λT. Thus at the wavelength of the maximum emission, $\lambda = \lambda_{max} = 4.83 \ \mu\text{m}$, and at $\lambda = 2.0$ and $6.0 \ \mu\text{m}$, one finds the monochromatic emissive power, with $T = 600°\text{K}$:

At $\lambda = 4.83 \ \mu\text{m}$:

$$\lambda T = 4.83 \times 600 = 2897.8 \ \mu\text{m-°K},$$

$$\text{Table A.12:} \quad \left(\frac{E_{b\lambda}}{\sigma T^5}\right) = 22.688 \times 10^{-5} \ (\mu\text{m-°K})^{-1},$$

$$E_{b\lambda} = 5.670 \times 10^{-8} \times (600)^5 \times 22.688 \times 10^{-5}$$

$$= 1000 \ \text{W/m}^2\text{-}\mu\text{m} \ (317 \ \text{Btu/h-ft}^2\text{-}\mu\text{m}).$$

At $\lambda = 2.0 \ \mu\text{m}$:

$$\lambda T = 2.0 \times 600 = 1200 \ \mu\text{m-°K},$$

$$\text{Table A.12:} \quad \left(\frac{E_{b\lambda}}{\sigma T^5}\right) = 1.646 \times 10^{-5} \ (\mu\text{m-°K})^{-1},$$

$$E_{b\lambda} = 72.57 \ \text{W/m}^2\text{-}\mu\text{m} \ (23.01 \ \text{Btu/h-ft}^2\text{-}\mu\text{m}).$$

At $\lambda = 6.0 \ \mu\text{m}$:

$$\lambda T = 6.0 \times 600 = 3600 \ \mu\text{m-°K},$$

$$\text{Table A.12:} \quad \left(\frac{E_{b\lambda}}{\sigma T^5}\right) = 20.432 \times 10^{-5} \ (\mu\text{m-°K})^{-1},$$

$$E_{b\lambda} = 5.670 \times 10^{-8} \times (600)^5 \times 20.432 \times 10^{-5}$$

$$= 900.8 \ \text{W/m}^2\text{-}\mu\text{m} \ (285.6 \ \text{Btu/h-ft}^2\text{-}\mu\text{m}).$$

The dependence of the monochromatic emissive power on wavelength is apparent. Note how much more rapidly $E_{b\lambda}$ falls off from the maximum (at $\lambda_{max} = 4.83 \ \mu\text{m}$) when the wavelength is reduced to $\lambda = 2.0 \ \mu\text{m}$ than when it is increased to $\lambda = 6.0 \ \mu\text{m}$.

(d) The fraction of the total emission occurring between two wavelengths, λ_1 and λ_2, is found from the fractional emission function given by Eqs. (9.29) and (9.30):

$$F_{\lambda 1-\lambda 2}(T) = F_{0-\lambda 2}(T) - F_{0-\lambda 1}(T).$$

The function giving the fraction of the emissive power occurring between $\lambda = 0$ and $\lambda = \lambda$, $F_{0-\lambda}(T)$, is tabulated in Table A.12. With the given surface temperature $T = 600°K$, the values of the F function at $\lambda = 2.0$ and $6.0\ \mu m$ are found to be:

$$\text{AT } \lambda_1 = 2.0\ \mu\text{m:} \quad \lambda_1 T = 2.0 \times 600 = 1200\ \mu\text{m-°K},$$

$$F_{0-\lambda 1}(T) = 0.00213.$$

$$\text{At } \lambda_2 = 6.0\ \mu\text{m:} \quad \lambda_2 T = 6.0 \times 600 = 3600\ \mu\text{m-°K},$$

$$F_{0-\lambda 2}(T) = 0.40360.$$

Thus the fraction of the emission between $\lambda_1 = 2.0\ \mu m$ and $\lambda_2 = 6.0\ \mu m$ is

$$F_{6.0\text{-}2.0}(T) = F_{0-\lambda 2}(T) - F_{0-\lambda 1}(T)$$
$$= 0.40360 - 0.00213$$
$$= 0.40147.$$

Calculations similar to those just illustrated in Example 9.3 reveal the dramatic effect of the surface temperature on the spectral distribution of blackbody radiation as shown in Table 9.1. The temperature of 300°K was chosen as representative of typically encountered ambient temperatures, and 5800°K will be shown later to approximate the emission of the sun. The effect of temperature on the total emissive power is profound in keeping with the fourth-power requirement of the Stefan–Boltzmann law. However, the effect of temperature on the spectral distribution is equally dramatic. Emission at typical ambient temperatures is seen to be predominately at the larger wavelengths and almost entirely in the infrared portion of the spectrum. As temperatures corresponding to the sun's emission are approached, it may be noted that significant portions occur in the ultraviolet and visible portions of the spectrum. It is worth noting that the wavelength of the maximum emission from the sun occurs almost exactly in the middle of the visible spectrum as sensed by the human eye. The last two rows of data, the fractions of the emission that occurs above and below a wavelength of about 4 μm, are particularly interesting as far as analyses associated with solar phenomena are concerned. Surfaces at typically ambient temperatures of about 300°K emit almost all their energy at wavelengths in excess of about 4 μm, while the sum emits almost all its energy at wavelengths less than 4 μm. Thus a cool surface subjected to solar irradiation will emit energy of a markedly different spectral character than that incident upon it. This "uncoupling" of the radiation may produce a significant difference between the emissive and solar-absorptive properties of real surfaces as will be illustrated later.

Table 9.1 Effect of Temperature on Blackbody Emission

	Surface Temperature, °K			
	300	**800**	**1600**	**5800**
Total emissive power, W/m^2	459.2	23,220	3.71×10^5	64.16×10^6
Wavelength of maximum emission, μm	9.66	3.62	1.81	0.500
Fraction of emission in the band				
Ultraviolet (5×10^{-3}–3.9×10^{-1} μm)	0.000	0.000	0.000	0.112
Visible light (3.9×10^{-1}–7.8×10^{-1} μm)	0.000	0.000	0.003	0.456
Infrared (7.8×10^{-1}–1×10^3 μm)	1.000	1.000	0.997	0.432
Fraction of emission				
Below $\lambda = 4$ μm	0.002	0.318	0.769	0.990
Above $\lambda = 4$ μm	0.998	0.682	0.231	0.010

9.4 RADIATION CHARACTERISTICS OF NONBLACK SURFACES

The monochromatic and total emissive powers of blackbody radiation as given by the Planck and Stefan–Boltzmann laws, together with the diffuse nature of blackbody radiation intensity, may be used as the ideal basis to which the radiative characteristics of real surfaces may be compared. The radiant characteristics of nonblack, real, surfaces differ from the ideal blackbody in several important ways. The monochromatic emissive power of a real surface at a given temperature differs from that of a blackbody at the same temperature both in amount and in spectral distribution. A nonblack surface exhibits an absorptivity less than unity, and its value may be dependent on the wavelength of the incident radiation. Thus a nonblack surface reflects, perhaps with spectral preference, some incident radiation, so that its surface radiosity consists of more than original emission.

Nonblack surfaces may also exhibit nondiffuse behavior, so that the intensity of emission or of reflected incident energy may not be constant as is the case for blackbody radiation. A proper, rigorous approach to the description of real surfaces would be that which first defines *monochromatic directional* properties and then proceeds to the *total* and *hemispherical* properties by appropriate integration over the spectrum and all directions. However, the presentation here will ignore possible directional dependencies by either defining simply the *hemispherical* properties or by assuming diffuseness when appropriate.

Emissivity—Monochromatic and Total

If $E_\lambda(\lambda, T)$ represents the hemispherical monochromatic emissive power of a nonblack surface maintained at a temperature T and measured at a particular wavelength λ, then the *hemispherical monochromatic emissivity* is defined as the ratio of $E_\lambda(\lambda, T)$ to the hemispherical monochromatic emissive power of a blackbody at the same T and λ:

$$\epsilon_\lambda(\lambda, T) = \frac{E_\lambda(\lambda, T)}{E_{b\lambda}(\lambda, T)}. \tag{9.31}$$

Note that no statement has been made as to whether the monochromatic emissivity, ϵ_λ, is greater or less than 1. That is, in this definition of ϵ_λ no assumption has been made as to whether or not a blackbody is a better or poorer emitter than a nonblack one. The fact that ϵ_λ must be less than 1 will be demonstrated later by use of Kirchhoff's law. It should be emphasized that ϵ_λ is a *hemispherical* emissivity, the emissive powers E_λ and $E_{b\lambda}$ being the sum of the energy emitted in all directions in the hemisphere above the surface.

Most real surfaces exhibit, to some degree, the characteristics of a "selective emitter" in that ϵ_λ is different for different wavelengths of the emitted energy. Also, ϵ_λ may be dependent on the surface temperature both in magnitude and spectral dependence, hence the notation $\epsilon_\lambda(\lambda, T)$; however, this temperature dependence may be small and is often ignored. Figures 9.9 and 9.10 show typical values of ϵ_λ, as functions of wavelength, for a number of surfaces. It may be noted that some surfaces exhibit dramatic variations of ϵ_λ with λ, while others show fairly constant values of ϵ_λ. Similar data are available in the literature for other surfaces and Refs. 1 through 4 may be consulted in this regard.

The hemispherical *total* emissivity is defined as the ratio of the total emissive power of a nonblack surface to that for a blackbody at the same temperature:

$$\epsilon(T) = \frac{E(T)}{E_b(T)}, \tag{9.32}$$

so that the emissive power of a nonblack surface is readily calculated when ϵ is known:

$$E(T) = \epsilon\sigma T^4. \tag{9.33}$$

The total emissivity may be expressed in terms of the monochromatic as follows:

$$\begin{aligned}\epsilon(T) = \frac{E(T)}{E_b(T)} &= \frac{\int_0^\infty E_\lambda(T)\, d\lambda}{\int_0^\infty E_{b\lambda}(\lambda, T)\, d\lambda} \\ &= \frac{\int_0^\infty \epsilon_\lambda(\lambda, T) E_{b\lambda}(\lambda, T)\, d\lambda}{\int_0^\infty E_{b\lambda}(\lambda, T)\, d\lambda} \\ &= \frac{\int_0^\infty \epsilon_\lambda(\lambda, T) E_{b\lambda}(\lambda, T)\, d\lambda}{\sigma T^4}.\end{aligned} \tag{9.34}$$

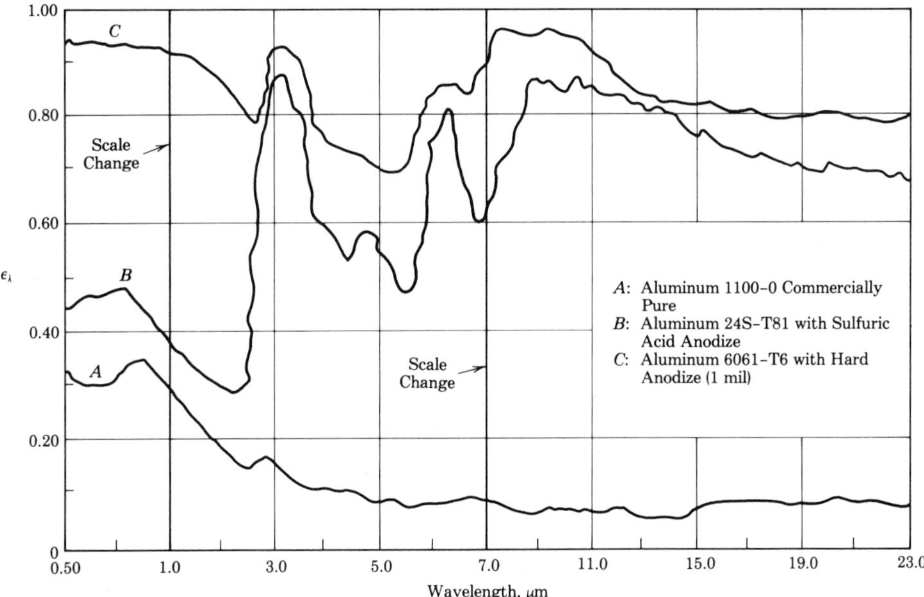

Figure 9.9. Monochromatic emissivity of various surfaces. (From D. K. Edwards, K. E. Nelson, R. D. Roddick, and J. T. Gier, "Basic Studies on the Use and Control of Solar Energy," *Univ. Calif. Dept. Eng. Rep. 60–93,* 1960.)

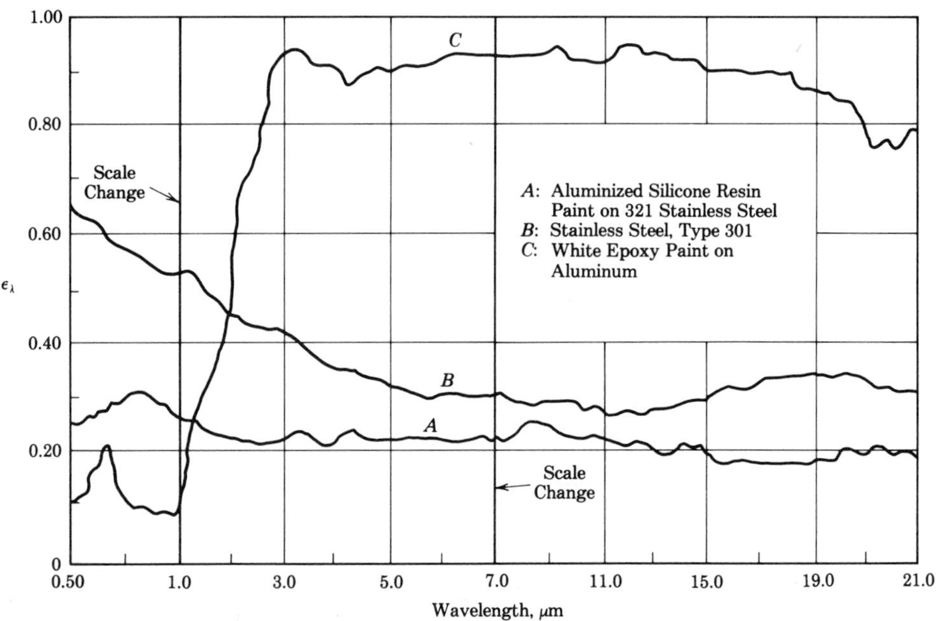

Figure 9.10. Monochromatic emissivity of various surfaces. (From D. K. Edwards, K. E. Nelson, R. D. Roddick, and J. T. Gier, "Basic Studies on the Use and Control of Solar Energy," *Univ. Calif. Dept. Eng. Rep. 60–93,* 1960.)

Since the monochromatic emissivity is a function of λ and T, Eq. (9.34) shows the total emissivity to depend on the surface temperature only. The total emissivity may be measured directly on the basis given in Eq. (9.32). Table A.10 lists values of the total emissivity for various surfaces, including the temperature dependency, and Fig. 9.11 illustrates graphically some typical data.

The total emissivity may be evaluated from the monochromatic emissivity by application of Eq. (9.34). If the $\epsilon_\lambda(\lambda, T)$ function is known, then Eq. (9.34) may be evaluated, probably numerically, using the Planck function in Eq. (9.25) or (9.28). If the ϵ_λ versus λ data are complex, as in Figs. 9.9 and 9.10, or known in

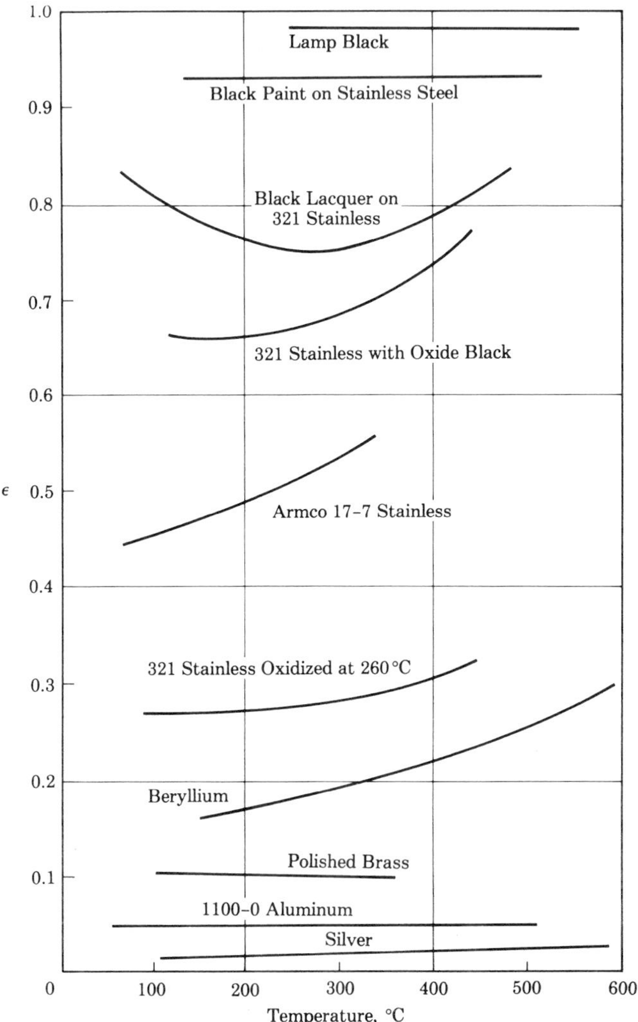

Figure 9.11. Total emissivity of various surfaces. (Based on data in Table A.10.)

numerical form, the integral of Eq. (9.34) may be expressed in terms of the function $F_{0-\lambda}$ given in Eq. (9.29) in a finite difference form. The spectrum of ϵ_λ versus λ may be divided into intervals ($\lambda_1, \lambda_2, \ldots, \lambda_i, \ldots, \lambda_n$) in which the monochromatic emissivity may be taken as approximately constant. If $\bar{\epsilon}_{\lambda_{i,i+1}}$ represents the average value of ϵ_λ between the wavelengths λ_i and λ_{i+1}, then Eq. (9.34) may be written in the form

$$\epsilon(T) = \sum \bar{\epsilon}_{\lambda_{i,i+1}}[F_{0-\lambda_{i+1}}(T) - F_{0-\lambda_i}(T)]. \tag{9.35}$$

Use of this relation is shown in Example 9.4.

A special type of nonblack surface, called the *ideal gray body,* is defined as one for which the monochromatic emissivity is independent of wavelength; i.e., the ratio of E_λ to $E_{b\lambda}$ is the same for all wavelengths of emitted energy at the given temperature. Thus, for the ideal gray surface, one readily obtains from Eq. (9.34):

$$\epsilon(T) = \epsilon_\lambda(T). \tag{9.36}$$

There are other rather profound simplifications that result from a surface exhibiting gray behavior, but these must be reserved until after the discussion of the property known as absorptivity. The assumption of a gray surface, ϵ_λ independent of λ, means that the curves of the monochromatic emissive power for a blackbody and a gray surface at the same temperature are affine to one another, there being no shift in the peak of the curve as illustrated in Fig. 9.12. Also shown in Fig. 9.12 is the monochromatic emissive power of a typical nonblack, nongray surface.

EXAMPLE 9.4

A nonblack surface exhibits a discontinuous monochromatic emissivity as shown in Fig. 9.13. If the surface is maintained at a temperature of 1500°K, find the total emissivity exhibited by the surface.

Solution. Since the specified variation of the monochromatic emissivity with wavelength is a step function with constant values of ϵ_λ in various ranges of the wavelength, the representation given in Eq. (9.35) for the total emissivity applies:

$$\epsilon(T) = \sum \bar{\epsilon}_{\lambda i,i+1}[F_{0-\lambda_{i+1}}(T) - F_{0-\lambda_i}(T)].$$

In this instance the spectrum is divided into three wavelength bands in which the monochromatic emissivity is constant as depicted in Fig. 9.13:

$$0 \le \lambda \le \lambda_1: \quad \bar{\epsilon}_{0,\lambda 1} = 0.2, \quad \lambda_1 = 2.0\ \mu\text{m}.$$

$$\lambda_1 \le \lambda \le \lambda_2: \quad \bar{\epsilon}_{\lambda 1,\lambda 2} = 0.8, \quad \lambda_1 = 2.0\ \mu\text{m}, \quad \lambda_2 = 5.0\ \mu\text{m}.$$

$$\lambda_2 \le \infty: \quad \bar{\epsilon}_{\lambda 2,\infty} = 0.4, \quad \lambda_2 = 5.0\ \mu\text{m}.$$

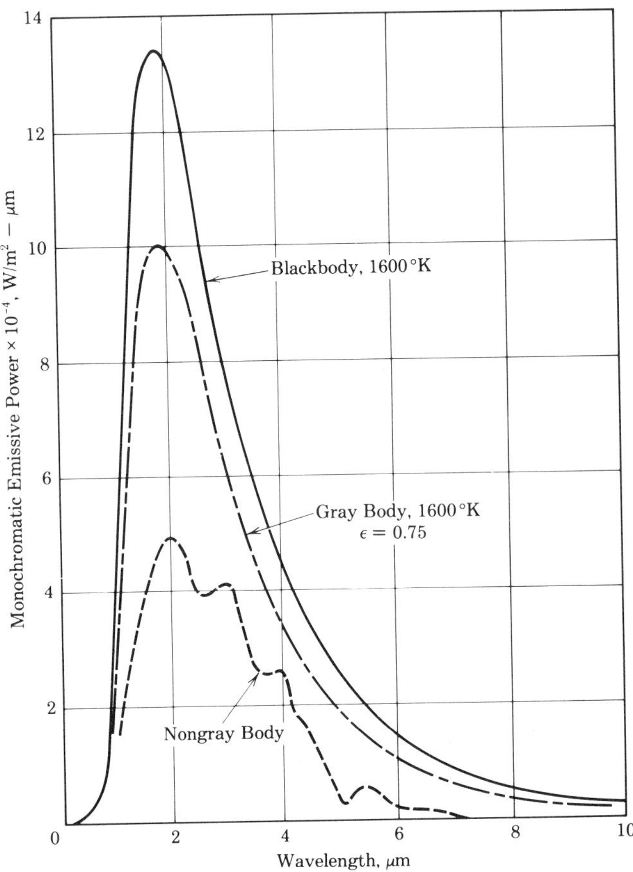

Figure 9.12. Monochromatic emissive power of a blackbody, a gray body, and a nongray body.

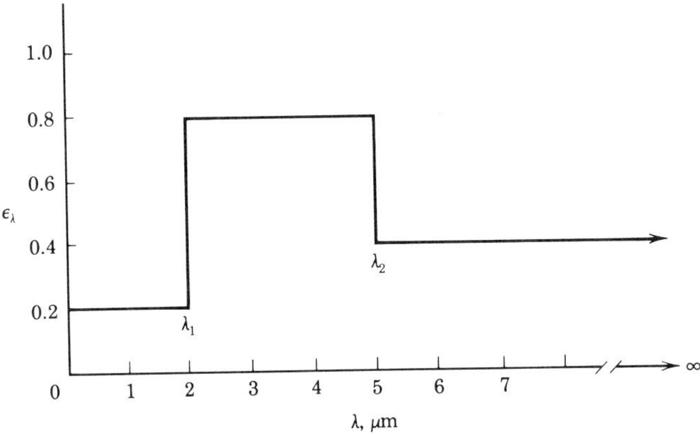

Figure 9.13. Example 9.4.

Thus Eq. (9.35) is

$$\epsilon = \bar{\epsilon}_{0,\lambda 1}[F_{0-\lambda 1}(T) - F_{0-0}(T)] + \bar{\epsilon}_{\lambda 1,\lambda 2}[F_{0-\lambda 2}(T) - F_{0-\lambda 1}(T)]$$
$$+ \bar{\epsilon}_{\lambda 2,\infty}[F_{0-\infty}(T) - F_{0-\lambda 2}(T)].$$

For $\lambda = 0$ the fractional function $F_{0-0}(T) = 0$ and as $\lambda \to \infty$ $F_{0-\infty}(T) = 1$. Thus

$$\epsilon = \bar{\epsilon}_{0,\lambda 1}[F_{0-\lambda 1} - 0] + \bar{\epsilon}_{\lambda 1,\lambda 2}[F_{0-\lambda 2}(T) - F_{0-\lambda 1}(T)] + \bar{\epsilon}_{\lambda 2,\infty}[1 - F_{0-\lambda 2}(T)].$$

The values of the function $F_{0-\lambda}(T)$ for $\lambda = \lambda_1 = 2.0$ μm, $\lambda = \lambda_2 = 5.0$ μm, and $T = 1500$°K are found from Table A.12:

$$\text{AT } \lambda_1 T = 2.0 \times 1500 = 3000 \ \mu\text{m-°K:} \quad F_{0-\lambda 1}(T) = 0.273,$$
$$\text{At } \lambda_2 T = 5.0 \times 1500 = 7500 \ \mu\text{m-°K:} \quad F_{0-\lambda 2}(T) = 0.834.$$

Thus

$$\epsilon = 0.2 \times 0.273 + 0.8(0.834 - 0.273) + 0.4(1 - 0.834)$$
$$= 0.570.$$

The total emissivity is dominated by the values of ϵ_λ beyond $\lambda = 2.0$ μm since the bulk of the emission from a blackbody at 1500°K occurs above this wavelength.

Absorptivity—Monochromatic and Total

One of the most significant ways in which real surfaces differ from the ideal blackbody is that real surfaces do not, generally, absorb all incident energy. Like emissivity, the absorptive character of a surface may exhibit both directional and spectral selectivity. Once again ignoring directional preferences by considering only the *hemispherical* irradiation incident upon a surface, G_λ, the *hemispherical monochromatic absorptivity* is defined as the fraction of this incident radiation that is absorbed:

$$\alpha_\lambda(\lambda, T) = \frac{[G_\lambda(\lambda)]_{abs}}{G_\lambda(\lambda)}. \tag{9.37}$$

In this definition G_λ is the irradiation, from all directions, incident on the surface at *a* particular wavelength and can be expressed in terms of the incident intensity by use of Eq. (9.20):

$$G_\lambda(\lambda) = \int_h I_{\lambda_i}(\lambda, \theta, \varphi) \cos\theta \, d\omega. \tag{9.38}$$

In the latter expression $I_{\lambda_i}(\lambda, \theta, \varphi)$ represents the incident *monochromatic* intensity, which, in general, is a function of direction as well as λ. Equation (9.37) presumes this function is known and the integral has been evaluated.

The notation $\alpha_\lambda(\lambda, T)$ in Eq. (9.37) is used to emphasize the fact that in addition to the obvious dependence of absorptivity on wavelength, experiment shows that it is also dependent on the surface temperature. It would appear that on the basis of the above definition, the monochromatic absorptivity α_λ is an additional surface property that would require experimental determination as a function of λ and T. However, as later discussion will show, Kirchhoff's law relates α_λ to ϵ_λ, defined previously. For the time being, however, α_λ will continue to be written as a separate entity.

The hemispherical *total* absorptivity is defined as the sum of the absorbed irradiation over all wavelengths divided by the total incidence. With this definition, the total absorptivity may be written in the following ways, using Eqs. (9.37) and (9.38):

$$\alpha(T, \text{source}) = \frac{\int_0^\infty [G_\lambda(\lambda)]_{\text{abs}}\, d\lambda}{\int_0^\infty G_\lambda(\lambda)\, d\lambda} = \frac{\int_0^\infty \alpha_\lambda(\lambda, T) G_\lambda(\lambda)\, d\lambda}{\int_0^\infty G_\lambda(\lambda)\, d\lambda} = \frac{\int_0^\infty \alpha_\lambda(\lambda, T) \int_h I_{\lambda_i}(\lambda, \theta, \varphi) \cos\theta\, d\omega\, d\lambda}{\int_0^\infty \int_h I_{\lambda_i}(\lambda, \theta, \varphi) \cos\theta\, d\omega\, d\lambda}. \tag{9.39}$$

The notation $\alpha(T, \text{source})$ is used to emphasize the fact that the total absorptivity depends not only on the absorbing surface temperature as implied in the function $\alpha_\lambda(\lambda, T)$, but also on the spectral and spatial characteristics of the incident radiation. In this sense the total absorptivity of a surface differs significantly from the total emissivity already defined. The total emissivity was seen to be a function of only the surface and its temperature; the total absorptivity is *also* a function of the source of the radiation incident on the surface. Thus, the total absorptivity is not a simple surface property that can be tabulated as can emissivity—a separate tabulation would have to be made for each source. This is normally not done except for a unique source, the sun, as noted in a later section.

It should be noted that a fundamental fact associated with the definition of the monochromatic and total absorptivities is that they are quantities necessarily less than or equal to 1.

One could continue in the line of reasoning presented for ϵ and α to define monochromatic and total values for the reflectivity and transmissivity of a surface. Reference 1 presents a comprehensive treatment of such formulations. However, most applications will be made to opaque surfaces so that $\tau = 0$, and the reflectivity may be deduced from

$$\rho_\lambda = 1 - \alpha_\lambda,$$

$$\rho = 1 - \alpha.$$

Kirchhoff's Law

The properties emissivity and absorptivity discussed in the foregoing may be related by what is known as Kirchhoff's law. It has already been established at the beginning of Sec. 9.3 that the irradiation field in an isothermal enclosure is equal to the emissive power of a blackbody at the temperature of the enclosure, i.e.,

$$G = E_b(T).$$

If a small nonblack surface is placed in the enclosure (small enough that it does not alter the irradiation field) with total emissivity ϵ and total absorptivity α, and allowed to come to the same temperature as the enclosure, then thermal equilibrium requires that the absorbed and emitted energies be equal:

$$\alpha G = E(T) = \epsilon E_b(T).$$

Since $G = E_b(T)$, as noted above, one has

$$\begin{aligned} \alpha E_b(T) &= \epsilon E_b(T), \\ \alpha &= \epsilon. \end{aligned} \tag{9.40}$$

Equation (9.40) is one form of the statement of Kirchhoff's law. Since $\alpha \leq 1$, then $\epsilon \leq 1$; that is, a blackbody, a perfect absorber, is also a perfect emitter.

It should be noted that the form of Kirchhoff's law just quoted in Eq. (9.40) is subject to the restrictions of *thermal equilibrium* in an *isothermal enclosure*. This also includes the condition of diffuseness since it was shown in Sec. 9.3 that the radiant field in an isothermal enclosure is diffuse. Extensions of Eq. (9.40) to other situations (as attractive as they may be) must be made with caution. Reference 1 presents an excellent and detailed exposition of Kirchhoff's law, including directional and spectral effects. In the case of *monochromatic* radiation, a derivation similar to that just shown yields the fact that

$$\alpha_\lambda(\lambda, T) = \epsilon_\lambda(\lambda, T) \tag{9.41}$$

as long as *either* the incident radiation is diffuse *or* the surface is such that α_λ and ϵ_λ have no directional dependencies. As long as these conditions are met, then known data for $\epsilon_\lambda(\lambda, T)$ such as are shown in Figs. 9.9 and 9.10 may be used to evaluate the total absorptivity from Eq. (9.39). Thus, additional physical data are not necessary; however, one still needs to know the spectral nature of the incident source.

The monochromatic form of Kirchhoff's law in Eq. (9.41) may be used with Eqs. (9.34) and (9.39) to deduce under what conditions that the total form of this law, Eq. (9.40), will be valid. Repeating Eq. (9.34) and rewriting Eq. (9.39) with the use of the monochromatic form of Kirchhoff's law:

$$\epsilon(T) = \frac{\int_0^\infty \epsilon_\lambda(\lambda, T) E_{b\lambda}(\lambda, T)\, d\lambda}{\int_0^\infty E_{b\lambda}(\lambda, T)\, d\lambda},$$

$$\alpha(T, \text{source}) = \frac{\int_0^\infty \epsilon_\lambda(\lambda, T) G_\lambda(\lambda)\, d\lambda}{\int_0^\infty G_\lambda(\lambda)\, d\lambda}. \tag{9.42}$$

Comparison of the two equations above shows that Kirchhoff's law in the total sense,

$$\alpha(T) = \epsilon(T), \tag{9.43}$$

occurs when *either* of the two following conditions is met:

1. If the incident irradiation on the receiving surface has a spectral distribution the same as that of the emission from a blackbody at the *same temperature*, i.e., $G_\lambda(\lambda) = E_{b\lambda}(\lambda)$.
2. If the receiving surface is an ideal *gray surface*, i.e., ϵ_λ is *not* a function of λ so that it may be moved from under the integral signs in Eqs. (9.34) and (9.42).

The first of the two conditions above is a restriction on the source of the irradiation, which must be that of a blackbody at the same temperature as the receiver. This was the case of the isothermal enclosure assumption leading to Eq. (9.40), but it is of little interest since thermal equilibrium seldom exists in heat transfer applications. The second condition, that of grayness, is a condition on the nature of the surface rather than the incident radiation. While there are numerous engineering applications in which the assumption of a gray surface may not be made (such as in the solar applications to be discussed next) there are many instances in which this idealization may be made. In such cases heat transfer calculations can be carried out using the total form of Kirchhoff's law, Eq. (9.43), even though thermal equilibrium does not exist. Such analyses are carried out extensively in Secs. 9.6 through 9.10.

9.5 SOLAR RADIATION AND SOLAR ABSORPTIVITY

Many applications of current engineering interest involving thermal radiation are ones in which radiation from the sun plays an important role. Many of these applications involve the use of surfaces for which the simplifying assumption of grayness cannot be made. Hence some characterization needs to be made of the solar radiation received at the earth.

Figure 9.14 illustrates the physical situation at hand. The sun, located at a distance from the earth equal to the earth's orbit radius r_e, radiates from its surface

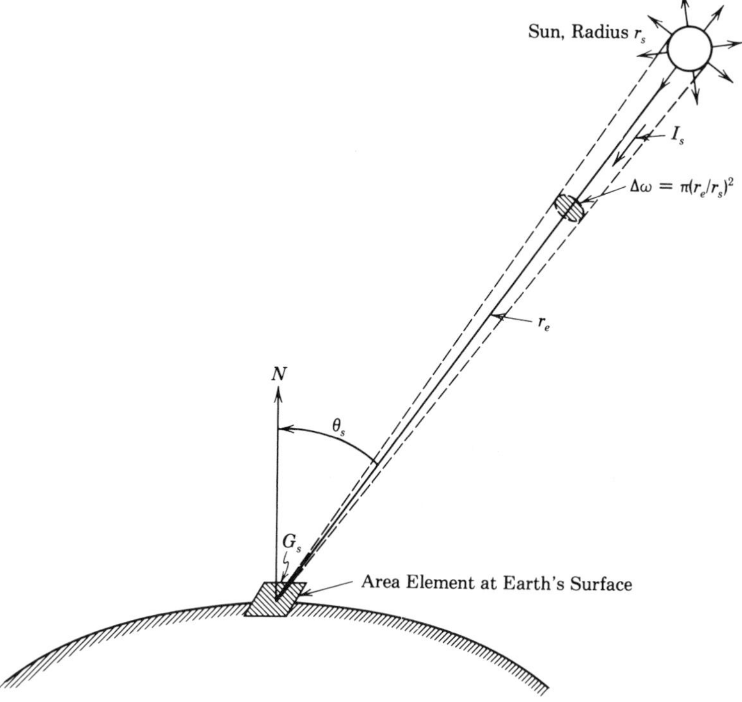

Figure 9.14. Geometric relations for solar irradiation.

in all directions. The radiation received at a surface element on the earth's surface appears to be virtually unidirectional, striking the surface at a solar angle of θ_s with the surface normal. As viewed from the surface, the sun subtends a solid angle $\Delta\omega$ which is

$$\Delta\omega = \frac{\pi r_s^2}{r_e^2},$$

where r_s is the sun's radius. Let the incident intensity from the sun, along the direction θ_s, be denoted by I_s (total, summed over all wavelengths) and let it be desired to find the total irradiation, G_s, falling on the surface. From the definition given in Eq. (9.20), the irradiation is

$$G_s = \int_h I_s(\theta, \varphi) \cos\theta \, d\omega,$$

where the integration is to be made over the hemisphere above the surface element. Because of the fact that all the energy incident on the element is contained in the solid angle $\Delta\omega$, the above integral is zero except in that solid angle. When the integration is taken in $\Delta\omega$, the polar angle θ varies slightly but may be taken as virtually constant and equal to θ_s, the polar angle of the center of the sun. Likewise, the solar intensity varies with θ, but because of the great distances involved, it may also be taken as constant in $\Delta\omega$. Thus the irradiation received at the surface is

$$G_s \cong I_s \cos\theta_s \,\Delta\omega$$
$$= I_s \cos\theta_s \left(\frac{\pi r_s^2}{r_e^2}\right).$$

The product $I_s(\pi r_s^2/r_e^2)$ is the irradiation the surface would receive were it placed normal to the sun's rays. Thus

$$G_s = G_{sn} \cos\theta_s,$$
$$G_{sn} = I_s\left(\frac{\pi r_s^2}{r_e^2}\right).$$

Spectral examination of the sun's radiation outside the earth's atmosphere indicates that its monochromatic intensity distributes among the wavelengths much as blackbody emission. There are some differences of course, and the earth's atmosphere further alters the spectral distribution by absorption and scattering. However, for present purposes, it is adequate to treat the sun as a blackbody emitter. Thus one may replace I_s in the equation above with the blackbody intensity, $I_b(T_s)$, which by use of the diffuseness of blackbody radiation given in Eq. (9.23), may be expressed in terms of the blackbody emissive power at the sun's apparent surface temperature, T_s:

$$I_s \cong I_b(T_s)$$
$$= \frac{E_b(T_s)}{\pi}.$$

Thus the sun's normal irradiation at the earth's surface is

$$G_{sn} = E_b(T_s)\frac{r_s^2}{r_e^2}$$
$$= \sigma T_s^4 \frac{r_s^2}{r_e^2}.$$

The normal irradiation of the sun received at the earth, G_{sn}, varies with the time of year since the earth's orbit is not circular. For our purposes here, it is sufficient to use the following observed yearly average value (just outside the atmosphere):

$$G_{sn} = 1396.9 \text{ W/m}^2 = 442.8 \text{ Btu/h-ft}^2.$$

This value may be used to estimate an equivalent solar temperature from the equation above if an average value of the earth's orbit radius $r_e = 149 \times 10^6$ km and the sun's radius $r_s = 0.70 \times 10^6$ km are used:

$$G_{sn} = \sigma T_s^4 \left(\frac{r_s}{r_e}\right)^2$$
$$1396.9 = 5.670 \times 10^{-8} \times T_s^4 \times \left(\frac{0.70}{149}\right)^2$$
$$T_s = 5780°\text{K} \cong 5800°\text{K (10,400 °R)}.$$

Thus, for many engineering applications it is adequate to represent the sun's radiation received at the earth's surface, outside the atmosphere, as

$$\text{blackbody radiation at } T_s = 5800°\text{K } (10{,}400°\text{R}),$$

$$G_{sn} = 1396.9 \text{ W/m}^2 \text{ (442.8 Btu/h-ft}^2\text{)}, \tag{9.44}$$

$$G_s = G_{sn} \cos \theta_s,\ \theta_s = \text{angle of incidence}.$$

Absorption by the earth's atmosphere depletes the normal component just calculated. The amount of depletion is dependent upon the location on the earth's surface, the time of year and day, as well as local atmospheric conditions. Also, scattering of the solar beam occurs in the atmosphere, so that a "diffuse" component is often significant in addition to the depleted direct beam. These details will not be treated here, and the reader may wish to consult Refs. 5 and 6 for more detailed information.

Solar Absorptivity

The discussion of Sec. 9.4 emphasized the fact that the total absorptivity of a surface depends on the spectral character of the incident irradiation as well as the nature of the receiving surface, while total emissivity is dependent on the surface alone. Since the sun represents a unique radiation source, values of the absorptivity of a surface exposed to the sun, the *solar absorptivity,* are unique surface properties that may be evaluated and tabulated. These values are necessary for use in the analysis of radiant heat transfer involving a solar phenomena. Unless it is gray, the solar absorptivity of a surface at a modest, earthbound temperature may differ significantly from its emissivity since the spectral distribution of the energy from the high temperature sun is vastly different from that of the emission at the lower temperature of the surface. If the surface is an ideal gray surface, then its emissivity and solar absorptivity will be the same.

Table A.11 quotes measured values of the solar absorptivity, α_s, for a number of surfaces, along with the emissivity. The ratio α_s/ϵ, a quantity whose significance will be seen shortly, is also tabulated, and a wide range of variation of α_s and α_s/ϵ is observed. Note, in particular, that α_s/ϵ may be much greater or much less than the gray case for which $\alpha_s/\epsilon = 1.0$.

The difference between α_s and ϵ, or the magnitude of α_s/ϵ, depends on the spectral dependence of the monochromatic emissivity $\epsilon_\lambda(\lambda, T)$. This dependence is best seen by returning to the definitions of the total emissivity and total absorptivity. Equation (9.34) for ϵ and its evaluation in terms of the partial integrals, $F_{0-\lambda}$, in Eq. (9.35) (as used in Example 9.3) are

$$\begin{aligned} \epsilon(T) &= \frac{\displaystyle\int_0^\infty \epsilon_\lambda(\lambda, T) E_{b\lambda}(\lambda, T)\, d\lambda}{\sigma T^4} \\ &= \Sigma\, \bar{\epsilon}_{\lambda_{i,i+1}} [F_{0-\lambda_{i+1}}(T) - F_{0-\lambda_i}(T)]. \end{aligned} \tag{9.45}$$

The solar absorptivity is given by Eq. (9.42) if the irradiation is taken to be that of blackbody at the solar temperature T_s. The resulting expression may also be written in terms of the function $F_{0-\lambda}$ over intervals in which ϵ_λ is taken constant:

$$\alpha_s(T) = \frac{\int_0^\infty \epsilon_\lambda(\lambda, T) E_{b\lambda}(\lambda, T_s)\, d\lambda}{\sigma T_s^4} = \Sigma\, \bar{\epsilon}_{\lambda_{i,i+1}} [F_{0-\lambda_{i+1}}(T_s) - F_{0-\lambda_i}(T_s)]. \tag{9.46}$$

Even though the function $\epsilon_\lambda(\lambda, T)$ is the same in both Eqs. (9.45) and (9.46), the values of the integrals will generally be different since the function $E_{b\lambda}$, or $F_{0-\lambda}$, will be different—being evaluated at the surface temperature T for ϵ or at the solar temperature T_s for α_s. The difference is illustrated in Example 9.5.

EXAMPLE 9.5

Find the solar absorptivity of the surface described in Example 9.4 and Fig. 9.13. Compare the result with the total emissivity found in Example 9.4.

Solution. As described in Example 9.4 and Fig. 9.13, the surface is a nongray one which exhibits a discontinuous monochromatic emissivity with respect to wavelength according to:

$$0 \le \lambda \le \lambda_1: \quad \bar{\epsilon}_{0,\lambda 1} = 0.2, \quad \lambda_1 = 2.0\ \mu\text{m}.$$

$$\lambda_1 \le \lambda \le \lambda_2: \quad \bar{\epsilon}_{\lambda 1,\lambda 2} = 0.8, \quad \lambda_1 = 2.0\ \mu\text{m}, \quad \lambda_2 = 5.0\ \mu\text{m}.$$

$$\lambda_2 \le \lambda < \infty: \quad \bar{\epsilon}_{\lambda 2,\infty} = 0.4, \quad \lambda_2 = 5.0\ \mu\text{m}.$$

Application of Eq. (9.46) over the wavelength intervals above (using, as in Example 9.4, the facts that $F_{0-0} = 0$, $F_{0-\infty} = 1$) gives the solar absorptivity to be

$$\alpha_s = \bar{\epsilon}_{0,\lambda 1}[F_{0-\lambda 1}(T_s) - 0] + \bar{\epsilon}_{\lambda 1,\lambda 2}[F_{0-\lambda 2}(T_s) - F_{0-\lambda 1}(T_s)] + \bar{\epsilon}_{\lambda 2,\infty}[1 - F_{0-\lambda 2}(T_s)].$$

The only difference between this expression for α_s and that for the total emissivity in Example 9.4 is that the functions involved are to be evaluated at the solar temperature T_s rather than the surface temperature. At $T_s = 5800°\text{K}$, as suggested in Eq. (9.44), and with $\lambda_1 = 2.0\ \mu\text{m}$, $\lambda_2 = 5.0\ \mu\text{m}$, Table A.12 gives:

$$\text{At } \lambda_1 T_s = 2.0 \times 5800 = 11{,}600\ \mu\text{m-°K}, \ F_{0-\lambda 1}(T_s) = 0.940,$$

$$\text{At } \lambda_2 T_s = 5.0 \times 5800 = 29{,}000\ \mu\text{m-°K}, \ F_{0-\lambda 2}(T_s) = 0.995.$$

Thus

$$\alpha_s = 0.2 \times 0.940 + 0.8(0.995 - 0.940) + 0.4(1 - 0.995) = 0.234.$$

The solar absorptivity is seen to be much smaller than the total emissivity $\epsilon = 0.570$ found in Example 9.4 at a surface temperature of 1500°K. This is due to the fact that at the solar temperature of $T_s = 5800$°K much more of the solar spectrum is concentrated in the spectral region where ϵ_λ is small (below, say, about $\lambda = 2.0\ \mu m$) than at 1500°K (as in Example 9.4).

The example just shown illustrated the calculation of the solar absorptivity when the sun's spectrum is represented as that from a blackbody at $T_s = 5800$°K. The same procedure is applied for an irradiation source that is the same as that from a blackbody at some other source temperature. The only difference is that one uses the source temperature instead of T_s. As long as the surface is nongray (i.e., $\epsilon_\lambda \neq$ constant) the total emissivity and the total absorptivity will be different as long as the surface temperature is different from the blackbody source. If the surface is an ideal gray one (i.e., $\epsilon_\lambda =$ constant), the emissivity and absorptivity will be the same, regardless of the temperatures, as the development of Kirchhoff's law in Sec. 9.4 showed.

Returning, now, to the discussion of solar absorptivity, the ratio α_s/ϵ, mentioned earlier, gives an indication of a surface's solar absorbing capability compared with its emitting capability. A large value of α_s/ϵ would be desirable for a solar collector surface since the surface should be able to absorb more energy than it would lose by reradiation. The reverse might be true for the surface of a spacecraft from which it might be desired to reject a net amount of heat in the presence of solar irradiation. These facts are sometimes expressed in terms of the equilibrium temperature that a surface would achieve if placed, adiabatically isolated, in the solar environment. In such a case the absorbed solar radiation would equal the emission, if any convection losses are ignored. Thus if T^* represents the equilibrium temperature of the surface,

$$\alpha_s G_s = E(T^*),$$

$$\alpha_s G_{sn} \cos\theta_s = \epsilon E_b(T^*) = \epsilon\, \sigma\, T^{*4}$$

or

$$T^* = \left(\frac{\alpha_s}{\epsilon}\,\frac{G_{sn}\cos\theta_s}{\sigma}\right)^{1/4}, \tag{9.47}$$

where θ_s is the angle between the surface normal and the sun's rays. Since G_{sn} and σ are fixed, the equilibrium temperature, T^*, depends on θ_s and the ratio α_s/ϵ. Large values of α_s/ϵ result in large T^*, while low values give low T^*.

In many applications the surface temperature may be fixed at a value different from T^*—as, for example, in a solar collector where it is desired to collect energy at a given temperature. In such a case, the *net* flux of energy *into* a surface maintained at temperature T is

$$\frac{q_{\text{net}}}{A} = \alpha_s G_{sn} \cos \theta_s - \epsilon \sigma T^4$$
$$= \alpha_s G_{sn} \cos \theta_s \left[1 - \left(\frac{T}{T^*} \right)^4 \right]. \tag{9.48}$$

For a specified collection temperature, T, the net flux absorbed would be positive only when a surface is chosen with α_s/ϵ large enough that $T^* > T$. Collection efficiency is improved the greater is T^*. Exactly the reverse is true if one desires, as perhaps in a spacecraft, to reject a net amount of heat in the presence of the sun.

The analysis above is made more complicated if simultaneous heat losses or gains to an ambient convecting fluid are included. Such problems are deferred to Chapter 11, where combined heat transfer modes are discussed.

EXAMPLE 9.6

A surface is placed normal to the sun's rays, outside the earth's atmosphere. Find the equilibrium temperature the surface would achieve if thermally isolated and the net heat flux absorbed by the surface if it is maintained at 130°C when the surface is composed of (a) silicon-coated aluminum foil; (b) anodized aluminum; (c) white epoxy paint on aluminum.

Solution. For the surface types specified, Table A.11 gives the solar absorptivity and total emissivity to be:

Silicon on Al: $\alpha_s = 0.522, \quad \epsilon = 0.12, \quad \alpha_s/\epsilon = 4.35.$

Anodized Al: $\alpha_s = 0.923, \quad \epsilon = 0.841, \quad \alpha_s/\epsilon = 1.10.$

White epoxy on Al: $\alpha_s = 0.248, \quad \epsilon = 0.882, \quad \alpha_s/\epsilon = 0.28.$

(a) The equilibrium temperature for zero heat flux is given by Eq. (9.47). Using $G_{sn} = 1396.9\ \text{W/m}^2$ and the foregoing properties of α_s and ϵ for silicon-coated aluminum foil, the result is

$$T^* = \left(\frac{\alpha_s}{\epsilon} \frac{G_{sn}}{\sigma} \right)^{1/4}$$
$$= \left(\frac{0.522}{0.12} \times \frac{1396.9}{5.67 \times 10^{-8}} \right)^{1/4} = 572°\text{K} = 299°\text{C}\ (570°\text{F}).$$

If this surface is maintained at 130°C, $T = 130 + 273 = 403°\text{K}$,† net heat is absorbed by the surface since $T < T^*$. Equation (9.48) gives the net heat absorption rate to be, for normal incidence,

†In all examples in this chapter absolute temperatures will be calculated as °K = °C + 273 or °R = °F + 460 rather than °K = °C + 273.15 or °R = °F + 459.67. Usually, the accuracy of the data given does not justify the more accurate definition.

$$\frac{q_{\text{net}}}{A} = \alpha_s G_{sn} \cos\theta_s \left[1 - \left(\frac{T}{T^*}\right)^4\right]$$

$$= 0.522 \times 1396.9 \times 1 \times \left[1 - \left(\frac{403}{572}\right)^4\right]$$

$$= 550 \text{ W/m}^2 \text{ (174 Btu/h-ft}^2\text{)}.$$

Similar calculations for the other two specified surfaces give:

(b) Anodized aluminum:

$$T^* = \left(\frac{\alpha_s}{\epsilon} \frac{G_{sn}}{\sigma}\right)^{1/4}$$

$$= \left(\frac{0.923}{0.841} \times \frac{1396.9}{5.67 \times 10^{-8}}\right)^{1/4} = 406°\text{K} = 133°\text{C (271°F)},$$

$$\frac{q_{\text{net}}}{A} = \alpha_s G_{sn} \cos\theta_s \left[1 - \left(\frac{T}{T^*}\right)^4\right]$$

$$= 0.923 \times 1396.9 \times 1 \left[1 - \left(\frac{403}{406}\right)^4\right]$$

$$= 37.7 \text{ W/m}^2 \text{ (11.9 Btu/h-ft}^2\text{)}.$$

(c) White epoxy on Al:

$$T^* = \left(\frac{\alpha_s}{\epsilon} \frac{G_{sn}}{\sigma}\right)^{1/4}$$

$$= \left(\frac{0.248}{0.882} \times \frac{1396.9}{5.67 \times 10^{-8}}\right)^{1/4} = 288°\text{K} = 15°\text{C (59°F)},$$

$$\frac{q_{\text{net}}}{A} = \alpha_s G_{sn} \cos\theta_s \left[1 - \left(\frac{T}{T^*}\right)^4\right]$$

$$= 0.248 \times 1396.9 \times 1 \left[1 - \left(\frac{403}{288}\right)^4\right]$$

$$= -982 \text{ W/m}^2 \text{ (311 Btu/h-ft}^2\text{)}.$$

Example 9.6 illustrates the effect of the magnitude of the ratio α_s/ϵ of a nongray surface on its performance in a solar environment. In case (a) the high value of α_s/ϵ led to a high value of T^*, greater than the surface temperature T, so that the surface *absorbed* a net heat flux. In case (c) the low α_s/ϵ led to the reverse result—a low T^* and a net rejection of heat while in the solar rays. A surface of the type of case (a) might be desirable for solar collector applications, while that in case (c) might prove useful in the passive thermal control of satellites or spacecraft. The surface in case (b), anodized aluminum, appears to be fairly neutral (almost gray) in the solar environment, at least for applications at the surface temperature used in the example.

The next example illustrates the spectral character a nongray surface must have in order to yield a high or low α_s/ϵ ratio.

EXAMPLE 9.7

Find the equilibrium temperature of a thermally isolated surface placed normal to the sun's irradiation, T^*, when the surface has a step-function form for its monochromatic emissivity, as suggested in Fig. 9.15. That is, there exists a "cutoff" value of the wavelength, $\lambda = \lambda_c$, below which the monochromatic emissivity has a constant value $\epsilon_{\lambda 1}$, and above which its value is a different constant, $\epsilon_{\lambda 2}$:

$$\text{For } 0 \leq \lambda \leq \lambda_c,\ \epsilon_\lambda = \epsilon_{\lambda 1}.$$

$$\text{For } \lambda_c \leq \lambda \leq \infty,\ \epsilon_\lambda = \epsilon_{\lambda 2}.$$

Find the equilibrium temperature T^* for (a) a surface which is a poor emitter in the short wavelengths and a good emitter in the long wavelengths, as in Fig. 9.15(a), with $\epsilon_{\lambda 1} = 0.1$, $\epsilon_{\lambda 2} = 0.9$, $\lambda_c = 4\ \mu\text{m}$; or (b) a surface which is a good emitter in the short wavelengths and a poor emitter in the long wavelengths, as in Fig. 9.15(b), with $\epsilon_{\lambda 1} = 0.9$, $\epsilon_{\lambda 2} = 0.1$, $\lambda_c = 4\ \mu\text{m}$.

Solution. The sought-for equilibrium temperature, T^*, is given by Eq. (9.47), for $\theta_s = 0°$:

$$T^* = \left(\frac{\bar{\alpha}_s}{\epsilon}\frac{G_{sn}}{\sigma}\right)^{1/4}. \tag{1}$$

The solar absorptivity, $\bar{\alpha}_s$, and the total emissivity, ϵ, of the surface are needed to evaluate T^*. The solar absorptivity of the surface is found by the methods illustrated in Example 9.5. Application of Eq. (9.46) over the two wavelength intervals below and above the cutoff wavelength λ_c gives

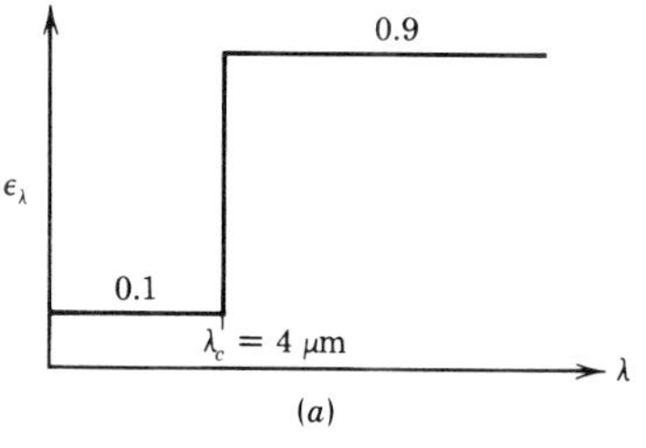

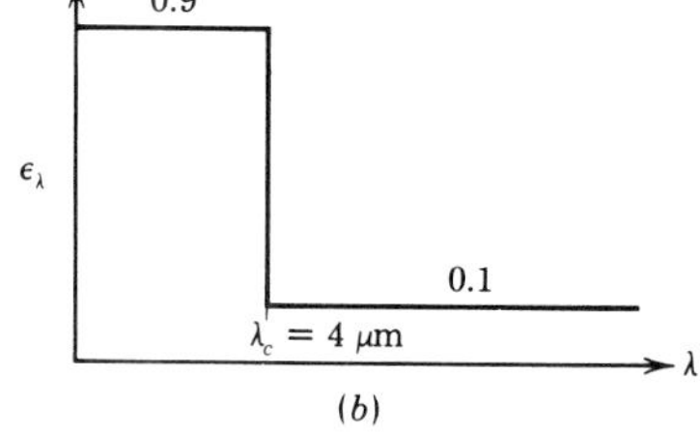

Figure 9.15. Example 9.7.

$$\begin{aligned}\alpha_s &= \Sigma\, \bar{\epsilon}_{\lambda i,i+1}[F_{0-\lambda_{i+1}}(T_s) - F_{0-\lambda_i}(T_s)] \\ &= \epsilon_{\lambda 1} F_{0-\lambda_c}(T_s) + \epsilon_{\lambda 2}[1 - F_{0-\lambda_c}(T_s)],\end{aligned} \tag{2}$$

when the facts that the fractional function $F_{0-\lambda} = 0$ for $\lambda = 0$ and $F_{0-\infty} = 1$ for $\lambda \to \infty$ are used. Thus the solar absorptivity may be found from known values of λ_1, λ_2, and λ_c along with the known solar temperature $T_s = 5800°\text{K}$. Thus α_s is independent of the surface temperature.

The total emissivity of the surface, however, must be evaluated at the yet-to-be-determined surface temperature, T^*. Application of Eq. (9.35), as in Example 9.4, at T^* gives

$$\begin{aligned}\epsilon &= \Sigma\, \bar{\epsilon}_{\lambda i,i+1}[F_{0-\lambda_{i+1}}(T^*) - F_{0-\lambda_i}(T^*) \\ &= \epsilon_{\lambda 1} F_{0-\lambda_c}(T^*) + \epsilon_{\lambda 2}[1 - F_{0-\lambda_c}(T^*)].\end{aligned} \tag{3}$$

With λ_c, $\epsilon_{\lambda 1}$, and $\epsilon_{\lambda 2}$ known, α_s may be found from Eq. (2); however, since ϵ is dependent on the surface temperature T^*, an iterative calculation for T^* and ϵ must be performed using Eqs. (1) and (3). This is illustrated for the two surface types specified for this example, as noted below.

(a) For the surface which is a poor emitter at low wavelengths and good at the longer [Fig. 9.15(a)], one has: $\lambda_c = 4\ \mu\text{m}$, $\epsilon_{\lambda 1} = 0.1$, $\epsilon_{\lambda 2} = 0.9$. Thus Eq. (2) gives the solar absorptivity of the surface (using the sun's temperature as $T_s = 5800°\text{K}$) to be

$$\lambda_c T_s = 4 \times 5800 = 23{,}200\ \mu\text{m-°K},$$

$$\text{Table A.12:}\quad F_{0-\lambda_c}(T_s) = 0.990,$$

$$\begin{aligned}\alpha_s &= \epsilon_{\lambda 1} F_{0-\lambda_c}(T_s) + \epsilon_{\lambda 2}[1 - F_{0-\lambda_c}(T_s)] \\ &= 0.1 \times 0.990 + 0.9(1 - 0.990) = 0.108.\end{aligned}$$

Equation (1) and (3) interrelate the unknown equilibrium temperature, T^*, and the total emissivity which is a function of T^*:

$$\begin{aligned}T^* &= \left(\frac{\alpha_s}{\epsilon}\frac{G_{sn}}{\sigma}\right)^{1/4} \\ &= \left(\frac{0.108}{\epsilon} \times \frac{1396.9}{5.67 \times 10^{-8}}\right)^{1/4}.\end{aligned}$$

and

$$\begin{aligned}\epsilon &= \epsilon_{\lambda 1} F_{0-\lambda_c}(T^*) + \epsilon_{\lambda 2}[1 - F_{0-\lambda_c}(T^*)] \\ &= 0.1 \times F_{0-\lambda_c}(T^*) + 0.9[1 - F_{0-\lambda_c}(T^*)].\end{aligned}$$

These two expressions must be solved iteratively. The final results are:

$$T^* = 233°\text{K, assumed},$$

$$\lambda_c T^* = 4 \times 233 = 932 \ \mu\text{m-°K},$$

$$F_{0-\lambda_c} = 0.00014 \quad \text{(Table A.12)},$$

$$\epsilon = 0.1 \times 0.00014 + 0.9(1 - 0.00014) = 0.900,$$

$$T^* = \left(\frac{0.107}{0.900} \times \frac{1396.9}{5.67 \times 10^{-8}}\right)^{1/4}$$

$$= 233°\text{K, calculated}$$

$$= -40°\text{C} \ (-40°\text{F}).$$

(b) For the second case of a surface which is a good emitter in the short wavelengths and a poor one in the longer [Fig. 9.15(b)], with $\lambda_c = 4$ μm, $\epsilon_{\lambda 1} = 0.9$, $\epsilon_{\lambda 2} = 0.1$, identical calculations may be carried out. Only the final results are shown:

$$\lambda_c T_s = 23{,}200 \ \mu\text{m-°K},$$

$$F_{0-\lambda_c}(T_s) = 0.990,$$

$$\alpha_s = 0.892$$

$$T^* = 577°\text{K, assumed},$$

$$\lambda_c T^* = 2308 \ \mu\text{m-°K},$$

$$F_{0-\lambda_c}(T^*) = 0.1216,$$

$$\epsilon = 0.197$$

$$T^* = 577°\text{K} = 340°\text{C} \ (579°\text{F}), \text{ calculated}.$$

In case (a) the surface had a very low value of ϵ_λ for wavelengths less than 4 μm, where most of the sun's spectrum is concentrated (as shown in Table 9.1), and a high ϵ_λ in the infrared region, where it does most of its emitting. Thus surface (a) is a poor solar absorber, $\alpha_s = 0.108$, but a good low-temperature emitter, $\epsilon = 0.900$. The resulting low value of $\alpha_s/\epsilon = 0.108/0.900 = 0.120$ leads to a low T^*, suggesting that the surface might be useful in a spacecraft application where it is desired to reject heat while in the direct rays of the sun. For case (b), exactly the reverse is true: $\alpha_s = 0.892$, $\epsilon = 0.197$, $\alpha_s/\epsilon = 4.53$. Such a surface might be desirable for a solar collector in which one wishes to absorb as much solar irradiation as possible, with the minimum of loss by reradiation.

9.6 DIFFUSE RADIANT EXCHANGE BETWEEN GRAY INFINITE PARALLEL PLANES

Much of the discussion of the preceding two sections, in particular Sec. 9.5 concerning solar phenomena, has emphasized the effects associated with a surface exhibiting nongray effects—ones in which Kirchhoff's law, that the total emissivity and absorptivity are the same, could not be applied. In this, and subsequent, sections the spectral behavior of a surface will not be emphasized so that the discussion may be devoted to some of the geometrical problems associated with radiant exchange. Thus in most of the remainder of this chapter, only gray surfaces will be treated, so that regardless of the nature of radiation incident on a surface, one may apply Kirchhoff's law for the total emissivity and absorptivity, Eq. (9.43):

$$\epsilon = \alpha,$$

and the emissive power of such a surface expressed by Eq. (9.40):

$$E = \epsilon E_b$$
$$= \epsilon \sigma T^4.$$

A general problem of radiant heat transfer that will be treated eventually is that of the exchange between a collection of such gray surfaces having arbitrary size and shape and located in an arbitrary way with respect to one another. One of the fundamental aspects of such a problem will be seen to be the geometric question of how much of the radiation leaving one surface is actually incident on another. In order to concentrate initially some of the physical processes taking place, however, the simplified case of the radiant exchange between the opposing faces of two parallel, infinite, gray planes will be treated first. In this instance one is assured of the fact that all the radiation leaving one surface is incident on the other. The assumption of infinite planes implies the neglection of edge effects in the case of finite surfaces and is often a justifiable simplification if the space between the planes is small compared to the size of the planes.

Consider, then, two infinite, parallel, gray planes, as suggested in Fig. 9.16, each of a known temperature on the opposing surfaces. The symbols T_1 and T_2 are used to denote the temperatures of the two planes. Even though the surfaces may not be black, knowing their temperatures is equivalent to knowing the associated blackbody emissive powers: $E_{b1} = \sigma T_1^4$, $E_{b2} = \sigma T_2^4$. Then the actual surface emissive powers are

$$E_1 = \epsilon_1 E_{b1} = \epsilon_1 \sigma T_1^4,$$

$$E_2 = \epsilon_2 E_{b2} = \epsilon_2 \sigma T_2^4.$$

This practice of using specified equivalent blackbody emissive power in the place of specified temperature, even when the surfaces are nonblack, will be used throughout the remainder of this chapter and has the advantage that the resultant heat exchange is expressible as a linear function of the E_b's rather than a nonlinear function of the T's.

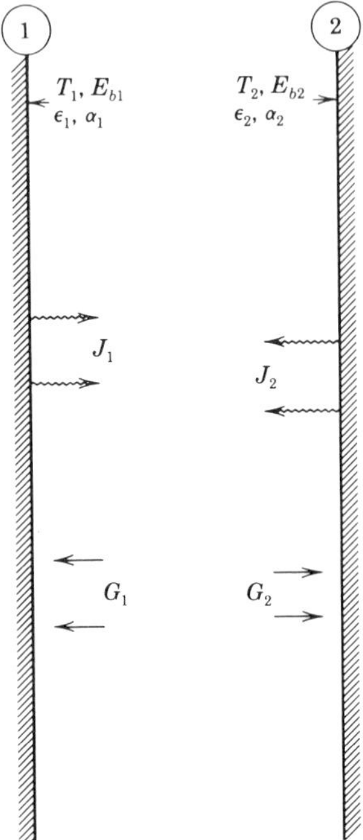

Figure 9.16. Infinite parallel gray planes.

Return now to the case depicted in Fig. 9.16. If the opposing faces of the two planes have specified temperatures, (i.e., E_b's) and known values of the surface properties ϵ_1, α_1, ϵ_2, α_2, let it be desired to find the *net* rate of heat flow *away* from each surface—i.e., the heat that must be supplied, or removed, to maintain the surfaces at the specified temperatures. One could determine this flow of energy by deducing how much of the emission from a surface is reflected back and absorbed through a multiplicity of interreflections (and partial absorptions) with the other surface. However, this energy flow may be deduced more simply by algebraic means. An elementary heat balance on, say, surface 1 for the *net* radiant energy leaving it, q_1, is

$$q_1 = A_1(J_1 - G_1),$$

where J_1 and G_1 represent the radiosity and irradiation, respectively, of that surface. Since the surfaces are infinite in extent, it only makes sense to work in terms of the heat flux, q_1/A_1. Also, since all the energy flowing away from surface 1 must end up at surface 2, no energy being lost out the ends, one has

$$\frac{q_1}{A_1} = -\frac{q_2}{A_2} = J_1 - G_1, \tag{9.49}$$

where q_2 represents the net radiant energy leaving A_2. Since the two planes are infinite in extent and parallel, the irradiation on one must be the radiosity of the other:

$$G_1 = J_2, \qquad G_2 = J_1. \tag{9.50}$$

In addition, the definition of radiosity, the energy leaving a surface gives, as in Eq. (9.6),

$$J = E + \rho G.$$

For an opaque surface, the reflectivity ρ is $\rho = 1 - \alpha$; however, if the surface is diffuse and *gray,* then Kirchhoff's law holds for the total properties so that $\alpha = \epsilon$, or

$$\begin{aligned} J &= E + (1 - \epsilon)G \\ &= \epsilon E_b + (1 - \epsilon)G. \end{aligned}$$

When applied to the two surfaces in question,

$$\begin{aligned} J_1 &= \epsilon_1 E_{b1} + (1 - \epsilon_1)G_1, \\ J_2 &= \epsilon_2 E_{b2} + (1 - \epsilon_2)G_2. \end{aligned} \tag{9.51}$$

Equations (9.50) and (9.51) are four equations in J_1, J_2, G_1, and G_2 and may be solved for J_1 and G_1 for use in Eq. (9.49):

$$\begin{aligned} J_1 &= \frac{\epsilon_1 E_{b1} + (1 - \epsilon_1)\epsilon_2 E_{b2}}{1 - (1 - \epsilon_1)(1 - \epsilon_2)}, \\ G_1 &= \frac{\epsilon_2 E_{b2} + (1 - \epsilon_2)\epsilon_1 E_{b1}}{1 - (1 - \epsilon_1)(1 - \epsilon_2)}. \end{aligned} \tag{9.52}$$

Equation (9.49) then yields the desired heat flows in terms of the specified properties and E_b's (i.e., temperature):

$$\begin{aligned} \frac{q_1}{A_1} = -\frac{q_2}{A_2} &= \frac{\epsilon_1\epsilon_2 E_{b1} - \epsilon_1\epsilon_2 E_{b2}}{1 - (1 - \epsilon_1)(1 - \epsilon_2)} \\ &= \frac{E_{b1} - E_{b2}}{(1/\epsilon_1) + (1/\epsilon_2) - 1} \\ &= \frac{\sigma(T_1^4 - T_2^4)}{(1/\epsilon_1) + (1/\epsilon_2) - 1}. \end{aligned} \tag{9.53}$$

If both surfaces are black, then $\epsilon_1 = \epsilon_2 = 1$ and Eq. (9.53) becomes simply

$$\frac{q_1}{A_1} = -\frac{q_2}{A_2} = E_{b1} - E_{b2}.$$

Since in Eq. (9.53) each of ϵ_1 and ϵ_2 is less than 1, the denominator is greater than 1. Thus the effect of the grayness of either surface is to reduce the heat flux from what it would be if the surfaces were black. This is the result of the fact that a nongray surface reflects back part of the radiation incident on it. For this reason, when it is desired to minimize the flow of radiant heat between two surfaces (such as in the air space in a building wall or between the walls of a vacuum bottle), one or both of the surfaces are usually coated with a material having a low ϵ (or high reflectivity, $\rho = 1 - \alpha = 1 - \epsilon$), such as aluminum.

EXAMPLE 9.8

Two opposed, parallel, infinite, gray planes are maintained at temperatures of 300°F and 400°F, respectively.

(a) What is the net heat flux away from each surface if one of them has an emissivity of 0.7 and the other an emissivity of 0.8? Does it matter which plane has which emissivity?
(b) What is the net heat flux if the planes are black instead of gray?
(c) Repeat part (a) if the temperature difference is doubled by raising the 400°F plane to 500°F.

Solution

(a) The two surface temperatures are $T_1 = 300 + 460 = 760$°R and $T_2 = 400 + 460 = 860$°R. In English units, Eq. (9.22) gives the Stefan–Boltzmann constant to be $\sigma = 0.1714 \times 10^{-8}$ Btu/h-ft^2-°R^4. Thus the equivalent black-body emissive powers are

$$E_{b1} = \sigma T_1^4 = 0.1714\left(\frac{T_1}{100}\right)^4 = 0.1714\left(\frac{760}{100}\right)^4 = 571.8 \text{ Btu/h-ft}^2,$$

$$E_{b2} = \sigma T_2^4 = 0.1714\left(\frac{T_2}{100}\right)^4 = 0.1714\left(\frac{860}{100}\right)^4 = 937.6 \text{ Btu/h-ft}^2.$$

Note that the values above are not the emissive powers of the surfaces, just those of black surfaces at the same temperatures. The actual emissive powers of the surfaces, which are not needed to solve this problem, are $E_1 = \epsilon_1 E_{b1}$ and $E_2 = \epsilon_2 E_{b2}$. However, it is not known which surface has which emissivity. The heat flux away from each surface is given by Eq. (9.53):

$$\frac{q_1}{A_1} = -\frac{q_2}{A_2} = \frac{E_{b1} - E_{b2}}{1/\epsilon_1 + 1/\epsilon_2 - 1}.$$

The way in which the surface emissivities enter the expression above shows that the result is independent of which plane has which emissivity. Thus for ϵ_1 or $\epsilon_2 = 0.7$ and ϵ_2 or $\epsilon_1 = 0.8$, one has

$$\frac{q_1}{A_1} = -\frac{q_2}{A_2} = \frac{571.8 - 937.6}{(1/0.7) + (1/0.8) - 1} = -217.9 \text{ Btu/h-ft}^2 \text{ (687.4 W/m}^2\text{)}.$$

(b) If the planes were black instead of gray, one simply puts $\epsilon_1 = \epsilon_2 = 1.0$ in the above, and finds

$$\frac{q_1}{A_1} = -\frac{q_2}{A_2} = E_{b1} - E_{b2}$$

$$= 571.8 - 937.6 = -365.8 \text{ Btu/h-ft}^2 \ (1154 \text{ W/m}^2).$$

The heat flux for black surfaces is seen to be significantly greater than when they are gray, in accord with the discussion above.

(c) If T_2 is changed to $T_2 = 500 + 460 = 960°\text{R}$, one has

$$E_{b2} = \sigma T_2^4 = 0.1714\left(\frac{T_2}{100}\right)^4 = 0.1714\left(\frac{960}{100}\right)^4 = 1455.8 \text{ Btu/h-ft}^2.$$

With the other data of part (a) unchanged, the heat flux becomes

$$\frac{q_1}{A_1} = -\frac{q_2}{A_2} = \frac{E_{b1} - E_{b2}}{1/\epsilon_1 + 1/\epsilon_2 - 1}$$

$$= \frac{571.8 - 1455.8}{(1/0.7) + (1/0.8) - 1} = 526.5 \text{ Btu/h-ft}^2 \ (1661 \text{ W/m}^2).$$

Doubling the temperature difference increases the heat flux by a factor of 2.4—illustrating the nonlinear aspects of radiant heat exchange.

Plane Radiation Shields

Parts (a) and (b) of Example 9.8 show the great influence that a surface emissivity different from unity has on the heat exchange between a pair of planes. This fact is often used to reduce the loss, or gain, of radiant energy from a surface by use of what is known as a radiation shield.

The radiant exchange for two planes, A_1 at T_1 and A_2 at T_2, has already been shown to be

$$\left(\frac{q_1}{A_1}\right)_0 = \mathscr{F}_{1-2}(E_{b1} - E_{b2}),$$

where (9.54)

$$\frac{1}{\mathscr{F}_{1-2}} = \frac{1}{\epsilon_1} + \frac{1}{\epsilon_2} - 1,$$

and the notation $(q_1/A_1)_0$ is used to signify the heat flux when nothing is placed between the planes.

Now let a third plane, a radiation shield (call it A_s with emissivity ϵ_s on each side), be placed between planes A_1 and A_2. Rather than being maintained at a specified temperature, let the shield be thermally isolated so that it comes to some equilibrium temperature T_s (or equivalent blackbody emissive power $E_{bs} = \sigma T_s^4$)

such that the radiant heat it exchanges with plane A_1 equals the negative of that it exchanges with A_2. Thus the shield A_s is neither a heat source nor sink, but simply a plane exchanging the same heat with each of its neighbors so that it assumes whatever temperature it must to satisfy this condition. The reflective (or emissive) properties of the shield determines this equilibrium temperature and, thereby, the net heat flow between A_1 and A_2. Application of Eq. (9.53) for the heat exchanged by A_s and A_1 gives

$$-\frac{q_{s-1}}{A_s} = \mathscr{F}_{1-s}(E_{b1} - E_{bs}),$$

$$\frac{1}{\mathscr{F}_{1-s}} = \frac{1}{\epsilon_1} + \frac{1}{\epsilon_s} - 1.$$

and that exchanged by A_s with A_2 is

$$\frac{q_{s-2}}{A_s} = \mathscr{F}_{s-2}(E_{bs} - E_{b2}),$$

$$\frac{1}{\mathscr{F}_{s-2}} = \frac{1}{\epsilon_s} + \frac{1}{\epsilon_2} - 1.$$

Equating the last two expressions for the shield heat flux gives the equilibrium blackbody emissive power of the shield, or its temperature, to be

$$E_{bs} = \frac{\mathscr{F}_{1-s}E_{b1} + \mathscr{F}_{s-2}E_{b2}}{\mathscr{F}_{1-s} + \mathscr{F}_{s-2}},$$

$$T_s^4 = \frac{\mathscr{F}_{1-s}T_1^4 + \mathscr{F}_{s-2}T_2^4}{\mathscr{F}_{1-s} + \mathscr{F}_{s-2}}. \tag{9.55}$$

The factors $\mathscr{F}_{1-s}$ and $\mathscr{F}_{s-2}$ are functions of only the emissivities of the surfaces and the shield as noted above. Since the heat flow away from surface A_1 with the shield in place is

$$\left(\frac{q_1}{A_1}\right)_s = -\frac{q_{s-1}}{A_s} = \mathscr{F}_{1-s}(E_{b1} - E_{bs}),$$

then Eq. (9.55) gives, after some algebra,

$$\left(\frac{q_1}{A_1}\right)_s = \frac{E_{b1} - E_{b2}}{1/\mathscr{F}_{1-s} + 1/\mathscr{F}_{s-2}}$$

$$= \frac{E_{b1} - E_{b2}}{1/\epsilon_1 + 2/\epsilon_s + 1/\epsilon_2 - 2}. \tag{9.56}$$

Comparison of Eqs. (9.54) and (9.56) shows that since $\epsilon_s < 1$, the heat flow with the shield in place, $(q_1/A_1)_s$, is always less than that with no shield, $(q_1/A_1)_0$. This reduction is the result of the fact that the equilibrium E_{bs} of the shield (or T_s) must lie between E_{b1} and E_{b2} (or T_1 and T_2) as required by Eq. (9.55). Thus the surface

A_1 now "sees" an environment of a higher temperature (if $T_1 > T_2$) than before, reducing the heat flow. The exact amount of the reduction of the heat flux depends on the relative magnitude of the emissivities ϵ_1, ϵ_2, and ϵ_s.

The amount of reduction in heat flow is best expressed as the ratio

$$\frac{(q_1/A_1)_s}{(q_1/A_1)_0} = \frac{1/\mathcal{F}_{1-2}}{1/\mathcal{F}_{1-s} + 1/\mathcal{F}_{s-2}} = \frac{1/\epsilon_1 + 1/\epsilon_2 - 1}{1/\epsilon_1 + 2/\epsilon_s + 1/\epsilon_2 - 2}. \tag{9.57}$$

The ratio above is clearly a number less than 1. How much this ratio is less than 1 depends on the relative magnitude of the term $2/\epsilon_s - 1$ with $(1/\epsilon_1) + (1/\epsilon_2) - 1$.

A feeling for the effectiveness of a shield is obtained by looking at the special case when all emissivities are the same. In this instance $\epsilon_1 = \epsilon_2 = \epsilon_s$, the above becomes

$$\frac{(q_1/A_1)_s}{(q_1/A_1)_0} = \frac{1}{2};$$

that is, a single shield of the same emissivity as the surfaces to be shielded reduces the heat flow by one-half. It may be shown that N shields of equal emissivities will reduce the heat flow according to

$$\frac{(q_1/A_1)_N}{(q_1/A_1)_0} = \frac{1}{N + 1}. \tag{9.58}$$

EXAMPLE 9.9

If the planes described in Example 9.8(a) are shielded by a thermally isolated plane with $\epsilon_s = 0.5$ on both sides, find (a) the equilibrium temperature of the shield; (b) the ratio of the heat flux between the two original planes with the shield in place to that with no shield. In this instance it *does* matter which surface has which emissivity. Hence take $\epsilon_1 = 0.7$ and $\epsilon_2 = 0.8$.

Solution. The specifications of Example 9.8 gave the following known data:

$$T_1 = 300°\text{F} = 760°\text{R}, \qquad E_{b1} = 571.8 \text{ Btu/h-ft}^2, \qquad \epsilon_1 = 0.7,$$

$$T_2 = 400°\text{F} = 860°\text{R}, \qquad E_{b2} = 937.6 \text{ Btu/h-ft}^2, \qquad \epsilon_2 = 0.8.$$

The transfer factors $\mathcal{F}_{1-s}$ and $\mathcal{F}_{s-2}$ defined in the foregoing are readily calculated:

$$\mathcal{F}_{1-s} = \left[\frac{1}{\epsilon_1} + \frac{1}{\epsilon_s} - 1\right]^{-1} = \left[\frac{1}{0.7} + \frac{1}{0.5} - 1\right]^{-1} = 0.4118,$$

$$\mathcal{F}_{s-2} = \left[\frac{1}{\epsilon_s} + \frac{1}{\epsilon_2} - 1\right]^{-1} = \left[\frac{1}{0.5} + \frac{1}{0.8} - 1\right]^{-1} = 0.4444.$$

(a) The equilibrium blackbody emissive power attained by the shield, so that it exchanges the same (in absolute value) heat with A_1 and A_2, is given by Eq. (9.55):

$$E_{bs} = \frac{\mathcal{F}_{1-s}E_{b1} + \mathcal{F}_{s-2}E_{b2}}{\mathcal{F}_{1-s} + \mathcal{F}_{s-2}}$$

$$= \frac{0.4118 \times 571.8 + 0.4444 \times 937.6}{0.4118 + 0.4444} = 486.6 \text{ Btu/h-ft}^2.$$

Thus the shield temperature is

$$E_{bs} = \sigma T_s^4,$$

$$486.6 = 0.1714\left(\frac{T_s}{100}\right)^4,$$

$$T_s = 730°\text{R} = 270°\text{F } (132°\text{C}).$$

The equilibrium temperature T_s lies between T_1 and T_2, as it should, but it is clearly not the arithmetic mean.

(b) The ratio of the heat flux with the shield in place to that with no shield is given by Eq. (9.57):

$$\frac{(q_1/A_1)_s}{(q_1/A_1)_0} = \frac{1/\epsilon_1 + 1/\epsilon_2 - 1}{1/\epsilon_1 + 2/\epsilon_s + 1/\epsilon_2 - 2}$$

$$= \frac{(1/0.7) + (1/0.8) - 1}{(1/0.7) + (2/0.5) + (1/0.8) - 2} = 0.359.$$

The shield is seen to reduce the heat flow to less than half its original value, due mainly to its low emissivity, or high reflectivity.

9.7 THE SHAPE FACTOR FOR DIFFUSE RADIANT EXCHANGE BETWEEN FINITE SURFACES

The simplicity of the problems just discussed in Sec. 9.6 was due primarily to the fact that all the radiation leaving one plane was incident on the other. Thus the radiosity of one surface was known to be the irradiation upon the other. In many applications, such as the radiant exchange between the several surfaces of a room, a furnace, a spacecraft, etc., this simplification does not hold. Thus it is now necessary to consider the general case of a collection of finite, nonparallel, nonplanar surfaces that are exchanging radiant energy.

The Shape Factor

Of basic importance is the *direct* exchange of energy between two surfaces—that fraction of the energy leaving one surface which strikes the second directly, none being considered which is transferred by reflection or reradiation from other surfaces that may be present. This direct flux of energy between one surface and another is expressed in terms of the *shape factor,* which is found by determining the exchange between differential area elements in each surface and then performing simultaneous integrations over both surfaces. When Lambert's law is presumed to hold, the result is dependent solely on the geometry of the problem.

The geometry of the situation is illustrated in Fig. 9.17. The two surfaces in question are denoted by A_1 and A_2. These surfaces are located arbitrarily in space, are of arbitrary shape, and are not necessarily plane. Elements of each surface are denoted by dA_1 and dA_2 with corresponding normals N_1 and N_2. The line connecting dA_1 to dA_2 has length r and makes polar angles θ_1 and θ_2 with the two normals N_1 and N_2, respectively. According to Eq. (9.12), the fraction of the flux leaving dA_1 which is intercepted by dA_2 is

$$df = I_1 \cos\theta_1 \, d\omega_{2-1},$$

in which I_1 represents the intensity of the radiation leaving dA_1 and $d\omega_{2-1}$ is the solid angle subtended by dA_2 at dA_1. Projecting dA_2 normal to the radial line r gives $d\omega_{2-1} = (dA_2 \cos\theta_2)/r^2$, so that

$$df = I_1 \frac{\cos\theta_1 \cos\theta_2}{r^2} dA_2.$$

Since df is the portion of the *flux* from dA_1 intercepted by dA_2, it may be written as

$$df = \frac{dq_{1\to 2}}{dA_1} = I_1 \frac{\cos\theta_1 \cos\theta_2}{r^2} dA_2,$$

where $q_{1\to 2}$ represents the energy leaving all of A_1 that strikes all of A_2. Thus $q_{1\to 2}$ is found by integrating the expression above over both the areas:

$$q_{1\to 2} = \int_{A_1} \int_{A_2} I_1 \frac{\cos\theta_1 \cos\theta_2}{r^2} dA_2 \, dA_1.$$

If the radiosity of the originating surface, J_1, is uniform over that surface, the total energy leaving it is $A_1 J_1$, so the *fraction* of the energy leaving A_1 which strikes

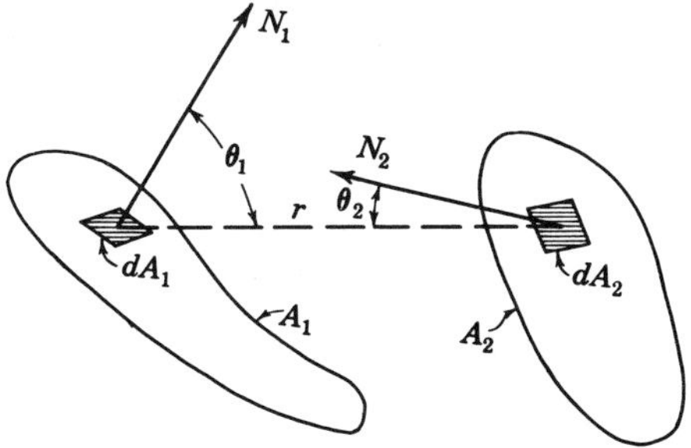

Figure 9.17

A_2 is

$$F_{1-2} = \frac{q_{1\to 2}}{A_1 J_1}. \tag{9.59}$$

If, in addition, the radiant energy leaving A_1 is *diffuse* (both the emissive and reflected portions), then I_1 is a constant and Eq. (9.19) states that $J_1 = \pi I_1$. Thus the two equations above give

$$F_{1-2} = \frac{1}{A_1}\int_{A_1}\int_{A_2} \frac{\cos\theta_1 \cos\theta_2}{\pi r^2}\, dA_2\, dA_1. \tag{9.60}$$

The factor F_{1-2} is called the diffuse *configuration factor* or *shape factor* of A_1 with respect to A_2—the fraction of the energy leaving A_1 which *directly* strikes A_2. As may be seen from its representation in Eq. (9.60), the shape factor for diffuse radiation is a purely geometric property of the two surfaces involved. The order of the subscripts on F_{1-2} is quite important since F_{2-1} will be used to represent the fraction of energy leaving A_2 which directly strikes A_1. In general, $F_{1-2} \neq F_{2-1}$, since reasoning similar to that given above yields

$$F_{2-1} = \frac{1}{A_2}\int_{A_2}\int_{A_1} \frac{\cos\theta_2 \cos\theta_1}{\pi r^2}\, dA_1\, dA_2. \tag{9.61}$$

Some Properties of the Shape Factor

Certain useful properties of the shape factor may be deduced from the definition given by Eqs. (9.59) and (9.60). These properties will be useful for the determination of shape factors for specific geometries and for the analysis of heat exchange between surfaces.

The Reciprocal Property. The most important property of the shape factor is readily deducible from Eq. (9.60) and (9.61). For two surfaces, A_1 and A_2, in general $F_{1-2} \neq F_{2-1}$; however, the following reciprocal relation exists:

$$A_1 F_{1-2} = A_2 F_{2-1}. \tag{9.62}$$

The Additive Property. If one of the two surfaces, say A_i, is divided into subareas $A_{i_1}, A_{i_2}, \ldots, A_{i_n}$ (so that $\Sigma A_{i_n} = A_i$), the application of Eqs. (9.59) and (9.60) will lead to the following relation:

$$A_i F_{i-j} = \sum_n A_{i_n} F_{i_n - j}. \tag{9.63}$$

If the reciprocal property of Eq. (9.62) is applied to Eq. (9.63),

$$\begin{aligned} A_j F_{j-i} &= \sum_n A_j F_{j-i_n}, \\ F_{j-i} &= \sum_n F_{j-i_n}. \end{aligned} \tag{9.64}$$

Thus if the transmitting surface is subdivided, the shape factor for that surface with respect to a receiving surface is not simply the sum of the individual shape factors, although the AF product is expressed by such a sum. The shape factor from a surface to a subdivided receiving surface *is* simply the sum of the individual shape factors.

If surface A_i is subdivided into n parts $(A_{i_1}, A_{i_j}, \ldots, A_{i_n})$ and surface A_j into m parts $(A_{j_1}, A_{j_2}, \ldots, A_{j_m})$, the reasoning above leads to

$$A_i F_{i-j} = \sum_n \sum_m A_{i_n} F_{i_n - j_m}. \tag{9.65}$$

These additive properties are useful in finding the shape factor for complex shapes by subdividing the surfaces into subsections for which the shape factor is either known or more simply evaluated. This is discussed in detail in Appendix C.

The Enclosure Property. If the interior surface of a completely enclosed space, such as a room, is subdivided into n parts—each having a finite area $A_1, A_2, \ldots, A_n$, then n equations of the following form may be written, one for each surface:

$$\sum_{j=1}^{n} F_{i-j} = 1, \qquad i = 1, 2, \ldots, n. \tag{9.66}$$

The representation above admits the shape factors $F_{1-1}, F_{2-2}, \ldots, F_{n-n}$ since some of the surfaces may "see" themselves if they are concave. That is, for example, the definition of the shape factor in Eq. (9.60) does not preclude the possibility that the two incremental areas, dA_1 and dA_2, may be on the same surface.

If a surface A_1 is completely enclosed by a second surface, A_2, and if A_1 does not see itself (i.e., it is convex) so that $F_{1-1} = 0$, then the enclosure property gives

$$F_{1-1} + F_{1-2} = 1,$$

$$F_{1-2} = 1.$$

For the enclosing surface, A_2, the shape factor with respect to the enclosed surface, F_{2-1}, is readily found from the above with the reciprocal property of Eq. (9.62), and since A_2 must surely see itself, the self-shape factor F_{2-2} is found from the enclosure property:

$$F_{2-1} = \frac{A_1}{A_2} F_{1-2} = \frac{A_1}{A_2},$$
$$F_{2-2} = 1 - F_{2-1} = 1 - \frac{A_1}{A_2}. \tag{9.67}$$

The Calculation of Shape Factors

The definition of the shape factor is given by

$$F_{1-2} = \frac{1}{A_1} \int_{A_1} \int_{A_2} \frac{\cos\theta_1 \cos\theta_2}{\pi r^2} \, dA_2 \, dA_1.$$

The two integrals in this equation are double integrals—each taken over a surface. Calculation of F_{1-2} for a specific geometrical setup will require that a coordinate system be established for each surface. Then the location of dA_1 and dA_2 on A_1 and A_2, respectively, may be expressed in terms of these coordinates. Likewise, the angles θ_1 and θ_2 and the connecting distance r may also be expressed in terms of the coordinate locations of dA_1 and dA_2. Finally, the area elements themselves must be expressed in terms of differentials of the coordinates—resulting in a twofold integral for each of the double integrals. Thus in terms of geometrical coordinates, the expression for F_{1-2} becomes, usually, a fourfold integral.

For even the simplest geometrical shapes the evaluation of the fourfold integral becomes quite involved. Since such integrations are principally mathematical in nature, the details of shape factor evaluation, for a selection of physically significant cases, are given in Appendix C. The resultant expressions for these cases are presented for convenience in Figs. 9.18 through 9.22. Figures 9.18 and 9.19 present the shape factors for directly opposed rectangles and perpendicular rectangles with a common edge; Fig. 9.20 gives F_{1-2} for parallel disks, and Fig. 9.21 presents F_{2-1} and F_{2-2} for coaxial cylinders of finite length. Two cases of importance in the present-day analysis of radiant exchange between a planetary body and elements of a spacecraft or satellite are given in Fig. 9.22—the shape factors between a large sphere and either a small plane element or a small sphere.

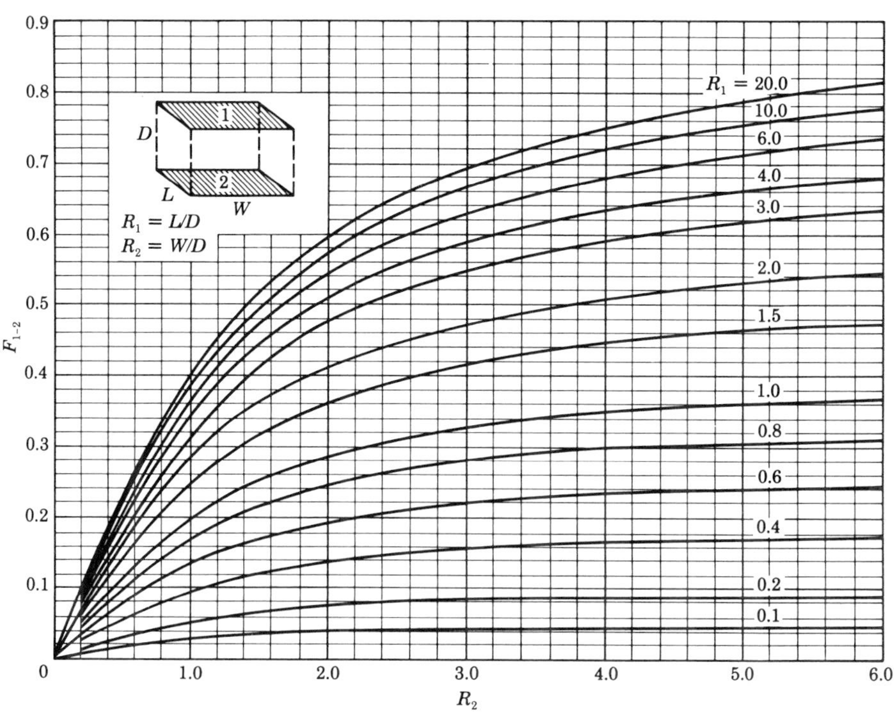

Figure 9.18. Shape factor for parallel, directly opposed, rectangles.

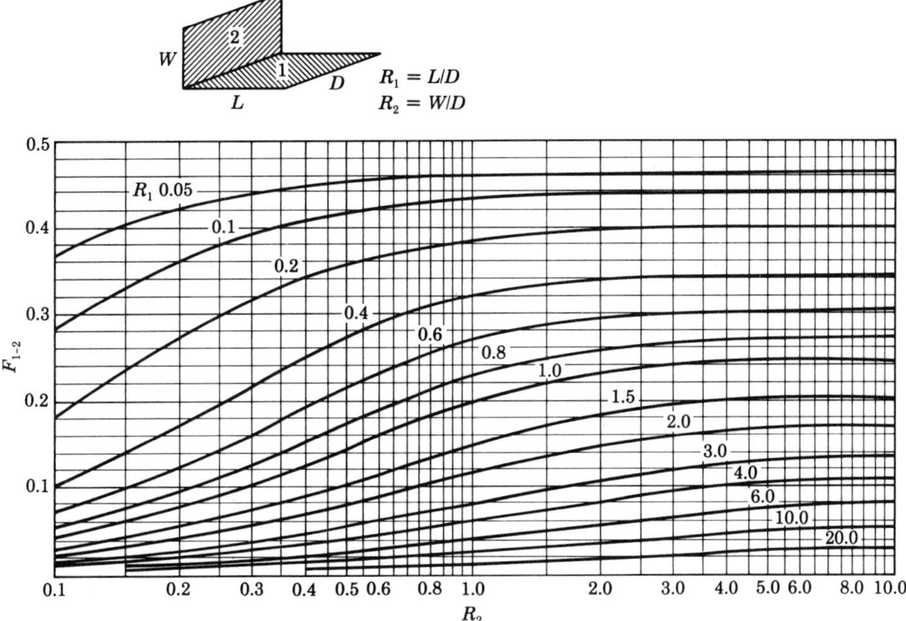

Figure 9.19. Shape factor for perpendicular rectangles with a common edge.

More complex shapes often may be reduced to simpler cases by the application of "shape factor algebra"—i.e., the use of some of the shape factor properties discussed in the preceding section, particularly the additive property. Appendix C illustrates the manner in which this reduction may be carried out. The material which follows in this chapter presumes that the reader is familiar with the methods of Appendix C and has mastered the methods of determining the shape factor for a given geometry.

Such mathematical methods as Stokes' theorem may be applied in some cases to reduce the double area integrals to contour integrals (Ref. 7). For many geometries, numerical methods utilizing digital computers may be advantageous (Ref. 8).

EXAMPLE 9.10

Find the shape factor F_{1-2} for the geometry shown in Fig. 9.23.

Solution. To reduce the desired shape factor to a form involving geometries for which F is already known, define the additional areas A_3, A_4, A_5, and A_6 as shown in Fig. 9.23. The given area A_1 may be subdivided into its three constituent parts, A_5 and two of the size A_6 (by symmetry):

$$A_1 = A_5 + A_6 + A_6$$

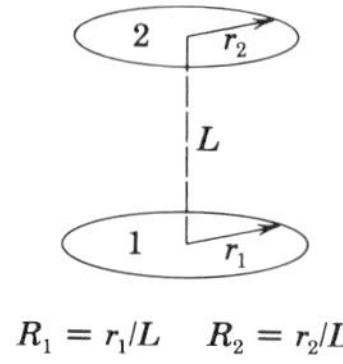

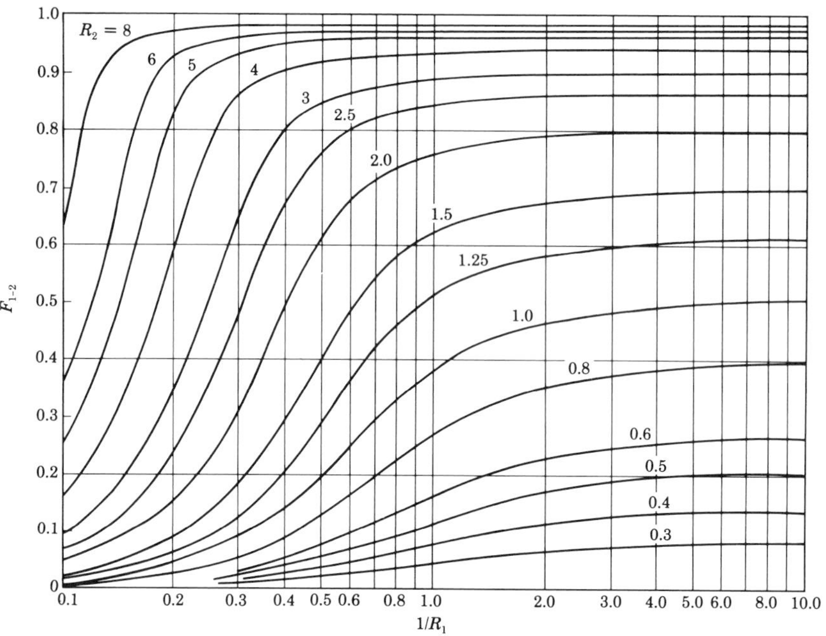

Figure 9.20. Shape factor for parallel concentric disks.

Then the additive property of Eq. (11.63) gives

$$A_1F_{1-2} = A_5F_{5-2} + 2A_6F_{6-2}.$$

The shape factor F_{6-2} is shown in Appendix C to be expressible in terms of the other areas involved as

$$A_6F_{6-2} = \tfrac{1}{2}[A_{(5,6)}F_{(5,6)-(2,3)} - A_5F_{5-2} - A_6F_{6-3}],$$

so that the expression for F_{1-2} above becomes

$$\begin{aligned} A_1F_{1-2} &= A_5F_{5-2} + 2 \times \tfrac{1}{2}[A_{(5,6)}F_{(5,6)-(2,3)} - A_5F_{5-2} - A_6F_{6-3}] \\ &= A_{(5,6)}F_{(5,6)-(2,3)} - A_6F_{6-3}. \end{aligned}$$

The shape factors on the right side of the equation above are both of the form of perpendicular rectangles with a common edge as shown in Fig. 9.19 or Eq. (C.3).

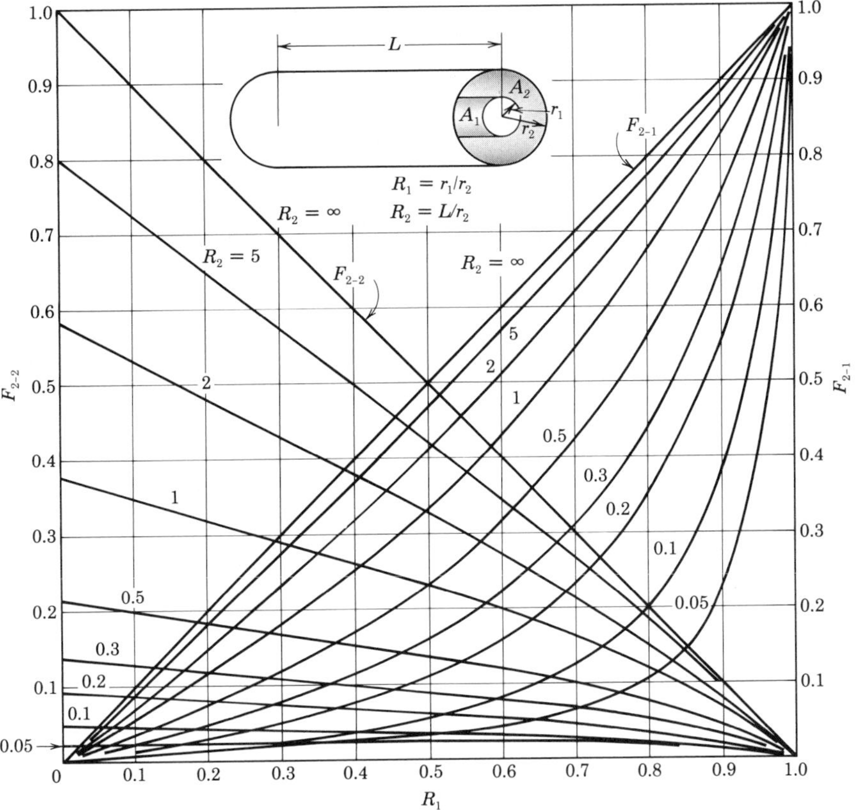

Figure 9.21. Shape factors for concentric cylinders of finite length.

For the factor $F_{(5,6)-(2,3)}$ the geometry of Fig. 9.19 gives

$$R_1 = \frac{L}{D} = \frac{12}{6 + 2} = 1.50, \qquad R_2 = \frac{W}{D} = \frac{3}{6 + 2} = \frac{3}{8},$$

so that Fig. 9.19 or, more accurately, Eq. (C.3) gives

$$F_{(5,6)-(2,3)} = 0.085.$$

For F_{6-3},

$$R_1 = \frac{L}{D} = \frac{12}{2} = 6.0, \qquad R_2 = \frac{3}{2} = 1.5$$

$$F_{6-3} = 0.051.$$

The areas needed are $A_1 = 10 \times 12 = 120$ m^2, $A_{(5,6)} = (6 + 2) \times 12 = 96$ m^2, $A_6 = 2 \times 12 = 24$ m^2, so

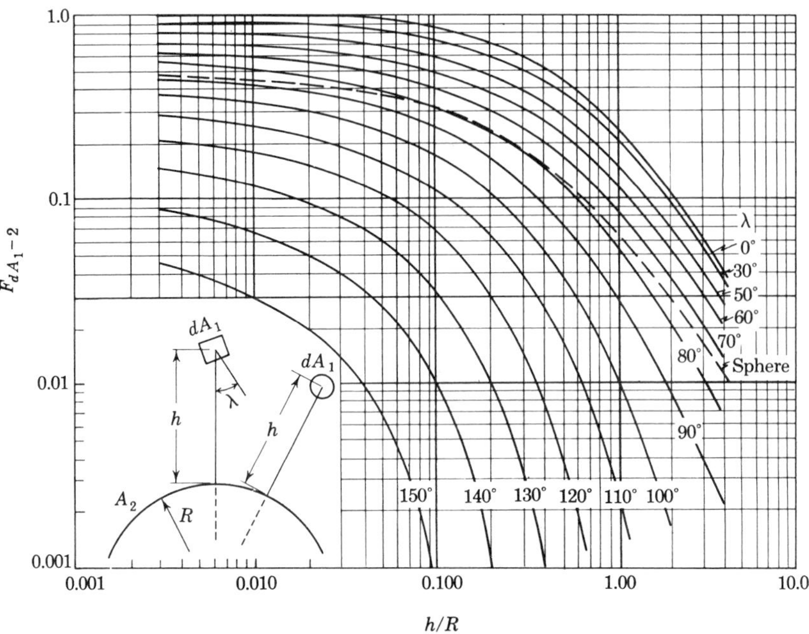

Figure 9.22. Shape factor for infinitesimal planes and spheres in the presence of a large sphere.

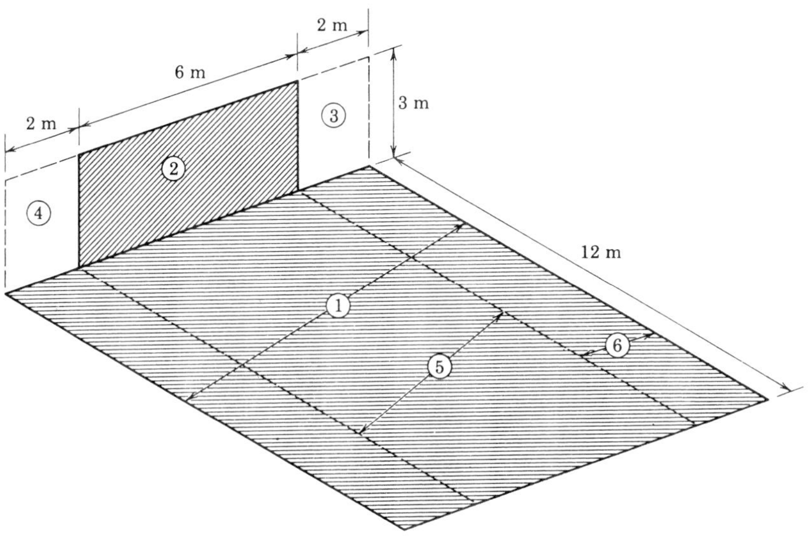

Figure 9.23. Example 9.10.

$$A_1F_{1-2} = A_{(5,6)}F_{(5,6)-(2,3)} - A_6F_{6-3}$$

$$120 \times F_{1-2} = 96 \times 0.085 - 24 \times 0.051$$

$$F_{1-2} = 0.058.$$

EXAMPLE 9.11

Figure 9.24 illustrates a furnace consisting of a parallelepiped having the dimensions: 10 m high, 20 m wide, 30 m long. The back wall of the furnace (10 m × 30 m) is to be treated as surface A_1. The top (20 m × 30 m) and front (10 m × 30 m) of the furnace are to be treated as a *single* surface A_2. The two end walls (10 m × 20 m) are to be taken together as another single surface A_3.

(a) Find the shape factor of the back wall with respect to the combined top and front, F_{1-2}.
(b) Find the shape factor of the combined top and front with respect to the combined ends, F_{2-3}.
(c) Since the surface comprising the combined top and front walls clearly sees itself, find the self-shape factor F_{2-2}.

Solutions. To reduce the problem of finding the desired shape factors to configurations involving only directly opposed rectangles or perpendicular rectangles having a common edge, as known in Figs. 9.18 and 9.19, it is useful to divide the areas specified as A_2 and A_3 into their constituent parts. The area designated as A_2 comprises the top and front wall of the furnace. Let A_a denote the top and A_b the

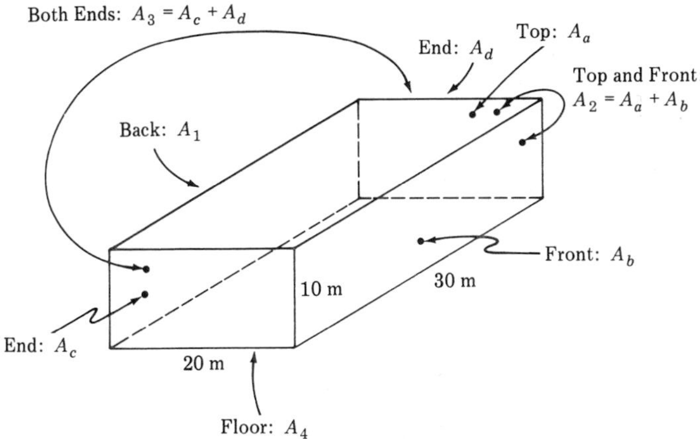

Figure 9.24. Example 9.11.

front wall, as indicated in Fig. 9.24. Then the geometry of Fig. 9.24 shows that

$$A_a = 20 \times 30 = 600 \text{ m}^2,$$

$$A_b = 10 \times 30 = 300 \text{ m}^2,$$

$$A_2 = A_a + A_b = 900 \text{ m}^2.$$

In like fashion, the area designated by A_3 comprises both ends of the furnace, so if A_c denotes one end and A_d the other,

$$A_c = 10 \times 20 = 200 \text{ m}^2,$$

$$A_d = 10 \times 20 = 200 \text{ m}^2,$$

$$A_3 = A_c + A_d = 400 \text{ m}^2.$$

The other surface area involved is the back wall of the furnace, A_1:

$$A_1 = 10 \times 30 = 300 \text{ m}^2.$$

(a) The shape factor of the back wall, A_1, with respect to the combined top and front wall, A_2 (i.e., F_{1-2}) may be found in terms of simpler elements by application of the additive property in Eq. (9.64) since $A_2 = A_a + A_b$:

$$A_1F_{1-2} = A_1F_{1-a} + A_1F_{1-b}.$$

The constituent shape factors F_{1-a} and F_{1-b} are of the known forms. For F_{1-a}, two perpendicular rectangles with a common edge, the geometry of Fig. 9.19 gives

$$R_1 = \frac{L}{D} = \frac{10}{30} = \frac{1}{3}, \qquad R_2 = \frac{W}{D} = \frac{20}{30} = \frac{2}{3},$$

so that Fig. 9.19 or, more accurately, Eq. (C.3) gives

$$F_{1-a} = 0.319.$$

For F_{1-b}, directly opposed rectangles, the geometry of Fig. 9.18 gives

$$R_1 = \frac{L}{D} = \frac{10}{20} = 0.500, \qquad R_2 = \frac{W}{D} = \frac{30}{20} = 1.50,$$

so that Fig. 9.18 or Eq. (C.2) gives

$$F_{1-b} = 0.146.$$

Thus the desired F_{1-2} is, from the above,

$$A_1F_{1-2} = A_1F_{1-a} + A_1F_{1-b}$$

$$F_{1-2} = F_{1-a} + F_{1-b}$$

$$= 0.319 + 0.146 = 0.465.$$

(b) The shape factor of the combined front and top, A_2, with respect to the combined ends, A_3, is similarly reduced to simpler configurations. This time, both A_2 and A_3 may be subdivided. First, dividing A_2 into its parts $A_2 = A_a + A_b$, the additive property of Eq. (9.63) gives

$$A_2F_{2-3} = A_aF_{a-3} + A_bF_{b-3}.$$

Since A_3 is divided into A_c and A_d ($A_3 = A_c + A_d$), the additive property of Eq. (9.64) may be applied to each of F_{a-3} and F_{b-3} in the above:

$$A_aF_{a-3} = A_aF_{a-c} + A_aF_{a-d},$$

$$A_bF_{b-3} = A_bF_{b-c} + A_bF_{b-d},$$

so that

$$A_2F_{2-3} = A_aF_{a-c} + A_aF_{a-d} + A_bF_{b-c} + A_bF_{b-d}.$$

Since A_c and A_d (the two ends) are identical in size and shape and are located symmetrically with each of A_a and A_b (the top and front), then $F_{a-c} = F_{a-d}$ and $F_{b-c} = F_{b-d}$. Thus the above reduces to

$$A_2F_{2-3} = 2(A_aF_{a-c} + A_bF_{b-c}).$$

Again, the constituent shape factors F_{a-c} and F_{b-c} are of known form—perpendicular rectangles with a common edge. Thus Fig. 9.19 or Eq. (C.3) give

For F_{a-c}:

$$R_1 = \frac{L}{D} = \frac{30}{20} = 1.5, \qquad R_2 = \frac{W}{D} = \frac{10}{20} = 0.5,$$

$$F_{a-c} = 0.103.$$

For F_{b-c}:

$$R_1 = \frac{L}{D} = \frac{30}{10} = 3.0, \qquad R_2 = \frac{20}{10} = 2.0,$$

$$F_{b-c} = 0.108.$$

So, finally, the desired shape factor F_{2-3} is, from the above,

$$A_2F_{2-3} = 2(A_aF_{a-c} + A_bF_{b-c})$$

$$900 \times F_{2-3} = 2(600 \times 0.103 + 300 \times 0.108)$$

$$F_{2-3} = 0.209.$$

(c) Since A_2, the combined top and front wall of the furnace does see itself, the shape factor F_{2-2} may similarly be found by using the additive principle. Dividing A_2 into $A_2 = A_a + A_b$ both as a receiver and originating surface, the additive property of Eq. (9.65) gives

$$A_2F_{2-2} = A_aF_{a-a} + A_aF_{a-b} + A_bF_{b-a} + A_bF_{b-b}.$$

Since both A_a and A_b are plane surfaces, $F_{a-a} = F_{b-b} = 0$. Also, the reciprocal property of Eq. (9.62) gives $A_bF_{b-a} = A_aF_{a-b}$. So the expression above reduces simply to

$$A_2F_{2-2} = 2A_bF_{b-a}.$$

Symmetry shows that $F_{b-a} = F_{1-a}$ (since A_1 and A_b are the same size and shape and located symmetrically with respect to A_a, the top). The shape factor F_{1-a} was already found in part (a) as $F_{1-a} = 0.319$. Thus with $F_{b-a} = F_{1-a} = 0.319$ and the given areas, the self-shape factor F_{2-2} is

$$A_2F_{2-2} = 2A_bF_{b-a}$$

$$900 \times F_{2-2} = 2 \times 300 \times 0.319$$

$$F_{2-2} = 0.213.$$

A considerable fraction of the energy leaving A_2 returns to itself.

The shape factor of the combined top and front wall, A_2, with respect to the back wall, A_1 (i.e., F_{2-1}), may readily be found from the known F_{1-2} and the reciprocal property:

$$F_{2-1} = \frac{A_1}{A_2}F_{1-2} = \frac{300}{900} \times 0.465 = 0.155.$$

With this value and the values already found for F_{2-2} and F_{2-3}, one could find the shape factor of A_2 with respect to the furnace floor (call it A_4 for convenience) by application of the enclosure rule [Eq. (9.66)] applied to A_2:

$$\sum_j F_{2-j} = 1$$

$$F_{2-1} + F_{2-2} + F_{2-3} + F_{2-4} = 1$$

$$0.155 + 0.213 + 0.209 + F_{2-4} = 1$$

$$F_{2-4} = 0.423.$$

In fact, noting the facts that $F_{1-1} = F_{4-4} = 0$ since those surfaces are plane, one can find all the remaining shape factors of the sixteen in the set $F_{i-j}(i, j = 1, 2, 3, 4)$ by application of only the enclosure rule and the reciprocal property without further reference to Figs. 9.18 and 9.19. Verification of this statement is left as an exercise for the reader.

9.8 RADIANT EXCHANGE AMONG BLACK SURFACES AND IN BLACK ENCLOSURES

The general problem of radiative exchange between a collection of surfaces, perhaps gray, of either specified temperature or specified heat flux will be treated in some detail in Sec. 9.9. A useful introduction to that problem is the special case in which the surfaces are black and of known temperature. In this instance one knows that the only energy leaving a surface is original emission, there being no reflection.

Consider, then, a system of two or more black surfaces, $A_1, A_2, \ldots, A_n$. Each of the surfaces has a known specified temperature, $T_1, T_2, \ldots, T_n$ or equivalently, known emissive power $E_{b1}, E_{b2}, \ldots, E_{bn}$. The surfaces may, or may not, form a complete enclosure. If they do not, imagine them to be situated in an irradiation-free environment. That is, any radiation leaving a surface that does not strike another in the collection may be presumed lost and no irradiation is incident upon a surface from any source that is not one of the black surfaces in the collection. Let it be desired to find the net rate of energy flow away from any one of the surfaces.

Since all the surfaces involved are black, the net energy flow away from a surface is the difference between its emissive power and the irradiation on it. For the ith surface of the collection

$$q_i = A_i E_{bi} - A_i G_i.$$

The energy falling on the surface (and absorbed), $A_i G_i$, comes from the other surfaces of the collection. Thus from the definition of the shape factor,

$$\begin{aligned} A_i G_i &= \sum_{j=1}^{n} (A_j J_j) F_{j-i}, \\ &= \sum_{j=1}^{n} A_j F_{j-i} E_{bj}. \end{aligned}$$

In the above J_j represents the radiosity of one of the surfaces in the collection and the fact that $J_j = E_{bj}$ for a black surface has been used. The summation is to be taken over *all* the surfaces in the collection, including the ith surface in the event it sees itself and $F_{i-i} \neq 0$. Thus the net heat flow away from A_i is

$$\begin{aligned} q_i &= A_i E_{bi} - \sum_j A_j F_{j-i} E_{bj} \\ &= A_i E_{bi} - \sum_j A_i F_{i-j} E_{bj} \qquad (9.68) \\ &= A_i \left(E_{bi} - \sum_j F_{i-j} E_{bj} \right). \end{aligned}$$

In arriving at Eq. (9.68) the general reciprocity of the shape factor, $A_i F_{i-j} = A_j F_{j-i}$, has been used. Since the geometry of the collection of surfaces is known, all the shape factors may be determined; thus Eq. (9.68) enables one to calculate the heat flow away from each surface, given all the E_b's.

If the system of surfaces forms a complete enclosure, one has from Eq. (9.66)

$$\sum_j F_{i-j} = 1$$

if $j = i$ is included in the summation. Then Eq. (9.68) takes the form

$$q_i = A_i\left(\sum_j F_{i-j}E_{bi} - \sum_j F_{i-j}E_{bj}\right)$$
$$= \sum_j A_iF_{i-j}(E_{bi} - E_{bj}). \tag{9.69}$$

In the equations above and those to follow, the abbreviated notation Σ_j has been introduced to indicate a summation to be taken over all surfaces in the enclosure, i.e., $j = 1, 2, \ldots, n$.

In a complete enclosure the net heat given up by a surface, say q_i, must be absorbed by the other surfaces of the enclosure. Thus the term $A_iF_{i-j}(E_{bi} - E_{bj})$ is often termed the "exchange" of energy between A_i and A_j, denoted by q_{ij}:

$$q_{ij} = A_iF_{i-j}(E_{bi} - E_{bj}),$$
$$q_i = \sum_j q_{ij}. \tag{9.70}$$

Clearly, for a complete enclosure energy conservation requires

$$\sum_i q_i = 0.$$

If a system of black surfaces does not form a complete enclosure, as described in Eq. (9.68), it may be treated as one, and Eq. (9.69) applied, by enclosing the system with a fictitious surface (call it A_s for "space") with $E_{bs} = 0$, $\epsilon_s = 1$.

EXAMPLE 9.12

Imagine a two-surface collection which has the same geometry specified for Example 9.10 in Fig. 9.23. Let the surface A_1 be maintained at a temperature of 40°C and the surface A_2 at 5°C. If both surfaces are ideal black surfaces and are located in a radiation-free environment, find the energy exchange between A_1 and A_2 and the net energy flow away from surface A_1.

Solution. At the specified surface temperatures the blackbody emissive powers are:

$$T_1 = 40 + 273 + 313\text{°K}, \qquad E_{b1} = \sigma T_1^4 = 5.67\left(\frac{313}{100}\right)^4$$
$$= 5.44.2 \text{ W/m}^2,$$
$$T_2 = 5 + 273 = 278\text{°K}, \qquad E_{b2} = \sigma T_2^4 = 5.67\left(\frac{278}{100}\right)^4$$
$$= 338.7 \text{ W/m}^2.$$

The data of Example 9.10 and Fig. 9.23 also give the surface areas noted below, and the analysis of Example 9.10 provides the needed shape factor F_{1-2}:

$$A_1 = 120 \text{ m}^2, \qquad A_2 = 18 \text{ m}^2, \qquad F_{1-2} = 0.058.$$

The heat exchange rate between A_1 and A_2 is found from (Eq. 9.70):

$$\begin{aligned} q_{12} &= A_1 F_{1-2}(E_{b1} - E_{b2}) \\ &= 120 \times 0.058(544.2 - 338.7) \\ &= 1430.3 \text{ W } (4880 \text{ Btu/h}). \end{aligned}$$

By Eq. (9.68) the net energy given up by surface A_1 is

$$q_1 = A_1\left(E_{b1} - \sum_j F_{1-j}E_{bj}\right).$$

Since there are two surfaces in the collection, $j = 1, 2$. However, $F_{1-1} = 0$ since A_1 is plane. Thus

$$\begin{aligned} q_1 &= A_1(E_{b1} - F_{1-1}E_{b1} - F_{1-2}E_{b2}) \\ &= A_1(E_{b1} - F_{1-2}E_{b2}) \\ &= 120(544.2 - 0.058 \times 338.7) \\ &= 62{,}950 \text{ W } (2.148 \times 10^5 \text{ Btu/h}). \end{aligned}$$

Note that considerably more heat is *lost* by A_1, q_1, than it exchanges with A_2, q_{12}. This is the result of the fact that most of A_1 sees "space" and only $F_{1-2} = 0.058$ sees A_2. That is, if one were to form an enclosure by adding an imaginary surface to simulate space (call it A_s), the shape factor of A_1 with respect to space is, by the enclosure rule and the obvious fact that $F_{1-1} = 0$,

$$\begin{aligned} F_{1-s} &= 1 - F_{1-1} - F_{1-2} \\ &= 1 - 0 - 0.058 = 0.942. \end{aligned}$$

Then the same result for the heat flow at A_1 may be found by applying the enclosure summation of Eq. (9.69):

$$\begin{aligned} q_1 &= \sum_j A_1 F_{1-j}(E_{b1} - E_{bj}) \\ &= A_1F_{1-1}(E_{b1} - E_{b1}) + A_1F_{1-2}(E_{b1} - E_{b2}) + A_1F_{1-s}(E_{b1} - E_{bs}), \end{aligned}$$

and since "space" has $E_{bs} = 0$:

$$\begin{aligned} q_1 &= A_1F_{1-2}(E_{b1} - E_{b2}) + A_1F_{1-s}E_{b1} \\ &= 120 \times 0.058(544.2 - 338.7) + 120 \times 0.942 \times 544.2 \\ &= 62{,}950 \text{ W}, \end{aligned}$$

as found earlier.

EXAMPLE 9.13

An outlet shoe store with a display window in the front is shown, with dimensions, in Fig. 9.25. The store is to be heated by making the floor a black radiant heating panel at 45°C. The glass window acts as a black plane at 10°C and the other walls and the ceiling act as black planes at 25°C. Find the net heat given up by the floor. What difference results if the ceiling height is raised to 4.5 m, the other dimensions remaining unchanged?

Solutions. If the floor is designated as surface A_1, the window as surface A_2, and the other walls and ceiling as A_w, the data given are:

$$T_1 = 45 + 273 = 318°\text{K}, \qquad A_1 = (6 + 2 + 2) \times 12 = 120 \text{ m}^2,$$

$$T_2 = 10 + 273 = 283°\text{K}, \qquad A_2 = 3 \times 6 = 18 \text{ m}^2,$$

$$T_w = 25 + 273 = 298°\text{K}.$$

Thus the various emissive powers are

$$E_{b1} = \sigma T_1^4 = 5.67 \times \left(\frac{T_1}{100}\right)^4 = 5.67\left(\frac{318}{100}\right)^4 = 579.8 \text{ W/m}^2,$$

$$E_{b2} = \sigma T_2^4 = 5.67 \times \left(\frac{T_2}{100}\right)^4 = 5.67\left(\frac{283}{100}\right)^4 = 363.7 \text{ W/m}^2,$$

$$E_{bw} = \sigma T_w^4 = 5.67 \times \left(\frac{T_w}{100}\right)^4 = 5.67\left(\frac{298}{100}\right)^4 = 447.1 \text{ W/m}^2.$$

The geometrical arrangement of the floor and window is identical to that of Example 9.10 and Fig. 9.23. Hence the shape factor F_{1-2} is known from that example. Since A_1 is plane $F_{1-1} = 0$, so that the shape factor between the floor and the remaining walls and ceiling may be found from the enclosure rule. Thus,

$$F_{1-2} = 0.058,$$

$$F_{1-w} = 1 - F_{1-1} - F_{1-2} = 1 - 0 - 0.058 = 0.942.$$

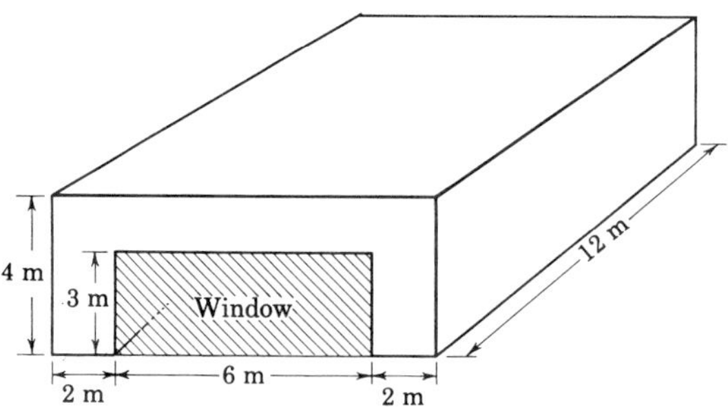

Figure 9.25. Geometry for Example 9.13.

Since there is a complete enclosure, the heat given up by the floor may be found from application of Eq. (9.69) to A_1:

$$q_1 = \sum_j A_1 F_{1-j}(E_{b1} - E_{bj}), \qquad j = 2, w$$

$$= A_1[F_{1-2}(E_{b1} - E_{b2}) + F_{1-w}(E_{b1} - E_{bw})]$$

$$= 120[0.058(579.8 - 363.7) + 0.942(579.8 - 363.7)]$$

$$= 16{,}500 \text{ W } (5.63 \times 10^5 \text{ Btu/h}).$$

Changing the ceiling height to 4.5 m instead of the 4 m shown in Fig. 9.25 does not alter the shape factor F_{1-2} nor $F_{1-w} = 1 - F_{1-2}$. Thus the result for q_1 will remain unchanged. In fact, any ceiling height greater than the window height will yield the same answer. All that changes is the relative amount of heat flowing from the floor to the walls or the ceiling. Since the walls and ceiling are treated as one surface at a given temperature, there is no change in the total heat lost by the floor.

Equations (9.69) and (9.70) suggest that for an enclosure of black surfaces, the radiant heat exchange may be represented by an electrical network analogy. In such an analogy, each surface is represented as a nodal point with its potential maintained at the blackbody emissive power, E_{bi}. Each node is connected to all other nodes through resistances which are $\mathcal{R}_{ij} = 1/A_i F_{i-j} = 1/A_j F_{j-i}$. Equation (9.70) notes that the current flow in each resistor is $(E_{bi} - E_{bj})/\mathcal{R}_{ij}$ and Eq. (9.69) expresses the fact that the current flow at any node equals the sum of the currents in the resistors connected to it. Figure 9.26 shows the equivalent network for Example 9.13. The equation used in Example 9.13 to determine the heat flow from the floor, q_1, could be written immediately from the network shown in Fig. 9.26 by summing current flows at the node representing A_1. One could, similarly, find the heat flows to the windows or the walls.

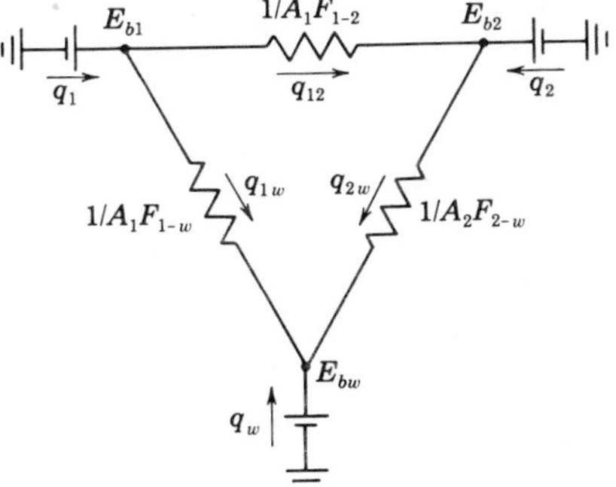

Figure 9.26. Electrical network analogy for Example 9.13.

9.9 RADIANT EXCHANGE IN ENCLOSURES OF DIFFUSE, GRAY SURFACES

The problem of the radiant exchange between parallel infinite planes considered in Sec. 9.6 was relatively simple since the shape factors between the surfaces were all unity, and the analysis could concentrate on the effects of multiple reflections due to grayness of the surfaces. The analysis just presented in Sec. 9.8, on the other hand, concentrated on the geometrical effects associated with finite-sized surfaces but avoided the complication of multiple interreflections by considering only black surfaces. Thus, the energy flow from one surface to another was known to be due to emissive power only, and all energy incident on a surface was absorbed there.

This section will consider (following the methods of Refs. 9 and 10) the general case in which the surfaces involved are of finite size so that $F_{i-j} \neq 1$. Also, the possibility that the surfaces, or some of them, may be gray will be accounted for. The analysis will presume that the collection of surfaces forms a complete enclosure. The radiant exchange between the various surfaces and windows of a room, between the surfaces in a furnace, etc., are examples of a complete enclosure. In some cases one may be interested in the exchange between a collection of surfaces that do not form an enclosure, but radiate at one another in an otherwise radiation-free environment—as, for example, between portions of a spacecraft in deep space. In such instances, the system may be made an enclosure by adding an enclosing surface with $E_b = 0$ and $\alpha = \epsilon = 1$. Thus the assumption of a complete enclosure will not be unduly restrictive.

All surfaces will be taken as diffuse emitters and reflectors that are gray so that the diffuse shape factor may be used and Kirchhoff's law in the total sense, $\alpha = \epsilon$, may be used. Even if a surface has a uniform temperature so that its emissive power is uniform, the irradiation upon it (and hence its radiosity of gray) may not be uniform, depending on how the various surfaces in that collection "see" one another. The analysis presented here will assume that E_b, G, and J of each surface is uniform. If for a given enclosure this is not the case, it will be presumed that the surfaces may be subdivided into smaller portions (thus increasing the number of surfaces involved) until the assumption can be made.

The entire analysis of this section presumes the existence of the steady state so that all quantities are time independent.

Basic Equations and the General Problem

For a thermally opaque surface the definition of radiosity, emissive power, and irradiation given in Sec. 9.2 led to Eq. (9.6):

$$
\begin{aligned}
J_i &= E_i + \rho_i G_i \\
&= E_i + (1 - \alpha_i) G_i \\
&= \epsilon_i E_{bi} + (1 - \alpha_i) G_i,
\end{aligned}
$$

and when Kirchhoff's law is applied,

$$J_i = \epsilon_i E_{bi} + (1 - \epsilon_i)G_i. \tag{9.71}$$

In Eq. (9.71) the subscript i denotes any one of the surfaces of the enclosure. A heat balance made on surface A_i shows that the heat flow *away* from that surface, q_i, is

$$q_i = A_i(J_i - G_i). \tag{9.72}$$

If Eq. (9.71) is used to eliminate G_i, the irradiation, one has

$$q_i = \frac{A_i\epsilon_i}{1 - \epsilon_i}(E_{bi} - J_i), \tag{9.73a}$$

$$E_{bi} = \frac{1 - \epsilon_i}{A_i\epsilon_i} q_i + J_i. \tag{9.73b}$$

The second form given in Eq. (9.73b) will be of use later; however, it is sufficient here to note that Eq. (9.73a) expresses the heat flow away from A_i in terms of its equivalent blackbody emissive power (i.e., temperature, since $E_{bi} = \sigma T_i^4$) and the radiosity leaving A_i.

A second form of the heat flow away from a surface in an enclosure may be obtained by considering its interaction with the other surfaces of the enclosure. If A_j represents another surface, and if n is the total number of surfaces in the enclosure, then the irradiation falling on A_i must come from these other surfaces:

$$A_iG_i = \sum_{j=1}^{n} F_{j-i}A_jJ_j.$$

In the above, the diffuse shape factor has been used, consistent with the assumption of diffuseness made earlier. The general reciprocity property gives

$$A_iG_i = \sum_{j=1}^{n} A_iF_{i-j}J_j. \tag{9.74}$$

Introduced into Eq. (9.72), Eq. (9.74) yields

$$q_i = A_jJ_i - \sum_j A_iF_{i-j}J_j, \tag{9.75}$$

where the abbreviated notation Σ_j means that the summation is to be made over *all* the surfaces in the enclosure—including A_i in the event A_i sees itself and $F_{i-i} \neq 0$. Equation (9.75) may be written in another form noting that for the enclosure $\Sigma_j F_{i-j} = 1$. Thus

$$\begin{aligned} q_i &= A_iJ_i \sum_j F_{i-j} - \sum_j A_iF_{i-j}J_j \\ &= \sum_j A_iF_{i-j}(J_i - J_j). \end{aligned} \tag{9.76}$$

Equations (9.75) and (9.76) give an alternative expression for q_i from that in Eq. (9.73), the heat flow being expressed in terms of the radiosity of a surface, J_i, and the radiosities of all the other surfaces of the enclosure. Equations (9.73a) and (9.76) may be equated to yield another important relation:

$$\frac{A_i\epsilon_i}{1 - \epsilon_i}(E_{bi} - J_i) = \sum_j A_i F_{i-j}(J_i - J_j). \tag{9.77}$$

Equations (9.73), (9.76), and (9.77) express the radiant exchange among surfaces of an enclosure in terms of the emissive powers and radiosities of the participating surfaces. These equations form the basis of the *radiosity method* of enclosure analysis. The general problem of an enclosure analysis may be stated as follows: Given an enclosure of surfaces of known geometry and known surface properties, and given the emissive power (i.e., temperature) of some of the surfaces and given the heat flux away from the others, let it be desired to find the heat flow away from the surfaces on which E_{bi} is specified and the temperature (i.e., E_{bi}) of the surfaces for which q_i is specified. In a general sense, this problem may be solved in the following way: For each of the surfaces on which E_{bi} is known, write an equation of the form of Eq. (9.77); on each of the surfaces on which q_i is known, write an equation of the form of Eq. (9.76). There results, then, n equations in the n unknown radiosities which can be solved for the J_i's. Then Eq. (9.73b) can be used to find E_{bi} on each surface for which q_i is known and Eq. (9.76) can be used to find q_i for each surface for which E_{bi} is known.

The actual carrying out of the general procedure just mentioned may involve complications such as some of the equations becoming indeterminate when a surface is black, etc.; however these matters will be treated later when detailed procedures are discussed. Also, the methods of actually implementing the procedure may differ depending on whether one is dealing with just a few surfaces or with a large number of surfaces or on whether one is seeking a closed-form analytic solution or a numerical answer for a particular situation. These different aspects will be treated in the material which follows.

As indicated in the foregoing discussion, conditions commonly imposed on surfaces in an enclosure include either the specification of a fixed surface temperature (or E_{bi}) or the specification of a known heat flow at the surface. In the first case one wishes to find the heat flow necessary to maintain the given emissive power and in the second case one seeks the resultant equilibrium temperature of the surface. While there are applications in radiant exchange analyses in which a surface heat flow of some known value is specified, the most commonly encountered case is that in which the flux is known to be zero—giving rise to the "adiabatic, reradiating" surface defined next. Most of the examples considered in what follows will be limited to cases in which the only heat flux specification applied is that of zero.

The Adiabatic, Reradiating Surface

In an enclosure, an *adiabatic, reradiating surface* is defined as one which is thermally isolated so that the net heat flow away from the surface is zero. Such a surface interacts radiatively with the other surfaces of the enclosure, absorbing and reflecting incident irradiation and reemitting the absorbed energy. Through this interaction it is allowed to come to an equilibrium emissive power (or temperature) compatible with its radiative environment so that its net heat flow is zero. Thus, for an adiabatic surface $J_i = G_i$. The radiation shield considered at the end of Sec. 9.6 was such an adiabatic, reradiating surface. The refractory walls in a furnace which serve to reflect or absorb and reradiate energy from the fire are also examples of adiabatic surfaces.

The enclosure problem to be treated in the remainder of this chapter will consist of collections of surfaces, some of which are of known temperature and some of which are adiabatic surfaces. It will be desired to find the equilibrium temperature of these adiabatic surfaces. Equation (9.73b) shows that for an adiabatic surface, the equilibrium temperature is readily found from its radiosity:

$$\sigma T_{ad}^4 = E_{bad} = J_{ad}, \tag{9.78}$$

where the subscript *ad* is used to denote an *adiabatic* surface. Equation (9.78) shows that the equilibrium temperature, or equivalent blackbody emissive power, is independent of the surface properties. Finding the radiosity of an adiabatic surface is tantamount to finding its temperature; hence the basic enclosure problem remains the same: Find the radiosities of all the surfaces in the enclosure.

The Network Analogy

The carrying out of the analysis of the radiation exchange in an enclosure may be facilitated by noting a generalization of the network analogy developed in Sec. 9.8 for black surfaces. Equations (9.73a) and (9.76) showed the net heat flow from a surface in an enclosure to be

$$q_i = \frac{A_i \epsilon_i}{1 - \epsilon_i}(E_{bi} - J_i) = \frac{E_{bi} - J_i}{(1 - \epsilon_i)/A_i\epsilon_i}$$

$$= \sum_j A_i F_{i-j}(J_i - J_j) = \sum_j \frac{J_i - J_j}{1/A_i F_{i-j}}.$$

As noted by Oppenheim in Ref. 11, the first of these equations suggests that the heat flow at a surface may be represented as the current flow in a resistor of magnitude $(1 - \epsilon_i)/A_i\epsilon_i$ connected between nodes, one of which is maintained at a potential equal to the blackbody emissive power of the surface, E_{bi}, and the other at a potential equal to the surface radiosity. The second equation suggests that the exchange between a surface and others in the enclosure may be represented as the current flows in resistors connecting the radiosity node to all other radiosity nodes through resistors of magnitude $1/A_iF_{i-j}$. Figure 9.27 depicts the representation of a surface maintained at a given E_{bi}.

If a surface is black, one knows that $E_{bi} = J_i$, so that the two nodes become one and no resistor connects them. If a surface is an adiabatic reradiator, the node is not maintained at a given E_{bi}, but is allowed to "float" and reach some equilibrium radiosity, in keeping with the concept discussed in the preceding section.

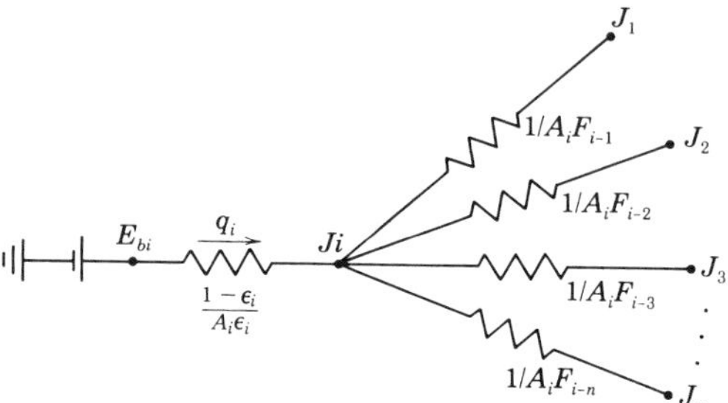

Figure 9.27. Electrical network analogy for a gray surface.

Figure 9.28 illustrates the network for an enclosure with a black surface of known temperature, A_1, two gray surfaces of known temperature, A_2 and A_3, and a reradiating surface, A_4.

The utility of the network analogy will be illustrated in the next section, in which enclosure analysis is discussed in some detail.

Direct Solution of Enclosures of Black, Gray, and Adiabatic Surfaces

The general concepts discussed in the foregoing will now be applied to the solution of specified enclosure problems. The details of implementation will differ depending on the magnitude of the particular problem at hand. If a rather large number of surfaces are involved so that the computations may require the use of a digital computer, matrix methods may be applied as described in the next section. However, for a modest number of surfaces in which one seeks the answer for a specific

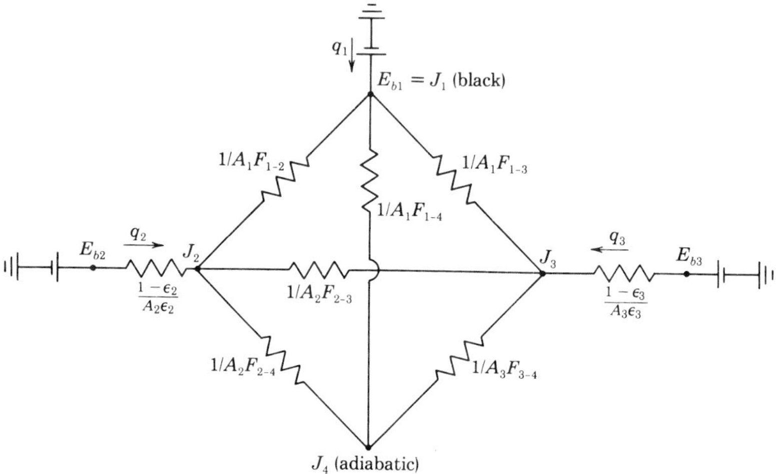

Figure 9.28. Network analogy for an enclosure of two gray surfaces, one black surface, and one adiabatic surface.

case or desires a closed-form solution, a more direct approach may be used. Such a direct solution is described here.

Consider, then, the case in which one is given an enclosure of n surfaces of known geometry. Some of the surfaces have specified temperature, or E_{bi}, and it is desired to find the heat flows at these surfaces. Call these surfaces "active surfaces" and number them $i = 1, 2, \ldots, n_1$, where $n_1 \leq n$. The active surfaces may be black or perhaps gray with known ϵ_i. Let the remaining surfaces, if any, be adiabatic surfaces with $q_i = 0$, and number them $i = n_1 + 1, n_1 + 2, \ldots, n$. It is desired to find the equilibrium temperatures, or E_{bi}, of these adiabatic surfaces. It is presumed that all the necessary shape factors, F_{i-j}, between all surfaces of the enclosure have been found according to the methods of Sec. 9.7. To find the desired heat flows at the n_1 active surfaces and the equilibrium temperatures at the $n - n_1$ adiabatic surfaces, it is necessary to find the n radiosities of all the surfaces. One proceeds as follows:

1. For each active surface write Eq. (9.77), the E_{bi}'s being known. If a particular active surface is black, Eq. (9.77) becomes indeterminate, but it is known that $J_i = E_{bi}$, effectively reducing the number of unknown J_i's. Thus one writes:

   For $i = 1, 2, \ldots, n_1;\ j = 1, 2, \ldots, n_1, \ldots, n$

   $$\text{If gray:} \quad \frac{A_i\epsilon_i}{1 - \epsilon_i}(E_{bi} - J_i) = \sum_j A_iF_{i-j}(J_i - J_j), \tag{9.79}$$

   $$\text{If black:} \quad J_i = E_{bi}.$$

2. For each adiabatic surface write Eq. (9.76) knowing $q_i = 0$:

   $$\text{For } i = n_1 + 1, \ldots, n;\ j = 1, 2, \ldots, n_1, \ldots, n \qquad \sum_j A_iF_{i-j}(J_i - J_j) = 0. \tag{9.80}$$

3. If one prefers, the analogous electrical network for the enclosure may be drawn, and steps 1 and 2 are equivalent to writing current balances at each of the nodes with unknown J_i. No balance is necessary at a black active node since $J_i = E_{bi}$ there.
4. Steps 1 and 2, or 3, yield n equations in the n unknown J_i's (or fewer when black active surfaces are present). Solve these equations for the unknown J_i's.
5. For each active surface, apply Eq. (9.76) [or Eq. (9.73a) if gray] to find the desired q_i. This is equivalent to finding the current flow at each active E_{bi} node if the network analogy is used. Thus

   For $i = 1, 2, \ldots, n_1;\ j = 1, 2, \ldots, n_1, \ldots, n$

   $$\text{If gray:} \quad q_i = \frac{A_i\epsilon_i}{1 - \epsilon_i}(E_{bi} - J_i), \tag{9.81}$$

   $$\text{If gray or black:} \quad q_i = \sum_j A_iF_{i-j}(J_i - J_j).$$

6. For each adiabatic surface find the equilibrium temperature according to Eq. (9.78):

$$\text{For } i = n_1 + 1, \ldots, n: \quad T_i = \left(\frac{E_{bi}}{\sigma}\right)^{1/4} = \left(\frac{J_i}{\sigma}\right)^{1/4} \tag{9.82}$$

In general, the procedure above involves solving a set of simultaneous equations equal to the number of gray active surfaces plus the number of adiabatic surfaces. If this number is small the procedure can be carried out by hand calculation, or even algebraically in some cases. If a large number of surfaces are present matrix solution, as described in the next section, may be desirable. Clearly, if there are no adiabatic surfaces present, one needs only to find the radiosities of the adiabatic surfaces. If the enclosure consists of only black active surfaces, the method degenerates to the simple case considered in Sec. 9.8.

EXAMPLE 9.14

For the same geometry of the shoe store considered in Example 9.13 and Fig. 9.25, let the floor, A_1, be a black surface at 45°C and let the window, A_2, be a black surface at 10°C. Denote the remaining walls and the ceiling by A_3 and in this instance presume that A_3 acts as a single adiabatic reradiating surface. Find (a) the net heat flow at the floor; (b) the net heat flow at the window; (c) the equilibrium temperature of the adiabatic surface, A_3.

Solution. The geometry of the system and the specified temperatures give the following data as evaluated in Example 9.13:

$$A_1 = 120\ \text{m}^2, \qquad T_1 = 45°\text{C} = 318°\text{K}, \qquad E_{b1} = 579.8\ \text{W/m}^2,$$

$$A_2 = 18\ \text{m}^2, \qquad T_2 = 10°\text{C} = 283°\text{K}, \qquad E_{b2} = 363.7\ \text{W/m}^2.$$

As determined in Example 9.10 and used in Example 9.13, the basic shape factor between the floor and the window is

$$F_{1-2} = 0.058.$$

The reciprocity principle yields

$$F_{2-1} = \frac{A_1}{A_2} F_{1-2} = \frac{120}{18} \times 0.058 = 0.387,$$

so that using the fact that $F_{1-1} = F_{2-2} = 0$ since A_1 and A_2 are plane, the enclosure rule gives

$$F_{1-3} = 1 - F_{1-1} - F_{1-2} = 1 - 0 - 0.058 = 0.942,$$

$$F_{2-3} = 1 - F_{2-1} - F_{2-2} = 1 - 0.387 - 0 = 0.613.$$

Thus the AF products required in the heat balance expressions (or the reciprocals of the resistances in the analogous network) are

$$A_1F_{1-2} = A_2F_{2-1} = 120 \times 0.058 = 6.96\ \text{m}^2,$$

$$A_1F_{1-3} = A_3F_{3-1} = 120 \times 0.942 = 113.0\ \text{m}^2,$$

$$A_2F_{2-3} = A_3F_{3-2} = 18 \times 0.613 = 11.0\ \text{m}^2.$$

One now proceeds to determine the radiosities of each of the three surfaces of the location as outlined in the foregoing discussion. Application of Eq. (9.79) at the two active surfaces, A_1 and A_2, is really unnecessary since they are black and it is known that their radiosities are equal to their respective emissive powers:

$$J_1 = E_{b1} = 579.8\ \text{W/m}^2,$$

$$J_2 = E_{b2} = 363.7\ \text{W/m}^2.$$

For the adiabatic surface, A_3, the heat balance of Eq. (9.80) gives

$$\sum_j A_3F_{3-j}(J_3 - J_j) = 0, \qquad j = 1, 2, 3$$

$$A_3F_{3-1}(J_3 - J_1) + A_3F_{3-2}(J_3 - J_2) + A_3F_{3-3}(J_3 - J_3) = 0.$$

The AF products have been noted earlier and the radiosities $J_1 = E_{b1}$ and $J_2 = E_{b2}$ are known as above. So

$$113.0(J_3 - 579.8) + 11.0(J_3 - 363.7) = 0,$$

$$J_3 = 560.63\ \text{W/m}^2.$$

Now that all surface radiosities have been found, the desired heat flows and temperatures are readily solved for.

(a) For surface A_1, Eq. (9.81) gives (using the second form since A_1 is black)

$$\begin{aligned} q_1 &= \sum_j A_1F_{1-j}(J_1 - J_j), \qquad j = 1, 2, 3 \\ &= A_1F_{1-2}(J_1 - J_1) + A_1F_{1-2}(J_1 - J_2) + A_1F_{1-3}(J_1 - J_3) \\ &= 0 + 6.96(579.8 - 363.7) + 113.0(579.8 - 560.63) \\ &= 3670\ \text{W}\ (1.252 \times 10^4\ \text{Btu/h}). \end{aligned}$$

This heat flow at A_1 is considerably less than that found in Example 9.13 since in this case the surrounding walls are an adiabatic reradiator which returns some of the energy to the floor. In Example 9.13 the walls were black and absorbed all the energy falling on them.

(b) Completely analogous to part (a), the heat flow at the window, A_2 is found from applying Eq. (9.81) there:

$$
\begin{aligned}
q_2 &= \sum_j A_2F_{2-j}(J_2 - J_j), \qquad j = 1, 2, 3 \\
&= A_2F_{2-1}(J_2 - J_1) + A_2F_{2-2}(J_2 - J_2) + A_2F_{2-3}(J_2 - J_3) \\
&= 6.96(363.7 - 579.8) + 0 + 11.0(363.7 - 560.63) \\
&= -3670 \text{ W } (1.252 \times 10^4 \text{ Btu/h}).
\end{aligned}
$$

The fact that $q_1 = -q_2$, as shown, results from the fact that the system of A_1, A_2, and A_3 forms a complete enclosure and A_3 is adiabatic.

(c) The equilibrium temperature of the adiabatic wall A_3 is found from its radiosity, $J_3 = 560.63$ W/m^2, as found above, and Eq. (9.82):

$$
\begin{aligned}
T_3 &= \left(\frac{E_{b3}}{\sigma}\right)^{1/4} = \left(\frac{J_3}{\sigma}\right)^{1/4} \\
&= \left(\frac{560.63}{5.67 \times 10^{-8}}\right)^{1/4} \\
&= 315°\text{K} = 42°\text{C } (108°\text{F}).
\end{aligned}
$$

In this elementary example the electrical network analogy may be applied to determine the same results from a more physical basis. Since the enclosure in question consists of two active black surfaces and one adiabatic reradiating surface, the corresponding network analogy is that shown in Fig. 9.29. There are two nodes of specified potential, $J_1 = E_{b1}$ and $J_2 = E_{b2}$, representing the black active surfaces and one floating node at an unknown potential J_3. The resistors connecting the three nodes comprise the reciprocals of the AF products found earlier. By adding the resistances in series and parallel shown in Fig. 9.29, one can show that the equivalent resistance between the nodes J_1 and J_2 is

$$
\left\{\left[\frac{1}{A_1F_{1-2}}\right]^{-1} + \left[\left(\frac{1}{A_1F_{1-3}}\right) + \left(\frac{1}{A_2F_{2-3}}\right)\right]^{-1}\right\}^{-1}.
$$

Thus the heat flow (or current) at A_1 or A_2 is the potential difference divided by the resistance above:

$$
q_1 = -q_2 = \frac{E_{b1} - E_{b2}}{\{A_1F_{1-2} + [(1/A_1F_{1-3}) + (1/A_2F_{2-3})]^{-1}\}^{-1}}.
$$

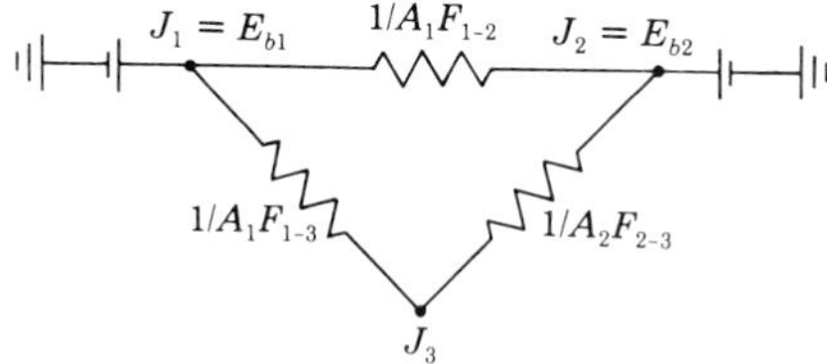

Figure 9.29. Network analogy for Example 9.14.

With the data given,

$$q_1 = -q_2 = \frac{579.8 - 363.7}{\{6.96 + [(1/113.0) + (1/11.0)]^{-1}\}^{-1}}$$

$$= 3670 \text{ W},$$

as found earlier.

EXAMPLE 9.15

A rectangle, A_1, 5 ft × 10 ft in size is maintained at $T_1 = 500°\text{F}$ and is gray with an emissivity of $\epsilon_1 = 0.7$. It is placed parallel and directly opposed to a second rectangle, A_2, of the same size, spaced 5 ft away. The second rectangle has a temperature of $T_2 = 900°\text{F}$ and it is also gray with $\epsilon_2 = 0.9$. Find the net heat flow away from each of the two surfaces (considering only the opposed faces) if (a) they are located in a radiation-free environment; (b) they are connected by a single adiabatic surface.

Solution. The two surfaces A_1 and A_2 are active gray surfaces of known temperature and emissive power. Both have the same area $A_1 = A_2 = 5 \times 10 = 50 \text{ ft}^2$. In English units the blackbody emissive power is given, with T in °R, by Eq. (9.22): $E_b = \sigma T^4 = 0.1714 \times 10^{-8} T^4$. Thus the specified data for the two active surfaces give

$$A_1 = 50 \text{ ft}^2, \quad \epsilon_1 = 0.7, \quad T_1 = 500°\text{F} = 960°\text{R}, \quad E_{b1} = 1456 \text{ Btu/h-ft}^2,$$

$$A_2 = 50 \text{ ft}^2, \quad \epsilon_2 = 0.9 \quad T_2 = 900°\text{F} = 1360°\text{R}, \quad E_{b2} = 5864 \text{ Btu/h-ft}^2.$$

Let the symbol A_3 denote a fictitious third surface that connects the other two, forming a complete enclosure. In part (a), in which A_1 and A_2 are treated as being isolated in a radiation-free space, A_3 will be taken as an active surface at $E_{b3} = J_3 = 0$ with $\epsilon_3 = 1.0$. In part (b) A_3 will be designated as an adiabatic reradiating surface, $q_3 = 0$, of unknown J_3 and with unspecified ϵ_3. The analogous electrical network for either case is shown in Fig. 9.30, where the two active surfaces are shown as maintained at known potentials, E_{b1} and E_{b2}, and the two options for the enclosing surface, A_3, indicated.

The shape factors necessary to find the resistances must now be found. For the two active surfaces, directly opposed parallel rectangles, Fig. 9.18 or Eq. (C.2) gives, from the geometry specified,

$$R_1 = \frac{L}{D} = \frac{10}{5} = 2.0, \qquad R_2 = \frac{W}{D} = \frac{5}{5} = 1.0, \qquad F_{1-2} = F_{2-1} = 0.286.$$

Since $A_1 = A_2$, F_{1-2} and F_{2-1} are necessarily the same. The remaining required shape factors, $F_{1-3}, = F_{2-3}$, are found by application of the enclosure rule, along

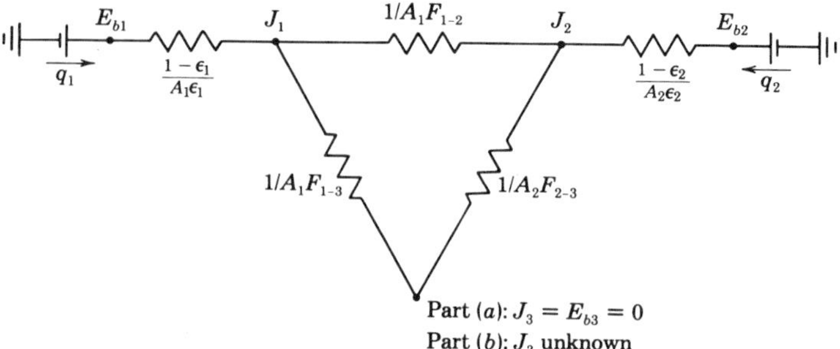

Figure 9.30. Network analogy for Example 9.15.

with the fact that $F_{1-1} = F_{2-2} = 0$:

$$F_{1-2} = F_{1-3} = 1 - F_{1-1} - F_{1-2},$$
$$= 1 - 0 - 0.286 = 0.714.$$

Now the factors required by the heat balance equations (i.e., the reciprocal resistors) may be calculated:

$$\frac{A_1\epsilon_1}{1 - \epsilon_1} = \frac{50 \times 0.7}{1 - 0.7} = 116.67 \text{ ft}^2,$$

$$\frac{A_2\epsilon_2}{1 - \epsilon_2} = \frac{50 \times 0.9}{1 - 0.9} = 450.0 \text{ ft}^2,$$

$$A_1F_{1-2} = A_2F_{2-1} = 50 \times 0.286 = 14.30 \text{ ft}^2,$$

$$A_1F_{1-3} = A_2F_{2-3} = A_3F_{3-1} = A_3F_{3-2} = 50 \times 0.714 = 35.70 \text{ ft}^2.$$

(a) In the case in which the surfaces are located in a radiation-free space, the enclosure is represented by simulating space with surface A_3 at $T_3 = E_{b3} = J_3 = 0$ and $\epsilon_3 = 1.0$. In this instance, then, no adiabatic surfaces are present, so heat balances need to be made at only the three active ones. Thus Eq. (9.79) applied at the two gray surfaces A_1 and A_2 gives

$$\frac{A_1\epsilon_1}{1 - \epsilon_1}(E_{b1} - J_1) = \sum_j A_1F_{1-j}(J_1 - J_j), \quad j = 1, 2, 3$$
$$= A_1F_{1-1}(J_1 - J_1) + A_1F_{1-2}(J_1 - J_2)$$
$$+ A_1F_{1-3}(J_1 - J_3),$$

$$\frac{A_2\epsilon_2}{1 - \epsilon_2}(E_{b2} - J_2) = \sum_j A_2F_{2-j}(J_2 - J_j), \quad j = 1, 2, 3$$
$$= A_2F_{2-1}(J_2 - J_1) + A_2F_{2-2}(J_2 - J_2)$$
$$+ A_2F_{2-3}(J_2 - J_3).$$

At the third surface, A_3, the second form of Eq. (9.79) applies since it is black, and one has the obvious result:

$$J_3 = E_{b3}.$$

Since $E_{b3} = 0$ to simulate space, the two equations at A_1 and A_2 reduce to

$$\frac{A_1\epsilon_1}{1 - \epsilon_1}(E_{b1} - J_1) = A_1F_{1-2}(J_1 - J_2) + A_1F_{1-3}J_1,$$

$$\frac{A_2\epsilon_2}{1 - \epsilon_2}(E_{b2} - J_2) = A_2F_{2-1}(J_2 - J_1) + A_2F_{2-3}J_2.$$

The various coefficients, along with E_{b1} and E_{b2}, have been calculated earlier, so the above reduce to two equations in the unknown radiosities J_1 and J_2:

$$116.67(1456 - J_1) = 14.30(J_1 - J_2) + 35.70J_1,$$

$$450.0(5864 - J_2) = 14.30(J_2 - J_1) + 35.70J_2,$$

from which one readily finds

$$J_1 = 1475 \text{ Btu/h-ft}^2, \qquad J_2 = 5319 \text{ Btu/h-ft}^2.$$

With the surface radiosities now known, one may find that heat flows at A_1 and A_2, both of which are gray, by application of Eq. (9.81):

$$q_1 = \frac{A_1\epsilon_1}{1 - \epsilon_1}(E_{b1} - J_1) = 116.67(1456 - 1475)$$

$$= -2217 \text{ Btu/h (650 W)},$$

$$q_2 = \frac{A_2\epsilon_2}{1 - \epsilon_2}(E_{b2} - J_2) = 450.0(5864 - 5319)$$

$$= 245{,}300 \text{ Btu/h (71,900 W)}.$$

Thus surface A_1 gains heat while A_2 loses heat. The heat flow to space could be found by applying Eq. (9.80) to A_3; however, energy conservation clearly requires that $q_3 = -(q_1 + q_2) = -243{,}080$ Btu/h.

This part of the example was relatively simple since the third surface, A_3, was of known temperature (0°R) and black. Thus only two unknown radiosities, at A_1 and A_2, had to be found. In the next case, however, A_3 is made an adiabatic surface of unknown temperature (or radiosity), so that three unknowns must be found.

(b) Now let the enclosing surface, A_3, be an adiabatic reradiating surface, the temperature of which must be found. In this case heat balances must be written at all three surfaces. Equation (9.79) is applied at the two active, gray ones, as before, and Eq. (9.80) is applied at the adiabatic surface, A_3. Omitting the obviously zero terms when $i = j$, one has

$$\frac{A_1\epsilon_1}{1-\epsilon_1}(E_{b1}-J_1) = A_1F_{1-2}(J_1-J_2) + A_1F_{1-3}(J_1-J_3),$$

$$\frac{A_2\epsilon_2}{1-\epsilon_2}(E_{b2}-J_2) = A_2F_{2-1}(J_2-J_1) + A_2F_{2-3}(J_2-J_3),$$

$$A_3F_{3-1}(J_3-J_1) + A_3F_{3-2}(J_3-J_2) = 0.$$

Introducing now the known E_b's and the calculated coefficients (noting that one does not need the shape factors F_{3-1} and F_{3-2} since the reciprocity relations $A_3F_{3-1} = A_1F_{1-3}$, $A_3F_{3-2} = A_2F_{2-3}$ apply), there results

$$116.67(1456 - J_1) = 14.30(J_1 - J_2) + 35.70(J_1 - J_3),$$

$$450.0(5864 - J_2) = 1430(J_2 - J_1) + 35.70(J_2 - J_3),$$

$$35.70(J_3 - J_1) + 35.70(J_3 - J_2) = 0.$$

This time a system of three equations in three unknown radiosities must be solved. The results are

$$J_1 = 2358 \text{ Btu/h-ft}^2, \qquad J_2 = 5630 \text{ Btu/h-ft}^2, \qquad J_3 = 3994 \text{ Btu/h-ft}^2.$$

Then for the two active surfaces A_1 and A_2 the heat flows are given by Eq. (9.81):

$$q_1 = \frac{A_1\epsilon_1}{1-\epsilon_1}(E_{b1}-J_1) = 116.67(1456 - 2358)$$

$$= -105{,}200 \text{ Btu/h } (30{,}800 \text{ W}).$$

$$q_2 = \frac{A_2\epsilon_2}{1-\epsilon_2}(E_{b2}-J_2) = 450.0(5864 - 5630)$$

$$= 105{,}200 \text{ Btu/h } (30{,}800 \text{ W}).$$

In this case $q_1 = -q_2$ since there are but two active surfaces and no heat is lost to the environment.

The equilibrium temperature of A_3, the adiabatic surface, is found from Eq. (9.82) and the radiosity J_3 just found:

$$T_3 = \left(\frac{J_3}{\sigma}\right)^{1/4} = \left(\frac{3994}{0.1714 \times 10^{-8}}\right)^{1/4} = 1236°\text{R} = 776°\text{F } (413°\text{C}).$$

Note that in parts (a) and (b) of this example, it was not necessary to find the shape factors of the adiabatic surface with respect to the active ones (i.e., F_{3-1} or F_{3-2}) since the required AF products were found from the reciprocity relations $A_3F_{3-1} = A_1F_{1-3}$ and $A_3F_{3-2} = A_2F_{2-3}$. The factors F_{1-3} and F_{2-3} were, in turn, found from the enclosure rule: $F_{1-3} = 1 - F_{1-1} - F_{1-2}$, $F_{2-3} = 1 - F_{2-1} - F_{2-2}$. Thus only the shape factor between the active surfaces, F_{1-2}, and their self-shape factors F_{1-1} and F_{2-2} (zero in this case) were needed. All this means that knowledge of the specific size and shape of the adiabatic surface A_3 was not necessary as long as it could be assumed that this surface could be represented as one at a single temperature. That is, any shape for A_3 could be assumed (as long as it does not interfere with the direct exchange between A_1 and A_2, F_{1-2}). Realistically, however, one could assume such a bizarre shape for A_3 that the assumption of a single representative temperature would be unrealistic.

EXAMPLE 9.16

Figure 9.31 depicts a furnace in the shape of a rectangular parallelepiped. The interior surfaces of the furnace, comprising a complete enclosure, exchange radiant energy. The furnace is 5 m high, 10 m wide, and 20 m long. The floor of the furnace, A_1, acts as a black active surface maintained at $T_1 = 200°C = 473°K$; the back wall (5 m × 20 m) acts as a gray active plane, A_2, maintained at $T_2 = 400°C = 673°K$ with $\epsilon_2 = 0.4$. The two 5 m × 10 m ends act, together, as a single adiabatic surface, A_3, and the remaining 10 m × 20 m ceiling together with the 5 m × 20 m front wall act as a second single adiabatic surface, A_4. Find the heat flow at each of the two active surfaces, A_1 and A_2, and the equilibrium temperature of the two adiabatic surfaces, A_3 and A_4.

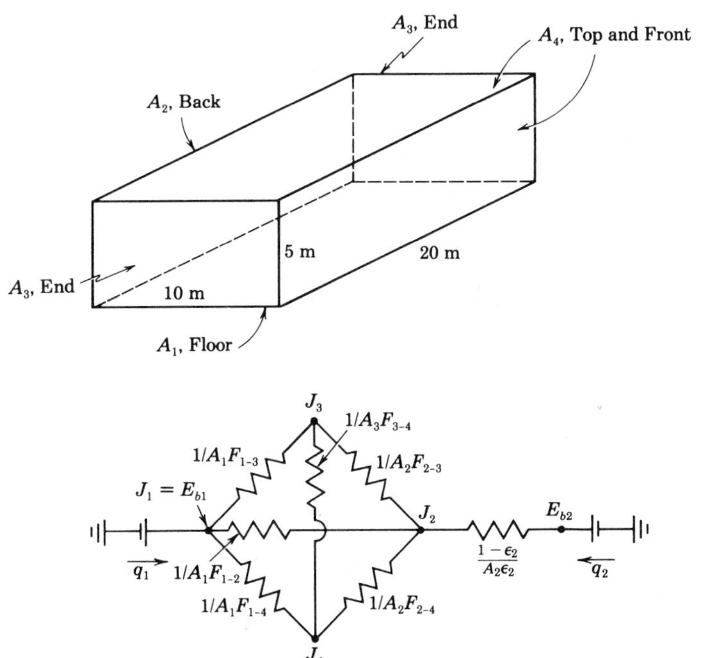

Figure 9.31. Geometry and network analogy for Example 9.16.

Solution. In this example, the enclosure system consists of four surfaces—two active ones of known temperature (one of which is black) and two adiabatic ones of known heat flux (zero). Since one of the active surfaces is black its radiosity is equal to its known emissive power. Thus there are only three unknown radiosities to be found (J_2, J_3, J_4), as in part (b) of Example 9.15. The analogous electrical network for this example is also shown in Fig. 9.31.

The given geometry, temperatures, and surface properties yield

$$A_1 = 10 \times 20 = 200 \text{ m}^2, \quad \epsilon_1 = 1.0, \quad T_1 = 473°\text{K}, \quad E_{b1} = \sigma T_1^4 = 2838 \text{ W/m}^2,$$

$$A_2 = 5 \times 20 = 100 \text{ m}^2, \quad \epsilon_2 = 0.4, \quad T_2 = 673°\text{K}, \quad E_{b2} = \sigma T_2^4 = 11{,}632 \text{ W/m}^2,$$

$$A_3 = (5 \times 10) \times 2 = 100 \text{ m}^2, \quad q_3 = 0,$$

$$A_4 = (5 \times 20) + (10 \times 20) = 300 \text{ m}^2, \quad q_4 = 0.$$

The various shape factors needed to carry out the analysis are found by the methods discussed in Sec. 9.7 and Appendix C. For this particular example involving surfaces consisting of a combination of simpler subdivisions, the methods outlined in Example 9.11 are particularly applicable. Rather than complicate the present example with such calculations, it is left as an exercise for the reader (Prob. 9.50) to show that the following shape factors may be found:

$$F_{1-1} = 0, \quad F_{1-2} = 0.167, \quad F_{1-3} = 0.158, \quad F_{1-4} = 0.675,$$

$$F_{2-2} = 0, \quad F_{2-3} = 0.168, \quad F_{2-4} = 0.498, \quad F_{3-4} = 0.484.$$

Thus the following parameters for the heat balance (or network analogy) may be calculated:

$$\frac{A_2\epsilon_2}{1 - \epsilon_2} = \frac{100 \times 0.4}{1 - 0.4} = 66.67 \text{ m}^2,$$

$$A_1F_{1-2} = A_2F_{2-1} = 200 \times 0.167 = 33.40 \text{ m}^2,$$

$$A_1F_{1-3} = A_3F_{3-1} = 200 \times 0.158 = 31.60 \text{ m}^2,$$

$$A_1F_{1-4} = A_4F_{4-1} = 200 \times 0.675 = 135.0 \text{ m}^2,$$

$$A_2F_{2-3} = A_3F_{3-2} = 100 \times 0.168 = 16.80 \text{ m}^2,$$

$$A_2F_{2-4} = A_4F_{4-2} = 100 \times 0.498 = 49.80 \text{ m}^2,$$

$$A_3F_{3-4} = A_4F_{4-3} = 100 \times 0.484 = 48.40 \text{ m}^2.$$

The procedure to follow in this case is basically that outlined in Examples 9.14 and 9.15. Heat balances of the form of Eqs. (9.79) and (9.80) are written for the four surfaces, resulting in four equations to be solved for the unknown radiosities. In this case there are only three unknown J's since that for the black active surface is known to be the given emissive power. Since the approach is much the same as in the two foregoing examples, less detail is given in the following analysis.

Equation (9.79), or current balances, at the two active surface nodes, A_1 and A_2, gives

$$J_1 = E_{b1},$$

$$\frac{A_2\epsilon_2}{1 - \epsilon_2}(E_{b2} - J_2) = A_2F_{2-1}(J_2 - J_1) + A_2F_{2-3}(J_2 - J_3) + A_2F_{2-4}(J_2 - J_4),$$

while Eq. (9.80), or current balances, at the two adiabatic surface nodes, A_3 and A_4, gives

$$A_3F_{3-1}(J_3 - J_1) + A_3F_{3-2}(J_3 - J_2) + A_3F_{3-4}(J_3 - J_4) = 0,$$

$$A_4F_{4-1}(J_4 - J_1) + A_4F_{4-2}(J_4 - J_2) + A_4F_{4-3}(J_4 - J_3) = 0.$$

The obviously zero values of $A_iF_{i-j}(J_i - J_i)$ when $i = j$ have been omitted. Introducing the known E_{b1}, E_{b2}, the calculated coefficients, and the fact that $J_1 = E_{b1}$, reduces the above to the following set of three equations in the unknown J_2, J_3, and J_4:

$$66.67(11{,}632 - J_2) = 33.40(J_2 - 2838) + 16.80(J_2 - J_3) + 49.80(J_2 - J_4),$$

$$31.60(J_3 - 2838) + 16.80(J_3 - J_2) + 48.40(J_3 - J_4) = 0,$$

$$135.0(J_4 - 2838) + 49.80(J_4 - J_2) + 48.40(J_4 - J_3) = 0.$$

The solution to these three equations is, along with the known fact that $J_1 = E_{b1}$,

$$J_1 = 2838, \quad J_2 = 6811, \quad J_3 = 4081, \quad J_4 = 3944 \text{ W/m}^2.$$

Then Eq. (9.81) applied at the two active surfaces gives the heat flows there [the alternative form being used for A_1 since it is black and $A_1\epsilon_1/(1 - \epsilon_1)$ is indeterminate]:

$$\begin{aligned} q_1 &= A_1F_{1-2}(J_1 - J_2) + A_1F_{1-3}(J_1 - J_3) + A_1F_{1-4}(J_1 - J_4) \\ &= 33.40(2838 - 6811) + 31.60(2838 - 4081) + 135.0(2838 - 3944) \\ &= -321.4 \text{ kW } (1.097 \times 10^6 \text{ Btu/h}), \end{aligned}$$

$$\begin{aligned} q_2 = \frac{A_2\epsilon_2}{1 - \epsilon_2}(E_{b2} - J_2) &= 66.67(11{,}632 - 6811) \\ &= 321.4 \text{ kW } (1.097 \times 10^6 \text{ Btu/h}). \end{aligned}$$

Here again, $q_1 = -q_2$ since there are only two active surfaces and no heat is lost to the surroundings.

The equilibrium temperature of the two adiabatic surfaces is determined from their radiosities according to Eq. (9.82):

$$T_3 = \left(\frac{J_3}{\sigma}\right)^{1/4} = \left(\frac{4081}{5.67 \times 10^{-8}}\right)^{1/4} = 518°\text{K} = 245°\text{C } (473°\text{F}),$$

$$T_4 = \left(\frac{J_4}{\sigma}\right)^{1/4} = \left(\frac{3944}{5.67 \times 10^{-8}}\right)^{1/4} = 514°\text{K} = 241°\text{C } (466°\text{F}).$$

The closeness of the values of the temperatures of the two adiabatic surfaces, T_3 and T_4 (or their radiosities J_3 and J_4), suggest that the two might have been treated as a single surface without much error and with a considerable saving in algebra since only two unknown radiosities would have to be found in such an instance. Examination of this possibility is left as an exercise for the reader.

Matrix Solution of Enclosures of Black, Gray, and Adiabatic Surfaces

The preceding examples illustrate the method that may be followed when the enclosure consists of only a few surfaces of unknown radiosity. If the number of gray active and adiabatic surfaces exceed about three, the solution for the unknown radiosities may become cumbersome unless matrix techniques are applied. In order to solve for the J's by matrix inversion, it is useful to recast the governing equations into forms useful for digital computer application and which do not become indeterminate when an active surface is black.

Equation (9.79) resulted from equating Eqs. (9.73a) and (9.76). However, for present purposes, it is desirable to use Eq. (9.75) in place of (9.76):

$$q_i = A_j J_i - \sum_j A_i F_{i-j} J_j.$$

If the Kronecker delta function δ_{ij} is introduced ($\delta_{ij} = 1$ when $i = j$, $\delta_{ij} = 0$ when $i \neq j$), the above is

$$\frac{q_i}{A_i} = \sum_j (\delta_{ij} - F_{i-j}) J_j. \tag{9.83}$$

When equated with Eq. (9.73a) some algebra gives

$$\frac{A_i \epsilon_i}{1 - \epsilon_i}(E_{bi} - J_i) = A_i \sum_j (\delta_{ij} - F_{i-j}) J_j$$

$$E_{bi} = \sum_j \frac{\delta_{ij} - (1 - \epsilon_i) F_{i-j}}{\epsilon_i} J_j. \tag{9.84}$$

Equations of the form of Eq. (9.84) may be written for each surface on which E_{bi} (i.e., T_i) is known, and equations of the form of Eq. (9.83) may be written for the adiabatic surfaces with $q_i/A = 0$. This set of equations may be solved for the unknown surface radiosities by matrix inversion routines normally found in most digital computing installations.

For the case of an enclosure of n surfaces, the first n_1 of which have specified E_{bi} and the remaining $n - n_1$ of which are adiabatic, the resulting set of n equations in the n radiosities may be represented by the matrix equation

$$[A][J] = [C]. \tag{9.85}$$

In Eq. (9.85), $[J]$ is the matrix of the radiosities,

$$[J] = \begin{bmatrix} J_1 \\ J_2 \\ \cdot \\ \cdot \\ \cdot \\ J_i \\ \cdot \\ \cdot \\ \cdot \\ J_n \end{bmatrix},$$

and $[C]$ is the matrix of the constant terms,

$$[C] = \begin{bmatrix} C_1 \\ C_2 \\ \cdot \\ \cdot \\ \cdot \\ C_i \\ \cdot \\ \cdot \\ \cdot \\ C_n \end{bmatrix}.$$

In the matrix $[C]$, Eqs. (9.84) and (9.83) show that

$$\begin{aligned} C_i &= E_{bi}, \qquad i = 1, 2, \ldots, n_1, \\ C_i &= 0, \qquad i = n_1 + 1, \ldots, n. \end{aligned} \tag{9.86}$$

The coefficient matrix $[A]$ is

$$[A] = \begin{bmatrix} a_{11} & a_{12} & \cdots & a_{1j} & \cdots & a_{1n} \\ a_{21} & a_{22} & \cdots & a_{2j} & \cdots & a_{2n} \\ \cdot \\ \cdot \\ \cdot \\ a_{i1} & a_{i2} & \cdots & a_{ij} & \cdots & a_{in} \\ \cdot \\ \cdot \\ \cdot \\ A_{n1} & a_{n2} & \cdots & a_{nj} & \cdots & a_{nn} \end{bmatrix},$$

in which the elements a_{ij} are formed, from Eqs. (9.83) and (9.84), according to the rule:

$$\text{For } i = 1, 2, \ldots, n_1; \qquad j = 1, 2, \ldots, n_1, \ldots, n$$

$$a_{ij} = \frac{\delta_{ij} - (1 - \epsilon_i)F_{i-j}}{\epsilon_i}. \tag{9.87}$$

$$\text{For } i = n_1 + 1, \ldots, n; \qquad j = 1, 2, \ldots, n_1, \ldots, n$$

$$a_{ij} = \delta_{ij} - F_{i-j}.$$

By inverting the matrix $[A]$, to obtain its inverse $[A]^{-1}$, the radiosities are found from

$$[J] = [A]^{-1}[C]. \tag{9.88}$$

Once the radiosities are known, the desired heat fluxes at the active surfaces are found from application of Eq. (9.83):

$$\text{For } i = 1, 2, \ldots, n_1; \quad j = 1, 2, \ldots, n_1, \ldots, n \tag{9.89}$$

$$\frac{q_i}{A_i} = \sum_j (\delta_{ij} - F_{i-j})J_j$$

and the equilibrium temperatures of the adiabatic surfaces from:

$$\text{For } i = n_1 + 1, \ldots, n \tag{9.90}$$

$$T_i = \left(\frac{J_i}{\sigma}\right)^{1/4}$$

Nothing in the foregoing procedure fails if one of the active surfaces is black, one obtaining $E_{bi} = J_i$ automatically.

The method just outlined may be extended to cases in which the heat fluxes at the "nonactive" surfaces are given values other than zero; however, this generalization is not presented here for brevity and to avoid complications which might result when the specified fluxes violate the energy conservation principle. Reference 9 may be consulted in regard to this general problem.

EXAMPLE 9.17

Repeat Example 9.16 using the matrix formulation.

Solution. For the enclosure described in Example 9.16, the following data were given for the two active surfaces, A_1 and A_2, and the two adiabatic surfaces, A_3

and A_4:

$$A_1 = 200 \text{ m}^2, \quad \epsilon_1 = 1.0, \quad T_1 = 473°\text{K}, \quad E_{b1} = 2838 \text{ W/m}^2,$$

$$A_2 = 100 \text{ m}^2, \quad \epsilon_2 = 0.4, \quad T_2 = 673°\text{K}, \quad E_{b2} = 11{,}632 \text{ W/m}^2,$$

$$A_3 = 100 \text{ m}^2, \quad q_3 = 0,$$

$$A_4 = 300 \text{ m}^2, \quad q_4 = 0.$$

Also, the shape factors were given:

$$F_{1-1} = 0, \quad F_{1-2} = 0.167, \quad F_{1-3} = 0.158, \quad F_{1-4} = 0.675,$$

$$F_{2-2} = 0, \quad F_{2-3} = 0.168, \quad F_{2-4} = 0.498, \quad F_{3-4} = 0.484.$$

The remaining eight shape factors of the 16 possible in a four-surface enclosure are also needed in this formulation. Six are found by applying the reciprocity principle:

$$F_{2-1} = \frac{A_1}{A_2} F_{1-2} = 0.334, \qquad F_{3-1} = \frac{A_1}{A_3} F_{1-3} = 0.316,$$

$$F_{4-1} = \frac{A_1}{A_4} F_{1-4} = 0.450, \qquad F_{3-2} = \frac{A_2}{A_3} F_{2-3} = 0.168,$$

$$F_{4-2} = \frac{A_2}{A_4} F_{2-4} = 0.166, \qquad F_{4-3} = \frac{A_3}{A_4} F_{3-4} = 0.161.$$

The remaining two are found by use of the enclosure rule:

$$F_{3-3} = 1 - F_{3-1} - F_{3-2} - F_{3-4} = 0.032,$$

$$F_{4-4} = 1 - F_{4-1} - F_{4-2} - F_{4-3} = 0.223.$$

With these data, the rules of Eq. (9.87) give the coefficient matrix $[A]$ to be

$$[A = \begin{bmatrix} \dfrac{1-(1-\epsilon_1)F_{1-1}}{\epsilon_1} & \dfrac{-(1-\epsilon_1)F_{1-2}}{\epsilon_1} & \dfrac{-(1-\epsilon_1)F_{1-3}}{\epsilon_1} & \dfrac{-(1-\epsilon_1)F_{1-4}}{\epsilon_1} \\[2ex] \dfrac{-(1-\epsilon_2)F_{2-1}}{\epsilon_2} & \dfrac{1-(1-\epsilon_2)F_{2-2}}{\epsilon_2} & \dfrac{-(1-\epsilon_2)F_{2-3}}{\epsilon_2} & \dfrac{-(1-\epsilon_2)F_{2-4}}{\epsilon_2} \\[2ex] -F_{3-1} & -F_{3-2} & 1-F_{3-3} & -F_{3-4} \\ -F_{4-1} & -F_{4-2} & -F_{4-3} & 1-F_{4-4} \end{bmatrix}$$

$$= \begin{bmatrix} 1 & 0 & 0 & 0 \\ -0.5010 & 2.500 & -0.2520 & -0.7470 \\ -0.3160 & -0.1680 & 0.9680 & -0.4840 \\ -0.4500 & -0.1660 & -0.1610 & 0.7770 \end{bmatrix}$$

The inverse of this matrix is

$$[A]^{-1} = \begin{bmatrix} 1 & 0 & 0 & 0 \\ 0.5482 & 0.4518 & 0.2118 & 0.5663 \\ 0.8587 & 0.1413 & 1.2187 & 0.8950 \\ 0.8742 & 0.1258 & 0.2978 & 1.5934 \end{bmatrix}$$

The constant matrix, from Eq. (9.86), is

$$[C] = \begin{bmatrix} E_{b1} \\ E_{b2} \\ 0 \\ 0 \end{bmatrix} = \begin{bmatrix} 2838 \\ 11{,}632 \\ 0 \\ 0 \end{bmatrix}$$

so that the radiosity matrix, by Eq. (9.88), is

$$[J] = \begin{bmatrix} J_1 \\ J_2 \\ J_3 \\ J_4 \end{bmatrix} = [A]^{-1}[C] = \begin{bmatrix} 2838 \\ 6811 \\ 4081 \\ 3944 \end{bmatrix}.$$

These are the same as the radiosities found in Example 9.16.

The heat flows at the two active surfaces are found from application of Eq. (9.89):

$$\begin{aligned} \frac{q_1}{A_1} &= \sum_j (\delta_{1j} - F_{1j})J_j, \qquad j = 1, 2, 3, 4 \\ &= (1 - F_{1-1}J_1) + (0 - F_{1-2})J_2 + (0 - F_{1-3})J_3 + (0 - F_{1-4})J_4 \\ &= (1 - 0)2838 - 0.167 \times 6811 - 0.158 \times 4081 - 0.675 \times 3944 \\ &= -1606 \text{ W/m}^2, \\ q_1 &= (-1606) \times 200 = -321.3 \text{ kW } (1.096 \times 10^6 \text{ Btu/h}), \end{aligned}$$

and

$$\begin{aligned} \frac{q_2}{A_2} &= \sum_j (\delta_{2j} - F_{2-j})J_j, \qquad j = 1, 2, 3, 4 \\ &= (0 - F_{2-1})J_1 + (1 - F_{2-2})J_2 + (0 - F_{2-3})J_3 + (0 - F_{2-4})J_4 \\ &= -0.334 \times 2838 + (1 - 0)6811 - 0.168 \times 4081 - 0.498 \times 3944 \\ &= 3213 \text{ W/m}^2, \\ q_2 &= 3213 \times 100 = 321.3 \text{ kW } (1.096 \times 10^6 \text{ Btu/h}) \end{aligned}$$

These results are the same as found in Example 9.16, as would be the temperatures of the adiabatic surfaces since this calculation here is identical to that in Example 9.16.

The coefficient matrix in this example, and its inverse, were rather simple in the first row since A_1 is a black surface. If A_1 had been gray, however, the first row would contain values other than 1 and 0, but still the inversion by digital computer would be no more complex. On the other hand, solution of this problem with A_1 gray by use of the direct approach would be cumbersome, indeed, since a set of four equations would have to be solved for four unknown radiosities. Thus, as mentioned earlier, when an enclosure of more than about three gray active and reradiating surfaces is involved, one would normally apply the matrix formulation.

9.10 SOME CLOSED-FORM SOLUTIONS FOR ENCLOSURES OF DIFFUSE, GRAY SURFACES

The methods outlined in Sec. 9.9 for the radiant exchange in enclosures of gray (or black) diffuse surfaces may be carried out algebraically when the number of unknown surface radiosities does not exceed two or three. In many instances the results may be written directly from examination of the corresponding electrical network analogy. The results of some of these analyses are summarized in the following and cover cases of some practical interest.

Two Gray Surfaces and a Single Adiabatic Surface

A case of some practical interest is that in which two active gray surfaces, of known E_b or temperature, are enclosed by a single adiabatic surface. The heat source and sink in a furnace enclosed by refractory walls is a typical example of such an enclosure. Figure 9.32 shows the electrical network for this case. The two active surfaces are denoted by A_1 and A_2 and the single adiabatic surface by A_r (the subscript r denoting the refractory surface). Summing the resistances in series and parallel between the nodes of specified E_{b1} and E_{b2} gives the heat flow from the two surfaces:

$$
\begin{aligned}
q_1 &= -q_2 \\
&= \frac{E_{b1} - E_{b2}}{\dfrac{1-\epsilon_1}{A_1\epsilon_1} + \left\{A_1F_{1-2} + \left[\left(\dfrac{1}{A_1F_{1-r}}\right) + \left(\dfrac{1}{A_2F_{2-r}}\right)\right]^{-1}\right\}^{-1} + \dfrac{1-\epsilon_2}{A_2\epsilon_2}}.
\end{aligned}
\tag{9.91}
$$

The equilibrium radiosity, and hence temperature, of the single adiabatic surface is found by applying Eq. (9.80) or summing currents at J_r:

$$
A_1F_{1-r}(J_r - J_1) + A_2F_{2-r}(J_r - J_2) = 0, \tag{9.92}
$$

$$
\sigma T_r^4 = J_r = \frac{A_1F_{1-r}J_1 + A_2F_{2-r}J_2}{A_1F_{1-r} + A_2F_{2-r}}.
$$

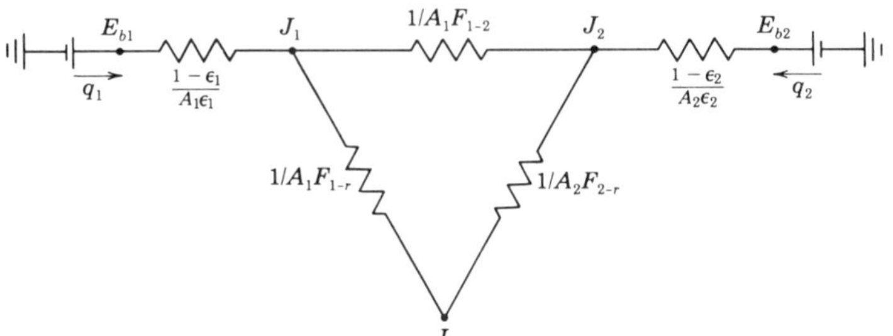

Figure 9.32. Network analogy for two gray surfaces and one adiabatic surface.

The J_1 and J_2 in Eq. (9.92) are found from Eq. (9.81) or the current flows between E_{b1} and J_1, and E_{b2} and J_2:

$$q_1 = \frac{A_1\epsilon_1}{1-\epsilon_1}(E_{b1} - J_1); \qquad J_1 = E_{b1} - \frac{1-\epsilon_1}{A_1\epsilon_1} q_1,$$
$$q_2 = -q_1 = \frac{A_2\epsilon_2}{1-\epsilon_2}(E_{b2} - J_2); \qquad J_2 = E_{b2} + \frac{1-\epsilon_2}{A_2\epsilon_2} q_1. \tag{9.93}$$

With q_1 known from Eq. (9.91), Eqs. (9.92) and (9.93) permit calculation of J_r or T_r.

If the active surfaces are black, Eq. (9.91) is correspondingly simplified, and the effect of the surface grayness is seen to be that of reducing the net heat flow. When A_1 and A_2 are black, the determination of J_r from Eq. (9.92) is simplifed since J_1 and J_2 are simply replaced by E_{b1} and E_{b2}.

The expressions above require the knowledge of three shape factors: F_{1-2}, F_{1-r}, F_{2-r}. These may be interrelated for some special cases as noted below.

Two Gray Convex Surfaces and a Single Adiabatic Surface

If the two active surfaces of the preceding case are plane or convex so that they do not "see" themselves and $F_{1-1} = F_{2-2} = 0$, then the two shape factors F_{1-r} and F_{2-r} may be eliminated by using the enclosure rule: $F_{1-r} = 1 - F_{1-2}$, $F_{2-r} = 1 - F_{2-1}$ and $A_1F_{1-2} = A_2F_{2-1}$. The resulting heat flow and equilibrium temperature become, after some algebra,

$$q_1 = -q_2 = \frac{E_{b1} - E_{b2}}{\dfrac{1-\epsilon_1}{A_1\epsilon_1} + \left[\dfrac{A_1 + A_2 - 2A_1F_{1-2}}{A_1A_2 - (A_1F_{1-2})^2}\right] + \dfrac{1-\epsilon_2}{A_2\epsilon_2}},$$
$$\sigma T_r^4 = J_r = \frac{A_1(1 - F_{1-2})J_1 + (A_2 - A_1F_{1-2})J_2}{A_1 + A_2 - 2A_1F_{1-2}}, \tag{9.94}$$
$$J_1 = E_{b1} - \frac{1-\epsilon_1}{A_1\epsilon_1} q_1; \qquad J_2 = E_{b2} + \frac{1-\epsilon_2}{A_2\epsilon_2} q_1,$$

In this instance only the single shape factor between the two active surfaces, F_{1-2}, is needed, and the configuration of the adiabatic surface is immaterial as long as it does not obstruct the view of A_1 and A_2. Again, the reduction to the case when the active surfaces are black is obvious.

Any Number of Black Surfaces and a Single Adiabatic Surface

If any enclosure consists of any number of black surfaces, say n_1 (all of specified E_{bi}) plus *one* adiabatic surface, it is left as an exercise to show that the heat flow from any one of the surfaces is

$$q_i = A_i \sum_{j=1}^{n_1} \left(F_{i-j} + \frac{F_{r-j}F_{i-r}}{1 - F_{r-r}} \right)(E_{bi} - E_{bj}), \tag{9.95}$$

and the equilibrium radiosity of the single adiabatic surface is

$$\sigma T_r^4 = J_r = \frac{\sum_{j=1}^{n_1} F_{r-j}E_{bj}}{1 - F_{r-r}}. \tag{9.96}$$

In Eqs. (9.95) and (9.96), F_{r-r} is the shape factor of the adiabatic surface with respect to itself.

Comparison of Eq. (9.95) with Eq. (9.69) for an enclosure of black surfaces only shows that the effect of an adiabatic surface is to enhance the heat flow between a pair of black active ones.

Two Convex Gray Surfaces in a Radiation-Free Space

Two gray surfaces, A_1 and A_2, located in an otherwise radiation-free space may be characterized as an enclosure in which the enclosing surface, A_3, is a black surface maintained with $E_{b3} = 0$. The corresponding network analogy is shown in Fig. 9.33. Current balances at nodes J_1 and J_2, or Eq. (9.79) yields

$$\frac{A_1\epsilon_1}{1 - \epsilon_1}(E_{b1} - J_1) = A_1F_{1-2}(J_1 - J_2) + A_1F_{1-3}(J_2 - 0),$$

$$\frac{A_2\epsilon_2}{1 - \epsilon_2}(E_{b2} - J_2) = A_2F_{2-1}(J_2 - J_1) + A_2F_{2-3}(J_2 - 0).$$

If the two active surfaces are convex so that $F_{1-1} = F_{2-2} = 0$, $F_{1-3} = 1 - F_{1-2}$, $F_{2-3} = 1 - F_{2-1}$, and if one uses $A_1F_{1-2} = A_2F_{2-1}$, solution of the two equations above will yield

$$J_1 = \frac{\epsilon_1E_{b1} + \epsilon_2(1 - \epsilon_1)F_{1-2}E_{b2}}{1 - (1 - \epsilon_1)(1 - \epsilon_2)(A_1/A_2)F_{1-2}^2}.$$

Thus since

$$q_1 = \frac{A_1\epsilon_1}{1 - \epsilon_1}(E_{b1} - J_1),$$

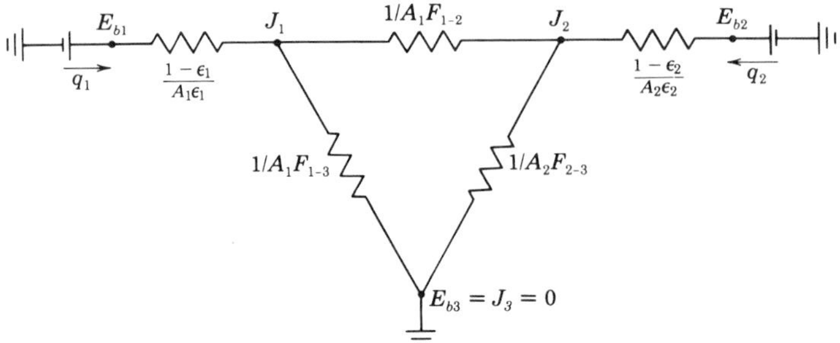

Figure 9.33. Network analogy for two gray surfaces in a radiation free space.

one has

$$q_1 = A_1\epsilon_1 \frac{[1 - (1 - \epsilon_2)(A_1/A_2)F_{1-2}^2]E_{b1} - \epsilon_2 F_{1-2} E_{b2}}{1 - (1 - \epsilon_1)(1 - \epsilon_2)(A_1/A_2)F_{1-2}^2}. \tag{9.97}$$

Note that in this instance, since energy is lost to space,

$$q_1 \neq -q_2.$$

The heat flow from surface A_2 may be obtained by interchanging the subscripts 1 and 2 throughout Eq. (9.97).

Two Gray Surfaces Only

If, as suggested in Fig. 9.34(a), an enclosure consists of only two gray active surfaces and nothing else, the network analogy readily yields

$$q_1 = -q_2 = \frac{E_{b1} - E_{b2}}{\dfrac{1 - \epsilon_1}{A_1\epsilon_1} + \dfrac{1}{A_1F_{1-2}} + \dfrac{1 - \epsilon_2}{A_2\epsilon_2}}. \tag{9.98}$$

If the surfaces are infinite parallel gray planes, $A_1 = A_2$, $F_{1-2} = 1$, and the result above reduces to that discussed in Sec. 9.6.

One Gray Surface Completely Enclosing a Second Convex Gray Surface

A case of considerable practical value is that in which one gray surface, A_2, completely encloses a second surface, A_1, which does not see itself, as depicted in Fig. 9.34(b). Concentric pipes, concentric spheres, objects in rooms, etc., are examples. In this instance one may use Eq. (9.98) with $F_{1-1} = 0$, $F_{1-2} = 1$:

$$\frac{q_1}{A_1} = \frac{E_{b1} - E_{b2}}{\dfrac{1}{\epsilon_1} + \dfrac{A_1}{A_2}\left(\dfrac{1 - \epsilon_2}{\epsilon_2}\right)}. \tag{9.99}$$

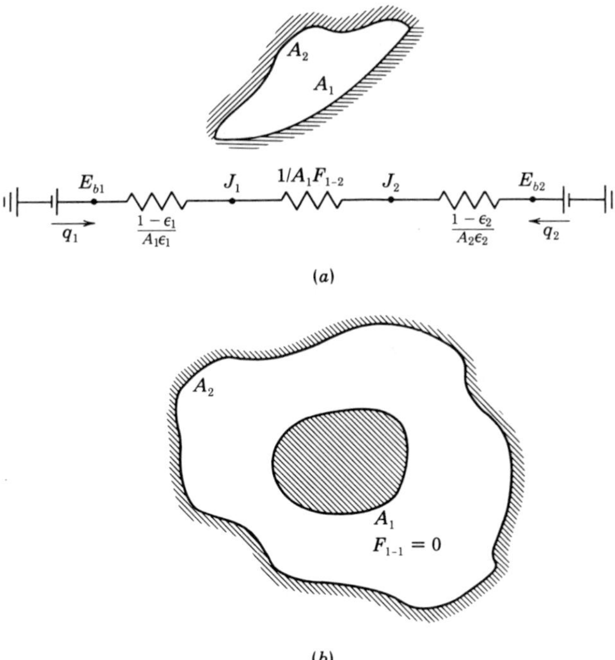

Figure 9.34. Enclosure of only two gray surfaces.

If, as may often be the case for heated bodies in a very *large* room, $A_1/A_2 \rightarrow 0$, then

$$\frac{q_1}{A_1} = \epsilon_1(E_{b1} - E_{b2}). \tag{9.100}$$

The heat loss from the small body is seen to be independent of the properties of the large room.

EXAMPLE 9.18

Repeat Example 9.14 using the closed-form solutions.

Solutions. Example 9.14 consisted of two black active surfaces

$$A_1 = 120 \text{ m}^2, \quad \epsilon_1 = 1.0, \quad T_1 = 318°\text{K}, \quad E_{b1} = 579.8 \text{ W/m}^2,$$

$$A_2 = 18 \text{ m}^2, \quad \epsilon_2 = 1.0, \quad T_1 = 283°\text{K}, \quad E_{b2} = 363.7 \text{ W/m}^2,$$

enclosed by a single adiabatic surface. The shape factor F_{1-2} between the two active surfaces was found to be

$$F_{1-2} = 0.058.$$

The surfaces A_1 and A_2 were both plane, so they do not see themselves ($F_{1-1} = F_{2-2} = 0$). Thus the results given in Eqs. (9.94) for the heat flow and the equilibrium temperature of the adiabatic surface apply. For the heat flow,

$$q_1 = -q_2 = \frac{E_{b1} - E_{b2}}{\dfrac{1 - \epsilon_1}{A_1\epsilon_1} + \left[\dfrac{A_1 + A_2 - 2A_1F_{1-2}}{A_1A_2 - (A_1F_{1-2})^2}\right] + \dfrac{1 - \epsilon_2}{A_2\epsilon_2}}$$

$$= \frac{A_1A_2 - (A_1F_{1-2})^2}{A_1 + A_2 - 2A_1F_{1-2}} \times (E_{b1} - E_{b2}),$$

since both A_1 and A_2 are black ($\epsilon_1 = \epsilon_2 = 1.0$). Thus for the data given,

$$q_1 = -q_2 = \frac{120 \times 18 - (120 \times 0.058)^2}{120 + 18 - 2(120 \times 0.058)}(579.8 - 363.7)$$

$$= 3678 \text{ W } (12{,}550 \text{ Btu/h}).$$

This value differs slightly from that found in Example 9.14 due to differences in round-off error in the two formulations.

The equilibrium radiosity of the adiabatic surface is, from Eq. (9.94),

$$J_r = \frac{A_1(1 - F_{1-2})J_1 + (A_2 - A_1F_{1-2})J_2}{A_1 + A_2 - 2A_1F_{1-2}}.$$

The two radiosities J_1 and J_2 are equal to the respective emissive powers since $\epsilon_1 = \epsilon_2 = 1.0$:

$$J_1 = E_{b1} - \frac{1 - \epsilon_1}{A_1\epsilon_1} q_1 = E_{b1},$$

$$J_2 = E_{b2} - \frac{1 - \epsilon_2}{A_2\epsilon_2} q_2 = E_{b2},$$

so, with the data given,

$$J_r = \frac{A_1(1 - F_{1-2})E_{b1} + (A_2 - A_1F_{1-2})E_{b2}}{A_1 + A_2 - 2A_1F_{1-2}}$$

$$= \frac{120(1 - 0.058)579.8 + (18 - 120 \times 0.058)363.7}{120 + 18 - 2 \times 120 \times 0.058}$$

$$= 560.57 \text{ W/m}^2.$$

This value, too, differs slightly from that found in Example 9.14. The temperature of the adiabatic surface is, then,

$$T_r = \left(\frac{J_r}{\sigma}\right)^{1/4} = \left(\frac{560.57}{5.67 \times 10^{-8}}\right)^{1/4}$$

$$= 315°\text{K} = 42°\text{C } (108°\text{F}).$$

EXAMPLE 9.19

Repeat Example 9.15 using the closed-form solutions.

Solution. Example 9.15 involved two active gray surfaces,

$$A_1 = 50 \text{ ft}^2, \quad \epsilon_1 = 0.7, \quad T_1 = 960°\text{R}, \quad E_{b1} = 1456 \text{ Btu/h-ft}^2,$$

$$A_2 = 50 \text{ ft}^2, \quad \epsilon_2 = 0.9, \quad T_2 = 1360°\text{R}, \quad E_{b2} = 5864 \text{ Btu/h-ft}^2,$$

with the shape factor

$$F_{1-2} = F_{2-1} = 0.286.$$

The surfaces are either (a) located in a radiation-free space or (b) enclosed by a single adiabatic reradiating surface.

(a) For the instance in which A_1 and A_2 are in a radiation-free environment, Eq. (9.97) applies since both A_1 and A_2 do not see themselves. Thus for surface A_1 the heat flow is

$$q_1 = A_1\epsilon_1 \frac{[1 - (1 - \epsilon_2)(A_1/A_2)F_{1-2}^2]E_{b1} - \epsilon_2 F_{1-2}E_{b2}}{1 - (1 - \epsilon_1)(1 - \epsilon_2)(A_1/A_2)F_{1-2}^2}$$

$$= 50 \times 0.7 \frac{[1 - (1 - 0.9)(50/50) \times (0.286)^2]1456 - 0.9 \times 0.286 \times 5864}{1 - (1 - 0.7)(1 - 0.9)(50/50) \times (0.286)^2}$$

$$= -2291 \text{ Btu/h } (671 \text{ W}).$$

For surface A_2, the heat flow is found by interchanging the 1's and 2's in Eq. (9.97):

$$q_2 = A_2\epsilon_2 \frac{[1 - (1 - \epsilon_1)(A_2/A_1)F_{2-1}^2]E_{b2} - \epsilon_1 F_{2-1}E_{b1}}{1 - (1 - \epsilon_2)(1 - \epsilon_1)(A_2/A_1)F_{2-1}^2}$$

$$= 50 \times 0.9 \frac{[1 - (1 - 0.7)(50/50) \times (0.286)^2]5864 - 0.7 \times 0.286 \times 1456}{1 - (1 - 0.9)(1 - 0.7)(50/50) \times (0.286)^2}$$

$$= 244{,}900 \text{ Btu/h } (71{,}780 \text{ W}).$$

Once again, these results differ slightly from those found in Example 9.15 because of differences in round-off error.

(b) When the two surfaces A_1 and A_2 are enclosed by a single adiabatic surface, Eqs. (9.94) apply since A_1 and A_2 are plane and do not see themselves. Thus the heat flow at the two surfaces is

$$q_1 = -q_2 = \frac{E_{b1} - E_{b2}}{\dfrac{1-\epsilon_1}{A_1\epsilon_1} + \dfrac{A_1 + A_2 - 2A_1F_{1-2}}{A_1A_2 - (A_1F_{1-2})^2} + \dfrac{1-\epsilon_2}{A_2\epsilon_2}}$$

$$= \frac{1456 - 5864}{\dfrac{1-0.7}{50 \times 0.7} + \dfrac{50 + 50 - 2 \times 50 \times 0.286}{50 \times 50 - (50 \times 0.286)^2} + \dfrac{1-0.9}{50 \times 0.9}}$$

$$= -105{,}200 \text{ Btu/h } (30{,}800 \text{ W}).$$

So the two active surface radiosities are

$$J_1 = E_{b1} - \frac{1-\epsilon_1}{A_1\epsilon_1} q_1 = 1456 - \frac{1-0.7}{50 \times 0.7}(-105{,}200)$$

$$= 2358 \text{ Btu/h-ft}^2,$$

$$J_2 = E_{b2} - \frac{1-\epsilon_2}{A_2\epsilon_2} q_2 = 5864 - \frac{1-0.9}{50 \times 0.9}(105{,}200)$$

$$= 5630 \text{ Btu/h-ft}^2.$$

Then the equilibrium radiosity and temperature of the adiabatic surface are

$$J_r = \frac{A_1(1 - F_{1-2})J_1 + (A_2 - A_1F_{1-2})J_2}{A_1 + A_2 - 2A_1F_{1-2}}$$

$$= \frac{50(1 - 0.286)2358 + (50 - 50 \times 0.286)5630}{50 + 50 - 2 \times 50 \times 0.286}$$

$$= 3994 \text{ Btu/h-ft}^2,$$

$$T_r = \left(\frac{J_r}{\sigma}\right)^{1/4} = \left(\frac{3994}{0.1714 \times 10^{-8}}\right)^{1/4}$$

$$= 1236°\text{R} = 776°\text{F } (413°\text{C}).$$

These results are the same as those in Example 9.15.

9.11 RADIATION IN THE PRESENCE OF ABSORBING AND EMITTING GASES

For other than the very broad defintions given in Sec. 9.2, all discussions in this chapter have been devoted to the characteristics of solid surfaces (emitting, absorbing and reflecting characteristics) and the exchange of radiation between solid surfaces separated by nonabsorbing media. The solid surfaces considered were taken as opaque to thermal radiation ($\tau = 0$) while the media between surfaces were taken as transparent and nonemitting ($\tau = 1$, $\epsilon = \alpha = 0$).

Elementary gases with symmetrical molecules are, indeed, transparent to thermal radiation. Many gases with more complex molecules, however, do emit and absorb thermal radiation—at least in certain wavelength bands. Water vapor, carbon dioxide, ammonia, and most hydrocarbons are examples of the latter.

Absorptivity and Emissivity of Gases

Figure 9.35 shows, as an example, the absorption characteristics of carbon dioxide. Similar data are available for other gases of engineering significance (Ref. 12). Examination of Fig. 9.35 reveals that the absorptivity depends upon the thickness of the gas layer as well as upon the wavelength of the incident radiation. The thermodynamic state of the gas is also a determining factor for the absorptivity. The dependence of the gas absorption on wavelength and thickness is described by the equation

$$dI_\lambda = -I_\lambda a_\lambda \, ds,$$

where I_λ represents the monochromatic intensity of the incident beam, s the path length of the beam, and a_λ the *absorption coefficient*. The absorption coefficient is dependent upon wavelength and the thermodynamic state of the gas. To a first approximation, a_λ varies linearly with pressure, at constant temperature. Thus, in a mixture of an absorbing and nonabsorbing gas the absorption coefficient should be proportional to the partial pressure of the absorbing gas.

An integration of the above equation yields

$$I_\lambda = I_{\lambda 0} e^{-a_\lambda s},$$

where $I_{\lambda 0}$ represents the intensity of the radiant beam as it enters the gas at $s = 0$. Since the reflection of thermal radiation at a gas-to-gas interface is generally negligible, the result above yields the following for the monochromatic transmissivity, emissivity, and absorptivity of a gas:

$$\tau_\lambda = e^{-a_\lambda s}, \qquad (9.101)$$

$$\alpha_\lambda = 1 - e^{-a_\lambda s}.$$

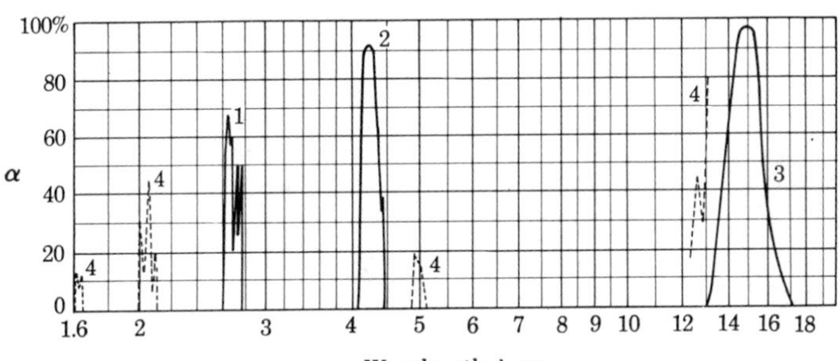

Figure 9.35. Absorptivity of carbon dioxide: (1) 5 cm thick; (2) 3 cm thick; (3) 6.3 cm thick; (4) 100 cm thick. (From E. R. G. Eckert and R. M. Drake, *Heat and Mass Transfer*, New York, McGraw-Hill, 1959. Used by permission.)

If Kirchhoff's law holds, than a monochromatic emissivity of a gas is

$$\epsilon_\lambda = 1 - e^{-a_\lambda s} = \frac{I_\lambda}{I_{b\lambda}}. \tag{9.102}$$

As indicated in Eq. (9.102), the emissivity of a gas has to be interpreted as the intensity of radiation arriving at a point, divided by the equivalent black body intensity. As such, the emissivity given above is associated with the radiation arriving at a point from a given direction and through a given thickness of gas. In order to account for radiant exchange between a gas mass and an element of its boundary surface, consideration must be made for *all* the radiation arriving at the element from *all* directions. The geometry of such a situation is depicted in Fig. 9.36.

Application of Eqs. (9.16), (9.17), and (9.31) shows that the hemispherical monochromatic emissivity of the entire gas mass with respect to dA is

$$\bar{\epsilon}_\lambda = \frac{1}{\pi}\int_0^{2\pi}\int_0^{\pi/2} (1 - e^{-a_\lambda s}) \sin\theta \cos\theta \, d\theta \, d\varphi.$$

The integration in the equation above can be performed only when s is related to θ and φ—meaning that the geometry of the gas mass must be known. Then a second integration must be performed, over the finite boundary surface.

In general, calculations like those just described are quite complex. However, in the case of a hemispherical mass of gas, the emissivity for radiation to the center of the base may be found easily. Hottel and Sarofim (Ref. 10) have shown that other shapes of practical interest may be reduced to equivalent hemispherical masses. The radii of these equivalent hemispheres are referred to as the *mean beam length*, customarily denoted by L_e. Table 9.2 gives recommended mean beam lengths for some shapes of interest. Since for the hemispherical shape, the mean beam length

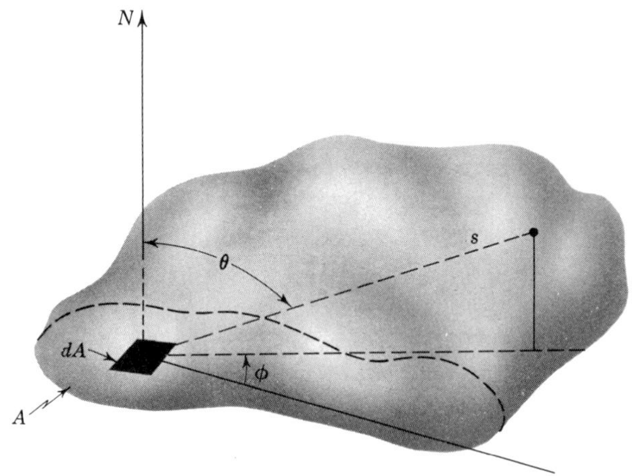

Figure 9.36

Table 9.2 Mean Beam Lengths for Various Gas Shapes

Shape	L_e
Sphere (radiation to surface)	0.65 × diameter
Circular cylinder, infinite length (radiation to curved surface)	0.95 × diameter
Circular cylinder, length = diameter (radiation to base center)	0.77 × diameter
Infinite planes (radiation to surfaces)	1.80 × distance between planes
Cube (radiation to surfaces)	0.66 × edge length
Arbitrary shape (radiation to surface)	≈ 3.6 × (volume/surface)

is constant, the associated value of $\bar{\epsilon}_\lambda$ may be integrated over all wavelengths, to obtain finally a total gas emissivity. Thus, the mean beam length becomes a useful engineering concept, giving the following simpler relation for emissive power of a given gas shape:

$$E = \bar{\epsilon}(L_e)\sigma T^4. \tag{9.103}$$

In Eq. (9.103), $\bar{\epsilon}$ represents the total gas emissivity described above. Quite apparently, $\bar{\epsilon}$ is a function of the gas composition, its thermodynamic state, *and* the mean beam length, L_e.

Values of $\bar{\epsilon}(L_e)$ may be found from experimental data such as those shown in Figs. 9.37 and 9.38 for carbon dioxide and water vapor mixed with air. As noted in the earlier discussion, the gas emissivity is dependent upon density, and in mixtures of emitting and nonemitting gases, $\bar{\epsilon}$ should depend on the partial pressure of the absorbing component—as Figs. 9.37 and 9.38 show.

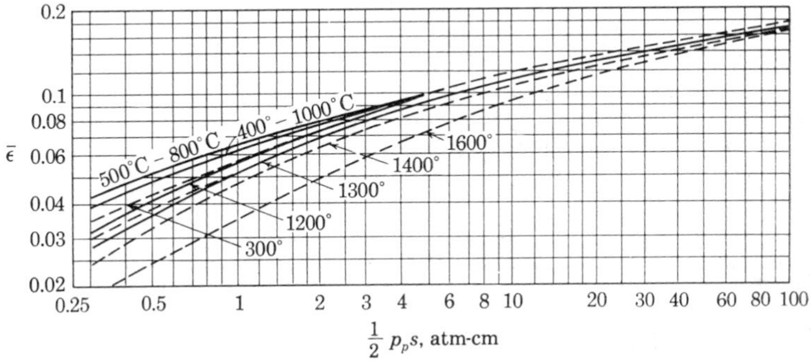

Figure 9.37. Emissivity of carbon dioxide in a mixture with air or nitrogen. The total pressure is 1 atm, and p_p represents the partial pressure of the carbon dioxide. (From E. R. G. Eckert and R. M. Drake, *Heat and Mass Transfer*, New York, McGraw-Hill, 1959. Used by permission.)

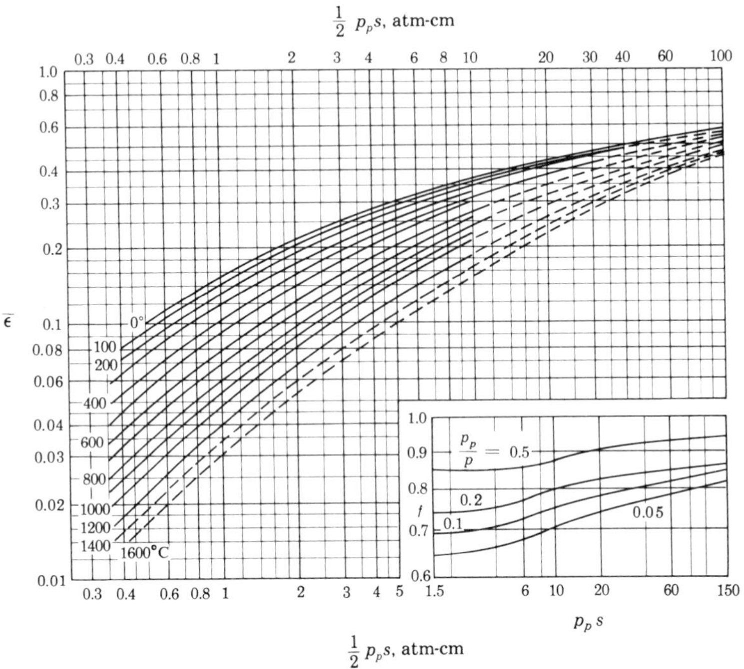

Figure 9.38. Emissivity of mixtures of water vapor and air or nitrogen for a total pressure of 1 atm, p_p representing the partial pressure of the water vapor. When p_p differs from 1, values from the large chart must be multiplied by f from the small chart. (From E. R. G. Eckert and R. M. Drake, *Heat and Mass Transfer,* New York, McGraw-Hill, 1959. Used by permission.)

EXAMPLE 9.20

Find the emissivity to be used for calculating the radiant exchange between a spherical mass of water vapor–air mixture and its surface if the mass has the diameter of 3 m, a temperature of 800°C, and a total pressure of 1 atm. The partial pressure of the water vapor is 0.2 atm.

Solution. For a diameter of 3 m, Table 9.2 gives the mean beam length to be $L_e = 0.65 \times 3 = 1.95$ m. Thus, $p_p L_e = 0.2 \times 1.95 \times 100 = 39$ atm-cm. At 800°C, $p_p/p = 0.2$, Fig. 9.38 gives

$$\bar{\epsilon} = 0.32 \times 0.83 = 0.27.$$

Radiation Exchange in Gas-Filled Enclosures

In the event that the enclosures in Secs. 9.8 and 9.9 are filled with an emitting and absorbing gas, the procedures given in those sections must be modified. The radiating gas must be treated as an individual "surface" itself. Since Eq. (9.103)

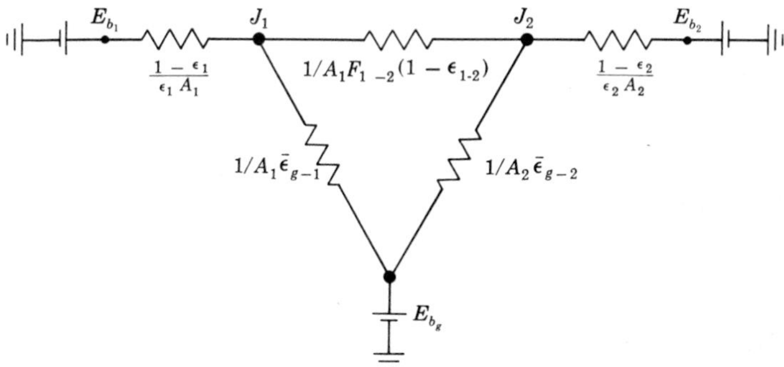

Figure 9.39. Network Analogy of two gray surfaces separated by an absorbing, emitting gas.

gives the emissive power of the gas relative to surface i as $\bar{\epsilon}_{g-i}E_{bg}$, the gas may be represented by a node maintained at a potential E_{bg} connected to each active radiating surface through a resistor of $1/A_i\bar{\epsilon}_{g-i}$. The direct radiant exchange between active surfaces, $A_iF_{i-j}(J_i - J_j)$, when no gas is present, must be decreased by the amount absorbed by the intervening gas. Thus the resistors connecting active nodes have to be of the form $1/A_iF_{i-j}(1 - \bar{\epsilon}_{i-j})$, in which $\bar{\epsilon}_{i-j}$ is the gas absorptivity evaluated for all rays traveling between A_i and A_j. This amounts to the evaluation of the equation for $\bar{\epsilon}$ over the geometrical angles admitted by F_{i-j}. This calculation can become quite complex. References 11 and 12 may be consulted in this respect. Figure 9.39 illustrates the equivalent network for a two-surface enclosure filled with an emitting and absorbing gas.

REFERENCES

1. Siegel, R., and J. R. Howell, *Thermal Radiation Heat Transfer,* 2nd ed., New York, McGraw-Hill, 1981.
2. Dunkle, R. V., "Thermal Radiation Tables and Applications" *Trans ASME,* Vol. 65, 1954, p. 549.
3. Kreith, Frank, *Radiation Heat Transfer,* Scranton, Pa., International Textbook, 1962.
4. Edwards, D. K., D. E. Nelson, R. D. Roddick, and J. T. Gier, "Basic Studies on the Use and Control of Solar Energy," *Univ. Calif. Dept. Eng. Rep. 60-93,* Los Angeles, 1960.
5. Duffie, J. A., and W. A. Beckman, *Solar Energy Thermal Processes,* New York, Wiley, 1974.
6. Kreith, Frank, and J. F. Kreider, *Principles of Solar Engineering*. Washington, D.C., Hemisphere, 1978.
7. Sparrow, E. M., "A New and Simpler Formulation for Radiative Angle Factors," *J. Heat Transfer, Trans. ASME,* Vol. 85, 1963, p. 81.

8. Toups, K. A., "Confac II, A General Computer Program for the Determination of Radiant Interchange Configuration and Form Factors," *Tech. Doc. Rep. FDL-TDR-64-43,* Air Force Flight Dynamics Laboratory, Wright-Patterson Air Force Base, Ohio, 1964.
9. Rohsenow, W. M., and J. P. Hartnett, eds., *Handbook of Heat Transfer,* New York, McGraw-Hill, 1973.
10. Hottel, H. C., and A. F. Sarofim, *Radiative Heat Transfer,* New York, McGraw-Hill, 1967.
11. Oppenheim, A. K., "Radiation Analysis by the Network Method," *Trans. ASME,* Vol. 78, 1956, p. 725.
12. McAdams, W. H., *Heat Transmission,* 3rd ed., New York, McGraw-Hill, 1954.

PROBLEMS

9.1 The accompanying figure depicts an emitting surface, dA_1, and two intercepting surfaces, dA_2 and dA_3. The radial lines connecting dA_1 to dA_2 and dA_3 both lie in the plane of the paper. The differential surface $dA_2 = 10\ \text{cm}^2$ is located at a polar angle $\theta_2 = 0°$ with respect to the normal to dA_1 (i.e., directly above dA_1) and it is oriented normal to the line connecting dA_1 and dA_2 (i.e., $\beta_2 = 0°$). The distance between dA_1 and dA_2, r_2, may assume different values. The differential surface $dA_3 = 8\ \text{cm}^2$ is located at a polar angle $\theta_3 = 30°$ with respect to the normal to dA_1, is located at a radial distance $r_3 = 1.2$ m, but its orientation with respect to r_3, β_3, may assume different values. If the *flux* of radiant energy leaving dA_1 is $f_1 = 500\ \text{W/m}^2$, find **(a)** the radiant energy received by dA_2 if (1) $r_2 = 1$ m,

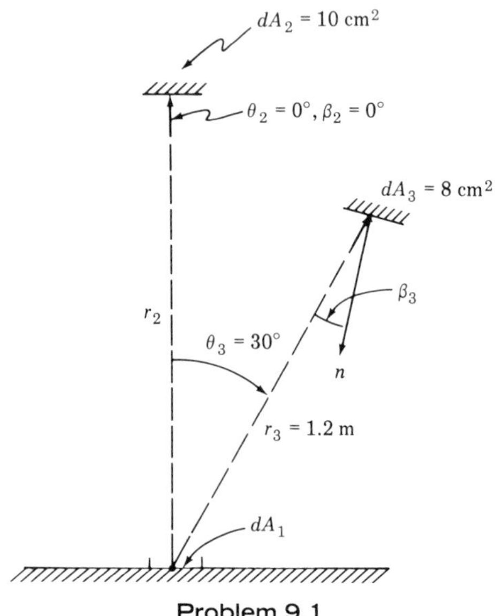

Problem 9.1

(2) $r_2 = 2$ m, or (3) $r_2 = 3$ m; and **(b)** the radiant energy received by dA_3 if (1) $\beta_3 = 0°$, (2) $\beta_3 = 30°$, or (3) $\beta_3 = 45°$. Assume that the emission from dA_1 is diffuse.

9.2 Two small areas, $dA_1 = 5\ cm^2$ and $dA_2 = 8\ cm^2$, are oriented as shown in the accompanying figure. A diffuse flux leaves dA_1 in the amount of $f_1 = 1000\ W/m^2$, and a diffuse flux of $f_2 = 1500\ W/m^2$ leaves dA_2.
(a) Find the solid angle subtended by dA_2 at dA_1 and that subtended by dA_1 at dA_2.
(b) Find the radiant energy intercepted by each surface.

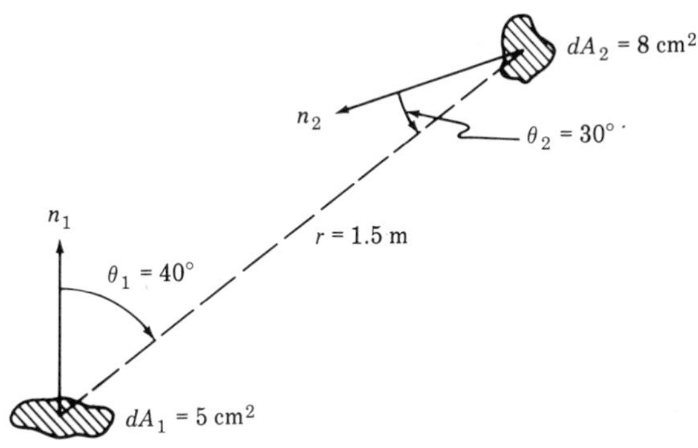

Problem 9.2

9.3 Find the fraction of the total hemispherical emissive power leaving a diffuse emitter that is contained in the directions **(a)** $\pi/4 \le \theta \le \pi/2$, $0 \le \varphi \le \pi$; and **(b)** $0 \le \theta \le \pi/4$, $0 \le \varphi \le \pi$.

9.4 Find the fraction of the total hemispherical emissive power leaving a diffuse emitter that is contained in the directions **(a)** $0 \le \theta \le \pi/6$, $0 \le \varphi \le 2\pi$; and **(b)** $0 \le \theta \le \pi/4$, $0 \le \varphi \le 2\pi$.

9.5 A hollow spherical cavity, 1 m in diameter, is evacuated and heated to an interior surface temperature of 500°K. How many watts of radiant energy is emitted from a hole 0.2 cm in diameter in the wall of the cavity if the cavity surface has an emissivity of **(a)** 0.9 and **(b)** 0.4?

9.6 The filament of an ordinary 100-W light globe may be approximated as a blackbody at 2900°K. What is the wavelength of the maximum monochromatic emission? What fraction of the emission is in the visible portion of the spectrum?

9.7 A blackbody is maintained at a temperature of 200°C. Find **(a)** the wavelength at which the maximum monochromatic emissive power occurs, **(b)** the value of the maximum monochromatic emissive power, **(c)** the total emissive power, and **(d)** the fraction of the radiation emitted between the wavelengths of 1.0 and 4.0 μm.

9.8 Repeat Prob. 9.7 if the blackbody temperature is 1100°C.

9.9 Verify the data given in Table 9.1 for the sun at a blackbody temperature of 5800°K.

9.10 A large cavity is evacuated and heated to an interior surface temperature of 750°K. What is the irradiation on a small test surface placed within the cavity if the cavity surface has an emissivity of **(a)** 0.9 or **(b)** 0.2?

9.11 A large cavity is evacuated and heated to a uniform interior surface temperature. If the radiant energy leaving a small hole in the cavity wall of 0.02 m^2 area is measured to be 80 W, find the temperature of the cavity walls.

9.12 Determine, and compare, the wavelength of the maximum blackbody emission from **(a)** the sun at 5800°K, **(b)** a light bulb filament at 2500°K, **(c)** a heated metal surface at 1600°K, and **(d)** the human skin at 305°K.

9.13 A blackbody surface is heated to 3000°R. Find the fraction of the emission that occurs in wavelength bands of **(a)** 0 to 1 μm, **(b)** 1 to 3 μm, **(c)** 3 to 6 μm, and **(d)** 6 to 12 μm.

9.14 A blackbody surface is heated to 2000°K. Find the fraction of the emission that occurs in the wavelength bands of **(a)** 0 to 1 μm, **(b)** 1 to 2 μm, **(c)** 2 to 5 μm, and **(d)** 5 to 15 μm.

9.15 For the surface of Example 9.4 and described in Fig. 9.13, find the total emissivity for a surface temperature of 1000°K. What is its total absorptivity with respect to a blackbody source at 1000°K and at 3000°K?

9.16 A surface has a monochromatic emissivity of 0.15 for all wavelengths less than or equal to 3.0 μm and 0.8 for all wavelengths greater than or equal to 3.0 μm. Find **(a)** the total emissivity of the surface for a surface temperature of 1000°K, and **(b)** the solar absorptivity of the surface.

9.17 If the surface in Prob. 9.16 is allowed to come to thermal equilibrium with the sun, find the equilibrium temperature if the sun's rays are **(a)** normal to the surface, and **(b)** inclined at an angle of 45° with the surface normal.

9.18 A surface has a monochromatic emissivity of 0.1 for all wavelengths below a "cutoff" value λ_c, and 0.9 for all wavelengths above λ_c. Find the solar absorptivity of the surface if λ_c is **(a)** 0 μm **(b)** 1 μm, **(c)** 2 μm, **(d)** 4 μm, and **(e)** 10 μm.

9.19 If the surface Prob. 9.18 is allowed to come to equilibrium with the sun's rays at normal incidence, find the equilibrium temperature of the surface for each of the cases.

9.20 Using the data of Fig. 9.10, make an approximate calculation of the solar absorptivity of white epoxy paint on aluminum.

9.21 The monochromatic emissivity of a surface varies with wavelength in the following way:

$$0 \le \lambda \le 0.5\ \mu\text{m}: \quad \epsilon_\lambda = 0.1,$$

$$0.5 \le \lambda \le 5\ \mu\text{m}: \quad \epsilon_\lambda = 0.5,$$

$$5 \le \lambda \le 15\ \mu\text{m}: \quad \epsilon_\lambda = 0.7,$$

$$15 \le \lambda \le \infty\ \mu\text{m}: \quad \epsilon_\lambda = 0.8.$$

Find the total emissivity of the surface when its temperature is **(a)** 1000°K, **(b)** 2000°K, and **(c)** 3000°K.

9.22 Find the total absorptivity of the surface described in Prob. 9.21 when it is subjected to a blackbody irradiation source at **(a)** 500°K, **(b)** 1000°K, or **(c)** 1500°K.

9.23 What is the solar absorptivity of the surface described in Prob. 9.21? If the surface is allowed to come to thermal equilibrium with the sun's rays at normal incidence, what is its temperature?

9.24 A surface has a monochromatic emissivity $\epsilon_{\lambda 1}$ for all wavelengths below 5 μm and $\epsilon_{\lambda 2}$ for all wavelengths in excess of 5 μm. Find the solar absorptivity and the equilibrium temperature T^*, for normal incidence, when **(a)** $\epsilon_{\lambda 1} = 0.15$, $\epsilon_{\lambda 2} = 0.85$; and **(b)** $\epsilon_{\lambda 1} = 0.85$, $\epsilon_{\lambda 2} = 0.15$.

9.25 For each of the two surface spectral distributions noted in Prob. 9.24, find net heat flux absorbed by the surface if it is placed normal to the sun's rays and maintained at **(a)** 100°C, or **(b)** 200°C. Note that T^* will be different from that found in Prob. 9.24 since the surface temperature is different.

9.26 Calculate the equilibrium temperature T^* for normal solar incidence on a surface composed of **(a)** type 410 stainless steel, **(b)** Ebanol on steel, **(c)** stainless steel with Armco black oxide, and **(d)** Titanox. Determine the heat flux from each surface when maintained at 200°C.

9.27 On clear cold nights tender vegetation may suffer freeze damage, even when the air temperature is above 32°F, due to radiant loss to the sky. On such clear nights the sky acts as a blackbody at about −40°F. By treating the vegetation surface (e.g., an orange tree leaf) as one with $\alpha = \epsilon = 1.0$, calculate the equilibrium temperature of the leaf when the ambient air is 40°F if it simultaneously exchanges radiant heat with the sky and convective heat with the air. For one estimate assume that the convective exchange is due to free convection only so that $h \approx 1$ Btu/h-ft²-°F. Then see what happens if a light wind exists so that $h \approx 10$ Btu/h-ft²-°F.

9.28 The air space in the wall of a house is sufficiently thin that the bounding surfaces may be treated as infinite parallel planes. If the inner surfaces of the outside brick

($\epsilon = 0.93$) is at 40°C and the opposite surface of the air space is building paper ($\epsilon = 0.90$) at 25°C, find the radiant heat flux through the space.

9.29 If in Prob. 9.28 a layer of aluminum foil ($\epsilon = 0.09$) is placed on top of the building paper, to what value is the radiant flux reduced if the foil assumes the temperature of the paper?

9.30 Two very large parallel planes are maintained at 150°F and 600°F, respectively. Find the radiant flux exchanged **(a)** if each plane is black; **(b)** if the 150°F plane is gray with $\epsilon = 0.9$, and the 600°F plane is gray with $\epsilon = 0.6$; **(c)** if the 150°F plane is gray with $\epsilon = 0.6$, and the 600°F plane is gray with $\epsilon = 0.9$; and **(d)** if the 150°F plane is gray with $\epsilon = 0.9$, and the 600°F plane is gray with $\epsilon = 0.1$.

9.31 Two parallel infinite planes each have $\epsilon = 0.095$. One plane is maintained at 90°C and the other at 350°C.

(a) What is the radiant heat exchanged between the planes?
(b) If a third plane of the same material is placed between the original two and allowed to come to equilibrium, what is the temperature of this third plane, and what is the radiant exchange between the two original surfaces?

9.32 Two infinite parallel planes, one at 400°C with $\epsilon = 0.8$ and the other at 150°C with $\epsilon = 0.3$ are to be shielded by placing a third plane ($\epsilon = 0.9$, each side) between them and allowing it to come to thermal equilibrium. Find the radiant flux between the planes before and after the insertion of the third plane, and find the equilibrium temperature of the shield.

9.33 Two large parallel planes, one at 500°F with $\epsilon = 0.4$ and the other at 150°F with $\epsilon = 0.8$, are to be shielded by the insertion of a third plane ($\epsilon = 0.15$ on each side) between them, allowing it to come to thermal equilibrium. Find the radiant exchange between the planes before and after the insertion of the shield, and find the equilibrium temperature of the shield.

9.34 As suggested in the accompanying figure, the two planes described in Prob. 9.33 are to be shielded by the insertion of two shields between them. The shield closest to the 500°F plane has an emissivity of 0.15 on each side and the other has $\epsilon = 0.25$ on each side. The two shields are allowed to come to thermal equilibrium. Find the equilibrium temperatures of the two shields and the radiant exchange between the two original planes before and after the shields are put in place.

9.35 As the accompanying figure depicts, the exposed surface of a spacecraft (approximated as a large plane) has a surface emissivity of 0.3 and a solar absorptivity of 0.75. The surface is maintained at 100°C and is subjected to solar irradiation at normal incidence. To protect the spacecraft from excessive heat gain from the sun a shield (or parasol) is placed, parallel to the surface, between the surface and the solar beam. The shield has an emissivity of 0.7 (on both sides) and a solar ab-

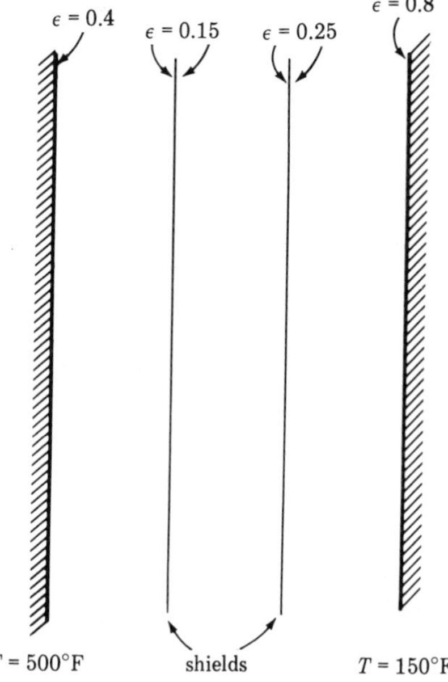

Problem 9.34

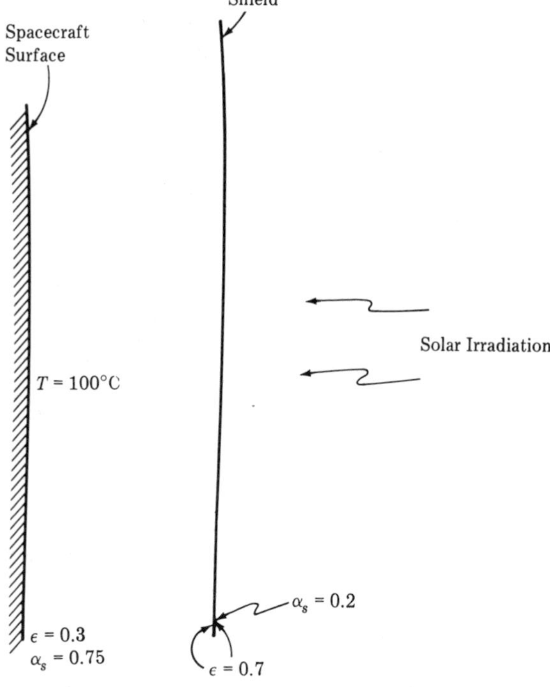

Problem 9.35

sorptivity of 0.2 on the side facing the sun. Find the net heat flux absorbed by the spacecraft surface, before and after the shield is put in place, and find the equilibrium temperature of the shield. Discuss the results.

9.36 The accompanying figure depicts two very long, concentric, cylinders. If the outer cylinder has a diameter twice that of the inner ($D_2 = 2D_1$), find the four shape factors F_{1-1}, F_{1-2}, F_{2-1}, and F_{2-2}.

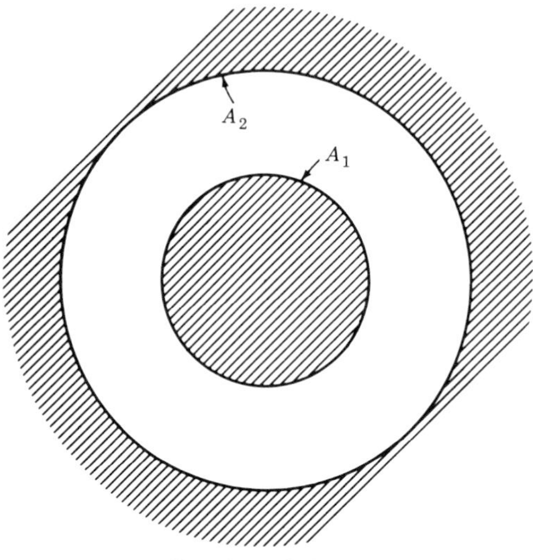

Problem 9.36

9.37 The accompanying figure depicts a very long drying oven that has the cross section of a plane capped by a semicylindrical roof. Find the four shape factors F_{1-1}, F_{1-2}, F_{2-1}, and F_{2-2}.

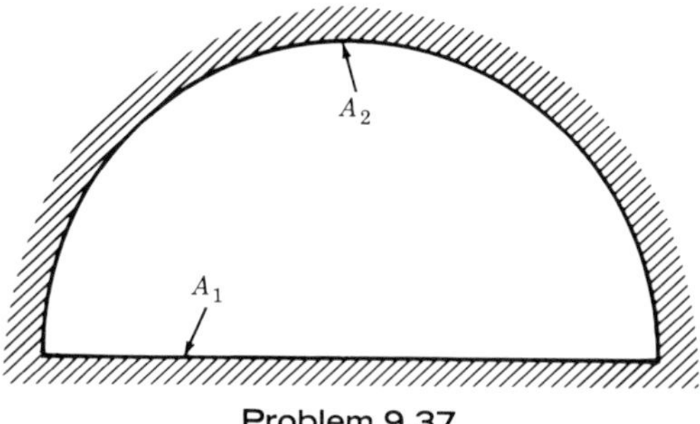

Problem 9.37

9.38–9.47 Find the shape factors, F_{1-2}, for the configurations shown in the accompanying figures.

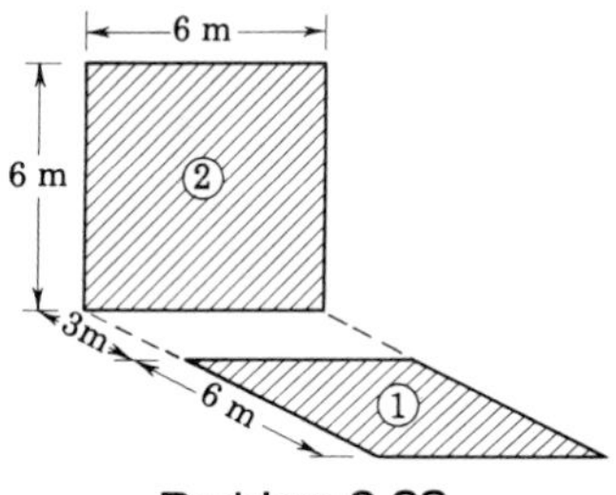

Problem 9.38

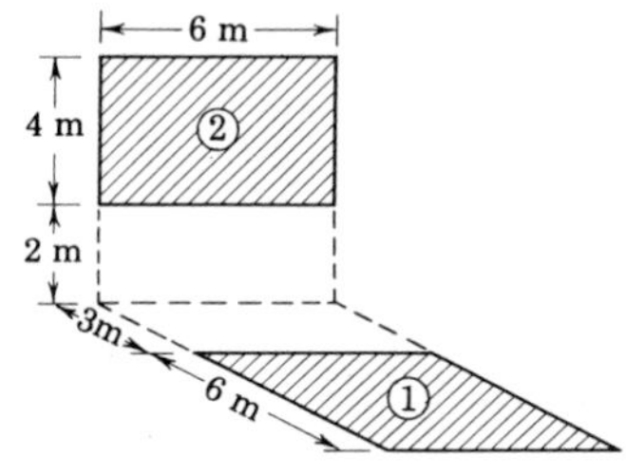

Problem 9.39

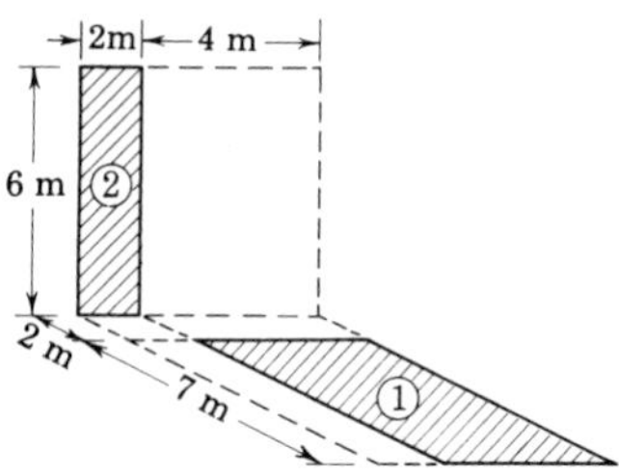

Problem 9.40

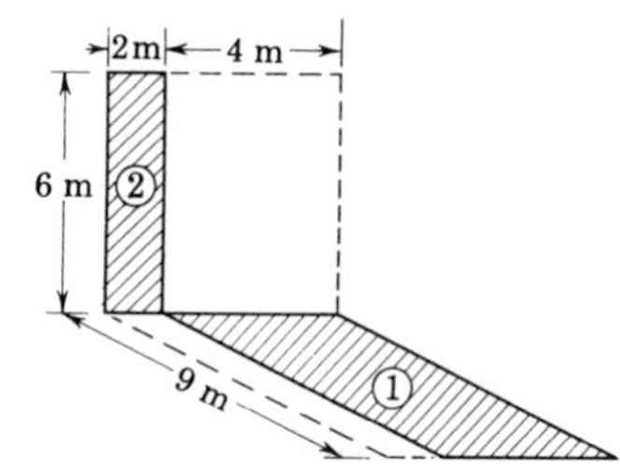

Problem 9.41

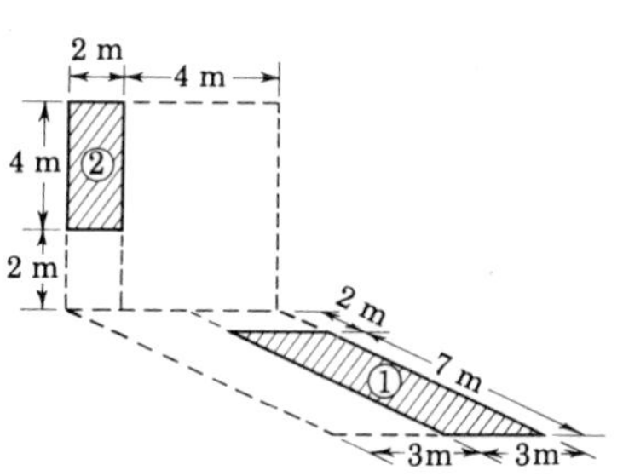

Problem 9.42

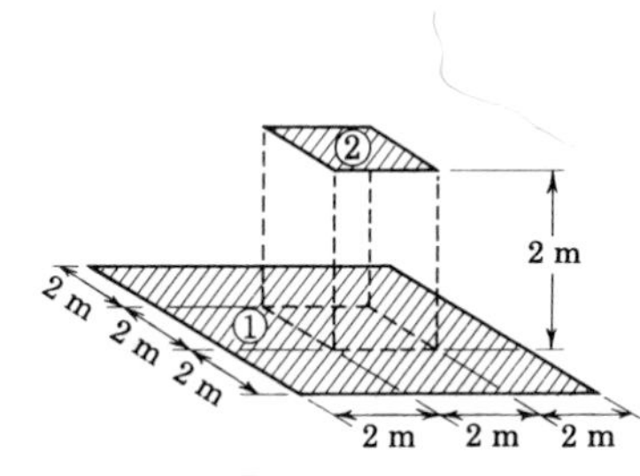

Problem 9.43

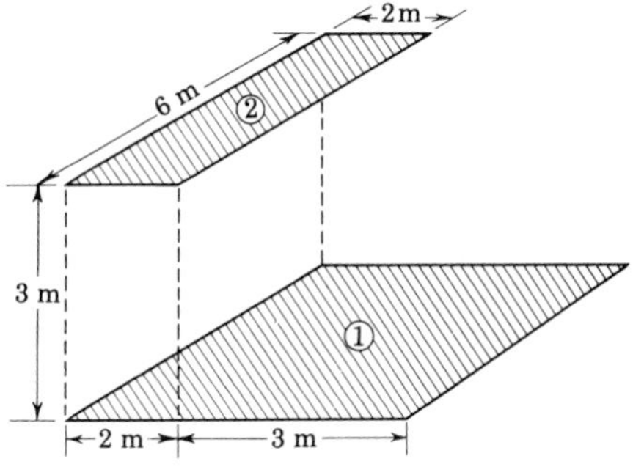

Problem 9.44

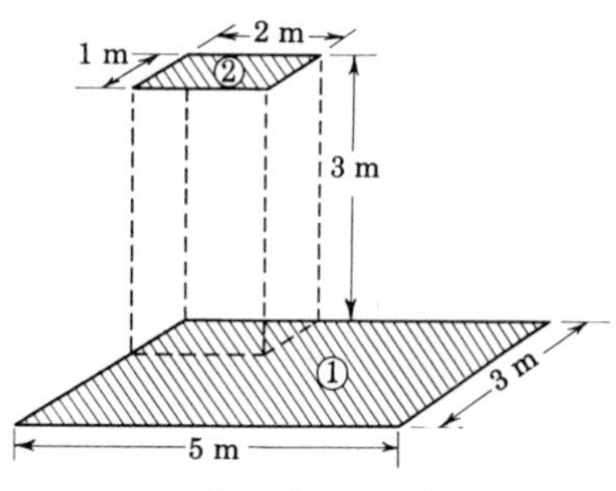

Problem 9.45

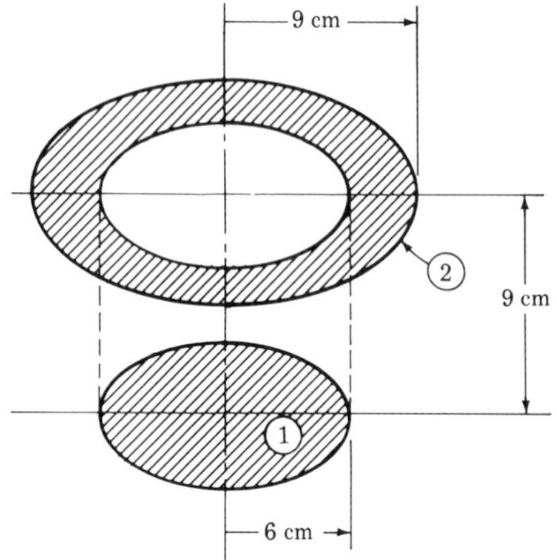

Problem 9.46

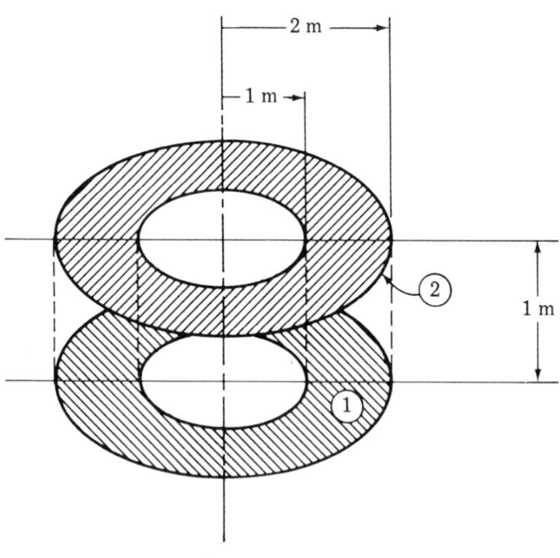

Problem 9.47

9.48 An enclosure consists of the inner surface of a 10-cm-ID cylinder 10 cm long (call it A_2), the outer surface of a 5-cm-OD cylinder of the same length and placed coaxially with the first (call it A_1), and the annular surfaces at each end (call them A_3 and A_4). Find the 16 shape factors: F_{i-j}, $i = 1, 2, 3, 4; j = 1, 2, 3, 4$.

9.49 Verify the statement made at the end of Example 9.11 that the remaining shape factors in the set of 16, $F_{i-j}(i, j = 1, 2, 3, 4)$, may be found by application of shape factor algebra and without further recourse to Fig. 9.18 or 9.19.

9.50 In Example 9.16, which considers the enclosure described in Fig. 9.31, the following shape factors are given: $F_{1-1} = 0$, $F_{1-2} = 0.167$, $F_{1-3} = 0.158$, F_{1-4}

$= 0.675$, $F_{2-2} = 0$, $F_{2-3} = 0.168$, $F_{2-4} = 0.498$, $F_{3-4} = 0.484$. Verify these values for the given geometry.

9.51 Two black rectangles, 2 m × 3 m, are parallel and directly opposed—spaced 2 m apart. If their temperatures are 250°C and 150°C, respectively, find the rate of radiant exchange between them and the rate at which the 250°C rectangle is losing energy. Assume that the surfaces are in a radiation-free environment.

9.52 Two black squares, 5 ft × 5 ft, are spaced 10 ft apart, parallel and directly opposed and placed in a radiation-free space. If their temperatures are 100°F and 300°F, respectively, find the rate of radiant exchange between them and the rate at which each square is losing energy.

9.53 A black rectangle, 2.5 m × 3.5 m, is maintained at 250°C. It is located normal to, and shares a common edge with, another black rectangle, 1.8 m × 3.5 m, at 500°C. What is the rate at which each rectangle is losing radiant energy if they are located in a radiation-free space?

9.54 For the geometry given in Prob. 9.39, let $T_1 = 300$°K and $T_2 = 400$°K. If both surfaces are black, what is the radiant heat flow away from each surface, assuming that no other surfaces are present?

9.55 As noted in the accompanying figure, a circular disk is parallel to and concentric with a circular ring. The disk, A_1, is black and maintained at 50°C. The ring, A_2, is also black but maintained at 500°C. Find the net radiative heat flow away from the facing sides of each surface.

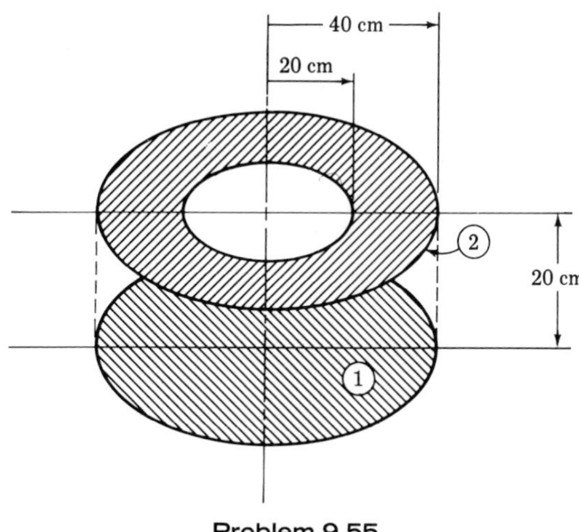

Problem 9.55

9.56 The semicylindrical roof of the drying oven in Prob. 9.37 has a diameter of 3 ft and is maintained at 1500°F. The floor of the oven is covered with a material to be dried and is kept at 212°F. If both surfaces are taken to be black, determine the drying rate, per foot of oven length.

9.57 A cubical room, 10 ft on a side, is composed of black surfaces. One wall is at 70°F and the opposite wall is at 90°F. All the other walls are at 80°F. What is the net radiant heat flow away from the 90°F wall?

9.58 The accompanying figure depicts an artist's studio with a skylight and door in one wall. The floor is to be used as a radiant heating panel and acts as a black plane at 70°C. The skylight and door act as black planes at 10°C, whereas the rest of the walls and ceiling act as black planes at 30°C. Find the net radiant energy given up by the floor.

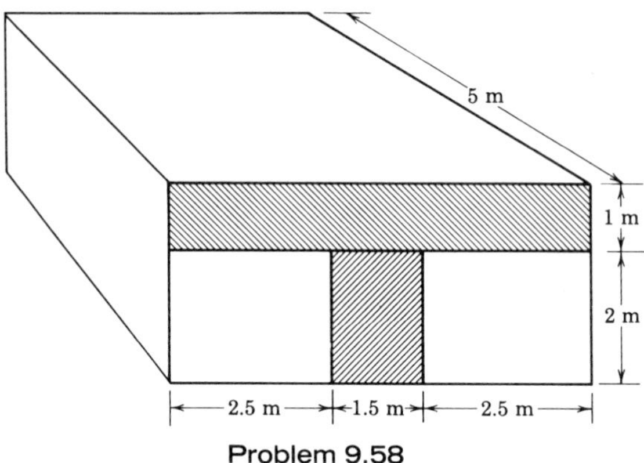

Problem 9.58

9.59 If for the geometry of Prob. 9.58, the floor acts as a black surface at 90°C, the door as a black surface at 40°C, and the skylight as a black plane at 5°C, find the net heat flow at each of these surfaces if the remaining walls act as a single adiabatic surface. What is the equilibrium temperature of the adiabatic surface?

9.60 Imagine a system of three black active surfaces. Surface 1 has $A_1 = 40\ \text{m}^2$, $T_1 = 500°\text{C}$; surface 2 has $A_2 = 20\ \text{m}^2$, $T_2 = 1000°\text{C}$; surface 3 has $A_3 = 15\ \text{m}^2$, $T_3 = 1600°\text{C}$. The following shape factors are known: $F_{1-1} = 0.0$, $F_{1-2} = 0.15$, $F_{1-3} = 0.10$, $F_{2-2} = 0.0$, $F_{2-3} = 0.0$, $F_{3-3} = 0.0$.

(a) If the three surfaces are located in a radiation-free space, find the net heat flow at each surface.

(b) If the three surfaces are enclosed by a fourth, adiabatic surface, find the equilibrium temperature of this surface and the heat flow at each active surface.

9.61 In Example 9.16, the values obtained for the radiosities (or temperatures) of the two adiabatic surfaces A_3 and A_4 suggests that not much error would result if these two surfaces had been represented as a *single* adiabatic surface. Repeat, then,

Example 9.16 by taking A_3 and A_4 to be a single adiabatic surface at a single temperature and determine the resultant heat flows at A_1 and A_2, comparing the results with those found in Example 9.16.

9.62 An enclosure consists of a rectangular parallelpiped 2 m × 4 m × 6 m. One of the 2 m × 4 m surfaces acts as a black surface at 100°C and the other acts as a black surface at 200°C. The other surfaces act as adiabatic surfaces; however, symmetry permits the treatment of the two 2 m × 6 m surfaces as one adiabatic surface and the two 4 m × 6 m surfaces as another. Find the equilibrium temperature of the two adiabatic surfaces and the heat flow at the two active surfaces.

9.63 The results of Prob. 9.62 should show that the two adiabatic surfaces have the same temperature. Show, perhaps with use of the electrical network analogy, that in *any* rectangular parallelpiped with two opposing active surfaces, *gray* or *black,* no error is incurred in finding the heat flow at the active surfaces if the other surfaces are taken to be a single adiabatic one as compared to treating them as individual adiabatic surfaces.

9.64 An enclosure consists of a parallelpiped 6 ft × 12 ft × 24 ft. One of the 12 × 24 surfaces (call it A_1) acts as a black surface at 500°F, and one of the 6 × 24 surfaces (call it A_2) acts as a black surface at 1000°F. The two 6 × 12 surfaces act as a single adiabatic surface (call it A_3) and the remaining 12 × 24 and 6 × 24 surfaces act as a second adiabatic surface (call it A_4). Find the equilibrium temperature of the two adiabatic surfaces and the net heat flow at the two active surfaces.

9.65 Let a complete enclosure be formed with the two surfaces of Prob. 9.55 by adding an adiabatic surface, A_3, to fill the circular hole in A_2 and by adding an adiabatic cylindric surface, A_4, connecting the outer edges of the two disks. The two original surfaces, A_1 and A_2, remain black and at the temperatures T_1 = 50°C, T_2 = 500°C. Determine **(a)** the net heat flow at A_1 and A_2, and **(b)** the equilibrium temperatures of the two adiabatic surfaces, A_3 and A_4.

9.66 The accompanying figure depicts a furnace of cylindrical shape, 3 ft in diameter, 12 ft long. One circular end is the heat source, A_1, taken to be a black surface maintained at 3000°F. The opposite end, the material to be heated, acts as another black surface, A_2, at 1000°F. The cylindric wall connecting the two ends is an adiabatic, refractory surface. To illustrate the effect of the uniform radiosity assumption used in enclosure analysis, determine the radiant exchange between A_1 and A_2 if the adiabatic surface is divided, longitudinally, into **(a)** one zone, **(b)** two equal zones, and **(c)** three equal zones.

9.67 An enclosure consists of a very long furnace with an equilateral triangular cross section as shown. Each side is 1 m wide. One surface is black and maintained at

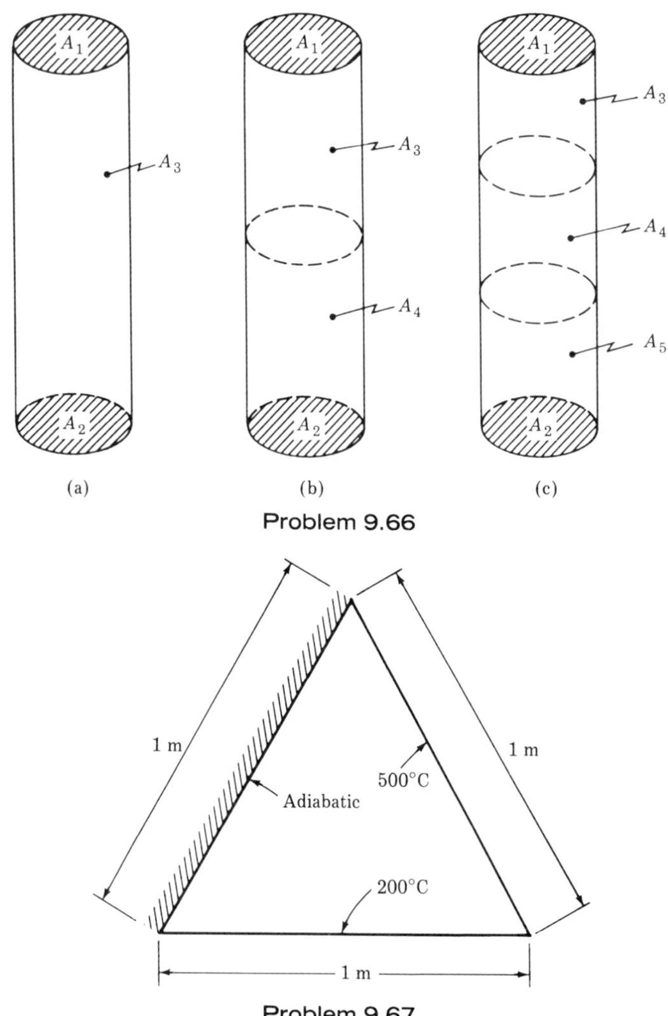

Problem 9.66

Problem 9.67

200°C, another is also black at 500°C, and the third is adiabatic. Find **(a)** the temperature of the adiabatic surface, and **(b)** the heat flow, per unit length, at the two active surfaces.

9.68 Repeat Prob. 9.53 if the 250°C rectangle is gray with $\epsilon = 0.8$ and the 500°C rectangle is gray with $\epsilon = 0.4$. All other data remain the same.

9.69 For the geometry of Prob. 9.39, let A_1 be gray with $\epsilon_1 = 0.2$, $T_1 = 300$°K and let A_2 be gray with $\epsilon_2 = 0.9$, $T_2 = 400$°K. The surfaces are in a radiation-free environment. Find the net heat flow at each surface.

9.70 For the geometry of Prob. 9.58, the floor acts as a black plane at 90°C, the skylight acts as a gray plane at 5°C with $\epsilon = 0.8$, and the door acts as a gray plane at 40°C with $\epsilon = 0.4$. The remaining walls act as a single adiabatic surface. Find the temperature of the adiabatic surface and the heat flow at each of the active surfaces.

9.71 Repeat Prob. 9.70, but let the floor be a gray plane with $\epsilon = 0.6$, all other data remaining the same.

9.72 An enclosure consists of an equilateral tetrahedron, each of the four faces of which is an equilateral triangle, 1 m on each side. One surface is gray with $\epsilon_1 = 0.2$, $T_1 = 200$°K; another is gray with $\epsilon_2 = 0.4$, $T_2 = 400$°K; a third is gray with $\epsilon_3 = 0.7$, $T_3 = 700$°K; the fourth surface is an adiabatic surface. Find the equilibrium temperature of the adiabatic surface.

9.73 Repeat Prob. 9.72 if the three active surfaces are black, all other data remaining unchanged.

9.74 For the data of Prob. 9.60, let A_1 remain black, but let A_2 be gray with $\epsilon_2 = 0.7$ and let A_3 be gray with $\epsilon_3 = 0.3$. All other data remain the same. Find the same information requested in parts (a) and (b).

9.75 The floor of a furnace acts as a plane at $T_1 = 1500$°F, $\epsilon_1 = 0.6$ and the ceiling acts as a plane at $T_2 = 800$°F, $\epsilon_2 = 0.8$. The furnace is 10 ft wide, 12 ft long, and 15 ft high. All the other surfaces act as a single adiabatic surface. What is the heat flow at the two active surfaces and what is the equilibrium temperature of the walls?

9.76 Repeat Prob. 9.66 for the cylindrical furnace if the heat source, surface A_1, is gray with $\epsilon_1 = 0.3$ and the heat sink, surface A_2, is also gray with $\epsilon_2 = 0.7$. All other data are unchanged. Use of the network analogy will permit the solution of this problem, using the results of Prob. 9.66, without recalculation of the radiosities of the adiabatic surfaces. This can be particularly useful when there are multiple adiabatic surfaces, as in parts (b) and (c).

9.77 The three surfaces shown in the accompanying figure are black and maintained at the temperatures: $T_1 = 500$°R, $T_2 = 800$°R, $T_3 = 1200$°R.

(a) If the three surfaces are in a radiation-free space, find the net heat flow away from each surface.

(b) If the three surfaces are enclosed by a fourth, adiabatic surface that does not interfere with the direct exchange among the three, find the temperature of this surface and the net heat flow away from each of the three active surfaces.

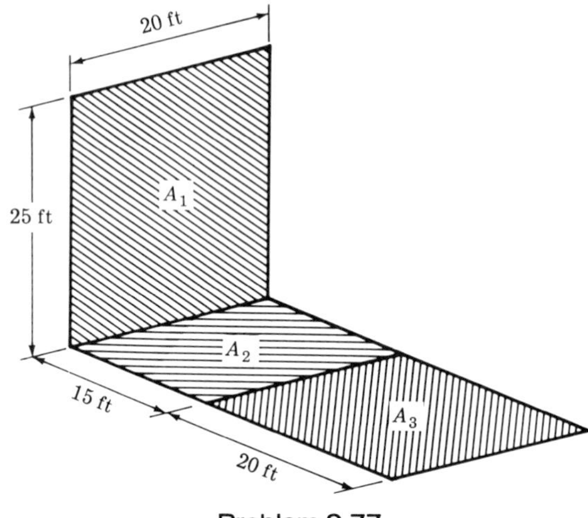

Problem 9.77

9.78 A furnace has the shape of a rectangular parallelpiped, 10 ft wide, 12 ft long, 15 ft high. The floor, 10 ft by 12 ft, acts as an active gray plane with $T_1 = 1200°F$, $\epsilon_1 = 0.6$. One of the 12-ft by 15-ft side walls acts as another active gray plane with $T_2 = 800°F$, $\epsilon_2 = 0.8$.

(a) If the remaining four surfaces act as a single adiabatic surface, find (1) the net heat flow at A_1, (2) the net heat flow at A_2, and (3) the equilibrium temperature of the adiabatic surface.

(b) If the remaining four surfaces act as a single *black* active surface at $T_3 = 300°F$, find the net heat flow at A_1, A_2, and A_3.

9.79 An enclosure consists of the interior surfaces of an equilateral tetrahedron, 2 m on each side. One of the surfaces is a gray active surface with $T_1 = 300°C$, $\epsilon_1 = 0.3$; another is a black active surface at $T_2 = 600°C$; a third surface acts as a black active one at $T_3 = 900°C$; the fourth surface is adiabatic. Determine the heat flux at each of the active surfaces and the temperature of the adiabatic surface.

9.80 The storage tank for liquid oxygen consists of an inner sphere, diameter $D_1 = 0.4$ m, containing the oxygen, placed concentrically within an outer sphere, $D_2 = 0.6$ m. The space between the two spheres is evacuated. The liquid oxygen maintains the inner sphere temperature at 100°K, and the outer sphere surface is at 300°K. Find the heat flow from the oxygen if the sphere surface emissivities are: **(a)** $\epsilon_1 = \epsilon_2 = 1.0$; **(b)** $\epsilon_1 = 0.1$, $\epsilon_2 = 1.0$; **(c)** $\epsilon_1 = 1.0$, $\epsilon_2 = 0.1$; and **(d)** $\epsilon_1 = 0.1$, $\epsilon_2 = 0.1$.

9.81 Two 1 m × 1 m squares are parallel, directly opposed, and located 2 m apart. One plane is at $T_1 = 100°C$ and the other $T_2 = 300°C$. Find the net heat flow at each surface if they are **(a)** black, unenclosed in a radiation-free space; **(b)** black, enclosed by a single adiabatic surface; **(c)** gray ($\epsilon_1 = 0.6$, $\epsilon_2 = 0.8$), in a radiation-free space; and **(d)** gray as in part **(c)**, enclosed by a single adiabatic surface.

9.82 Two disks, each 3 ft in diameter, directly opposed and spaced 6 ft apart, are maintained at $T_1 = 300°F$ and $T_2 = 600°F$. Find the net heat flow at each disk if they are **(a)** black, unenclosed and in radiation-free space; **(b)** black, enclosed by a single adiabatic surface; **(c)** gray ($\epsilon_1 = 0.4$, $\epsilon_2 = 0.7$), in a radiation-free space; and **(d)** gray as in part (c), enclosed by a single adiabatic surface.

9.83 Two parallel, directly opposed, square planes (2 m × 2 m) are spaced 2 m apart. One plane ($\epsilon_1 = 0.6$) is at $T_1 = 200°C$ and the other ($\epsilon_2 = 0.7$) is at $T_2 = 400°C$. What is the heat flow at each plane **(a)** if they are in a radiation-free environment; and **(b)** if they are enclosed by a single adiabatic surface?

9.84 Two parallel, directly opposed square planes (3 m × 3 m) are spaced 2 m apart. One plane is maintained at $T_1 = 400°C$, $\epsilon_1 = 0.8$ and the other at $T_2 = 150°C$, $\epsilon_2 = 0.3$. A third plane of the same size with $\epsilon_2 = 0.9$ on each side is placed equidistant between the first two and allowed to come to thermal equilibrium, with the same temperature on each side. The entire system is in a radiation-free space. Find the net heat flow from each of the two original planes before and after the third is inserted.

9.85 A cylindrical cavity is 0.5 m in diameter and 0.5 m long. The ends are closed by circular disks. One end acts as a gray plane with $T_1 = 100°C$, $\epsilon_1 = 0.8$; the other end acts as a gray plane with $T_2 = 300°C$, $\epsilon_2 = 0.4$.

(a) If the wall of the cavity is an adiabatic surface, find its temperature and the net heat flow at each active surface.

(b) If the wall of the cavity is a third gray surface at $T_3 = 200°C$, $\epsilon_3 = 0.6$, find the net heat flow at each surface.

9.86 Two infinitely long cylinders are placed coaxially. The outer cylinder has an inside diameter of 6 in., an emissivity of 0.4, and is maintained at 400°F. The inner cylinder has an outside diameter of 2 in., an emissivity of 0.2, and is maintained at 800°F. A third cylindrical shield (very thin so that its temperature is the same on both sides) is placed concentrically midway between the first two and allowed to come to thermal equilibrium. It has an emissivity of 0.6 on both sides. Find **(a)** the net heat flow, per foot of length, from each cylinder before the shield is put in place; **(b)** the net heat flow, per foot of length, from each cylinder after the shield is put in place; and **(c)** the equilibrium temperature of the shield.

9.87 A cylindrical cavity has a diameter of 10 cm and a length of 15 cm. It is closed on one end by a circular disk. The other end is capped with a circular disk provided with a 5-cm-diameter concentric hole. The hole acts as a black plane at 0°K. Find the radiant loss out of the hole **(a)** if all interior surfaces are black at 600°K;

(b) if the base of the cavity (i.e., the 10-cm-diameter end) is gray at 600°K, $\epsilon = 0.5$, and the other surfaces are adiabatic; and **(c)** if all interior surfaces are gray at 600°K, $\epsilon = 0.5$.

9.88 A piece of steel (30 cm × 100 cm × 300 cm) is heated to a temperature of 1000°C and then placed in a large room with walls at 40°C. The emissivity of the steel is 0.5. What is the initial rate of radiant heat loss from the steel as it cools?

9.89 A nominal 3-in. bare wrought iron pipe with a surface temperature of 300°F passes through a large furnace at 800°F. What is the radiant loss from the pipe, per foot of length?

9.90 The accompanying figure depicts an annealing furnace in which the "fire" acts as a gray plane ($\epsilon = 0.7$, 2500°F). The steel, shielded so that it does not "see" the fire, acts as a gray surface at 1500°F with $\epsilon = 0.8$. All other surfaces are refractory (adiabatic) surfaces. The furnace is 10 ft deep into the plane of the paper. Even though the fire and steel do not see each other, find the rate of radiant heat transfer between them.

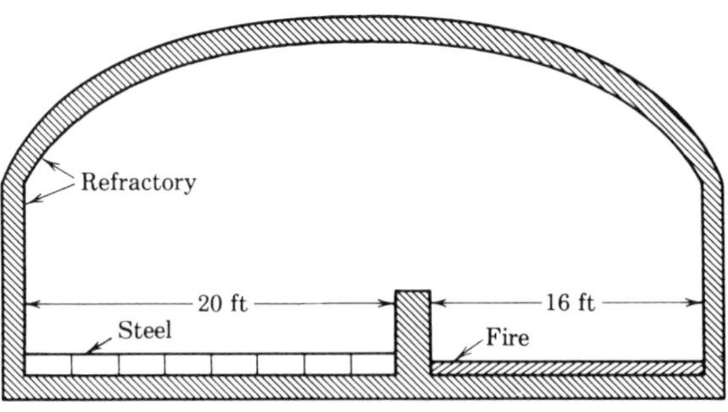

Problem 9.90

9.91 Verify the algebra leading from Eqs. (8.91), (9.92), and (9.93) to Eq. (9.94).

9.92 Derive Eq. (9.97).

CHAPTER 10

Heat Transfer by Combined Conduction and Convection—Heat Exchangers

10.1 INTRODUCTORY REMARKS

The title of this chapter would seem to imply that all the foregoing chapters have dealt with situations in which only one mode of heat transfer was taking place. Strictly speaking, this is not true. The previous chapters on the various aspects of conduction and convection certainly emphasized one or the other of these modes; however, in many instances both mechanisms were occurring simultaneously.

In much of the work concerning conduction, particularly Chapters 2 through 4, convection was included as a boundary condition in the solution of the conduction equation. In these cases it was presumed that the surface heat transfer coefficient, h, was known; and from this, and other given facts, much could be deduced about the conduction process within a given body—including the determination of the temperature of the solid surface in contact with the convecting fluid. On the other hand, all of Chapters 5 through 8, which considered the hydrodynamical and thermodynamical aspects of convection, presumed knowledge of the surface temperature for determination of the heat transfer coefficient.

In real problems encountered in engineering practice, however, conduction may be taking place within the interior of a solid body while convection takes place with ambient fluids in contact with one or more of its surfaces. In such instances neither the surface heat transfer coefficients nor the surface temperatures are known, and it becomes necessary to solve the conduction and convection problems simultaneously. Imagine, for instance, a hot fluid of given temperature and hydrodynamic conditions separated by a wall (of known geometry and composition) from a cold fluid of known conditions. Based on the fluid temperatures, the heat flow through the wall (and, hence, the surface temperatures of the wall) can be found only if the heat transfer coefficients at the two surfaces are known along with the wall geometry and thermal properties. However, the heat transfer coefficients can generally be found only when the surface temperatures are known. Thus a simultaneous, perhaps iterative solution must be carried out to find the surface temperature *and* the heat transfer coefficients.

The situation cited in the preceding paragraph, that of a wall separating two convecting fluids of different temperatures, is encountered frequently in engineering applications. The wall may be of any geometry (i.e., plane, cylindrical, etc.) and may consist of composite layers. In particular, the analysis of the performance of a heat exchanger (a device to transfer heat between two fluids without mixing) presents such a problem and is considered in some detail in later sections of this chapter.

However, other situations also occur wherein a simultaneous solution of conduction and convection must be carried out—such as the analysis of heat flow in an extended surface on which the surface heat transfer coefficient is not known a priori. It is clearly impossible to generalize sufficiently in order to establish a solution technique that will apply in all instances. Hence the next section presents the approach that may be used in two particular cases with the hope that the reader may gain sufficient experience to be able to handle other instances requiring simultaneous solution of conduction and convection problems.

10.2 EXAMPLES OF ITERATIVE SOLUTION OF COMBINED CONDUCTION AND CONVECTION

This section will illustrate by the use of two examples the methodology of the simultaneous, iterative solution that may be applied in cases of combined convection and conduction. In both examples the problem of heat transfer between two convecting fluids separated by a wall will be considered, and in both cases a cylindrical-shaped wall will be used. In the first case forced convection will occur inside the cylinder and free convection will exist at the outer surface. In the second case forced convection and condensation occur at the surfaces. These cases are simply examples; any of the convective processes of Chapters 5 through 8 can be occurring at the boundaries—forced convection, free convection, dissipative flow, etc. Similarly, the separating wall may be of any geometry for which a conduction solution can be obtained (e.g., a plane wall), and the exposed surfaces may or may not be equipped with an array of extended surfaces.

In general, then, one must ascertain two surface heat transfer coefficients and two surface temperatures. In the first example, however, a special case will be considered in which one of the surface heat transfer coefficients is independent of temperature.

One Coefficient Independent of Temperature

As a first example, then, consider the case of a pipe (perhaps insulated) through which a hot fluid is flowing at a specified bulk temperature and velocity. Let the pipe be horizontal and exposed to still, atmospheric air at a known temperature so that there is heat loss by free convection at the outer surface. Both the inside pipe surface and the outer exposed surface temperatures are unknown. However, let it be assumed that the flow inside the pipe is turbulent and that the inner surface temperature is such that the conditions specified for determination of the inside heat transfer coefficient by Eq. (6.31) are satisfied. In this fortunate instance, the heat transfer coefficient can be found directly from the known, bulk, fluid temperature and no iterative solution for it is required. At the outer surface, however, the free convection coefficient is dependent on the unknown exposed surface temperature. This temperature is dependent on the heat flow through the pipe, which, in turn, is dependent on the heat transfer coefficient. The dependence of h on the surface temperature is implied through the complex dependence of the air thermal properties on temperature as well as through the correlations of Chapter 7, which involve the difference between the surface and the air temperature. Hence one cannot, generally, solve directly for the temperature, and some iterative procedure must be established.

First, a trial value of the outer surface temperature, t_s, is assumed. Based on this assumption and the ambient air temperature, t_o, the free convection coefficient at the outside surface, h_o, can be calculated. The heat flow, per unit of outside surface area A_o, is then

$$\frac{q}{A_o} = h_o(t_s - t_o).$$

Since the heat transfer coefficient at the inner surface is known, as is the geometry and thermal conductivity of the pipe and insulation, the overall coefficient, U_o, based on the outer surface area can be calculated in accordance with the methods of Sec. 2.6. Then the heat flow may also be expressed in terms of the overall temperature difference between the outside fluid at t_o and the inside fluid at t_i:

$$\frac{q}{A_o} = U_o(t_i - t_o).$$

Equating these last two expressions gives

$$h_o(t_s - t_o) = U_o(t_i - t_o) \tag{10.1}$$

from which the surface temperature t_s may be calculated. This computed result is compared with the initial assumption, a revised value of t_s is chosen, and the calculations are repeated. This process is continued until a satisfactory agreement between the assumed value and the derived value of t_s is achieved. The extent to which this iterative process is carried out depends on the accuracy desired or the accuracy that can be justified by the given data.

The revised value of t_s used on the second, and subsequent, iterations need not necessarily be the derived value from the previous iteration. Indeed, depending on the problem at hand, the convergence to an acceptable answer may be hastened by choosing a new value of t_s between the original one and the derived one. Because of the complex way in which the surface temperature enters the calculation of h_o

through the dependence of the fluid properties on t_s, it is difficult to formulate a general rule for choosing t_s in subsequent iterations, and the experience gained by working examples is essential. Obviously, the iterative process is shortened if the initial assumption of the surface temperature is close to the correct value. Again, previous experience is invaluable in this regard. In instances in which free convection into atmospheric air is occuring it is useful to remember that the heat transfer coefficient is often very close to 1 Btu/h-ft^2-°F $\approx$ 6 W/m^2-°C. These values may be used as an initial assumption (rather than t_s) to start the iterative process. These points are best illustrated by example.

The following examples are presented in some detail to illustrate the methodology involved. However, the reader should note that modern computer methods may be applied to facilitate considerably the solution of such iterative problems—provided that a capability of generating the fluid thermophysical properties is available.

EXAMPLE 10.1

Steam at 2000 kN/m^2 pressure and 300°C flows with a velocity of 5 m/s in a 6-in. schedule 40 pipe (k = 45 W/m-°C) which is covered with 3.5 cm of magnesia insulation (k = 0.075 W/m-°C). The pipe is positioned horizontally in a room in which the ambient air is at atmospheric pressure and 20°C. Find the temperature of the outer insulation surface, the overall heat transfer coefficient, and the rate of heat loss per unit length of pipe.

Solution. The notation to be used is shown in Fig. 10.1. Table B.1 yields the following diameters:

$$D_2 = 15.41 \text{ cm}, \qquad D_3 = 16.83 \text{ cm}, \qquad D_4 = 23.83 \text{ cm}.$$

The data given also yield

$$t_1 = 300°\text{C} \qquad t_5 = 20°\text{C},$$

$$k_{23} = 45 \text{ W/m-°C}, \qquad k_{34} = 0.075 \text{ W/m-°C}.$$

First the inside heat transfer coefficient, h_{12}, is determined. For steam at 2000 kN/m^2, 300°C, Table A.5 gives

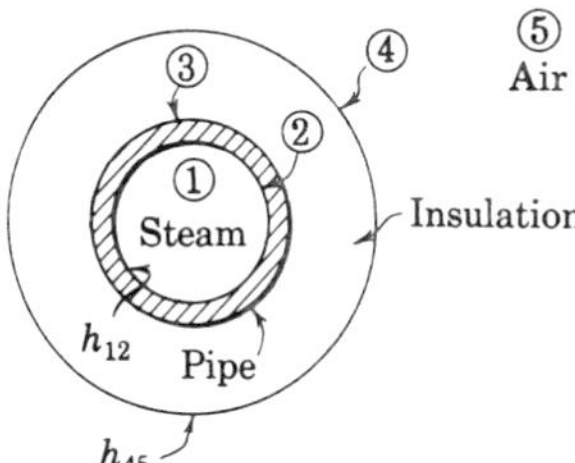

Figure 10.1

$$\rho = 7.9681 \text{ kg/m}^3, \qquad \mu = 20.10 \times 10^{-6} \text{ kg/m-s}$$

$$k = 45.95 \times 10^{-3} \text{ W/m-°C}, \qquad \text{Pr} = 1.02.$$

Thus the Reynolds number is

$$\text{Re}_D = \frac{vD_2\rho}{\mu} = \frac{5 \times 0.1451 \times 7.9681}{20.10 \times 10^{-6}} = 3.05 \times 10^5.$$

So the flow is turbulent and Eq. (6.31) may be applied if it is assumed that $|t_1 - t_2| < 60°\text{C}$. The validity of this assumption will have to be verified later. Thus, since the steam is being cooled, Eq. (6.31) gives

$$\text{Nu}_D = 0.023\text{Re}_D^{0.8}\text{Pr}^{0.3}$$

$$= 0.023 \times (3.05 \times 10^5)^{0.8} \times (1.02)^{0.3} = 564.9,$$

$$h_{12} = \text{Nu}_D \frac{k}{D_2} = 564.9 \times \frac{0.04595}{0.1541} = 168.5 \text{ W/m}^2\text{-°C}.$$

The coefficient h_{12} is not dependent on the unknown surface temperature t_2 and may thus be taken as constant in the subsequent calculations. Note that one might have used Eq. (6.33) instead of Eq. (6.31) to find h_{12} since the surface temperature is not needed in that equation for gases.

To obtain a starting estimate for the outside surface temperature, t_4, let h_{45} (the heat transfer coefficient at that surface) be taken to be approximately 6 W/m²-°C, a value typical for free convection to atmospheric air. Thus, according to Eq. (2.31), the overall heat transfer coefficient for the pipe, based on the outside surface area, U_4, is

$$U_4 = \left[\frac{r_4}{r_2 h_{12}} + \frac{r_4 \ln (r_3/r_2)}{k_{23}} + \frac{r_4 \ln (r_4/r_3)}{k_{34}} + \frac{1}{h_{45}}\right]^{-1}$$

$$= \left[\frac{0.2383}{0.1541 \times 168.5} + \frac{0.2383}{2 \times 45} \ln\left(\frac{0.1683}{0.1541}\right) + \frac{0.2383}{2 \times 0.075} \ln\left(\frac{0.2383}{0.1683}\right) + \frac{1}{6}\right]^{-1}$$

$$= (0.00918 + 0.00023 + 0.5525 + 0.1667)^{-1} = 1.373 \text{ W/m}^2\text{-°C}.$$

It is immediately noticed that the controlling resistances in this case are those of the insulation and the outside coefficient—the resistance of the inside coefficient and the pipe wall being rather small. In some instances it may be justified to neglect these latter two resistances, but they are retained here for the sake of completeness. The fact that the inside coefficient has little effect on U_4 means that it is not important whether Eq. (6.31) or (6.33) is used to find h_{12}; hence the simpler, Eq. (6.31), was employed.

Now the heat flow, per unit of exposed surface, may be written in two ways—based on the overall temperature difference $(t_1 - t_5)$ and the surface-to-air difference $(t_4 - t_5)$, as in Eq. (10.1), to find an estimate of t_4. Thus, using the assumed value for $h_{45} = 6\ \text{W/m}^2\text{-°C}$:

$$\frac{q}{A_4} = h_{45}(t_4 - t_5) = U_4(t_1 - t_5)$$

$$6.0(t_4 - 20) = 1.373(300 - 20)$$

$$t_4 = 84°\text{C}.$$

This first estimate of the outer surface temperature may now be used to calculate an improved value for the outside heat transfer coefficient, h_{45}, by applying the free convection correlations of Chapter 7. With this estimate of $t_4 = 84°\text{C}$ and the specified ambient air temperature $t_5 = 20°\text{C}$, one has

$$\Delta t = t_4 - t_5 = 64°\text{C}. \qquad t_m = \frac{t_4 + t_5}{2} = 52°\text{C} \qquad \beta = \frac{1}{T_5} = \frac{1}{293.15°\text{K}}.$$

Table A.6 gives for air at $t_m = 52°\text{C}$:

$$\nu = 18.12 \times 10^{-6}\ \text{m}^2/\text{s}, \qquad k = 27.95 \times 10^{-3}\ \text{W/m-°C}, \qquad \text{Pr} = 0.709.$$

Thus the Grashof and Rayleigh numbers are

$$\text{Gr}_D = \frac{D_4^3 g\beta\,\Delta t}{\nu^2} = \frac{(0.2383)^3 \times 9.8 \times (1/293.15) \times 64}{(18.12 \times 10^{-6})^2} = 8.818 \times 10^7,$$

$$\text{Ra}_D = \text{Gr}_D \times \text{Pr} = 6.252 \times 10^7.$$

For a horizontal cylinder, Eq. (7.32) gives

$$\text{Nu}_D = \left\{0.60 + 0.387\text{Ra}_D^{1/6}\left[1 + \left(\frac{0.559}{\text{Pr}}\right)^{9/16}\right]^{-8/27}\right\}^{-2}$$

$$= \left\{0.60 + 0.387(6.252 \times 10^7)^{1/6}\left[1 + \left(\frac{0.559}{0.709}\right)^{9/16}\right]^{-8/27}\right\}^{2}$$

$$= 49.00,$$

$$h_{45} = \text{Nu}_D\frac{k}{D_4} = 49.00 \times \frac{27.95 \times 10^{-3}}{0.2383} = 5.75\ \text{W/m}^2\text{-°C}.$$

This value of h_{45} is close to the originally assumed value of $6\ \text{W/m}^2\text{-°C}$. It may be used to revise U_4 and find a further improved value of the surface temperature. Since only the outside heat transfer coefficient has changed, the other resistances found in calculating the overall coefficient U_4, above, may be used to revise this quantity and generate another, improved, estimate for t_4:

$$U_4 = \left(0.00918 + 0.00023 + 0.5525 + \frac{1}{5.75}\right)^{-1} = 1.355\ \text{W/m}^2\text{-°C},$$

$$h_{45}(t_4 - t_5) = U_4(t_1 - t_5)$$

$$5.75(t_4 - 20) = 1.355(300 - 20)$$

$$t_4 = 86°\text{C}.$$

For most engineering calculations this value of $t_4 = 86°C$ is probably close enough to the immediately preceding value of 84°C to terminate the procedure. However, just to complete the process for illustrative purposes, one may carry out one additional iteration with $t_4 = 85.5°C$. The results are summarized below:

$$t_4 = 85.5°\text{C}, \qquad t_m = 52.8°\text{C}, \qquad \Delta t = 65.5°\text{C},$$

$$\text{Gr}_D = 8.955 \times 10^7,$$

$$\text{Ra}_D = 6.349 \times 10^7,$$

$$\text{Nu}_D = 49.23,$$

$$h_{45} = 5.78 \text{ W/m}^2\text{-°C},$$

$$U_4 = 1.357 \text{ W/m}^2\text{-°C},$$

$$t_4 = 85.7°\text{C}.$$

Thus the desired answers may be taken to be

$$t_4 = 85.6°\text{C (186.1°F)},$$

$$U_4 = 1.357 \text{ W/m}^2\text{-°C (0.239 Btu/h-ft}^2\text{-°F)},$$

$$\frac{q}{L} = \pi D_4 U_4 (t_1 - t_5)$$

$$= 284.5 \text{ W/m (295.8 Btu/h-ft)}.$$

It will be recalled that Eq. (6.31) was used to obtain the value of the inside pipe surface heat transfer coefficient, h_{12}. The advantage of using this particular correlation for turbulent pipe flow was the fact that knowledge of the inside pipe temperature, t_2, is not required—eliminating the necessity of performing a simultaneous iterative calculation for that temperature. However, the restrictions placed on Eq. (6.31) require that $|t_1 - t_2| < 60°C$. Thus one needs to verify that this condition is met by determining t_2:

$$q = A_2 h_{12}(t_1 - t_2) = A_4 U_4 (t_1 - t_5),$$

$$q/L = \pi D_2 h_{12}(t_1 - t_2) = \pi D_4 U_4 (t_1 - t_5),$$

$$0.1541 \times 168.5(300 - t_2) = 0.2383 \times 1.357(300 - 20)$$

$$t_2 = 297°\text{C}.$$

thus $|t_1 - t_2| = 3°C < 60°C$, and the application of Eq. (6.31) is justified.

As was pointed out earlier, the above example is just that—an *example*. The hydrodynamic conditions existing at the two surfaces could be any of those described in Chapters 5 through 8 and the separating wall could be of some geometry other than cylindrical. The basic procedure, however, remains the same—a surface temperature is assumed to find the unknown h, and this h is used to find the surface temperature, which is then revised. The agreement between the assumed and derived temperatures required to terminate the calculation depends on the situation at hand. Usually, the accuracy of the given data does not justify the carrying of the solution as far as was done in Example 10.1.

Two Coefficients Dependent on Surface Temperatures

Example 10.1 was simplified in that the inside surface temperature did not directly influence the calculation of the heat transfer coefficient there. Had the flow in the pipe been laminar or had the conditions on Eq. (6.31) not been met so that Eq. (6.32) must be applied, then knowledge of t_2 would be necessary for the determination of h_{12}. Then a double iterative process would have to be performed, as noted in the next example.

EXAMPLE 10.2

A horizontal condenser tube ($\frac{3}{4}$ in. OD, 16 gage) has steam at 5 psia condensing on its outer surface. Water at 85°F flows through the tube with a velocity of 5 ft/s. If the tube is made of stainless steel with $k = 10$ Btu/h-ft-°F, find the temperature of the tube surface and the overall heat transfer coefficient.

Solution. The Steam Tables show the saturation temperature and latent heat for steam at 5 psia to be 162°F and 1000.9 Btu/lb$_m$, respectively. These data, along with the problem specifications and Table B.2, yield the following known information when the notation of Fig. 10.2 is used:

$$t_1 = 85°\text{F}, \qquad v_1 = 5 \text{ ft/s},$$

$$t_4 = 162°\text{F}, \qquad h_{fg} = 1000.9 \text{ Btu/lb}_m,$$

$$D_2 = 0.620 \text{ in.}, \qquad D_3 = 0.750 \text{ in.}, \qquad k_{23} = 10 \text{ Btu/h-ft-°F}.$$

The procedure to be followed here is not vastly different from that in Example

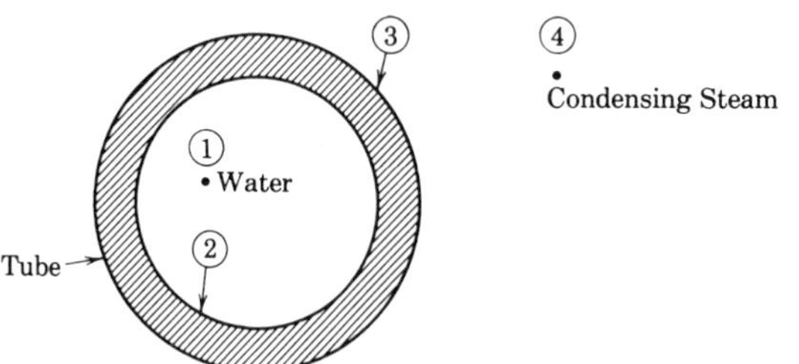

Figure 10.2. Example 10.4.

10.1. An initial trial value of the inside heat transfer coefficient, h_{12}, is obtained. With this value held fixed, an iterative calculation is performed to determine the outside surface temperature and coefficient, t_3 and h_{34}. Then the value of t_2, the inside surface temperature, is found, and the trial value of h_{12} is revised. Using the revised h_{12}, the iterative determination of t_3 and h_{34} is repeated. The process is continued, alternating between revisions of h_{12} and the iterative determination of h_{34}, until a satisfactory agreement of all assumed and calculated temperatures is obtained.

1. *Initial estimate of* h_{12}. First it will be assumed that $|t_1 - t_2| < 6°\text{C}$ so that Eq. (6.31) may be used for h_{12}. This turns out not to be true, and a different correlation must be used; however, it provides a good starting point. Thus with $t_1 = 85°\text{F}$, Table A.3 gives, for water,

$$\nu = 0.03150 \text{ ft}^2/\text{h}, \qquad k = 0.3549 \text{ Btu/h-ft-°F}, \qquad \text{Pr} = 5.52.$$

Thus

$$\text{Re}_D = \frac{v_1 D_2}{\nu} = \frac{5 \times 3600 \times (0.62/12)}{0.03150} = 2.95 \times 10^4.$$

The flow in the tube is turbulent and Eq. (6.31) gives

$$\begin{aligned} \text{Nu}_D &= 0.023\text{Re}_D^{0.8}\text{Pr}^{0.4} \\ &= 0.023(2.95 \times 10^4)^{0.8} \times (5.52)^{0.4} = 171.65, \end{aligned}$$

$$\begin{aligned} h_{12} = \text{Nu}_D \frac{k}{D_2} &= 171.65 \times \frac{0.3549}{(0.62/12)} \\ &= 1179.1 \text{ Btu/h-ft}^2\text{-°F}. \end{aligned}$$

2. *Iteratively find* t_3 *and* h_{34}.

(a) *Assume* that $t_3 = 135°\text{F}$. Then $t_m = (135 + 162)/2 = 148.5°\text{F}$ and Table A.3 gives, for the steam condensate,

$$c_{pl} = 1.00 \text{ Btu/lb}_\text{m}\text{-°F}, \qquad \rho_l = 61.22 \text{ lb}_\text{m}/\text{ft}^3,$$

$$\mu_l = 1.054 \text{ lb}_\text{m}/\text{ft-h}, \qquad k_l = 0.3758 \text{ Btu/h-ft-°F},$$

Thus application of Eq. (8.20) for condensation on a horizontal cylinder, with $\Delta t = 162 - 135 = 27°\text{F}$ and $\rho_l(\rho_l - \rho_v) \approx \rho_l^2$, gives

$$\begin{aligned} \text{Ja} &= \frac{c_{pl}\,\Delta t}{h_{\text{fg}}} \\ &= \frac{1.00 \times 27}{1000.9} = 0.0270, \end{aligned}$$

$$\begin{aligned} h'_{\text{fg}} &= h_{\text{fg}}(1 + 0.68\text{Ja}) \\ &= 1000.9(1 + 0.68 \times 0.0270) = 1019.3 \text{ Btu/lb}_\text{m}), \end{aligned}$$

$$\mathrm{Nu}_D = 0.728\left(\frac{g\rho_l^2 h'_{fg} D^3}{\mu_l k_l\,\Delta t}\right)^{1/4}$$

$$= 0.728\left[\frac{(32.2)(3600)^2(61.22)^2(1019.3)(0.75/12)^3}{1.054 \times 0.3785 \times 27}\right]^{1/4}$$

$$= 317.4$$

$$h_{34} = \mathrm{Nu}_D \frac{k}{D_3} = 317.4 \times \frac{0.3785}{0.75/12}$$

$$= 1922.2 \text{ Btu/h-ft}^2\text{-°F}.$$

(b) *Find the overall coefficient.* For the geometry of Fig. 10.2, the overall heat transfer coefficient based on the outside surface area is

$$U_3 = \left[\frac{r_3}{r_2 h_{12}} + \frac{r_3 \ln (r_3/r_2)}{k_{23}} + \frac{1}{h_{34}}\right]^{-1}$$

$$= \left[\frac{0.75}{0.62 \times 1179.1} + \frac{0.75 \ln (0.75/0.62)}{2 \times 12 \times 10} + \frac{1}{1922.2}\right]^{-1}$$

$$= 467.1 \text{ Btu/h-ft}^2\text{-°F}.$$

(c) *Verify the assumed t_3.* The value of t_3 compatible with the current values of h_{12} and h_{34} can be calculated from

$$\frac{q}{A_3} = h_{34}(t_3 - t_4) = U_3(t_1 - t_4)$$

$$1922.2(t_3 - 162) = 467.1(85 - 162)$$

$$t_3 = 143\text{°F}.$$

This calculated value of t_3 is significantly different from the assumed 135°F. Thus, a new assumption and calculation must be performed.

(d) *Assume a new t_3 = 144°F.* Repeat the calculations of steps (a) through (c). The results are summarized below.

$$t_3 = 144\text{°F}, \qquad t_m = 153\text{°F}, \qquad \Delta t = 18\text{°F},$$

$$c_{pl} = 1.00 \text{ Btu/lb}_m\text{-°F}, \qquad \rho_l = 61.13 \text{ lb}_m/\text{ft}^2,$$

$$\mu_l = 1.018 \text{ lb}_m/\text{ft-h}, \qquad k_l = 0.3797 \text{ Btu/h-ft-°F},$$

$$\mathrm{Ja} = 0.0180,$$

$$h'_{fg} = 1013.1 \text{ Btu/lb}_m,$$

$$\mathrm{Nu}_D = 353.2,$$

$$h_{34} = 2145.8 \text{ Btu/h-ft}^2\text{-°F},$$

$$U_3 = 479.2 \text{ Btu/h-ft}^2\text{-°F},$$

$$t_3 = 144.8\text{°F}.$$

The latter value is sufficiently close to the assumed $t_3 = 144$°F to proceed.

3. *Verify the trial value of h_{12}.* Now that the values of h_{34} and t_3 consistent with the value of h_{12} found in part 1 have been determined, a value of t_2 may be calculated and h_{12} reevaluated. Find the inside tube surface temperature from

$$q = A_2 h_{12}(t_1 - t_2) = A_3 h_{34}(t_3 - t_4)$$

$$q/\pi L = D_2 h_{12}(t_1 - t_2) = D_3 h_{34}(t_3 - t_4)$$

$$0.62 \times 1179.1(85 - t_2) = 0.75 \times 2145.8 \times (144 - 162)$$

$$t_2 = 124\text{°F}.$$

This value of t_2 yields $|t_1 - t_2| = 40\text{°F} = 22\text{°C}, > 6\text{°C}$, which exceeds the limits imposed on Eq. (6.31) which was used to estimate h_{12}. So one must start again, now with a more rational value of t_2.

4. *Assume t_2 and find h_{12}.* Based on the above, now assume that $t_2 = 124$°F. In this case $|t_1 - t_2| > 6$°C, so Eq. (6.33) must be used. [Equation (6.32) could also be used.] At $t_1 = 85$°F, the properties and Re_D found in step 1 still apply. However, one now needs also the viscosity of the tube water at both the bulk and surface temperatures, t_1 and t_2. Table A.3 gives

$$\text{At } t_1 = 85\text{°F:} \quad \mu = 1.959 \text{ lb}_\text{m}\text{/ft-h},$$

$$\text{At } t_2 = 124\text{°F:} \quad \mu_s = 1.301 \text{ lb}_\text{m}\text{/ft-h}.$$

At the bulk temperature $t_1 = 85$°F, the calculations of step 1 gave $\text{Re}_D = 2.95 \times 10^4$ and $\text{Pr} = 5.52$. Thus Eq. (6.33) gives

$$\frac{f}{8} = \tfrac{1}{8}(1.82 \log_{10}\text{Re}_D - 1.64)^{-2}$$

$$= \tfrac{1}{8}[1.82 \log_{10}(2.95 \times 10^4) - 1.64]^{-2} = 0.00294,$$

$$\text{Nu}_D = \frac{(f/8)\text{Re}_D\text{Pr}}{1.07 + 12.7\sqrt{f/8}(\text{Pr}^{2/3} - 1)}\left(\frac{\mu}{\mu_s}\right)^{0.11}$$

$$= \frac{0.00294 \times 2.95 \times 10^4 \times 5.52}{1.07 + 12.7\sqrt{0.00294}(5.52^{2/3} - 1)}\left(\frac{1.959}{1.301}\right)^{0.11}$$

$$= 198.1,$$

$$h_{12} = \text{Nu}_D \frac{k}{D_2} = 198.1 \times \frac{0.3549}{0.620/12}$$

$$= 1361.0 \text{ Btu/h-ft}^2\text{-°F}.$$

5. *Repeat step 2*. One must now repeat the iterative calculation for t_3 and h_{34}. However, a very good starting point is the last iteration given in part (d) of step 2—in fact, no iteration is necessary. For an assumed $t_3 = 144°F$ (as found last), one has the identical calculations for h_{34}. Thus, $h_{34} = 2145.8$ Btu/h-ft²-°F. However, U_3 must be recalculated since h_{12} has changed. Thus

$$h_{12} = 1307.1 \text{ Btu/h-ft}^2\text{-°F},$$

$$h_{34} = 2145.8 \text{ Btu/h-ft}^2\text{-°F},$$

$$U_3 = \left[\frac{r_3}{r_2 h_{12}} + \frac{r_2 \ln (r_3/r_2)}{k_{23}} + \frac{1}{h_{34}}\right]^{-1} = 512.9 \text{ Btu/h-ft}^2\text{-°F},$$

and the verification of t_3 yields

$$h_{34}(t_3 - t_4) = U_3(t_1 - t_4)$$

$$2145.8(t_3 - 162) = 512.9(85 - 162)$$

$$t_3 = 144°\text{F, as assumed.}$$

6. *Verify the assumed* t_2. Now, one repeats step 3 to check the assumed t_2 of 124°F. Thus

$$D_2 h_{12}(t_1 - t_2) = D_3 h_{34}(t_3 - t_4)$$

$$0.62 \times 1307.1(85 - t_2) = 0.75 \times 2145.8(144 - 162)$$

$$t_2 = 121°\text{F}.$$

This value for the inside tube surface temperature, $t_2 = 119°F$, is probably not close enough to the assumed 124°F. Thus another iteration should be carried out. If steps 4, 5, and 6 are repeated (including another iterative calculation of h_{34} as in step 2) with t_2 assumed to be 120°F, the following results are obtained:

$$t_2 = 120°\text{F, assumed,}$$

$$h_{12} = 1355.7 \text{ Btu/h-ft}^2\text{-°F},$$

$$t_3 = 144°\text{F, assumed,}$$

$$h_{34} = 2145.8 \text{ Btu/h-ft}^2\text{-°F},$$

$$U_3 = 512.0 \text{ Btu/h-ft}^2\text{-°F},$$

$$t_3 = 144°\text{F, calculated,}$$

$$t_2 = 120°\text{F, calculated.}$$

The calculated temperatures t_3 and t_2 agree with the assumed values, so that results above are taken to be the desired answers.

The procedures outlined in the two preceding examples will be used later in this chapter. The two fluids in a heat exchanger are separated by a wall (plane or cylindrical, with or without fins) and thus calculations of the foregoing sort must be carried out. These calculations may be further complicated by the fact that the local fluid temperatures also change as the fluids move through the heat exchanger.

10.3 HEAT EXCHANGERS—THE VARIOUS TYPES AND SOME OF THEIR GENERAL CHARACTERISTICS

The process of exchanging heat between two different fluid streams is one of the most important and frequently encountered processes found in engineering practice. Boilers, condensers, water heaters, automobile radiators, air heating or cooling coils, etc., are examples of processes in which heat is exchanged between a "hot" and a "cold" fluid. (The terms "hot" and "cold" are used in a relative sense.) The modern petrochemical industry, energy generating plants, etc., are based on innumerable processes involving the use of devices to exchange heat between two fluid streams without physically mixing them. Such devices are generally termed *heat exchangers*. This section will be devoted to descriptions of the more commonly encountered forms of heat exchangers and some of their general characteristics.

Ordinary heat exchangers may be divided into two general classes depending on the relative orientation of the flow direction of the two fluid streams. If the two streams cross one another in space—usually at right angles—the exchanger is termed a *cross-flow* heat exchanger. An automobile radiator or the cooling unit in an air conditioning duct are simple examples of cross-flow heat exchangers.

The second general class of heat exchangers comprise those in which the two streams move in parallel directions in space. The usual *shell-and-tube* heat exchanger is the most frequently encountered form of such an exchanger, and the concentric pipe exchanger (or *double-tube* exchanger) is also of this type. The term *parallel-flow* is not applied to this general class but is reserved for a special subclass as described in the following section.

Shell-and-Tube or Concentric Pipe Heat Exchangers

The simple form of the concentric pipe heat exchanger is depicted in Fig. 10.3. In the form shown in Fig. 10.3(a), the two fluids enter at the same end of the device and flow in the same direction to the other end. Such an arrangement is termed a *parallel-flow* heat exchanger. Figure 10.3(b) depicts an alternative arrangement for the same device in which the flow direction of the fluid in the inner pipe has been reversed. In this instance the two fluid streams travel in parallel paths but in opposite directions, and the arrangement is termed a *counterflow* heat exchanger. A concentric pipe heat exchanger may be limited in the amount of available surface area for the transfer of heat between the two fluid streams. The transfer area may be made greater by increasing the number of inner pipes as is done in the shell-and-tube arrangement described next.

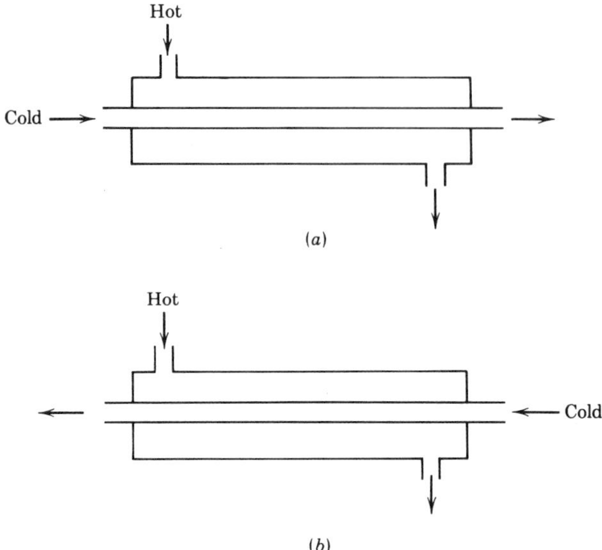

Figure 10.3. Schematic drawing of a concentric pipe heat exchanger; (a) parallel-flow; and (b) counterflow.

The "classical" form of shell-and-tube heat exchanger is indicated by the simplified drawing in Fig. 10.4(a). The exchanger consists of a bundle of tubes (usually many more than shown in Fig. 10.4) secured at either end in tube sheets which are large drilled plates into which the tubes are either rolled or soldered. The entire tube bundle is placed inside a closed shell, which seals around the tube sheets to form the two domains for the hot and cold fluids. One fluid circulates through the tubes—entering and leaving via the water boxes formed between the tube sheets and shell heads. The other fluid flows around the outside of the tubes—in the space between the tube sheets and enclosed by the outer shell. Baffle plates are located normal to the tube bundle to ensure thorough distribution of the shell side fluid.

Figure 10.4(a) is to be interpreted only as a simplified illustration of the principal components of a typical shell-and-tube exchanger. There are many refinements in the design and construction of a heat exchanger which are not shown. Such factors as thermal expansion stresses, tube fouling due to contaminated fluids, ease of assembly and disassembly, size, weight, etc., must be provided for in a detailed design of a heat exchanger.

The heat exchanger shown in Fig. 10.4(a) is designated as having one shell pass and one tube pass since both the shell side fluid and the tube side fluid make a single traverse through the heat exchanger. Schematically, a one-shell-pass, one-tube-pass heat exchanger will be indicated as shown in Figs. 10.4(b) and 10.4(c). Figure 10.4(b) indicates the flow arrangement for a parallel-flow, one-shell-pass, one-tube-pass heat exchanger. As in the concentric pipe exchanger, the fluids enter and leave at the same ends of the exchanger.

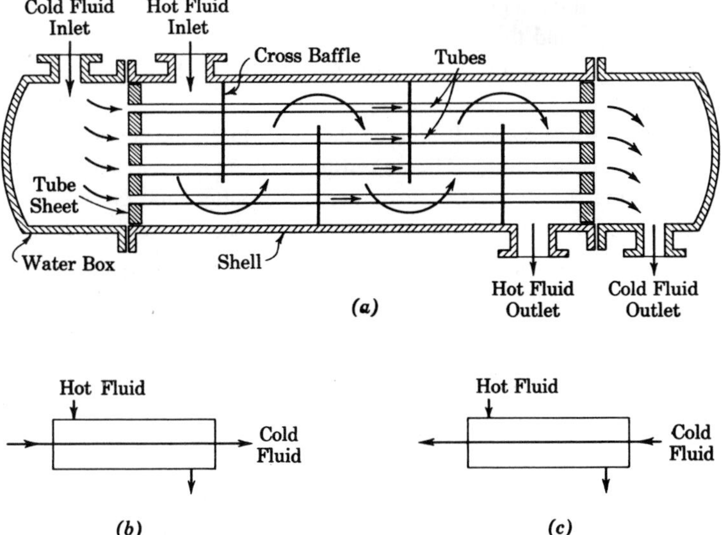

Figure 10.4. Schematic drawing of (a) one-shell-pass, one-tube-pass heat exchanger; (b) parallel-flow; and (c) counterflow.

Figure 10.4(c) indicates, schematically, a counterflow, one-shell-pass, one-tube-pass heat exchanger in which the two fluid streams flow in opposite directions through the exchanger.

Although they are formally quite similar, the parallel-flow and counterflow heat exchangers differ greatly in the manner in which the fluid temperatures vary as the fluids pass through. This difference is illustrated in the diagrams in Fig. 10.5, which plot the temperature of the fluid streams against the distance along the heat exchanger and apply whether the exchangers are of the concentric pipe or shell-and-tube configuration.

As indicated in Fig. 10.5, the parallel-flow exchanger brings the hottest portion of the hot fluid into communication with the coldest portion of the cold fluid. This provides a large thermal *potential difference* initially, but as the fluids move through the exchanger and transfer heat, one to the other, the temperature difference drops rapidly. The temperatures of the two streams approach one another asymptotically, and it would theoretically require an infinitely long heat exchanger to raise the cold fluid to the temperature of the hot fluid.

On the other hand, the counterflow exchanger places the hottest portion of the hot fluid in communication with the hottest portion of the cold fluid—and the coldest portion of the hot fluid in communication with the coldest portion of the cold fluid. This provides a more nearly constant temperature difference between the two fluids all along the heat exchanger and allows the possibility of heating the cold fluid to a temperature greater than the outlet temperature of the hot fluid. In general, the counterflow arrangement results in a greater rate of heat exchange than in the corresponding parallel-flow case.

Figure 10.5 also illustrates the importance of another parameter—the product of the mass flow rate and the specific heat of each of the fluids. If $\dot{m}$ is the mass flow rate and c_p is the specific heat of the fluids, the product, $\dot{m}c_p$, is called the *heat capacity rate*.

The subscript c will be used to denote the cold fluid, and the subscript h will be used to denote the hot fluid. The subscript i will be used to indicate the inlet fluid conditions and o will be used to indicate the outlet fluid conditions. If one uses these subscripts and the definition of the heat capacity rate, an overall energy balance of the heat exchanger gives the total heat transferred between the fluids, q, expressed in two ways:

$$q = \dot{m}_c c_{p_c}(t_{c_o} - t_{c_i}) = \dot{m}_h c_{p_h}(t_{h_i} - t_{h_o}), \tag{10.2}$$

$$\frac{\dot{m}_c c_{p_c}}{\dot{m}_h c_{p_h}} = \frac{t_{h_i} - t_{h_o}}{t_{c_o} - t_{c_i}}.$$

The sketches of Fig. 10.5 show how the relative variation of the two fluid temperatures through the heat exchanger is influenced by whether $\dot{m}_c c_{p_c}$ is greater or less than $\dot{m}_h c_{p_h}$.

An important aspect of the particular flow arrangement chosen is the maximum possible heat that could be transferred, expressed in terms of the input parameters: $\dot{m}_c c_{p_c}$, $\dot{m}_h c_{p_h}$, t_{c_i}, t_{h_i}. This maximum possible heat transfer is that which would result if the exchanger is infintely long, or has an infinite surface area for heat exchange. In particular, for counterflow, examination of the sketches in Fig. 10.5 shows that this limiting condition is determined by whether $\dot{m}_c c_{p_c}$ is greater or less than $\dot{m}_h c_{p_h}$. When $\dot{m}_c c_{p_c} > \dot{m}_h c_{p_h}$, the maximum possible heat transfer is determined by the fact that the hot fluid can be cooled to the temperature of the cold fluid inlet. That is,

For $\dot{m}_c c_{p_c} > \dot{m}_h c_{p_h}$:

$$t_{h_o} \to t_{c_i},$$

$$\begin{aligned} q_{\max} &= \dot{m}_h c_{p_h}(t_{h_i} - t_{h_o}) \\ &= \dot{m}_h c_{p_h}(t_{h_i} - t_{c_i}). \end{aligned}$$

For the reverse case, however, the limit is determined as the cold fluid is heated to the inlet temperature of the hot fluid:

For $\dot{m}_c c_{p_c} < m_h c_{p_h}$:

$$t_{c_o} \to t_{h_i},$$

$$\begin{aligned} q_{\max} &= \dot{m}_c c_{p_c}(t_{c_o} - t_{c_i}) \\ &= \dot{m}_c c_{p_c}(t_{h_i} - t_{c_i}). \end{aligned}$$

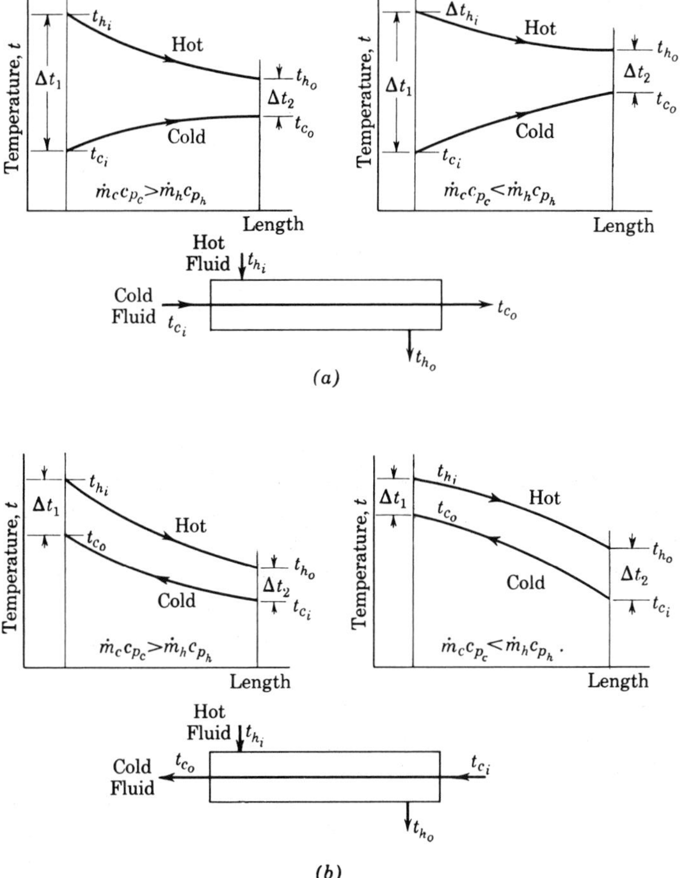

Figure 10.5. Temperature variations in (a) parallel-flow; and (b) counterflow heat exchangers.

Thus, for the counterflow exchanger, the two sets of equations above show that in terms of the inlet parameters, the maximum possible heat exchange is determined by the stream with the smaller heat capacity rate:

$$q_{\max,\text{ctr}} = (\dot{m}c_p)_{\min}(t_{h_i} - t_{c_i}). \tag{10.3}$$

The limiting case is, however, quite different for the parallel-flow arrangement. Regardless of the relative sizes of the two streams, the limiting heat transfer is determined by the fact that the two fluid streams approach the same outlet temperature, the weighted average of the inlet streams:

$$t_{h_o} \rightarrow t_{c_o} \rightarrow \frac{\dot{m}_c c_{p_c} t_{c_i} + \dot{m}_h c_{p_h} t_{h_i}}{\dot{m}_c c_{p_c} + \dot{m}_h c_{p_h}}. \tag{10.4}$$

Thus the maximum possible heat transfer is, after some algebra:

$$q_{\max} = \dot{m}_c c_{p_c}(t_{c_o} - t_{c_i})$$

$$= \frac{1}{\dfrac{1}{\dot{m}_c c_{p_c}} + \dfrac{1}{\dot{m}_h c_{ph}}}(t_{h_i} - t_{c_i}).$$

So, for parallel flow, the above may be written in the following way regardless of which stream has the larger heat capacity rate:

$$q_{\max,\text{par}} = \frac{(\dot{m}c_p)_{\min}(t_{h_i} - t_{c_i})}{1 + [(\dot{m}c_p)_{\min}/(\dot{m}c_p)_{\max}]}. \tag{10.5}$$

Comparison of Eqs. (10.3) and (10.5) shows that for given inlet conditions the counterflow arrangement always has a greater potential for heat transfer than does the parallel-flow arrangement. For this reason the parallel-flow configuration is rarely used in practice unless special conditions demand it or unless multiple passes are used, as described later. Also, since counterflow provides the best potential for heat transfer, its limiting case, as expressed in Eq. (10.3), is used for the basis of comparison (or nondimensionalization) of heat exchanger performance as described later.

It is not necessary for all shell-and-tube heat exchangers to be of the one-shell-pass, one-tube-pass form just discussed. In fact, space limitations frequently demand other arrangements, such as shown in Fig. 10.6. In Fig. 10.6(a) a simplified drawing of another heat exchanger is shown. This is the same in all respects as the exchanger depicted in Fig. 10.4(a) except that a pass partition has been added in the water box at one end and the tube fluid outlet has been relocated. This arrangement causes the tube fluid to make one pass through half of the tubes, reverse its direction of flow, and make a second pass through the remaining half of the tubes. This type of exchanger is referred to as a *one-shell-pass, two-tube-pass* heat exchanger and is typical of many steam condensers. This arrangement is shown schematically in Fig. 10.6(b) and it is apparent that the exchanger is neither a parallel-flow nor counterflow heat exchanger—it is often called a parallel-counterflow heat exchanger.

Many other possible flow arrangements exist and are used. By including longitudinal baffles within the heat exchanger shell, it is possible to cause the shell side fluid to make two or more passes through the shell. Proper positions of pass partitions in the tube fluid circuit will provide as many tube passes per shell pass as may be desired. A two-shell-pass, four-tube-pass heat exchanger is depicted schematically in Fig. 10.6(c).

In any of the multiple pass arrangements, as in Fig. 10.6, some portion of the exchanger is in parallel flow. Thus, such exchangers have a heat transfer potential less than that of pure counterflow—leaving the counterflow exchanger as the ideal maximum.

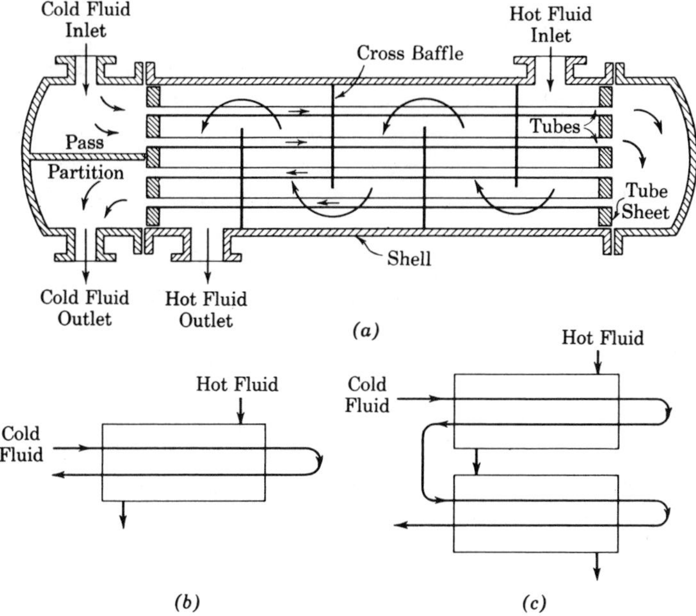

Figure 10.6. Multiple pass heat exchangers: (a), (b) one-shell-pass and two-tube-passes; (c) two-shell-passes, four-tube-passes.

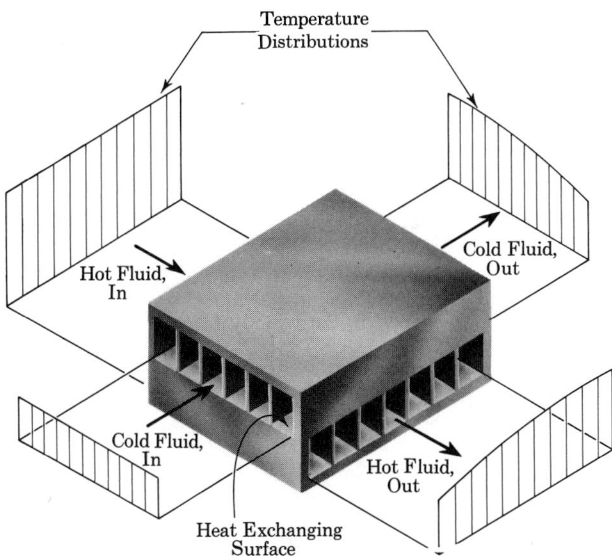

Figure 10.7. Temperature variation of unmixed fluids in a cross-flow heat exchanger.

Cross-Flow Heat Exchangers

As defined earlier, a cross-flow heat exchanger is one in which the two fluid streams are oriented at an angle to one another. For purposes of discussion, Fig. 10.7 depicts a simplified plate type of cross-flow heat exchanger. As is often the case in the cross-flow heat exchangers used in aircraft or spacecraft applications, Fig. 10.7 shows the separating plate, or wall, to be provided with fins. Although simple in principle, the analysis of this type of heat exchanger is extremely complex, since the fluid temperatures may vary both in the direction of flow and normal to the direction of flow if no fluid mixing is allowed in the normal direction. This produces inlet and outlet fluid temperature distributions as indicated in Fig. 10.7—resulting in the fact that the fluid temperature difference varies with the two coordinate fluid stream directions. The designations ''mixed'' and ''unmixed'' are used to denote whether or not mixing of either fluid is allowed in the direction transverse to the direction of the flow stream. The exchanger illustrated in Fig. 10.7 is, as noted, one in which there is no mixing in either fluid stream. Other cross-flow heat exchangers may allow for mixing in one or both of the fluid streams.

EXAMPLE 10.3

In a one-shell-pass, one-tube-pass heat exchanger, the cold fluid is water (c_{p_c} = 4.18 kJ/kg-°C) entering at 40°C with a flow rate of 1000 kg/h. The hot fluid is oil (c_{p_h} = 2.22 kJ/kg-°C) entering at 150°C and flowing at the rate of 1500 kg/h. If the heat exchanger becomes infinitely large, find the maximum heat that can be transferred between the two streams and find the outlet temperatures of the fluids for (a) parallel flow; (b) counterflow.

Solutions. The stated data gives the following inlet specifications for the two streams:

$$\text{Cold:}\quad \dot{m}_c = 1000 \text{ kg/h}, \qquad c_{p_c} = 4.18 \text{ kJ/kg-°C}, \qquad t_{c_i} = 40\text{°C}.$$

$$\text{Hot:}\quad \dot{m}_h = 1500 \text{ kg/h}, \qquad c_{p_h} = 2.22 \text{ kj/kg-°C}, \qquad t_{h_i} = 150\text{°C}.$$

The capacity rates and which is larger are then readily found:

$$\dot{m}_c c_{p_c} = 1000 \times \frac{1}{3600} \times 4.18 = 1.611 \text{ kW/°C} = (\dot{m}c_p)_{\text{max}},$$

$$\dot{m}_h c_{p_h} = 1500 \times \frac{1}{3600} \times 2.22 = 0.925 \text{ kW/°C} = (\dot{m}c_p)_{\text{min}}.$$

(a) For the parallel-flow arrangement, the maximum heat that could be transferred for an infintely large exchanger is given by Eq. (10.5):

$$q_{\max,\text{par}} = \frac{(\dot{m}c_p)_{\min}(t_{h_i} - t_{c_i})}{1 + [(\dot{m}c_p)_{\min}/(\dot{m}c_p)_{\max}]}$$

$$= \frac{0.925(150 - 40)}{1 + (0.925/1.611)}$$

$$= 56.63 \text{ kW } (1.932 \times 10^5 \text{ Btu/h}).$$

The two fluids leave the heat exchanger at the same temperature, as given by Eq. (10.4):

$$t_{h_o} = t_{c_o} \rightarrow \frac{\dot{m}_c c_{p_c} t_{c_i} + \dot{m}_h c_{p_h} t_{h_i}}{\dot{m}_c c_{p_c} + \dot{m}_h c_{p_h}}$$

$$= \frac{1.611 \times 40 + 0.925 \times 150}{1.611 + 0.925}$$

$$= 88.8°\text{C } (192°\text{F}).$$

(b) For the counterflow arrangement the maximum possible heat transfer is considerably greater, as given by Eq. (10.3):

$$q_{\max,\text{ctr}} = (\dot{m}c_p)_{\min}(t_{h_i} - t_{c_i})$$

$$= 0.925(150 - 40)$$

$$= 101.8 \text{ kW } (3.473 \times 10^5 \text{ Btu/h}).$$

Since the heat exchange is dominated by the cold fluid (i.e., $\dot{m}_c c_{p_c} > \dot{m}_h c_{p_h}$), the maximum heat exchange above is limited by the fact that the hot fluid outlet is cooled to the temperature of the cold fluid inlet:

$$t_{h_o} \rightarrow t_{c_i} = 40°\text{C } (104°\text{F}).$$

The outlet temperature of the cold fluid is then determined from the basic heat balance on the exchanger in Eq. (10.2):

$$\dot{m}_c c_{p_c}(t_{c_o} - t_{c_i}) = \dot{m}_h c_{p_h}(t_{h_i} - t_{h_o})$$

$$1.611(t_{c_o} - 40) = 0.925(150 - 40)$$

$$t_{c_o} = 127.6°\text{C } (261.7°\text{F}).$$

While an infinitely large heat exchanger may be unrealistic, the calculations above do illustrate the inherent advantage of the counterflow arrangement over the parallel-flow pattern. For the data of this example the counterflow arrangement has the potential of transferring nearly twice as much heat as in the parallel-flow case.

10.4 THE OVERALL HEAT TRANSFER COEFFICIENT OF A HEAT EXCHANGER

One of the basic parameters used to rate a heat exchanger and used to calculate the performance of a given heat exchanger is the overall heat transfer coefficient. This coefficient depends on the geometric configuration of the wall separating the two fluids and the convective heat transfer coefficients on each side. The separating surface may, or may not, be equipped with a fin array to enhance the exchange of heat. Figure 10.8 depicts a separating wall in which t_c and t_h denote the local cold and hot fluid temperatures, h_c and h_h the heat transfer coefficients on each side, A_c and A_h the total exposed surface area of each side, and $\mathcal{R}_w$ the total resistance of the wall exclusive of the fins. The overall coefficient for this geometry, U, is then defined as that quantity which yields the total rate of local heat transfer between the cold and hot fluids according to

$$q = UA(t_h - t_c).$$

Clearly, the value of U depends on which of the surface areas, A_c or A_h, is used. According to the discussion of Secs. 2.6 and 2.11, the UA product in the equation above is

$$\frac{1}{UA} = \frac{1}{A_c\eta_c h_c} + \mathcal{R}_w + \frac{1}{A_h\eta_h h_h}. \tag{10.6}$$

In Eq. (10.6) η_c and η_h represent the total surface effectiveness given in Eq. (2.56):

$$\eta_c = 1 - \frac{A_{f,c}}{A_c}(1 - \kappa_c),$$

$$\eta_h = 1 - \frac{A_{f,h}}{A_h}(1 - \kappa_h),$$

where $A_{f,c}$ and $A_{f,h}$ are the areas of *fins only* on the cold and hot sides, respectively. In the above expressions the κ's represent the fin efficiency which can be found

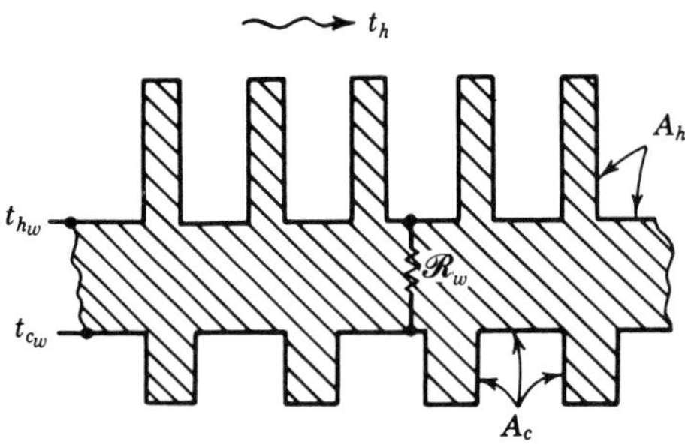

Figure 10.8

from the fin geometry (i.e., rectangular, triangular, etc.) according to the methods of Sec. 2.10. Thus, given the fin shapes and the surface geometries, the η's in Eq. (10.6) can be found, and are presumed known in what follows.

For shell-and-tube heat exchangers, the separating wall has a cylindrical geometry (outside radius r_o, inside radius r_i) and the wall resistance is known to be $\mathcal{R}_w = \ln (r_o/r_i)/2\pi Lk$, with L being the tube length and k its thermal conductivity. In shell and tube exchangers, the overall U is invariably based on the outside surface area, which is also almost always the hot surface, A_h. Thus for a shell-and-tube exchanger, the overall U is found from

$$\frac{1}{U} = \frac{A/A_c}{\eta_c h_c} + \frac{A/2\pi L}{k} \ln\left(\frac{r_o}{r_i}\right) + \frac{1}{\eta_h h_h}. \tag{10.7}$$

The A without a subscript is the *total* exposed surface area on the hot, outer surface, including fins, if any. This total exposed surface is another important performance parameter for a heat exchanger, as will be seen subsequently. Most shell-and-tube exchangers do not have fins at the inner tube surface, so that $A_c = 2\pi L r_i$ and $\eta_c = 1$:

$$\frac{1}{U} = \frac{A/2\pi L r_i}{h_c} + \frac{A/2\pi L}{k} \ln\left(\frac{r_o}{r_i}\right) + \frac{1}{\eta_h h_h}. \tag{10.8}$$

In many instances the tubes are bare on the outside as well, so that the familiar relation results:

$$\frac{1}{U} = \frac{r_o/r_i}{h_c} + \frac{r_o}{k} \ln\left(\frac{r_o}{r_i}\right) + \frac{1}{h_h}. \tag{10.9}$$

For cross-flow heat exchangers, the situation is somewhat different. The wall resistance is $\mathcal{R}_w = \Delta x/A_w k$, in which Δx and A_w represent the thickness and the bare, unfinned area of the plane separating wall. Since there is no obvious preferred surface, such as the outside surface of a shell and tube exchanger, one may base the overall coefficient on the total exposed surface at the cold side *or* the hot side:

$$\begin{aligned} \frac{1}{U_c} &= \frac{1}{\eta_c h_c} + \left(\frac{A_c}{A_w}\right)\frac{\Delta x}{k} + \frac{A_c/A_h}{\eta_h h_h}, \\ \frac{1}{U_h} &= \frac{A_h/A_c}{\eta_c h_c} + \left(\frac{A_h}{A_w}\right)\frac{\Delta x}{k} + \frac{1}{\eta_h h_h}. \end{aligned} \tag{10.10}$$

Depending on which U is used, the exchanger size is specified by the corresponding total area, A_c or A_h. Most cross-flow exchangers will have fins on both the hot and cold sides.

The preceding discussion presumed the heat transfer surfaces involved to be clean. In practice, however, it is not uncommon for the surface to become contaminated with deposits due to impurities in the fluids, chemical reaction, or the buildup of scale by biological action. Such surface contamination can significantly influence the overall heat transfer coefficient by the introduction of additional thermal re-

Table 10.1 Selected Values of Heat Exchanger Fouling Factors*

Fluid	R_f m^2-°C/W	R_f h-ft^2-°F/Btu
Seawater, below 50°C (125°F)	0.00009	0.0005
Seawater, above 50°C (125°F)	0.00018	0.001
Treated boiler feedwater	0.00018	0.001
Steam, non-oil bearing	0.00009	0.0005
Refrigerant vapor, oil bearing	0.00035	0.002
Refrigerant liquid	0.00018	0.001
Natural gas	0.00018	0.001

*From Ref. 2.

sistances. The magnitude of the additional resistances due to such fouling is, of course, dependent on a great number of factors such as the nature of the fluid, the length of service, etc. For general-purpose use, experimental values of the "fouling factor" have been determined for a number of situations and some are quoted in Table 10.1. Additional data are available in Ref. 2. These fouling factors, R_f, are *unit* thermal resistances and thus their application must account for the size of the surface area to which they apply. Since either the hot or cold surface may become fouled, Eq. (10.6) must be modified in the following way:

$$\frac{1}{(UA)_{\text{dirty}}} = \frac{1}{(UA)_{\text{clean}}} + \frac{1}{A_c(1/R_{fc})} + \frac{1}{A_h(1/R_{fh})},$$

in which R_{fc} and R_{fh} are the fouling factors for the cold and hot surfaces, respectively. Thus the U_c or U_h expressions in the foregoing should be modified according to

$$\frac{1}{(U_h)_{\text{dirty}}} = \frac{1}{(U_h)_{\text{clean}}} + \frac{A_h}{A_c} R_{fc} + R_{fh},$$
$$\frac{1}{(U_c)_{\text{dirty}}} = \frac{1}{(U_c)_{\text{clean}}} + R_{fc} + \frac{A_c}{A_h} R_{fh}. \tag{10.11}$$

The additions above may be made to any of Eqs. (10.7) through (10.10); however, the corresponding special forms are not written here for the sake of brevity.

In what follows, it is presumed that the expressions of this section have been applied to evaluate the overall U for a given heat exchanger—based on whichever total surface area has been taken to characterize the size of the exchanger. As a guide to the reader, Table 10.2 shows typical values encountered in practice for the overall heat transfer coefficient. A wide range of values is apparent, depending on the particular application.

While the foregoing discussion has concentrated on the definition and calculation of the overall coefficient, U, for a heat exchanger, one must remember that it is

Table 10.2 Representative Values of Heat Exchanger Overall Heat Transfer Coefficients

Situation	U W/m²-°C	U Btu/h-ft²-°F
Air to air, cross-flow	2–10	10–50
Air to water, cross-flow	5–15	25–80
Water to oil, shell and tube	20–60	110–350
Water to water, shell and tube	150–300	850–1700
Steam condenser, shell and tube	200–1000	1100–5600

the product of U and the total available transfer area, A, that is the measure of the heat transferring capability of a given heat exchanger. That is, for given fluid temperatures and flow rates, the product UA represents the heat-exchanging capacity of an exchanger, per unit temperature difference, and is the fundamental parameter to be used in comparing different heat exchangers. The next two examples illustrate this point.

EXAMPLE 10.4

Figure 10.9 illustrates a $\frac{3}{4}$-in. 16-gage heat exchanger tube equipped with 20 equally spaced straight fins of rectangular profile placed longitudinally along the tube. The fins are 0.75 in. long in the radial direction and are 0.05 in. thick. Both the tube and the fins are made of steel with $k = 13$ Btu/h-ft-°F. The fluid flowing inside the tube produces a heat transfer coefficient of 200 Btu/h-ft²-°F at the inner tube surface. The fluid flowing past the finned surface on the outside of the tube produces

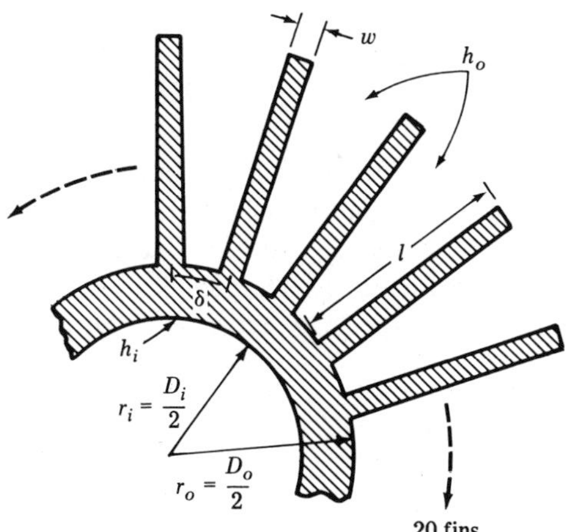

Figure 10.9. Example 10.4.

a heat transfer coefficient of 45 Btu/h-ft^2-°F on all the exposed surfaces. Find the overall heat transfer coefficient, based on the total outside exposed surface area, of a heat exchanger consisting of such tubes. Determine, also, the *UA* product for the heat exchanger, per unit length of tube.

Solution. Table B.2 gives the dimensions for a $\frac{3}{4}$-in. 16-gage tube. These dimensions and the other specified quantities yield the following, using the notation given in Fig. 10.9:

$$\text{Tube:} \quad D_i = 0.620 \text{ in.}, \quad D_o = 0.750 \text{ in.}$$

$$\text{Find:} \quad l = 0.75 \text{ in.}, \quad w = 0.05 \text{ in.}, \quad N = 20 \text{ fins.}$$

$$\delta = \frac{\pi D_o}{N} = \pi \times \frac{0.750}{20} = 0.1178 \text{ in.}$$

$$\text{Thermal quantities: } k = 23 \text{ Btu/h-ft-°F}, \quad h_i = 200 \text{ Btu/h-ft}^2\text{-°F}, \quad h_o = 45 \text{ Btu/h-ft}^2\text{-°F}.$$

To find the fin efficiency and the overall heat transfer coefficient, one needs two additional geometric parameters, the total outer exposed surface area (per unit length of tube) and the ratio of the exposed fin area to the total area. Per unit of tube length, the total exposed surface area is, in terms of the given geometry,

$$\frac{A}{L} = \pi D_o + N(2l)$$

$$= \pi\left(\frac{0.750}{12}\right) + 20\left(2 \times \frac{0.75}{12}\right) = 2.6964 \text{ ft}^2/\text{ft}.$$

For an array of straight fins of rectangular profile (length, l, thickness w, spacing δ), Eq. (2.57) and Fig. 2.18 show that the ratio of the exposed fin surface to the total exposed surface is given by

$$\frac{A_f}{A} = \frac{2l + w}{2l + \delta} = \frac{2 \times 0.750 + 0.05}{2 \times 0.750 + 0.1178} = 0.9581.$$

Before proceeding to the determination of the overall U from Eq. (10.8) it is necessary first to determine the surface effectiveness of the fin array, which, in turn, requires finding the individual fin efficiency. The fins are exposed to the outside fluid through the heat transfer coefficient h_o. Thus application of the principles developed in Sec. 2.10 gives the fin efficiency according to Eq. (2.55):

$$ml = l\sqrt{\frac{2h_o}{kw}} = \frac{0.750}{12}\sqrt{\frac{2 \times 45}{23(0.05/12)}} = 1.9153,$$

$$\kappa = \frac{1}{ml}\tanh ml = \frac{1}{1.9153}\tanh(1.9153) = 0.4999.$$

Thus Eq. (2.56) gives the surface effectiveness of the fin array:

$$\eta = 1 - \frac{A_f}{A}(1 - \kappa)$$

$$= 1 - 0.9581(1 - 0.4999) = 0.5209.$$

Since there are no extended surfaces at the inner tube surface, the form given in Eq. (10.8) for the overall heat transfer coefficient based on the outer exposed area applies:

$$U = \left[\frac{(A/L)}{2\pi r_i h_i} + \frac{(A/L)}{2\pi k}\ln\left(\frac{r_o}{r_i}\right) + \frac{1}{\eta h_o}\right]^{-1}$$

$$= \left[\frac{2.6964}{2\pi\left(\dfrac{0.620/2}{12}\right)200} + \frac{2.6964}{2\pi \times 23}\ln\left(\frac{0.750}{0.620}\right) + \frac{1}{0.5209 \times 45}\right]^{-1}$$

$$= 7.74 \text{ Btu/h-ft}^2\text{-°F } (43.9 \text{ W/m}^2\text{-°C}).$$

Per unit of tube length the UA product is, then,

$$\frac{UA}{L} = 7.74 \times 2.6964$$

$$= 20.86 \frac{\text{Btu/h-°F}}{\text{ft}}\left(36.10 \frac{\text{W/°C}}{\text{m}}\right).$$

EXAMPLE 10.5

If the heat exchanger tube considered in Example 10.4 has no fins on the outer surface (i.e., it is a plain bare tube) and the same inside and outside heat transfer coefficients exist, find the overall heat transfer coefficient based on the outside surface area and the UA product. Compare these results with those found for the finned tube in Example 10.4.

Solution. The total exposed surface area, per unit of tube length, is now

$$\frac{A}{L} = \pi D_o = \pi\left(\frac{0.75}{12}\right) = 0.1964 \text{ ft}^2\text{/ft}.$$

The overall coefficient is, by either Eq. (10.8) or (10.9),

$$U = \left[\frac{r_o/r_i}{h_i} + \frac{r_o}{k}\ln\left(\frac{r_o}{r_i}\right) + \frac{1}{h_o}\right]^{-1}$$

$$= \left[\frac{0.750/0.620}{200} + \frac{0.750/2}{12 \times 23}\ln\left(\frac{0.750}{0.620}\right) + \frac{1}{45}\right]^{-1}$$

$$= 35.05 \text{ Btu/h-ft}^2\text{-°F } (199.0 \text{ W/m}^2\text{-°C}).$$

The overall U is seen to be nearly five times as great as that for the finned tube. However, this does not imply an improved heat exchanger performance since the total available surface area for heat transfer has been significantly reduced. In fact, the UA product, per unit of tube length, is now

$$\frac{UA}{L} = 35.05 \times 0.1964$$

$$= 6.88 \frac{\text{Btu/h-°F}}{\text{ft}} \left(11.9 \frac{\text{W/°C}}{\text{m}} \right).$$

This value is less than one-third of that for the finned tube. Thus, under the specified service conditions as given by h_o and h_i, the finned tube heat exchanger would perform significantly better than the bare tube one.

10.5 A SIMPLE HEAT EXCHANGER—FLOW PAST AN ISOTHERMAL SURFACE

The analysis of the performance of a heat exchanger is customarily done in one of two formulations—the "mean-temperature-difference" formulation and the "effectiveness" formulation. These two approaches are treated in some detail in the following sections; however, it is instructive to see the motivation behind each approach and their equivalence by first examining a rather simple special case.

Consider, then, the flow of a fluid past an isothermal surface of known temperature and surface area. Figure 10.10 depicts such a case in which a fluid of known inlet temperature, t_{f_i}, flows at a known mass flow rate, $\dot{m}$, in a tube of known circumference, C, and length, L. The specific heat of the fluid, c_p, is known as is the temperature of the tube surface, t_s, and the heat transfer coefficient at that surface, h. Let it be desired to find the outlet temperature of the fluid, t_{f_o}, and the total heat transferred to it, q. For convenience of discussion the surface temperature is taken as greater than that of the fluid. The fluid temperature increases with distance along the tube, x, and at $x = L$, the exit fluid temperature is t_{f_o}. In any incremental portion of the tube the heat transfer to the fluid is expressed in two ways, in terms of a heat balance on the fluid and in terms of the exchange between the fluid and the surface:

$$\begin{aligned} dq &= \dot{m}c_p dt_f \\ &= hC\,dx(t_s - t_f). \end{aligned}$$

For the entire tube length where $A = LC$ is the total surface area, integrated forms of the equations above are

$$q = \dot{m}c_p(t_{f_o} - t_{f_i}) \tag{10.12}$$

$$= hLC(t_s - t_f)_m = hA(t_s - t_f)_m. \tag{10.13}$$

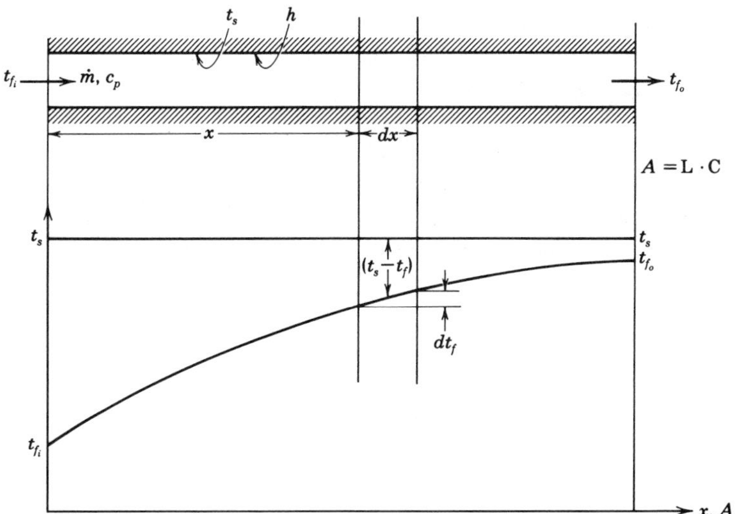

Figure 10.10. Temperature variation in an isothermal passage.

The term $(t_s - t_f)_m$ represents some suitable mean value of the surface-to-fluid temperature difference which varies from $t_s - t_{f_i}$ at the tube entrance to $t_s - t_{f_o}$ at the exit. It is t_{f_o} and q which are sought, given t_{f_i}, $\dot{m}$, c_p, h, A, and t_s.

The situation just described and depicted in Fig. 10.10 is identical to that discussed for flow in a pipe at the end of Sec. 6.4 and depicted in Fig. 6.2. Hence, those results will simply be quoted here, with appropriate notational changes. In Sec. 6.4 it was shown that the fluid temperature increases with x according to Eq. (6.41) and the mean value of the temperature difference is the logarithmic mean given in Eq. (6.43):

$$\frac{t_f - t_s}{t_{fi} - t_s} = \exp\left(-\frac{hCx}{\dot{m}c_p}\right), \tag{10.14}$$

$$\frac{t_{fo} - t_s}{t_{fi} - t_s} = \exp\left(-\frac{hA}{\dot{m}c_p}\right), \tag{10.15}$$

$$(t_s - t_f)_m = \frac{(t_s - t_{fi}) - (t_s - t_{fo})}{\ln\left(\dfrac{t_s - t_{fi}}{t_s - t_{fo}}\right)}. \tag{10.16}$$

Given t_{fi}, t_s, $\dot{m}$, c_p, h, and A (i.e., given the geometry of the "exchanger," the surface temperature, and the fluid inlet conditions) it is desired to determine the exchanger performance as determined by the total heat transferred and the fluid outlet temperature, q and t_{f_o}. In the *mean-temperature-difference* method these parameters are determined as follows:

1. Find t_{fo} from Eq. (10.15).
2. Find the mean temperature difference from Eq. (10.16).
3. Find q from Eq. (10.12).

An exactly equivalent method can be formulated by considering the maximum possible heat that could be transferred to the fluid if the tube (i.e., "exchanger") were infinitely long. In such an instance it is obvious that the maximum heat transfer possible occurs when the fluid is heated to the surface temperature, that is, $t_{fo} \to t_s$, so that Eq. (10.12) gives

$$q_{\max} = \dot{m}c_p(t_s - t_{fi}), \tag{10.17}$$

a quantity calculable from the data given. Then let the "effectiveness" of a finite tube, ϵ, be defined

$$\epsilon = \frac{q}{q_{\max}} = \frac{\dot{m}c_p(t_{fo} - t_{fi})}{\dot{m}c_p(t_s - t_{fi})} = 1 - \frac{t_s - t_{fo}}{t_s - t_{fi}}.$$

From Eq. (10.15), the effectiveness is then

$$\epsilon = 1 - \exp\left(-\frac{hA}{\dot{m}c_p}\right).$$

The quantity $hA/\dot{m}c_p$ is a special case of what is termed "number of transfer units," or NTU:

$$\text{NTU} = \frac{hA}{\dot{m}c_p}, \qquad \epsilon = 1 - \exp(-\text{NTU}). \tag{10.18}$$

The *effectiveness* method of heat exchanger analysis determines the performance parameters q and t_{fo} in terms of the same input data (t_{fi}, t_s, $\dot{m}$, c_p, h, A) used in the mean-temperature-difference method:

1. Find NTU $= hA/\dot{m}c_p$.
2. Find the effectiveness, ϵ, from Eq. (10.18).
3. Find $q_{\max}$ from Eq. (10.17).
4. Find $q = \epsilon q_{\max}$ and t_{fo} from Eq. (10.12).

Clearly, the two approaches, the mean-temperature-difference approach and the effectiveness approach, are exactly equivalent. The same equations are used, but cast in different forms. The first method concentrates on expressing performance in terms of the mean difference between the fluid and the surface as the fluid proceeds through the "exchanger." The second method expresses performance in terms of the maximum heat that *could* be transferred under the stated conditions. The equivalence of the two formulations is easy to see because of the simple case chosen—that of a single fluid and a surface of uniform, known, temperature. In the general case of the heat exchangers described in Sec. 10.4, there are *two* fluids, both of varying temperature, and the separating surface varies in temperature along the flow path. Thus the problem of describing the heat exchanger performance will become more complicated; however, two basic representations are possible, in terms of a mean temperature difference or in terms of an effectiveness based on maximum possible performance. These two representations are described for the more complex exchangers in the next two sections.

10.6 HEAT EXCHANGER MEAN TEMPERATURE DIFFERENCES

For the case of a heat exchanger involving two fluids the concept of the mean temperature difference introduced in Sec. 10.5 must be broadened. Let it be assumed that the geometry of a heat exchanger is completely known. That is, if it is a shell and tube exchanger the dimensions of tubes, the number of tubes, the arrangement of the tubes and the shell passes, and the configuration of any fins are all known. If it is a cross-flow exchanger, comparable data are known. Then, in any incremental portion of the exchanger in which an increment of transfer surface area, dA, is exposed between the two fluids, the incremental heat transferred is

$$dq = U\,dA\,\Delta t. \tag{10.19}$$

In Eq. (10.19) Δt is the temperature difference between the two fluids in the increment dA, and U is the overall heat transfer coefficient (based on the area being used) between the two fluids. One seeks to integrate Eq. (10.19) to give the total heat transferred in the entire heat exchanger in the form

$$q = (U\,\Delta t)_m A,$$

in which A is the heat exchanger total surface area and $(U\,\Delta t)_m$ is a suitable mean value of the $U\,\Delta t$ product. For most cases to be treated here, let it be assumed that an appropriate average value for U can be found that applies throughout the exchanger. Then one seeks a mean temperature difference, Δt_m, so that

$$q = UA\,\Delta t_m. \tag{10.20}$$

Equations (10.19) and (10.20) imply the existence of a mean temperature difference defined as

$$\Delta t_m = \frac{1}{A}\int_0^A (\Delta t)\,dA. \tag{10.21}$$

To determine Δt_m, then, one needs to determine how the temperature of each fluid varies as it flows through the exchanger, so that the difference, Δt, may be expressed as a function of position (i.e., as a function of surface area) and Eq. (10.21) evaluated.

In the case of the elementary exchanger considered in the preceding section, this temperature difference was easily determined. The determination of the more complicated tube-pass and shell-pass configurations will be described in the next few sections.

Parallel-Flow and Counterflow Heat Exchangers

First consider the one-tube-pass, one-shell-pass heat exchangers described in Sec. 10.3. The heat exchanger may be of either the parallel-flow or counterflow arrangement. The discussion which follows considers the parallel flow case, but the discussion should make it apparent that the results apply equally well to the counterflow case. The following assumptions are made:

1. The average overall U is constant throughout the exchanger.
2. If either fluid undergoes a phase change (e.g., as in a condenser), then this phase change occurs throughout the heat exchanger—not just part of it—resulting in a constant fluid temperature.
3. The specific heat and mass flow rate, and hence the heat capacity rate, of each fluid are constant.
4. No heat is lost to the surroundings.
5. There is no conduction in the direction of flow in either the fluids or the tube walls.
6. At any cross section in the heat exchanger, each of the fluids may be characterized by a single temperature—i.e., perfect transverse mixing in each fluid is presumed.

The notation used in Sec. 10.3 will be used here. That is, the subscripts h and c refer to the hot and cold fluids, respectively, whereas the subscripts i and o represent inlet and outlet conditions. The subscripts 1 and 2 denote the left and right ends of the exchanger. Thus for the parallel flow device shown in Fig. 10.5(a) the temperature differences at the two ends are

$$\Delta t_1 = t_{h_i} - t_{c_i},$$

$$\Delta t_2 = t_{h_o} - t_{c_o};$$

whereas for a counterflow exchanger, as in Fig. 10.5(b),

$$\Delta t_1 = t_{h_i} - t_{c_o},$$

$$\Delta t_2 = t_{h_o} - t_{c_i}.$$

The four fluid terminal temperatures t_{c_i}, t_{c_o}, t_{h_i}, t_{h_o} are presumed known.

Now, for the parallel-flow heat exchanger depicted in Fig. 10.11, the heat

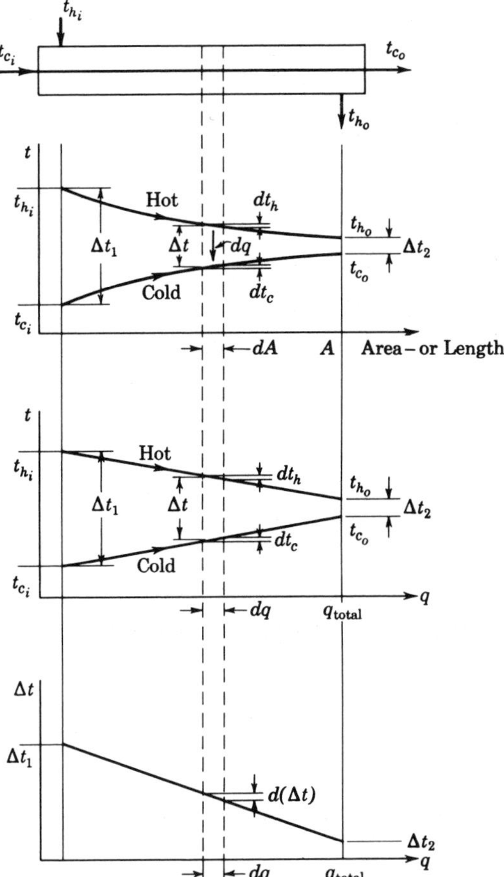

Figure 10.11

exchanged in an incremental length of the exchanger (of surface area dA) may be expressed in the following three ways:

$$dq = U\,dA\,\Delta t, \tag{10.22}$$

$$dq = \dot{m}_c c_{p_c}\,dt_c, \tag{10.23}$$

$$dq = -\dot{m}_h c_{p_h}\,dt_h. \tag{10.24}$$

The difference between the two fluid temperatures in the increment dA is denoted by $\Delta t = t_h - t_c$, and dt_c and dt_h represents the changes in the cold and hot streams as they pass through the increment.

Equations (10.23) and (10.24) show that if one were to plot the fluid temperatures against the amount of heat transferred up to the section under consideration, the resulting curve would be a straight line as illustrated in Fig. 10.11. Thus the temperature difference, Δt, is also a straight line when plotted against q—also shown in Fig. 10.11. This fact is true whether or not the exchanger is a parallel-flow one or a counterflow one. Thus Δt is a linear function of q—varying from Δt_1 where $q = 0$ at the left end to Δt_2 where $q = q_{\text{total}}$ at the right end. Thus

$$\frac{d(\Delta t)}{dq} = \frac{\Delta t_2 - \Delta t_1}{q_{\text{total}}}.$$

Since Eq. (10.22) gives $dq = U\, dA\, \Delta t$,

$$\frac{d(\Delta t)}{\Delta t} = \frac{U(\Delta t_2 - \Delta t_1)}{q_{\text{total}}}\, dA. \tag{10.25}$$

Integration of the above between $A = 0$, $\Delta t = \Delta t_1$ and $A = A$, $\Delta t = \Delta t_2$ gives

$$\ln \Delta t_2 - \ln \Delta t_1 = U \frac{\Delta t_2 - \Delta t_1}{q_{\text{total}}} A$$

or

$$q_{\text{total}} = UA \frac{\Delta t_1 - \Delta t_2}{\ln (\Delta t_1/\Delta t_2)},$$

with A representing the total surface area. The latter expression, when compared with Eq. (10.20), yields the desired mean temperature difference:

$$\Delta t_m = \Delta t_{\text{lm}} = \frac{\Delta t_1 - \Delta t_2}{\ln (\Delta t_1/\Delta t_2)}, \tag{10.26}$$

$$q = UA\, \Delta t_{\text{lm}}. \tag{10.27}$$

Again, the logarithmic mean temperature difference is seen to apply, hence the symbol Δt_{lm}. The logarithmic mean is the mean of the temperature differences at the two ends of the heat exchanger, and while the same relation appears to apply to both the parallel-flow and counterflow cases, it is important to note that Δt_1 and Δt_2 are different in the two instances. In terms of the four fluid terminal temperatures, the log-mean temperature differences for the parallel-flow and counterflow cases are

$$\Delta t_{\text{lm,par}} = \frac{(t_{h_i} - t_{c_i}) - (t_{h_o} - t_{c_o})}{\ln \left(\dfrac{t_{h_i} - t_{c_i}}{t_{h_o} - t_{c_o}}\right)}, \tag{10.28}$$

$$\Delta t_{\text{lm,ctr}} = \frac{(t_{h_i} - t_{c_o}) - (t_{h_o} - t_{c_i})}{\ln \left(\dfrac{t_{h_i} - t_{c_o}}{t_{h_o} - t_{c_i}}\right)}. \tag{10.29}$$

For a condenser, or any other heat exchanger in which a hot fluid maintains a constant temperature $t_h = t_{h_o} = t_{h_i}$, the above two mean temperature differences become the same, as one would expect, and become equal to the mean temperature difference deduced for the case of an isothermal surface in Sec. 10.5.

EXAMPLE 10.6

In a one-tube-pass, one-shell-pass exchanger, the hot fluid enters at $t_{h_i} = 425°\text{C}$ and leaves $t_{h_o} = 260°\text{C}$, while the cold fluid enters at $t_{c_i} = 40°\text{C}$ and leaves at $t_{c_o} = 150°\text{C}$. Find the log-mean temperature difference if the heat exchanger is arranged for (a) parallel flow; (b) counterflow.

Solution.

(a) For the parallel-flow arrangement the temperature difference between the hot and cold fluids at each end of the exchanger are

$$\Delta t_1 = t_{h_i} - t_{c_i} = 425 - 40 = 385°\text{C},$$

$$\Delta t_2 = t_{h_o} - t_{c_o} = 260 - 150 = 110°\text{C},$$

and the mean temperature difference is, by Eq. (10.26) or (10.28),

$$\Delta t_m = \frac{\Delta t_1 - \Delta t_2}{\ln (\Delta t_1/\Delta t_2)}$$

$$= \frac{385 - 110}{\ln (385/110)} = 219.5°\text{C (395.1°F)}.$$

(b) For the counterflow arrangement, the fluid temperature differences at each end are different:

$$\Delta t_1 = t_{h_i} - t_{c_o} = 425 - 150 = 275°\text{C},$$

$$\Delta t_2 = t_{h_o} - t_{c_i} = 260 - 40 = 220°\text{C},$$

so that the mean temperature difference is

$$\Delta t_m = \frac{\Delta t_1 - \Delta t_2}{\ln (\Delta t_1/\Delta t_2)}$$

$$= \frac{275 - 220}{\ln (275/220)} = 246.5°\text{C (443.7°F)}.$$

The mean temperature difference for counterflow is 12% greater than that for parallel flow.

Multiple Pass Shell-and-Tube Heat Exchangers. Cross-Flow Heat Exchangers

The foregoing analysis and the derived log-mean temperature difference applies only to heat exchangers having just one shell pass and one tube pass. In the event that a multiple pass heat exchanger is under consideration, it is necessary to obtain

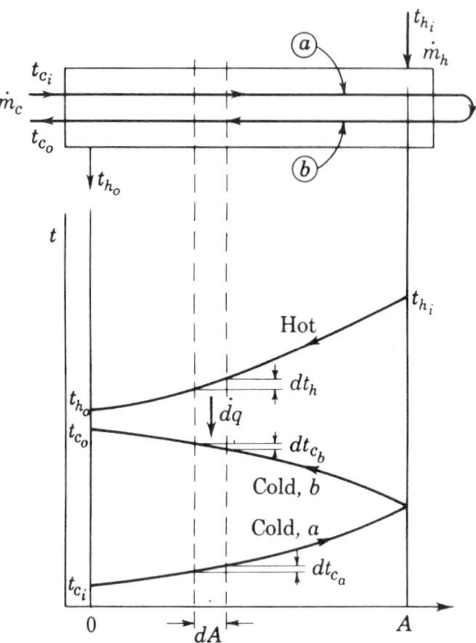

Figure 10.12. Temperature variation in a one-shell-pass, two-tube-pass heat exchanger.

a different expression for the mean temperature difference, dependent on the arrangement of the shell and tube passes.

Consider, for instance, the one-shell-pass, two-tube-pass heat exchanger discussed earlier and shown again in Fig. 10.12. The subscripts h and c are used as in the above example, and again the subscripts i and o will denote inlet and outlet fluid conditions. The subscripts a and b will be used to denote the first and second tube passes, respectively.

For an incremental length of the heat exchanger—involving a total tube surface area of dA—the following expression may be written for the heat transferred:

$$
\begin{aligned}
dq &= \dot{m}_h c_{ph}\, dt_h, \\
dq &= \dot{m}_c c_{pc}(dt_{ca} - dt_{cb}), \qquad (10.30) \\
dq &= U\, dA[(t_h - t_{ca}) + (t_h - t_{cb})].
\end{aligned}
$$

These three equations may be combined to eliminate any two of the three unknown temperatures—say t_{ca} and t_{cb}—resulting in a differential equation in t_h. This differential equation may be solved along with the following fact, derived from the overall heat balance of the exchanger given in Eq. (10.2):

$$q = \dot{m}_h c_{ph}(t_{h_i} - t_{h_o}) = \dot{m}_c c_{pc}(t_{co} - t_{ci}).$$

The following expression for the total heat transfer then results:

$$q = UA\, \Delta t_m,$$

in which the mean temperature difference, Δt_m, is

$$\Delta t_m = \frac{\sqrt{(t_{h_i} - t_{h_o})^2 + (t_{c_o} - t_{c_i})^2}}{\ln\left[\dfrac{(t_{h_i} + t_{h_o}) - (t_{c_i} + t_{c_o}) + \sqrt{(t_{h_i} - t_{h_o})^2 + (t_{c_o} - t_{c_i})^2}}{(t_{h_i} + t_{h_o}) - (t_{c_i} + t_{c_o}) - \sqrt{(t_{h_i} - t_{h_o})^2 + (t_{c_o} - t_{c_i})^2}}\right]}. \tag{10.31}$$

The actual algebra leading to this expression has been omitted due to its length and complexity.

Similar expressions may be developed for each individual shell and tube pass arrangement—or, in a much more complex analysis, for cross-flow heat exchangers. This is done in some detail in Ref. 1. In each case the mean temperature difference is expressed in terms of the four terminal fluid temperatures: t_{h_i}, t_{h_o}, t_{c_i}, and t_{c_o}. The expressions for the mean temperature difference may be simplified (see Ref. 2) by introducing the following dimensionless ratios:

$$\text{Capacity ratio:} \quad R = \frac{\dot{m}_c c_{pc}}{\dot{m}_h c_{ph}} = \frac{t_{h_i} - t_{h_o}}{t_{c_o} - t_{c_i}}. \tag{10.32}$$

$$\text{Effectiveness:} \quad P = \frac{\dot{m}_c c_{pc}(t_{c_o} - t_{c_i})}{\dot{m}_c c_{pc}(t_{h_i} - t_{c_i})} = \frac{t_{c_o} - t_{c_i}}{t_{h_i} - t_{c_i}}. \tag{10.33}$$

The capacity ratio, R, is so named since it is the ratio of $\dot{m}_c c_{p_c}$ to $\dot{m}_h c_{p_h}$, above. The effectiveness is the ratio of the total heat transferred to the cold fluid to the heat that would be transferred to the cold fluid if it were raised to the inlet temperature of the hot fluid. With these definitions it is possible to relate the mean temperature difference to P, R, and the log-mean temperature difference calculated for *counterflow*. That is, one may define a function of P and R, call it $F(P, R)$, so that

$$\Delta t_m = F(P, R)\, \Delta t_{\text{lm,ctr}}. \tag{10.34}$$

Here $\Delta t_{\text{lm,ctr}}$ is given by Eq. (10.29), and the function $F(R, P)$ is then determined by the shell and tube pass arrangement of the particular heat exchanger.

For instance, in the case just discussed of a one-shell-pass, two-tube-pass heat exchanger, Eqs. (10.31) through (10.34) may be combined with Eq. (10.29) to obtain

$$F(P, R) = \frac{\sqrt{R^2 + 1}}{R - 1} \cdot \frac{\ln\left(\dfrac{1 - P}{1 - PR}\right)}{\ln\left[\dfrac{2 - P(R + 1 - \sqrt{R^2 + 1})}{2 - P(R + 1 + \sqrt{R^2 + 1})}\right]}. \tag{10.35}$$

With F thus defined, it may be considered to be a *correction factor* to be applied to the usual, and readily calculated, log-mean temperature difference for counterflow according to Eq. (10.29). It can be noted in Eq. (10.35) that a solution for F does not exist for all possible combinations of R and P. The reason that there are certain areas of impossible solutions is due to the definition of the effectiveness P.

The effectiveness is defined as the ratio of $\dot{m}_c c_{p_c}(t_{c_o} - t_{c_i})$, the actual heat transferred, to $\dot{m}_c c_{p_c}(t_{h_i} - t_{c_i})$. In Sec. 10.3 it was noted that the *maximum* possible heat that could be transferred in a *counterflow* heat exchanger with *infinite* surface area was equal to $\dot{m}_c c_{p_c}(t_{h_i} - t_{c_i})$ for $R < 1$. This quantity is the denominator in the definition of the effectiveness [see Eq. (10.33)]. The two-tube-pass heat exchanger described by Eq. (10.35) exchanges some heat in parallel flow as well as counterflow. For this reason there are certain values of the effectiveness which cannot be achieved—even for an exchanger of infinite area. The maximum effectiveness would, of course, be dependent on the capacity ratio R.

In addition to the limitations just described, it is economically poor practice to employ a heat exchanger for conditions under which $F < 0.75$. A value of F less than about 0.75 results in a mean temperature difference, Δt_m, sufficiently small that an uneconomically large heat exchanger would probably result. This criterion further restricts the maximum effectiveness of a heat exchanger. For applications in which the specified values of the terminal temperatures are such that the associated values of P and R give no solution for F, or if F is less than 0.75, some other shell and tube arrangement should be employed. Some other arrangements will be discussed below.

Values of F for a one-shell-pass, two-tube-pass heat exchanger have been computed from Eq. (10.35) by Bowman et al. (Ref. 3). The results are shown in Fig. 10.13. Similar results have been obtained by similar analyses for a number of shell and tube arrangements and for various cross-flow heat exchangers. These results may be found in Ref. 2, and some of them are presented here. Figure 10.14 gives F for a heat exchanger with two shell passes and two tube passes per shell pass. Figure 10.15 presents the results of Bowman et al. (Ref. 3) for a cross-flow heat exchanger in which there is no mixing of either fluid stream in a direction transverse to their direction of flow. The outlet fluid temperature in this case is interpreted as the "mixed mean" outlet temperature of the fluids—i.e., transverse mixing being allowed after exit from the heat exchanger.

Another important item to be noted about this manner of presentation of the mean temperature difference of a multiple pass heat exchanger is the unfortunate shape of the curves shown in Figs. 10.13 through 10.15. As is readily apparent, the curves are difficult to read with accuracy as P increases (depending on R). This is particularly true for $R > 1$ as $P \rightarrow 1/R$, since F becomes undefined. However, as has been recently pointed out by Shamsundar (Ref. 4), half the F versus P curves are redundant since it may be shown from the definitions of P and R that the following reciprocity relation holds:

$$F(P, R) = F(PR, 1/R). \tag{10.36}$$

Thus the curves for $R > 1$ are uniquely related to those for $R < 1$ and only one set is actually needed. Consequently, in the regions of the charts of Figs. 10.13 through 10.15 for $R > 1$ where good accuracy is difficult to obtain, the above reciprocity relation may be applied and the curves for $R < 1$ used instead.

One final note concerning the F factor might be made. The presentation above concentrated on heat exchangers with multiple tube and shell passes. It could be

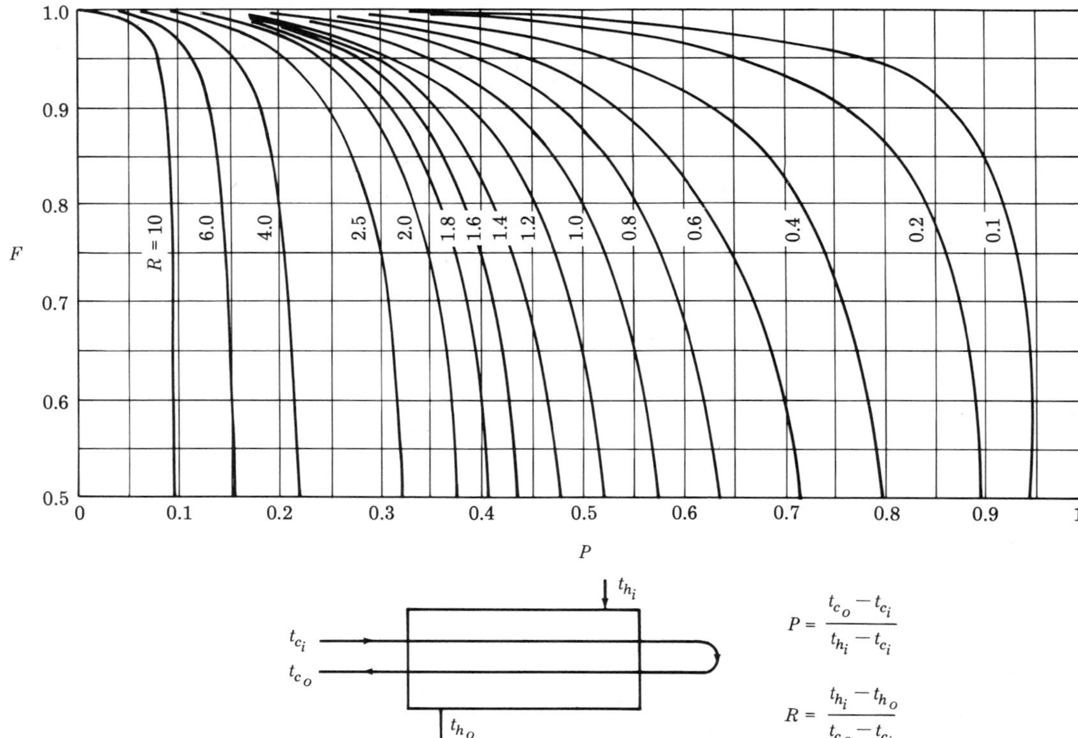

Figure 10.13. Correction factor F for a one-shell-pass, two-tube-pass heat exchanger.

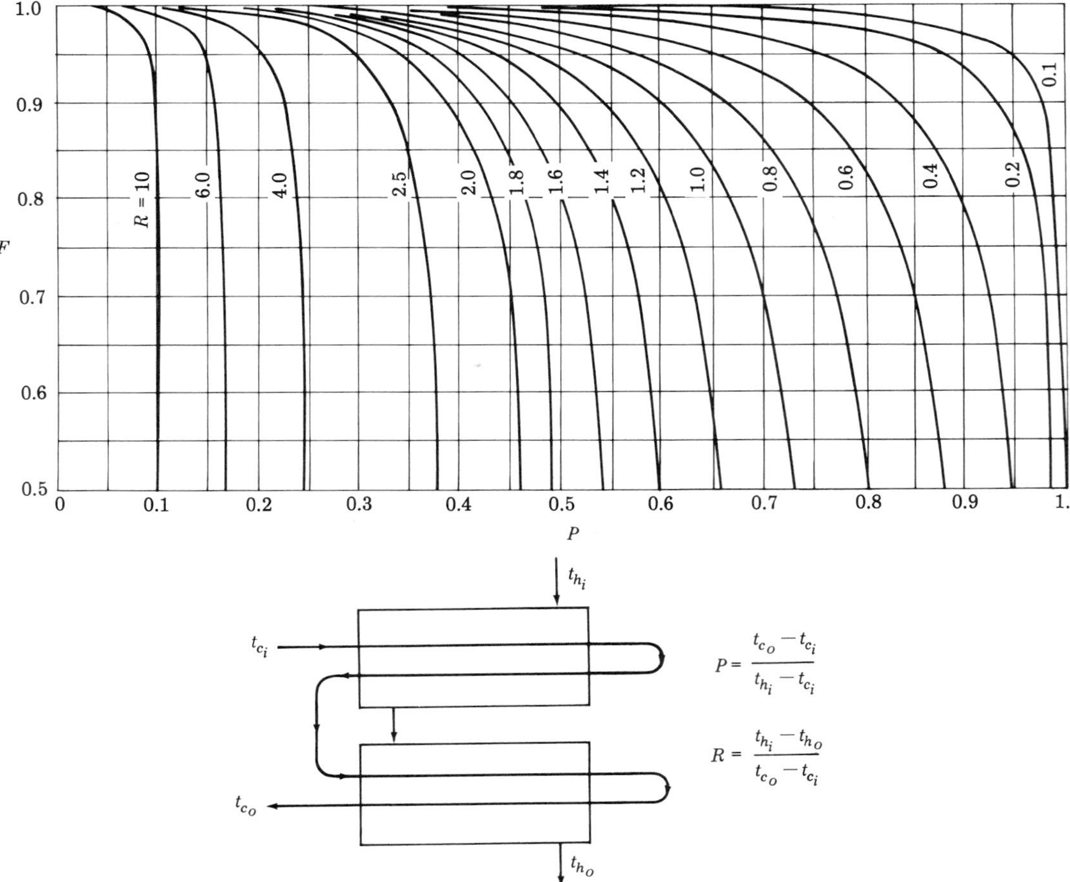

Figure 10.14. Correction factor F for a two-shell-pass, four-tube-pass heat exchanger.

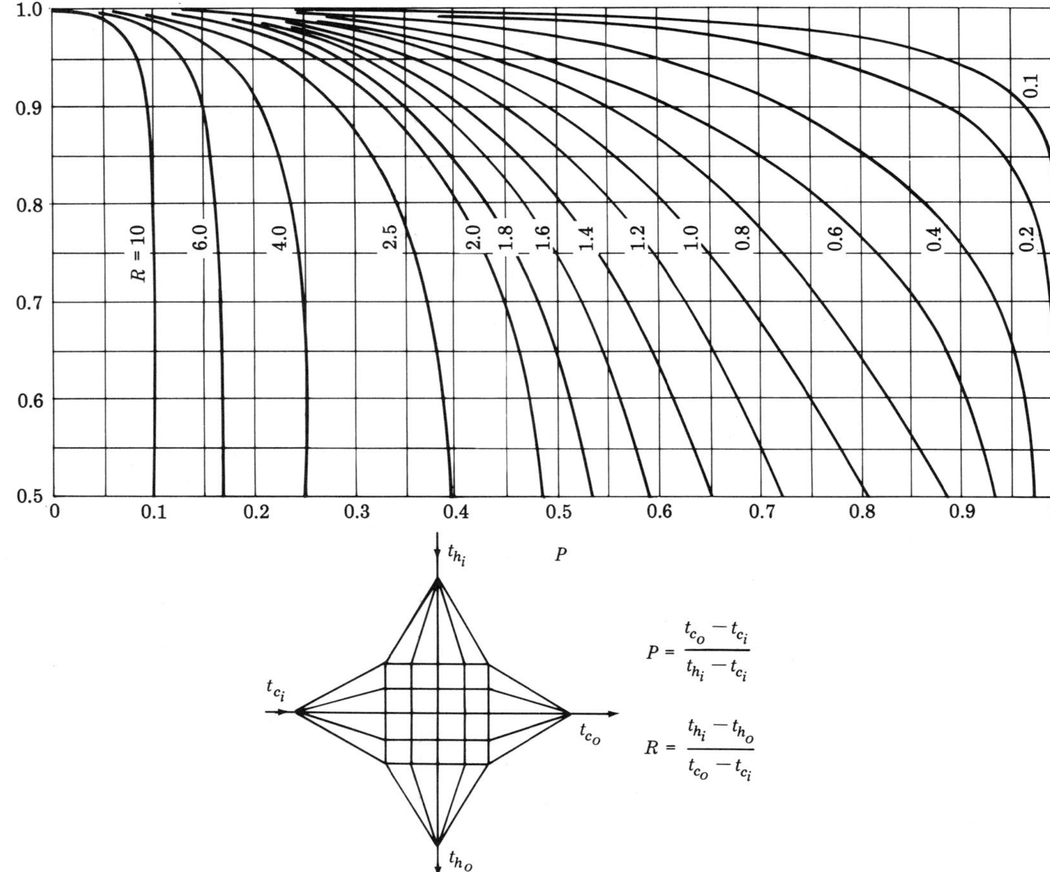

Figure 10.15. Correction factor F for a cross-flow heat exchanger with both fluids unmixed.

applied as well to the one-shell-pass, one-tube-pass exchangers discussed earlier. Clearly, from the definition of F, it is equal to 1 for the simple counterflow exchanger; however, this is not so for the parallel-flow case. One could derive the $F(P, R)$ relation for a simple parallel-flow exchanger (this is left as an exercise in one of the problems at the end of the chapter), but this is usually not done since Δt_m for this case is so easily calculated from Eq. (10.26) or (10.28).

EXAMPLE 10.7

A one-shell-pass, two-tube-pass heat exchanger is used to heat 20,000 lb_m/h of water ($c_p = 1.0$ Btu/lb_m-°F) from 100°F to 250°F by the use of oil ($c_p = 0.667$ Btu/lb_m-°F) flowing at the rate of 15,000 lb_m/h and entering at 600°F. The total exposed surface area for heat exchange is $A = 50 \text{ ft}^2$. Find the rate of heat transfer between the two fluids and the overall heat transfer coefficient for the heat exchanger, based on the exposed surface area.

Solution. The data given include:

$$\dot{m}_c = 20{,}000 \text{ lb}_\text{m}/\text{h}, \quad c_{p_c} = 1.0 \text{ Btu/lb}_\text{m}\text{-°F}, \quad t_{c_i} = 100\text{°F}, \quad t_{c_o} = 250\text{°F},$$

$$\dot{m}_h = 15{,}000 \text{ lb}_\text{m}/\text{h}, \quad c_{p_h} = 0.667 \text{ Btu/lb}_\text{m}\text{-°F}, \quad t_{h_i} = 600\text{°F}.$$

The heat transfer rate between the two fluids and the associated outlet temperature of the hot fluid may be found from the overall exchanger heat balance as given in Eq. (10.2):

$$q = \dot{m}_c c_{p_c}(t_{c_o} - t_{c_i}) = \dot{m}_h c_{p_h}(t_{h_i} - t_{h_o})$$

$$= 20{,}000 \times 1.0 \times (250 - 100) = 15{,}000 \times 0.667 \times (600 - t_{h_o})$$

$$q = 3.0 \times 10^6 \text{ Btu/h } (8.79 \times 10^5 \text{ W}),$$

$$t_{h_o} = 300\text{°F } (149\text{°C}).$$

With the outlet temperature of the hot fluid now known, the mean temperature difference of the heat exchanger may be found. To apply Eq. (10.34) for this purpose, one first needs to find the log mean for counterflow as given by Eq. (10.29):

$$\Delta t_{\text{lm,ctr}} = \frac{(t_{h_i} - t_{c_o}) - (t_{h_o} - t_{c_i})}{\ln\left(\dfrac{t_{h_i} - t_{c_o}}{t_{h_o} - t_{c_i}}\right)}$$

$$= \frac{(600 - 250) - (300 - 100)}{\ln\left(\dfrac{600 - 250}{300 - 100}\right)}$$

$$= 268\text{°F}.$$

This value has to be corrected to account for the one-shell-pass, two-tube-pass flow pattern as required by Eq. (10.34). Thus the capacity ratio and effectiveness defined in Eqs. (10.32) and (10.33) are

$$R = \frac{t_{h_i} - t_{h_o}}{t_{c_o} - t_{c_i}} = \frac{\dot{m}_c c_{pc}}{\dot{m}_h c_{ph}}$$

$$= \frac{600 - 300}{250 - 100} = 2.0,$$

$$P = \frac{t_{c_o} - t_{c_i}}{t_{h_i} - t_{c_i}}$$

$$= \frac{250 - 100}{600 - 100} = 0.30.$$

For the one-shell-pass, two-tube-pass heat exchanger, the correction factor to be applied to $\Delta t_{\text{lm,ctr}}$ is given as a function of P and R, just found, in Fig. 10.13 or, more accurately, by Eq. (10.35):

$$F(P, R) = 0.883.$$

Thus the true mean temperature difference is, by Eq. (10.34),

$$\Delta t_m = F(P, R)\, \Delta t_{\text{lm,ctr}}$$

$$= 0.883 \times 268 = 236.6°\text{F}.$$

The total heat exchange, given by Eq. (10.2) in terms of the fluid heat balance, is also expressed in terms of the mean temperature difference and in terms of the overall heat transfer coefficient according to Eq. (10.20):

$$q = UA\, \Delta t_m.$$

Thus for the given exchanger area $A = 50$ ft^2 and $q = 3.0 \times 10^6$ Btu/h found previously, the required overall heat transfer coefficient of the heat exchanger is

$$3.0 \times 10^6 = U \times 50 \times 236.6$$

$$U = 253.6 \text{ Btu/h-ft}^2\text{-°F } (1440 \text{ W/m}^2\text{-°C}).$$

10.7 HEAT EXCHANGER EFFECTIVENESS AND THE NUMBER OF TRANSFER UNITS

One difficulty associated with the mean-temperature-difference formulation for heat exchanger performance is that Δt_m cannot be found unless all four fluid terminal temperatures (t_{c_i}, t_{h_i}, t_{c_o}, t_{h_o}) are known. In certain design problems these temperatures are known and one may proceed as in Sec. 10.6. However, if one wishes to estimate the performance (i.e., q) of a given exchanger subject to given *inlet* conditions, then the outlet temperatures are not known until q is determined. Thus, an iterative calculation would appear to be necessary.

If the performance equations are recast in the terms of an effectiveness parameter as suggested in Sec. 10.5, this difficulty can be circumvented. While this is not the only reason for use of the effectiveness formulation, it is an important one. The concept of heat exchange "effectiveness" and the associated "number of transfer units" was first suggested by Nusselt and has been developed extensively by Kays and London (Ref. 5).

The effectiveness of a heat exchanger, consistent with the concept introduced in Sec. 10.5, is defined as the actual heat transferred by the exchanger divided by the maximum possible heat that could be transferred, i.e.,

$$\epsilon = \frac{q}{q_{\max}}.$$

The discussion of Sec. 10.3 showed that for an infinite transfer area the most heat that could be transferred is that in counterflow, and further, that this maximum is determined by which fluid stream has the least capacity rate. That is, for given flow rates and fluid inlet temperatures, the maximum heat that could be transferred under any circumstances is, according to Eq. (10.3),

$$q_{\max} = \dot{m}_c c_{p_c}(t_{h_i} - t_{c_i}), \quad \dot{m}_c c_{p_c} < \dot{m}_h c_{p_h},$$

$$q_{\max} = \dot{m}_h c_{p_h}(t_{h_i} - t_{c_i}), \quad \dot{m}_c c_{p_c} > \dot{m}_h c_{p_h}.$$

The actual heat transfer is found from a heat balance on either the cold or the hot stream: $q = \dot{m}_c c_{p_c}(t_{c_o} - t_{c_i}) = \dot{m}_h c_{p_h}(t_{h_i} - t_{h_o})$. To make the effectiveness dependent on only the terminal temperatures, the first form for q is used when $\dot{m}_c c_{p_c} < \dot{m}_h c_{p_h}$ and the second when the reverse is true. Thus the following definition is used for heat exchanger effectiveness:

$$\epsilon = \begin{cases} \dfrac{t_{c_o} - t_{c_i}}{t_{h_i} - t_{c_c}}, & \dot{m}_c c_{p_c} < \dot{m}_h c_{p_h}, \\[2ex] \dfrac{t_{h_i} - t_{h_o}}{t_{h_i} - t_{c_i}}, & \dot{m}_c c_{p_c} > \dot{m}_h c_{p_h}. \end{cases} \tag{10.37}$$

The relative thermal "size" of the two fluid streams is obviously an important parameter, and in order to make it a quantity that is always less than unity, the *capacity ratio* is defined as

$$C_R = \frac{(\dot{m}c_p)_{\min}}{(\dot{m}c_p)_{\max}}. \tag{10.38}$$

The product UA represents the heat exchanging capacity of an exchanger, per degree of temperature difference. This thermal "size" of an exchanger can be nondimensionalized by referring it to the storage capacity of one of the fluid streams. Since the stream with the smaller heat capacity rate is the one which limits the maximum heat transfer, it is used in the nondimensionalization. Thus, similar to the definition made for the elementary case in Sec. 10.5, the *number of transfer units* of a heat exchanger, NTU, is defined:

$$\mathrm{NTU} = \frac{UA}{(\dot{m}c_p)_{\min}}. \tag{10.39}$$

The three parameters just defined (ϵ, C_R, and NTU) do not involve any new quantities not used in the mean-temperature-difference formulation but simply constitute a different combination of the same ones. The utility of the ϵ–C_R-NTU formulation lies in the fact that the effectiveness, ϵ, can be written as an explicit function of C_R and NTU for a given heat exchanger. Both C_R and NTU involve only the fluid flow rates, heat capacities, the heat exchanger size, and the overall coefficient. Thus, the effectiveness, ϵ, can be found without knowing any temperatures, and since ϵ involves only *one* of the outlet temperatures, that temperature may be found explicitly when the inlet temperatures are given.

The following sections give the $\epsilon = \epsilon(C_R, \mathrm{NTU})$ functions for several heat exchanger configurations. The actual determination of this function is discussed for a parallel-flow heat exchanger.

The Parallel-Flow Exchanger

For any heat exchanger one may write the heat transferred in the following three ways:

$$q = UA\,\Delta t_m,$$

$$q = \dot{m}_c c_{p_c}(t_{c_o} - t_{c_i}),$$

$$q = \dot{m}_h c_{p_h}(t_{h_i} - t_{h_o}),$$

or, equivalently,

$$\dot{m}_c c_{p_c}(t_{c_o} - t_{c_i}) = \dot{m}_h c_{p_h}(t_{h_i} - t_{h_o}), \tag{10.40}$$

$$UA\,\Delta t_m = \dot{m}_h c_{p_h}(t_{h_i} - t_{h_o}). \tag{10.41}$$

For the parallel-flow heat exchanger, Eq. (10.28) for Δt_m may be introduced into Eq. (10.41) to yield

$$\ln\left(\frac{t_{h_i} - t_{c_i}}{t_{h_o} - t_{c_o}}\right) = \left[\frac{(t_{h_i} - t_{c_i}) - (t_{h_o} - t_{c_o})}{t_{h_i} - t_{h_o}}\right]\frac{UA}{\dot{m}_h c_{p_h}}$$

$$= \left(1 + \frac{t_{c_o} - t_{c_i}}{t_{h_i} - t_{h_o}}\right)\frac{UA}{\dot{m}_h c_{p_h}}.$$

Equation (10.40) may be introduced to give

$$\ln\left(\frac{t_{h_i} - t_{c_i}}{t_{h_o} - t_{c_o}}\right) = \left(1 + \frac{\dot{m}_h c_{p_h}}{\dot{m}_c c_{p_c}}\right)\frac{UA}{\dot{m}_h c_{p_h}}$$

or

$$\frac{t_{h_o} - t_{c_o}}{t_{h_i} - t_{c_i}} = \exp\left[-\left(\frac{1}{\dot{m}_c c_{p_c}} + \frac{1}{\dot{m}_h c_{p_h}}\right)UA\right]. \tag{10.42}$$

The latter equation may be solved with Eq. (10.40) for either of the fluid outlet temperatures, t_{c_o} or t_{h_o}.

Solving first for t_{c_o}, Eqs. (10.40) and (10.42) yield

$$1 - \left(1 + \frac{\dot{m}_c c_{p_c}}{\dot{m}_h c_{p_h}}\right)\frac{t_{c_o} - t_{c_i}}{t_{h_i} - t_{c_i}} = \exp\left[-\left(1 + \frac{\dot{m}_c c_{p_c}}{\dot{m}_h c_{p_h}}\right)\frac{UA}{\dot{m}_c c_{p_c}}\right].$$

In this form, one sees that if $\dot{m}_c c_{p_c} < \dot{m}_h c_{p_h}$, then the temperature difference ratio on the left side of the equation is the effectiveness defined in Eq. (10.37), $\dot{m}_c c_{p_c}/\dot{m}_h c_{p_h}$ is C_R in Eq. (10.38), and $UA/\dot{m}_c c_{p_c}$ is the NTU of Eq. (10.37). Thus

$$1 - (C_R + 1)\epsilon = \exp\left[-(C_R + 1)\text{NTU}\right]. \tag{10.43}$$

If, on the other hand, Eqs. (10.40) and (10.42) are solved for t_{h_o}, the hot fluid outlet, one has

$$1 - \left(1 + \frac{\dot{m}_h c_{p_h}}{\dot{m}_c c_{p_c}}\right)\frac{t_{h_i} - t_{h_o}}{t_{h_i} - t_{c_i}} = \exp\left[-\left(1 + \frac{\dot{m}_h c_{p_h}}{\dot{m}_c c_{p_c}}\right)\frac{UA}{\dot{m}_h c_{p_h}}\right],$$

so that if $\dot{m}_h c_{p_h} < \dot{m}_c c_{p_c}$ there results

$$1 - (C_R + 1)\epsilon = \exp\left[-(C_R + 1)\text{NTU}\right].$$

The latter expression is identical with Eq. (10.43), which is precisely why the parameters ϵ, C_R, and NTU were defined in the way they were. For either relative size of the hot and cold streams, $\dot{m}_c c_{p_c} < \dot{m}_h c_{p_h}$ or $\dot{m}_c c_{p_c} > \dot{m}_h c_{p_h}$, then

$$\epsilon = \frac{1 - \exp\left[-(C_R + 1)\text{NTU}\right]}{C_R + 1},$$

$$\text{NTU} = \frac{-\ln\left[1 - (C_R + 1)\epsilon\right]}{C_R + 1}. \tag{10.44}$$

Equation (10.44) is the sought-for relation giving ϵ as a function of C_R and NTU. Given the heat exchanger size and overall U, the fluid flow rate and heat capacities, and the fluid inlet temperatures, Eq. (10.44) yields whichever fluid outlet temperature is involved in ϵ—depending on which form of Eq. (10.37) applies. The remaining fluid outlet temperature is then simply obtained from the heat balance of Eq. (10.40). Also quoted in Eq. (10.44) is its inversion, expressing NTU as a function of C_R and ϵ. This form is sometimes useful for design calculations.

Equation (10.44) is easy to evaluate; however, it is also presented graphically in Fig. 10.16. It is useful to note that for the case in which one fluid changes phase, as in a condenser, then $t_{h_o} = t_{h_i} = t_h$ is constant. Thus $\dot{m}_h c_{p_h} \to \infty$ and $C_R \to 0$. In this case the first form for ϵ in Eq. (10.37) applies and

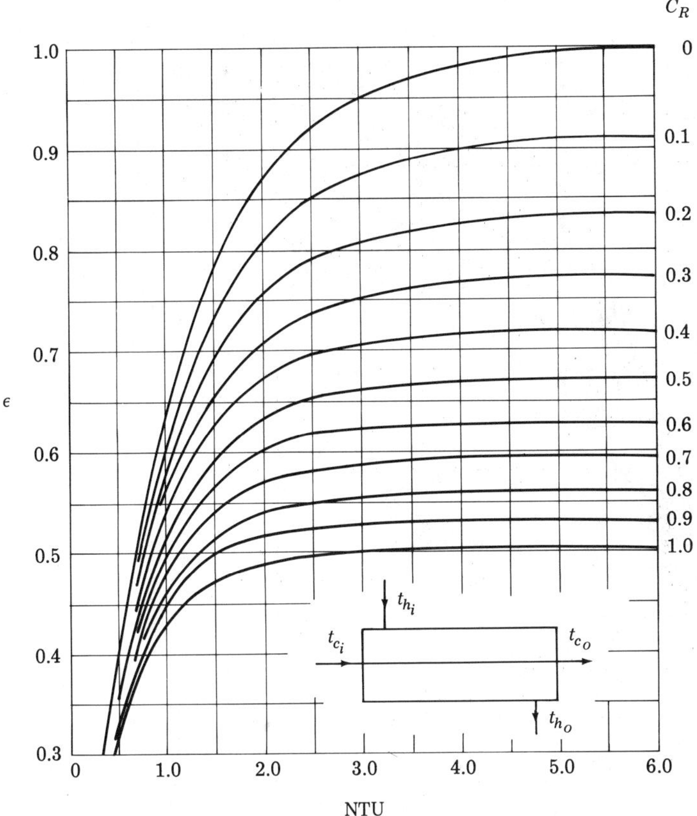

Figure 10.16. Effectiveness-NTU relationship for a parallel-flow heat exchanger.

$$\epsilon = \frac{t_{c_o} - t_{c_i}}{t_{h_i} - t_{c_i}} = 1 - \exp(-\text{NTU}), \qquad (10.45)$$

$$\text{NTU} = -\ln(1 - \epsilon).$$

The Counterflow Exchanger

An analogous derivation may be carried out for the counterflow heat exchanger to show that

$$\epsilon = \frac{1 - \exp[-(1 - C_R)\text{NTU}]}{1 - C_R \exp[-(1 - C_R)\text{NTU}]}, \qquad (10.46)$$

$$\text{NTU} = \frac{\ln[(1 - \epsilon)/(1 - \epsilon C_R)]}{C_R - 1}.$$

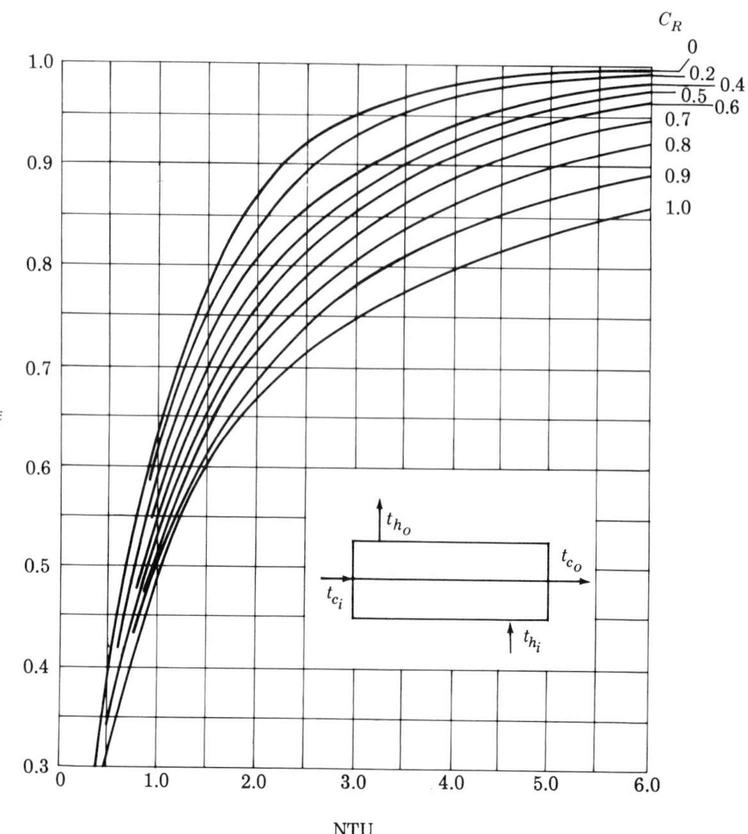

Figure 10.17. Effectiveness-NTU relationship for a counterflow heat exchanger.

Equation (10.46) is shown graphically in Fig. 10.17. In the case of a condenser with $C_R \to 0$, Eq. (10.45) is again obtained. This must be so since the distinction between a parallel-flow and a counterflow heat exchanger does not apply when one fluid remains at a fixed temperature. The form of Eq. (10.45) is, for the same reason, identical to that found in Sec. 10.5, Eq. (10.18), for the elementary isothermal wall case.

Equation (10.46) for the counterflow exchanger becomes indeterminate when the two fluid streams are the same "size" thermally, or as $C_R \to 1$. One may readily show that in this case Eq. (10.46) becomes

$$\epsilon = \frac{\text{NTU}}{1 + \text{NTU}}, \qquad \text{NTU} = \frac{\epsilon}{1 - \epsilon}. \tag{10.47}$$

The One-Shell-Pass, Two-Tube-Pass Exchanger

For the one-shell-pass, two-tube-pass heat exchanger depicted in Fig. 10.6, Ref. 3 shows the effectiveness to be

$$\epsilon = 2\left[(1 + C_R) + (1 + C_R^2)^{1/2}\frac{1 + \exp[-(1 + C_R^2)^{1/2}\text{NTU}]}{1 - \exp[-(1 + C_R^2)^{1/2}\text{NTU}]}\right]^{-1}, \tag{10.48}$$

$$\text{NTU} = -(1 + C_R^2)^{-1/2} \ln\left[\frac{2/\epsilon - 1 - C_R - (1 + C_R^2)^{1/2}}{2/\epsilon - 1 - C_R + (1 + C_R^2)^{1/2}}\right].$$

Equation (10.48) is awkward to use, so the graphical display in Fig. 10.18 may be used.

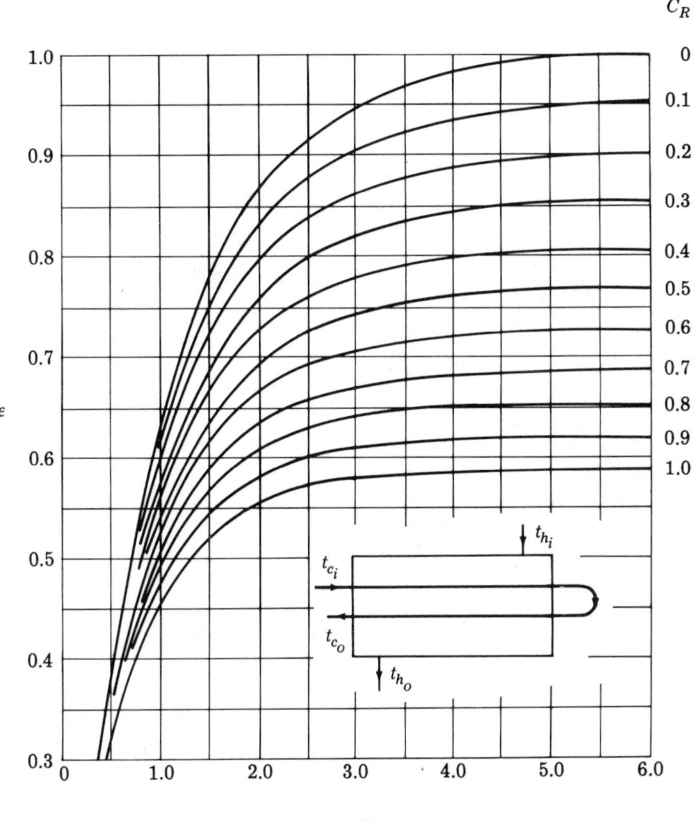

Figure 10.18. Effectiveness-NTU relationship for a one-shell-pass, two-tube-pass heat exchanger.

Other Heat Exchanger Configurations

Relations and graphical plots of the $\epsilon = \epsilon(C_R, \text{NTU})$ function for a number of other heat exchanger configurations, particularly multiple pass cases for both shell-and-tube and cross-flow exchangers have been determined and reported in the literature. Reference 5 should be consulted for these results.

Figure 10.19 presents the results for a heat exchanger with two shell passes and two tube passes per shell pass.

Determination of the ϵ–C_RNTU relation in cross flow is complex since the fluid temperatures are dependent on two spatial variables. An analytic result for the case in which both fluids are unmixed may be found in Ref. 6; however, this result is a complicated double infinite series. The results of this analysis are displayed in Fig. 10.20. An approximation that may be used when C_R is close to unity is:

$$\epsilon \cong 1 - \exp\{C_R \text{NTU}^{0.22}[\exp(-C_R \text{NTU}^{0.78}) - 1]\}.$$

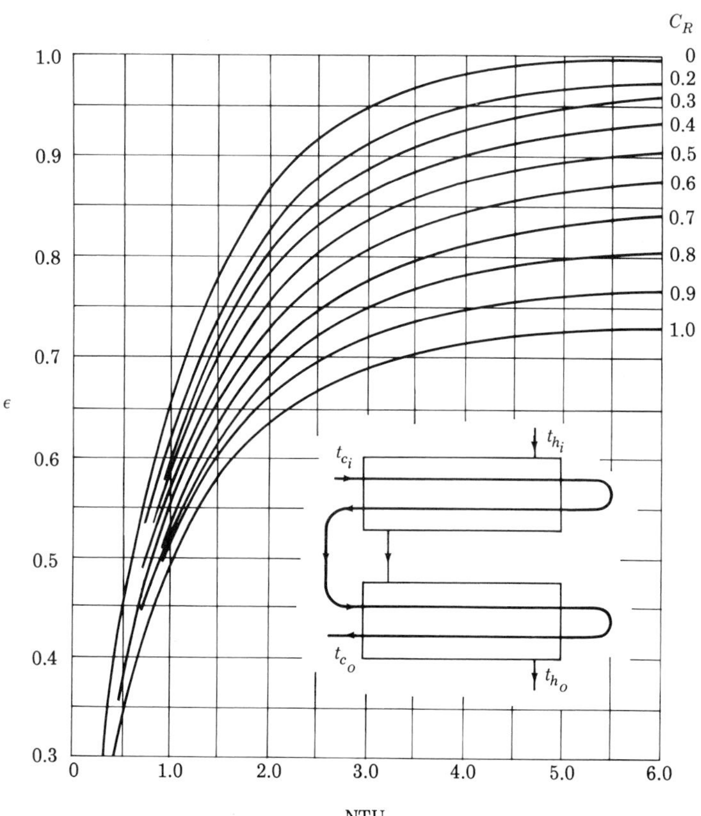

Figure 10.19. Effectiveness-NTU relationship for a two-shell-pass, four-tube-pass heat exchanger.

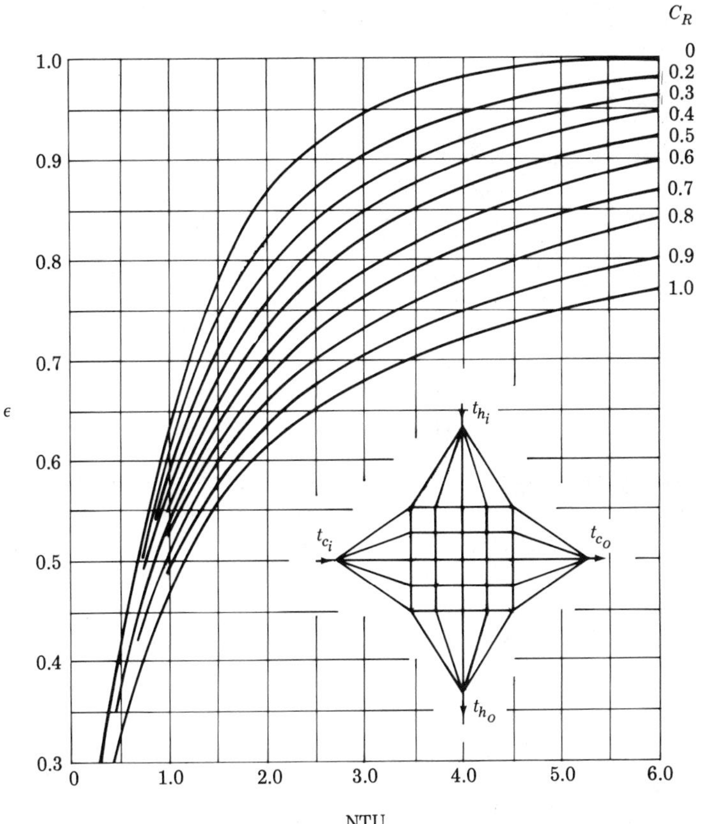

Figure 10.20. Effectiveness-NTU relationship for a cross-flow heat exchanger with both fluids unmixed.

EXAMPLE 10.8

Repeat Example 10.7 using the ϵ–C_R–NTU formulation rather than the mean temperature difference.

Solution. The data given in Example 10.7 and the calculated hot fluid outlet temperature were

$$\dot{m}_c = 20{,}000 \text{ lb}_\text{m}/\text{h}, \quad c_{p_c} = 1.0 \text{ Btu/lb}_\text{m}\text{-°F}, \quad t_{c_i} = 100\text{°F}, \quad t_{c_o} = 250\text{°F},$$

$$\dot{m}_h = 15{,}000 \text{ lb}_\text{m}/\text{h}, \quad c_{p_h} = 0.667 \text{ Btu/lb}_\text{m}\text{-°F}, \quad t_{h_i} = 600\text{°F}, \quad t_{h_o} = 300\text{°F}.$$

For these data the capacity rates of the two fluid streams are

$$\dot{m}_c c_{p_c} = 20{,}000 \times 1.0 = 20{,}000 \text{ Btu/h-°F} = (\dot{m}c_p)_{\text{max}},$$

$$\dot{m}_h c_{p_h} = 15{,}000 \times 0.667 = 10{,}000 \text{ Btu/h-°F} = (\dot{m}_p)_{\text{min}}.$$

The hot fluid clearly has the minimum capacity rate and thus is the stream that limits the heat exchanger performance. The capacity ratio, from Eq. (10.38), is then

$$C_R = \frac{(\dot{m}c_p)_{\min}}{(\dot{m}c_p)_{\max}} = \frac{10{,}000}{20{,}000} = 0.50.$$

Since $\dot{m}_c c_{p_c} > \dot{m}_h c_{p_h}$ in this instance, the defining relation for the effectiveness is the second of the two given in Eq. (10.37):

$$\epsilon = \frac{t_{h_i} - t_{h_o}}{t_{h_i} - t_{c_i}}$$

$$= \frac{600 - 300}{600 - 100} = 0.60.$$

With the capacity ratio, C_R, and effectiveness, ϵ, known, the number of transfer units (i.e., the dimensionless value of the *UA* product) is given by the appropriate ϵ–C_R–NTU relation. In this case of a one-shell-pass, two-tube-pass exchanger, one may use Fig. 10.18; however, Eq. (10.48) gives the result more accurately:

$$\text{NTU} = -(1 + C_R^2)^{-1/2} \ln \left[\frac{2/\epsilon - 1 - C_R - (1 + C_R^2)^{1/2}}{2/\epsilon - 1 - C_R + (1 + C_R^2)^{1/2}}\right]$$

$$(1 + C_R^2)^{1/2} = (1 + 0.5^2)^{1/2} = 1.118$$

$$\text{NTU} = -\frac{1}{1.118} \ln \left[\frac{2/0.6 - 1 - 0.5 - 1.118}{2/0.6 - 1 - 0.5 + 1.118}\right]$$

$$= 1.267.$$

The number of transfer units, NTU, yields the *UA* product from its definition in Eq. (10.39):

$$\text{NTU} = \frac{UA}{(\dot{m}c_p)_{\min}}.$$

In this case the minimum capacity rate is that of the hot fluid: $(\dot{m}c_p)_{\min} = \dot{m}_h c_{p_h} = 10{,}000$ Btu/h-°F. Thus

$$1.267 = \frac{UA}{10{,}000}$$

$$UA = 12{,}670 \text{ Btu/h-°F}.$$

Since the surface area was given as $A = 50$ ft^2, the overall U is

$$U = \frac{12{,}670}{50} = 253.6 \text{ Btu/h-ft}^2\text{-°F } (1440 \text{ W/m}^2\text{-°C}),$$

as in Example 10.7.

10.8 CALCULATION OF THE PERFORMANCE OF A GIVEN HEAT EXCHANGER

Two general types of problems are normally encountered in dealing with heat exchanger calculations. The first consists of the determination of the *performance* of a given heat exchanger. That is, one wishes to find the heat exchanged between two fluid streams (and their outlet temperatures) when given the complete physical description of the exchanger and the inlet conditions of the streams. The *design* problem consists of determining the physical description of a heat exchanger that will accomplish a desired amount of heat transfer between the two given streams. Clearly, the design problem is more complex than the performance problem and has no unique answer. Most design procedures consist of making a great number of performance calculations; hence the performance problem will be discussed first.

The performance problem, then, presumes that the complete physical description of the heat exchanger is known. That is, one knows the flow arrangement (i.e., whether it is a cross-flow exchanger, a shell-and-tube exchanger, etc.; how many passes of each fluid there are; whether it is parallel-flow, counterflow, etc.) and one also knows the physical dimensions involved (i.e., tube size and length, the size of the separating wall and any fins if cross-flow, etc.). The question then is: Given the above and given the inlet fluid conditions (i.e., $\dot{m}_c$, c_{p_c}, t_{c_i}, $\dot{m}_h$, c_{p_h}, t_{h_i}), what are the rate of heat transfer and the fluid outlet temperatures (i.e., q, t_{c_o}, t_{h_o})? Three relations are needed to find these three performance quantities. Two relations come immediately from the heat exchanger heat balance:

$$\begin{aligned} q &= \dot{m}_c c_{p_c}(t_{c_o} - t_{c_i}), \\ q &= \dot{m}_h c_{p_h}(t_{h_i} - t_{h_o}). \end{aligned} \tag{10.49}$$

The third relation to be used to complete the problem depends on whether the mean temperature difference or the effectiveness formulation is to be used. In terms of the mean temperature difference,

$$q = UA\ \Delta t_m, \tag{10.50}$$

in which Δt_m depends on the four terminal fluid temperatures and the fluid flow arrangement. In terms of the effectiveness one has

$$\epsilon = \epsilon(C_R, \text{NTU}), \tag{10.51}$$

in which the form of the effectiveness function depends on the flow arrangement and the definitions in Eqs. (10.37) through (10.39).

In either formulation one needs the *UA* product for the exchanger. The surface area is known from the given geometry. However, in general the overall U depends on the fluid temperatures and the hydrodynamic conditions on each side of the exchanger. Thus the determination of U has to be done, iteratively perhaps, simultaneous with the determination of the fluid outlet temperatures. In order to separate the question of the determination of t_{c_o} and t_{h_o} and the determination of

U, a special subcase of the performance problem will be discussed first in which it is presumed U is known. Such knowledge might be the result of extensive experience or it might be the result of some previous calculation in a longer iterative scheme to be described later.

Performance Calculation—Known U

If one presumes U to be known, then the only unknowns of the performance problem are q, t_{c_o}, and t_{h_o}. Once any one of these three are known, Eqs. (10.49) readily yield the other two. In the use of the mean-temperature-difference formulation, the Δt_m is a function (often complex) of the *known* t_{c_i}, t_{h_i}, and the unknown t_{c_o}, t_{h_o}. Thus an iterative calculation would be necessary. However, the definition of the effectiveness in Eq. (10.37) involves only one of the unknown outlets and since ϵ can be expressed as a function of C_R and NTU, one may solve directly for t_{c_o} or t_{h_o} without iteration. Both approaches are illustrated in the following example for the sake of completeness. Clearly, the effectiveness method would be preferred; however, the advantage of this method vanishes in the general problem when U is not known.

EXAMPLE 10.9

A one-shell-pass, two-tube-pass heat exchanger has a total surface area of $A = 5$ m^2, and its overall heat transfer coefficient is known to be $U = 1400$ W/m^2-°C. The hot fluid is water entering the shell side at 315°C with a flow rate of 4500 kg/h. The cold fluid is also water, entering the tubes at 40°C with a flow rate of 9000 kg/h. Find the fluid outlet temperatures and the rate of heat transfer between the two streams using (a) the mean temperature difference formulation; (b) the ϵ–C_R–NTU formulation. For simplicity of calculation ignore the dependence of the water c_p on temperature and use $c_p = 4.187$ kJ/kg-°C for both streams.

Solution. For the data stated one has

$$A = 5 \text{ m}^2, \qquad U = 1400 \text{ W/m}^2\text{-°C},$$

and the two streams inlet conditions are

$$\dot{m}_c = 9000 \text{ kg/h}, \qquad c_{p_c} = 4.187 \text{ kJ/kg-°C}, \qquad t_{c_i} = 40\text{°C},$$

$$\dot{m}_h = 4500 \text{ kg/h}, \qquad c_{p_h} = 4.187 \text{ kJ/kg-°C}, \qquad t_{h_i} = 315\text{°C}.$$

Thus the capacity rates are calculated as noted below, and the hot stream is seen to have the smaller value:

$$\dot{m}_c c_{p_c} = 9000 \times 4.187 \times \frac{1000}{3600} = 10{,}468 \text{ W/°C} = (\dot{m}c_p)_{\max},$$

$$\dot{m}_h c_{ph} = 4500 \times 4.187 \times \frac{1000}{3600} = 5234 \text{ W/°C} = (\dot{m}c_p)_{\min}.$$

(a) To find the outlet fluid temperatures using the mean temperature difference formulation, one equates the heat balances of Eq. (10.49) to Eq. (10.50):

$$q = \dot{m}_c c_{pc}(t_{c_o} - t_{c_i}) = \dot{m}_h c_{ph}(t_{h_i} - t_{h_o}) = UA\ \Delta t_m.$$

Since the capacity rates, the inlet temperatures, and the UA product are known, the fluid outlets, t_{h_o} and t_{c_o} can be found if the mean temperature difference, Δt_m, can be determined. However, Δt_m depends on t_{h_o} and t_{c_o} (as well as t_{h_i} and t_{c_i}), so that an iterative calculation is required.

As a first assumption, take $t_{h_o} = 175$°C. Then Eq. (10.49) gives the corresponding cold fluid outlet temperature:

$$q = \dot{m}_c c_{pc}(t_{c_o} - t_{c_i}) = \dot{m}_h c_{ph}(t_{h_i} - t_{h_o})$$

$$10{,}468(t_{c_o} - 40) = 5234(315 - 175)$$

$$t_{c_o} = 110\text{°C}.$$

The mean temperature difference depends on the log mean for counterflow and the function $F(P, R)$. Equations (10.29), (10.32), and (10.33) give

$$\Delta t_{\text{lm,ctr}} = \frac{(t_{h_i} - t_{c_o}) - (t_{h_o} - t_{c_i})}{\ln\left(\dfrac{t_{h_i} - t_{c_o}}{t_{h_o} - t_{c_i}}\right)}$$

$$= \frac{(315 - 110) - (175 - 40)}{\ln\left(\dfrac{315 - 110}{175 - 40}\right)} = 167.6\text{°C},$$

$$R = \frac{\dot{m}_c c_{pc}}{\dot{m}_h c_{ph}} = \frac{10{,}468}{5234} = 2.0,$$

$$P = \frac{t_{c_o} - t_{c_i}}{t_{h_i} - t_{c_i}} = \frac{110 - 40}{315 - 40} = 0.255.$$

For a one-shell-pass, two-tube-pass exchanger the correction factor to be applied to $\Delta t_{\text{lm,ctr}}$ is found from Fig. 10.13 or, more accurately, Eq. 10.35:

$$F(P, R) = 0.938.$$

Thus the mean temperature difference, by Eq. (10.34), is

$$\Delta t_m = F(P, R)\ \Delta t_{\text{lm,ctr}}$$

$$= 0.938 \times 167.6 = 157.2\text{°C}.$$

By equating Eqs. (10.49 and (10.50) one may find a revised value of the hot fluid outlet:

$$q = UA\,\Delta t_m = \dot{m}_h c_{p_h}(t_{h_i} - t_{h_o})$$

$$1400 \times 5 \times 157.2 = 5234(315 - t_{h_o})$$

$$t_{h_o} = 104.8°\text{C}.$$

This calculated value of t_{h_o} is considerably less than the assumed value of 175°C, so another calculation must be performed with a revised estimate of t_{h_o}. Following the same procedure as above, the following final, verified results are obtained:

$$t_{h_o} = 146.5°\text{C, assumed},$$

$$t_{c_o} = 124.3°\text{C},$$

$$\Delta t_{\text{lm,ctr}} = 144.5°\text{C},$$

$$R = 2.0, \qquad P = 0.306, \qquad F(P, R) = 0.872,$$

$$\Delta t_m = 126.0°\text{C},$$

$$t_{h_o} = 146.5°\text{C, calculated}.$$

Thus verified, the desired outlet temperatures are

$$t_{h_o} = 146.5°\text{C } (295.7°\text{F}),$$

$$t_{c_o} = 124.3°\text{C } (255.7°\text{F}),$$

and the heat transfer rate is

$$q = \dot{m}_h c_{p_h}(t_{h_i} - t_{h_o})$$

$$= 5234(315 - 146.5) \times 10^{-3} = 881.0\text{ kW } (3.01 \times 10^6\text{ Btu/h}).$$

(b) The use of the ϵ–C_R–NTU formulation requires no iteration since the effectiveness involves only one of the unknown outlet temperatures. From the data given, the capacity ratio, C_R, and the number of transfer units, NTU, are readily calculated from their definitions in Eqs. (10.38) and (10.39):

$$C_R = \frac{(\dot{m}c_p)_{\min}}{(\dot{m}c_p)_{\max}} = \frac{5234}{10{,}468} = 0.50,$$

$$\text{NTU} = \frac{UA}{(\dot{m}c_p)_{\min}} = \frac{1400 \times 5}{5234} = 1.337.$$

Figure 10.18 or, more accurately, Eq. (10.48) gives the effectiveness for a one-shell-pass, two-tube-pass exchanger to be

$$\epsilon = 2\left[(1 + C_R) + (1 + C_R^2)^{1/2}\frac{1 + \exp[-(1 + C_R^2)^{1/2}\text{NTU}]}{1 - \exp[-(1 + C_R^2)^{1/2}\text{NTU}]}\right]^{-1}$$

$$(1 + C_R^2)^{1/2} = (1 + 0.5^2)^{1/2} = 1.118$$

$$\epsilon = 2\left[(1 + 0.5) + 1.118\frac{1 + \exp[-1.118 \times 1.337]}{1 - \exp[-1.118 \times 1.337]}\right]^{-1} = 0.6127.$$

Since $\dot{m}_c c_{pc} > \dot{m}_h c_{ph}$, the second definition of the effectiveness in Eq. (10.37) applies, so the hot fluid outlet may be calculated directly:

$$\epsilon = \frac{t_{h_i} - t_{h_o}}{t_{h_i} - t_{c_i}}$$

$$0.6127 = \frac{315 - t_{h_o}}{315 - 40}$$

$$t_{h_o} = 146.5°\text{C } (295.7°\text{F}).$$

The cold fluid outlet and the heat transfer rate are then found from the heat balance of Eq. (10.49):

$$q = \dot{m}_c c_{pc}(t_{c_o} - t_{c_i}) = \dot{m}_h c_{ph}(t_{h_i} - t_{h_o})$$

$$= 10{,}468(t_{c_o} - 40) = 5234(315 - 146.5)$$

$$t_{c_o} = 124.3°\text{C } (255.7°\text{F})$$

$$q = 881.9 \text{ kW } (3.01 \times 10^6 \text{ Btu/h}).$$

In this instance of known U the ϵ–C_R–NTU formulation is clearly preferred since no iterative calculation is required; however, as noted earlier, when U is not known a priori, this advantage disappears as the next example will show.

In this example it was presumed that the c_p of the fluids was known at the outset. If the fluids have c_p's that are strongly dependent on temperature, it may be necessary to assume representative values, find the resulting fluid outlet temperatures, and then revise the chosen c_p values based on the average fluid temperatures.

Performance Calculation—Unknown *U*

As noted earlier, the overall coefficient, U, is generally not known. In reality, U depends on the temperatures of the two fluids. As shown in Examples 10.1 and 10.2, an iterative calculation for U is often required even when the two fluid temperatures are known. In addition, in the case of a heat exchanger only the *inlet* temperatures of the fluids are known. For a parallel-flow heat exchanger one might estimate U at one end of the exchanger; however, if one is dealing with, say, a counterflow exchanger or if, as is more often the case, one wishes to find a representative average U based on the *mean* temperatures of the two fluids, one needs knowledge of the fluid *outlet* temperature before U can be found. Thus a doubly iterative calculation must be made in which fluid outlet temperatures are assumed, an overall U calculated (iteratively, perhaps) based on these assumptions, the outlet temperatures calculated by either of the methods of Example 10.9, and then these results are compared with the initial assumptions and revised accordingly. Since assumptions must be made for t_{c_o} and t_{h_o} to find U, there is no particular advantage of the ϵ–NTU method over the Δt_m method.

Even with some knowledge of a representative mean temperature for each fluid stream, determination of the overall U involves finding the heat transfer coefficients between each stream and the separating wall. For shell-and-tube exchangers, the relations of Chapter 6 can be used to find the h to be applied at the inside tube surface. The method of estimating h on the outside tube surface depends on the mechanism taking place there (forced convection, condensing, etc.) and many of the relations of Chapter 6 or 8 may be applied. If the tubes are equipped with fins, or if one is dealing with a cross-flow heat exchanger which almost invariably involves a complicated finned surface, then some knowledge of the heat transfer correlation applicable for the particular geometry is necessary. In order to establish the method an example will first be shown in which the correlations of Chapters 6 and 8 are applied. Then typical data for more complex surface geometries and an associated second example will be presented.

EXAMPLE 10.10

A steam condenser consists of a one-shell-pass, two-tube-pass heat exchanger. Saturated steam at a pressure of 15 kN/m^2 (approximately 4.5 in. of mercury) is to be condensed on the outside of the tubes by cooling water flowing through the tubes. The tubes are $\frac{5}{8}$-in. 18-gage brass ($k = 110$ W/m-°C) tubes, 1.85 m long, and there are no fins at either tube surface. There are 129 tubes in each of the two tube passes and the water enters the tubes with an inlet temperature of 20°C. The average water velocity in the tubes is to be 1.4 m/s. Estimate the average overall heat transfer coefficient for the condenser, the outlet temperature of the cooling water, and the mass rate at which steam is condensed. Presume that the tubes of the condenser are in the horizontal position.

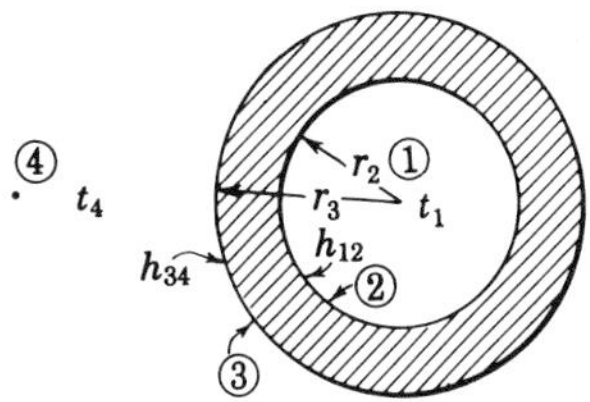

Figure 10.21

Solution. Since the physical geometry of the condenser is completely specified, certain important geometric parameters may be calculated. The notation to be used in reference to the tube geometry is shown in Fig. 10.21. For a $\frac{5}{8}$-in. 18-gage tube, Table B.2 gives

$$D_2 = 1.339 \text{ cm},$$

$$D_3 = 1.588 \text{ cm},$$

$$\text{surface area (per unit tube length)} = 0.04989 \text{ m}^2/\text{m},$$

$$\text{cross-sectional area of a tube} = 0.0001408 \text{ m}^2.$$

The total cross-sectional area available for cooling water flow is that presented in each pass of 129 tubes:

$$A_x = 0.0001408 \times 129 = 0.0182 \text{ m}^2.$$

The total exposed surface area for heat transfer between the cooling water and the condensing steam is, however, that provided by *both* tube passes, 1.85 m long each:

$$A = 0.04989 \times 1.85 \times 129 \times 2 = 23.81 \text{ m}^2.$$

At the condensing steam pressure of 15 kN/m^2, the Steam Tables give the saturation temperature and latent heat to be

$$t_{\text{sat}} = 53.97 \approx 54°\text{C},$$

$$h_{\text{fg}} = 2373.1 \text{ kJ/kg}.$$

These data will be needed in later calculations.

The inlet temperature of the cold, tube-side fluid is given as $t_{c_i} = 20°\text{C}$. The outlet temperature of the cooling water, t_{c_o}, is one of the desired answers; however, it is also needed in order to estimate the overall heat transfer coefficient for the condenser. To estimate the heat transfer coefficient at the inside tube surface (and later to find an average tube surface temperature for the determination of the outside condensing heat transfer coefficient), one must assume an outlet cold fluid temperature, t_{c_o}, and later verify or revise this assumption. Likewise, a trial value of the outlet hot fluid temperature would generally be needed also, and could be found from the assumed t_{c_o} and an overall heat balance on the exchanger. In the present special case of a condenser, however, it is known that the hot fluid is of constant temperature, so such a calculation need not be made and one takes $t_4 = t_{\text{sat}} = t_{h_i} = t_{h_o} = 54°\text{C}$. Based, then, on an average cold fluid temperature (obtained from the given t_{c_i} and the assumed t_{c_o}) and an average hot fluid temperature (known in this case), one calculates the heat transfer coefficients at both tube surfaces, finds the corresponding overall U, and then verifies (or modifies) the assumed t_{c_o} according to the methods of Example 10.9. Thus one proceeds as follows:

1. *Assume that* $t_{c_o} = 35°C$. This assumption is based on the experience that cooling water temperature increases in steam condensers are typically of the order of 15 to 25°C.

   (a) *Determine the inside tube coefficient* h_{12}. This coefficient is based, in part, on the average cold fluid temperature: $t_1 = (t_{c_o} + t_{c_i})/2 = (35 + 20)/2 = 27.5°C$. Subsequent calculations reveal (as in Example 10.2) that the inside tube surface temperature is such that the conditions of Eq. (6.31) are not met and one must use Eqs. (6.32) or (6.33) which involve knowledge of t_2. However, subsequent calculations also show that the temperature drop through the tube wall is negligibly small so that one may take $t_2 \approx t_3$. Thus one need not perform iterative calculations for both t_2 and t_3 as was done in Example 10.2. So, anticipating the results above, and showing the *last* iteration only, *assume* that $t_2 = 46°C$. With $t_1 = 27.5°C$, as above, then Table A.3 gives $\mu = 0.8443 \times 10^{-3}$ kg/m-s, $\mu_s = 0.5864 \times 10^{-3}$ kg/m-s, $\nu = 0.8473 \times 10^{-6}$ m²/s, $k = 0.6113$ W/m-°C, Pr = 5.78. Thus Eq. (6.33) gives

$$\mathrm{Re}_D = \frac{U_m D_2}{\nu} = \frac{1.4 \times 0.01339}{0.8473 \times 10^{-6}} = 2.212 \times 10^4,$$

$$\frac{f}{8} = \tfrac{1}{8}(1.82 \log_{10} \mathrm{Re}_D - 1.64)^{-2}$$

$$= \tfrac{1}{8}[1.82 \log_{10} (2.212 \times 10^4) - 1.64]^{-2} = 0.00318,$$

$$\mathrm{Nu}_D = \frac{(f/8)\mathrm{Re}_D \mathrm{Pr}}{1.07 + 12.7\sqrt{f/8}(\mathrm{Pr}^{2/3} - 1)}\left(\frac{\mu}{\mu_s}\right)^{0.11}$$

$$= \frac{0.00318 \times 2.212 \times 10^4 \times 5.78}{1.07 + 12.7\sqrt{0.00318}(5.78^{2/3} - 1)}\left(\frac{0.8443}{0.5864}\right)^{0.11}$$

$$= 159.2,$$

$$h_{12} = \mathrm{Nu}_D \frac{k}{D_2} = 159.2 \times \frac{0.6113}{0.01339} = 7267.0 \text{ W/m}^2\text{-°C}.$$

   (b) *Determine* h_{34}. The heat transfer coefficient on the outside tube surface, h_{34}, is that due to the steam condensation there in accord with the relations developed in Sec. 8.4. Consistent with the foregoing assumption of $t_2 = 46°C$, assume that $t_3 = 46°C$. Thus with $t_4 = 54°C$, $\Delta t = 54 - 46 = 8°C$, $t_m = (54 + 46)/2 = 50°C$, Table A.3 gives $c_{pl} = 4.182$ kJ/kg-°C, $\rho_l = 988.0$ kg/m³, $\mu_l = 0.5471 \times 10^{-3}$ kg/m-s, $k_l = 0.6405$ W/m-°C. Thus Eq. (8.20) gives

$$\mathrm{Ja} = \frac{c_{pl}\,\Delta t}{h_{fg}} = \frac{4.182 \times 8}{2373.1} = 0.0141,$$

$$h'_{fg} = h_{fg}(1 + 0.68\mathrm{Ja})$$

$$= 2373.1(1 + 0.68 \times 0.141) = 2396 \text{ kJ/kg},$$

$$\text{NU}_D = 0.728\left[\frac{g\rho_l^2 h'_{fg} D^3}{\mu_l k_l \, \Delta t}\right]^{1/4}$$

$$= 0.728\left[\frac{9.8 \times (988.0)^2 \times 2396 \times 10^3 \times (0.01588)^3}{0.5471 \times 10^{-3} \times 0.6405 \times 8}\right]^{1/4}$$

$$= 309.7,$$

$$h_{34} = \text{Nu}_D \frac{k}{D_3} = 309.7 \times \frac{0.6405}{0.01588} = 12{,}490 \text{ W/m}^2\text{-°C}.$$

(c) *Determine* U. For the geometry in question, the overall U is

$$U = \left[\frac{r_3}{r_2 h_{12}} + \frac{r_3 \ln (r_3/r_2)}{k_{23}} + \frac{1}{h_{34}}\right]^{-1}$$

$$= \left[\frac{0.01588}{0.01339 \times 7267} + \frac{0.01588 \ln (0.01588/0.01339)}{2 \times 110} + \frac{1}{12{,}490}\right]^{-1},$$

$$= 3913 \text{ W/m}^2\text{-°C}.$$

(d) *Verify* t_2 *and* t_3. The trial values of t_2 and t_3 may now be checked:

$$q/\pi L = D_2 h_{12}(t_1 - t_2) = D_3 \, h_{34}(t_3 - t_4) = D_3 \, U(t_1 - t_4),$$

$$0.01339 \times 7267(27.5 - t_2) = 0.01588 \times 12{,}490(t_3 - 54)$$

$$= 0.01588 \times 3913(27.5 - 54),$$

$$t_2 = 44.4\text{°C}, \qquad t_3 = 45.7\text{°C}.$$

These values are sufficiently close to the assumed value of 46°C and verify the presumption that the tube wall temperature drop is negligible. Had these computed values differed significantly from the assumed, steps (a), (b), and (c) should be repeated. Only the last iteration is shown here.

2. *Verify the assumed* t_{co}. The verification of the assumed cold fluid outlet may be made using either the Δt_m or ϵ method. Both are shown for sake of completeness.

Based on the cold fluid average temperature, $t_1 = 27.5$°C, $\rho = 996.4$ kg/m^3, $c_{pc} = 4.180$ kJ/kg-°C. Thus

$$\dot{m}_c = A_x \rho v = 0.0182 \times 996.4 \times 1.4 = 25.39 \text{ kg/s}$$

$$\dot{m}_c c_{pc} = 25.39 \times 4.180 = 106.1 \text{ kW/°C}.$$

For application of the mean-temperature-difference formulation to find the total heat transferred there is no need to find the capacity ratio R and the effectiveness P since the hot fluid is of constant temperature. In this instance the factor F is 1.0 and the mean temperature difference is the log mean as given by Eq. (10.29):

$$\Delta t_m = \Delta t_{\text{lm,ctr}} = \frac{(t_h - t_{c_i}) - (t_h - t_{c_o})}{\ln\left(\dfrac{t_h - t_{c_i}}{t_h - t_{c_o}}\right)}.$$

Based on the given $t_{c_i} = 20°\text{C}$, $t_h = 54°\text{C}$ and the assumed $t_{c_o} = 35°\text{C}$, one has

$$\Delta t_m = \frac{(54 - 20) - (54 - 35)}{\ln\left(\dfrac{54 - 20}{54 - 35}\right)} = 25.78°\text{C},$$

so that the heat transfer rate is

$$q = UA\ \Delta t_m.$$

The total transfer area was found to be $A = 23.81\ \text{m}^2$ at the outset of this example. With $U = 3913\ \text{W/m}^2\text{-}°\text{C}$ and $\Delta t_m = 25.78°\text{C}$, based on the assumed $t_{c_o} = 35°\text{C}$,

$$\begin{aligned} q &= 3913 \times 23.81 \times 25.78 \times 10^{-3} \\ &= 2402\ \text{kW}. \end{aligned}$$

A heat balance on the cold fluid stream, using the $\dot{m}_c c_{p_c}$ just found above, yields a revised value for t_{c_o}:

$$\begin{aligned} q &= \dot{m}_c c_{p_c}(t_{c_o} - t_{c_i}) \\ 2402 &= 106.1(t_{c_o} - 20) \\ t_{c_o} &= 42.6°\text{C}. \end{aligned}$$

If, on the other hand, one wishes to use the ϵ–NTU method to revise t_{c_o}, it should be noted that since the hot fluid is changing phase at constant temperature,

$$\dot{m}_h c_{p_h} \to \infty = (\dot{m}c_p)_{\max},$$

$$\dot{m}_c c_{p_c} = 106.1\ \text{kW/°C} = (\dot{m}c_p)_{\min}.$$

Thus the capacity rate and NTU are

$$C_R = \frac{(\dot{m}c_p)_{\min}}{(\dot{m}c_p)_{\max}} = 0,$$

$$\text{NTU} = \frac{UA}{(\dot{m}c_p)_{\min}} = \frac{3913 \times 23.81}{106.1 \times 1000} = 0.878.$$

Thus the effectiveness is given by any of Eq. (10.44), (10.45), or (10.46):

$$\epsilon = 1 - \exp(-\text{NTU})$$
$$= 1 - \exp(-0.878) = 0.584.$$

Then from Eq. (10.37), since $\dot{m}_c c_{p_c} < \dot{m}_h c_{p_h}$,

$$\epsilon = \frac{t_{c_o} - t_{c_i}}{t_{h_i} - t_{c_i}}$$

$$0.584 = \frac{t_{c_o} - 20}{54 - 20}$$

$$t_{c_o} = 39.9°\text{C}.$$

This result is slightly different from that obtained by the mean temperature difference method since Δt_m is based on an incorrect assumption for t_{c_o}. This difference in results by the two methods will disappear when a verified t_{c_o} is achieved.

3. *Modify the assumed* t_{c_o}, *and repeat steps 1 and 2*. By either method the calculated t_{c_o}, above, differs significantly from the assumed value of 35°C. Hence revised calculations are in order. The process is not as extensive as one might fear since the values of $t_3 \approx t_2$ found in the first iteration usually work well in the second. If one assumes a new value of $t_{c_o} = 40°\text{C}$, a repetition of steps 1 and 2 yields the following results:

$$t_{c_o} = 40°\text{C, assumed},$$

$$t_1 = \frac{t_{c_i} + t_{c_o}}{2} = 30°\text{C},$$

$$t_2 = 46°\text{C, assumed},$$

$$h_{12} = 7403\ \text{W/m}^2\text{-°C},$$

$$t_3 = 46°\text{C, assumed},$$

$$t_4 = 54°\text{C},$$

$$h_{34} = 12{,}490\ \text{W/m}^2\text{-°C},$$

$$U = 3959\ \text{W/m}^2\text{-°C},$$

$$t_2 = 45.2°\text{C, calculated},$$

$$t_3 = 46.4°\text{C, calculated},$$

$$\Delta t_m = 22.5°\text{C},$$

$$\dot{m}_c c_{p_c} = 106.0\ \text{kW/°C},$$

$$C_R = 0,$$

$$\text{NTU} = 0.889,$$

$$\epsilon = 0.589,$$

$$t_{c_o} = 40.0°\text{C, calculated.}$$

4. The calculated outlet cold fluid temperature agrees with the assumed value so that this may be taken to be the desired result. The heat transfer rate and the rate of steam condensed are then readily found:

$$t_{c_o} = 40°\text{C } (104°\text{F}),$$

$$q = \dot{m}_c c_{p_c}(t_{c_o} - t_{c_i})$$

$$= 106.0(40 - 20) = 2120 \text{ kW } (7.23 \times 10^6 \text{ Btu/h}),$$

$$\dot{m} = \frac{q}{h_{\text{fg}}} = \frac{2120}{2373.1}$$

$$= 0.893 \text{ kg/s} = 3216 \text{ kg/h } (7090 \text{ lb}_\text{m}/\text{h}).$$

The example above utilized heat transfer correlations developed in Chapters 6 and 8 for the calculation of the heat transfer coefficients at the two surfaces of the heat exchanger tubes. For other surface configurations or other heat transfer processes it is necessary to have appropriate correlations for the particular situations at hand. This is particularly true for finned surfaces for which analytical solutions are generally not available. Data of this sort are available in various sources in the heat transfer literature, and the reader is referred to sources such as Refs. 2, 5, and 7.

As *examples* of the kind of information available, Figs. 10.22 and 10.23, taken from Ref. 5, are shown. Figure 10.22 shows information for plates with straight fins (in a triangular pitch for ease of manufacture) and Fig. 10.23 shows data for a bank of finned circular tubes. Similar data are available for a vast variety of configurations and these are shown simply to indicate typical presentations. Shown are correlations for both the surface heat transfer coefficients and the friction factor for use in estimating pressure losses. Also, various geometric parameters (such as fin area, total area, etc.) are given. Because of the complexity of the geometry involved, the dimensionless parameters used are modified in form somewhat from those used in Chapters 5 through 8. The Reynolds number is based on the *hydraulic diameter* introduced in Eq. (6.44):

$$D_h = \frac{4 \times \text{flow area}}{\text{wetted perimeter}}$$

and the *mass* velocity, G, introduced in Eq. (5.12) or (8.24):

$$G = \frac{\text{mass flow rate}}{\text{flow area}}. \tag{10.52}$$

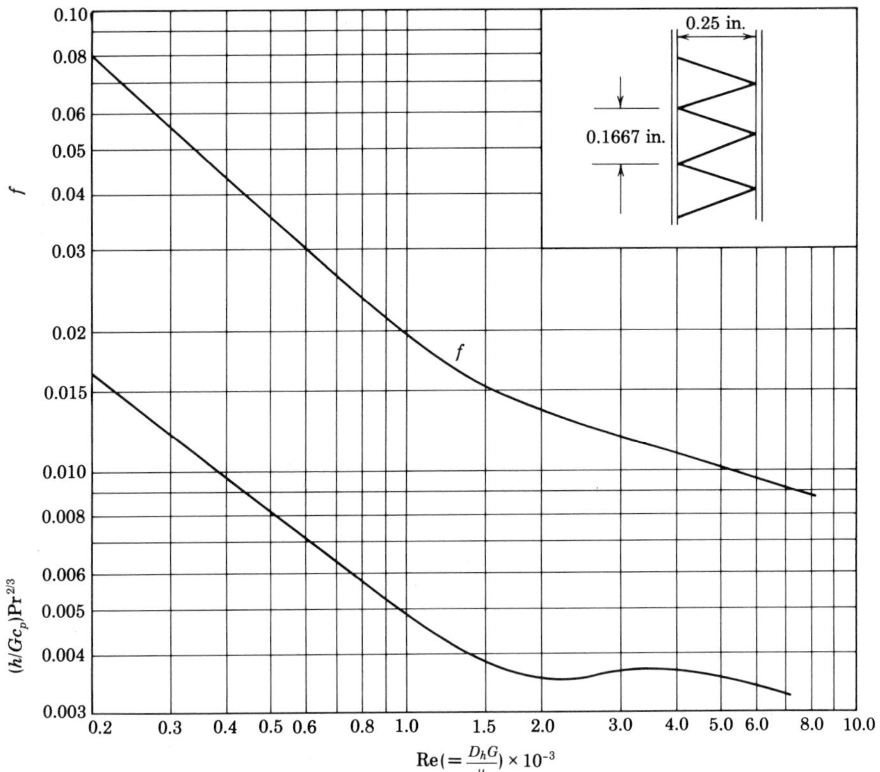

Fin pitch = 12.0 per in.; Fin thickness = 0.006 in.
Plate spacing = 0.250 in.; Flow hydraulic diamter = 0.009412 ft
Fin area/total surface area = 0.773
Total surface area/volume between plates = 392.7 $\mathrm{ft^2/ft^3}$
Free flow area/normal area between plates = 0.9240

Figure 10.22. Heat transfer and friction data for a plate-fin heat exchanger surface. (From W. M. Kays and A. L. London, *Compact Heat Exchangers*, 2nd ed., New York, McGraw-Hill, 1964. Used by permission.)

Thus the Reynolds number is expressed

$$\mathrm{Re} = \frac{U\rho D_h}{\mu} = D_h \frac{G}{\mu}. \tag{10.53}$$

The available flow area for the calculation of G is shown in the figures presented as is the hydraulic diameter.

Instead of expressing the heat transfer coefficient in terms of the Nusselt number, it is given in terms of

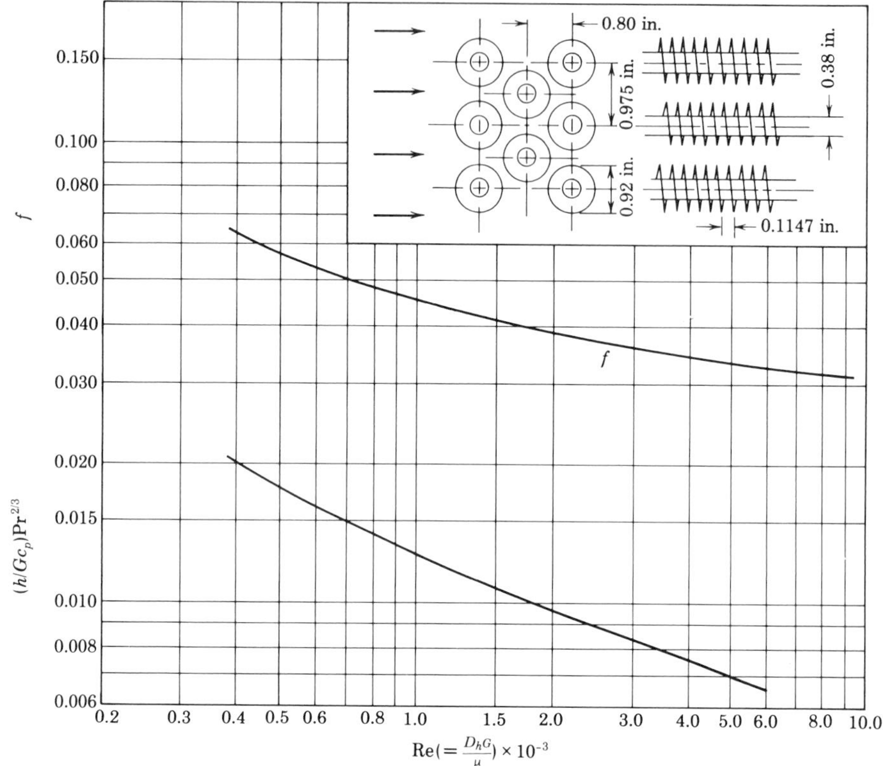

Fin pitch = 8.72 per in.; Fin thickness = 0.018 in.
Tube OD = 0.38 in.; Flow hydraulic diameter = 0.01288 ft
Fin area/total surface area = 0.910
Total surface area/total core volume = 163 $\mathrm{ft^2/ft^3}$
Free flow area/frontal area = 0.524

Figure 10.23. Heat transfer and friction data for a finned-tube heat exchanger core. (From W. M. Kays and A. L. London, *Compact Heat exchangers*, 2nd ed., New York, McGraw-Hill, 1964. Used by permission.)

$$\left(\frac{\mathrm{Nu}}{\mathrm{Re}}\right)\mathrm{Pr}^{-1/3} = \left(\frac{hD_h}{k}\frac{\mu}{D_hG}\right)\mathrm{Pr}^{-1/3} = \left(\frac{h}{Gc_p}\right)\mathrm{Pr}^{2/3}. \tag{10.54}$$

Likewise, the friction factor is modified to the form

$$f = \frac{\tau_0}{\frac{1}{2}\rho U^2} = \frac{2\rho^2\tau_0}{G^2}. \tag{10.55}$$

Use of these parameters is no different from those used previously as shown in the next example. This example is that of a typical cross-flow heat exchanger using an array of plate-fin surfaces. This example also illustrates the approach to be taken when both fluid temperatures vary through the heat exchanger—unlike Example 10.10, in which one fluid was changing phase at constant temperature.

EXAMPLE 10.11

A compact, cross-flow, aircraft heat exchanger is depicted in Fig. 10.24. The overall dimensions are shown. The basic construction of the exchanger is of the plate-fin type. Plates 0.012 in. thick, spaced 0.25 in. apart, form the separating walls between the two fluids. The surface of each plate is covered with an array of longitudinal fins, 0.006 in. thick, placed in a triangular pitch with 12 fins per inch. The material of both the fins and the plates has a $k = 15$ Btu/h-ft-°F. By alternating the direction of the fins on successive plates, a multiple cross-flow pattern is created as shown. The heat exchanger is 6 in. wide normal to the cold fluid flow direction and 4 in. wide normal to the hot fluid flow direction. Appropriate manifolding (not shown) distributes the two fluid streams into the exchanger and collects the exiting streams. The fin array need not be the same in the two flow directions as assumed here, but heat transfer data for each type would be needed.

Let there be 12 flow passages in each flow direction; hence, the total height of the stack is 6.30 in. when the plate thickness is accounted for. (Actually, there probably would be one more passage in one direction than the other to provide symmetry at the ends of the stack; however, for simplicity of discussion imagine

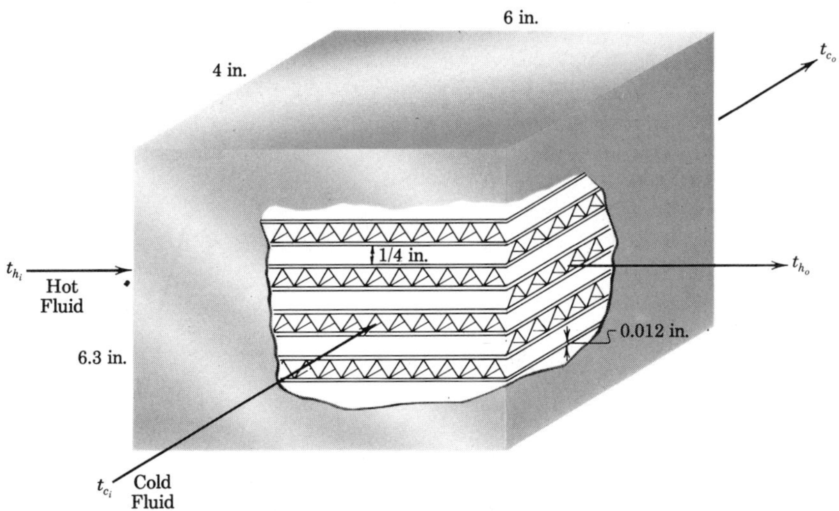

Figure 10.24. Compact cross-flow heat exchanger.

there to be an equal number in each direction, representing the last passage on each end as a half-passage.)

The cold fluid is air entering at $t_{c_i} = 0°\text{F}$ and flowing at the rate of $\dot{m}_c = 800$ $\text{lb}_\text{m}/\text{h}$. The hot fluid consists of combustion gases (to be approximated with the transport properties of air) entering at $t_{h_i} = 800°\text{F}$ and flowing at the rate of $\dot{m}_h =$ 200 $\text{lb}_\text{m}/\text{h}$. Estimate the mixed outlet temperature of the two fluid streams and the rate of heat transfer between them.

Solution. The plate-fin configuration conforms to that in Fig. 10.22. Thus the following geometric parameters apply:

$$D_h = 0.009412 \text{ ft},$$

$$\frac{\text{total area}}{\text{volume between plates}} = 392.7 \text{ ft}^2/\text{ft}^3,$$

$$\frac{\text{fin area}}{\text{total area}} = 0.773$$

$$\frac{\text{flow area}}{\text{normal area between plates}} = 0.9240.$$

For either fluid stream (with 12 passages) the volume between plates is $(12 \times 0.25 \times 4 \times 6)/1728 = 0.04167 \text{ ft}^3$. Thus the total surface area on both the hot and cold sides is

$$A_c = A_h = 0.04167 \times 392.7 = 16.325 \text{ ft}^2.$$

The free-flow area for each of the streams is

$$A_{x_c} = \frac{12 \times (0.25 \times 6) \times 0.9240}{144} = 0.1155 \text{ ft}^2,$$

$$A_{x_h} = \frac{12 \times (0.25 \times 4) \times 0.9240}{144} = 0.0770 \text{ ft}^2.$$

Thus the mass velocity in each stream is

$$G_c = \frac{\dot{m}_c}{A_{x_c}} = \frac{800}{0.1155} = 6926.4 \text{ lb}_\text{m}/\text{h-ft}^2,$$

$$G_h = \frac{\dot{m}_h}{A_{x_h}} = \frac{200}{0.077} = 2597.4 \text{ lb}_\text{m}/\text{h-ft}^2.$$

The other known data are

$$t_{c_i} = 0°\text{F} \qquad \dot{m}_c = 800 \text{ lb}_\text{m}/\text{h},$$

$$t_{h_i} = 800°\text{F}, \qquad \dot{m}_h = 200 \text{ lb}_\text{m}/\text{h}.$$

The same basic approach is used as in the preceding example. Trial values of t_{c_o} and t_{h_o} are chosen. Based on average cold and hot fluid temperatures, values of the heat transfer coefficients on each fluid surface, h_c and h_h, are then found. Then an overall U and the outlet fluid temperatures can be calculated and compared with the trial values. Repetition of this process eventually yields acceptable values for t_{c_o} and t_{h_o}. Only the last, verified, iteration is given here:

1. *Assume that* $t_{c_o} = 140°\text{F}$. Since $t_{c_i} = 0°\text{F}$, an average cold fluid temperature of $(0 + 140)2 = 70°\text{F}$ results for which $c_{p_c} = 0.2403$ Btu/lb$_m$-°F. If, for the moment, one takes $c_{p_h} \approx c_{p_c}$, then a basic heat balance gives $800(140 - 0) = 200(800 - t_{h_o})$, or $t_{h_o} = 240°\text{F}$. This leads to an average t_h of $(800 + 246)/2 = 520°\text{F}$ for which $c_{p_h} \cong 0.248$ Btu/lb$_m$-°F. Repeating this calculation, the following consistent assumptions of t_{c_o} and t_{h_o} are made:

$$t_{c_o} = 140°\text{F},$$

$$t_{h_o} = 256°\text{F}.$$

2. *Determine* h_c. Based on the assumed t_{c_o} the cold side heat transfer coefficient can be found using $t_c = (t_{c_i} + t_{c_o})/2 = 70°\text{F}$. Table A.6 gives $c_{p_c} = 0.2403$ Btu/lb$_m$-°F, Pr $= 0.713$, $\mu = 4.409 \times 10^{-2}$ lb$_m$/ft-h. Equation (12.53) for Re gives

$$\text{Re} = \frac{D_h G_c}{\mu} = \frac{0.009412 \times 6926.4}{4.409 \times 10^{-2}} = 1478.$$

Figure 10.22 presents the heat transfer correlation for the plate-fin geometry in question. Thus, Fig. 10.22 at Re = 1478 gives

$$\frac{h_c}{G_c c_{p_c}} \text{Pr}^{2/3} = 0.0038,$$

so that

$$h_c = 7.92 \text{ Btu/h-ft}^2\text{-°F}.$$

3. *Determine* h_h. With $t_{h_i} = 800°\text{F}$ and t_{h_o} assumed to be 256°F, an average hot fluid temperature of 528°F results for which $c_{p_h} = 0.2483$ Btu/lb$_m$-°F, Pr $= 0.698$, $\mu = 6.905 \times 10^{-2}$ lb$_m$/ft-h. With $G_h = 2597.4$ lb$_m$/h-ft^2, calculations similar to the above give

$$\text{Re} = \frac{D_h G_h}{\mu} = 354,$$

$$\frac{h_h}{G_h c_{p_h}} \text{Pr}^{2/3} = 0.0105 \quad \text{(Fig. 10.22)},$$

$$h_h = 8.61 \text{ Btu/h-ft}^2\text{-°F}.$$

4. *Determine the effectiveness of both surfaces.* The plate fins have a thickness of $w = 0.006$ in. Their length (half-length between plates) is $L = [(0.25)^2 + (0.1667/2)^2]^{1/2}/2 = 0.1318$ in. According to Eq. (2.55), the fin efficiency is $\kappa = (\tanh mL)/mL$ with $mL = L\sqrt{2h/kw}$. The surface effectiveness is, then, according to Eq. (2.56): $\eta = 1 - (A_f/A)(1 - \kappa)$. The ratio A_f/A is 0.773, for both sides, as given by Fig. 10.22. Thus one has

$$\text{Cold side:}\quad mL = (0.1318/12)\sqrt{2 \times 7.92/15 \times (0.006/12)} = 0.5048,$$

$$\kappa_c = 0.9228,$$

$$\eta_c = 0.940.$$

$$\text{Hot side:}\quad mL = (0.1318/12)\sqrt{2 \times 8.61/15 \times (0.006/12)} = 0.5263,$$

$$\kappa_h = 0.9169,$$

$$\eta_h = 0.936.$$

5. *Determine the overall U.* The overall U based on the hot side surface is, by Eq. (10.10),

$$U_h = \left[\frac{A_h/A_c}{\eta_c h_c} + \left(\frac{A_h}{A_w}\right)\frac{\Delta x}{k} + \frac{1}{\eta_h h_h}\right]^{-1}.$$

The unfinned plate area (for 23 separating plates) is $A_w = (4 \times 6 \times 23)/144 = 3.833$ ft^2. Thus

$$U_h = \left(\frac{16.325/16.325}{0.940 \times 7.92} + \frac{16.325}{3.833} \times \frac{0.012/12}{15} + \frac{1}{0.936 \times 8.61}\right)^{-1}$$

$$= 3.87 \text{ Btu/h-ft}^2\text{-°F}.$$

Since $A_c = A_h$ in this instance, then $U_c = U_h$.

6. *Verify the fluid outlet temperatures.* From the given mass flow rates and the specific heats found in steps 2 and 3:

$$\dot{m}_c c_{p_c} = 800 \times 0.2403 = 192.24 \text{ Btu/h-°F},$$

$$\dot{m}_h c_{p_h} = 200 \times 0.2483 = 49.66 \text{ Btu/h-°F}.$$

Thus

$$C_R = \frac{49.66}{192.24} = 0.258,$$

$$\text{NTU} = \frac{UA}{(\dot{m}c_p)_{\min}} = \frac{3.87 \times 16.325}{49.66} = 1.27,$$

so that Fig. 10.20 gives $\epsilon = 0.68$. Then by Eq. (10.37) with $\dot{m}_h c_{p_h} < \dot{m}_c c_{p_c}$,

$$\epsilon = \frac{t_{h_i} - t_{h_o}}{t_{h_i} - t_{c_i}}$$

$$0.68 = \frac{800 - t_{h_o}}{800 - 0}$$

$$t_{h_o} = 256°\text{F}.$$

This agrees with the assumption, so no further iteration is necessary and t_{c_o} = 140°F as assumed, also. Then since $q = \dot{m}_h c_{p_h}(t_{h_i} - t_{h_o})$, the final desired results are

$$t_{c_o} = 140°\text{F } (60°\text{C}),$$

$$t_{h_o} = 256°\text{F } (124°\text{C}),$$

$$q = 27{,}000 \text{ Btu/h } (7910 \text{ W}).$$

10.9 HEAT EXCHANGER DESIGN

The discussion of the preceding section was limited to the calculation of the performance of a given heat exchanger of known geometry. That is, the total surface area and the flow pass arrangement were known, and it was desired to determine the outlet temperatures of the fluids for various inlet conditions.

The design problem, that of selecting the geometry of a heat exchanger to accomplish a desired transfer of heat between two fluids of specified terminal temperatures, is more complex inasmuch as there is no single answer to the problem. Several different heat exchangers may be able to satisfy stated conditions equally well from the thermal point of view. The ultimate decision as to which of several possible designs should be used usually depends on many other factors—such as cost, space requirements, operating expenses, personal taste of the designer, etc. Designs are also often limited by the desirability of adhering to certain standard practices—such as the use of standard tube sizes, standard fin arrays, etc.

If the flow rates, specific heats, and terminal temperatures of two heat exchanging fluids are specified, one may easily calculate the total heat to be transferred, the capacity ratio (R or C_R), and the effectiveness (P or ϵ). Then selection of the basic flow pattern readily yields Δt_m or the NTU from which the necessary UA product is found. The question becomes, then, what heat exchanger will have the proper area and overall coefficient to satisfy this need?

For a shell-and-tube exchanger, the tube size and the fluid velocity determine the number of tubes, per pass, needed to accommodate the desired flow. The number of tube passes and their length determine the total surface, A. These same dimensions, and the fluid velocities, and temperatures, serve to establish U. Thus one seeks a synthesis of all these factors to find the combination to give the desired UA. A similar reasoning applies to compact cross-flow exchangers.

In order to separate some of these parameters, two sets of examples will be shown. First it will be assumed that U is known, and the effect of flow pass arrangement and tube size will be shown. Then a case in which U is simultaneously found will be discussed.

EXAMPLE 10.12

A shell-and-tube heat exchanger is to be constructed with 1-in. 12-gage tubes. The cold fluid, flowing in the tubes, is water flowing at the rate of 18,000 kg/h and is to be heated from $t_{c_i} = 35°C$ to $t_{c_o} = 65°C$. The hot shell-side fluid is also water and flows at the rate of 12,800 kg/h, entering at 100°C. The tube water flows at an average velocity of 0.3 m/s and the overall heat transfer coefficient of the exchanger is known to be 1600 W/m²-°C. For simplicity, assume that the specific heat of each fluid is constant at 4.18 kJ/kg-°C. Determine the number of tubes, per pass, and the required length of the tubes—examining various tube-pass possibilities but limiting the design to one shell pass.

Solution. Table B.2 gives the following data for 1-in. 12-gage tubes:

$$\frac{\text{surface area}}{\text{unit length}} = 0.0798 \text{ m}^2/\text{m, per tube,}$$

$$\text{cross-sectional area} = 0.0003098 \text{ m}^2\text{, per tube.}$$

For the given $t_{c_i} = 35°C$, $t_{c_o} = 65°C$, an average of 50°C gives $\rho = 988.0 \text{ kg/m}^3$ (Table A.3), so that the number of tubes needed, per pass, to accommodate the 18,000-kg/h flow is

$$\dot{m} = 18{,}000 = N\rho A_x v = N \times 988.0 \times 0.0003098 \times 0.3 \times 3600,$$

$$N = 54.5.$$

Thus 55 tubes per pass should be used. The capacity rates of the two streams are

$$\dot{m}_c c_{p_c} = 18{,}000 \times \frac{4.18}{3600} = 20.90 \text{ kW/°C},$$

$$\dot{m}_h c_{p_h} = 12{,}800 \times \frac{4.18}{3600} = 14.86 \text{ kW/°C}.$$

The hot fluid outlet is given by

$$q = \dot{m}_c c_{p_c}(t_{c_o} - t_{c_i}) = \dot{m}_h c_{p_h}(t_{h_i} - t_{h_o})$$

$$20.90(65 - 35) = 14.86(100 - t_{h_o})$$

$$t_{h_o} = 57.8°C.$$

The problem specifications indicate that only one shell pass should be used. Investigate, first, the use of one tube pass. In this instance the use of a parallel-flow exchanger is ruled out since $t_{h_o} < t_{c_o}$. Thus the only one-shell-pass, one-tube-pass arrangement possible is that for counterflow. Since $\dot{m}_h c_{p_h} < \dot{m}_c c_{p_c}$, the definition of the effectiveness and capacity ratio given in Eqs. (10.37) and (10.38) give

$$C_R = \frac{(\dot{m}c_p)_{\min}}{(\dot{m}c_p)_{\max}} = \frac{14.86}{20.90} = 0.711,$$

$$\epsilon = \frac{t_{h_i} - t_{h_o}}{t_{h_i} - t_{c_i}} = \frac{100 - 57.8}{100 - 35} = 0.649.$$

Figure 10.17 may be used to find the required NTU, but Eq. (10.46) is more accurate:

$$\text{NTU} = \frac{\ln\left[(1 - \epsilon)/(1 - \epsilon C_R)\right]}{C_R - 1} = 1.481,$$

and since $\text{NTU} = UA/(\dot{m}c_p)_{\min}$,

$$\text{NTU} = 1.481 = \frac{UA}{14.86},$$

$$UA = 22.02 \text{ kW/°C}.$$

One could, as well, find Δt_m from Eq. (10.29):

$$\Delta t_m = \Delta t_{\text{lm,ctr}} = \frac{(t_{h_i} - t_{c_o}) - (t_{h_o} - t_{c_i})}{\ln\left(\dfrac{t_{h_i} - t_{c_o}}{t_{h_o} - t_{c_i}}\right)}$$

$$= \frac{(100 - 65) - (57.8 - 35)}{\ln\left(\dfrac{100 - 65}{57.8 - 35}\right)} = 28.47\text{°C},$$

and UA from

$$q = UA\,\Delta t_m = \dot{m}_c c_{p_c}(t_{c_o} - t_{c_i})$$

$$UA \times 28.47 = 20.90(65 - 35)$$

$$UA = 22.02 \text{ kW/°C}.$$

In this example U is known to be 1600 W/m^2-°C = 1.6 kW/m^2-°C, so the necessary exchanger surface area is

$$A = \frac{22.02}{1.6} = 13.76 \text{ m}^2.$$

The total surface area is determined by the tube length, L, so that for N tubes

$$A = N \times L \times (\text{surface area/unit length/tube}),$$

$$13.76 = 55 \times L \times 0.0798,$$

$$L = 3.13 \text{ m}.$$

Thus a counterflow, one-shell-pass, one-tube-pass exchanger with 55 tubes approximately 3 m long will satisfy the stated requirements.

If, because of space limitations or other constraints, the tube length calculated above results in a heat exchanger considered to be too long, one might examine the possibility of using a one-shell-pass, two-tube-pass arrangement. In this instance, the effectiveness and capacity ratio remain $\epsilon = 0.649$, $C_R = 0.711$. Figure 10.18 is now applicable, but it is difficult to read the required NTU with much accuracy—a value of NTU of about 2.5 seeming to be indicated. Using the Δt_m–P–R formulation does not give much better results, the definitions of R and P giving

$$R = \frac{\dot{m}_c c_{pc}}{\dot{m}_h c_{ph}} = \frac{20.90}{14.86} = 1.41,$$

$$P = \frac{t_{c_o} - t_{c_i}}{t_{h_i} - t_{c_i}} = \frac{65 - 35}{100 - 35} = 0.462.$$

Figure 10.13 for the F correction to be applied to $\Delta t_{\text{lm,ctr}}$ is likewise difficult to read with accuracy. The implication of either of these facts is that such an arrangement is not an effective use of the exchanger surface area and should probably be avoided. Proceeding, however, to show the method, one may apply Eq. (10.48) to find NTU with better accuracy than use of Fig. 10.18. Thus, for $\epsilon = 0.649$, $C_R = 0.711$, this equation yields

$$\text{NTU} = -(1 + C_R^2)^{-1/2} \ln \left[\frac{2/\epsilon - 1 - C_R - (1 + C_R^2)^{1/2}}{2/\epsilon - 1 - C_R + (1 + C_R^2)^{1/2}} \right] = 2.359$$

$$\text{NTU} = \frac{UA}{\dot{m}_h c_{ph}}$$

$$2.359 = \frac{UA}{14.86}$$

$$UA = 35.05 \text{ kW/°C},$$

$$A = 21.9 \text{ m}^2.$$

[Similarly, Eq. (10.35) will give $F = 0.620$, an undesirably low value.] While the surface area required is greater than the one-tube-pass case, there are twice as many tubes to provide the area:

$$A = 21.9 = 2 \times 55 \times L \times 0.0798,$$

$$L = 2.5 \text{ m}.$$

The reduction in length is not much, and the increased number of tubes (i.e., twice) probably results in a higher cost of manufacture. Thus, this second option should probably not be used. The marginal improvement of the second design is the result of the fact, as noted earlier, that the surface area is not being utilized very effectively as evidenced by the low value of the temperature difference correction factor F.

Example 10.12 emphasized the effect of the flow pass arrangement on heat exchanger design. Altering other parameters will also affect the solution. The fluid velocity in the tubes affects the number of tubes needed to pass the tube side fluid and altering it will significantly modify the results. The fluid velocity also determines to some degree the overall U, assumed known above, as well as the pumping requirements to move the fluid. Likewise, the tube size may be varied to alter performance. It is not possible to quantify all these effects to produce general rules for design. However, the following example illustrates some of these effects, as well as showing the procedure involved when, as is actually the case, the overall U is not known. This example, as the above, is presented for a shell-and-tube exchanger. Analogous procedures would apply for compact cross-flow devices.

EXAMPLE 10.13

It is desired to design a feedwater heater to supply a boiler. The inlet water is at a temperature of $t_{c_i} = 70°\text{F}$, flows at the rate of 20,000 $\text{lb}_\text{m}/\text{h}$, and is to be heated to 150°F. The feedwater is to be heated in a horizontal shell-and-tube heat exchanger by condensing saturated steam at 20 psia. Because of space limitations, the tubes should not exceed 6 ft in length. Determine the configuration of a heat exchanger to accomplish these needs.

Solution. The Steam Tables give, at 20 psia, $t_\text{sat} = 228°\text{F}$. Using 1 $\text{Btu/lb}_\text{m}\text{-°F}$ for the specific heat of the tube water, the data give

$$\dot{m}_c c_{p_c} = 20{,}000 \text{ Btu/h-°F}, \quad t_{c_i} = 70°\text{F}, \quad t_{c_o} = 150°\text{F},$$

$$\dot{m}_h c_{ph} \to \infty, \quad t_{h_i} = 228°\text{F}, \quad t_{h_o} = 228°\text{F}.$$

Since the shell fluid is of constant temperature, the number of tube passes does not affect the mean temperature difference or the ϵ–NTU–C_R relation to be used (since $C_R \to 0$). Thus the required UA product can be found in advance, and one does not have to be concerned with the effect of the flow pass arrangement as in the preceding example. For the data given either Eq. (10.28) or (10.29) gives the mean temperature difference:

$$\Delta t_m = \frac{(t_h - t_{c_i}) - (t_h - t_{c_o})}{\ln\left(\dfrac{t_h - t_{c_i}}{t_h - t_{c_o}}\right)}$$

$$= \frac{(228 - 70) - (228 - 150)}{\ln\left(\dfrac{228 - 70}{228 - 150}\right)} = 113.3°\text{F},$$

and $q = \dot{m}_c c_{p_c}(t_{c_o} - t_{c_i}) = UA\,\Delta t_m$ yields

$$UA = \frac{20{,}000(150 - 70)}{113.3} = 14{,}120 \text{ Btu/h-°F}.$$

Use of the definition of $\epsilon = (t_{c_o} - t_{c_i})/(t_{h_i} - t_{c_i}) = 0.5063$ and Eq. (10.45) for $C_R = 0$ gives NTU $= 0.7059$ and the same result for UA. Thus the design process devolves into finding configurations and flow velocities that will yield the UA above.

As a starting assumption, pick 1-in. 16-gage brass ($k = 64$ Btu/h-ft-°F) tubes. To keep pressure losses to a minimum, select a low water velocity in the tubes, say $v = 1$ ft/s. Using the notation of Fig. 10.21, or Example 10.10, Table B.2 gives the geometric data for the tubes:

$$D_2 = 0.870 \text{ in.}, \qquad D_3 = 1.00 \text{ in.},$$

$$\text{cross section} = 0.004128 \text{ ft}^2,$$

$$\text{surface area} = 0.2618 \text{ ft}^2/\text{ft}.$$

The number of tubes per pass is established by the need to pass 20,000 lb_m/h at a velocity of 1 ft/s. For a mean tube water temperature of 110°F, $\rho = 61.86$ $\text{lb}_\text{m}/\text{ft}^3$. Thus

$$\dot{m}_c = N\rho A_x v,$$

$$N = \frac{20{,}000}{61.86 \times 1 \times 3600 \times 0.004128} = 21.8,$$

so 22 tubes per pass must be provided.

Determination of the overall U will enable one to establish the total area needed and thus the required tube length. The overall U is found by obtaining the inside and outside heat transfer coefficients, h_{12} and h_{34}. These coefficients are found by the methods outlined in some detail in both Examples 10.2 and 10.10. The tube surface temperatures t_2 and t_3 need to be assumed, h_{12} found by application of the forced convection relation of Eq. (6.33), and h_{34} found by use of the condensing formula of Eq. (8.20). Then the assumed values of t_2 and t_3 are verified, modified appropriately, and the process repeated until satisfactory agreement is obtained. As in Example 10.10, one can probably neglect the tube wall resistance and let $t_2 \approx t_3$, so that an iteration on only one temperature need be made. Since these calculations are identical to those in Example 10.10 (or 10.2), they are not shown here. The results of this iterative calculation are

$$t_2 \approx t_3 = 212°\text{F},$$

$$h_{12} = 407 \text{ Btu/h-ft}^2\text{-°F}$$

$$h_{34} = 2303 \text{ Btu/h-ft}^2\text{-°F}.$$

Thus the overall U may be found:

$$U = \left[\frac{r_3}{r_2 h_{12}} + \frac{r_3 \ln (r_3/r_2)}{k_{23}} + \frac{1}{h_{34}}\right]^{-1}$$

$$= \left[\frac{1.0}{0.87 \times 407} + \frac{1.0 \ln (1.0/0.87)}{2 \times 12 \times 64} + \frac{1}{2303}\right]^{-1} = 299 \text{ Btu/h-ft}^2\text{-°F}.$$

Then the required surface area is

$$A = \frac{14{,}120}{299} = 47.2 \text{ ft}^2.$$

For one tube pass, the required tube length will be, for 22 tubes per pass,

$$L = \frac{A}{N \times (\text{surface area/length})}$$

$$= \frac{47.2}{22 \times 0.2618} = 8.2 \text{ ft}.$$

This exceeds the 6-ft maximum imposed in the specifications, so two tube passes must be provided. Since in this case the Δt_m does not change when two tube passes are used, the length is $L = 8.2/2 = 4.1$. As a summary, then, one has for a first design:

1 shell pass
2 tube passes
22 tubes per pass, 1-in.-OD 16-gage 4.1-ft-long tubes
1 ft/s water velocity
$h_{12} = 407$ Btu/h-ft^2-°F — Design 1
$h_{34} = 2303$ Btu/h-ft^2-°F
$U = 299$ Btu/h-ft^2-°F
$\Delta t_m = 113.3$°F
$A = 47.2$ ft^2

This design does not allow for other factors such as tube fouling, shell side fluid distribution, etc. Nonetheless, these calculations illustrate the method and the effect of certain parameters.

Since the length of a single pass in the calculation above was not too much greater than the imposed maximum of 6 ft, one might hope to avoid the use of two passes by increasing U and thus reducing the required surface area. An increased tube water velocity should serve to increase U, and indeed it does, but the number of tubes required per pass is reduced—thus giving fewer tubes to provide the area.

To show these effects, increase the water velocity to 1.5 ft/s, leaving all other quantities the same. With the same size tubes, the final results are summarized below:

Design 2:

1 shell pass
2 tube passes
15 tubes per pass, 1-in.-OD 16-gage 4.7-ft-long tubes
1.5 ft/s water velocity
h_{12} = 563 Btu/h-ft^2-°F
h_{34} = 2173 Btu/h-ft^2-°F
U = 386 Btu/h-ft^2-°F
Δt_m = 113.3°F
A = 36.6 ft^2

The overall U is, indeed, greater and A is smaller. However, only 15 tubes are required per pass, to allow the tube water to flow at the rate of 20,000 lb$_m$/h. Thus the tube length in a single pass becomes greater, 9.3 ft, so that two tube passes are still required. The second design requires fewer total tubes than the first, so its cost of manufacture would be less. However, the pressure loss required to pump the tube water in the second design would be of the order of 2.5 times that in the first design since this loss varies, approximately, as the square of the water velocity. Thus operating costs associated with the second design would be expected to be greater than the first.

The effect of tube size can be seen by returning to a water velocity of 1 ft/s, as in the first design, but using instead $\frac{3}{4}$-in. 16-gage tubes. In this case, calculations yield

Design 3:

1 shell pass
1 tube pass
43 tubes per pass, $\frac{3}{4}$-in.-OD 16-gage 5.4-ft-long tubes
1 ft/s water velocity
h_{12} = 437 Btu/h-ft^2-°F
h_{34} = 2610 Btu/h-ft^2-°F
U = 308 Btu/h-ft^2-°F
Δt_m = 113.3°F
A = 45.8 ft^2

This design requires only one tube pass, but the total number of tubes and their length are comparable to those required in the first design. The pressure loss in the tubes would be of the same order as that in the first case, so one may take Design 1 and Design 3 as approximately equivalent.

The three designs given above are certainly not the only ones possible for the stated specifications. Others exist, and many should be examined and then compared through, probably, a detailed economic analysis to ascertain which should be adopted as the final design. More details of typical design procedures are available in sources such as Refs. 5, 8, and 9.

10.10 CLOSURE

The emphasis of this chapter has been on the *methodology* of the treatment of problems of combined conduction and convection. Section 10.2 concentrated on the simultaneous solution of these two heat transfer mechanisms. Iterative, and sometimes double-iterative, solutions were found to be necessary. In Sections 10.8 and 10.9, in which heat exchanger performance analysis and heat exchanger design were discussed, multiple iterative solutions were again found to be necessary. The solution techniques were discussed in some detail in order to emphasize the fundamental origin of the conduction solutions and the convective correlations being used.

Quite apparently, the actual implementation of the solution methods described in this chapter can be greatly facilitated by the use of modern digital computing equipment. As noted earlier, the successful implementation of any computing system to the solution of these problems requires the capability of generating the necessary physical property data for the large number of fluids and solids that might be involved. Many modern engineering design offices have available computing systems with such capability.

REFERENCES

1. Jakob, M., *Heat Transfer,* Vol. 2., New York, Wiley, 1957.
2. *Standards of Tubular Exchange Manufacturers Association,* 6th ed., New York, Tubular Exchanger Manufacturers Association, Inc., 1978.
3. Bowman, R. A., A. C. Mueller, and W. M. Nagle, "Mean Temperature Difference in Design," *Trans. ASME,* Vol. 63, 1940, p. 283.
4. Shamsundar, N., "A Property of the Log-Mean Temperature-Difference Correction Factor," *Mech. Eng. News,* Vol. 19, No. 3, 1982, p. 14.
5. Kays, W. M., and A. L. London, *Compact Heat Exchangers,* 2nd ed., New York, McGraw-Hill, 1964.
6. Mason, J. L., "Heat Transfer in Cross-Flow," *Proc. Appl. Mech.,* 2nd U.S. Natl. Congr. 1954, p. 801.
7. Rohsenow, W. M., and J. P. Hartnet, eds., *Handbook of Heat Transfer,* New York, McGraw-Hill, 1975.
8. Kern, D. Q., *Process Heat Transfer,* New York, McGraw-Hill, 1950.
9. Kern, D. Q., and A. D. Kraus, *Extended Surface Heat Transfer,* New York, McGraw-Hill, 1972.

PROBLEMS

10.1 Steam at 4000 kN/m^2, 450°C, flows at a velocity of 7.5 m/s through a 6-in. schedule 40 pipe which is covered with 3.5 cm of 85% magnesia insulation. The pipe is located horizontally in a room in which the ambient air temperature is 25°C. Find the temperature of the outer surface of the insulation, the overall heat transfer coefficient, and the heat loss, per meter of length.

10.2 Steam at 600°F, 800 psia, flows at 20 ft/s through a horizontal 5-in. schedule 40 pipe covered with 1 in. of insulation (k = 0.5 Btu/h-ft-°F). The pipe is exposed to atmospheric stagnant air at 100°F. Find the temperature of the outer surface of the insulation and the overall heat transfer coefficient.

10.3 A 1-in. 16-gage condenser tube is made of brass. Steam at 20 kN/m^2 is condensing on the tube surface, in a horizontal position, and water at 20°C flows with a velocity of 1.2 m/s through the tube. Find the tube surface temperature and the overall heat transfer coefficient.

10.4 A $\frac{3}{4}$-in. 18-gage horizontal condenser tube has steam at 5 psia condensing on its outer surface and cooling water at 80°F flowing through it with a velocity of 2 ft/s. Find the overall heat transfer coefficient **(a)** if the tube is brass with k = 64 Btu/h-ft-°F, and **(b)** if the tube is stainless steel with k = 10 Btu/h-ft-°F.

10.5 A bare 10-in. schedule 40 steel pipe passes horizontally through a room of still air at 85°C. The pipe carries steam at 3000 kN/m^2, 550°C, flowing with a velocity of 100 m/s. Find the overall heat transfer coefficient and the pipe surface temperature.

10.6 Steam at 500 kN/m^2, 450°C, flows at a velocity of 5 m/s through a horizontal 5-in. schedule 80 pipe covered with 7.5 cm of 85% magnesia insulation. The pipe is exposed to still atmospheric air at 35°C. Find the overall heat transfer coefficient and the outer insulation temperature.

10.7 The lubricating oil listed in Table A.4 flows at 0.3 m/s through a 1-in. 16-gage brass tube 2 m long. The oil is at 50°C and the outer tube surface is exposed, horizontally, to atmospheric air at 20°C by free convection. Find the outer tube surface temperature.

10.8 The lubricating oil listed in Table A.4 flows at 1.5 ft/s through a horizontal 1-in. 16-gage brass tube 7 ft long. The oil temperature is 120°F and the tube is exposed to atmospheric air at 70°F. Find the tube wall temperature.

10.9 The lubricating oil listed in Table A.4 flows with a velocity of 0.5 m/s through a copper tube (OD = 3.75 cm, ID = 3.25 cm) that is 3.5 m long. The bulk temperature of the oil is 140°C. Atmospheric pressure air at 30°C flows over the outside of the tube, normal to it, with a freestream velocity of 20 m/s. Find **(a)** the tube surface temperature, and **(b)** the overall heat transfer coefficient between the oil and the air. Note clearly any simplifying assumptions made and justify these assumptions by appropriate calculations.

10.10 The lubricating oil listed in Table A.4 flows at a temperature of 300°F and with an average velocity of 1.2 ft/s through a $1\frac{1}{2}$-in. 12-gage copper tube that is 12 ft long. Atmospheric air at 90°F flows normal to the outside of the tube with a freestream velocity of 60 ft/s. Assume that the temperature drop through the tube wall is negligible. Determine **(a)** the tube surface temperature, and **(b)** the rate of heat transfer from the tube.

10.11 Superheated steam at 2000 kN/m^2, 300°C, flows with a velocity of 6 m/s through a 6-in. schedule 40 steel pipe. Atmospheric air at 20°C flows normal to the cylinder with a velocity of 12 m/s. Estimate the outer pipe surface temperature.

10.12 A cast iron rod, 2 cm in diameter and 25 cm long, protrudes horizontally from a heat source at 150°C into still atmospheric air at 30°C. Estimate the heat loss from the rod.

10.13 In a one-shell-pass, one-tube-pass heat exchanger, the cold fluid (water with c_{pc} = 4.18 kJ/kg-°C) enters at 50°C and flows at the rate of 900 kg/h. The hot fluid (also water with c_{ph} = 4.18 kJ/kg-°C) enters at 370°C and flows at the rate of 1260 kg/h. If the heat exchanger becomes infinitely large, find the maximum heat that can be transferred and the outlet temperatures of the fluids **(a)** for parallel flow, and **(b)** for counterflow.

10.14 Repeat Prob. 10.13 if the cold fluid flow rate is 1260 kg/h and the hot fluid flow rate is 900 kg/h, the other data remaining the same.

10.15 The cooler for the lubricating oil for a large gas turbine consists of a one-shell-pass, one-tube-pass heat exchanger. The oil (c_p = 0.5 Btu/lb$_m$-°F) flows through the tubes at the rate of 7000 lb$_m$/h, entering at 320°F. The coolant is water flowing in the shell at the rate of 9000 lb$_m$/h, entering at 80°F. Find the maximum possible heat transfer rate (i.e., for a very large heat exchanger) and the outlet temperatures of the two fluids if the flow arrangement **(a)** is parallel flow, and **(b)** is counter-flow.

10.16 Repeat Prob. 10.15 if the oil flow rate is 9000 lb$_m$/h and the water flow rate is 7000 lb$_m$/h, all other data remaining the same.

10.17 Water at a bulk temperature of 25°C flows with a velocity of 1.5 m/s in a brass steam condenser tube 6 m long. The outside and inside tube diameters are 1.58

cm and 1.34 cm, respectively. The condensing steam on the outer tube surface produces a heat transfer coefficient of 12,000 W/m^2-°C there. Calculate the overall heat transfer coefficient for the exchanger.

10.18 A heat exchanger brass tube (OD = 3.34 cm, ID = 2.5 cm) has outside and inside surface heat transfer coefficients of 2000 W/m^2-°C and 1200 W/m^2-°C, respectively, when the surfaces are clean. Find the overall heat transfer coefficient for the heat exchanger **(a)** when the surfaces are clean, and **(b)** when the surfaces are dirty, each having a fouling factor of R_f = 0.00018 m^2-°C/W.

10.19 A 1-in. 14-gage heat exchanger tube is equipped with 20 equally spaced straight fins of uniform thickness placed longitudinally along the tube. The fins are 2.5 cm long in the radial direction and are 0.16 cm thick. Both the tube and the fins are made of steel with k = 45 W/m-°C. The inside and outside surface heat transfer coefficients are 1130 and 255 W/m^2-°C, respectively. What is the overall heat transfer coefficient for the exchanger, based on the outer exposed surface area? Compare this result with that for the same tube, without fins, subjected to the same heat transfer coefficients.

10.20 The separating surface of a cross-flow heat exchanger is a plate 0.03 cm thick. On the hot fluid side, there is an array of straight uniform thickness fins 0.3 cm long, 0.015 cm thick, spaced 0.2 cm on centers, and the heat transfer coefficient there is 45 W/m^2-°C. On the cold fluid side of the plate there is also an array of straight uniform thickness fins of the same dimensions but spaced 0.13 cm on centers and the heat transfer coefficient is 54 W/m^2-°C. What is the overall heat transfer coefficient based on the exposed surface area where h = 45 W/m^2-°C if the fins and the plate have a thermal conductivity of 40 W/m-°C?

10.21 In a one-shell-pass, one-tube-pass heat exchanger, the hot fluid enters at 425°C and leaves at 315°C. The cold fluid enters at 40°C and leaves at 260°C. Find the mean temperature difference for **(a)** parallel flow, and **(b)** counterflow.

10.22 Find the mean temperature difference in a one-shell-pass, two-tube-pass heat exchanger for the temperatures noted in Prob. 10.21.

10.23 In a heat exchanger, the cold fluid enters at 110°F and leaves at 500°F while the hot fluid enters at 750°F and leaves at 600°F. Find the mean temperature difference for **(a)** parallel-flow, **(b)** counterflow, and **(c)** one-shell-pass, two-tube-pass.

10.24 In a one-shell-pass, one-tube-pass heat exchanger the cold fluid enters at 40°C and leaves at 200°C while the hot fluid enters a 370°C and leaves at 150°C. Find the mean temperature difference.

10.25 In a cross-flow heat exchanger, the cold fluid enters at 40°C and leaves at 150°C while the hot fluid enters at 425°C and leaves at 260°C. Find the mean temperature difference.

10.26 In terms of the parameters R and P defined in Eqs. (10.32) and (10.33), show that the F correction factor defined in Eq. (10.34) for a parallel-flow heat exchanger is

$$F = \frac{R+1}{R-1} \frac{\ln\left(\dfrac{1-P}{1-PR}\right)}{\ln\left[\dfrac{1}{1-P(R+1)}\right]}.$$

10.27 Equation (10.44) for the effectiveness of a parallel-flow exchanger was derived from Eqs. (10.40) and (10.41) by introduction of Eq. (10.28). Follow the same procedure to derive Eq. (10.46) from Eqs. (10.40) and (10.41) by introducing Eq. (10.29).

10.28 Show that Eq. (10.47) is the limiting form for Eq. (10.46) as $C_R \to 1$.

10.29 The cold fluid of a heat exchanger increases in temperature from 15°C to 30°C as it passes through. The hot fluid temperature drops from 65°C to 40°C.
(a) Find the capacity ratio, R, and the effectiveness, P, used in the Δt_m formulation.
(b) Find the capacity ratio C_R and the effectiveness, ϵ, used in the ϵ–NTU formulation.
(c) If the exchanger has one shell pass and one tube pass, find, if possible, the NTU and Δt_m for (1) parallel flow, or (2) counterflow.
(d) If the exchanger has one shell pass and two tube passes, find the NTU and Δt_m.

10.30 Repeat Prob. 10.29 if the cold fluid temperatures are $t_{c_i} = 70°\text{F}$, $t_{c_o} = 120°\text{F}$ and the hot fluid temperatures are $t_{h_i} = 175°\text{F}$, $t_{h_o} = 140°\text{F}$.

10.31 The overall heat transfer coefficient for a one-shell-pass, two-tube-pass heat exchanger is known to be 1700 W/m^2-°C. The shell side fluid ($c_p \cong 4.18$ kJ/kg-°C) flows at the rate of 9100 kg/h, entering at 260°C. The tube-side fluid ($c_p = 3.55$ kJ/kg-°C) flows at the rate of 27,300 kg/h, entering at 30°C. For a total surface area of 9.5 m^2, find the rate of heat transfer between the two fluids and their outlet temperatures **(a)** using the Δt_m formulation, and **(b)** using the ϵ–NTU formulation.

10.32 A one-shell-pass, one-tube-pass, counterflow heat exchanger uses 6800 kg/h of water ($c_p = 4.18$ kJ/kg-°C) entering at 25°C to cool 18,000 kg/h of oil ($c_p = 2.9$ kJ/kg-°C) entering at 100°C. The total exchanger surface area is 14 m^2 and the overall coefficient is $U = 370$ W/m^2-°C. Find the rate of heat exchange and the outlet temperature of the two fluids by either the Δt_m or ϵ–NTU formulation.

10.33 A one-shell-pass, two-tube-pass heat exchanger has a known overall heat transfer coefficient of 280 Btu/h-ft^2-°F. The shell-side fluid is water ($c_p \cong 1.0$ Btu/lb$_m$-°F) flowing at the rate of 18,000 lb$_m$/h and entering at 500°F. The tube-side fluid is oil ($c_p = 0.85$ Btu/lb$_m$-°F) flowing at the rate of 50,000 lb$_m$/h and entering at 80°F. For a total surface area of 125 ft^2, find the outlet fluid temperatures **(a)** using the Δt_m formulation, and **(b)** using the ϵ–NTU formulation.

10.34 A one-shell-pass, one-tube-pass heat exchanger operates in counterflow. The exchanger uses 6500 kg/h of water ($c_p = 4.18$ kJ/kg-°C) entering at 25°C to cool 13,000 kg/h of oil ($c_p = 2.09$ kJ/kg-°C) entering at 100°C. The total surface area of the exchanger is 16 m^2 and the overall heat transfer coefficient is 350 W/m^2-°C. Find the outlet temperatures of the two fluids using both the Δt_m and ϵ–NTU formulations.

10.35 A steam condenser is made of $\frac{3}{4}$-in. 14-gage brass tubes, 5 ft long. There are two tube passes with 115 tubes per pass, and the cooling water enters at 70°F with a velocity of 5 ft/s. The inside tube heat transfer coefficient is known to be 1140 Btu/h-ft^2-°F and the outside tube surface condensing heat transfer coefficient is known to be 1250 Btu/h-ft^2-°F. The steam is being condensed at 5 psia. Find **(a)** the outlet temperature of the cooling water, and **(b)** the lb$_m$/h of steam condensed.

10.36 A steam condenser is made of $\frac{5}{8}$-in. 14-gage brass tubes, 1.8 m long. There are two tube passes with 125 tubes per pass. Cooling water enters at 24°C and flows with an average tube velocity of 1.37 m/s. The inside tube heat transfer coefficient is known to be 6735 W/m^2-°C and the outside tube surface condensing heat transfer coefficient is known to be 11,340 W/m^2-°C. Find the outlet water temperature and the kg/h of steam condensed if the steam is saturated at 14 kN/m^2 pressure.

10.37 A cross-flow heat exchanger has a surface configuration with heat transfer coefficients as described in Prob. 10.20. The fluid passages are 0.6 cm wide, so the fin length, by symmetry, is 0.3 cm as given. The hot fluid flows at a rate of 45 kg/h ($c_p = 1$ kJ/kg-°C), entering at 300°C while the cold fluid flows at the rate of 55 kg/h ($c_p = 1.2$ kJ/kg-°C), entering at 100°C. The total exposed surface area on the hot side is 1.5 m^2. Find the fluid outlet temperatures.

10.38 A one-shell-pass, one-tube-pass, counterflow heat exchanger is used to cool oil ($c_p = 1.811$ kJ/kg-°C) entering the shell at 120°C and flowing at the rate of 15,000 kg/h by use of cooling water ($c_p = 4.18$ kJ/kg-°C) entering the tubes at 30°C and flowing at the rate of 6500 kg/h. The tubes are brass, 1-in. OD, 14 gage. The heat transfer coefficient at the inside tube surface is known to be 700 W/m^2-°C and that on the outside tube surface is known to be 750 W/m^2-°C. The total exposed surface area of the tubes is 20 m^2. Find the outlet temperatures of the two fluids and the total rate of heat exchange.

10.39 A one-shell-pass, two-tube-pass, heat exchanger is to be used to cool ethylene glycol, flowing at the rate of 1 kg/s and entering at 85°C, with water entering at 15°C and flowing at the rate of 2 kg/s. The overall heat transfer coefficient for the exchanger is 500 W/m^2-°C and the total exposed tube surface area is 10 m^2. Find the total rate of heat transfer and the outlet temperatures of the two fluids.

10.40 The lubicating oil in an engine is cooled by air in a cross-flow heat exchanger, both fluids being unmixed. The air, at atmospheric pressure, enters at 30°C and flows at the rate of 0.55 kg/s. The oil flows at the rate of 0.025 kg/s and enters at 75°C. The overall heat transfer coefficient, based on the hot side, is 53 W/m^2-°C and the surface area there is 1 m^2. Find the exit temperature of the oil.

10.41 Water flows at an average velocity of 0.3 m/s through a 1-in. schedule 40 steel pipe. The pipe is bare and passes horizontally through a large room, where it loses heat by free convection to the ambient air at 20°C. If the water enters the pipe at 95°C and the pipe is 150 m long, estimate the outlet temperature of the water.

10.42 Water flows at the rate of 2.5 gpm through a 1-in. schedule 40 steel pipe. The pipe is bare and passes horizontally through a large room, where it loses heat by free convection to the ambient air at 70°F. If the water enters the pipe at 190°F and is 500 ft long, estimate the outlet temperature of the water.

10.43 A horizontal steam condenser is condensing steam at a pressure of 14 kN/m^2. There are two tube passes with 180 tubes per pass, 2.4 m long. The tubes are brass, $\frac{5}{8}$-in., 16-gage, and tube water velocity is 1 m/s. If the cooling water enters at 20°C, find the outlet water temperture nd the kg/h of steam condensed.

10.44 A horizontal steam condenser has two tube passes, 210 tubes per pass, of $\frac{3}{4}$-in. 16-gage brass tubes 2.75 m long. Cooling water is supplied at 25°C with an average tube velocity of 1.5 m/s. What is the outlet water temperature if the steam is being condensed at a pressure of 10 kN/m^2?

10.45 A one-shell-pass, two-tube-pass steam condenser is made of $\frac{3}{4}$-in. 18-gage brass tubes, 8 ft long. There are 220 tubes per pass and the average tube water velocity is 5 ft/s. If the condenser is in a horizontal position, the water inlet at 70°F, and the condensing pressure is 1.5 psia, find the outlet water temperature and the lb_m/h of steam condensed.

10.46 A compact cross-flow heat exchanger is constructed of a plate-fin array with the same plate thickness, plate spacing, fin spacing, and fin dimensions as that considered in Example 10.11. The plates are, however, 8 in. × 8 in. square, and there are 20 flow passages in each direction. The cold fluid is gaseous carbon dioxide at atmospheric pressure entering at 50°F and flowing at the rate of 1200 lb_m/h. The hot fluid is air, at atmospheric pressure, entering at 600°F and flowing at the rate of 900 lb_m/h. Determine the outlet temperature of the two streams and the rate of heat transfer between them. Take the plate and fin thermal conductivity to be 15 Btu/h-ft-°F.

10.47 The heating coil in a ventilation system duct consists of a bank of finned tubes of the configuration shown in Fig. 10.23. The duct is 2 ft × 2 ft in cross section. Thus the tubes are 2 ft long and, as the geometry in Fig. 10.23 shows, the 2-ft height allows for a bank of tubes 25 high in the first row, 24 in the second, and 25 in the third (last) row. The tubes have an ID of 0.28 in., OD of 0.38 in., and are equipped with fins (k = 95 Btu/h-ft°-F) of the dimensions shown. According to the dimensions shown, the thickness from the front face of the tube bank to the back face is 2.52 in. Hot water is the heating medium, and enters at one end of the tubes at 120°F. The water velocity is 2 ft/s in in the tubes. Air flows in the duct at a volume rate of 2000 cfm in an unobstructed cross section. If the air approaching the heating coil is at 50°F (1 atm pressure), what is the temperature of the air leaving the coil?

10.48 In a counterflow heat exchanger, 400,000 lb_m/h of water ($c_p \cong 1$ Btu/lb_m-°F) is heated from 250°F to 310°F by the use of hot gases entering at 750°F anbd leaving at 500°F. If the exchanger has a total surface area of 22,000 ft^2, find the overall U. Use both the Δt_m and ϵ–NTU formulations.

10.49 A heat exchanger is used to heat 3600 kg/h of water ($c_p \cong 4.18$ kJ/kg-°C) from 40°C to 175°C by cooling another stream of water flowing at 4550 kg/h and entering at 315°C. The overall heat transfer coefficient for the exchanger is 1760 W/m^2-°C. Using both the Δt_m and ϵ–NTU formulations, find the required surface area if the heat exchanger is **(a)** parallel-flow, **(b)** counterflow, and **(c)** one-shell-pass, two-tube-pass.

10.50 A one-shell-pass, two-tube-pass heat exchanger is used to heat 27,000 kg/h of water from 85°C to 100°C by the use of steam condensing at 345 kN/m^2. The overall heat transfer coefficient is known to be 2800 W/m^2-°C. If there are 30 1-in.-OD tubes per pass, how long must the tubes be?

10.51 A counterflow heat exchanger is used to heat 2160 kg/h of water (c_p = 4.18 kJ/kg-°C) from 35°C to 90°C by the use of oil (c_p = 2.1 kJ/kg-°C) flowing at the rate of 3240 kg/h with an inlet temperature of 175°C. If the overall heat transfer coefficient is 425 W/m^2-°C, find the surface area required.

10.52 A cross-flow heat exchanger is used to heat water from 35°C to 85°C, flowing at the rate of 3 kg/s, with atmospheric pressure air being cooled from 225°C to

100°C. If the overall heat transfer coefficient is 210 W/m^2-°C, find the surface area required.

10.53 A two-shell-pass, four-tube pass, heat exchanger is used to heat 1.2 kg/s of water from 20°C to 80°C by using 2.2 kg/s of oil (c_p = 2.1 kJ/kg-°C) entering at 160°C. If the overall heat transfer coefficient of the heat exchanger is 300 W/m^2-°C, find the required surface area.

10.54 The combustion air supplied to a boiler furnace is to be preheated in a cross-flow heat exchanger by use of the hot exhaust gases from the furnace. The hot gases (c_p = 1.075 kJ/kg-°C) flow at the rate of 15 kg/s and enter the exchanger at 825°C while the air (c_p = 1.007 kJ/kg-°C) flows at the rate of 10 kg/s, entering at 25°C. If the overall heat transfer coefficient is estimated to be 100 W/m^2-°C, find the required surface area for the heat exchanger to heat the air to 575°C.

10.55 A shell-and-tube heat exchanger is to be made of brass $\frac{3}{4}$-in. 18-gage tubes. The cold fluid is water, flowing at the rate of 33,000 lb$_m$/h, entering the tubes at 85°F and heated to an exit temperature of 140°F. The hot fluid, also water, flows at the rate of 22,000 lb$_m$/h through the shell, entering at 220°F. The tube water velocity is to be 1.3 ft/s, and the overall heat transfer coefficient for the heat exchanger is estimated to be 310 Btu/h-ft^2-°F. It is desired to design the exchanger, requiring the minimum number of passes in each fluid, that satisfies the foregoing conditions. Space limitations require, further, that the tube length be less than 6 ft. Find **(a)** the number of tubes per pass; **(b)** the required number of passes for each fluid and the flow arrangement (i.e., 1 shell pass–1 tube pass, parallel flow; 1 shell pass–1 tube pass, counterflow; 1 shell pass–2 tube passes; etc.; and **(c)** the tube length required.

10.56 A water-to-water, shell-and-tube heat exchanger is to be made of $\frac{3}{4}$-in. 18-gage tubes. Cold water enters the tubes at a flow rate of 15,000 kg/h and is heated from 30°C to 60°C. The hot fluid, also water, flows at the rate of 10,000 kg/h, entering at 105°C. The tube water velocity is to be 0.4 m/s and the overall heat transfer coefficient for the exchanger is 1750 W/m^2-°C. Find the number of tubes per pass and the required length of the tubes if the exchanger is **(a)** parallel-flow; **(b)** counterflow; **(c)** one-shell-pass, two-tube-pass; and **(d)** two-shell-passes, four-tube-passes. Discuss the relative merits of each design.

10.57 A water-to-water heat exchanger is to be made of $\frac{3}{4}$-in. 16-gage tubes. The tube water enters at 90°F leaves at 120°F and flows at the rate of 30,000 lb$_m$/h. The shell water flows at the rate of 20,000 lb$_m$/h, entering at 200°F. The overall heat transfer coefficient is 250 Btu/h-ft^2-°F. The exchanger is to have one shell pass, and the tube water velocity is to be 1 ft/s. If the length of the tubes is not to exceed 8 ft, find **(a)** the number of tubes per pass, **(b)** the number of tube passes, and **(c)** the length of the tubes.

10.58 A one-shell-pass, one-tube-pass heat exchanger is made of 60 tubes ($\frac{5}{8}$ in., 14 gage). Water enters the shell side at 150°C and leaves at 40°C, flowing at the rate of 9000 kg/h. The tube water enters at 25°C and leaves at 65°C. The inside and outside tube surface heat transfer coefficients are 1700 and 8500 W/m^2-°C, respectively. Find the required length of the tubes.

10.59 A steam condenser is made of brass tubes, $\frac{3}{4}$ in., 14 gage. There are 150 tubes per pass. The cooling water flows at the rate of 1.52 m/s, entering at 25°C and leaving at 50°C. The condensing heat transfer coefficient is known to be 10,200 W/m^2-°C and the inside tube surface coefficient is 6800 W/m^2-°C. Steam is being condensed at 77°C. The tubes are not to exceed 3 m in length. Find the number of tube passes and the length of the tubes.

10.60 It is desired to design a steam condenser to condense 10,200 kg/h of saturated steam at 35 kN/m^2 pressure. The condenser is to be supplied with cooling water at 26°C and the water temperature rise is to be 14°C. The tube water velocity is to be 2.1 m/s. The tubes are selected to be $\frac{7}{8}$-in. 16-gage brass (2.23 cm OD, 1.89 cm ID) and are not to exceed 2.5 m in length. Find the number of tube passes and the required tube length.

10.61 Repeat Prob. 10.60 if the tube water velocity is changed to 1.8 m/s.

10.62 A heat exchanger is to be designed to heat 6800 kg/h of water from 20°C to 38°C by the use of saturated steam condensing at 240 kN/m^2. The tubes are to be 1-in. 12-gage brass and are not to exceed 2 m in length. The tube water velocity is to be 0.45 m/s. Determine the number of tubes per pass, the number of tube passes, and the length of the tubes.

10.63 Design a steam condenser (i.e., find the number of tubes per pass, number of passes, and tube length) that will condense 20,000 lb$_m$/h of steam at 5 psia. It is desired to use tubes with 0.875 in. OD, 0.745 in. ID. The tube water velocity is 7 ft/s and the tubes are not to exceed 12 ft in length. The tube water is to enter at 75°F and leave at 100°F.

10.64 Design a feedwater heater to heat 1.5 kg/s of water from 20°C to 55°C by the use of steam condensing at 200 kN/m^2. Set the maximum tube length at 2.5 m.

10.65 A cross-flow heat exchanger is to be designed using the plate-fin configuration of Example 10.11. The same plate thickness, spacing, fin dimensions and thermal conductivity are to be used; however, the plate dimensions in the directions transverse to the fluid streams and the number of flow passages for each fluid are to be determined. It is desired to heat a stream of air flowing at the rate of 1000 lb$_m$/h from 20°F to 160°F by the use of another airstream being cooled from 700°F to 200°F.

CHAPTER 11

Additional Cases of Combined Heat Transfer

11.1 INTRODUCTORY REMARKS

The examples of combined heat transfer considered in Chapter 10 were restricted to cases in which only the conduction and convection modes were present, and primary emphasis was placed on the application of these cases to the analysis and design of heat exchangers. This chapter will be devoted to other examples of combined heat transfer, some of which will include thermal radiation as one of the simultaneous modes taking place. Applications of these principles for the estimation of the thermometric errors will be illustrated. Some recent engineering applications involving multimode heat transfer will be discussed briefly. The analysis of a solar collector involves all of the basic heat transfer modes—conduction, convection, and radiation. The so-called *heat pipe* involves the combined effects of conduction, convection, vaporization, and condensation.

11.2 SIMULTANEOUS CONVECTION AND RADIATION—THE RADIATION COEFFICIENT

The instances of heat transfer from a body surface to its surroundings treated thus far have been limited to cases in which the sole mechanism taking place at the surface was either pure convection (Chapters 5 through 8) or pure radiation (Chapter 9). In many cases of practical interest both the convective and radiative mechanisms occur simultaneously and the conditions may be such that the heat transfer by each of these mechanisms may be of the same order of magnitude. When the convecting fluid surrounding a surface is radiatively nonabsorbing and nonemitting, one may treat the convection and radiation mechanisms as independent.

The surface heat transfer by convection, q_c, is represented by the familiar relation

$$q_c = Ah_c(T_s - T_a), \tag{11.1}$$

in which h_c is the *convective* heat transfer coefficient as determined by the methods of Chapters 5 through 8, T_s is the surface temperature, and T_a represents the temperature of the ambient fluid surroundings. In Chapter 9, the radiant heat transfer from a surface, q_r, was shown to be of the form

$$\begin{aligned} q_r &= A\mathcal{F}(E_{bs} - E_{be}) \\ &= A\sigma\mathcal{F}(T_s^4 - T_e^4). \end{aligned} \tag{11.2}$$

In Eq. (11.2), T_e is the temperature of the radiation environment seen by the surface and $\mathcal{F}$ is a function (sometimes complex) of the surface radiation properties (ϵ, α, etc.) *and* the geometry of the radiative environment. The exact form of $\mathcal{F}$ depends on the particular situation at hand, but all the results of Chapter 9 (particularly the analytic results of Sec. 9.10) may be reduced to this form.

Often Eq. (11.2) is linearized into the form

$$q_r = Ah_r(T_s - T_a) \tag{11.3}$$

by introduction of the *radiation coefficient* h_r:

$$h_r = \sigma\mathcal{F}\frac{T_s^4 - T_e^4}{T_s - T_a}.$$

If, in particular, $T_e = T_a$, the above becomes

$$h_r = \sigma\mathcal{F}(T_s^2 + T_a^2)(T_s + T_a). \tag{11.4}$$

The radiation coefficient was first introduced in Eq. (8.35) in connection with film boiling. The linearization given by Eq. (11.3) is deceptive in that the radiation coefficient, h_r, is a strong function of temperature.

The use of h_r allows one to express, formally, the combined heat flow from the surface as

$$\begin{aligned} q &= q_c + q_r \\ &= A(h_c + h_r)(T_s - T_a) \\ &= Ah_{cr}(T_s - T_a), \end{aligned} \tag{11.5}$$

in which $h_{cr} = h_c + h_r$ is termed the *combined coefficient*.

Some applications of these concepts will now be presented.

Combined Heat Loss from a Completely Enclosed Body

As a first example, imagine a body completely enclosed in a large room. The convective coefficient h_c is determined by the hydrodynamic conditions imposed—

free convection, forced convection, etc., and let it be presumed that h_c has been appropriately determined.

For a body with surface temperature T_s and surface emissivity ϵ_s, completely enclosed by a large room at temperature T_a, Eq. (9.100) gives the radiant heat transfer to be

$$\begin{aligned} q &= \epsilon_s(E_{bs} - E_{ba}) \\ &= \sigma\epsilon_s(T_s^4 - T_a^4), \end{aligned}$$

that is, the $\mathcal{F}$ in Eq. (11.2) is simply $\mathcal{F} = \epsilon_s$. Then the associated radiation coefficient is, by Eq. (11.4),

$$h_r = \sigma\epsilon_s(T_s^2 + T_a^2)(T_s + T_a). \qquad (11.6)$$

The relations above may be used to estimate the combined convective and radiative losses from bodies in large enclosures as illustrated in the next example.

EXAMPLE 11.1

A horizontal oxidized wrought iron pipe (outside diameter D = 16 cm) has a surface temperature of t_s = 100°C and passes through a large room in which heat loss occurs to the ambient air by free convection and to the walls of the room by radiation. Both the ambient air and the walls are T_a = 20°C = 293°K. Find the combined heat loss, per unit of surface area, by convection and radiation.

Solution. The convective coefficient is found by the methods of Chapter 7 for free convection. At a mean film temperature of $t_m = (t_s + t_a)/2 = (100 + 20)/2 = 60$°C, Table A.6 gives the following air properties:

$$\nu = 18.90 \times 10^{-6}\ \text{m}^2/\text{s}, \quad k = 28.52 \times 10^{-3}\ \text{W/m-°C}, \quad \text{Pr} = 0.708.$$

The coefficient of expansion of the air is evaluated at the ambient air temperature T_a = 20°C = 293°K:

$$\beta = \frac{1}{T_a} = \frac{1}{293°\text{K}}.$$

Thus with $\Delta t = t_s - t_a = 100 - 20 = 80$°C, Eq. (7.32) for free convection around horizontal cylinders gives the convective coefficient:

$$\text{Gr}_D = \frac{D^3 g\beta\,\Delta t}{\nu^2} = \frac{(0.16)^3 \times 9.8 \times (1/293) \times 80}{(18.90 \times 10^{-6})^2} = 3.07 \times 10^7,$$

$$\text{Ra}_D = \text{Gr}_D\text{Pr} = 3.07 \times 10^7 \times 0.708 = 2.17 \times 10^7,$$

$$\text{Nu}_D = \left\{0.60 + 0.387\text{Ra}_D^{1/6}\left[1 + \left(\frac{0.559}{\text{Pr}}\right)^{9/16}\right]^{-8/27}\right\}^2$$

$$\mathrm{Nu}_D = \left\{0.60 + 0.387(2.17 \times 10^7)^{1/6}\left[1 + \left(\frac{0.559}{0.708}\right)^{9/16}\right]^{-8/27}\right\}^2 = 35.59,$$

$$h_c = \mathrm{Nu}_D \frac{k}{D} = 35.59 \times \frac{0.02852}{0.16} = 6.34\ \mathrm{W/m^2\text{-}°C}.$$

The radiation coefficient for the radiant exchange between the pipe and the walls is given by Eq. (11.6) if the room is taken to be very large when compared with the pipe:

$$h_r = \sigma\epsilon_s(T_s^2 + T_a^2)(T_s + T_a).$$

The absolute values of the pipe surface temperature and the ambient walls are

$$T_s = 100 + 273 = 373°\mathrm{K},$$

$$T_a = 20 + 273 = 293°\mathrm{K}.$$

For oxidized wrought iron, the pipe surface, Table A.10 suggests that the surface emissivity is $\epsilon_s = 0.94$. Thus the relation above for the radiation coefficient yields

$$\begin{aligned} h_r &= \sigma\epsilon_s(T_s^2 + T_a^2)(T_s + T_a) \\ &= 5.67 \times 10^{-8} \times 0.94(373^2 + 293^2)(373 + 293) \\ &= 7.99\ \mathrm{W/m^2\text{-}°C}. \end{aligned}$$

Thus the combined coefficient for both modes of heat transfer and the surface heat flux are

$$h_{cr} = h_c + h_r = 6.34 + 7.99 = 14.33\ \mathrm{W/m^2\text{-}°C}\ (2.53\ \mathrm{Btu/h\text{-}ft^2\text{-}°F}),$$

$$\begin{aligned} \frac{q}{A} = h_{cr}(T_s - T_a) &= 14.33(373 - 293) \\ &= 1146\ \mathrm{W/m^2}\ (363\ \mathrm{Btu/h\text{-}ft^2}). \end{aligned}$$

Note that even for the modest surface temperature used in this example, the convective and radiation coefficients are of the same order of magnitude ($h_c = 6.34$, $h_r = 7.99$ W/m²-°C). Hence the effects of radiant loss are significant and should not be neglected as is often done. If the surface temperature were increased to, say, $T_s = 200°\mathrm{C} = 473°\mathrm{K}$, identical calculations will reveal that the convective coefficient increases only modestly to $h_c = 7.81$ W/m²-°C, while the radiative coefficient increases by a factor of about 1.6 to $h_r = 12.6$ W/m²-°C. Thus the combined coefficient becomes $h_{cr} = 20.4$ W/m²-°C, emphasizing, again, the importance of including radiative heat transfer in such calculations.

The method of calculation shown in Example 11.1 forms the basis for the various tables and charts available (see, for example, Ref. 1) in the literature which give values of the combined coefficient for pipes as a function of the diameter and temperature difference. The results for insulated pipes are substantially the same as those illustrated above for a bare pipe (except that lower surface temperatures are usually involved) because of the fact that the emissivity of the canvas insulation wrapping is about the same as that of oxidized steel.

An instance of heat transfer by a combination of all three modes is encountered, for example, in the case of heat loss from a pipe (or wall) when the surface temperature is not specified as it was in the above example. If, instead, the bulk temperature of the fluid flowing inside the pipe is specified, one must account for the inside convective film coefficient and the thermal resistance of the pipe and insulation (if any) as well as the combined coefficient. The determination of the rate of heat loss from the pipe then necessitates a trial and error solution just like those presented in Sec. 10.2, for cases where radiation was neglected, except that the computation of h_{cr} illustrated above replaces the step of the calculations wherein the outside convective coefficient was computed. The iterative solution of such problems converges less rapidly when the thermal radiation is included than when it is neglected, because the radiation coefficient is much more sensitive to changes in the surface temperature than is the convective coefficient.

Combined Convection and Radiation in Air Spaces

Another example of combined convection and radiation heat transfer is that found in the case of air spaces between two surfaces of different temperature. Such air spaces are found between the walls of a building, between glass layers in windows, or between the absorbing surface of a solar collector and its glass cover plate. The air spaces may be inclined or in a vertical position.

If the two surface temperatures are denoted by T_h and T_c (for "hot" and "cold," respectively) the convective transfer between them is

$$\frac{q_c}{A} = h_c(T_h - T_c)$$

where the convective coefficient is that due to free convection as found from the correlations given in Eqs. (7.38) through (7.43). The radiative transfer may be modeled as that between two infinite parallel planes as given by Eq. (9.53):

$$\frac{q_r}{A} = \frac{\sigma(T_h^4 - T_c^4)}{(1/\epsilon_h) + (1/\epsilon_c) - 1}.$$

Even when the surfaces are glass, the above representation is probably valid since glass is nearly opaque to low temperature radiation in the long (infrared) part of the spectrum. If *solar* radiation is present, other considerations must be made, as illustrated later in the chapter.

For the representation above the radiation coefficient is then defined as

$$\frac{q_r}{A} = h_r(T_h - T_c), \qquad (11.7)$$

$$h_r = \frac{\sigma(T_h^2 + T_c^2)(T_h + T_c)}{(1/\epsilon_h) + (1/\epsilon_c) - 1},$$

so that the combined heat transfer is

$$\begin{aligned} q &= (h_r + h_c)(T_h - T_c) \\ &= h_{cr}(T_h - T_c). \end{aligned} \tag{11.8}$$

Given the thermal conditions and the surface emissivities, the combined effects of convection and radiation may be deduced. Calculations such as those given in the next example form the basis of an air space conductance values quoted in sources such as Ref. 2. Representations such as the above are also used in the thermal analysis of solar collectors as shown later in this chapter.

EXAMPLE 11.2

A vertical air space exists between the inner and outer walls of a building. The space between the walls is 1.5 in. wide and it is 3 ft high. One of the bounding surfaces of the air space is at $T_h = 95°\text{F} = 555°\text{R}$ and the other is at $T_c = 65°\text{F} = 525°\text{R}$. Find the combined heat transfer coefficient due to convection and radiation between the two surfaces of the air space if the emissivity of each is (a) 0.9; (b) 0.333.

Solution. The convective heat transfer coefficient between the two surfaces may be found from the correlation for free convection in rectangular enclosures as given in Eq. (7.39). At the average temperature for the air space of $t_{av} = (t_h + t_c)/2 = (95 + 65)/2 = 80°\text{F}$, Table A.6 gives for air,

$$\nu = 0.6086 \text{ ft}^2/\text{h}, \qquad k = 1.511 \times 10^{-2} \text{ Btu/h-ft-°F}, \qquad \text{Pr} = 0.711,$$

and the coefficient of expansion at that temperature is

$$\beta = \frac{1}{T_{av}} = \frac{1}{80 + 460} = \frac{1}{540°\text{R}}.$$

The data given also yield the temperature potential and aspect ratio for the space:

$$\Delta t = t_h - t_c = 95 - 65 = 30°\text{F},$$

$$L = 1.5 \text{ in.} = 0.125 \text{ ft}, \qquad H = 3.0 \text{ ft},$$

$$\text{Ar} = \frac{H}{L} = \frac{3.0}{0.125} = 24.$$

Since the air space is vertical, the free convective exchange between the two bounding surfaces is given by Eq. (7.39):

$$\text{Gr}_L = \frac{L^3 g\beta\,\Delta t}{\nu^2} = \frac{(0.125)^3 \times 32.2 \times (3600)^2 \times (1/540) \times 30}{(0.6086)^2}$$

$$= 1.223 \times 10^5,$$

$$\text{Ra}_L = \text{Gr}_L\text{Pr} = 1.223 \times 10^5 \times 0.711 = 8.692 \times 10^4,$$

$$\text{Nu}_1 = 0.0605\text{Ra}_L^{1/3} = 0.0605(8.692 \times 10^4)^{1/3} = 2.68,$$

$$\text{Nu}_2 = \left\{1 + \left[\frac{0.104\text{Ra}_L^{0.293}}{1 + (6310/\text{Ra}_L)^{1.36}}\right]^3\right\}^{1/3}$$

$$= \left\{1 + \left[\frac{0.104 \times (8.692 \times 10^4)^{0.293}}{1 + (6310/8.692 \times 10^4)^{1.36}}\right]^3\right\}^{1/3} = 2.87,$$

$$\text{Nu}_3 = 0.242\left(\frac{\text{Ra}_L}{\text{Ar}}\right)^{0.272}$$

$$= 0.242\left(\frac{8.692 \times 10^4}{24.0}\right)^{0.272} = 2.25.$$

The applicable Nusselt number is the largest of the three above, $\text{Nu}_2 = 2.87$. Thus the convective coefficient is

$$h_c = \text{Nu}_2 \frac{k}{L}$$

$$= 2.87 \times \frac{0.01511}{0.125} = 0.347 \text{ Btu/h-ft}^2\text{-°F}.$$

For the radiative heat transfer coefficient between the two surfaces of the air-space, one may apply Eq. (11.7) if the space is small compared with the size of the surfaces:

$$h_r = \frac{\sigma(T_h^2 + T_c^2)(T_h + T_c)}{(1/\epsilon_h) + (1/\epsilon_c) - 1}.$$

The temperatures are given: $T_h = 555$°R, $T_c = 525$°R, and in English units $\sigma = 0.1714 \times 10^{-8}$ Btu/h-ft^2-°R^4. Thus h_r is easily found once the surface emissivities are specified.

(a) For $\epsilon_h = \epsilon_c = 0.9$, the radiative coefficient is

$$h_r = \frac{\sigma(T_h^2 + T_c^2)(T_h + T_c)}{(1/\epsilon_h) + (1/\epsilon_c) - 1}$$

$$= \frac{0.1714 \times 10^{-8}(555^2 + 525^2)(555 + 525)}{(1/0.9) + (1/0.9) - 1}$$

$$= 0.884 \text{ Btu/h-ft}^2\text{-°F}.$$

Thus the combined convective and radiative coefficient is

$$h_{cr} = h_c + h_r$$

$$= 0.347 + 0.884 = 1.23 \text{ Btu/h-ft}^2\text{-°F } (6.98 \text{ W/m}^2\text{-°C}).$$

Note, once more, that the radiative exchange is of the same order of magnitude as the convective exchange.

(b) If the surface emissivities are reduced to $\epsilon_h = \epsilon_c = 0.333$, the radiation coefficient and the combined coefficient are correspondingly reduced:

$$h_r = \frac{\sigma(T_h^2 + T_c^2)(T_h + T_c)}{(1/\epsilon_h) + (1/\epsilon_c) - 1}$$

$$= \frac{0.1714 \times 10^{-8}(555^2 + 525^2)(555 + 525)}{(1/0.333) + (1/0.333) - 1}$$

$$= 0.216 \text{ Btu/h-ft}^2\text{-°F},$$

$$h_{cr} = h_c + h_r$$

$$= 0.347 + 0.216 = 0.56 \text{ Btu/h-ft}^2\text{-°F } (3.18 \text{ W/m}^2\text{-°C}).$$

11.3 THERMOCOUPLE LEAD ERROR IN SURFACE TEMPERATURE MEASUREMENTS

Some of the principles of the various modes of heat transfer may be combined to analyze, approximately, the error that may be expected when the temperature of a body or fluid is measured with one of the usual thermometric devices—a thermometer or a thermocouple. Since the introduction of a temperature-measuring device into a system in which heat transfer is taking place will alter the thermal conditions existing there, one must expect that the device will indicate a value of the sought-after temperature that differs from the temperature that would actually exist had no measurement been made. This error is due to the thermal effects of the measuring device itself on the system and is not a calibration error of the thermometer. Some typical cases will be discussed in the following sections.

If a thermocouple is attached to the surface of a solid body in the manner illustrated in Fig. 11.1(a), heat will be conducted along the thermocouple leads and dissipated into the surroundings by convection from the wires. This conduction of heat sets up temperature gradients within the body in the vicinity of the point of attachment—a condition that would not exist had the thermocouple not been attached. The thermocouple will read a value corresponding to the local depression in the surface temperature at the point of attachment.

Since thermocouple wires are usually long and thin, the conduction of heat along them—and its eventual dissipation into the surroundings—may be modeled as heat flow along a spine of uniform cross section as discussed in Sec. 2.9 in connection with the concept of an extended surface. To simplify the system under consideration, let the thermocouple be represented as a single wire of radius R and thermal conductivity k_t. Further, let h represent the heat transfer coefficient between the wire surface and the ambient air at temperature t_a. Then if t_t is used to denote the depressed temperature of the surface under the point of attachment of the thermocouple, the wire may be thought of as a constant cross section spine main-

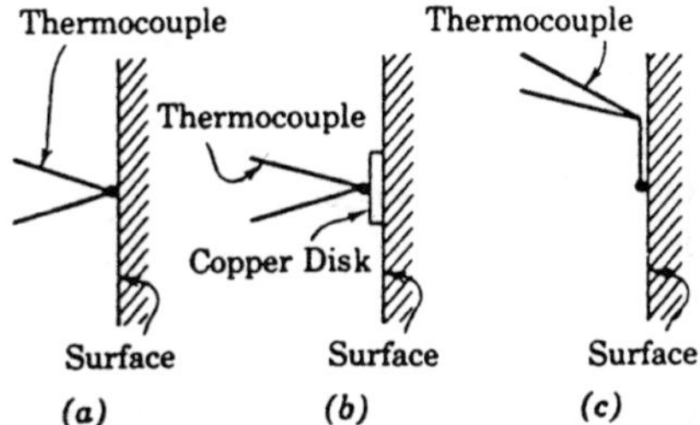

Figure 11.1. Measurement of surface temperatures by the use of thermocouples.

tained at t_t on one end and exchanging heat through its surface with air at t_a. Solutions for the heat flow in such a geometry were discussed in Sec. 2.9 for various boundary conditions; however, the simple case in which the fin, or spine, is taken to be infinitely long would appear to be applicable in the present case since one would normally expect a thermocouple wire to be very long compared with its diameter. In this instance the heat flow into the wire at its point of attachment may be found from application of Eq. (2.56):

$$q = k_t \sqrt{\frac{hC}{k_t A}} \times A(t_t - t_a).$$

Here C and A are the perimeter and cross-sectional areas of the wire, respectively. Thus

$$q = \pi\sqrt{2k_t R^3 h} \times (t_t - t_a). \tag{11.9}$$

The heat noted in Eq. (11.9) must be supplied from within the body. The body may be represented as a semi-infinite solid maintained at a temperature t_s at points far removed from the point of the thermocouple attachment. The point of attachment may be approximated as an area of radius R maintained at temperature t_t. Thus t_s is the true surface temperature, t_t is the temperature indicated by the thermocouple, and $(t_s - t_t)$ is the error introduced by the attachment of the thermocouple.

One of the classical solutions of the mathematical theory of heat conduction is that for a semi-infinite solid of a uniform temperature, say t_s, at points far removed from a heat sink on the surface which has a finite radius, say R, and a temperature t_t. The solution is not given here but may be found in Ref. 3. The rate of heat flow into the sink is given by this solution to be

$$q = 4Rk_s(t_s - t_t). \tag{11.10}$$

In Eq. (11.10) k_s is used to denote the thermal conductivity of the surface material. The result quoted in Eq. (11.10) may also be obtained from the tabulation of conduction shape factors given in Table 3.1. Combination of Eqs. (11.9) and (11.10) gives the measured error to be

$$t_s - t_t = \frac{(\pi/k_s)\sqrt{k_1 Rh/8}(t_s - t_a)}{1 + (\pi/k_s)\sqrt{k_t Rh/8}}.$$

Now, h is determined by the free convection relations of Chapter 7. Generally, the diameter of a thermocouple wire is quite small, and the average temperature difference between the wire and the ambient air would not be expected to be very great. Hence one may estimate h by assuming that the Grashof, or Rayleigh, number for the free convection around the wire is approximately zero. In such an instance, the correlation for free convection on a horizontal cylinder given in Eq. (7.32) yields $\mathrm{Nu}_D = 0.36 = h(2R)/k_a$, k_a being the thermal conductivity of the ambient air. Thus the relation above gives the estimated error to be

$$\begin{aligned} t_s - t_t &= \frac{0.47(k_a k_t/k_s^2)^{1/2}}{1 + 0.47(k_a k_t/k_s^2)^{1/2}}(t_s - t_a) \\ &= \frac{t_s - t_a}{1 + [0.47(k_a k_t/k_s^2)^{1/2}]^{-1}}. \end{aligned} \tag{11.11}$$

This expression gives an approximation to the magnitude of the error that might be expected in the measurement of a surface temperature by a thermocouple when it is attached in the manner shown in Fig. 11.1(a). Equation (11.11) is only an approximation in view of the simple model chosen and the simplifying assumptions made. It is useful for estimating the magnitude of the error involved but is certainly not accurate enough to be used to correct observed readings. It is interesting to note that the estimated error is independent of the wire diameter as long as it is small compared with the other dimensions involved. The greater the conductivity of the thermocouple wire, k_t, the greater is the error since heat will be more readily conducted away from the solid surface. Similarly, the smaller the value of the thermal conductivity of the solid, k_s, the greater is the error since the body has greater difficulty giving up the heat taken away by the wire—increasing the local temperature depression. In some instances, the magnitude of the error involved may be quite significant, as the following example shows.

EXAMPLE 11.3

A furnace wall is made of chrome brick and has an exposed surface temperature of $t_s = 200$°C. The surface is exposed to ambient air at $t_a = 40$°C. Estimate the error that may be expected to be indicated if the surface temperature is to be measured by a thermocouple made of constantan (60% Cu, 40% Ni).

Solution. The thermal conductivities of the brick surface, the thermocouple, and the ambient air may be estimated by use of Tables A.2, A.1, and A.6:

$$k_s = 2.32, \qquad k_t = 26, \qquad k_a = 0.0326 \text{ W/m-°C}.$$

With $t_s = 200$°C and $t_a = 40$°C, Eq. (11.11) estimates the error to be

$$t_s - t_t = \frac{t_s - t_a}{1 + [0.047(k_a k_t/k_s^2)^{1/2}]^{-1}}$$

$$= \frac{200 - 40}{1 + \{0.47[0.0326 \times 26/(2.32)^2]^{1/2}\}^{-1}}$$

$$= 25°\text{C } (45°\text{F}).$$

The estimate of error is seen to be rather significant. Materials were chosen to illustrate a "worst case." The wire has a large conductivity which allows it to withdraw heat readily from the surface. The surface, on the other hand, has a low conductivity so that it has difficulty supplying the heat and the surface temperature is depressed locally rather dramatically. Had the surface been made of a better conductor (say, iron with $k_s = 62$ W/m-°C), similar calculations give a much lower error:

$$t_s - t_t = 1.1°\text{C } (2.0°\text{F}).$$

The error in the reading of a thermocouple attached to a surface may be reduced by one of the schemes indicated in Fig. 11.1(b) or (c). Figure 11.1(b) shows the thermocouple attached to a thin disk of material of high thermal conductivity (say copper). The disk is then placed in contact with the surface. The arrangement is advantageous since it provides a large area at the surface from which the heat conducted away by the thermocouple is withdrawn, producing a less severe local depression in the surface temperature.

If the leads of the thermocouple are placed along the surface for a short distance before being led away, as shown in Fig. 11.1(c), the junction of the thermocouple does not lie at the point of departure of the leads from the surface—the point at which the greatest local depression in temperature is likely to occur.

11.4 THERMOMETER WELL ERRORS DUE TO CONDUCTION

The temperature of a fluid flowing in a closed conduit, or pipe, is often measured by the use of a thermometer or thermocouple placed in a thermometer well inserted into the fluid stream. Such an arrangement is indicated schematically in Fig. 11.2. The sketch of Fig. 11.2 presumes that the fluid in the pipe is hotter than the ambient conditions outside the pipe (although the reverse might be true), so that the pipe wall temperature, t_0, is less than the bulk temperature of the fluid, t_f. It is the temperature t_f that is desired to be measured by the temperature measuring device inserted into the thermometer well. However, the presence of the well, extending from the pipe wall at t_0 into the fluid at $t_f > t_0$, establishes a path for heat conduction, resulting in a temperature gradient along the walls of the well as suggested in Fig. 11.2. The thermometer or thermocouple inserted into the well measures the temperature at the end of the well, t_t, which may differ from t_f by some amount

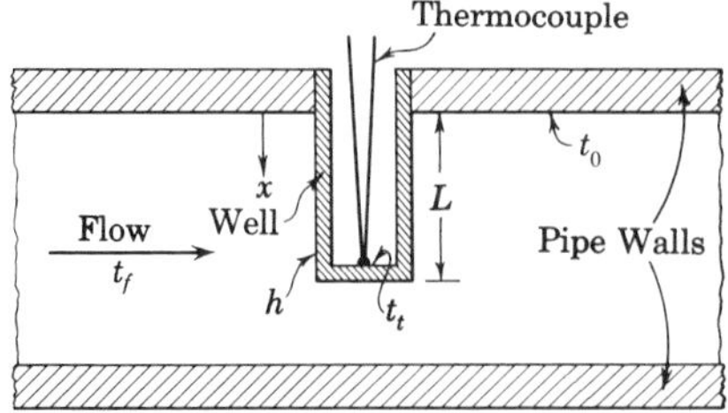

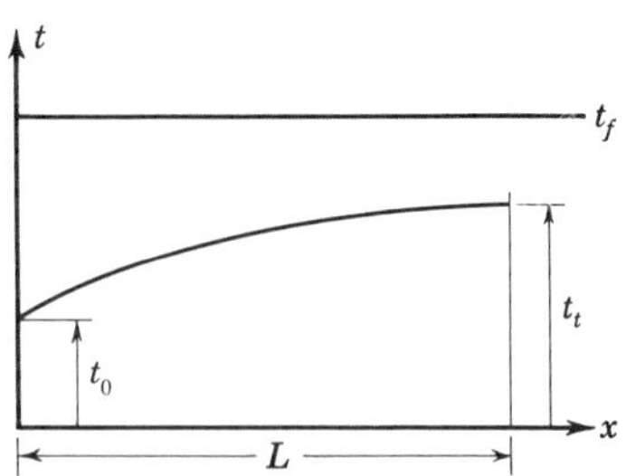

Figure 11.2

dependent on the gradient established along the well. It is desired to estimate the resultant error, $t_t - t_f$.

If the thermometer well is assumed to have good thermal contact with the pipe wall, it may be treated as a spine of uniform cross section and finite length L, maintained at t_0 at one end and extending into an ambient fluid at t_f. Such a geometry was examined in Sec. 2.9 in connection with extended surfaces. Neglecting heat flow out of the end of the spine, one may apply Eq. (2.51) at $x = L$ to give the temperature difference at the measurement end, the error $t_t - t_f$, to be

$$t_t - t_f = \frac{t_0 - t_f}{\cosh mL}, \qquad mL = \sqrt{\frac{hC}{kA}}\,L. \tag{11.12}$$

In Eq. (11.12), L represents the well length and A is the cross-sectional area for heat conduction in the well wall, $(\pi/4)(D_o^2 - D_i^2)$ if cylindric with outside and inside diameters D_o and D_i, respectively. The quantity C is the exposed perimeter of the well exterior, πD_o, since in such installations the well interior is usually filled with insulation or a stagnant liquid to minimize heat loss at the inner surface.

Equation (11.12) permits one to estimate the error associated with the use of a thermometer well if its geometry is specified and if one knows h, the heat transfer coefficient at the well surface. For a cylindrically shaped well placed normal to the flow in the pipe, the correlations given in Eq. (6.48) or (6.49) for forced convection in cross flow over a cylinder may be used to estimate h. These correlations, as is usually the case, require the knowledge of the surface temperature of the cylinder. In the application considered here the surface temperature varies along the well (from t_0 to t_t), so some representative mean value must be selected. In view of the many simplifying assumptions already made, a simple arithmetic average of t_0 and t_t is probably all that is justified [although one could use the concept of fin efficiency to calculate an average surface temperature from $\kappa = (t_{av} - t_f)/(t_0 - t_f) = (1/mL)\tanh mL$]. In either event an iterative calculation would be required since neither t_t nor h is known and the determination of one involves the other. This type of calculation is illustrated in the next example.

The estimate of the error in the temperature reading given by Eq. (11.12) involves knowledge of the pipe surface temperature t_0. This, itself, could possibly require an iterative calculation also, as discussed in Sec. 10.2. In any case, the error of Eq. (11.12) is directly proportional to the difference $(t_0 - t_f)$, so the error may be reduced by insulating the outside pipe surface in the vicinity of the thermometer well in order to reduce this difference. Also, Eq. (11.12) shows that any means of making mL as great as possible, such as increasing the length L, will reduce the error.

EXAMPLE 11.4

A thermometer well is inserted in a pipe, normal to the direction of flow, as depicted in Fig. 11.2. The well is of cylindric shape (outside diameter $D_o = 0.5$ cm, inside diameter $D_i = 0.25$ cm, length $L = 6$ cm) and is made of steel with $k = 30$ W/m-°C. The pipe has a diameter of 12 cm, and carries atmospheric pressure air flowing with a velocity $U = 5$ m/s and a temperature $t_f = 250$°C. Heat losses to the surroundings reduces the pipe wall temperature below that of the flowing air, and this temperature is observed to be $t_0 = 150$°C. Estimate the temperature observed by a thermocouple attached to the end of the well, t_t. That is, estimate the error in the observation of t_f, $(t_t - t_f)$.

Solution. The following calculations will require knowledge of two geometric properties of the thermometer well, the exposed perimeter, C, and the cross-sectional area for conduction, A. These are

$$C = \pi D_o = \pi \times 0.005 = 0.01571 \text{ m},$$

$$A = \frac{\pi}{4}(D_o^2 - D_i^2) = \frac{\pi}{4}(0.005^2 - 0.0025^2)$$

$$= 1.4726 \times 10^{-5} \text{ m}^2.$$

The temperature indicated by the thermocouple, t_t, or the measurement error, $t_t - t_f$, may be found from Eq. (11.12). However, use of Eq. (11.12) requires the determination of the heat transfer coefficient for the air as it flows over the thermometer well. Determination of this coefficient, h, requires, in addition to the state of the flowing air, a representative surface temperature for the well. This temperature varies from t_0 at the pipe wall to t_t at the end of the well. While an integrated average surface temperature could be deduced from the temperature distribution along the well (i.e., the fin efficiency κ) the approximate model being used probably does not justify such an elaborate method; hence the average surface temperature of the well will simply be taken as the mean of those at the two ends: $t_s = (t_0 + t_t)/2$. Even so, an iterative calculation is required since h depends on t_s (and thus t_t) and t_t depends on h. So one must assume a value for t_t; find h from the associated value of t_s and the appropriate convection correlation; calculate a resultant t_t from this value of h using Eq. (11.12); and then revise the estimate of t_t accordingly. The process is repeated until the assumed and calculated values are in sufficient agreement. For this example the results are shown below.

1. *Assume t_t and determine h.* As an initial estimate it will be assumed that

$$t_t = 230°\text{C}.$$

The mean surface temperature of the well is then

$$t_s = \frac{t_0 + t_t}{2} = \frac{150 + 230}{2} = 190°\text{C}.$$

Use of the correlation for cross flow over a cylinder requires properties of the fluid to be evaluated at the mean film temperature, the mean of the flowing air temperature $t_f = 250°\text{C}$ and t_s just found:

$$t_m = \frac{t_s + t_f}{2} = \frac{190 + 250}{2} = 220°\text{C}.$$

At this temperature, Table A.6 gives, for atmospheric pressure air,

$$\nu = 37.08 \times 10^{-6}\ \text{m}^2/\text{s}, \quad k = 0.03908\ \text{W/m-°C}, \quad \text{Pr} = 0.699.$$

Thus since the air velocity in the pipe is known to be $U = 5$ m/s, Eq. (6.49) for cross flow over a cylinder gives [Eq. (6.48) could also be used with somewhat different results]

$$\text{Re}_D = \frac{UD_o}{\nu} = \frac{5 \times 0.005}{37.08 \times 10^{-6}} = 674,$$

$$\text{Nu}_D = 0.3 + 0.62\text{Re}_D^{1/2}\text{Pr}^{1/3}[1 + (0.4/\text{Pr})^{2/3}]^{-1/4} \times [1 + (\text{Re}_D/2.82 \times 10^5)^{5/8}]^{4/5}$$

$$= 0.3 + 0.62(674)^{1/2}(0.699)^{1/3}[1 + (0.4/0.699)^{2/3}]^{-1/4} \times [1 + (674/2.82 \times 10^5)^{5/8}]^{4.5}$$

$$= 13.1,$$

$$h = \text{Nu}_D \frac{k}{D_o} = 13.1 \times \frac{0.03908}{0.005} = 102.1\ \text{W/m}^2\text{-°C}.$$

2. *Determine a revised t_t.* With this initial estimate of the heat transfer coefficient, Eq. (11.12) may be applied to determine an improved value of the end temperature (remembering that the well conductivity and length are known to be $k = 30$ W/m-°C, $L = 6$ cm $= 0.06$ m):

$$mL = \sqrt{\frac{hC}{kA}}L = \left(\frac{102.1 \times 0.01571}{30 \times 1.4726 \times 10^{-5}}\right)^{1/2} \times 0.06 = 3.6153$$

$$t_t - t_f = \frac{t_0 - t_f}{\cosh mL} = \frac{150 - 250}{\cosh(3.6153)} = -5.4°\text{C}$$

$$t_t = (t_t - t_f) + t_f = -5.4 + 250 = 244.6°\text{C}.$$

This value of $t_t = 244.6°\text{C}$ is in considerable variance with the assumed value of 230°C. Thus a revised estimate must be made, and the calculations repeated.

3. *Assume a new t_t.* A revised value of t_t is now chosen, and all the calculations of steps 1 and 2 repeated. With much less detail the results are:

$$t_t = 245°\text{C, assumed,}$$

$$t_s = \frac{t_0 + t_t}{2} = 197.5°\text{C},$$

$$t_m = \frac{t_s + t_f}{2} = 224°\text{C},$$

$$\nu = 37.59 \times 10^{-6}\ \text{m}^2/\text{s}, \quad k = 0.03933\ \text{W/m-°C}, \quad \text{Pr} = 0.699,$$

$$\text{Re}_D = \frac{UD_o}{\nu} = 665,$$

$$\text{Nu}_D = 0.3 + 0.62\text{Re}_D^{1/2}\text{Pr}^{1/3}[1 + (0.4/\text{Pr})^{2/3}]^{-1/4} \times [1 + \text{Re}_D/2.82 \times 10^5)^{5/8}]^{4/5}$$

$$= 12.97$$

$$h = \text{Nu}_D \times \frac{k}{D_o} = 102.0\ \text{W/m}^2\text{-°C},$$

$$mL = \sqrt{\frac{hC}{kA}}L = 3.6136,$$

$$t_t - t_f = \frac{t_0 - t_f}{\cosh mL} = -5.4°\text{C},$$

$$t_t = 244.6°\text{C, calculated.}$$

The agreement between the assumed and calculated values of t_t is quite good and the final results are then:

Indicated temperature: $t_t = 245°\text{C}$ (473°F).

Temperature error: $(t_t - t_f) = -5°\text{C}$ (9°F).

For the stated conditions of this problem, the error incurred, 5°C, is probably quite significant. The error may be reduced by insulating the pipe surface to reduce the difference $t_0 - t_f$ to which the error is proportional (the temperatures t_s and t_m would also change, altering h and mL somewhat). One could also reduce the error by increasing the denominator of Eq. (11.12). The most direct way of doing this is to increase the well length L. The reader may verify the fact that if L is increased from 6 cm to 7 cm, with the other data unchanged, the error is reduced to $(t_t - t_f) = -3°\text{C}$. There is, of course, a limit to this approach since the pipe diameter was given as 12 cm. Thus, increasing L much more than 7 cm will move the sensing portion of the thermometer well out of the central core of the airflow in the pipe. Alternatively, the well could be made longer by inserting it at an angle with the pipe axis—however, a different correlation for the heat transfer coefficient h would have to be sought.

11.5 RADIATION EFFECTS IN THE MEASUREMENT OF GAS TEMPERATURES

If the temperature of a high-temperature gas stream is to be measured by the insertion of a thermometer or thermocouple, the effects of the radiant exchange between the pipe walls and the temperature-sensing element may be of such an order of magnitude as to influence markedly the value indicated.

Conduction Effects Negligible

The effects of thermal radiation on the measurement of gas temperatures is best illustrated by first examining the case in which conduction along the thermometer or thermocouple is negligible. Such might be the case when the thermocouple probe has a sufficiently long length as, perhaps, illustrated in Fig. 11.3. With conduction neglected, the equilibrium temperature of the thermocouple is determined by a balance between the radiative transfer with the pipe walls and the convective transfer with the flowing gas stream. On a unit area basis, this balance is

$$-h_c(T_t - T_f) = \epsilon\sigma(T_t^4 - T_0^4), \tag{11.13}$$

where T_f is the fluid temperature, T_t the thermocouple surface temperature, T_0 the pipe wall temperature, h_c the convective heat transfer coefficient, and ϵ the emissivity of the thermocouple housing. The expression on the right side of Eq. (11.13) is based on Eq. (9.100) for the radiant exchange between a surface (the thermo-

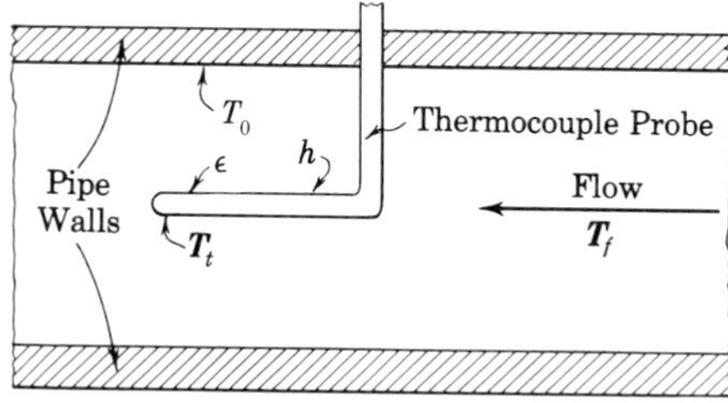

Figure 11.3

couple) completely enclosed by a second, much larger, surface (the pipe walls). The term $(T_t - T_f)$ on the left side of Eq. (11.13) constitutes the error in the gas temperature indicated by the thermocouple.

The convective heat transfer coefficient between the gas stream and the thermocouple housing, h_c, is found by the methods of Chapter 6, depending on the geometry of the thermocouple. Since the correlations for forced convection in external flow are generally based on the mean film temperature [$(T_t + T_f)/2$ in this case], it is necessary to perform an iterative solution of Eq. (11.13) in order to estimate the error indicated by the thermocouple reading, $T_t - T_f$.

EXAMPLE 11.5

High-temperature atmospheric-pressure air flows in a large duct with a velocity of 20 ft/s. The duct wall temperature is 300°F. A thermocouple probe (diameter 0.5 in.) with a surface emissivity of $\epsilon = 0.5$ is placed normal to the flow. If the effects of conduction are negligible, estimate the air temperature when the thermocouple reads 1000°F.

Solution. Since the heat balance of Eq. (11.13) expresses the fluid temperature T_f in terms of the convective coefficient, which, in turn, is dependent on T_f, an iterative calculation is necessary. For the sake of brevity, only the final, verified, iteration is given here.

Thus *assume* that the temperature of the flowing gas is $T_f = 1300°\text{F} = 1760°\text{R}$. Then the mean film temperature for determination of the convective heat transfer coefficient is, using the known thermocouple surface temperature $T_t = 1000°\text{F} = 1460°\text{R}$,

$$t_m = \frac{t_f + t_t}{2} = \frac{1300 + 1000}{2} = 1150°\text{F}.$$

At this temperature, Table A.6 gives the following air properties:

$$\nu = 3.840 \text{ ft}^2/\text{h}, \qquad k = 0.03586 \text{ Btu/h-ft-°F}, \qquad \text{Pr} = 0.706.$$

For the given air velocity $U = 20$ ft/s and probe diameter $D = 0.5$ in., the heat transfer coefficient of the probe surface is given by Eq. (6.49) for cross flow normal to a cylinder:

$$\text{Re}_D = \frac{UD}{\nu} = \frac{20 \times 3600 \times 0.5/12}{3.840} = 781,$$

$$\begin{aligned}\text{Nu}_D &= 0.3 + 0.62\text{Re}_D^{1/2}\text{Pr}^{1/3}[1 + (0.4/\text{Pr})^{2/3}]^{-1/4} \\ &\qquad \times [1 + (\text{Re}_D/2.82 \times 10^5)^{5/8}]^{4/5} \\ &= 0.3 + 0.62(781)^{1/2}(0.706)^{1/3}[1 + (0.4/0.706)^{2/3}]^{-1/4} \\ &\qquad \times [1 + (781/2.82 \times 10^5)^{5/8}]^{4/5}\end{aligned}$$

$$\mathrm{Nu}_D = 14.1,$$

$$h_c = \mathrm{Nu}_D \frac{k}{D} = 14.1 \times \frac{0.03586}{0.5/12} = 12.1 \text{ Btu/h-ft}^2\text{-}^\circ\text{F}.$$

Equation (11.13) may now be used to calculate the gas temperature T_f for comparison with the assumed value, using the known values for the pipe wall temperature, $T_0 = 300°\text{F} = 760°\text{R}$, and the thermocouple surface temperature, $T_t = 1000°\text{F} = 1460°\text{R}$:

$$h_c(T_f - T_t) = \epsilon\sigma(T_t^4 - T_0^4)$$

$$12.1(T_f - 1460) = 0.5 \times 0.1714 \times 10^{-8}[(1460)^4 - (760)^4]$$

$$T_f = 1758°\text{R} = 1298°\text{F}.$$

This calculated value of $T_f = 1298°\text{F}$ is sufficiently close to the assumed value of 1300°F to take the final value for the estimate of the gas temperature and the error to be

$$T_f = 1298°\text{F}\ (703°\text{C}),$$

$$T_t - T_f = -298°\text{F}\ (166°\text{C}).$$

For the high gas temperature and relatively low pipe wall temperature used in this example, the error in the thermocouple reading due to thermal radiation is seen to be rather large.

The radiation error involved in the measurement of the temperature of a very hot gas may be reduced considerably by the use of radiation shields. The use of plane radiation shields was discussed in Sec. 9.6, and it was noted in that case that the use of one shield reduced the radiant energy exchange to one-half of its original, two shields reduced it to one-third, etc. The same sort of advantage can be deduced for a radiation shield around a thermocouple housing. If T_{sh} denotes the equilibrium temperature of a radiation shield placed around a thermocouple, and if one assumes

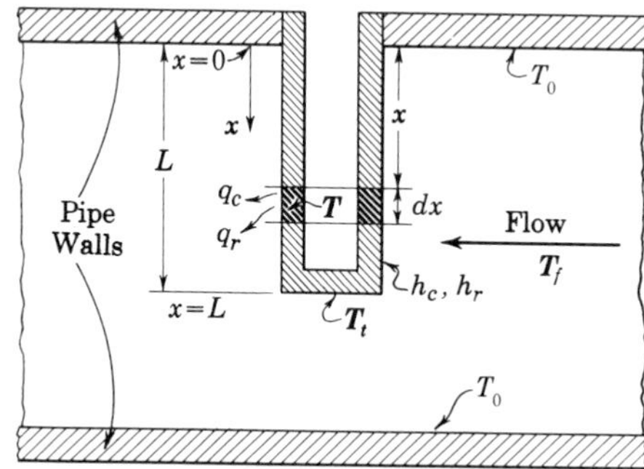

Figure 11.4

that the shield is large compared to the thermocouple but small compared to the duct walls, the equilibrium temperature is given by

$$\sigma\epsilon_t(T_t^4 - T_{sh}^4) = \sigma\epsilon_{sh}(T_{sh}^4 - T_0^4).$$

If one can say that the emissivities of the thermocouple and the shield are practically the same,

$$T_{sh}^4 = \frac{T_0^4 + T_t^4}{2}.$$

Thus Eq. (11.13) may be replaced by

$$\begin{aligned} h_c(T_f - T_t) &= \sigma\epsilon(T_{sh}^4 - T_0^4) \qquad (11.14) \\ &= \frac{\sigma\epsilon}{2}(T_t^4 - T_0^4). \end{aligned}$$

The error is reduced to one-half of its former value. It is not difficult to show, as in the case of plane radiation shields, that if N shields are placed around the thermocouple probe, the error will be reduced to $1/(N + 1)$ of its value when no shield is used.

Conduction Effects Not Negligible

Section 11.4 considered the error involved in the use of a thermometer well in a gas stream when the effects of radiation were neglected. The discussion above considered the case in which conduction errors were negligible and radiant effects were dominant. This section will consider the case of the error likely to be involved in the measurement of the temperature of a fluid stream by means of a thermometer well when all three modes of heat transfer must be taken into account.

Figure 11.4 illustrates the simplified system assumed to represent the thermometer well. The temperatures of the pipe surface, flowing gas, and thermometer well end are denoted by T_s, T_f, and T_t, respectively. By assuming no losses in the interior of the well, T_t will be the temperature indicated by the thermometer or thermocouple.

The temperature of the well surface varies, in some manner, from its value at the pipe wall, T_0, to the indicated temperature, T_t, at the end. In fact, it is this variation that one needs to find in order to predict T_t in terms of the convective and radiative environments "seen" by the well.

Let, as shown in Fig. 11.4, x be the local coordinate measured along the thermometer well—from $x = 0$ at the pipe wall, where the temperature is T_0 to $x = L$, the well length, where the temperature is T_t. At some intermediate location, $x = x$, let the local well surface temperature be denoted by T. Then the convective exchange between the flowing gas at T_f and the well surface, per unit area, is

$$\frac{q_c}{A} = h_c(T - T_f),$$

h_c being the convective heat transfer coefficient determined according to the methods of Chapter 6.

The radiative exchange from the well surface is to an environment at the pipe wall temperature T_0. Presuming that the thermometer well is small compared to the enclosing pipe, the radiation coefficient may be taken to be represented by Eq. (9.100) as was done earlier in Sec. 11.2 and Eq. (11.6):

$$h_r = \epsilon\sigma(T^2 + T_0^2)(T + T_0), \tag{11.15}$$

where ϵ is the surface emissivity. Then the radiative exchange at the well surface is

$$\frac{q_r}{A} = h_r(T - T_0),$$

and the combined exchange is

$$\frac{q_c + q_r}{A} = h_c(T - T_f) + h_r(T - T_0). \tag{11.16}$$

Equation (11.16) indicates that the well surface sees two environments for the transfer of heat: a fluid at T_f through the coefficient h_c, and the radiative environment at T_0 through the coefficient h_r. This two environment situation may be expressed in terms of a single equivalent environment at a weighted environment temperature, T_e, defined as

$$T_e = \frac{h_c T_f + h_r T_0}{h_c + h_r}. \tag{11.17}$$

Then the combined heat exchange of Eq. (11.16) may be expressed in terms of this single equivalent temperature and the combined coefficient $(h_c + h_r)$:

$$\frac{q_c + q_r}{A} = (h_c + h_r)(T - T_e). \tag{11.18}$$

With these definitions the thermometer well may be characterized as a spine of uniform cross section, A, and uniform perimeter, C, maintained at a base $(x = 0)$ temperature of T_0 and exposed to an environment at T_e through a linear coefficient $h_c + h_r$. These conditions are identical to those used to develop Eq. (2.53) in Sec. 2.9 [or as just applied in Eq. (11.12) of Sec. 11.4] with T_e replacing t_f and $(h_c + h_r)$ replacing h. Thus the temperature distribution in the well is given by

$$m = \sqrt{\frac{(h_c + h_r)C}{kA}}, \tag{11.19}$$

$$\frac{T - T_e}{T_0 - T_e} = \frac{\cosh m(L - x)}{\cosh mL}. \tag{11.20}$$

At the end of the well, $x = L$, the temperature indicated by the thermocouple or thermometer inserted into it, T_t, is given by

$$\frac{T_t - T_e}{T_0 - T_e} = \frac{1}{\cosh mL}. \tag{11.21}$$

Equation (11.21) may be used as it stands to estimate the indicated temperature, T_t, and how it differs from the correct fluid temperature, T_f, or one may introduce Eq. (11.17) to find the indicated error, $T_t - T_f$:

$$\frac{T_t - T_f}{T_0 - T_f} = 1 - \frac{h_c}{h_c + h_r}\left(1 - \frac{1}{\cosh mL}\right). \tag{11.22}$$

Given the well geometry, the flow conditions of the fluid, and the pipe wall temperature one may use Eqs. (11.21) and (11.22) to determine T_t or $T_t - T_f$ after appropriately finding the convective coefficient h_c. As in the preceding cases, an iterative solution is often required since h_c is normally based on an average well surface temperature. As done in Example 11.4, this average surface temperature may usually be simply taken as the mean of T_0 and T_t, $(T_0 + T_t)/2$, with T_t being unknown.

In applying Eqs. (11.21) and (11.22) it must be remembered that the parameter m is defined differently than in the usual case, as in Eq. (11.19), with the combined coefficient $h_c + h_r$ being used.

EXAMPLE 11.6

Repeat Example 11.4 by including the effects of radiant exchange with the pipe wall. Let the emissivity of the thermometer well surface be $\epsilon = 0.5$.

Solution. In the statement of Example 11.4, the gas flowing in the pipe, atmospheric-pressure air, has a temperature of

$$T_f = 250°\text{C} = 523°\text{K},$$

and a velocity of

$$U = 5 \text{ m/s}.$$

The pipe walls are at a temperature of

$$T_0 = 150°\text{C} = 423°\text{K}.$$

The specific dimensions and thermal properties of the thermometer well gave the following necessary parameters:

$$D_o = 0.5 \text{ cm} = 0.005 \text{ m},$$

$$L = 6 \text{ cm} = 0.06 \text{ m},$$

$$C = 0.01571 \text{ m},$$

$$A = 1.4726 \times 10^{-5} \text{ m}^2,$$

$$k = 30 \text{ W/m-°C},$$

$$\epsilon = 0.5.$$

As in Example 11.4, both the convective coefficient and the radiative coefficient will be based on an average surface temperature, which will be taken as the mean of those at each end of the thermometer well: $T_s = (T_0 + T_t)/2$. Thus the determination of T_t by Eq. (11.21) will involve knowing T_t to find h_c and h_r. Thus an iterative calculation for T_t is necessary in the same manner as in Example 11.4. For brevity, only the final, verified, iteration is given here.

Thus *assume* that the temperature indicated by the thermocouple is

$$T_t = 237\text{°C} = 510\text{°K, assumed.}$$

Then the mean temperature of the thermometer well surface is

$$T_s = \frac{T_t + T_0}{2} = \frac{237 + 150}{2} = 193.5\text{°C} = 466.5\text{°K}.$$

The convective geometry for the flow past the well is that of cross flow normal to a cylinder. For these purposes, the fluid properties are evaluated at the mean film temperature:

$$T_m = \frac{T_s + T_f}{2} = \frac{193.5 + 250}{2} = 222\text{°C}.$$

At this temperature Table A.6 gives, for atmospheric-pressure air,

$$\nu = 37.34 \times 10^{-6} \text{ m}^2\text{/s}, \quad k = 0.03921 \text{ W/m-°C}, \quad \text{Pr} = 0.699.$$

Then for the stated air velocity $U = 5$ m/s and well diameter $D_o = 0.005$ m, Eq. (6.49) for cross flow over a cylinder gives the convective coefficient to be

$$\text{Re}_D = \frac{UD_o}{\nu} = \frac{5 \times 0.005}{37.34 \times 10^{-6}} = 669.5,$$

$$\begin{aligned}
\text{Nu}_D &= 0.3 + 0.62\text{Re}_D^{1/2}\text{Pr}^{1/3}[1 + (0.4/\text{Pr})^{2/3}]^{-1/4} \\
&\qquad \times [1 + (\text{Re}_D/2.82 \times 10^5)^{5/8}]^{4/5} \\
&= 0.3 + 0.62(669.5)^{1/2}(0.699)^{1/3}[1 + (0.4/0.699)^{2/3}]^{-1/4} \\
&\qquad \times [1 + (669.5/2.82 \times 10^5)^{5/8}]^{4/5} \\
&= 13.01,
\end{aligned}$$

$$h_c = \text{Nu}_D \frac{k}{D_o} = 13.01 \times \frac{0.03921}{0.005} = 102.1 \text{ W/m}^2\text{-°C}.$$

The radiative coefficient is given by Eq. (11.15), evaluated at the mean surface temperature:

$$h_r = \epsilon\sigma(T_s^2 + T_0^2)(T_s + T_0)$$

$$\begin{aligned} h_r &= 0.5 \times 5.67 \times 10^{-8}(466.5^2 + 423^2)(466.5 + 423) \\ &= 10.0\ \text{W/m}^2\text{-°C}. \end{aligned}$$

Thus the equivalent environment temperature, T_e, defined by Eq. (11.17), is

$$\begin{aligned} T_e &= \frac{h_c T_f + h_r T_0}{h_c + h_r} \\ &= \frac{102.1 \times 250 + 10.0 \times 150}{102.1 + 10.0} = 241.1°\text{C} = 514.1°\text{K}. \end{aligned}$$

Then the temperature indicated at the well end is given by Eqs. (11.19) and (11.21) when one applies the geometric and thermal parameters noted above:

$$\begin{aligned} mL &= \sqrt{\frac{(h_c + h_r)C}{kA}}\, L \\ &= \left[\frac{(102.1 + 10.0) \times 0.01571}{30 \times 1.4726 \times 10^{-5}}\right]^{1/2} \times 0.06 = 3.7883, \end{aligned}$$

$$\frac{T_t - T_e}{T_0 - T_e} = \frac{1}{\cosh mL}$$

$$\frac{T_t - 514.1}{423 - 514.1} = \frac{1}{\cosh (3.7883)}$$

$$T_t = 510°\text{K} = 237°\text{C, calculated.}$$

This calculated value of T_t agrees with the assumed value, so one can terminate the calculation procedure. Had the calculated value differed significantly from the assumed, a new assumption for T_t would have to be made and the entire calculation outlined above repeated.

In this example, the final value for the indicated temperature, T_t, and the error, $T_t - T_f$, are

$$T_t = 237°\text{C } (459°\text{F}),$$

$$T_t - T_f = 237 - 250 = -13°\text{C } (23°\text{F}).$$

Reference to Example 11.4 will show that the radiant effect is to increase the measurement error indicated by the thermocouple. As illustrated above, the provision of a radiation shield for the thermometer well can reduce the effects of radiation.

11.6 HEAT TRANSFER FROM RADIATING FINS

One of the most important problems of modern space technology is that of *thermal control*—the maintenance of a desired temperature level within a spacecraft (manned

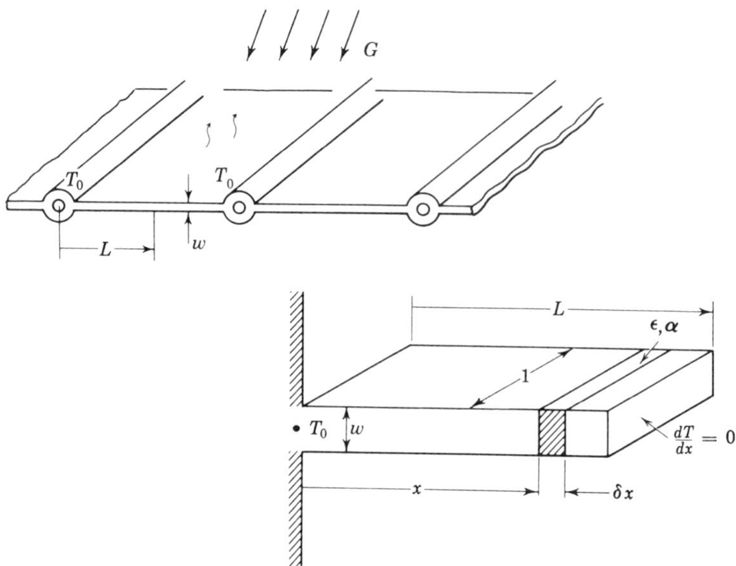

Figure 11.5. Heat rejection space radiator.

or unmanned) or satellite. Such spacecraft are subjected to heat inputs of both internal and external sources. Internal sources consist of electrically generated heat and metabolically generated heat, while the external sources consist of thermal irradiation from the sun, planetary bodies, etc. If a net gain in heat is experienced, some means of heat rejection must be provided in order that thermal control be maintained.

In outer space, the only means of rejecting heat is that of thermal radiation. For this purpose *heat rejection space radiators* have been developed. Many varieties of such systems have been developed (see Ref. 4). A typical space radiator configuration is illustrated in Fig. 11.5. Internally generated heat within the spacecraft is conveyed to the radiator by a circulating coolant (e.g., a water-glycol mixture). The coolant circulates in tubes (usually connected in a parallel circuit) which are joined by finned surfaces. Heat from the coolant conducts into the fin and is dissipated by radiation to the surroundings. The finned surface may be subjected to an external irradiation from a source such as the sun. The basic question to be answered is: Given a radiating fin of known geometry, maintained at a known base (i.e., tube temperature) and subjected to a certain external radiation, what is the rate at which heat is dissipated by the fin to the environment?

The solution for one such situation will be given here. If the fluid in adjacent radiator tubes is at the same temperature, the fin may be represented by the model suggested in Fig. 11.5—a straight fin of length L and uniform thickness w maintained, at one end, at a known temperature T_0. The other end of the fin has the condition $dT/dx = 0$ imposed by symmetry. The fin material has thermal conductivity k, and the exposed surface has a total emissivity ϵ. The surface is exposed to an external irradiation G (perhaps solar) to which it exhibits a total absorptivity α. Since such radiators are often made an integral part of the spacecraft skin, only one surface of the fin is considered to be radiatively active.

An approach similar to that developed in Sec. 2.9 for a fin with a convective boundary condition may be used here. If the conduction in the fin is taken to be one-dimensional, a heat balance taken on an element dx in length yields the following relation (for a unit depth):

$$-kw\,\frac{dT}{dx} = -\,kw\,\frac{dT}{dx} - kw\,\frac{d^2T}{dx^2}\,\delta x + \cdots + \epsilon\sigma T^4\,\delta x - \alpha G\,\delta x.$$

No convective loss at the fin surface is included. For $\delta x \to 0$, the following differential equation must be satisfied at each point:

$$\frac{d^2T}{dx^2} - \frac{\epsilon\sigma}{kw}\left(T^4 - \frac{\alpha G}{\epsilon\sigma}\right) = 0.$$

The nonhomogeneous term is recognized as the equilibrium temperature, or equivalent sink temperature, defined in Eq. (9.47). This temperature is the equilibrium temperature the surface would achieve if isolated in the irradiation G. Thus with

$$T_s^4 = \frac{\alpha G}{\epsilon\sigma}, \tag{11.23}$$

the differential equation for the temperature distribution in the fin is

$$\frac{d^2T}{dx^2} - \frac{\epsilon\sigma}{kw}\,(T^4 - T_s^4) = 0. \tag{11.24}$$

The solution of Eq. (11.24) is best discussed in terms of the following dimensionless variables:

$$\theta = \frac{T}{T_0},$$

$$\theta_s = \frac{T_s}{T_0}, \tag{11.25}$$

$$\xi = \frac{x}{L},$$

$$\lambda = \frac{\epsilon\sigma T_0^3 L^2}{kw}.$$

Introduction of these definitions into Eq. (11.24) yields

$$\frac{d^2\theta}{d\xi^2} - \lambda(\theta^4 - \theta_s^4) = 0. \tag{11.26}$$

The boundary conditions to be satisfied are

$$\begin{aligned} &\text{At } \xi = 0 (x = 0): \quad T = T_0,\ \theta = 1. \\ &\text{At } \xi = 1 (x = L): \quad \frac{dT}{dx} = \frac{d\theta}{d\xi} = 0. \end{aligned} \tag{11.27}$$

The nonlinearity of Eq. (11.26) makes its solution in closed form impossible, although the first integral may readily be written

$$\frac{d\theta}{d\xi} = -\sqrt{\tfrac{2}{5}\lambda(\theta^5 - 5\theta\theta_s^4) + C}, \tag{11.28}$$

C being a constant of integration.

A complete solution of the foregoing system, Eqs. (11.27) and (11.28), may be accomplished only by numerical means. Rather than seek the solution for the temperature distribution, the principal item of interest is the heat dissipated by the fin:

$$q = -kw\left(\frac{dT}{dx}\right)_{x=0} = -\frac{kwT_0}{L}\left(\frac{d\theta}{d\xi}\right)_{\xi=0}. \tag{11.29}$$

As in the case of the convective fins of Chapter 2, the heat dissipation is best expressed in terms of a fin efficiency (see Sec. 2.10), which is the ratio of q to the heat that *would* be dissipated if the whole fin was maintained at T_0:

$$\begin{aligned} \kappa &= \frac{q}{(\epsilon\sigma T_0^4 - \alpha G)L} \\ &= \frac{q}{\epsilon\sigma L T_0^4(1 - \theta_s^4)}. \end{aligned} \tag{11.30}$$

Thus when eqs. (11.29) and (11.30) are combined,

$$\kappa = -\frac{1}{\lambda(1 - \theta_s^4)}\left(\frac{d\theta}{d\xi}\right)_{\xi=0}. \tag{11.31}$$

When Eq. (11.28) is solved, subject to the conditions of Eqs. (11.27), the results may be applied to Eq. (11.31) to yield the fin efficiency in the following form:

$$\kappa = fn(\theta_s, \lambda). \tag{11.32}$$

The parameter θ_s measures the relative magnitude of the fin base temperature and the irradiation of the environment, while λ represents a combination of the thermal parameters of the fin. Lieblein, in Ref. 5, performed the numerical integration referred to above. The results are shown graphically in Fig. 11.6. By use of Fig. 11.6, along with the definitions of κ, θ_s, and λ given in Eqs. (11.23), (11.25), and (11.30), the heat rejected by the radiator may be determined.

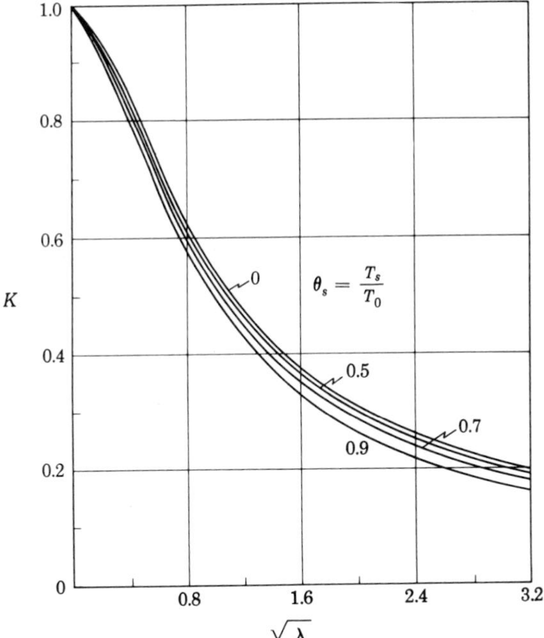

Figure 11.6. Efficiency of a radiating fin of uniform thickness. (from S. Lieblein, *NASA Tech. Note D-196*, National Aeronautics and Space Administration, Washington, D.C., 1959.)

EXAMPLE 11.7

A space radiator is constructed of fins with a half-width of $L = 3$ in. and a thickness $w = 0.0150$ in. The fin material is an aluminum alloy with $k = 65$ Btu/h-ft-°F, and the coating applied to the exposed surface has an emissivity and total absorptivity of $\epsilon = 0.5$, $\alpha = 0.1$. If the fin base is maintained at a temperature of $T_0 = 120°\text{F} = 580°\text{R}$ find the heat rejected, per foot of width, when the external irradiation is $G = 100$ Btu/h-ft^2.

Solution. The equivalent sink temperature of the irradiation environment, with respect to the exposed surface properties, is given by Eq. (11.23):

$$T_s^4 = \frac{\alpha G}{\epsilon\sigma}$$

$$T_s = \left(\frac{0.1 \times 100}{0.5 \times 0.1714 \times 10^{-8}}\right)^{1/4}$$

$$= 328.7°\text{R}.$$

Thus the dimensionless sink temperature, θ_s, and the conduction parameter, λ, are by use of Eq. (11.25):

$$\theta_s = \frac{T_s}{T_0} = \frac{328.7}{580} = 0.567,$$

$$\lambda = \frac{\epsilon\sigma T_0^3 L^2}{kw}$$

$$= \frac{0.5 \times 0.1714 \times 10^{-8} \times (580)^3 \times (3/12)^2}{65 \times 0.0150/12}$$

$$= 0.1286.$$

At these values of θ_s and λ, Fig. 11.6 gives the fin efficiency to be

$$\kappa = 0.82.$$

So Eq. (11.30) gives the fin heat loss, per foot of width, to be

$$q = \kappa\epsilon\sigma L T_0^4(1 - \theta_s^4)$$

$$= 0.82 \times 0.5 \times 0.1714 \times 10^{-8} \times (3/12) \times (580)^4(1 - 0.567^4)$$

$$= 17.83 \text{ Btu/h-ft } (17.14 \text{ W/m}).$$

11.7 THE FLAT-PLATE SOLAR COLLECTOR

An application of considerable current interest which involves all three modes of heat transfer is the collection of solar energy for commercial or domestic uses. One form of device used for solar collection is the flat-plate solar collector such as depicted in Fig. 11.7. Such devices usually consist of a flat absorber plate on which solar energy falls and is partly absorbed. Some of the absorbed energy is reradiated by the absorber plate and that which is not is removed from the surface by a coolant fluid in tubes attached to the plate.

Some energy is also lost from the absorbing surface by convection to whatever fluid surrounds the plate, usually the ambient air. In order to reduce the energy lost both by reradiation from the absorber and by convection to the ambient, most solar collectors are also provided with a transparent cover, often glass, which is opaque to the long-wave reemitted radiation, thus returning it to the absorber. The cover also serves to protect the absorbing surface from air movement, thus reducing the convective loss there to that by free convection. Some advanced designs provide an evacuated space between the absorbing plate and the cover to further reduce convective losses.

Heat removal from the absorbing surface is accomplished by circulating a fluid in tubes attached to the surface as suggested in Fig. 11.7. The tubes may be attached to the back side of the plate, as shown, or may be made intergrally with the plate

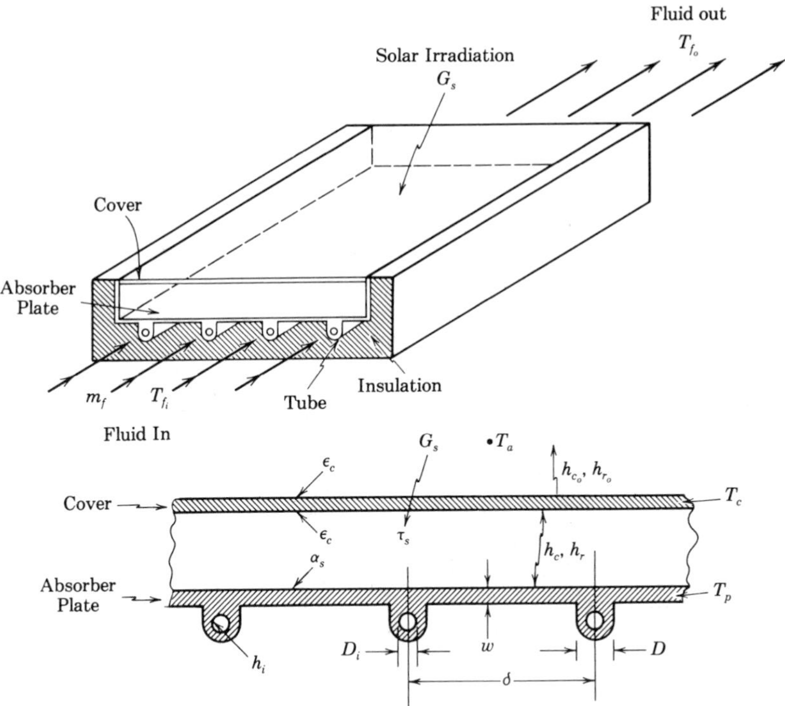

Figure 11.7. Flat plate solar collector.

in a form quite analogous to that illustrated in Fig. 11.6 for a space radiator. The analysis differs only slightly for the two configurations, so that shown in Fig. 11.7 will be used here. In this instance, then, the heat is transferred to the fluid by conduction in the plate surface between the tubes. For the geometry shown in Fig. 11.7, the tubes act as heat sinks with fins attached. The fin length is $(\delta - D)/2$ if δ is the on-center tube spacing and D is the outside tube diameter.

Let the solar irradiation falling on the collector be denoted by G_s. This amount depends on the orientation of the collector with respect to the sun, the time of year and day, the longitude and latitude on the earth's surface where the collector is located, etc. Let it be assumed that this quantity has been determined and is known. The cover of the collector should be as transparent to the incident solar radiation as possible; although some energy may be lost to the environment by reflection from the cover. The solar radiation that passes through the cover is mostly absorbed at the absorbing surface, but some may be reflected there also. Let $(\tau_s\alpha_s)$ represent the apparent solar transmissivity–absorbtivity product for the cover-plate array that accounts for all these effects. Typically, $\tau_s \approx 0.9$ and α_s is close to the solar absorbtivity of the collector plate. In any event $(\tau_s\alpha_s)G_s$ represents the solar energy absorbed at the absorbing surface.

In addition to the solar beam just discussed, the absorber plate and cover exchange radiation which results from their own emission. The cover radiates to the environment, and both the cover and the plate are subjected to convective heat transfer. All of these energy flows interact to establish equilibrium temperatures for the cover and the absorbing plate when the tube fluid temperature is known. The tube fluid, of course, varies in temperature as it proceeds through the collector and one wishes to find, among other things, the temperature of the fluid leaving the collector (given its inlet) so that the amount of useful energy absorbed can be found. The variation of the fluid temperature along the length of the collector implies a variation in the temperatures of the cover and the absorber plate, also in the longitudinal direction.

The analysis to follow will attempt to determine the useful heat collected (i.e., find the outlet fluid temperature) when one is given the inlet fluid temperature, the fluid flow rate, the various geometrical and thermal parameters for the collector (i.e., cover, absorber, tubes), and the ambient environment and solar irradiation. In this analysis it will be assumed that the ambient air is at a temperature, T_a, and that the radiative environment to which the cover radiates is at the same temperature. Further, it will be assumed that there exist representative average temperatures for the cover and the absorber plate, T_c and T_p, respectively, which take into account any longitudinal variation such as discussed above. These representative temperatures, T_c and T_p, will be assumed known for the present; however, later it will be pointed out how they may be determined. The analysis here will also assume that the collector is well insulated on the back side of the absorber plate so that no heat is lost to the environment there.

There are two convective exchanges to be accounted for—that at the outer cover surface and that between the inner cover surface and the absorber plate. The coefficient at the outer cover surface will be denoted by h_{co}, and may be determined when T_a, T_c, and the airflow conditions past the cover are known. The coefficient between the inner cover surface and the absorber plate will be denoted by h_c and is found by application of the air-space correlations of Eqs. (7.38) through (7.43), involving knowledge of T_p and T_c. Often solar collectors are tilted toward the sun, so the inclined air-space correlations are particularly important.

Radiation coefficients must also be found for the outer cover and between the cover and the absorber. For the outer cover one may once again use the result of Eq. (9.100) to obtain the radiation coefficient there, h_{ro}, as

$$h_{ro} = \sigma\epsilon_c(T_c^2 + T_a^2)(T_c + T_a), \tag{11.33}$$

in which ϵ_c is the emissivity of the cover plate material. The radiation exchange, due to surface emission, between the inner surface of the cover and the absorber plate is normally in the low-temperature long-wave portion of the spectrum to which the cover plate is thermally opaque. Hence the radiation coefficient used in Eq. (11.7) may be applied here:

$$h_r = \frac{\sigma(T_p^2 + T_c^2)(T_p + T_c)}{(1/\epsilon_c) + (1/\epsilon_p) - 1}. \tag{11.34}$$

In Eq. (11.34) ϵ_p represents the emissivity of the absorber plate and is generally different from its solar absorptivity, α_s above, if the surface is not gray.

With the foregoing four coefficients (h_{co}, h_c, h_{ro}, h_r) known, an effective cover-absorber overall coefficient, U_c, may be found:

$$\frac{1}{U_c} = \frac{1}{h_{co} + h_{ro}} + \frac{1}{h_c + h_r}, \tag{11.35}$$

and expresses the combined conductance due to convection and long-wave radiation between the absorber plate at T_p and the ambient environment at T_a. The total heat flow, per unit of collector surface area, is the sum of that due to the mechanisms just discussed and that absorbed from the solar beam.

The *net* heat flow into the absorber plate, q_p, per unit of plate area, A_p, is the difference between the solar energy absorbed,

$$(\tau_s\alpha_s)G_s,$$

and that returned to the surroundings by convection and radiation through the cover,

$$U_c(T_p - T_a).$$

Thus

$$\begin{aligned} \frac{q_p}{A_p} &= (\tau_s\alpha_s)G_s - U_c(T_p - T_a) \\ &= (\tau_s\alpha_s)G_s + U_c(T_a - T_p). \end{aligned} \tag{11.36}$$

The environment seen by the absorber plate consists of two parts—the solar beam, G_s, and the ambient air, T_a. These two sinks may be reduced to one by defining an *equivalent* environment temperature, T'_a, so that

$$T'_a = T_a + \frac{(\tau_s\alpha_s)G_s}{U_c}, \tag{11.37}$$

$$\frac{q_p}{A_p} = U_c(T'_a - T_p). \tag{11.38}$$

The heat flow in Eq. (11.36) or (11.38) must eventually go into the fluid circulating in the tubes. The solar collector may be characterized as a heat exchanger in which the "cold" fluid is the collecting fluid entering at T_{f_i} and leaving at T_{f_o}, and the "hot" fluid is one which remains at a constant temperature equal to the equivalent environment temperature, T'_a. Thus Eq. (10.45) gives the fluid temperature rise to be

$$\frac{T_{f_o} - T_{f_i}}{T'_a - T_{f_i}} = 1 - \exp\left(-\frac{\overline{U}A_p}{\dot{m}_f c_{pf}}\right) \tag{11.39}$$

in which $\overline{U}$ is some overall coefficient for the collector (between the fluid and the environment) to be discussed shortly and $\dot{m}_f$ and c_{pf} are the flow rate and specific heat of the collector fluid.

In Eq. (11.39), $\overline{U}$ is the overall coefficient for the collector, between the fluid at T_f and the environment at T_a'. If h_i represents the heat transfer coefficient applying at the inside tube surface (as given by the appropriate forced convection correlation of Chapter 6) and if A_i is the total *inside* surface area of all the tubes, then the principles of Sec. 2.11 give

$$\frac{1}{\overline{U}A_p} = \frac{1}{h_i A_i} + \frac{1}{A_p U_c \eta},$$

$$\frac{1}{\overline{U}} = \frac{1}{(h_i(A_i/A_p)} + \frac{1}{U_c \eta}.$$

In arriving at the expression for $\overline{U}$ above, the thermal resistance of the tube wall has been neglected and η is used to denote the effectiveness of the plate surface (i.e., "fins") between adjacent tubes. To avoid interrupting the current line of reasoning, detailed discussion of the determination of the effectiveness η will be delayed until later, so let it be treated as known for the present.

If, as noted in Fig. 11.7, δ represents the spacing between the tubes, the ratio of the exposed plate area to the inside tube area is $(A_p/A_i) = \delta/\pi D_i$ when D_i is used to denote the inside tube diameter. Thus the equation above becomes

$$\frac{1}{\overline{U}} = \frac{\delta}{\pi D_i h_i} + \frac{1}{U_c \eta}. \tag{11.40}$$

With $\overline{U}$ now determined in terms of h_i and the cover coefficient U_c [a function of h_{co}, h_{ro}, h_c, and h_r as given by Eq. (11.35)], the total heat gained by the fluid (i.e., the total useful heat collected) may be determined from

$$q = \dot{m}_f c_{pf}(T_{f_o} - T_{f_i}). \tag{11.41}$$

Substitution of Eqs. (11.39) and (11.37) into (11.41) gives

$$q = \frac{\dot{m}_f c_{pf}}{U_c}[(\tau_s \alpha_s)G_s - U_c(T_{f_i} - T_a)]\left[1 - \exp\left(-\frac{\overline{U}A_p}{\dot{m}_f c_{pf}}\right)\right] \tag{11.42}$$

Equation (11.42) constitutes the desired result. It expresses the useful heat absorbed in a collector by the circulating fluid when the fluid inlet conditions are known, the collector geometry is specified, and the environment (solar and ambient) is known. One proceeds as follows: Values for the average cover and absorber plate temperatures are assumed, i.e., T_c and T_p. The convective coefficients h_{co} and h_c are found using these temperatures and the applicable convection correlations. The radiation coefficients are found from Eqs. (11.33) and (11.34) so that the cover-absorber coefficient U_c can be found from Eq. (11.35). Knowledge of the flow conditions in the collector tubes then permits calculation of the overall coefficient $\overline{U}$ of the collector from Eq. (11.40), and Eq. (11.42) then yields their desired heat collected. The assumed value of T_p is verified by substituting the just-found q into Eq. (11.38). Then the assumed value of T_c is verified by use of the identity $(h_{co} + h_{ro})(T_c - T_a) = (h_c + h_r)(T_p - T_c)$. If necessary T_p and T_c can be revised and the entire calculation repeated.

The calculation above requires the determination of the effectiveness of the absorber plate surface. This is found by first finding the fin efficiency for the portion of the plate between the tubes. Equation (2.58) shows this to be

$$\kappa = \frac{\tanh ml'}{ml'}, \tag{11.43}$$

where l' is the apparent fin length between tubes and $m = \sqrt{hC/kA}$. For the geometry for Fig. 11.7, and the fact that only one side of the plate is subjected to heat transfer, one has for Eq. (11.43):

$$ml' = \sqrt{\left(\frac{U_c}{kw}\right)}\frac{\delta - D}{2}. \tag{11.44}$$

In Eq. (11.44) δ is the tube spacing, D the tube outside diameter, and w the plate thickness. The surface effectiveness, η, is then by Eq. (2.59):

$$\begin{aligned} \eta &= 1 - \frac{A_f}{A_p}(1 - \kappa) \\ &= 1 - \left(1 - \frac{D}{\delta}\right)(1 - \kappa) \\ &= \frac{D}{\delta} + \kappa\left(1 - \frac{D}{\delta}\right). \end{aligned} \tag{11.45}$$

Practitioners in solar energy collectors often rewrite the formulation given in the foregoing by defining the *collector efficiency factor, F'*, and the *collector heat removal factor, F_R*, as follows:

$$F' = \frac{\overline{U}}{U_c} = \frac{1/U_c}{\dfrac{A_p}{A_i h_i} + \dfrac{1}{U_c\eta}} = \frac{1/U_c}{\delta\left[\dfrac{1}{\pi D_i h_i} + \dfrac{1}{U_c[D + \kappa(\delta - D)]}\right]}, \tag{11.46}$$

$$F_R = \frac{\dot{m}_f c_{pf}}{A_p U_c}\left[1 - \exp\left(-\frac{A_p U_c F'}{\dot{m}_f c_{pf}}\right)\right] \tag{11.47}$$

In terms of these expressions, Eq. (11.42) for the useful heat collected is

$$q = A_p F_R[(\tau_s\alpha_s)G_s - U_c(T_{fi} - T_a)]. \tag{11.48}$$

The formulation given by Eqs. (11.46) through (11.48) is exactly the same as that leading to Eq. (11.42).

EXAMPLE 11.8

A flat-plate solar collector panel, 1 m × 2 m, consists of a lower absorbing plate maintained at 70°C with a glass cover plate at 10°C. The air in the space between the plates is at atmospheric pressure and the surfaces are spaced 8 cm apart. The

1-m dimension of the collector is inclined at an angle of 40° with the horizontal while the 2-m dimension is in the horizontal plane (i.e., parallel to the ground). The long-wave emissivity of the glass cover is 0.9 on both sides and that for the absorber plate is 0.3. The ambient air is at 0°C, and this may also be taken to be the temperature of the radiative environment. Airflow, due to wind, over the collector cover produces a convective heat transfer coefficient $h_{co} = 24.5$ W/m^2-°C on the outside of that surface. Determine the cover–absorber overall coefficient, U_c.

Solution. The data given include the following temperatures of the absorber plate, T_p, cover plate, T_c, and the environment, T_a:

$$T_p = 70°\text{C} = 343°\text{K},$$

$$T_c = 10°\text{C} = 283°\text{K},$$

$$T_a = 0°\text{C} = 273°\text{K}.$$

The emissivities of the absorber and cover plates are given as

$$\epsilon_p = 0.3,$$

$$\epsilon_c = 0.2.$$

Equation (11.35) will be used to calculate the cover–absorber overall coefficient. Hence one needs values for the convective and radiative coefficients between the absorber and cover plate, h_c and h_r, as well as the convective and radiative coefficients for exchange between the outside cover surface and the environment, h_{co} and h_{ro}. Each of these four coefficients will now be determined.

The convective coefficient for the free convection exchange in the space between the absorber and cover surfaces, h_c, is that obtained according to the correlations given for free convection in inclined rectangular enclosures by Eqs. (7.38) through (7.43). Referring to the geometry of such an inclined enclosure given in Fig. 7.5, in the present instance the length of the inclined surface is $H = 1$ m and the enclosure spacing is $L = 8$ cm. The lower, heated absorber plate (inclined at 40° with the horizontal) is at $T_h = T_p = 70$°C, while the upper, cold surface is the cover at $T_c = 10$°C. These geometric and thermal conditions are identical to those specified in Example 7.7. Hence the calculations of that example yield the following value for h_c:

$$h_c = 2.78 \text{ W/m}^2\text{-°C}.$$

The radiative transfer between the plates is given by Eq. (11.34), which, with the data specified above, gives

$$h_r = \frac{\sigma(T_p^2 + T_c^2)(T_p + T_c)}{(1/\epsilon_p) + (1/\epsilon_c) - 1}$$

$$h_r = \frac{5.67 \times 10^{-8}(343^2 + 283^2)(343 + 283)}{(1/0.3) + (1/0.9) - 1}$$

$$= 2.04 \text{ W/m}^2\text{-°C}.$$

The radiative coefficient between the outer cover surface at T_c and the environment at T_a is given by Eq. (11.33):

$$h_{ro} = \sigma\epsilon_c(T_c^2 + T_a^2)(T_c + T_a)$$

$$= 5.67 \times 10^{-8} \times 0.9(283^2 + 273^2)(283 + 273)$$

$$= 4.39 \text{ W/m}^2\text{-°C}.$$

The convective coefficient at the outer cover surface, h_{co}, due to air movement there is determined by the flow conditions prevailing on that surface. In this example, this coefficient is given as being known:

$$h_{co} = 24.5 \text{ W/m}^2\text{-°C}.$$

With these four coefficients now known, the effective conductance between the absorber plate and the environment is given by the cover–absorber overall coefficient, U_c, as shown in Eq. (11.35):

$$\frac{1}{U_c} = \frac{1}{h_{co} + h_{ro}} + \frac{1}{h_c + h_r}$$

$$= \frac{1}{24.5 + 4.39} + \frac{1}{2.78 + 2.04}$$

$$U_c = 4.1 \text{ W/m}^2\text{-°C (0.72 Btu/h-ft}^2\text{-°F)}.$$

EXAMPLE 11.9

A flat-plate solar collector is constructed in the geometrical configuration depicted in Fig. 11.7. The absorber is 3 ft wide and 6 ft long in the direction of flow in the tubes. The absorber plate and the tubes are made of aluminum with $k = 85$ Btu/h-ft-°F. In reference to the notation used in Fig. 11.7, the tubes have an outside diameter $D = 0.5$ in., an inside diameter $D_i = 0.4$ in., and are spaced $\delta =$ 6.0 in. on centers. The absorber plate thickness is $w = 0.03$ in. The cover–absorber plate combination has an apparent solar transmissivity–absorptivity product of $(\tau_s\alpha_s) = 0.80$. The fluid circulating in the tubes to carry away the absorbed heat is water ($c_{pf} = 1.0$ Btu/lb$_m$-°F) flowing at the rate of $\dot{m}_f = 160$ lb$_m$/h and entering the tubes with a temperature of $T_{fi} = 110$°F. The heat transfer coefficient at the inside tube surface may be calculated from the given tube geometry and water flow rate, but to minimize the details of this example, take it to be known as $h_i = 370$ Btu/h-ft^2-°F. At the particular time of day and day of the year, the solar irradiation falling on the absorber is known to be $G_s = 230$ Btu/h-ft^2 and the ambient atmosphere temperature is $T_a = 70$°F.

As indicated in the foregoing discussion, the cover–absorber overall coefficient U_c depends on the average cover and absorber plate temperatures as shown in Example 11.8. These temperatures are not known a priori, so one would have to perform an iterative calculation to find U_c. However, again to minimize the detail here, presume that the cover–absorber coefficient is already known to be U_c = 2.0 Btu/h-ft²-°F.

Given all the information above, estimate the net heat absorbed by the circulating tube water and determine the outlet temperature of this water.

Solution. The heat absorbed will be found from the relation developed in Eq. (11.42), which requires knowledge of the overall collector coefficient, $\overline{U}$. This coefficient, in turn, requires knowing the effectiveness of the plate array between the tubes. To find the latter quantity one first finds the parameter ml' given by Eq. (11.44):

$$ml' = \sqrt{\frac{U_c}{kw}}\,\frac{\delta - D}{2}$$

$$= \sqrt{\frac{2.0}{85 \times 0.03/12}}\,\frac{6.0 - 0.5}{12 \times 2} = 0.7031.$$

The efficiency of plate between the tubes is, then, by Eq. (11.43),

$$\kappa = \frac{\tanh ml'}{ml'}$$

$$= \frac{\tanh (0.7031)}{0.7031} = 0.8624,$$

so that the surface effectiveness of the entire plate array is, from Eq. (11.45),

$$\eta = \frac{D}{\delta} + \kappa\left(1 - \frac{D}{\delta}\right)$$

$$= \frac{0.5}{6.0} + 0.8624\left(1 - \frac{0.5}{6.0}\right) = 0.8739.$$

The overall heat transfer coefficient for the entire collector (i.e., between the fluid and the environment) is then found from Eq. (11.40):

$$\overline{U} = \left[\frac{\delta}{\pi D_i h_i} + \frac{1}{U_c \eta}\right]^{-1}$$

$$= \left[\frac{6.0}{\pi \times 0.4 \times 370} + \frac{1}{2.0 \times 0.8739}\right]^{-1}$$

$$= 1.71 \text{ Btu/h-ft}^2\text{-°F}.$$

The determination of the heat collected by Eq. (11.42) requires the parameter $\overline{U}A_p/\dot{m}_f c_{pf}$, the NTU of the collector heat exchanger. The plate area A_p is known from the given collector dimensions,

$$A_p = 3.0 \times 6.0 = 18.0 \text{ ft}^2,$$

so that

$$\frac{\overline{U}A_p}{\dot{m}_f c_{pf}} = \frac{1.71 \times 18.0}{160 \times 1.0} = 0.1924.$$

The absorbed heat may now be found from Eq. (11.42):

$$q = \frac{\dot{m}_f c_{pf}}{U_c}[(\tau_s\alpha_s)G_s - U_c(T_{f_i} - T_a)]\left[1 - \exp\left(-\frac{\overline{U}A_p}{\dot{m}_f c_{pf}}\right)\right]$$

$$= \frac{160 \times 1.0}{2.0}[0.80 \times 230 - 2.0(110 - 70)][1 - \exp(-0.1924)]$$

$$= 1456 \text{ Btu/h } (427 \text{ W}).$$

This amount absorbed is seen to be about one-third of that falling on the collector from the sun: $G_s A_p = 230 \times 18 = 4140$ Btu/hr. The remaining energy is either reradiated to the environment or lost by convection from the surface.

The outlet water temperature is now readily calculated from the heat balance on the water stream given by Eq. (11.41):

$$q = \dot{m}_f c_{pf}(T_{f_o} - T_{f_i})$$

$$1456 = 160 \times 1.0(T_{f_o} - 110)$$

$$T_{f_o} = 110°\text{F } (48°\text{C}).$$

One could obtain the same results using the collector efficiency factor, F', and the collector heat removel factor, F_R, given in Eqs. (11.46) and (11.47):

$$F' = \frac{1/U_c}{\delta\left[\dfrac{1}{\pi D_i h_i} + \dfrac{1}{U_c[D + \kappa(\delta - D)]}\right]}$$

$$= \frac{1/2.0}{6.0\left[\dfrac{1}{\pi \times 0.4 \times 370} + \dfrac{1}{2.0[0.5 + 0.8624(6.0 - 0.5)]}\right]}$$

$$= 0.8546,$$

$$F_R = \frac{\dot{m}_f c_{pf}}{A_p U_c}\left[1 - \exp\left(-\frac{A_p U_c F'}{\dot{m}_f c_{pf}}\right)\right]$$

$$F_R = \frac{160 \times 1.0}{18.0 \times 2.0}\left[1 - \exp\left(\frac{18.0 \times 2.0 \times 0.8546}{160 \times 1.0}\right)\right]$$

$$= 0.7775.$$

Then Eq. (11.48) gives the heat collected to be

$$q = A_p F_R[(\tau_s \alpha_s)G_s - U_c(T_{fi} - T_a)]$$

$$= 18.0 \times 0.7775[0.80 \times 230 - 2.0(110 - 70)]$$

$$= 1456 \text{ Btu/h},$$

as before.

11.8 THE HEAT PIPE

The heat pipe is a recent (Refs. 6 and 7) development which utilizes the high latent heat of evaporation, or condensation, together with the phenomenon of capillary pumping to transfer very high heat fluxes without the addition of external work. As a result, the device, seemingly inert, can transport heat fluxes many times that which can be conducted by solid metal bars of equal cross section.

The basic schematic diagram of a typical heat pipe is shown in Fig. 11.8. Structurally, the device consists of a closed vessel, or pipe, of circular cross section, a capillary wick, and a transport fluid. The wick materials are usually wire screen, woven cloth, ceramic materials, or narrow grooves machined into the pipe wall. At one end, the evaporator, heat is added at some temperature, vaporizing the liquid fluid from the wick material. The vaporized fluid flows down the central core of the pipe to the somewhat-cooler, lower-pressure end, termed the *condenser*. At the condenser end, the vapor is condensed back to a liquid with the release of the associated latent heat. The condensed liquid is "pumped" back to the evaporator section by the action of surface tension in the capillary structure of the wick material. The evaporator and condenser sections are normally separated by an adiabatic section. Because of the relatively low pressure drop in the flowing vapor, the difference between the evaporator temperature (where the heat is added) and the condensor temperature (when the heat is rejected) is relatively small—at least compared with the latent heat transported. These facts lead to a very high apparent conductivity between the two ends of the pipe.

The nature of the working fluid has a significant effect on the performance of a heat pipe, as do the properties of the wick material and the relative orientation of the pipe in the gravity field. Working fluids such as water, ammonia, alcohol, molten metals, and nitrogen are used—the choice being based, in part, on the desired working temperature of the device.

In space applications in which no gravity forces act, the heat pipe may be used to transport heat over relatively large distances by lengthening the adiabatic section.

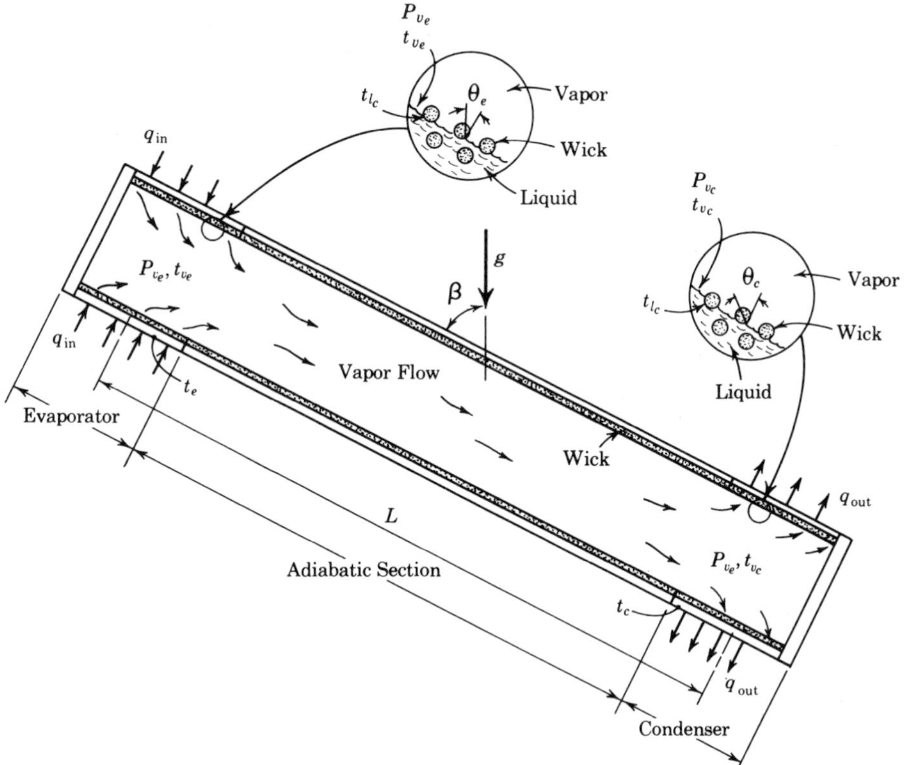

Figure 11.8. Schematic drawing of a typical heat pipe.

Some of the governing parameters of heat pipe performance may be obtained in the following simplified analysis. The basic transport mechanism involved is that of capillary pumping in the wick. The driving potential for this process is the capillary pressure difference which results from the surface tension at a vapor–liquid interface. Figure 11.9 shows the basic physical mechanism. By assuming that the vapor–liquid interface occurs in a circular tube of radius, r; that the liquid wets the tube wall with a contact angle θ; and that the surface tension at the interface is σ, a simple balance of forces shows that the pressure of the vapor exceeds that of the liquid by an amount

$$\Delta P = \frac{2\pi r\sigma \cos \theta}{\pi r^2} = \frac{2\sigma \cos \theta}{r}. \tag{11.49}$$

Using the notation of Fig. 11.8, one finds that the pressure of the liquid at the condensing interface in the condenser, P_{l_c}, exceeds that at the evaporative interface

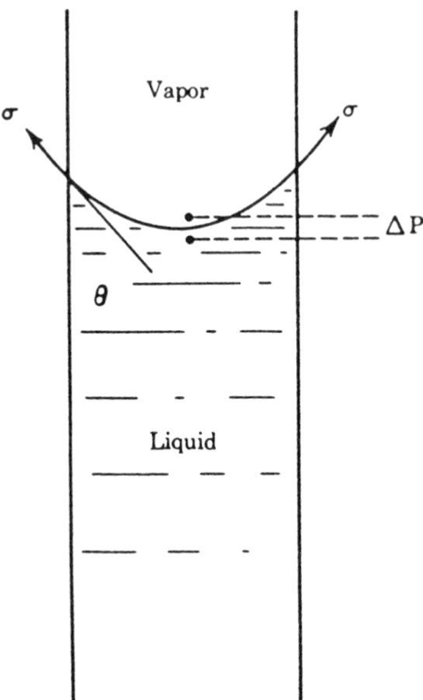

Figure 11.9

in the evaporator, P_{l_e}, by an amount determined by the difference in the vapor pressure (P_{v_c} and P_{v_e}), the difference in the interfacial pressures, and the hydrostatic head imposed by gravity (if any):

$$\begin{aligned} P_{l_c} - P_{l_e} &= (P_{v_c} - P_{v_e}) - (\Delta P_c - \Delta P_e) - g\rho_l L \cos\beta \\ &= (P_{v_c} - P_{v_e}) + 2\sigma\left(\frac{\cos\theta_e}{r} - \frac{\cos\theta_c}{r}\right) - g\rho_l L \cos\beta. \end{aligned} \tag{11.50}$$

In Eq. (11.50) the subscripts l and v denote liquid and vapor and the subscripts c and e denote condenser and evaporator. The angle β specifies the orientation of the pipe relative to the gravity vector and, for simplicity, L denotes a mean distance between the condenser and evaporator sections. It has been presumed that the wick material can be represented as being composed of circular capillaries of radius r. The symbols θ_e and θ_c represent the liquid contact angles in the evaporator and condenser. Also in Eq. (11.50) it has been presumed that the evaporator is located higher in the gravity field than the condenser—hence the negative sign on the last term.

The pressure difference given by Eq. (11.50) is the driving force to transport the liquid from the condenser back to the evaporator through the wick material. The wick, naturally, offers a resistance to this flow. Darcy's law may be used in this instance:

$$\dot{m} = (P_{l_c} - P_{l_e})\,\frac{\rho_l A K}{\mu_l L},$$

in which $\dot{m}$ is the mass flow rate of liquid, A is the cross-sectional area of wick material available for the liquid flow, and K is the permeability of the wick. Thus the flow rate in the heat pipe is

$$\dot{m} = \frac{AK}{L}\frac{\sigma\rho_l}{\mu_l}\left[\frac{P_{v_c} - P_{v_e}}{\sigma} + \frac{2}{r}(\cos\theta_e - \cos\theta_c) - \frac{g\rho_l L\cos\beta}{\sigma}\right].$$

For many applications, certainly adequate for this simple analysis, one may neglect the difference between the vapor pressures at each end of the pipe ($P_{v_c} \approx P_{v_e}$). Thus the mass flow is

$$\dot{m} = \frac{AK}{L}\frac{\sigma\rho_l}{\mu_l}\left[\frac{2}{r}(\cos\theta_e - \cos\theta_c) - \frac{g\rho_l L}{\sigma}\cos\beta\right], \tag{11.51}$$

and the heat flux between the condenser and the evaporator is

$$q = \frac{AK}{L}\frac{\sigma\rho_l h_{fg}}{\mu_l}\left[\frac{2}{r}(\cos\theta_e - \cos\theta_c) - \frac{g\rho_l L}{\sigma}\cos\beta\right], \tag{11.52}$$

where h_{fg} is the latent heat of vaporization.

In order to produce a positive flow rate, and thus heat flux, $(\cos\theta_e - \cos\theta_c)$ must be positive. As the power input, q, to the heat pipe increases and the rates of vaporization and condensation increase, the apparent contact angles in the evaporator and condenser must change so that $(\cos\theta_e - \cos\theta_c)$ becomes more positive. The maximum flow rate and maximum heat flux are given when $\cos\theta_e \to 1$ and $\cos\theta_c \to 0$:

$$q_{max} = \frac{2AK}{rL}\frac{\sigma\rho_l h_{fg}}{\mu_l}\left(1 - \frac{g\rho_l Lr}{2\sigma}\cos\beta\right). \tag{11.53}$$

Although Eq. (11.53) may be lacking in some respects for the prediction of actual heat pipe performance, it does permit the identification of significant parameters. It is seen that q_{max} is the product of three factors. The first, $2AK/rL$, is a grouping of wick properties. The third, $g\rho_l Lr\cos\beta/2\sigma$, is a grouping of wick and liquid properties which represents the relative importance of gravitational and surface tension forces. In space, where $g = 0$, this is 0. The second term is quite significant in that it is a grouping of liquid properties called the *liquid transport factor:*

$$N_l = \frac{\rho_l h_{fg}\sigma}{\mu_l}.$$

The designer wishes to choose this factor to be as large as possible. It is limited on one extreme by the liquid freezing point, where $\mu_l \to \infty$, and the other extreme by the critical point, where $h_{fg} \to 0$. Figure 11.10 presents values of N_l for some typical heat pipe fluids. It is apparent that the desired operating range temperature has a profound effect in the choice of fluid to be used. Also, it may be noticed that the heat pipes operating at high temperatures using, say, liquid metals, may transport heat fluxes of several orders of magnitude greater than those at lower temperatures.

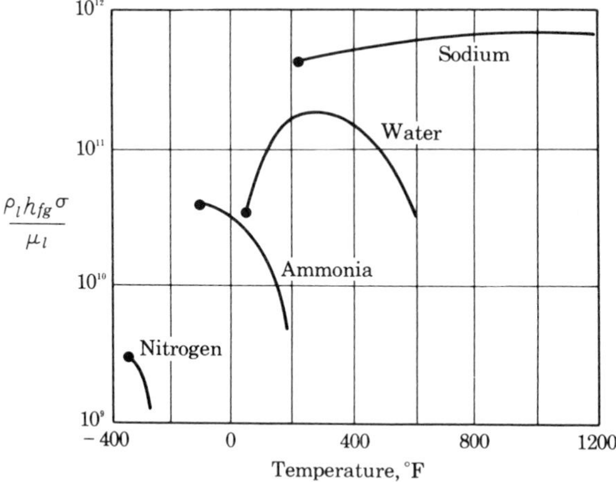

Figure 11.10. Heat pipe liquid transport factor for several fluids.

The maximum limitation on heat flux given by Eq. (11.53) was imposed by the maximum pumping capability of the wick material. Other considerations may override this limitation. One is associated with the fact that in practice a finite temperature difference is required between the evaporator, T_{v_e}, and the outside evaporator wall, T_e. This difference is required to drive the heat from the heat source into the heat pipe. A typical temperature distribution through a heat pipe is shown in Fig. 11.11. The overall temperature drop is seen to exceed considerably the vapor temperature drop. For high heat loads the required difference, $t_e - t_{v_e}$, may be large enough that local superheating of the liquid will occur within the wick ma-

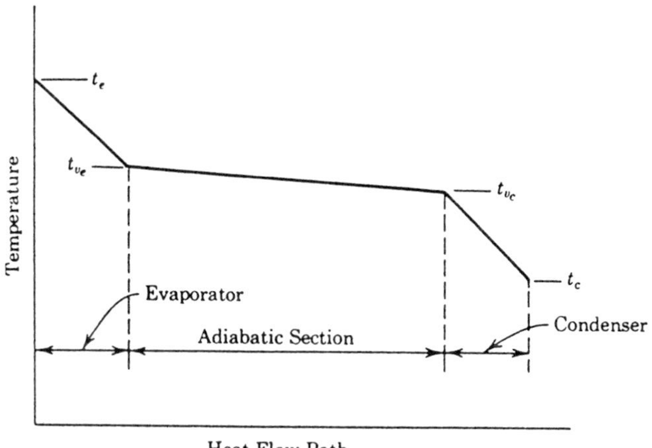

Figure 11.11. Temperature profile of a typical heat pipe.

terial. If this superheating is great enough, nucleate boiling may be initiated within the wick with a subsequent degradation of its pumping capability. Thus the heat pipe may become "boiling-limited" rather than "pumping-limited," as discussed above.

REFERENCES

1. Marks, L. S., ed., *Mechanical Engineers' Handbook,* 8th ed., New York, McGraw-Hill, 1978.
2. *ASHRAE Handbook of Fundamentals,* Atlanta, American Society of Heating, Refrigeration, and Air-Conditioning Engineers, 1981.
3. Gröber, H., and S. Erk, *Die Grundgesetze der Wärmeübertragung,* 3rd ed., rev. by U. Grigull, Berlin, Springer-Verlag, 1961.
4. Mackay, D. B., *Design of Space Power Plants,* Englewood Cliffs, N.J., Prentice-Hall, 1963.
5. Lieblein, S., "Analysis of Temperature Distribution and Radiant Heat Transfer Along a Rectangular Fin of Uniform Thickness," *NASA Tech. Note D-196,* Washington, D.C., Nov. 1959.
6. Cotter, T. P., "Theory of Heat Pipes," *Los Alamos Sci, Lab. Rep. LA-3246-MS,* Los Alamos, N.M., 1965.
7. Feldman, K. T., and G. H. Whiting, "The Heat Pipe," *Mech. Eng.,* Feb. 1967, p. 30.

PROBLEMS

11.1 A bare, oxidized, wrought iron pipe (8-in. schedule 40) is placed horizontally in still air at 20°C. Its surface temperature is 95°C. Find the combined heat transfer coefficient due to convection and radiation, and find the heat loss from 30 m of the pipe.

11.2 Repeat Prob. 11.1 if the pipe temperature raised to 465°C.

11.3 Steam at 800°F, 500 psia flows through a 6-in. schedule 40 horizontal steel pipe with a velocity of 10 ft/s. The pipe is covered with 1 in. of 85% magnesia insulation. The emissivity of the outer insulation surface is 0.75. The pipe is located in a large room where the ambient temperature of the walls and still air is 70°F. Accounting for both convective and radiative losses, find the temperature of the outer insulation surface and the overall heat transfer coefficient.

11.4 Steam at 4000 kN/m^2, 450°C, flows with a velocity of 7.5 m/s through a 6-in. schedule 40 pipe which is covered with 3.5 cm of 85% magnesia insulation (emissivity = 0.8). The pipe is located horizontally in a large room in which the ambient air and walls are at 25°C. Accounting for both convection and radiation, find the temperature of the outer insulation surface, the overall heat transfer coefficient, and the heat loss, per meter of pipe. Compare the results with those found in Prob. 10.1, in which radiation was neglected.

11.5 For the oil flowing in a tube described in Prob. 10.8, assume that the brass tube is oxidized with $\epsilon = 0.3$ and that the radiative environment is also at 70°F. Repeat the problem, accounting for the combined effects of convection and radiation, and compare the results.

11.6 A 100-W light bulb delivers about 10% of its energy in the visible spectrum. This energy and about 2% of the remaining radiation from the filament passes through the glass of the bulb. Make an engineering estimate of the surface temperature of such a light bulb.

11.7 A vertical air space is 2 m high and 6 cm wide. One bounding surface is maintained at 35°C and the other at 15°C. The emissivity of the two surfaces is 0.65. Find the heat flux through the air space **(a)** if radiation is neglected, and all heat transfer is by convection; **(b)** if the space is evacuated and *only* radiation takes place; and **(c)** if both the effects of radiation and convection are accounted for.

11.8 The air space in the wall of a house is approximately 3.75 in. wide and 8 ft high. The outer surface of the air space is at 95°F and has an emissivity of 0.8. The inner surface is at 70°F. Find the heat flux through the wall by combined convection and radiation if the emissivity of the inner surface is **(a)** 0.8 or **(b)** 0.2.

11.9 A solar collector panel, 6 ft × 3 ft, has its 6-ft dimension inclined at an angle of 60° with the horizontal. The lower collector surface is at 150°F and has an emissivity of 0.2. The upper cover plate, spaced 1.5 in. away, is at 90°F and has an emissivity of 0.8. Find the heat transferred through the air space between the two surfaces by **(a)** convection, **(b)** radiation, and **(c)** combined convection and radiation if the air space is at (1) 1 atm pressure, or (2) 0.5 atm pressure.

11.10 The temperature of a surface made of stainless steel (18% Cr, 8% Ni) is measured by attaching a thermocouple made of copper. The thermocouple extends horizontally into air at 25°C. If the thermocouple indicates 250°C, estimate the temperature of the surface.

11.11 The temperature of a surface at 300°C is to be measured with a thermocouple extending horizontally into ambient air at 25°C. Find the magnitude of the error of the reading indicated by the thermocouple **(a)** if the surface is made of iron and the thermocouple is made of (1) constantan or (2) iron, and **(b)** if the surface is made of magnisite brick and the thermocouple is made of (1) constantan or (2) iron.

11.12 A thermocouple is attached to a steel surface and extends horizontally into still air at 75°F. If the thermocouple is made of constantan and indicates a temperature of 350°F, what is the temperature of the surface?

11.13 A thermometer well (0.6 cm OD, 0.3 cm ID) is made of brass and extends into a pipe normal to the direction of flow. The pipe has an inside diameter of 15 cm and carries air at 315°C, 700 kN/m^2, flowing with a velocity of 4.6 m/s. The thermometer well is 10 cm long and the pipe wall temperature is 175°C. Neglecting the effects of radiation, find the temperature indicated by a thermometer in contact with the bottom of the well.

11.14 A thermometer well (0.25 in OD, 0.125 in ID) is made of brass and extends a length of 6 in. normal to the direction of flow in an 8-in. schedule 40 steel pipe. The pipe carries air at 20 psia, flowing at the rate of 1000 lb_m/h. The pipe is bare and is located horizontally in a room of atmospheric air at 100°F. Neglecting the effects of radiation, estimate the temperature of the air flowing in the pipe if a thermometer inserted in the well reads 700°F.

11.15 A thermometer well (0.5 cm OD, 0.25 cm ID) is made of stainless steel (18% Cr, 8% Ni) and extends 5 cm into a 30-cm-diameter duct through which air at a pressure of 1.2 atm flows at the rate of 0.5 kg/s. The air flows normal to the thermometer well, and the duct wall temperature is 120°C. Estimate the temperature of the air when the thermometer reads 200°C, neglecting radiation.

11.16 A thermocouple ($\epsilon = 0.8$) is used to measure the temperature of a hot gas flowing in a large duct. The convective coefficient between the thermocouple and the hot gas is 100 W/m^2-°C. If the duct walls are at 25°C, estimate the gas temperature when the thermocouple reads 325°C. Assume that the thermocouple is sufficiently long to neglect the effects of conduction.

11.17 Neglecting the effects of conduction, estimate the error in the reading given by a thermocouple ($\epsilon = 0.7$) placed in a large duct carrying air at 800°C if the duct walls are at 325°C. Assume that the convective heat transfer coefficient between the thermocouple and the air is 120 W/m^2-°C.

11.18 Repeat the calculation of Prob. 11.17 if the thermocouple is shielded by a material of the same emissivity. Assume that the shield is large compared to the thermocouple but small compared to the duct walls. Assume also that the convective heat transfer coefficient remains the same.

11.19 A bare thermocouple ($\epsilon = 0.8$) is 0.25 cm in diameter, and it is placed in a flue duct carrying combustion gases (approximated as air) at 340°C. The duct walls are at 120°C. The thermocouple is placed normal to the steam in the duct, but sufficiently long that conduction may be neglected. The gas in the duct is at atmospheric pressure and flows with a velocity of 7.6 m/s. What error in the thermocouple reading results?

11.20 A glass thermometer 0.25 in. in diameter is inserted normal to the flow of air in a duct. The thermometer extends for a distance of 6 in. into the duct, which is 1 ft in diameter and has a wall surface temperature of 100°F. The air in the duct is at 14.7 psia and flows at a velocity of 50 ft/s. If the thermometer reads 350°F, estimate the air temperature.

11.21 A thermocouple well (0.6 cm OD, 0.3 cm ID) is made of steel ($\epsilon = 0.5$) and extends 15 cm into a 30-cm-diameter duct, normal to the flow. The duct carries atmospheric air flowing with a velocity of 6 m/s. The duct is also steel with $\epsilon = 0.5$ and has a surface temperature of 95°C. The thermocouple inserted into the well reads 315°C. Accounting for conduction effects in the well, estimate the air temperature **(a)** if radiation is neglected, and **(b)** if radiation is not neglected.

11.22 If the surface emissivity of the thermometer well given in Prob. 11.13 is 0.5, repeat the problem but include the effects of radiation.

11.23 If the surface emissivity of the thermometer well given in Prob. 11.15 is 0.7, repeat the problem but include the effects of radiation.

11.24 Verify that Eq. (11.28) is the first integral of Eq. (11.26).

11.25 Repeat Example 11.7 for the case of zero-irradiation environment, all other data remaining the same.

11.26 For the radiating fin described in Sec. 11.6, if the length L becomes very great, one might assume that $\theta_L \rightarrow \theta_s$ as well as $(d\theta/d\xi)_L = 0$. Under this circumstance, show that the efficiency is

$$\kappa_\infty = \frac{[(2/5\lambda)(1 - 5\theta_s^4 + 4\theta_s^5)]^{1/2}}{1 - \theta_s^4}.$$

11.27 A flat-plate solar collector panel, 2 m × 1 m, consists of a lower absorbing plate at 70°C with a glass cover plate at 30°C. The air in the space between the surfaces is at atmospheric pressure and the surfaces are spaced 5 cm apart. The 2 m dimension of the collector is inclined at an angle of 60° with the horizontal. The long-wave emissivity of the glass cover plate is 0.9 on both sides and that for the absorber surface is 0.2. The ambient air is at 20°C and this is also the temperature of the radiative environment. Wind blowing over the collector cover produces a convective heat transfer coefficient of 17 W/m^2-°C on the outside of that surface. Find the cover–absorber overall coefficient, U_c.

11.28 A flat solar collector, such as depicted in Fig. 11.7, is 1.05 m wide and 3 m long in the direction of flow. The absorber plate and tubes are made of aluminum.

The tubes are 1.2 cm OD, 1.1 cm ID, spaced 15 cm on centers, and the absorber plate is 0.05 cm thick. The cover–absorber overall heat transfer coefficient is known to be $U_c = 6 \text{ W/m}^2\text{-°C}$. The cover–absorber combination has an apparent solar transmissivity–absorptivity product of $(\tau_s\alpha_s) = 0.81$. Water enters the tubes at 57°C and flows at the rate of 0.02 kg/s with an inside tube heat transfer coefficient of 1200 $\text{W/m}^2\text{-°C}$. The solar irradiation incident on the collector is $G_s = 730 \text{ W/m}^2$ and the ambient air temperature is 20°C. Find **(a)** the water outlet temperature; **(b)** the useful energy collected; **(c)** the collector efficiency factor, F'; and **(d)** the heat removal factor, F_R.

11.29 A flat-plate solar collector panel, 6 ft × 3 ft, consists of a lower absorbing plate at 150°F with a glass cover plate at 90°F. The air in the space between the surfaces is at atmospheric pressure and the surfaces are spaced 1.5 in. apart. The 6-ft dimension of the collector is inclined at an angle of 60° with the horizontal. The long-wave emissivity of the glass cover plate is 0.8 on both sides and that of the absorber surface is 0.2. The ambient air is at 75°F and this is also the temperature of the radiative environment. Wind blowing over the collector cover produces a convective heat transfer coefficient of 3.5 $\text{Btu/h-ft}^2\text{-°F}$ on the outside of that surface. Find the cover–absorber overall coefficient, U_c.

11.30 A flat solar collector, such as depicted in Fig. 11.7, is 30 in. wide and 5 ft long in the direction of flow. The absorber plate and tubes are made of an aluminum alloy with $k = 80$ Btu/h-ft-°F. The tubes are 0.45 in OD and 0.35 in. ID, spaced 5 in. on centers. The absorber plate is 0.02 in. thick. The cover–absorber overall heat transfer coefficient is known to be 1.8 $\text{Btu/h-ft}^2\text{-°F}$. The cover–absorber combination has an apparent solar transmissivity–absorptivity product of $(\tau_s\alpha_s) = 0.85$. Water enters the tubes at 115°F and flows at the rate of 140 $\text{lb}_\text{m}\text{/h}$ with an inside tube heat transfer coefficient of 350 $\text{Btu/h-ft}^2\text{-°F}$. The solar irradiation incident on the collector is known to be 245 Btu/h-ft^2 and the ambient air temperature is 70°F. Find **(a)** the outlet water temperature; **(b)** the useful energy collected; **(c)** the collector efficiency factor, F'; and **(d)** the heat removal factor, F_R.

APPENDIX A

Tables of the Thermal Properties of Substances

Table A.1 Thermal Properties of Metals*

Metal	Properties at 20°C				Thermal Conductivity, k, W/m-°C									
	ρ kg/m³	c_p kJ/kg-°C	k W/m-°C	$\alpha \times 10^5$ m²/s	−100°C	0°C	100°C	200°C	300°C	400°C	600°C	800°C	1000°C	1200°C
Aluminum														
Pure	2,707	0.896	204	8.418	215	202	206	215	228	249				
Al–Cu (Duralumin) 94–96 Al, 3–5 Cu, trace Mg	2,787	0.883	164	6.676	126	159	182	194						
Al–Mg (Hydronalium) 91–95 Al, 5–9 Mg	2,611	0.904	112	4.764	93	109	125	142						
Al–Si (Silumin) 87 Al, 13 Si	2,659	0.871	164	7.099	149	163	175	185						
Al–Si (Silumin, copper bearing) 86.5 Al, 12.5 Si, 1 Cu	2,659	0.867	137	5.933	119	137	144	152	161					
Al–Si (Alusil) 78–80 Al, 20–22 Si	2,627	0.854	161	7.172	144	157	168	175	178					
Al–Mg–Si, 97 Al, 1 Mg, 1 Si, 1 Mn	2,707	0.892	177	7.311		175	189	204						
Lead	11,373	0.130	35	2.343	36.9	35.1	33.4	31.5	29.8					
Iron														
Pure	7,897	0.452	73	2.026	87	73	67	62	55	48	40	36	35	36
Wrought iron (C < 0.5%)	7,849	0.46	59	1.626		59	57	52	48	45	36	33	33	33
Cast iron (C ≈ 4%)	7,272	0.42	52	1.703										
Steel (C max. ≈ 1.5%)														
Carbon steel, C ≈ 0.5%	7,833	0.465	54	1.474		55	52	48	45	42	35	31	29	31
1.0%	7,801	0.473	43	1.172		43	43	42	40	36	33	29	28	29
1.5%	7,753	0.486	36	0.970		36	36	36	35	33	31	28	28	29
Nickel steel, Ni ≈ 0%	7,897	0.452	73	2.026	87	73	67	62	55	48	40	36	35	36
10%	7,945	0.46	26	0.720										
20%	7,993	0.46	19	0.526										
30%	8,073	0.46	12	0.325										

40%	8,169	0.46	10	0.279										
50%	8,266	0.46	14	0.361										
60%	8,378	0.46	19	0.493										
70%	8,506	0.46	26	0.666										
80%	8,618	0.46	35	0.872										
90%	8,762	0.46	47	1.156										
100%	8,906	0.448	90	2.276										
Invar, Ni = 36%	8,137	0.46	10.7	0.286										
Chrome steel, Cr = 0%	7,897	0.452	73	2,026	87	73	67	62	55	48	40	36	35	36
1%	7,865	0.46	61	1.665		62	55	52	47	42	36	33	33	
2%	7,865	0.46	52	1.443		54	48	45	42	38	33	31	31	
5%	7,833	0.46	40	1.110		40	38	36	36	33	29	29	29	
10%	7,785	0.46	31	0.867		31	31	31	29	29	28	28	29	
20%	7,689	0.46	22	0.635		22	22	22	22	24	24	26	29	
30%	7,625	0.46	19	0.542										
Cr–Ni (chrome–nickel);														
15 Cr, 10 Ni	7,865	0.46	19	0.526										
18 Cr, 8 Ni (V2A)	7,817	0.46	16.3	0.444		16.3	17	17	19	19	22	26	31	
20 Cr, 15 Ni	7,833	0.46	15.1	0.415										
25 Cr, 20 Ni	7,865	0.46	12.8	0.361										
Ni-Cr (nickel–chrome);														
80 Ni, 15 Cr	8,522	0.46	17	0.444										
60 Ni, 15 Cr	8,266	0.46	12.8	0.333										
40 Ni, 15 Cr	8,073	0.46	11.6	0.305										
20 Ni, 15 Cr	7,865	0.46	14.0	0.390		14.0	15.1	15.1	16.3	17	19	22		
Cr–Ni–Al; 6 Cr, 1.5 Al, 0.55 Si (Sicromal 8)	7,721	0.490	22	0.594										
24 Cr, 2.5 Al, 0.55 Si (Sicromal 12)	7,673	0.494	19	0.501										
Manganese steel, Mn = 0%	7,897	0.452	73	2.026	87	73	67	62	55	48	40	36	35	36
1%	7,865	0.46	50	1.388										
2%	7,865	0.46	38	1.050		36	36	36	36	35	33			
5%	7,849	0.46	22	0.637										
10%	7,801	0.46	17	0.483										

Table A.1 (cont.)

Metal	Properties at 20°C				Thermal Conductivity, k, W/m-°C									
	ρ kg/m³	c_p kJ/kg-°C	k W/m-°C	$\alpha \times 10^5$ m²/s	−100°C	0°C	100°C	200°C	300°C	400°C	600°C	800°C	1000°C	1200°C
Tungsten steel,														
W = 0%	7,897	0.452	73	2.026	87	73	67	62	55	48	40	36	35	36
1%	7,913	0.448	66	1.858										
2%	7,961	0.444	62	1.763		62	59	54	48	45	36			
5%	8,073	0.435	54	1.525										
10%	8,314	0.419	48	1.391										
20%	8,826	0.389	43	1.249										
Silicon steel, Si = 0%	7,897	0.452	73	2.026	87	73	67	62	55	48	40	36	35	36
1%	7,769	0.46	42											
2%	7,673	0.46	31											
5%	7,417	0.46	19											
Copper														
Pure	8,954	0.3831	386	11.234	407	386	379	374	369	363	353			
Aluminum bronze; 95 Cu, 5 Al	8,666	0.410	83	2.330										
Bronze; 75 Cu, 25 Sn	8,666	0.343	26	0.859										
Red brass; 85 Cu, 9 Sn, 6 Zn	8,714	0.385	61	1.804		59	71							

Brass; 70 Cu, 30 Zn	8,522	0.385	111	3.412	88		128	144	147	147				
German silver, 62 Cu, 15 Ni, 22 Zn	8,618	0.394	24.9	0.733	19.2		31	40	45	48				
Constantan; 60 Cu, 40 Ni	8,922	0.410	22.7	0.612	21		22.2	26						
Magnesium														
Pure	1,746	1.013	171	9.708	178	171	168	163	157					
Mg–Al (electrolytic); 6–8 Al, 1–2 Zn	1,810	1.00	66	3.605		52	62	74	83					
Mg–Mn; 2% Mn	1,778	1.00	114	6.382	93	111	125	130						
Molybdenum	10,220	0.251	123	4,790	138	125	118	114	111	109	106	102	99	92
Nickel														
Pure (99.9%)	8,906	0.4459	90	2.266	104	93	83	73	64	59				
Impure (99.2%)	8,906	0.444	69	1.747		69	64	59	55	52	55	62	67	69
Ni–Cr; 90 Ni, 10 Cr	8,666	0.444	17	0.444		17.1	18.9	20.9	22.8	24.6				
80 Ni, 20 Cr	8,314	0.444	12.6	0.343		12.3	13.8	15.6	17.1	18.9	22.5			
Silver														
Purest	10,524	0.2340	419	17.004	419	417	415	412						
Pure (99.9%)	10,524	0.2340	407	16.563	419	410	415	374	362	360				
Tin pure	7,304	0.2265	64	3.884	74	65.9	59	57						
Tungsten	19,350	0.1344	163	6.271		166	151	142	133	126	112	76		
Zinc, pure	7,144	0.3843	112.2	4.106	114	112	109	106	100	93				

*From E. R. G. Eckert and R. M. Drake, *Analysis of Heat and Mass Transfer*, New York, copyright McGraw-Hill, 1972. Used by permission.

Table A.2 Thermal Properties of Some Nonmetals*

Substance	c_p kJ/kg-°C	ρ kg/m³	t °C	k W/m-°C	$\alpha \times 10^7$ m²/s
Structural Materials					
Asphalt			20	0.74 (a)	
Bakelite	1.59 (b)	1273 (b)	20	0.232 (b)	1.14
Bricks					
Common	0.84 (d)	1602 (d)	20	0.69 (a)	5.2
Face		2050 (d)	20	1.32 (a)	
Carborundum brick			600	18.5 (a)	
			1400	11.1 (a)	
Chrome brick	0.84 (d)	3011 (d)	200	2.32 (a)	9.2
			550	2.47 (a)	9.8
			900	1.99 (a)	7.9
Diatomaceous earth (fired)			204	0.24 (a)	
			871	0.31 (a)	
Fire-clay brick (burnt 1330°C)	0.96 (d)	2050 (d)	500	1.04 (a)	5.3
			800	1.07 (a)	5.4
			1100	1.09 (a)	5.5
Fire-clay brick (burnt 1450°C)	0.96 (d)	2323 (d)	500	1.28 (a)	5.7
			800	1.37 (a)	6.1
			1100	1.40 (a)	6.3
Fire-clay brick (Missouri)	0.96 (d)	2643 (d)	200	1.00 (a)	3.9
			600	1.47 (a)	5.8
			1400	1.77 (a)	7.0
Magnesite	1.13 (d)		204	3.81 (a)	
			650	2.77 (a)	
			1204	1.90 (a)	
Cement, portland		1500 (a)		0.29 (a)	
Cement, mortar			24	1.16 (a)	
Concrete	0.88 (b)	1900–2300 (b)	20	0.81–1.40 (b)	4.0–8.4
Concrete, cinder			24	0.76 (a)	
Glass, plate	0.84 (b)	2700 (b)	20	0.76 (b)	3.2

Glass, borosilicute		2225 (b)	30	1.09 (b)	
Plaster, gypsum	0.84 (d)	1440 (d)	21	0.48 (a)	4.0
Plaster, metal lath			21	0.47 (a)	
Plaster, wood lath			21	0.28 (a)	
Stone					
Granite	0.82 (d)	2640 (d)		1.73–3.98 (a)	8.0–18.3
Limestone	0.91 (d)	2480 (d)	100–300	1.26–1.33 (a)	5.6–5.9
Marble	0.81 (b)	2500–2700 (b)	20	2.77 (b)	12.7–13.7
Sandstone	0.71 (b)	2160–2300 (b)	20	1.63–2.08 (b)	10.0–13.6
Wood, cross grain					
Balsa		140 (a)	30	0.055 (a)	
Cypress		465 (d)	30	0.097 (a)	
Fir	2.72 (d)	417 (b)	24	0.11 (a)	0.96
Oak	2.39 (d)	480–600 (b)	30	0.17 (a)	1.3
Yellow pine	2.81 (d)	640 (d)	24	0.16 (a)	0.91
White pine		430 (d)	30	0.11 (a)	
Wood, radial					
Fir	2.72 (b)	417–421 (b)	20	0.14 (b)	1.2
Oak	2.39 (b)	480–600 (b)	20	0.17–0.21 (b)	1.2–1.8
Insulating Materials					
Asbestos		469 (b)	−200	0.074 (b)	
			0	0.155 (b)	
Asbestos		577 (b)	0	0.151 (b)	
			100	0.192 (b)	
			200	0.208 (b)	
			400	0.223 (b)	
Asbestos		697 (b)	−200	0.156 (b)	
			0	0.234 (b)	
Asbestos cement				2.08 (a)	
Asbestos cement board			20	0.744 (a)	
Asbestos sheet			51	0.166 (a)	
Asbestos felt (40 laminations per inch)			38	0.057 (a)	
			149	0.069 (a)	
			260	0.083 (a)	

Table A.2 (Cont.)

Substance	c_p kJ/kg-°C	ρ kg/m³	t °C	k W/m-°C	α × 10⁷ m²/s
Asbestos felt (20 laminations per inch)			38	0.078 (a)	
			149	0.095 (a)	
			260	0.113 (a)	
Asbestos, corrugated (4 plies per inch)			38	0.087 (a)	
			93	0.100 (a)	
			149	0.119 (a)	
Balsam wool		32 (a)	32	0.040 (a)	
Cardboard, corrugated				0.064 (a)	
Celotex			32	0.048 (a)	
Corkboard		160 (b)	30	0.043 (b)	
Cork, expanded scrap	1.88 (b)	45–119 (b)	20	0.036 (b)	1.6–4.3
Cork, ground		151 (b)	30	0.043 (b)	
Diatomaceous earth (powdered)		160 (e)	93	0.050 (e)	
			204	0.066 (e)	
			316	0.083 (e)	
Diatomaceous earth (powdered)		224 (e)	93	0.057 (e)	
			204	0.068 (e)	
			316	0.080 (e)	
Diatomaceous earth (powdered)		288 (e)	93	0.069 (e)	
			204	0.078 (e)	
			316	0.085 (e)	
Felt, hair		131 (c)	−7	0.0410 (c)	
			38	0.0466 (c)	
			93	0.0537 (c)	
Felt, hair		183 (c)	−7	0.0367 (c)	
			38	0.0440 (c)	
			93	0.0518 (c)	
Felt, hair		205 (c)	−7	0.0403 (c)	
			38	0.0454 (c)	
			93	0.0511 (c)	
Fiber insulating board		237 (b)	21	0.049 (b)	
Glass wool		24 (c)	−7	0.0376 (c)	
			38	0.0542 (c)	
			93	0.0753 (c)	

Glass wool		64 (c)	−7	0.0310 (c)	
			38	0.0414 (c)	
			93	0.0549 (c)	
Glass wool		96 (c)	−7	0.0282 (c)	
			38	0.0377 (c)	
			93	0.0499 (c)	
Kapok			30	0.035 (a)	
Magnesia, 85%		270 (c)	38	0.068 (a)	
			93	0.071 (a)	
			149	0.074 (a)	
			204	0.080 (a)	
Rock wool		64 (c)	−7	0.0260 (c)	
			38	0.0388 (c)	
			93	0.0549 (c)	
Rock wool		128 (c)	−7	0.0296 (c)	
			38	0.0345 (c)	
			93	0.0518 (c)	
Rock wool		192 (c)	−7	0.0317 (c)	
			38	0.0391 (c)	
			93	0.0486 (c)	
Miscellaneous Materials					
Aerogel, silica		136 (b)	120	0.023 (b)	
Clay	0.88 (b)	1458 (b)	20	1.28 (b)	10.0
Coal, anthracite	1.26 (b)	1200–1500 (b)	20	0.26 (b)	1.4–1.7
Coal, powdered	1.30 (b)	737 (b)	30	0.116 (b)	1.2
Cotton	1.30 (b)	80 (b)	20	0.059 (b)	5.7
Earth, coarse	1.84 (b)	2050 (b)	20	0.52 (b)	1.4
Ice	1.93 (b)	913 (b)	0	2.22 (b)	12.6
Rubber, hard		2000 (b)	0	0.151 (b)	
Sawdust			24	0.059 (a)	
Silk	1.38 (b)	58 (b)	20	0.036 (b)	4.5

*Adapted from (a) A. I. Brown and S. M. Marco, *Introduction to Heat Transfer,* 3rd ed., New York, McGraw-Hill, 1958; (b) E. R. G. Eckert, *Introduction to the Transfer of Heat and Mass,* New York, McGraw-Hill, 1950; (c) R. H. Heilman, *Ind. Eng. Chem.*, Vol. 28, 1936, p. 782; (d) L. S. Marks, *Mechanical Engineers' Handbook,* 6th ed., New York, McGraw-Hill, 1958; (e) R. Calvert, *Diatomaceous Earth,* Chemical Catalog Company, Inc., 1930; (f) H. F. Norton, *J. Am. Ceramic Soc.*, Vol. 10, 1957, p. 30.

Table A.3-SI Properties of Saturated Water (SI Units)*

t °C	c_p kJ/kg-°C	ρ kg/m^3	$\mu \times 10^3$ kg/m-s	$\nu \times 10^6$ m^2/s	k W/m-°C	$\alpha \times 10^7$ m^2/s	$\beta \times 10^3$ 1/°K	Pr
0	4.218	999.8	1.791	1.792	0.5619	1.332	−0.0853	13.45
5	4.203	1000.0	1.520	1.520	0.5723	1.362	0.0052	11.16
10	4.193	999.8	1.308	1.308	0.5820	1.389	0.0821	9.42
15	4.187	999.2	1.139	1.140	0.5911	1.413	0.148	8.07
20	4.182	998.3	1.003	1.004	0.5996	1.436	0.207	6.99
25	4.180	997.1	0.8908	0.8933	0.6076	1.458	0.259	6.13
30	4.180	995.7	0.7978	0.8012	0.6150	1.478	0.306	5.42
35	4.179	994.1	0.7196	0.7238	0.6221	1.497	0.349	4.83
40	4.179	992.3	0.6531	0.6582	0.6286	1.516	0.389	4.34
45	4.182	990.2	0.5962	0.6021	0.6347	1.533	0.427	3.93
50	4.182	998.0	0.5471	0.5537	0.6405	1.550	0.462	3.57
55	4.184	985.7	0.5043	0.5116	0.6458	1.566	0.496	3.27
60	4.186	983.1	0.4668	0.4748	0.6507	1.581	0.529	3.00
65	4.187	980.5	0.4338	0.4424	0.6553	1.596	0.560	2.77
70	4.191	977.7	0.4044	0.4137	0.6594	1.609	0.590	2.57
75	4.191	974.7	0.3783	0.3881	0.6633	1.624	0.619	2.39
80	4.195	971.6	0.3550	0.3653	0.6668	1.636	0.647	2.23
85	4.201	968.4	0.3339	0.3448	0.6699	1.647	0.675	2.09
90	4.203	965.1	0.3150	0.3264	0.6727	1.659	0.702	1.97
95	4.210	961.7	0.2978	0.3097	0.6753	1.668	0.728	1.86
100	4.215	958.1	0.2822	0.2945	0.6775	1.677	0.755	1.76
120	4.246	942.8	0.2321	0.2461	0.6833	1.707	0.859	1.44
140	4.282	925.9	0.1961	0.2118	0.6845	1.727	0.966	1.23
160	4.339	907.3	0.1695	0.1869	0.6815	1.731	1.084	1.08
180	4.411	886.9	0.1494	0.1684	0.6745	1.724	1.216	0.98
200	4.498	864.7	0.1336	0.1545	0.6634	1.706	1.372	0.91
220	4.608	840.4	0.1210	0.1439	0.6483	1.674	1.563	0.86
240	4.770	813.6	0.1105	0.1358	0.6292	1.622	1.806	0.84
260	4.991	783.9	0.1015	0.1295	0.6059	1.549	2.130	0.84
280	5.294	750.5	0.0934	0.1245	0.5780	1.455	2.589	0.86
300	5.758	712.2	0.0858	0.1205	0.5450	1.329	3.293	0.91
320	6.566	666.9	0.0783	0.1174	0.5063	1.156	4.511	1.02
340	8.234	610.2	0.0702	0.1151	0.4611	0.918	7.170	1.25
360	16.138	526.2	0.0600	0.1139	0.4115	0.485	21.28	2.35

*c_p, ρ, μ, β computed from equations recommended in *ASME Steam Tables,* 3rd ed., New York, Am. Soc. Mech. Engrs., 1977. k computed from equation recommended in J. Kestin, "Thermal Conductivity of Water and Steam," *Mech. Eng.*, Aug. 1978, p. 47.

Table A.3-ENGL Properties of Saturated Water (English Units)*

t °F	c_p Btu/lb$_m$-°F	ρ lb$_m$/ft^3	μ lb$_m$/ft-h	$\nu \times 10^2$ ft^2/h	k Btu/h-ft-°F	$\alpha \times 10^3$ ft^2/h	$\beta \times 10^3$ 1/°R	Pr
32	1.008	62.41	4.333	6.943	0.3247	5.163	−0.0474	13.45
40	1.004	62.42	3.742	5.994	0.3300	5.264	−0.0023	11.39
50	1.001	62.41	3.163	5.069	0.3363	5.381	0.0456	9.42
60	1.000	62.37	2.715	4.354	0.3421	5.485	0.0862	7.94
70	0.999	62.31	2.361	3.789	0.3475	5.584	0.121	6.79
80	0.998	62.22	2.075	3.335	0.3525	5.677	0.153	5.88
90	0.998	62.12	1.842	2.965	0.3572	5.762	0.181	5.15
100	0.998	62.00	1.648	2.659	0.3616	5.844	0.206	4.55
110	0.998	61.86	1.486	2.402	0.3656	5.921	0.230	4.06
120	0.998	61.71	1.348	2.185	0.3693	5.994	0.253	3.65
130	0.999	61.55	1.231	2.000	0.3728	6.062	0.274	3.30
140	1.000	61.38	1.129	1.840	0.3760	6.127	0.294	3.00
150	1.000	61.19	1.041	1.701	0.3789	6.190	0.313	2.75
160	1.001	60.99	0.964	1.580	0.3815	6.251	0.331	2.53
170	1.002	60.79	0.896	1.474	0.3839	6.303	0.349	2.34
180	1.003	60.57	0.835	1.379	0.3861	6.356	0.366	2.17
190	1.004	60.34	0.782	1.296	0.3880	6.407	0.383	2.02
200	1.006	60.11	0.734	1.221	0.3897	6.448	0.400	1.89
210	1.007	59.86	0.691	1.154	0.3912	6.487	0.416	1.78
220	1.009	59.61	0.652	1.094	0.3924	6.527	0.432	1.68
230	1.010	59.35	0.617	1.039	0.3934	6.567	0.448	1.58
240	1.012	59.08	0.585	0.990	0.3943	6.592	0.464	1.50
250	1.014	58.80	0.556	0.945	0.3949	6.624	0.480	1.43
260	1.016	58.52	0.529	0.904	0.3953	6.648	0.497	1.36
270	1.019	58.22	0.505	0.867	0.3955	6.667	0.513	1.30
280	1.022	57.92	0.483	0.834	0.3956	6.680	0.530	1.25
290	1.025	57.61	0.462	0.803	0.3954	6.697	0.547	1.20
300	1.027	57.30	0.444	0.774	0.3950	6.711	0.565	1.15
350	1.050	55.59	0.369	0.663	0.3905	6.693	0.663	0.99
400	1.078	53.66	0.316	0.589	0.3816	6.596	0.784	0.89
450	1.125	51.46	0.277	0.538	0.3681	6.360	0.946	0.85
500	1.192	48.94	0.246	0.502	0.3501	6.001	1.183	0.84
550	1.302	45.97	0.219	0.476	0.3269	5.463	1.569	0.87
600	1.516	42.31	0.193	0.457	0.2978	4.643	2.316	0.99

*c_p, ρ, μ, β computed from equations recommended in *ASME Steam Tables,* 3rd ed., New York, Am. Soc. Mech. Engrs., 1977. k computed from equation recommended in J. Kestin, "Thermal Conductivity of Water and Steam," *Mech. Eng.*, Aug. 1978, p. 47

Table A.4 Properties of Saturated Liquids*

t °C	ρ kg/m³	c_p kJ/kg-°C	$\nu \times 10^6$ m²/s	k W/m-°C	$\alpha \times 10^7$ m²/s	Pr	$\beta \times 10^3$ 1/°K
			Sulfur Dioxide, SO_2				
−50	1560.84	1.3595	0.484	0.242	1.141	4.24	
−40	1536.81	1.3607	0.424	0.235	1.130	3.74	
−30	1520.64	1.3616	0.371	0.230	1.117	3.31	
−20	1488.60	1.3624	0.324	0.225	1.107	2.93	
−10	1463.61	1.3628	0.288	0.218	1.097	2.62	
0	1438.46	1.3636	0.257	0.211	1.081	2.38	
10	1412.51	1.3645	0.232	0.204	1.066	2.18	
20	1386.40	1.3653	0.210	0.199	1.050	2.00	1.94
30	1359.33	1.3662	0.190	0.192	1.035	1.83	
40	1329.22	1.3674	0.173	0.185	1.019	1.70	
50	1299.10	1.3683	0.162	0.177	0.999	1.61	
			Methyl Chloride, CH_3Cl				
−50	1052.58	1.4759	0.320	0.215	1.388	2.31	
−40	1033.35	1.4826	0.318	0.209	1.368	2.32	
−30	1016.53	1.4922	0.314	0.202	1.337	2.35	
−20	999.39	1.5043	0.309	0.196	1.301	2.38	
−10	981.45	1.5194	0.306	0.187	1.257	2.43	
0	962.39	1.5378	0.302	0.178	1.213	2.49	
10	942.36	1.5600	0.297	0.171	1.166	2.55	
20	923.31	1.5860	0.293	0.163	1.112	2.63	
30	903.12	1.6161	0.288	0.154	1.058	2.72	
40	883.10	1.6504	0.281	0.144	0.996	2.83	
50	861.15	1.6890	0.274	0.133	0.921	2.97	
			Ammonia, NH_3				
−50	703.69	4.463	0.435	0.547	1.742	2.60	
−40	691.68	4.467	0.406	0.547	1.775	2.28	
−30	679.34	4.476	0.387	0.549	1.801	2.15	
−20	666.69	4.509	0.381	0.547	1.819	2.09	
−10	653.55	4.564	0.378	0.543	1.825	2.07	
0	640.10	4.635	0.373	0.540	1.819	2.05	
10	626.16	4.714	0.368	0.531	1.801	2.04	
20	611.75	4.798	0.359	0.521	1.775	2.02	2.45
30	596.37	4.890	0.349	0.507	1.742	2.01	
40	580.99	4.999	0.340	0.493	1.701	2.00	
50	564.33	5.116	0.330	0.476	1.654	1.99	
			Carbon Dioxide, CO_2				
−50	1156.34	1.84	0.119	0.0855	0.4021	2.96	
−40	1117.77	1.88	0.118	0.1011	0.4810	2.46	
−30	1076.76	1.97	0.117	0.1116	0.5272	2.22	
−20	1032.39	2.05	0.115	0.1151	0.5445	2.12	
−10	983.38	2.18	0.113	0.1099	0.5133	2.20	
0	926.99	2.47	0.108	0.1045	0.4578	2.38	
10	860.03	3.14	0.101	0.0971	0.3608	2.80	
20	772.57	5.0	0.091	0.0872	0.2219	4.10	14.00
30	597.81	36.4	0.080	0.0703	0.0279	28.7	

Table A.4 Properties of Saturated Liquids*

t °C	ρ kg/m^3	c_p kJ/kg-°C	$\nu \times 10^6$ m^2/s	k W/m-°C	$\alpha \times 10^7$ m^2/s	Pr	$\beta \times 10^3$ 1/°K
			Dichlorodifluoromethane (Freon-12), CCl_2F_2				
−50	1546.75	0.8750	0.310	0.067	0.501	6.2	2.63
−40	1518.71	0.8847	0.279	0.069	0.514	5.4	
−30	1489.56	0.8956	0.253	0.069	0.526	4.8	
−20	1460.57	0.9073	0.235	0.071	0.539	4.4	
−10	1429.49	0.9203	0.221	0.073	0.550	4.0	
0	1397.45	0.9345	0.214	0.073	0.557	3.8	
10	1364.30	0.9496	0.203	0.073	0.560	3.6	
20	1330.18	0.9659	0.198	0.073	0.560	3.5	
30	1295.10	0.9835	0.194	0.071	0.560	3.5	
40	1257.13	1.0019	0.191	0.069	0.555	3.5	
50	1215.96	1.0216	0.190	0.067	0.545	3.5	
			Lubricating Oil (Approx. SAE 50)				
0	899.12	1.796	4280	0.147	0.911	47100	
20	888.23	1.880	900	0.145	0.872	10400	0.70
40	876.05	1.964	240	0.144	0.834	2870	
60	864.04	2.047	83.9	0.140	0.800	1050	
80	852.02	2.131	37.5	0.138	0.769	490	
100	840.01	2.219	20.3	0.137	0.738	276	
120	828.96	2.307	12.4	0.135	0.710	175	
140	816.94	2.395	8.0	0.133	0.686	116	
160	805.89	2.483	5.6	0.132	0.663	84	
			Glycerine, $C_3H_5(OH)_3$				
0	1276.03	2.261	8310	0.282	0.983	84700	
10	1270.11	2.319	3000	0.284	0.965	31000	
20	1264.02	2.386	1180	0.286	0.947	12500	0.50
30	1258.09	2.445	500	0.286	0.929	5380	
40	1252.01	2.512	220	0.286	0.914	2450	
50	1244.96	2.583	150	0.287	0.893	1630	
			Ethylene Glycol, $C_2H_4(OH_2)$				
0	1130.75	2.294	57.53	0.242	0.934	615	
20	1116.65	2.382	19.18	0.249	0.939	204	0.65
40	1101.43	2.474	8.69	0.256	0.939	93	
60	1087.66	2.562	4.75	0.260	0.932	51	
80	1077.56	2.650	2.98	0.261	0.921	32.4	
100	1058.50	2.742	2.03	0.263	0.908	22.4	

*From E. R. G. Eckert and R. M. Drake, *Analysis of Heat and Mass Transfer*, New York, copyright McGraw-Hill, 1972 used by permission.

Table A.5-SI Properties of Steam (SI Units)*

Temperature, ↓ °C		Pressure, kN/m^2(kPa) (Sat. Temp., °C) 100 (99.63)	500 (151.84)	1000 (179.88)	1500 (198.29)	2000 (212.37)	4000 (250.33)	6000 (275.55)	8000 (294.97)	10,000 (310.96)	12,000 (324.65)	14,000 (336.64)	16,000 (347.33)	18,000 (356.96)
Saturated vapor	c_p	2.027	2.329	2.595	2.820	3.025	3.788	4.613	5.600	6.770	8.158	10.15	13.46	18.22
	ρ	0.5904	2.6690	5.1469	7.5955	10.047	20.101	30.828	42.507	55.432	70.035	87.106	107.61	132.70
	$\mu \times 10^6$	12.26	14.08	15.07	15.72	16.22	17.61	18.61	19.49	20.33	21.21	22.20	23.37	24.87
	$k \times 10^3$	24.75	31.03	35.40	38.76	41.65	51.28	59.99	68.96	78.96	90.89	106.3	127.6	159.5
	Pr	1.00	1.06	1.10	1.14	1.18	1.30	1.43	1.58	1.74	1.90	2.12	2.47	2.84
150°C	c_p	1.987												
	ρ	0.5165												
	$\mu \times 10^6$	14.19												
	$k \times 10^3$	28.80												
	Pr	0.98												
200°C	c_p	1.979	2.161	2.446	2.800									
	ρ	0.4603	2.3532	4.8563	7.5538									
	$\mu \times 10^6$	16.18	16.07	15.93	15.79									
	$k \times 10^3$	33.37	34.24	36.06	38.75									
	Pr	0.96	1.01	1.08	1.14									
250°C	c_p	1.990	2.088	2.233	2.406	2.608								
	ρ	0.4156	2.1078	4.2965	6.5793	8.9727								
	$\mu \times 10^6$	18.22	18.15	18.07	17.99	17.91								
	$k \times 10^3$	38.28	38.81	39.69	40.82	42.22								
	Pr	0.95	0.98	1.02	1.06	1.11								
300°C	c_p	2.010	2.066	2.145	2.235	2.337	2.876	3.715	5.232					
	ρ	0.3790	1.9136	3.8763	5.8927	7.9681	16.997	27.666	41.214					
	$\mu \times 10^6$	20.29	20.25	20.20	20.15	20.10	19.93	19.80	19.74					
	$k \times 10^3$	43.49	43.89	44.49	45.17	45.95	50.14	56.67	67.27					
	Pr	0.94	0.95	0.97	1.00	1.02	1.14	1.30	1.53					
350°C	c_p	2.037	2.071	2.118	2.169	2.225	2.504	2.883	3.404	4.175	5.413	7.420	11.06	
	ρ	0.3483	1.7542	3.5407	5.3610	7.2172	15.050	23.684	33.391	44.601	58.116	75.682	102.34	
	$\mu \times 10^6$	22.37	22.35	22.32	22.29	22.27	22.18	22.14	22.13	22.18	22.33	22.64	23.35	

	$k \times 10^3$	48.97	49.32	49.79	50.31	50.86	53.55	57.14	61.96	68.57	78.05	92.89	120.8	
	Pr	0.93	0.94	0.95	0.96	0.97	1.04	1.12	1.22	1.35	1.55	1.81	2.14	
400°C	c_p	2.067	2.090	2.120	2.152	2.186	2.346	2.550	2.806	3.127	3.545	4.116	4.933	6.131
	ρ	0.3223	1.6203	3.2628	4.9281	6.6170	13.629	21.107	29.146	37.867	47.429	58.050	70.055	83.948
	$\mu \times 10^6$	24.45	24.44	24.43	24.41	24.40	24.38	24.38	24.42	24.49	24.61	24.79	25.05	25.43
	$k \times 10^3$	54.71	55.02	55.43	55.87	56.33	58.41	60.96	64.08	67.94	72.75	78.84	86.69	97.09
	Pr	0.92	0.93	0.93	0.94	0.95	0.98	1.02	1.07	1.13	1.20	1.29	1.43	1.61
450°C	c_p	2.099	2.116	2.138	2.160	2.183	2.286	2.408	2.554	2.728	2.935	3.184	3.490	3.877
	ρ	0.2999	1.5059	3.0275	4.5651	6.1191	12.507	19.193	26.216	33.623	41.467	49.813	58.736	68.327
	$\mu \times 10^6$	26.52	26.52	26.52	26.51	26.52	26.53	26.57	26.63	26.72	26.85	27.01	27.21	27.46
	$k \times 10^3$	60.69	60.98	61.35	61.73	62.14	63.92	65.97	68.36	71.12	74.33	78.06	82.43	87.56
	Pr	0.92	0.92	0.92	0.93	0.93	0.95	0.97	1.00	1.02	1.06	1.10	1.15	1.22
500°C	c_p	2.132	2.146	2.162	2.179	2.197	2.270	2.353	2.446	2.551	2.672	2.810	2.968	3.151
	ρ	0.2805	1.4069	2.8251	4.2549	5.6962	11.582	17.670	23.978	30.525	37.333	44.426	51.831	59.578
	$\mu \times 10^6$	28.57	28.58	28.58	28.59	28.60	28.64	28.71	28.79	28.90	29.03	29.18	29.36	29.57
	$k \times 10^3$	66.89	67.16	67.50	67.86	68.22	69.80	71.56	73.54	75.75	78.23	81.01	84.11	87.59
	Pr	0.91	0.91	0.92	0.92	0.92	0.93	0.94	0.96	0.97	0.99	1.01	1.04	1.06
550°C	c_p	2.166	2.177	2.191	2.205	2.219	2.276	2.337	2.403	2.475	2.554	2.640	2.735	2.839
	ρ	0.2634	1.3203	2.6489	3.9859	5.3313	10.799	16.411	22.172	28.092	34.181	40.448	46.904	53.561
	$\mu \times 10^6$	30.61	30.62	30.63	30.64	30.65	30.72	30.80	30.90	31.02	31.15	31.30	31.47	31.67
	$k \times 10^3$	73.30	73.55	73.87	74.20	74.53	75.96	77.54	79.26	81.15	83.22	85.47	87.94	90.64
	Pr	0.91	0.91	0.91	0.91	0.91	0.92	0.93	0.94	0.95	0.96	0.97	0.98	0.99
600°C	c_p	2.201	2.210	2.221	2.233	2.245	2.292	2.341	2.392	2.446	2.503	2.563	2.627	2.694
	ρ	0.2483	1.2439	2.4939	3.7501	5.0125	10.125	15.341	20.664	26.096	31.643	37.308	43.097	49.013
	$\mu \times 10^6$	32.62	32.63	32.64	32.66	32.68	32.76	32.86	32.96	33.09	33.22	33.37	33.54	33.73
	$k \times 10^3$	79.89	80.13	80.43	80.73	81.05	82.37	83.80	85.34	87.01	88.81	90.75	92.83	95.07
	Pr	0.90	0.90	0.90	0.90	0.91	0.91	0.92	0.92	0.93	0.94	0.94	0.95	0.96
650°C	c_p	2.235	2.243	2.253	2.262	2.272	2.313	2.354	2.396	2.439	2.484	2.530	2.577	2.626
	ρ	0.2348	1.1759	2.3564	3.5414	4.7312	9.5366	14.418	19.376	24.412	29.529	34.728	40.010	45.379
	$\mu \times 10^6$	34.60	34.61	34.63	34.65	34.68	34.77	34.87	34.99	35.11	35.25	35.40	35.56	35.74
	$k \times 10^3$	86.65	86.88	87.16	87.45	87.74	88.97	90.28	91.69	93.20	94.81	96.52	98.34	100.28
	Pr	0.89	0.89	0.90	0.90	0.90	0.90	0.91	0.91	0.92	0.92	0.93	0.93	0.94

Note: c_p in kJ/kg-°C, ρ in kg/m^3, μ in kg/m-s, k in W/m-°C.

*c_p, ρ, μ computed from equations recommended in *ASME Steam Tables*, 3rd ed., New York, Am. Soc. Mech. Engrs., 1977. k computed from equation recommended in J. Kestin, "Thermal Conductivity of Water and Steam," *Mech. Eng.*, Aug. 1978, p. 47.

Table A.5-ENGL Properties of Steam (English Units)*

Temperature, ↓ °F		Pressure, psia (Sat. Temp., °F): 14.696 (212.00)	100 (327.82)	200 (381.80)	400 (444.60)	600 (486.20)	800 (518.21)	1000 (544.58)	1200 (567.19)	1400 (587.07)	1600 (604.87)	1800 (621.02)	2000 (635.80)	2500 (668.11)
Saturated vapor	c_p	0.4844	0.5820	0.6610	0.7924	0.9172	1.051	1.202	1.373	1.566	1.779	2.029	2.360	3.910
	ρ	0.03732	0.2257	0.4372	0.8614	1.2991	1.7576	2.2424	2.7590	3.3137	3.9146	4.5757	5.3119	7.6711
	$\mu \times 10^2$	2.970	3.513	3.768	4.070	4.277	4.447	4.599	4.742	4.882	5.026	5.177	5.342	5.871
	$k \times 10^2$	1.432	1.900	2.196	2.632	2.998	3.345	3.693	4.059	4.456	4.899	5.414	6.029	8.442
	Pr	1.00	1.08	1.13	1.23	1.31	1.40	1.50	1.60	1.72	1.83	1.94	2.09	2.72
300°F	c_p	0.4748												
	ρ	0.03276												
	$\mu \times 10^2$	3.421												
	$k \times 10^2$	1.659												
	Pr	0.98												
400°F	c_p	0.4729	0.5365	0.6364										
	ρ	0.02884	0.2026	0.4238										
	$\mu \times 10^2$	3.957	3.920	3.876										
	$k \times 10^2$	1.952	2.032	2.200										
	Pr	0.96	1.04	1.12										
500°F	c_p	0.4761	0.5078	0.5542	0.6806	0.8664								
	ρ	0.02579	0.1790	0.3670	0.7787	1.2589								
	$\mu \times 10^2$	4.508	4.486	4.461	4.413	4.367								
	$k \times 10^2$	2.271	2.315	2.388	2.605	2.949								
	Pr	0.95	0.98	1.04	1.15	1.28								
600°F	c_p	0.4820	0.4992	0.5225	0.5817	0.6594	0.7601	0.8964	1.094	1.396				
	ρ	0.02333	0.1609	0.3270	0.6774	1.0575	1.4763	1.9465	2.4902	3.1487				
	$\mu \times 10^2$	5.065	5.052	5.039	5.013	4.991	4.974	4.962	4.958	4.966				
	$k \times 10^2$	2.610	2.643	2.690	2.807	2.961	3.162	3.424	3.775	4.267				
	Pr	0.94	0.95	0.98	1.04	1.11	1.20	1.30	1.44	1.62				

700°F	c_p	0.4945	0.4997	0.5129	0.5439	0.5821	0.6284	0.6838	0.7507	0.8335	0.9398	1.081	1.273	2.133
	ρ	0.02131	0.1464	0.2961	0.6061	0.9323	1.2775	1.6448	2.0386	2.4640	2.9282	3.4416	4.0197	5.9542
	$\mu \times 10^2$	5.624	5.618	5.611	5.600	5.592	5.587	5.586	5.591	5.601	5.618	5.644	5.682	5.870
	$k \times 10^2$	2.968	2.996	3.033	3.116	3.212	3.326	3.458	3.615	3.802	4.030	4.311	4.669	6.248
	Pr	0.93	0.94	0.95	0.98	1.01	1.06	1.10	1.16	1.23	1.31	1.42	1.55	2.00
800°F	c_p	0.4977	0.5046	0.5130	0.5316	0.5531	0.5778	0.6063	0.6389	0.6761	0.7189	0.7685	0.8269	1.031
	ρ	0.01961	0.1344	0.2709	0.5509	0.8409	1.1417	1.4546	1.7810	2.1222	2.4802	2.8569	3.2547	4.3616
	$\mu \times 10^2$	6.182	6.180	6.178	6.176	6.176	6.180	6.187	6.197	6.210	6.228	6.251	6.279	6.378
	$k \times 10^2$	3.344	3.369	3.401	3.471	3.550	3.638	3.737	3.849	3.975	4.117	4.278	4.460	5.037
	Pr	0.92	0.93	0.93	0.95	0.96	0.98	1.00	1.03	1.05	1.09	1.12	1.16	1.31
900°F	c_p	0.5065	0.5115	0.5176	0.5304	0.5442	0.5594	0.5762	0.5947	0.6152	0.6378	0.6628	0.6905	0.7749
	ρ	0.01816	0.1242	0.2500	0.5061	0.7688	1.0384	1.3153	1.6001	1.8933	2.1956	2.5075	2.8299	3.6868
	$\mu \times 10^2$	6.736	6.737	6.738	6.743	6.749	6.758	6.769	6.783	6.800	6.820	6.843	6.869	6.951
	$k \times 10^2$	3.736	3.760	3.788	3.850	3.917	3.990	4.069	4.156	4.250	4.353	4.465	4.587	4.942
	Pr	0.91	0.92	0.92	0.93	0.94	0.95	0.96	0.97	0.98	1.00	1.02	1.03	1.09
1000°F	c_p	0.5155	0.5195	0.5242	0.5339	0.5440	0.5547	0.5660	0.5780	0.5909	0.6047	0.6194	0.6352	0.6801
	ρ	0.01691	0.1155	0.2321	0.4686	0.7096	0.9551	1.2055	1.4609	1.7215	1.9875	2.2593	2.5371	3.2595
	$\mu \times 10^2$	7.284	7.287	7.291	7.300	7.311	7.323	7.338	7.355	7.374	7.395	7.418	7.444	7.519
	$k \times 10^2$	4.144	4.165	4.192	4.247	4.307	4.371	4.439	4.511	4.589	4.672	4.761	4.855	5.120
	Pr	0.91	0.91	0.91	0.92	0.92	0.93	0.94	0.94	0.95	0.96	0.97	0.97	1.00
1100°F	c_p	0.5245	0.5278	0.5317	0.5396	0.5476	0.5559	0.5644	0.5732	0.5824	0.5919	0.6018	0.6122	0.6402
	ρ	0.01583	0.1080	0.2168	0.4367	0.6597	0.8859	1.1153	1.3481	1.5844	1.8242	2.0677	2.3149	2.9499
	$\mu \times 10^2$	7.825	7.830	7.835	7.848	7.861	7.877	7.893	7.912	7.932	7.954	7.978	8.003	8.075
	$k \times 10^2$	4.565	4.585	4.609	4.661	4.715	4.772	4.832	4.896	4.963	5.034	5.109	5.188	5.404
	Pr	0.90	0.90	0.90	0.91	0.91	0.92	0.92	0.93	0.93	0.94	0.94	0.94	0.96
1200°F	c_p	0.5335	0.5364	0.5396	0.5463	0.5529	0.5597	0.5666	0.5736	0.5807	0.5880	0.5955	0.6031	0.6230
	ρ	0.01487	0.1014	0.2034	0.4090	0.6169	0.8270	1.0393	1.2541	1.4711	1.6906	1.9125	2.1368	2.7087
	$\mu \times 10^2$	8.359	8.364	8.371	8.386	8.401	8.418	8.437	8.456	8.477	8.500	8.524	8.549	8.618
	$k \times 10^2$	4.998	5.017	5.040	5.087	5.137	5.189	5.244	5.301	5.361	5.424	5.490	5.558	5.743
	Pr	0.89	0.89	0.90	0.90	0.90	0.91	0.91	0.91	0.92	0.92	0.92	0.93	0.93

Note: c_p in Btu/lb$_m$-°F, ρ in lb$_m$/ft^3, μ in lb$_m$/ft-h, k in Btu/h-ft-°F.

*c_p, ρ, μ computed from equations recommended in *ASME Steam Tables,* 3rd ed., New York, Am. Soc. Mech. Engrs., 1977. k computed from equation recommended in J. Kestin, "Thermal Conductivity of Water and Steam," *Mech. Eng.*, Aug. 1978, p. 47.

Table A.6-SI Properties of Dry Air at Atmospheric Pressure (SI Units)*

t °C	c_p kJ/kg-°C	ρ kg/m³	$\mu \times 10^6$ kg/m-s	$\nu \times 10^6$ m²/s	$k \times 10^3$ W/m-°C	Pr
−50	1.0064	1.5819	14.63	9.25	20.04	0.735
−40	1.0060	1.5141	15.17	10.02	20.86	0.731
−30	1.0058	1.4518	15.69	10.81	21.68	0.728
−20	1.0057	1.3944	16.20	11.62	22.49	0.724
−10	1.0056	1.3414	16.71	12.46	23.29	0.721
0	1.0057	1.2923	17.20	13.31	24.08	0.718
10	1.0058	1.2467	17.69	14.19	24.87	0.716
20	1.0061	1.2042	18.17	15.09	25.64	0.713
30	1.0064	1.1644	18.65	16.01	26.38	0.712
40	1.0068	1.1273	19.11	16.96	27.10	0.710
50	1.0074	1.0924	19.57	17.92	27.81	0.709
60	1.0080	1.0596	20.03	18.90	28.52	0.708
70	1.0087	1.0287	20.47	19.90	29.22	0.707
80	1.0095	0.9996	20.92	20.92	29.91	0.706
90	1.0103	0.9721	21.35	21.96	30.59	0.705
100	1.0113	0.9460	21.78	23.02	31.27	0.704
110	1.0123	0.9213	22.20	24.10	31.94	0.704
120	1.0134	0.8979	22.62	25.19	32.61	0.703
130	1.0146	0.8756	23.03	26.31	33.28	0.702
140	1.0159	0.8544	23.44	27.44	33.94	0.702
150	1.0172	0.8342	23.84	28.58	34.59	0.701
160	1.0186	0.8150	24.24	29.75	35.25	0.701
170	1.0201	0.7966	24.63	30.93	35.89	0.700
180	1.0217	0.7790	25.03	32.13	36.54	0.700
190	1.0233	0.7622	25.41	33.34	37.18	0.699
200	1.0250	0.7461	25.79	34.57	37.81	0.699
210	1.0268	0.7306	26.17	35.82	38.45	0.699
220	1.0286	0.7158	26.54	37.08	39.08	0.699
230	1.0305	0.7016	26.91	38.36	39.71	0.698
240	1.0324	0.6879	27.27	39.65	40.33	0.698
250	1.0344	0.6748	27.64	40.96	40.95	0.698
260	1.0365	0.6621	27.99	42.28	41.57	0.698
270	1.0386	0.6499	28.35	43.62	42.18	0.698
280	1.0407	0.6382	28.70	44.97	42.79	0.698
290	1.0429	0.6268	29.05	46.34	43.40	0.698
300	1.0452	0.6159	29.39	47.72	44.01	0.698
310	1.0475	0.6053	29.73	49.12	44.61	0.698
320	1.0499	0.5951	30.07	50.53	45.21	0.698
330	1.0523	0.5853	30.41	51.95	45.84	0.698
340	1.0544	0.5757	30.74	53.39	46.38	0.699

Table A.6-SI Properties of Dry Air at Atmospheric Pressure (SI Units)*

t °C	c_p kJ/kg-°C	ρ kg/m³	$\mu \times 10^6$ kg/m-s	$\nu \times 10^6$ m²/s	$k \times 10^3$ W/m-°C	Pr
350	1.0568	0.5665	31.07	54.85	46.92	0.700
360	1.0591	0.5575	31.40	56.31	47.47	0.701
370	1.0615	0.5489	31.72	57.79	48.02	0.701
380	1.0639	0.5405	32.04	59.29	48.58	0.702
390	1.0662	0.5323	32.36	60.79	49.15	0.702
400	1.0686	0.5244	32.68	62.31	49.72	0.702
410	1.0710	0.5167	32.99	63.85	50.29	0.703
420	1.0734	0.5093	33.30	65.39	50.86	0.703
430	1.0758	0.5020	33.61	66.95	51.44	0.703
440	1.0782	0.4950	33.92	68.52	52.01	0.703
450	1.0806	0.4882	34.22	70.11	52.59	0.703
460	1.0830	0.4815	34.52	71.70	53.16	0.703
470	1.0854	0.4750	34.82	73.31	53.73	0.703
480	1.0878	0.4687	35.12	74.93	54.31	0.704
490	1.0902	0.4626	35.42	76.57	54.87	0.704
500	1.0926	0.4566	35.71	78.22	55.44	0.704
510	1.0949	0.4508	36.00	79.87	56.01	0.704
520	1.0973	0.4451	36.29	81.54	56.57	0.704
530	1.0996	0.4395	36.58	83.23	57.13	0.704
540	1.1020	0.4341	36.87	84.92	57.68	0.704
550	1.1043	0.4288	37.15	86.63	58.24	0.704
560	1.1066	0.4237	37.43	88.35	58.79	0.705
570	1.1088	0.4187	37.71	90.07	59.33	0.705
580	1.1111	0.4138	37.99	91.82	59.87	0.705
590	1.1133	0.4090	38.27	93.57	60.41	0.705
600	1.1155	0.4043	38.54	95.33	60.94	0.705
610	1.1177	0.3997	38.81	97.11	61.47	0.706
620	1.1198	0.3952	39.09	98.89	62.00	0.706
630	1.1219	0.3908	39.36	100.69	62.52	0.706
640	1.1240	0.3866	39.62	102.50	63.03	0.707
650	1.1260	0.3824	39.89	104.32	63.55	0.707

*ρ computed from ideal gas law. c_p, μ, ν, k computed from equations recommended in *Thermophysical Properties of Refrigerants,* New York, ASHRAE, 1976.

Table A.6-ENGL Properties of Dry Air at Atmospheric Pressure (English Units)*

t °F	c_p Btu/lb$_m$-°F	$\rho \times 10^2$ lb$_m$/ft^3	$\mu \times 10^2$ lb$_m$/ft-h	ν ft^2/h	$k \times 10^2$ Btu/h-ft-°F	Pr
−100	0.2407	11.029	3.230	0.2929	1.045	0.744
−80	0.2405	10.448	3.380	0.3235	1.099	0.739
−60	0.2404	9.925	3.526	0.3552	1.153	0.735
−40	0.2403	9.452	3.669	0.3882	1.206	0.731
−20	0.2402	9.022	3.809	0.4222	1.258	0.727
0	0.2402	8.630	3.947	0.4574	1.310	0.724
20	0.2402	8.270	4.082	0.4936	1.361	0.720
40	0.2402	7.939	4.215	0.5309	1.412	0.717
60	0.2403	7.633	4.345	0.5692	1.462	0.714
80	0.2403	7.350	4.473	0.6086	1.511	0.711
100	0.2405	7.088	4.599	0.6489	1.557	0.710
120	0.2406	6.843	4.723	0.6902	1.602	0.709
140	0.2408	6.615	4.845	0.7324	1.648	0.708
160	0.2409	6.401	4.965	0.7756	1.693	0.707
180	0.2412	6.201	5.083	0.8197	1.737	0.706
200	0.2414	6.013	5.199	0.8647	1.781	0.705
220	0.2417	5.836	5.314	0.9105	1.824	0.704
240	0.2419	5.669	5.427	0.9573	1.867	0.703
260	0.2422	5.512	5.539	1.0049	1.910	0.702
280	0.2426	5.363	5.649	1.0533	1.952	0.702
300	0.2429	5.222	5.757	1.1026	1.995	0.701
320	0.2433	5.088	5.864	1.1527	2.036	0.701
340	0.2437	4.960	5.970	1.2036	2.078	0.700
360	0.2441	4.839	6.075	1.2552	2.119	0.700
380	0.2445	4.724	6.178	1.3077	2.160	0.699
400	0.2450	4.614	6.280	1.3609	2.201	0.699
420	0.2455	4.509	6.380	1.4149	2.242	0.699
440	0.2460	4.409	6.480	1.4697	2.282	0.698
460	0.2465	4.313	6.578	1.5252	2.322	0.698
480	0.2470	4.221	6.676	1.5814	2.362	0.698
500	0.2476	4.133	6.772	1.6384	2.402	0.698
520	0.2481	4.049	6.867	1.6960	2.441	0.698
540	0.2487	3.968	6.961	1.7544	2.480	0.698
560	0.2493	3.890	7.055	1.8134	2.519	0.698
580	0.2499	3.815	7.147	1.8732	2.558	0.698

Table A.6-ENGL Properties of Dry Air at Atmospheric Pressure (English Units)*

t °F	c_p Btu/lb$_m$-°F	$\rho \times 10^2$ lb$_m$/ft^3	$\mu \times 10^2$ lb$_m$/ft-h	ν ft^2/h	$k \times 10^2$ Btu/h-ft-°F	Pr
600	0.2505	3.743	7.238	1.9336	2.597	0.698
620	0.2511	3.674	7.329	1.9948	2.635	0.698
640	0.2517	3.607	7.418	2.0566	2.623	0.699
660	0.2523	3.543	7.507	2.1190	2.707	0.700
680	0.2530	3.481	7.595	2.1821	2.743	0.701
700	0.2536	3.421	7.682	2.2459	2.778	0.701
720	0.2542	3.363	7.768	2.3103	2.814	0.702
740	0.2549	3.307	7.854	2.3753	2.851	0.702
760	0.2555	3.252	7.939	2.4409	2.887	0.702
780	0.2561	3.200	8.023	2.5072	2.924	0.703
800	0.2568	3.149	8.106	2.5741	2.961	0.703
820	0.2574	3.100	8.189	2.6416	2.998	0.703
840	0.2580	3.052	8.270	2.7098	3.035	0.703
860	0.2587	3.006	8.352	2.7785	3.072	0.703
880	0.2593	2.961	8.432	2.8478	3.108	0.703
900	0.2600	2.917	8.512	2.9177	3.145	0.704
920	0.2606	2.875	8.592	2.9882	3.182	0.704
940	0.2612	2.834	8.670	3.0593	3.218	0.704
960	0.2618	2.794	8.748	3.1310	3.254	0.704
980	0.2625	2.755	8.826	3.2032	3.290	0.704
1000	0.2631	2.718	8.903	3.2760	3.326	0.704
1020	0.2637	2.681	8.979	3.3494	3.361	0.704
1040	0.2643	2.645	9.055	3.4234	3.397	0.705
1060	0.2649	2.610	9.130	3.4978	3.432	0.705
1080	0.2655	2.576	9.205	3.5729	3.466	0.705
1100	0.2661	2.543	9.279	3.6485	3.501	0.705
1120	0.2667	2.511	9.353	3.7246	3.535	0.706
1140	0.2672	2.480	9.426	3.8013	3.569	0.706
1160	0.2678	2.449	9.499	3.8785	3.602	0.706
1180	0.2684	2.419	9.571	3.9562	3.635	0.706
1200	0.2689	2.390	9.643	4.0345	3.668	0.707

*ρ computed from ideal gas law. c_p, μ, ν, k computed from equations recommended in *Thermophysical Properties of Refrigerants,* New York, ASHRAE, 1976.

Table A.7 Properties of Gases at Atmospheric Pressure*

t °C	c_p kJ/kg-°C	$\mu \times 10^6$ kg/m-s	$k \times 10^3$ W/m-°C	Pr
		Ammonia (NH_3)		
−20	2.251	8.64	19.96	0.974
0	2.194	9.32	21.82	0.938
20	2.167	10.02	23.83	0.912
40	2.164	10.73	25.98	0.894
60	2.177	11.45	28.23	0.883
80	2.203	12.18	30.59	0.877
100	2.236	12.91	33.05	0.874
120	2.274	13.65	35.59	0.872
140	2.314	14.39	38.21	0.871
160	2.353	15.13	40.91	0.870
180	2.390	15.87	43.68	0.869
200	2.425	16.62	46.51	0.867
220	2.457	17.37	49.41	0.864
240	2.488	18.13	52.37	0.861
260	2.519	18.88	55.39	0.859
		Carbon Dioxide (CO_2)		
−50	0.7813	11.34	11.05	0.80
0	0.8277	13.74	14.57	0.78
50	0.8730	16.05	18.58	0.75
100	0.9167	18.27	22.24	0.75
150	0.9581	20.40	26.31	0.74
200	0.9969	22.42	30.25	0.74
250	1.033	24.37	33.96	0.74
300	1.062	26.24	38.16	0.73
400	1.114	29.75	46.59	0.71
500	1.158	33.01	54.14	0.71
600	1.196	36.08	60.60	0.71
		Helium (He)		
−100	5.1931	13.75	104.5	0.68
−50	5.1931	16.01	123.5	0.67
0	5.1931	18.03	142.3	0.66
50	5.1931	20.95	160.2	0.68
100	5.1931	23.15	177.7	0.68
150	5.1931	25.23	194.8	0.67
200	5.1931	27.22	211.5	0.67
300	5.1931	31.01	243.3	0.66
400	5.1931	34.58	272.9	0.65
500	5.1931	37.96	300.4	0.65
		Nitrogen (N_2)		
−100	1.044	11.35	15.89	0.745
−50	1.043	14.07	20.08	0.731
0	1.041	16.58	24.04	0.718
50	1.041	18.89	27.59	0.713
100	1.043	21.05	30.86	0.711

Table A.7 (Cont.)

t °C	c_p kJ/kg-°C	$\mu \times 10^6$ kg/m-s	$k \times 10^3$ W/m-°C	Pr
		Nitrogen (N_2)		
150	1.047	23.07	34.00	0.710
200	1.053	24.97	37.02	0.710
300	1.069	28.40	42.71	0.711
400	1.091	31.55	47.92	0.719
500	1.116	34.46	52.85	0.728
600	1.140	37.16	57.50	0.737
		Oxygen (O_2)		
0	0.9159	19.15	24.49	0.716
50	0.9239	21.95	28.55	0.710
100	0.9348	24.57	32.26	0.712
150	0.9479	27.02	35.83	0.715
200	0.9627	29.33	39.33	0.718
300	0.9944	33.64	46.23	0.724
400	1.0249	37.55	52.73	0.730
500	1.0492	41.16	58.74	0.735
600	1.0693	44.52	64.54	0.738
700	1.0864	47.68	70.25	0.738
		Freon-12 (CCl_2F_2)		
−20	0.5525	10.70	7.36	0.804
0	0.5748	11.52	8.34	0.794
20	0.5956	12.33	9.35	0.785
40	0.6150	13.12	10.39	0.777
60	0.6331	13.90	11.44	0.769
80	0.6500	14.66	12.51	0.758
100	0.6656	15.41	13.60	0.754
120	0.6801	16.14	14.70	0.747
140	0.6936	16.86	15.81	0.740
160	0.7060	17.56	16.92	0.733
180	0.7175	18.25	18.05	0.726
200	0.7282	18.93	19.17	0.719
		Freon-22 ($CHClF_2$)		
−40	0.5675	10.13	4.04	1.42
0	0.6156	11.85	9.39	0.777
40	0.6625	13.54	11.79	0.761
80	0.7076	15.20	14.19	0.758
120	0.7503	16.81	16.59	0.760
160	0.7900	18.39	18.99	0.765
200	0.8259	19.92	21.39	0.769
240	0.8575	21.42	23.79	0.772
280	0.8841	22.89	26.19	0.773
320	0.9051	24.31	28.59	0.770

Table A.7 (Cont.)

t °C	c_p kJ/kg-°C	$\mu \times 10^6$ kg/m-s	$k \times 10^3$ W/m-°C	Pr
		Butane (normal, C_4H_{10})		
20	1.6860	7.41	15.41	0.811
40	1.7781	7.89	17.27	0.812
60	1.8692	8.37	19.25	0.813
80	1.9591	8.85	21.31	0.813
100	2.0478	9.32	23.45	0.814
120	2.1351	9.80	25.64	0.816
140	2.2210	10.27	27.87	0.818
160	2.3052	10.74	30.12	0.822
180	2.3877	11.20	32.38	0.826
200	2.4684	11.66	34.62	0.832
220	2.5472	12.12	36.85	0.838
240	2.6238	12.58	39.04	0.845
260	2.6983	13.03	41.19	0.855
		Ethane (C_2H_6)		
−100	1.4252	5.62	8.00	1.00
−50	1.5260	7.14	12.66	0.861
0	1.6791	8.65	18.30	0.794
50	1.8660	10.12	24.87	0.759
100	2.0713	11.53	31.96	0.747
150	2.2825	12.87	39.49	0.744
200	2.4904	14.15	47.36	0.744
250	2.6890	15.40	55.50	0.746
300	2.8750	16.58	63.83	0.747
350	3.0485	17.71	72.44	0.745
400	3.2125	18.80	81.03	0.745
		Methane (CH_4)		
0	2.1702	10.32	30.78	0.727
20	2.2193	10.98	33.39	0.721
40	2.2724	11.62	36.06	0.732
60	2.3290	12.25	38.73	0.737
80	2.3889	12.86	41.55	0.740
100	2.4515	13.46	44.38	0.744
120	2.5166	14.05	47.25	0.748
140	2.5839	14.62	50.88	0.742
160	2.6529	15.18	54.61	0.737
180	2.7233	15.72	58.37	0.734
200	2.7948	16.26	62.14	0.731
220	2.8670	16.79	65.91	0.730
240	2.9396	17.30	69.68	0.730
260	3.0121	17.81	73.44	0.730

Table A.7 (Cont.)

t °C	c_p kJ/kg-°C	$\mu \times 10^6$ kg/m-s	$k \times 10^3$ W/m-°C	Pr
		Propane (C_3H_8)		
0	1.5956	7.55	15.27	0.789
20	1.6779	8.07	17.41	0.778
40	1.7628	8.60	19.60	0.773
60	1.8496	9.12	21.84	0.773
80	1.9375	9.64	24.12	0.775
100	2.0260	10.16	26.44	0.779
120	2.1144	10.67	28.78	0.784
140	2.2022	11.18	31.15	0.790
160	2.2889	11.68	33.52	0.797
180	2.3742	12.17	35.92	0.805
200	2.4577	12.66	38.31	0.812
220	2.5391	13.14	40.72	0.819
240	2.6183	13.61	43.12	0.826

*Tabulated values computed from equations recommended in *Thermophysical Properties of Refrigerants*, New York, ASHRAE, 1976. The values of c_p for helium, Freon-12, and Freon-22 are for the ideal gas state as $p \rightarrow 0$; they may be used at one atmosphere without serious error.

Table A.8 Critical Properties

Substance	Formula	Molecular Weight	T_c °K	p_c kN/m²(kPa)
Air	—	28.96	132.4	3,770
Ammonia	NH_3	17.03	405.5	11,280
Butane	C_4H_{10}	58.124	425.2	3,800
Carbon dioxide	CO_2	44.01	304.2	7,390
Ethane	C_2H_6	30.070	305.5	4,880
Freon-12	CCl_2F_2	120.91	384.7	4,010
Freon-22	$CHC1F_2$	86.48	369.2	4,980
Helium	He	4.003	5.3	230
Methane	CH_4	16.043	191.1	4,640
Nitrogen	N_2	28.013	126.2	3,390
Oxygen	O_2	31.999	154.8	5,080
Propane	C_3H_8	44.097	370.0	4,260

Table A.9 Properties of Some Liquid Metals*

Metal	t °C	$\rho \times 10^{-3}$ kg/m³	$\mu \times 10^{4}$ kg/m-s	k W/m-°C	c_p kJ/kg-°C	$\nu \times 10^{7}$ m²/s	$\alpha \times 10^{5}$ m²/s	Pr
Bismuth	316	10.011	16.04	16.4	0.1444	1.602	1.134	0.0141
	427	9.867	13.48	15.6	0.1495	1.366	1.058	0.0129
	538	9.739	11.08	15.6	0.1545	1.138	1.037	0.0110
	649	9.611	9.18	15.6	0.1595	0.955	1.018	0.0094
	760	9.467	8.06	15.6	0.1645	0.851	1.002	0.0085
Lead	371	10.540	24.22	16.1	0.1591	2.298	0.960	0.024
	454	10.444	19.92	15.6	0.1549	1.907	0.964	0.020
	538	10.348	16.82	15.4	0.1549	1.625	0.961	0.017
	621	10.236	15.58	15.1	0.1549	1.522	0.952	0.016
	704	10.140		14.9				
Mercury	10	13.567	16.12	8.13	0.1382	1.188	0.434	0.027
	93	13.359	12.07	10.4	0.1382	0.904	0.563	0.016
	149	13.231	10.25	11.6	0.1382	0.775	0.634	0.012
	204	13.087	10.13	12.5	0.1340	0.774	0.713	0.011
	316	12.847	8.60	14.0	0.1340	0.669	0.813	0.0082
Potassium	149	0.8073	3.75	45.0	0.795	4.65	7.01	0.0066
	260	0.7801	2.41	42.7	0.795	3.40	6.89	0.0049
	427	0.7417	1.82	39.5	0.754	2.45	7.06	0.0035
	593	0.7016	1.50	35.7	0.754	2.14	6.75	0.0032
	704	0.6744	1.36	33.1	0.754	2.02	6.51	0.0031

Sodium	93	0.9291	6.95	86.2	1.38	7.48	6.72	0.011
	204	0.9018	4.42	80.3	1.34	4.90	6.65	0.0074
	371	0.8602	2.89	72.3	1.30	3.36	6.47	0.0052
	538	0.8201	2.03	65.4	1.26	2.48	6.33	0.0039
	704	0.7785	1.82	59.7	1.26	2.32	6.09	0.0038
Zinc	454	6.904	32.08	58.3	0.498	4.65	1.70	0.027
	538	6.856	26.54	57.5	0.486	3.87	1.72	0.022
	649	6.760	20.92	56.8	0.473	3.09	1.78	0.017
	816	6.536	16.87	56.4	0.448	2.58	1.93	0.013
55% Bi,	149	10.524		9.05	0.146		0.589	
44.5% Pb	288	10.348	17.61	10.73	0.146	1.70	0.710	0.024
(eutectic)	371	10.236	15.34	11.86	0.146	1.50	0.794	0.019
	593	9.931						
	649	9.835						
44% K,	93	0.8874	5.75	25.6	1.130	6.48	2.55	0.025
56% Na	204	0.8618	3.56	26.5	1.093	4.13	2.82	0.015
	371	0.8217	2.31	27.5	1.055	2.81	3.17	0.0089
	538	0.7817	1.78	28.4	1.038	2.28	3.50	0.0065
	704	0.7401	1.61	28.9	1.043	2.18	3.74	0.0058

*Adapted from M. Jakob, *Heat Transfer,* Vol. II, New York, Wiley, 1957, p. 564. Original data are contained in *Liquid Metals Handbook,* 2nd ed. (rev.) *NAVEXOS p. 733* (rev.), Atomic Energy Commission, Dept. of the Navy, Washington, D.C., Jan. 1954.

Table A.10 Emissivities of Various Surfaces*

Surface Description	Emissivity, ϵ/T, °C
Metals and metal plating	
Aluminum	
Foil, bright side, as received (1)	0.04/371
75S-T alloy, weathered (1)	0.16/816
1100-0, commercially pure (1)	0.05/93; 0.05/204; 0.05/316; 0.05/427
1100-0, commercially pure, oxidized at 316°C (1)	0.04/93; 0.04/204; 0.05/316; 0.05/427
24S-T81, with chromic acid anodize (1)	0.17/149; 0.17/168; 0.17/185; 0.17/204
24S-T81, with H_2SO_4 anodize	0.85/149; 0.82/168; 0.80/185; 0.78/427
Beryllium (1)	0.16/149; 0.21/371; 0.26/482; 0.30/593
Beryllium, anodized (1)	0.90/149; 0.88/371; 0.85/482; 0.82/593
Brass	
Highly polished (2)	0.030/277
Polished (8)	0.10/38; 0.10/316
Rolled plate (3)	0.06/22
Chromium	
Polished (6)	0.08/38; 0.36/1093
0.1-mil-thick plate on 0.5-mil nickel on 321 stainless steel (1)	0.12/93; 0.13/171; 0.14/249; 0.15/327; 0.15/399
Gold, evaporated on fiber glass (1)	0.05/93; 0.05/149; 0.075/260
Gold, coated on stainless steel (1)	0.09/93; 0.09/171; 0.11/254; 0.15/316; 0.14/399
Iron	
Cast, polished (5)	0.21/200
Cast, oxidized (5)	0.64/199; 0.78/593
Plate, completely rusted (3)	0.69/19
Wrought, polished (7)	0.28/38; 0.28/249
Wrought, oxidized (7)	0.94/21; 0.94/360
Platinum on polished steel (1)	0.13/93; 0.15/171; 0.14/249; 0.15/327; 0.15/399
Silver	
Pure, polished (2)	0.020/227; 0.032/627
5-mil silver plate on 0.5-mil nickel on 321 stainless steel (1)	0.06/93; 0.06/171; 0.06/249; 0.07/327; 0.08/399
Silver plate on stainless steel (1)	0.11/93; 0.11/204; 0.11/316; 0.13/427
Steel	
Polished plate (4)	0.066/100
Rough plate (8)	0.94/38; 0.97/371
Stailness, type 301 (1)	0.14/93; 0.15/204; 0.16/316; 0.18/427
Stainless, type 321, oxidized at 260°C (1)	0.27/93; 0.27/204; 0.28/316; 0.32/427
Stainless, type 321, with black oxide (1)	0.66/93; 0.66/204; 0.69/316; 0.76/427
Inconel X, oxidized at 1052°C (1)	0.61/93; 0.79/171; 0.81/249; 0.86/327; 0.81/399
Tungsten filament (9)	0.39/3316

Table A.10 Emissivities of Various Surfaces*

Surface Description	Emissivity, ϵ/T, °C
Paints and Lacquers	
Aluminized silicon resin paint on 321 stainless steel, baked at 316°C (1)	0.20/93; 0.20/204; 0.21/316; 0.22/427
Black lacquer on iron (3)	0.88/24
Black lacquer, flat black (8)	0.96/38; 0.98/93
Black, heat resistant, on 321 stainless (1)	0.81/93; 0.76/204; 0.76/316; 0.80/427
Black, high heat Dixon 208, on stainless (1)	0.93/149; 0.92/260; 0.93/371; 0.93/482; 0.93/593
International orange on 2024-T4A1(1)	0.72/93; 0.68/204; 0.53/316
White enamel, fused on iron (3)	0.90/19
White acrylic resin paint on aluminum (1)	0.90/93; 0.87/204
Miscellaneous	
Asbestos board (3)	0.96/23
Asbestos paper (8)	0.93/38; 0.94/371
Brick	
Red, rough (3)	0.93/21
Building (10)	0.45/1000
Fire clay (10)	0.75/1000
Magnesite refractory (10)	0.38/1000
Concrete tile (10)	0.63/1000
Glass (3)	0.94/22
Graphite, pressed (11)	0.98/249; 0.98/510
Lampblack, thick coat on iron (3)	0.97/20
Plaster, rough (7)	0.91/10; 0.91/88
Roofing paper (3)	0.91/21

*When more than one value is given, linear interpolation is permissible. Compiled from data given in (1) D. K. Edwards, K. E. Nelson, R. D. Roddick, and G. T. Gier, "Basic Studies on the Use and Control of Solar Energy," *Univ. Calif. Dept. Eng. Rep. 60-93,* Los Angeles, 1960; and from data given in W. H. McAdams, *Heat Transmission,* 3rd ed., New York, McGraw-Hill, 1954, as compiled by H. C. Hottel, from (2) H. Schmidt and E. Furthmann, *Mitt. Kaiser-Wilhelm-Inst. Eisenforsch.,* Abhandl., Vol. 109, 1928, p. 225; (3) E. Schmidt, *Gesundh.-Ing.,* Beiheft 20, Reihe 1, 1927; (4) B. T. Barnes, W. E. Forsythe, and E. Q. Adams, *J. Opt. Soc. Am.,* Vol. 37, 1947, p. 804; (5) C. F. Randolph and M. J. Overholtzer, *Phys. Rev.,* Vol. 2, 1913, p. 144; (6) E. O. Hurlbert, *Astrophys. J.,* Vol. 42, 1915, p. 205; (7) F. Wamsler, *Z.V.D.I.,* Vol. 55, 1911, p. 599; *Mitt. Forsch.,* Vol. 98, 1911, p. 1; (8) R. H. Heilmann, *Trans. ASME,* Vol. 32, 1944, p. 239; (9) C. Zwikker, *Arch. Neerl. Sci.,* Vol. 9, 1925, p. 207; (10) M. W. Thring, *Sciences of Flames and Furnaces,* London, Chapman & Hall, 1952; (11) M. Pirani, *J. Sci. Instrum.,* Vol. 16, 1939.

Table A.11 Solar Absorptivity of Various Surfaces*

Surface Description	Solar Absorptivity α_s	Infrared Emissivity 250°K ϵ	250°K α_s/ϵ	308°K ϵ	308°K α_s/ϵ	556°K ϵ	556°K α_s/ϵ
Aluminum foil, coated with 10-μm silicon	0.522			0.12	4.35	0.12	4.35
Silicon solar cell, 1 mm thick	0.938			0.316	2.97	0.497	1.89
Chromium plate 0.1 mil thick on 0.5-mil nickel on 321 stailness steel	0.778			0.150	5.18	0.182	4.27
Stainless steel, type 410	0.764			0.130	5.88	0.180	4.24
Titanium 75A, heated 454°C 300 h	0.798			0.211	3.78	0.294	2.72
Titanium C-110, heated 427°C 100 h	0.524			0.162	3.24	0.202	2.59
Titanium, anodized	0.515	0.866	0.59			0.835	0.62
Ebanol C on copper	0.908			0.11	8.25		
Ebanol S on steel	0.848			0.10	8.48		
Aluminum, 6061-T-6, 1 mil anodize	0.923	0.841	1.10			0.847	1.09
Iconel X, oxidized	0.898	0.711	1.26			0.809	1.11
Stainless steel 301, with Armco black oxide	0.891	0.746	1.19			0.756	1.18
Graphite, on sodium silicate on polished aluminum	0.960	0.908	1.06			0.930	1.03
Glass, 3 mils on silicon solar cell	0.925	0.843	1.10			0.877	1.05
Titanox, 2 mils on black paint	0.154	0.885	0.17			0.905	0.17
Flat black epoxy paint on aluminum	0.951	0.888	1.07			0.924	1.03
White epoxy paint on aluminum	0.248	0.882	0.28			0.912	0.27

*Compiled from data given in D. K. Edwards, R. D. Roddick, and J. T. Gier, "Basic Studies on the Use and Control of Solar Energy," *Univ. Calif. Dept. Eng. Rept. 60-93*, Los Angeles, 1960.

Table A.12 Radiation Functions

λT μm-°K	$\frac{E_{b\lambda}}{\sigma T^5} \times 10^5$ (μm-°K)$^{-1}$	$F_{0-\lambda}(T)$	λT μm-°K	$\frac{E_{b\lambda}}{\sigma T^5} \times 10^5$ (μm-°K)$^{-1}$	$F_{0-\lambda}(T)$	λT μm-°K	$\frac{E_{b\lambda}}{\sigma T^5} \times 10^5$ (μm-°K)$^{-1}$	$F_{0-\lambda}(T)$
500	0.00000672	0	3000	22.627	0.27323	5500	10.340	0.69088
600	0.0003269	0	3100	22.450	0.29578	5600	9.939	0.70102
700	0.004650	0	3200	22.178	0.31810	5700	9.553	0.71077
800	0.03114	0.000016	3300	21.827	0.34011	5800	9.182	0.72013
900	0.1275	0.000087	3400	21.411	0.36173	5900	8.827	0.72914
1000	0.3723	0.000321	3500	20.942	0.38291	6000	8.486	0.73779
1100	0.8550	0.000911	3600	20.432	0.40360	6100	8.159	0.74611
1200	1.646	0.00213	3700	19.891	0.42376	6200	7.845	0.75411
1300	2.774	0.00432	3800	19.327	0.44337	6300	7.544	0.76181
1400	4.223	0.00779	3900	18.748	0.46241	6400	7.256	0.76920
1500	5.934	0.01285	4000	18.160	0.48087	6500	6.980	0.77632
1600	7.827	0.01972	4100	17.568	0.49873	6600	6.716	0.78317
1700	9.811	0.02853	4200	16.976	0.51600	6700	6.463	0.78976
1800	11.799	0.03934	4300	16.389	0.53268	6800	6.220	0.79610
1900	13.716	0.05211	4400	15.809	0.54878	6900	5.988	0.80220
2000	15.501	0.06673	4500	15.240	0.56430	7000	5.766	0.80808
2100	17.113	0.08305	4600	14.681	0.57926	7100	5.553	0.81373
2200	18.524	0.10089	4700	14.136	0.59367	7200	5.348	0.81918
2300	19.720	0.12003	4800	13.606	0.60754	7300	5.153	0.82443
2400	20.698	0.14026	4900	13.091	0.62089	7400	4.966	0.82949
2500	21.465	0.16136	5000	12.591	0.63373	7500	4.786	0.83437
2600	22.031	0.18312	5100	12.108	0.64608	8000	3.995	0.85625
2700	22.412	0.20536	5200	11.642	0.65795	8500	3.354	0.87457
2800	22.626	0.22789	5300	11.191	0.66936	9000	2.832	0.88999
2900	22.692	0.25056	5400	10.758	0.68034	9500	2.404	0.90304

Table A.12 (Cont.)

λT μm-°K	$\frac{E_{b\lambda}}{\sigma T^5} \times 10^5$ (μm-°K)$^{-1}$	$F_{0-\lambda}(T)$	λT μm-°K	$\frac{E_{b\lambda}}{\sigma T^5} \times 10^5$ (μm-°K)$^{-1}$	$F_{0-\lambda}(T)$	λT μm-°K	$\frac{E_{b\lambda}}{\sigma T^5} \times 10^5$ (μm-°K)$^{-1}$	$F_{0-\lambda}(T)$
10000	2.052	0.91416	15000	0.5399	0.96893	35000	0.0247	0.99695
10500	1.761	0.92367	15500	0.4821	0.97149	40000	0.0149	0.99792
11000	1.518	0.93185	16000	0.4317	0.97377	45000	0.0095	0.99852
11500	1.315	0.93892	16500	0.3877	0.97581	50000	0.0063	0.99890
12000	1.145	0.94505	17000	0.3492	0.97765	55000	0.0044	0.99917
12500	1.000	0.95041	18000	0.2853	0.98081	∞	0	1.00000
13000	0.8779	0.95509	19000	0.2353	0.98341			
13500	0.7733	0.95921	20000	0.1958	0.98555			
14000	0.6837	0.96285	25000	0.0869	0.99217			
14500	0.6066	0.96607	30000	0.0441	0.99529			

APPENDIX B

Miscellaneous Tables

Table B.1 Dimensions of Standard Pipe Sizes

Nominal Size	Outside Diameter		Outside Surface Area per Length		Schedule No.	Inside Diameter		Inside Cross-Sectional Area	
in.	in.	cm	ft^2/ft	m^2/m		in.	cm	ft^2	m^2
$\frac{1}{2}$	0.840	2.134	0.2199	0.06704	40	0.622	1.580	0.002110	0.0001961
					80	0.546	1.387	0.001626	0.0001511
1	1.315	3.340	0.3443	0.1049	40	1.049	2.664	0.006002	0.0005574
					80	0.957	2.431	0.004995	0.0004642
2	2.375	6.033	0.6218	0.1895	40	2.067	5.250	0.02330	0.002165
					80	1.939	4.925	0.02051	0.001905
3	3.500	8.890	0.9163	0.2793	40	3.068	7.793	0.05134	0.004770
					80	2.900	7.366	0.04587	0.004261
4	4.500	11.43	1.178	0.3591	40	4.026	10.226	0.08840	0.008213
					80	3.826	9.718	0.07984	0.007417
5	5.563	14.13	1.456	0.4439	40	5.047	12.82	0.1389	0.01291
					80	4.813	12.23	0.1263	0.01175
6	6.625	16.83	1.734	0.5287	40	6.065	15.41	0.2006	0.01865
					80	5.761	14.63	0.1810	0.01681
8	8.625	21.91	2.258	0.6683	40	7.981	20.27	0.3474	0.03227
					80	7.625	19.37	0.3171	0.02947
10	10.750	27.31	2.814	0.8580	40	10.020	25.45	0.5476	0.05087
					80	9.564	24.29	0.4989	0.04634
12	12.750	32.39	3.338	1.0176	40	11.938	30.32	0.7773	0.07220
					80	11.376	28.90	0.7058	0.06560

Table B.2 Dimensions of Standard Tubing

Nominal Size in.	Outside Diameter		Outside Surface Area per Length		Gage	Inside Diameter		Inside Cross-Sectional Area	
	in.	cm	ft^2/ft	m^2/m		in.	cm	ft^2	m^2
$\frac{1}{2}$	0.500	1.270	0.1309	0.03990	18	0.402	1.021	0.0008814	0.00008187
					16	0.370	0.940	0.0007467	0.00006940
					14	0.334	0.848	0.0006084	0.00005648
$\frac{5}{8}$	0.625	1.588	0.1636	0.04989	18	0.527	1.339	0.001515	0.0001408
					16	0.495	1.257	0.001336	0.0001241
					14	0.459	1.166	0.001149	0.0001068
$\frac{3}{4}$	0.750	1.905	0.1963	0.05985	18	0.652	1.656	0.002319	0.0002154
					16	0.620	1.575	0.002097	0.0001948
					14	0.584	1.483	0.001860	0.0001727
1	1.000	2.540	0.2618	0.07980	18	0.902	2.291	0.004438	0.0004122
					16	0.870	2.210	0.004128	0.0003836
					14	0.834	2.118	0.003794	0.0003523
					12	0.782	1.986	0.003335	0.0003098
					10	0.732	1.589	0.002922	0.0002714
$1\frac{1}{4}$	1.250	3.175	0.3272	0.09975	18	1.152	2.926	0.007238	0.0006724
					16	1.120	2.845	0.006842	0.0006357
					14	1.084	2.753	0.006409	0.0005953
					12	1.032	2.621	0.005809	0.0005395
					10	0.982	2.494	0.005260	0.0004885
$1\frac{1}{2}$	1.500	3.810	0.3927	0.1197	13	1.310	3.327	0.009360	0.0008694
					12	1.282	3.256	0.008964	0.0008326
					11	1.260	3.200	0.008659	0.0008042
					10	1.232	3.129	0.008278	0.0007690
					8	1.170	2.972	0.007460	0.0006937
2	2.000	5.080	0.5236	0.1596	13	1.810	4.597	0.01787	0.001660
					12	1.782	4.526	0.01732	0.001609
					11	1.760	4.470	0.01690	0.001569
					10	1.732	4.399	0.01632	0.001520
					8	1.670	4.242	0.01521	0.001413

APPENDIX C

Determination of the Shape Factor

The shape factor, F_{1-2}, was introduced in Sec. 9.7 as the fraction of diffuse radiation leaving one surface, A_1, which directly strikes a second surface, A_2. This definition led to the following expression for the shape factor,

$$F_{1-2} = \frac{1}{A_1} \int_{A_2} \int_{A_1} \frac{\cos \theta_1 \cos \theta_2 \, dA_1 \, dA_2}{\pi r^2}. \tag{C.1}$$

The geometrical variables θ_1 θ_2, r, A_1, and A_2 are illustrated in Fig. 9.17. The shape factor is seen to be a purely geometrical quantity—dependent only on the size, shape, and relative orientation of the two surfaces.

The discussions of Sec. 9.7, and succeeding sections, emphasized the importance of this shape factor in the calculation of radiant exchange. Whether dealing with individual surfaces, enclosures, etc., the shape factor entered into almost every consideration. Thus it is important to have at hand numerical values of the shape factor for geometric configurations of practical significance. Some of the fundamental configurations will be considered here. References 1, 2, 3, and 4 may be consulted for more extensive tabulations of the shape factor and for a great variety of cases. In addition to consideration of basic configurations, application may be made of the reciprocal and additive properties of the shape factor, discussed in Sec. 9.6 to calculate F for other, more complex, configurations.

C.1 THE SHAPE FACTOR FOR FINITE PARALLEL, OPPOSED RECTANGLES

Of very great application in engineering is the configuration of two equal, parallel, directly opposed rectangles. Figure C.1 shows two rectangles $W \times L$ in size and spaced a distance D apart. By arbitrarily denoting the upper area by A_1 and the

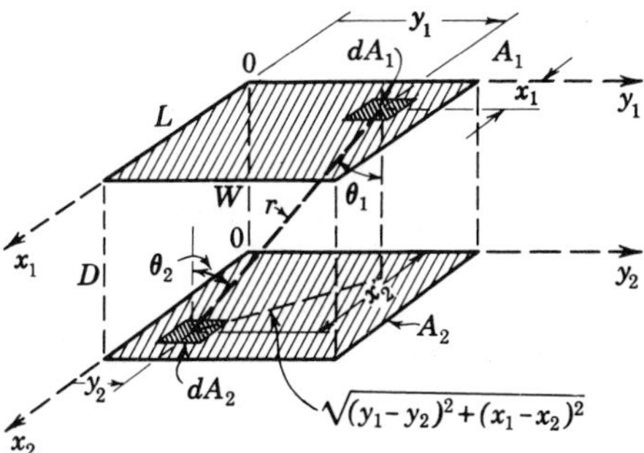

Figure C.1

lower by A_2, the two coordinate systems (x_1, y_1) and (x_2, y_2) may be chosen as shown. Selecting area elements dA_1 and dA_2 on each surface, one finds that their centers have coordinates (x_1, y_1) and (x_2, y_2), respectively. The distance r between dA_1 and dA_2 is then given by

$$r^2 = D^2 + (x_1 - x_2)^2 + (y_1 - y_2)^2.$$

The cosines of the two angles θ_1 and θ_2 are identical and equal to $\cos\theta_1 = \cos\theta_2 = D/r$. Equation (C.1) for the shape factor F_{1-2} (or F_{2-1} in this case of equal areas) gives

$$F_{1-2} = \frac{1}{A_1}\int_{A_1}\int_{A_2}\frac{\cos\theta_1\cos\theta_2\,dA_1\,dA_2}{\pi r^2}$$

$$= \frac{1}{WL}\int_0^W\int_0^L\int_0^W\int_0^L\frac{D^2\,dx_1\,dy_1\,dx_2\,dy_2}{\pi[D^2 + (x_1 - x_2)^2 + (y_1 - y_2)^2]^2}.$$

The result of the integration of the equation above may be written in terms of the dimensionless ratios W/D and L/D. The result is given below with the following symbols introduced for simplicity:

$$R_1 = \frac{L}{D},$$

$$R_2 = \frac{W}{D},$$

$$F_{1-2} = \frac{1}{\pi}\left[\frac{1}{R_1R_2}\ln\frac{(1 + R_1^2)(1 + R_2^2)}{(1 + R_1^2 + R_2^2)} - \frac{2}{R_1}\tan^{-1}R_2 - \frac{2}{R_2}\tan^{-1}R_1 + 2\sqrt{1 + \frac{1}{R_1^2}}\tan^{-1}\frac{R_2}{\sqrt{1 + R_1^2}} + 2\sqrt{1 + \frac{1}{R_2^2}}\tan^{-1}\frac{R_1}{\sqrt{1 + R_2^2}}\right]. \tag{C.2}$$

For rapid determination of the shape factor expressed by this equation. Fig. 9.18 presents F_{1-2} as a function of $R_1 = L/D$ and $R_2 = W/D$.

C.2 THE SHAPE FACTOR FOR PERPENDICULAR RECTANGLES HAVING A COMMON EDGE

Another configuration of particular engineering importance is that of two perpendicular rectangles with a common edge. Such a configuration exists between the walls of a room (or furnace) and the floor or ceiling.

The approach to the evaluation of the shape factor is identical to that used in the preceding section. Figure C.2 depicts a rectangle, call it A_1 of dimensions $D \times L$ located normal to A_2 with dimensions $D \times W$. The dimension D is, then, the length of the common edge. Figure C.2 shows the coordinate system selected, and the various quantities needed to evaluate F_{1-2} are indicated. Without discussion, the shape factor is expressed by

$$r^2 = y_1^2 + y_2^2 + (x_1 - x_2)^2,$$

$$\cos \theta_1 = \frac{y_2}{r}, \qquad \cos \theta_2 = \frac{y_1}{r}.$$

Thus the definition of F_{1-2} in Eq. (C.1) gives

$$F_{1-2} = \frac{1}{A_1} \int_{A_1} \int_{A_2} \frac{\cos \theta_1 \cos \theta_2 \, dA_1 \, dA_2}{\pi r^2}$$

$$= \frac{1}{LD} \int_0^W \int_0^D \int_0^L \int_0^D \frac{y_1 y_2 \, dx_1 \, dy_1 \, dx_2 \, dy_2}{\pi [y_1^2 + y_2^2 + (x_1 - x_2)^2]^2}.$$

In terms of the two dimensionless parameters,

$$R_1 = \frac{L}{D},$$

$$R_2 = \frac{W}{D},$$

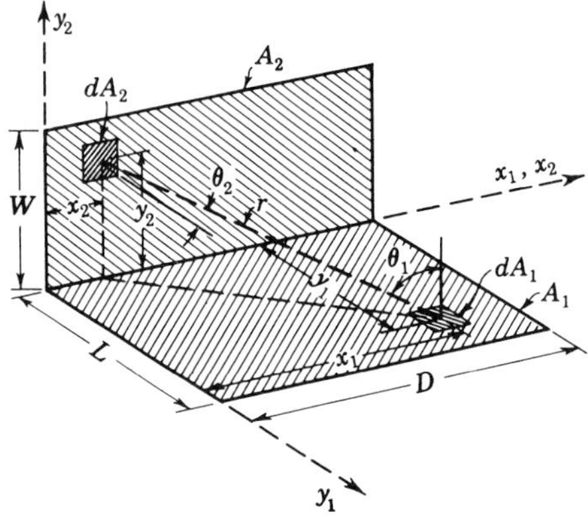

Figure C.2

integration gives

$$F_{1-2} = \frac{1}{\pi R_1}\left\{R_1 \tan^{-1}\frac{1}{R_1} + R_2 \tan^{-1}\frac{1}{R_2} - \sqrt{R_1^2 + R_2^2}\tan^{-1}\frac{1}{\sqrt{R_1^2 + R_2^2}}\right.$$

$$+ \frac{1}{4}\ln\left[\frac{(1 + R_1^2)(1 + R_2^2)}{(1 + R_1^2 + R_2^2)}\left(\frac{R_2^2(1 + R_1^2 + R_2^2)}{(1 + R_2^2)(R_1^2 + R_2^2)}\right)^{R_2^2}\right.$$

$$\left.\left.\times\left(\frac{R_1^2(1 + R_1^2 + R_2^2)}{(1 + R_1^2)(R_1^2 + R_2^2)}\right)^{R_1^2}\right]\right\} \tag{C.3}$$

The results of this expression are presented graphically in Fig. 9.19. As should be intuitively apparent, Fig. 9.19 shows that F_{1-2} must always be less than 0.5 for the configuration under consideration.

EXAMPLE C.1

Find the diffuse shape factor for (a) two square planes, 5 m × 5 m, parallel and directly opposed and spaced 2.5 m apart; (b) a square plane, 5 m × 5 m located normal to another rectangle, 5 m × 2.5 m, with the 5 m side in common.

Solution

(a) For the two parallel directly opposed 5 m × 5 m squares, spaced 2.5 m apart, the geometry of Fig. C.1 applies with

$$W = 5 \text{ m}, \quad L = 5 \text{ m}, \quad D = 2.5 \text{ m}.$$

Thus the ratios R_1 and R_2 defined in Eq. (C.2) are

$$R_1 = \frac{L}{D} = \frac{5}{2.5} = 2.0, \qquad R_2 = \frac{W}{D} = \frac{5}{2.5} = 2.0,$$

so that Fig. 9.18 or, more accurately, Eq. (C.2) gives the fraction of the radiation leaving one square that strikes the other, F_{1-2}, to be

$$F_{1-2} = 0.415.$$

(b) For a 5 m × 5 m square, A_1, perpendicular to a 5 m × 2.5 m rectangle, A_2, with the 5 m side in common, the geometry of Fig. C.2 applies with

$$D = 5 \text{ m}, \qquad L = 5 \text{ m}, \qquad W = 2.5 \text{ m}.$$

Thus the ratios R_1 and R_2 in Eq. (C.3) are

$$R_1 = \frac{L}{D} = \frac{5}{5} = 1.0, \qquad R_2 = \frac{W}{D} = \frac{2.5}{5} = 0.50.$$

Then Fig. 9.19 or, more accurately, Eq. (C.3) gives the shape factor from the square to the rectangle to be

$$F_{1-2} = 0.146.$$

C.3 OTHER CONFIGURATIONS DERIVABLE FROM PERPENDICULAR RECTANGLES WITH A COMMON EDGE

The two basic configurations just discussed, two directly opposed parallel rectangles or two perpendicular rectangles having a common edge, may be used to deduce the shape factors for other configurations involving rectangles that lie in planes that are parallel or perpendicular for which the conditions that they be directly opposed or have a common edge are not met. Appropriate application of the additive rule for shape factors, as given in Eq. (9.65), may reduce such cases that that of finding a number of shape factors of the basic forms just discussed.

As examples, some cases derivable from that for perpendicular rectangles having a common edge will be considered. The same principles illustrated for this instance may be applied to other basic configurations.

Case 1. Let it be desired to find the shape factor F_{1-2} for the configuration shown in Fig. C.3(a). Here the two areas involved, A_1 and A_2, are in perpendicular planes, but one of them, A_2, is displaced from the common edge. Define the fictitious area A_3 as shown in the figure and let A_2 and A_3 together comprise a single area, call it $A_{(2,3)}$. The additive principle, Eq. (9.65), applied to the shape factor between A_1 and $A_{(2,3)}$ gives

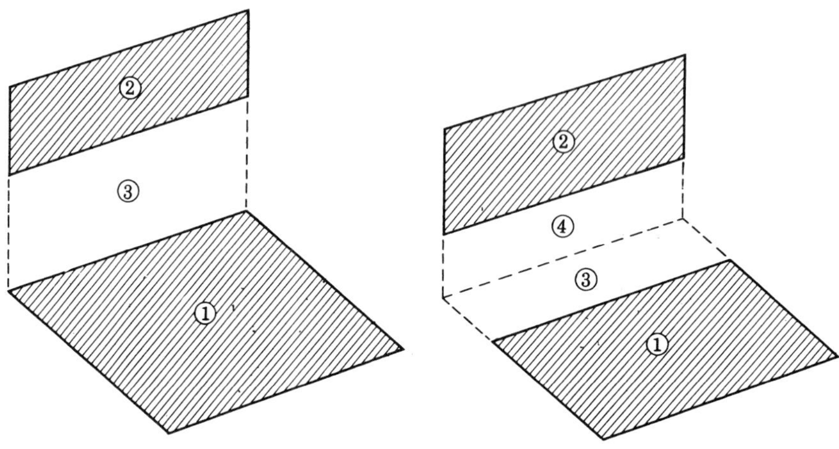

Figure C.3

$$A_i F_{i-j} = \sum_n \sum_m A_{i_n} F_{i_n - j_m},$$

$$A_1 F_{1-(2,3)} = A_1 F_{1-2} + A_1 F_{1-3}, \tag{C.4}$$

$$F_{1-2} = F_{1-(2,3)} - F_{1-3}.$$

Both $F_{1-(2,3)}$ and F_{1-3} are of the form given by Eq. (C.3) or Fig. 9.19—that of perpendicular rectangles with a common edge. Thus the desired F_{1-2} may be found by using the already known configuration twice.

To obtain F_{2-1} for this configuration, one could use the reciprocal property [Eq. (9.62)] or one could apply the additive property to $F_{(2,3)-1}$:

$$A_{(2,3)} F_{(2,3)-1} = A_2 F_{2-1} + A_3 F_{3-1}, \tag{C.5}$$

$$F_{2-1} = \frac{A_{(2,3)} F_{(2,3)-1} - A_3 F_{3-1}}{A_2}.$$

Again, the desired result is found by evaluating the basic configuration of Fig. 9.19 or Eq. (C.3) twice.

Case 2. Figure C.3(b) involves two perpendicular rectangles, A_1 and A_2, both of which are displaced from the line of common intersection. In this instance one defines two additional areas A_3 and A_4, as shown, so that combined areas

$$A_{(1,3)} = A_1 + A_3, \qquad A_{(2,4)} = A_2 + A_4$$

may be defined. Then the shape factor between the two combined areas may be divided according to the additive rule:

$$A_{(1,3)} F_{(1,3)-(2,4)} = A_{(1,3)} F_{(1,3)-4} + A_{(1,3)} F_{(1,3)-2}.$$

The quantity $A_{(1,3)} F_{(1,3)-2}$ may be further subdivided by the additive principle to obtain

$$A_{(1,3)} F_{(1,3)-(2-4)} = A_{(1,3)} F_{(1,3)-4} + (A_1 F_{1-2} + A_3 F_{3-2}).$$

The last term in this equation, $A_3 F_{3-2}$, is of the form considered in Case 1, and the result given by Eq. (C.4), with appropriate change of subscripts, is $A_3 F_{3-2} = A_3 F_{3-(2,4)} - A_3 F_{3-4}$. Thus one finally has

$$A_{(1,3)} F_{(1,3)-(2,4)} = A_{(1,3)} F_{(1,3)-4} + A_1 F_{1-2} + A_3 F_{3-(2,4)} - A_3 F_{3-4}, \tag{C.6}$$

$$A_1 F_{1-2} = A_{(1,3)} F_{(1,3)-(2,4)} + A_3 F_{3-4} - A_3 F_{3-(2,4)} - A_{(1,3)} F_{(1,3)-4}.$$

Once again, the desired shape factor F_{1-2} has been expressed in terms of several (this time, four) configurations of the form given in Eq. (C.3) or Fig. 9.19.

Case 3. The determination of the shape factor F_{1-2} for the configuration shown in Fig. C.4(a) is a little more involved than the above two cases. If one defines the fictitious areas A_3 and A_4 as shown, the additive principle gives

$$F_{1-2} = \frac{1}{A_1} [A_{(1,3)} F_{(1,3)-(2,4)} - A_1 F_{1-4} - A_3 F_{3-2} - A_3 F_{3-4}]. \tag{C.7}$$

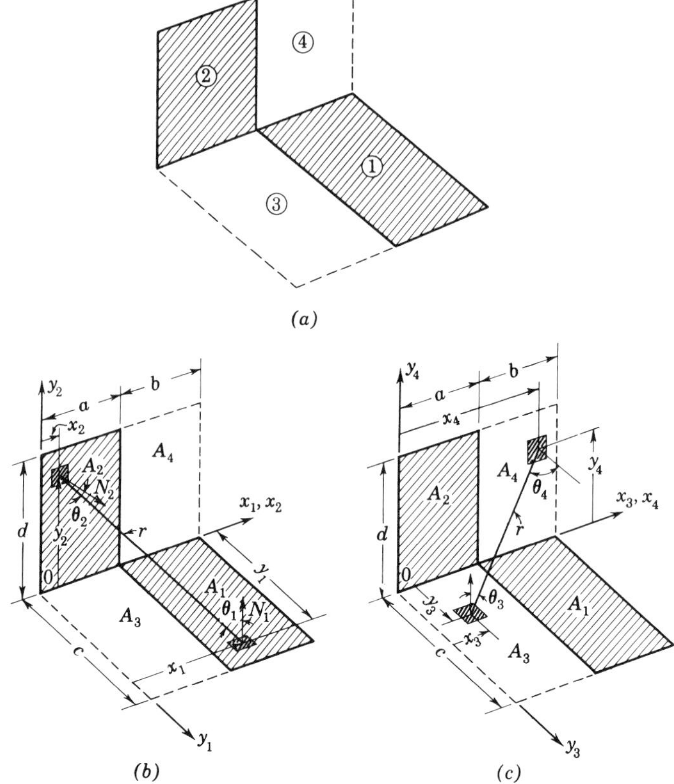

Figure C.4

One notes now that the shape factor F_{3-4} required in this expression is for a configuration of the same form as that for F_{1-2}—which is what is sought. This dilemma may be solved by developing another reciprocal relation of considerable importance.

The integral representation of the two shape factors F_{1-2} and F_{3-4} must be formed. Figure C.4(b) and (c) shows the dimensions and coordinates used for this purpose. The integrands of these expressions will be the same as for the basic configuration discussed in the preceding section, only the limits of the integration will be different. The resulting expressions for F_{1-2} and F_{3-4} are

$$A_1F_{1-2} = \frac{1}{\pi}\int_0^d\int_0^a\int_0^c\int_a^{a+b} \frac{y_1y_2\,dx_1\,dy_1\,dx_2\,dy_2}{[y_1^2 + y_2^2 + (x_1 - x_2)^2]^2},$$

$$A_3F_{3-4} = \frac{1}{\pi}\int_0^d\int_a^{a+b}\int_0^c\int_0^a \frac{y_3y_4\,dx_3\,dy_3\,dx_4\,dy_4}{[y_3^2 + y_4^2 + (x_3 - x_4)^2]^2}.$$

These two integrals are of identical form except for the order of integration. Since the nature of the integrand permits the interchange of the order of the integration, the following reciprocal formula is obtained for the configuration of Fig. C.4(a):

$$A_1F_{1-2} = A_3F_{3-4}.$$

This reciprocal relation then enables one to obtain F_{1-2} since Eq. (C.7) now reduces to

$$F_{1-2} = \frac{1}{2A_1}[A_{(1,3)}F_{(1,3)-(2,4)} - A_1F_{1-4} - A_3F_{3-2}]. \quad \text{(C.8)}$$

This involves only shape factors obtainable from Fig. 9.19 or Eq. (C.3).

C.4 GENERAL RELATIONS FOR PERPENDICULAR AND PARALLEL RECTANGLES

The methods of the foregoing sections may be applied to obtain expressions for the shape factor for perpendicular and parallel rectangles oriented in rather general ways. The results obtained by Hamilton and Morgan (Ref. 1) are given below. These relations will enable one to calculate shape factors between any portions (windows, doors, walls, etc.) of a parallelepiped structure.

Perpendicular Rectangles. Figure C.5(a) illustrates a general orientation of two rectangles located in perpendicular planes. The following reciprocal relations exist [see Fig. C.5(a) for the explanation of the subscripts used to denote the various areas involved]:

$$A_1F_{1-3'} = A_3F_{3-1'} = A_{3'}F_{3'-1} = A_{1'}F_{1'-3}. \quad \text{(C.9)}$$

The shape factor, $F_{1-3'}$, is given by

$$\begin{aligned} A_1F_{1-3'} = K_{1-3'} = \tfrac{1}{2}[&K_{(1,2,3,4,5,6)^2} - K_{(2,3,4,5)^2} - K_{(1,2,5,6)^2} + K_{(4,5,6)^2} \\ &- K_{(4,5,6)-(1',2',3',4',5',6')} - K_{(1,2,3,4,5,6)-(4',5',6')} \\ &+ K_{(1,2,5,6)-(5',6')} + K_{(2,3,4,5)-(4',5')} + K_{(5,6)-(1',2',5',6')} \\ &+ K_{(4,5)-(2',3',4',5')} + K_{(2,5)^2} - K_{(2,5)-5'} - K_{(5,6)^2} \\ &- K_{(4,5)^2} - K_{5-(2',5')} + K_{5^2}]. \end{aligned} \quad \text{(C.10)}$$

In Eq. (C.10) K_{m-n} is used to symbolize $K_{m-n} = A_mF_{m-n}$, and $K_{(m)^2} = A_mF_{m-m'}$.

Parallel Rectangles. Similar relations for opposed parallel rectangles are given below, using the notation depicted in Fig. C.5(b). The following reciprocal relations exist:

$$A_1F_{1-9'} = A_3F_{3-7'} = A_9F_{9-1'} = A_7F_{7-3'}. \quad \text{(C.11)}$$

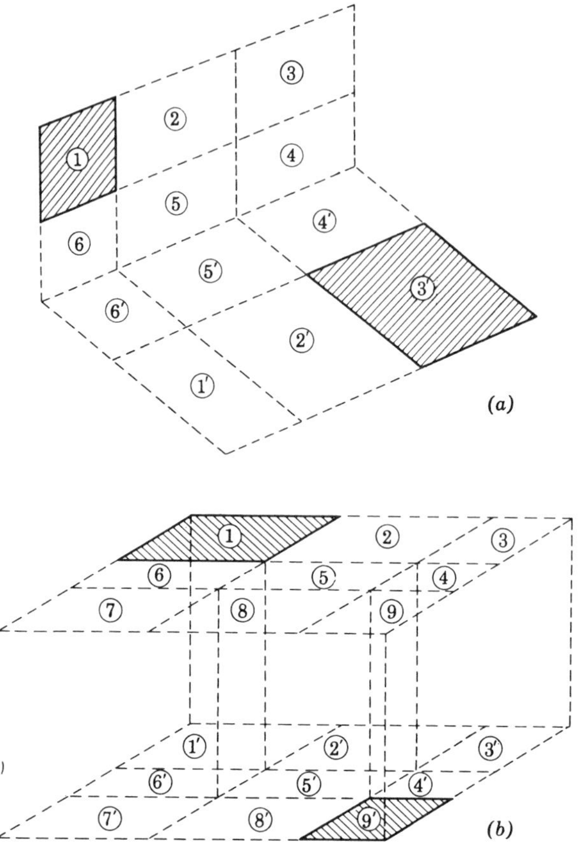

Figure C.5

The shape factor $F_{1-9'}$ is given by

$$
\begin{aligned}
A_1F_{1-9'} = K_{1-9'} = \tfrac{1}{4}\{ & K_{(1,2,3,4,5,6,7,8,9)^2} - K_{(1,2,5,6,7,8)^2} - K_{(2,3,4,5,8,9)^2} \\
& - K_{(1,2,3,4,5,6)^2} + K_{(1,2,5,6)^2} + K_{(2,3,4,5)^2} + K_{(4,5,8,9)^2} \\
& - K_{(4,5)^2} - K_{(5,8)^2} - K_{(5,6)^2} - K_{(4,5,6,7,8,9)^2} \\
& + K_{(5,6,7,8)^2} + K_{(4,5,6)^2} + K_{(2,5,8)^2} - K_{(2,5)^2}\}. \qquad \text{(C.12)}
\end{aligned}
$$

C.5 OTHER CONFIGURATIONS OF INTEREST

The number of configurations which may be considered is nearly limitless. However, three other situations of practical value will be presented here. No derivations are given; only the analytical results are presented, with corresponding graphical displays given in the main body of the text.

Directly Opposed Disks. For two disks, radii r_1 and r_2, placed L apart in parallel planes and having collinear centers

$$F_{1-2} = \tfrac{1}{2}[x - \sqrt{x^2 - 4(R_2/R_1)^2}],$$

$$x = 1 + \frac{1 + R_2^2}{R_1^2}, \tag{C.13}$$

$$R_1 = \frac{r_1}{L}, \qquad R_2 = \frac{r_2}{L}.$$

Figure 9.20 presents this shape factor in graphical form.

Concentric Cylinders. Two concentric cylinders of equal length L form a configuration of considerable engineering interest. If A_1 represents the outer surface of the inner cylinder of radius r_1, and A_2 represents the inner surface of the outer cylinder of radius r_2, the various shape factors involved may be expressed in terms of the ratios $R_1 = r_1/r_2$ and $R_2 = L/r_2$. Equations for F_{2-1} and F_{2-2} (the self-shape factor of the outer cylinder) in terms of R_1 and R_2 are reported in Ref. 2 but are not quoted here for the sake of brevity. These shape factors are displayed graphically in Fig. 9.21.

Small Spheres or Planes Irradiated by a Large Sphere. Present-day applications in space technology require knowledge of the amounts of heat acquired by satellites and spacecraft from planetary bodies. In most instances the size of the receiving body may be taken to be quite small compared to that of the planetary body. Figure 9.22 illustrates the geometrical parameters involved in consideration of the radiant exchange between a large planetary body and a small sphere or plane element. The small receiving surface is denoted by dA_1 and the large emitting one by A_2. In the case of the plane element (which might be part of a larger space structure), the direction of the surface normal with respect to the line drawn to the planetary center, λ, must be known. Because the expressions for the shape factor F_{dA_1-2} cannot be written in simple closed forms, only the results of numerical integrations are given in Fig. 9.22 (Ref. 5) for the two cases.

One must note that the shape factors given are valid only if the large sphere has a uniform radiosity. In the event that only part of the sphere is radiating (as might be the case for reflected solar radiation) one must take this fact into account. Results of such calculations are given in Refs. 3 and 5.

REFERENCES

1. Hamilton, D. C., and W. R. Morgan, "Radiant Interchange Configuration Factors," *NACA Tech. Note 2836,* Dec. 1952.
2. Siegel, R., and J. R. Howell, *Thermal Radiation Heat Transfer,* 2nd ed., New York, McGraw-Hill, 1981.

3. Stevenson, J. A., and J. C. Grafton, "Radiation Heat Transfer Analysis for Space Vehicles," *Rep. ASD 61–119, Part I,* Flight Accessories Laboratory, Aeronautical Systems Division, Wright-Patterson Air Force Base, Ohio, Dec. 1961.
4. Plamondon, J. A., "Numerical Determination of Radiation Configuration Factors for Some Common Geometrical Situations," *Tech. Rep. 32–127,* Jet Propulsion Laboratory, California Institute of Technology, 1961.
5. Clark, L. G., and E. C. Anderson, "Geometric Shape Factors for Planetary-Thermal and Planetary-Reflected Radiation Incident upon Spinning and Non-spinning Spacecraft," *NASA Tech. Note D-2835,* May 1965.

APPENDIX D

SI-Unit Conversion Table for Heat Transfer Calculations

The discussion presented in Section 1.8 describes the SI and English engineering systems of units. This appendix presents conversion factors between these two systems to facilitate computations and the use of the various tables of physical properties. These conversions are limited to those quantities occurring commonly in heat transfer calculations (i.e., no conversions are given for electrical quantities, etc.).

Dimension	English Units	SI Units
Acceleration	1 ft/s^2 1 ft/h^2	$= 3.0480 \times 10^{-1}$ m/s^2 $= 2.3519 \times 10^{-8}$ m/s^2
Area	1 ft^2 1 $in.^2$	$= 9.2903 \times 10^{-2}$ m^2 $= 6.4516 \times 10^{-4}$ m^2
Conductance, thermal	1 Btu/h-ft^2-°F	= 5.6784 W/m^2-°C
Conductivity, thermal	1 Btu/h-ft-°F	= 1.7308 W/m-°C
Density	1 lb_m/ft^3	$= 1.6018 \times 10$ kg/m^3
Diffusivity, thermal	1 ft^2/s 1 ft^2/h	$= 9.2903 \times 10^{-2}$ m^2/s $= 2.5806 \times 10^{-5}$ m^2/s
Energy	1 Btu 1 kW-h 1 ft-lb_f 1 hp-h	= 1.0551 kJ $= 3.6000 \times 10^3$ kJ $= 1.3558 \times 10^{-3}$ kJ $= 2.6845 \times 10^3$ kJ
Force	1 lb_f	= 4.4482 N
Heat	1 Btu	= 1.0551 kJ
Heat flow rate	1 Btu/s 1 Btu/h	$= 1.0551 \times 10^3$ W $= 2.9308 \times 10^{-1}$ W
Heat flux (unit area) (unit length)	 1 Btu/h-ft^2 1 Btu/h-ft	 = 3.1546 W/m^2 $= 9.6152 \times 10^{-1}$ W/m
Heat generation rate (unit mass) (unit volume)	 1 Btu/h-lb_m 1 Btu/h-ft^3	 $= 6.4612 \times 10^{-1}$ W/kg $= 1.0350 \times 10$ W/m^3
Heat transfer coefficient	1 Btu/h-ft^2-°F	= 5.6784 W/m^2-°C
Latent heat	1 Btu/lb_m	= 2.3260 kJ/kg
Length	1 ft 1 μm 1 in. 1 mile	$= 3.0480 \times 10^{-1}$ m $= 1.0000 \times 10^{-6}$ m $= 2.5400 \times 10^{-2}$ m $= 1.6093 \times 10^3$ m
Mass	1 lb_m	$= 4.5359 \times 10^{-1}$ kg
Mass flow rate	1 lb_m/s 1 lb_m/h	$= 4.5359 \times 10^{-1}$ kg/s $= 1.2600 \times 10^{-4}$ kg/s
Mass flux	1 lb_m/s-ft^2 1 lb_m/h-ft^2 1 lb_m/s-$in.^2$ 1 lb_m/h-$in.^2$	= 4.8824 kg/s-m^2 $= 1.3562 \times 10^{-3}$ kg/s-m^2 $= 7.0362 \times 10^2$ kg/s-m^2 $= 1.9545 \times 10^{-1}$ kg/s-m^2
Momentum, linear	1 lb_m-ft/s 1 lb_m-ft/h	$= 1.3825 \times 10^{-1}$ kg-m/s $= 3.8404 \times 10^{-5}$ kg-m/s
Power	1 Btu/s 1 ft-lb_f/s 1 Btu/h 1 hp	$= 1.0551 \times 10^3$ W = 1.3558 W $= 2.9308 \times 10^{-1}$ W $= 7.4570 \times 10^2$ W

Dimension	English Units	SI Units
Pressure	1 lb_f/ft^2	$= 4.7880 \times 10^{-2}$ kN/m^2
	1 $lb_f/in.^2$	$= 6.8948$ kN/m^2
	1 standard atmosphere	$= 1.0133 \times 10^2$ kN/m^2
	1 in. water	$= 2.4909 \times 10^{-1}$ kN/m^2
	1 ft water	$= 2.9891$ kN/m^2
	1 in. mercury	$= 3.3866$ kN/m^2
Resistance, thermal		
(total)		$= 1.8956$°C/W
(unit)	1 h-°F/Btu	$= 1.7611 \times 10^{-1}$
	1 h-ft^2-°F/Btu	m^2-°C/W
Specific energy	1 Btu/lb_m	$= 2.3260$ kJ/kg
	1 ft-lb_f/lb_m	$= 2.9891 \times 10^{-3}$ kJ/kg
Specific heat	1 Btu/lb_m-°F	$= 4.1868$ kJ/kg-°C
Specific volume	1 ft^3/lb_m	$= 6.2428 \times 10^{-2}$ m^3/kg
Surface tension	1 lb_f/in.	$= 1.7513 \times 10^2$ N/m
Temperature	°R	°K $= \frac{5}{9} \times$ °R
	°F	°C $= \frac{5}{9} \times$ (°F $-$ 32)
Temperature difference	1°F(°R)	$= \frac{5}{9}$°C(°K)
Time	1 h	$= 3.6000 \times 10^3$ s
	1 min	$= 6.0000 \times 10$ s
Velocity	1 ft/s	$= 3.0480 \times 10^{-1}$ m/s
	1 ft/h	$= 8.4667 \times 10^{-5}$ m/s
	1 mph	$= 4.4704 \times 10^{-1}$ m/s
Viscosity, dynamic	1 poise (g/cm-s)	$= 1.0000 \times 10^{-1}$ kg/m-s (N-s/m^2)
	1 lb_m/ft-s	$= 1.4882$ kg/m-s
	1 lb_m/ft-h	$= 4.1338 \times 10^{-4}$ kg/m-s
	1 lb_f-s/in.^{2}	$= 6.8947 \times 10^3$ kg/m-s
	1 lb_f-h/ft^2	$= 1.7237 \times 10^5$ kg/m-s
Viscosity, kinematic	1 stoke (cm^2/s)	$= 1.0000 \times 10^{-4}$ m^2/s
	1 ft^2/s	$= 9.2903 \times 10^{-2}$ m^2/s
	1 ft^2/h	$= 2.5806 \times 10^{-5}$ m^2/s
Volume	1 ft^3	$= 2.8317 \times 10^{-2}$ m^3
	1 in.^{3}	$= 1.6387 \times 10^{-5}$ m^3
Volume flow rate	1 ft^3/s	$= 2.8317 \times 10^{-2}$ m^3/s
	1 ft^3/min	$= 4.7195 \times 10^{-4}$ m^3/s
	1 ft^3/h	$= 7.8658 \times 10^{-6}$ m^3/s

Index